VERHANDELINGEN DER KONINKLIJKE NEDERLANDSE
AKADEMIE VAN WETENSCHAPPEN, AFD. NATUURKUNDE

TWEEDE REEKS, DEEL 62

CHECK LIST OF EUROPEAN POLYPORES

M. A. DONK

NORTH-HOLLAND PUBLISHING COMPANY — AMSTERDAM-LONDON — 1974

Isbn 7204–82550

Library of Congress
Catalogue Card Number: 73–77662

Aangeboden in de vergadering
van 31 maart 1973

SYNOPSIS OF SECTIONS

PREFACE

This publication was written for the purpose of taking stock of what has been published so far on European polypores. Considerable time was spent in surveying the literature. Once I realized that a really exhaustive survey would take too much of my time I halted it rather abruptly; during the last few years I limited myself as far as this project was concerned to editing the collected data and implementing the manuscript with such additions and corrections as had been encountered in the daily routine.

I wish to make it quite clear that from the start I never had any intention of going beyond the very first steps necessary for the preparation of a monograph. No taxonomic revision at either the specific or any other level was aimed at. The taxonomic basis presented in the main section of this publication is not a critical one; it simply reflects what judging from the experience gained during many years of intermittent attention to polypores in general I thought to be reasonably up-to-date. Many innovations, or at any rate alterations, could have been worked into the arrangement of the species but I deliberately restricted myself to an arrangement that I regard as a somewhat conservative presentation of what had previously been accomplished in this field—in Europe. "In Europe" is added to emphasize one of the principal weaknesses of the taxonomic arrangement I adopted. This arrangement has been built up by authors who for the most part have only a negligible critical knowledge of the polypores growing outside Europe, a continent relatively poor in species of this group. It is about time that European taxonomists ceased thinking that the taxonomy of the polypores can be worked out completely with knowledge of only a small percentage of species of the group as a whole. This is not to say that the 'European' scheme of classification will be of no use at all in accommodating extra-European species; even though it is inadequate, at the same time it appears to be a useful frame both for reference and for correction and additions. Except for notable exceptions in connection with cultural characters, mainly of species from temperate North America, it must be admitted in all fairness that mycologists outside Europe have scarcely even begun to take a share in advancing the classification of the polypores.

In several instances I could have entered fresh results, but this again did not agree with the primary intention. What was more important was to deliver a basis for future work by consulting and checking the literature, and also to reassess the various components that make up the more complex protologues of revalidated names. During these activities I encountered a complaint dating from as long as 150 years ago that revealed that a failing common to many a modern author is of long standing:

"The practice of accurately examining the synonyms, instead of copying them from other authors, has been too long neglected" (Purton, App. Midl. Fl. 335. 1821). This failing has culminated in a number of errors and puzzles in references in some of the much used European publications that is really stupendous.

The group of fungi indicated in the title of this book as the 'polypores' coincides almost completely with what remains of the Friesian conception of the "Polyporei" after the exclusion of the Boleti and the meruloid and cyphelloid elements and after the addition of a sprinkling of species taken from such genera as *Irpex*, *Lenzites*, and *Sistotrema*. Nowadays almost without exception specialists agree that the polypores, even with the foregoing restrictions, are still a strongly mixed group of similar life forms based on a few features only, the principal ones being the presence of holobasidia and a more or less tubulose hymenophore with sterile edges of the tube-dissepiments.

At present I am prepared to distribute the polypores over seven families, of which one, the Polyporaceae sensu stricto, contains the large majority of the genera.[1] The restricted family of Polyporaceae is still a thoroughly heterogeneous assemblage and includes a number of artificial genera (at any rate as delimited on this Check list) that have not as yet been broken up satisfactorily: *Albatrellus*, *Fomitopsis*, *Poria*, *Trametes*, and *Tyromyces*. Some of the genera of the Hymenochaetaceae (like *Inonotus* and *Phellinus*) also represent groups that are too broadly conceived.

As to the composition of the Check list proper, it will be seen that in the separate entries of the taxa the purely nomenclative are kept apart from the rest, which aims at providing a key to the current conception and further knowledge about each of the taxa. Only generic and specific names are taken into consideration (save for indispensable exceptions). The generic names are mostly treated more summarily, partly in anticipation of a review of the genera that is in preparation. Most of the pertinent nomenclative information has already been dealt with in a series of previously published papers primarily concerned with this kind of information (in particular, Donk, *1960, 1962*).

The appended "Notes" differ in tenor from those that preceded them in a series of separately published papers (Donk, *1966a, 1966b, 1967, 1969a–c, 1971a–c, e, 1972a–c*). While the earlier notes attempted to clarify some of the conclusions that entered into the shaping of the Check list, the notes that are offered in the present publication for the most part draw attention to various problems requiring further study.

The "List of omitted names" has expanded to unexpected length. It contains mainly nomina dubia and not validly published names but also such names as have been reduced to the synonymy of the 'admitted'

---

[1] Compare also M. A. Donk (1964), A conspectus of the families of Aphyllophorales, *in* Persoonia **3**: 199–324.

names on the Check list proper but that for one reason or another had to drop out. The "Index" makes it possible to trace presumed or now rejected connections of the 'omitted' names with those 'admitted'.

Fossils have not been taken into consideration. A few papers dealing with them are listed in the bibliography (Bondarcev, *1960*; Kirchheimer, *1942*).

This registration of names is not intended to assign to them any status under the "Code" other than the one they had before this publication appeared.

ACKNOWLEDGEMENTS.—Since much of the composition of this book consisted of checking literature, a special word of thanks is extended to Librarians. Of these Mr. L. Vogelenzang and his staff of the Rijksherbarium, Leiden, deserve first mention. Other libraries consulted were those of the British Museum (Natural History), London; of the Herbarium, Royal Botanic Gardens, Kew; of the New York Botanical Garden; of the National Fungus Collections, Beltsville. During my stay in the U.S.A., while enjoying a National Science Foundation Senior Scientist Fellowship at the Department of Botany, University of Tennessee, I received extensive help from the Librarians of that Department. To all of them and to the institutes I express my gratitude.

It is not feasible to mention here all of the many mycologists and herbaria from which I received specimens or information. In many cases they have been mentioned individually in the appropriate places in this and former publications. However, it would be negligent were I not to make particular mention of Dr. J. L. Lowe, Syracuse, U.S.A. and Dr. H. Jahn, Heiligenkirchen, Germany.

I am indebted, as usual, to the constant help received from Dr. Elizabeth Helmer van Maanen for improving the English text of the introductory chapters, the "Notes", and the "List of omitted names".

Note—Long after the manuscript of this work had been accepted for publication, it became apparent that some of the technicalities proposed by the late author could not be realized. The undersigned takes responsibility for possible ambiguity resulting from economy in printing.

Dr. Donk left no instructions as to what should be included in the index of his work. Here again the necessity to economize dictated the adoption of an arrangement as explained on p. 452.

R. A. Maas Geesteranus

METHOD OF PRESENTATION

Europe

'Europe' is accepted in the traditional sense, without Greenland, but including Caucasia.

Generic names

Generic names published before 1962 are listed without the usual references. Variant spellings are not mentioned. For these and other nomenclative details the series "The generic names proposed for Hymenomycetes" [2] should be consulted, particularly "The generic names proposed for Polyporaceae" (Donk, *1960, 1962*). Instead of the usual bibliographic citations references to the treatment of the generic names in this series are added between square brackets.

Example.—"PIPTOPORUS P. Karst. / 1881 [1960 (Pe 1): 257]. — Lectotype: *Polyporus betulinus* (Bull.) per Fr." is an abbreviated form of
PIPTOPORUS P. Karst. 1881 (for place of publication and other nomenclative details, see Donk *in* Persoonia 1: 257. 1960). — Lectotype (selection discussed by Donk, l.c.): *Polyporus betulinus* (Bull.) per Fr.

Specific names

This Check list distinguishes between four kinds of specific names: (i) the *basionym* and (ii) the *recombinations* of its epithet, as well as (iii) the corresponding *new name(s)*, viz. substitute name(s) (name changes) replacing the basionym, including its epithet. These recombinations and substitute names form the *isonyms* of the (ultimate) basionym.

The last category is (iv) the *non-isonymous synonyms* of a correct name, viz. names that when published were not (or not primarily) intended to replace a previously published name. Some of these may later prove to have based on the same type as another name, in which case they become obligate synonyms (typonyms).

Of a *correct* name the specific epithet is printed in bold face type, followed by the author's citation and the date of publication. Then, separated by a dash, come the basionym and/or the recombinations, as well as substitute names (preceded by the symbol ≡), in so far as they are devalidated names or have been validly published (provided no

---

[2] Parts I–IX, XII, XIII were brought together in a facsimile edition to which an "Index" was added (Verlag J. Cramer, 1966). "The generic names proposed for Agaricaceae" (*in* Beih. Nova Hedw. **5**. 1962) makes up Part XI of the series. "The generic names proposed for Polyporaceae" is Part X (1960) and Part XIV (1962); of the former there was a facsimile edition (Verlag J. Cramer).

qualification to the contrary is added); each is likewise followed by the author's citation and the date of publication.

Non-isonymous synonyms form separate entries. These entries are arranged in chronological order according to the date of the first-published *specific* combination—validly published or devalidated. Where nomina anamorphosium are listed these come after the names based on the perfect states.

Then, preceded by the indication "M.", there follows a selection of misapplications in such cases as these were thought to be worth mentioning at all.

REFERENCES

There are various kinds of references. Besides an occasional reference of a more customary type, one of these kinds consists of references comprising only a date not printed in italics. These references are not further elucidated or taken into consideration in the following explanation.

The dates are presumed to be the correct ones. Sometimes, where additional information has shown that a correction was necessary, they replace the dates in the publications themselves. In the reference '1959 (Ll 21): 105' the date in the publication referred to ["1958"] is not mentioned.

(i) REFERENCES CONSISTING OF A DATE NOT PRINTED IN ITALICS AND FOLLOWED BY ADDITIONAL INFORMATION.—In connection with this category a distinction is made between 'individually cited publications' (a) and 'serials' (b). The former are cited by date or by a date and a strongly reduced title, and where necessary the number of the volume, fascicle, &c., followed by the indication of such items as page, plate, figure, or, in the cases of exsiccati of series with 'printed' labels, of number ("No."). Titles of 'serials', including periodicals and journals, are abbreviated to no more than three letters and are usually followed by the number of the series (except for the first series) and the volume, all between brackets. In other respects the same pattern as in (a) is adhered to. Where alternative pages are mentioned, the second is that of the reprint.

The abbreviated titles are listed and elucidated by their more usual, less strongly abbreviated form in the Chapter "Strongly reduced bibliographic references".

EXAMPLES.—
(ia): Bull. 1786: *pl. 312* = Bull., Herb. Fr. *pl. 312.* 1786.
(ia): Bull. 1791 H.: 348 = Bull., Hist. Champ. Fr. 348. 1791.
(ib): Bres. 1897 (AAR III 3): 78 = Bres. *in* Atti I.R. Acad. Agiati [Rovereto] III 3: 78. 1897.

(ii) REFERENCES TO TITLES LISTED IN THE "BIBLIOGRAPHY OF SPECIAL LITERATURE."—These are in the form of *dates printed in italics.*

Authors cited in combination with references in the form of year-dates in italics are not always entered on the list of abbreviations of authors' names. Their initials can be traced from the "Bibliography of special literature".

I wish to emphasize that every reference not followed by '(n.v.)' has been checked. This also applies to those references that consist of a date only ('1846'), i.e. without any further indication of place of publication.

COMPOSITION OF ENTRIES

To each entry of a correct specific name or of a non-isonymous specific synonym at least one reference to a description is added. If there is not more than one such reference this indicates that I know of no improved description or illustration. Usually this one reference is to the original description or the protologue of the name. If for some reason the original account was thought somehow to be useful a reference to it is given separately, following the nomenclative information.

The one or more references following the nomenclative information about an entry and separated from it by a dash (—), are such as I consider to be of a certain importance to the knowledge of the taxon or to its re- cognition. These are arranged in chronological order and usually refer to the more representative descriptions and illustrations of the taxon, and occasionally also to notes on other subjects, such as nomenclature, distribution, and cytology. The descriptions and notes referred to are not necessarily reliable. They may for instance have been drawn up for a too-inclusively conceived taxon. They may even represent misconceptions that have so far not been recognized as such. Sometimes they contain only a minor addition to previous knowledge of the taxon, but in that case very little is known about the latter and the information may con- ceivably be of some use to future workers.

These references are nearly always followed by a generic or specific name between brackets. This was added to indicate the specific or infra- specific name under which the matter referred to was published, usually without repetition of the corresponding epithet. In cases where the same name would have followed two or more consecutive references it has been placed only after the last of the series, and where it is the same for all references as the name given at the beginning of the entry it has been deleted. In the case of homonyms the swung dash (∼) avoids repetition in full of the preceding name minus the author's citation.

EXAMPLE of an entry of a correct name.—
"**grisea** (Peck) Bond. & S. 1951 (17–19). — *Polyporus* Peck 1873, 1874 (U.S.A., New York), not ∼ Bres. 1912, not ∼ (Wint.) Pilát (n.v.p.); *Scutiger* Murrill 1903. — Lloyd 1911 (LMW 3, 0.): 78 *fs. 499, 500*; Lowe 1942: 35; Overh. 1953: 228 *pl. 168, pl. 130 fig.* (*Polyporus*); Malenç. 1956 (BmF 71): 272 (*Boletopsis*)" is to be read as follows:
Boletopsis grisea (Peck) Bond. & S. 1941 (see notes **17–19**). — Synonyms: *Polyporus*

griseus Peck 1873, 1874 (basionym; type locality, U.S.A., New York State), not *Polyporus griseus* Bres. 1912, not *Polyporus griseus* (Wint.) Pilát (not validly published); *Scutiger griseus* (Peck) Murrill 1903. — Description, illustrations, &c.: Lloyd, Mycol. Writ. 3 (Syn. Sect. Ovinus Polyp.): 78 *fs. 499, 500*. 1911 (as *Polyporus griseus*); Overh., Polyporac. U.S., Alaska & Canada 228 *pl. 168, pl. 130 unnumbered figure* (as *Polyporus griseus*); Malenç. *in* Bull. Soc. mycol. Fr. 71: 272 (as *Boletopsis griseus*).

A reference will often be found at the end of the first member of an entry. This is to the author who reduced the name to synonymy. He may not have been the first to do so. There are various reasons why the citation of a later author may be preferable; he may have seen the type or have recently studied it microscopically. If such a reference fails to mention the taxon to which a name is reduced this indicates that the name had already been reduced to the name considered to be correct (in one of its manifestations: basionym or one of its isonyms). In other cases the name of the taxon is mentioned specifically.

EXAMPLES of entries of non-isonymous synonyms:—
[**Chaetoporus collabens** . . .]
 "*Polyporus emollitus* (Fr.) Cooke 1878; fide Bres. 1897 (AAR III 3): 82 = *Poria blytii* [sensu Bres.]." is to be read as follows:
 Polyporus emollitus (Fr.) Cooke 1878; fide Bres. *in* Atti I.R. Accad. Agiati, Rovereto III 3: 82. 1897 = *Poria blyttii* in Bresadola's conception in the cited work.
[**Coriolus hirsutus** . . .]
 "*Polyporus cinereus* Lév. 1846 . . ., not ∼ Schw. 1832; fide Bres. 1916 (Am 14): 223." should be read as follows:
 Polyporus cinereus Lév. 1846 . . ., not *Polyporus cinereus* Schw. 1832; fide Bres. in Annls mycol. **14**: 223. 1916 = *Polyporus hirsutus*.

NOTES

Numbers in bold-face type between brackets refer to the remarks assembled in the section "Notes".

SPECIAL LITERATURE

The references listed under this caption and following the generic name (and its synonymy) are to the titles in the Section "Bibliography of special literature". With a few exceptions these items deal *exclusively* with the subject, or part of the subject, for which they are cited.

ABBREVIATIONS AND MISCELLANEOUS

The following list does not mention the abbreviations of authors' names and titles of publications. These will be found in the Sections "Strongly reduced bibliographic references" and "Abbreviations of authors' names". The abbreviations below do not, however, include many of those that are in common use.

c.	*circa*
d.n.	devalidated name
descr.	description
f., *fs.*, *fig.*	figure(s); *fig.*, unnumbered figure
Ind.	Index
M.	misapplication
No.	number, numéro, &c., especially used in connection with exsiccati series.
nom. altern.	nomen alternativum
nom. anam.	nomen anamorphosis
nom. conf.	nomen confusum
nom. monstr.	nomen monstrositatis
nom. nud.	nomen nudum
nom. prov.	nomen provisorium
n.v.	*non vidi* (not seen)
n.v.p.	not validly published
pl., *pls.*, *plate*	plate(s); *plate*, unnumbered plate
ref.	reference
repr.	description reproduced in/by ...
sp., spp.	species (singular, plural)
tpl., *tpls.*	text-plate(s)
→	= & see ...
(O)	refers to the "List of omitted names".

Plates or figures not simultaneously published with the corresponding name are placed between square brackets.

In an attempt to correlate other authors' nomenclature with the nomenclature adopted in this publication, additions between square brackets are used freely. This avoids a search for the correct name via the "Index". It should be emphasized, however, that the use of the square bracket implies that the additions reflect my personal view and are not necessarily correct.

EXAMPLES.—"*Merulius* [*Phlebia*] *rufus*" means that the species called *Merulius rufus* by the author mentioned in connection with it I regard as belonging to *Phlebia*. — "*Polyporus subsquamosus* [sensu Bres. = *Boletopsis griseus*]" means that in my opinion the taxon Bresadola called *Polyporus subsquamosus* belongs to *Boletopsis grisea*.

The qualification of 'extra-European' often implies that the species involved has not been otherwise dealt with elsewhere in this publication.

CHECK LIST OF EUROPEAN POLYPORES

ABORTIPORUS Murrill

1904 [1960 (Pe 1): 175]. — Holotype: *Boletus distortus* Schw. — Cf. Donk 1971 (PNA 74): 1.

Daedalea Quél. 1886, emend. Pat. 1900: 95, not ∼ Pers. per Fr. 1821 [cf. 1960 (Pe 1): 205, in obs.]. — [≡ *Daedalea* Pers. sensu Quél. 1886: 184, excl. of type.] — Lectotype: *Daedalea biennis* (Bull.) per Fr.

Irpicium Bref. 1912 [1960 (Pe 1): 231]. — Monotype: *Irpicium ulmicola* Bref.

Heteroporus Lázaro 1916 [1960 (Pe 1: 223]. — Lectotype: *Daedalea biennis* (Bull.) per Fr.

M.—*Daedalea* Pers. sensu Quél. 1886: 184 ("Pers.") emend. Pat. 1900:95 → *Daedalea* Quél.

Special literature.—Boudier, *1888*; O. Fidalgo, *1959b, 1964, 1969*; Graff, *1939*; Schulzer von Müggenburg, *1874, 1880a*; de Seynes, *1888a, 1888b*.

biennis (Bull. per Fr.) Sing. 1944 **(1, 2)**. — *Boletus* Bull. 1789 (France) (d.n.); *Sistotrema* Pers. 1801 (d.n.); *Hydnum* DC. 1805 (d.n.); *Daedalea* (Bull.) per Fr. 1821; *Hydnum* Mérat 1821; *Sistotrema* S. F. Gray 1821; *Boletus* Purt. 1821; *Polyporus* Fr. 1838; *Heteroporus* Lázaro 1916; *Phaeolus* Pilát 1935. — Bull. 1789: *pl. 449 f. 1*; 1791 H.: 333; Sow. 1799: *pl. 191* [referred to *Polyporus rufescens* by Fr. 1838: 433] (*Boletus*); Berk. 1836: 130; Gillet 1874–90 P.: *pl. 265/476* (*Daedalea*); Seyn. 1888 P.: 55 *pl. 5 fs. 4–16, pl. 6* (*Polyporus*); Bourd. & G. 1928: 526; Konr. & M. 1932 I. 5: *pl. 447* (*Daedalea*); Donk 1933: 177; Pilát 1936–7 (ACE 3): 115 *pls. 48–52*; D. Reid 1958 (TBS 41): 434 *f. 15, pl. 23 f. 2*: Fid. 1969 (Ri 4): 127 *fs. 1–17, 50*, in part: var. *biennis* (*Heteroporus*). — Sensu Sow. → *Daedalea sowerbei* Fr. (**0**), a not validly published syn. of the present sp.

Sistotrema rufescens Pers. 1801 (Germany) (d.n.). — *Hydnum* Poir. 1808 (d.n.), not ∼ Schaeff. 1774 (d.n.); *Polyporus* (Pers.) per Fr. 1821; *Sistotrema* Pers. 1825; *Daedalea* Secr. 1833, not ∼ (Lázaro) Sacc. & Trott. apud Trott. 1925; *Trametes* Fr. 1849 (n.v.p.[3]), Otth 1860; *Polystictus* P. Karst. 1882; *Spongiosus* Torrend 1924. — Pers. 1803 I.p.: 14 *pl. 6* (*Sistotrema*); Berk. 1860: 238; Fr. 1874: 529 (*Polyporus*); Bres. 1931 (BIm 20): *pl. 958* (*Polyporus biennis* var.). — Sensu Rostk. 1828 (StP 4): 17 *pl. 7* = *Onnia*

---

[3] When Fries (1849: 322) published the new combination *Trametes rufescens* he gave no description and only one reference, "B. n. 130." This is apparently an error to be corrected into 'B. 130' = 'Berk. 1836: 130' where *Daedalea biennis* is described but without any direct or indirect reference to *Polyporus rufescens* (Pers.) ex Fr. In my opinion this makes the name *Trametes rufescens* of Fries a nomen nudum, with a reference to a taxonomic synonym rather than to the 'basionym'.

sp., fide Fr. 1832[Ind.]: 148 = *Polyporus [Onnia] tomentosus*; sensu Buller 1922 = *Abortiporus distortus* (extra-European) (**O**).

Poria terrestris Pers. 1805 (France?) (d.n.); fide Bourd. & G. 1925 (BmF 41): 153 (forma). — *Boletus* DC. 1815 (d.n.), misapplied; *Polyporus* (Pers.) per Fr. 1821, in part; *Physisporus* Chev. 1826; *Boletus* Spreng. 1827; *Poria* Cooke 1886, Quél. 1886, misapplied, not ∼ Bourd. & G. 1925. — Pers. 1805 I.p.: 35 *pl. 16 f. 1* (*Poria*); sensu Bourd. & G. 1925 (BmF 41): 153 & 1928: 577 (*Daedalea biennis* f.); & cf. Donk 1967 (Pe 5): 112. — Sensu DC. → *Poria mollicula*; sensu Bres. 1897 → *Rigidoporus sanguinolentus*.

Boletus rugosus Sow. 1815: *pl. 422* (England) (d.n.), not ∼ Jacq. 1774 (d.n.), not ∼ Fr. 1835; fide Fid. 1969 (Ri 4): 117, 129, perhaps a doubtful interpretation. — *Polyporus* (Sow.) per Pers. 1825, not ∼ Bl. & Nees 1826: Fr. 1828, not ∼ Trog 1844; not ∼ (Jacq.) per E. Krause 1928; *Cladomeris* Big. & Guill. 1909 (syn.: n.v.p.).

Daedalea albida Purt. 1821: 253 *pl. 38* (England), not ∼ Fr. 1821: 338, not ∼ Schw. 1822; fide Berk. 1836: 131 & Graff 1939 (M 31): 468, 470.

Sistotrema lobatum Desm. 1823: 19 (France). — *Hydnum* Duby 1830.

Polyporus laciniatus Pers. 1825, not ∼ Velen. 1922; fide Bourd. & G. 1928: 730. — *Daedalea* Pat. 1900; [≡ "*Agaricus Tubae Fallopianae instar laciniatus* Tou. J.R.H. 562" sensu Batt. 1755: 66 *pl. 32 f. B* (Italy)].

Polyporus acanthoides Rostk. 1848, misapplied, not ∼ (Bull. per Mérat) Fr. 1838; fide Donk 1971 (PNA 74): 2. — [≡ *Polyporus acanthoides* (Bull. per Mérat) Fr. sensu Fr. 1838: 488 (Sweden)]. — Sensu Rostk. → *Meripilus giganteus*.

Sistotrema carneum Bon. 1857 (BZ 15): 237 *pl. 5 f. B* (Germany), not ∼ Ehrenb. 1818 (d.n.), not ∼ Fr. 1818 (d.n.); fide Fid. 1969 (Ri 4): 120, 130, & Donk. — Eeden 1868–72 (Fb 14): *pl. 1095*; Luc. & Gillot 1885 (Rm 7): 41.

Polyporus occultus Lasch in Rab. 1858 Kl. II: No. 617 [repr. 1858 (BZ 16): 79 & 1858 (Fl 41): 388] (Germany); fide Donk 1971 (PNA 74): 3.

Irpex radicatus Fuck. 1870 (Jna 23–24): 23 (Germany); fide Fid. 1969 (Ri 4): 120, 130, & Donk. — *Xylodon* O.K. 1898 ("*radicalis*"). — Velen. 1922: 740 *f. 133 at left*.

Polyporus heteroporus Fr. apud Quél. 1872 (MMb II 5): 274/257 (France), not ∼ Mont. 1841; fide Quél. 1888: 374. — *Merisma* Gillet 1877; *Trametes* Cooke & Q. 1878, not ∼ (Mont.) Bres. 1912; *Daedalea* Pat. 1900.

Sistotrema notarisii Terracc. 1872: 234 (n.v.) [repr. Sacc. 1916: 1109] (Italy); cf. Sacc., l.c., "È forse affine a *Polyp. biennis?*"

? *Sistotrema strobi* Terracc. 1872: 234 (n.v.) [repr. Sacc. 1916: 1109] (Italy).

? *Trametes terrei* B. & Br. 1876 (AM IV 17): 136 [repr. 1876 (H. 15): 58] (England); fide Bres. 1916 (Am 14): 240, "Specimen abortivum, probaliter *Pol. biennis* Bull." — *Antrodia* P. Karst. 1879. — V.s.: "terryi", "terreyi".

Polyporus sericellus Sacc. 1876 (NGi 8): 163 (Italy), not ∼ Lév. 1846;

fide Seyn. 1888 P.: 53, 56. — *Daedalea* Pat. 1900; = *Polyporus saccardoi* Cooke & Q. 1878. — Sacc. 1877 F.d.: *f. 106*; Seyn. 1888 P.: 46 *pl. 5 f. 8.*, descr. of type; Sacc. 1916: 957.

Daedalea polymorpha S. Schulz. 1880 (Yugoslavia, Slavonia). — S. Schulz. 1880 (ÖbZ 30): 144 *plate.*

Daedalea incarnato-albida Chod. & Mart. 1889 (BGe 5): 221 [repr. 1891 (BbC 1): 100] (Switzerland); fide C. E. Mart. 1899 (BGe 7): 194.

Irpicium ulmicola Bref. 1912 (Germany); fide Donk 1960 (Pe 1): 231. — Bref. 1912 U. 15: 143 *pl. 5 fs. 15–21, pl. 6 fs. 1–4.*

Daedalea capitata (Quél.) Big. & Guill. 1913. — *Daedalea biennis* var. Quél. 1888: 374 (France). — Bull. 1789: *pl. 499 f. 1 at left* (*Boletus biennis*).

? *Daedalea aquosa* Velen. 1922: 689 [see Pilát 1948: 261 for Latin translation] (Czechoslovakia); cf. Pilát 1937 (ACE 3): 116.

Irpex hydniformis Velen. 1922: 741 *f. 133 at right* [see Pilát 1948: 271 for Latin translation] (Czechoslovakia); fide Pilát 1937 (ACE 3): 116, & Donk.

Fibrillaria subterranea Pers. 1822: 53 (Germany) (nom. anam.); fide Seyn. 1888 P.: 56 ("mycélium"). — Sensu Tul. 1851 F.h.: 2 *pl. 21 f. 12*, ascribed to *Daedalea quercina* as a state; Seyn. 1888 P.: 44 *pl. 5 f. 10.*

Ceriomyces terrestris S. Schulz. 1874 (VW 24): 451 (Yugoslavia, Slavonia) (nom. anam.); fide Bres. apud Wint. 1884 (H 23): 169 & Seyn. 1888 P.: 56 (conidial state). — Seyn. 1888 P.: 41 *pl. 5 fs. 4, 5, 7, 14–16, pl. 6 fs. 6, 14*

Ptychogaster alveolatus Boud. 1888 (BmF 4): lv *pl. 3* (France) (nom. anam.); fide Boud., op. cit., p. lvi (conidial state). — *Ceriomyces* Sacc. 1891. — Fid. 1969 (Ri 4): 145, in obs., *fs. 7, 17.*

Ptychogaster lindtneri Pilát 1937 (BmF 53): 86 (Yugoslavia) (nomen anam.); fide Pilát 1951 (Sbč 12): 63 (forma). — *Leptoporus* Pilát 1937. — Pilát 1937–8 (ACE 3): 166, 207 *f. 62, pl. 125* (*Leptoporus*).

M.—*Boletus albidus* Schaeff. sensu With. 1796: 321; fide Fid. 1969 (Ri 4): 116, 128. — Fid., l.c., incorrectly thought that With. referred to *Boletus albus* Schaeff.

M.—*Boletus acanthoides* Bull. sensu Fr. 1838: 448 (*Polyporus*); fide Fid. 1959 (BTC 60): 134, & Donk 1971 (PNA 74): 2.

M.—*Polyporus pes-caprae* Pers. sensu Payer 1850: *f. 488* ("*Systotrema*"); fide Fid. 1969 (Ri 4): 130.

M.—*Hydnum compactum* Pers. (**O**) senzu Inz. 1867 (GSP 3): 190 *pl. 10, pl. 11 f. 2*; 1869 F.s. 1: 52 *pl. 5, pl. 6 f. 2*; fide Bres. 1931 (BIm 20): *pl. 957* ("forma vegeta").

M.—*Hydnum crispum* Schaeff. (**O**) sensu Inz. 1867 (GSP 3): 192 *pl. 11 f. 1*; 1869 F.s. 1: 54 *pl. 6 f. 1*; fide Bres. 1931 (BIm 20): pl. 957 ("forma exsiccata rufescens").

M.—*Boletus distortus* Schw. sensu Pilát 1934 (BbC 52): 75 (*Phaeolus biennis* f.), as to European collections. — Cf. (**2**).

AGARICUM [Mich.] Maratti

1822: 455. — *Agaricum* Mich. 1729: 117 (pre-Linnaean name) ≡ *Agaricus* Tourn. 1694, 1700 (pre-Linnaean name) [1960 (Pe 1): 180] ≡ *Agaricon* [Tourn.] Adans. 1763 (d.n. [1960 (Pe 1: 180] ≡ *Agaricus* [Tourn.] Lam. 1783, Juss. 1789 (d.n.) [1960 (Pe 1): 181, in obs.]; ≡ *Agaricus* [Tourn.] Rafin. 1830 [1962 (Pe 2): 202], typonym, not *Agaricus* L. 1753 (d.n.) per Fr. 1821 (Agaricaceae), not ∼ Murrill 1905 (Polyporaceae). — Lectotype: "*Agaricus, sive Fungus Laricis* C.B. pin. 375". — Cf. Donk 1971 (PNA 74): 25.

Agarico-pulpa Paul. 1791 (d.n.) [1960 (Pe 1): 176, 178; 1962 (Pe 2): 202]. — Lectotype: *Agarico-pulpa officinalis* Paul.

Laricifomes Kotl. & P. 1957 [1960 (Pe 1): 232]. — Holotype: *Polyporus officinalis* (Vill.) per Fr.

SPECIAL LITERATURE.—Borzini, *1941*; fon Branke, *1896* (n.v.); Christison, *1874*; Faull, *1916, 1919*; Harz, *1868*; Henry, *1924*; Igmándy, *1956*; Kartavenko, *1958*; H. Lohwag, *1938*; K. Lohwag, *1948*; Martinez, *1943*; Muraškinskij, *1927*; Pilát, *1927*; Rubel, *1778*; Schmieder, *1886*; Schreiber, *1925*; Snell, *1921*; Stanley-Brown, *1891*; Teixeira, *1958b*; Thoms & Vogelsang, *1907*.

officinale (Vill. per Fr.) Donk 1971. — *Boletus* Vill. 1789 (d.n.); *Polyporus* (Vill.) per Fr. 1821; *Piptoporus* P. Karst. 1882; *Cladomeris* Quél. 1886; *Ungulina* Pat. 1900; *Fomes* A. Ames 1913; *Placodes* Rick. 1918; *Placoderma* Ulbr. 1928; *Fomitopsis* Bond. & S. 1941; *Laricifomes* Kotl. & P. 1957; [≡ *Agaricus* Mattioli 1554 (n.v.) (Italy); ≡ *Agaricus sive fungus laricis* C. Bauh. 1623]; ≡ *Boletus laricis* Rubel 1778 (d.n.) per Roques 1821, typonym, not ∼ Schleich. 1821 (n.v.p.); *Polyporus* Delle-Chiaje 1824 (n.v.), Duby 1830; *Fomes* Murrill 1903; *Cladomeris* Big. & Guill. 1909 (syn.: n.v.p.); ≡ *Agaricus laricis* Lam. 1783 (d.n.), typonym; ≡ *Boletus officinalis* Batsch 1783 (d.n.), typonym; ≡ *Boletus agaricum* All. 1785 (d.n.) per Pollini 1824, typonym; ≡ *Agarico-pulpa officinalis* Paul. 1793 (d.n.) (≡ *Boletus laricis* Rubel); ≡ *Boletus purgans* J. F. Gmel. 1792 (d.n.) (≡ *Boletus laricis* Rubel); *Agaricum* Paul. 1812–35 (generic name n.v.p.?); *Cladomeris* Big. & Guill. 1909 (syn.: n.v.p.). — Mattioli 1554: 485 (n.v.) (*Agaricus*); Mich. 1727: 117 *pl. 61 Ordo III f. 1* (*Agaricum, sive Fungus Laricis*); Rubel 1778 (MaJ 1): 172 *pls. 19, 20*, descr. by Wulf.; Bull. 1786; *pl. 296*; 1791 H.: 353 (*Boletus laricis*); Pers. 1801: 531 (*Boletus purgans*); Harz 1868 (BSM 41): 5 *pls. 1, 2* (*Polyporus officinalis*); Murrill 1903 (BTC 30): 230 (*Fomes laricis*); Faull *1916* (*Fomes o.*); Boyce 1923 (BUS 1163): 3 *pl. 6* (*Fomes laricis*); Bourd. & G. 1928: 607 (*Ungulina o.*); Bres. 1931 (BIm 20): *pl. 989*; Pilát 1940 (ACE 3): 355 *fs. 153, 158, pl. 237 f. b, pl. 238, pl. 239 f. b*; Overh. 1953: 48 *pl. 59 f. 355, pl. 60 f. 360, pl. 110 f. 605, pl. 126 fig.*; Kawam. 1954 I. 1: 168 *f. 145* (*Fomes o.*); Malenç. 1955 (BmF 71): 281 (*Ungulina o.*); Lowe 1957 F.: 83 *f. 65* (*Fomes o.*); S. Ito 1955: 301 *f. 220*; Kreisel 1961: 145 *f. 97* (*Fomitopsis o.*); A. Teix. *1958b* (*Laricifomes o.*).

Fomes albogriseus Peck 1903 (U.S.A., Michigan); fide Overh. 1953: 48 & Lowe 1957 F.: 83. — Overh. 1915 (WUS 3[1]): 57.

ALBATRELLUS S. F. Gray

1821 [1960 (Pe 1): 182]. — Lectotype: *Boletus albidus* Pers.

Polyporus P. Karst. 1881, in part, not ∼ [Mich.] Fr. 1821, not ∼ (Pers.) per S. F. Gray 1821 [1960 (Pe 1): 263, in obs.]. — [≡ *Polyporus* [Mich.] Fr. sensu P. Karst. 1881 (Rm 3 / No. 9): 17, excl. of type]. — Lectotype: *Polyporus ovinus* (Schaeff.) per Fr.

Caloporus Quél. 1886, not ∼ P. Karst. 1881 [1960 (Pe 1): 193]. — *Lectotype*: *Polyporus ovinus* (Schaeff.) per Fr.

Ceriomyces Pat. 1887: 137, in part. — [*Ceriomyces* Quél. sensu Pat. 1887, excl. of type]. — Lectotype: *Polyporus hirtus* Quél. — Cf. Donk 1960 (Pe 1): 197.

Scutiger Paul. per Murrill 1903 [1960 (Pe 1): 176, 179, 279]. — *Scutiger* Paul. 1808 (d.n.), 1812–35 (n.v.p.). — Lectotype: *Scutiger tuberosus* Paul.

Ovinus (Lloyd) Torrend 1920 [1960 (Pe 1): 249. — [≡ *Polyporus* stirps *Polypori ovini* Fr. 1851]. — *Polyporus* sect. *Ovini* Cooke 1885. — ≡ *Polyporus* sect. *Ovinus* Lloyd 1911. — Lectotype: *Polyporus ovinus* (Schaeff.) per Fr.

M.—*Polyporus* [Mich.] Fr. sensu P. Karst. 1881 → *Polyporus* P. Karst.

M.—*Ceriomyces* Quél. sensu Pat. 1887, in part → *Ceriomyces* Pat.

SPECIAL LITERATURE.—Lloyd, *1911*; Pilát, *1931*.

Albatrellus syringae: Malençon, *1966* (*Polyporus peckianus*); Niemelä, *1970*.

Albatrellus similis: Pouzar, *1966b*.

confluens (A. & S. per Fr.) Kotl. & P. 1957. — *Boletus* A. & S. 1805 (Germany) (d.n.), not ∼ Schum. 1803; *Polyporus* (A. & S.) per Fr. 1821; *Boletus* Lenz 1840; *Merisma* Gillet 1878; *Polypilus* P. Karst. 1881; *Cladomeris* Quél. 1886; *Caloporus* Quél. 1888; *Scutiger* Bond. & S. 1941. — Fr. 1862 S.S.: 18 *pl. 24*; Bres. 1899 F.m.: 104 *pl. 96*; Lloyd 1911 (LMW 3, O.): 81; Gramb. 1913 P.H. 2: 20 *pl. 20*; Bourd. & G. 1928: 521; Bres. 1931 (BIm 20): *pl. 970*; Shope 1931 (AMo 18): 351 *pl. 28 f. 1* (*Polyporus*); Pilát 1936 (ACE 3): 22 *f. 5, pls. 6, 7* (*Caloporus*); Overh. 1953: 216 *pl. 29 fs. 176, 177, pl. 129 fig.* (*Polyporus*); Pouz. 1966 (Fgp 1): 360 (*Albatrellus*). — Sensu Rostk. → *Ischnoderma benzoinum*.

Boletus aurantius Schaeff. 1774: 79 [*pls. 109, 110*] (Germany) (d.n.), not ∼ Pers. 1801 (d.n.; error for *B. aurantiacus* Bull.?), not ∼ Fr. 1838 ("Sow."; syn.: n.v.p.); fide Fr. 1874: 539, & Donk. — *Polyporus* (Schaeff.) per Trog 1832; Secr. 1833 (as a sp. of *Boletus*: n.v.p.), in part: as to var. A. — Trog 1832 (Fl 15): 554 (*Polyporus*).

Polyporus *pachypus* Pers. 1825: 47 (Switzerland, not ∼ Mont. 1842; fide Fr. 1874: 525 = *Polyporus politus*; fide Lloyd 1910 (LMW 3): 466 (forma).

Boletus artemidorus Lenz 1831 (Germany); fide Fr. 1838: 447, & Lenz 1840: 96. — *Polyporus* Fr. 1838 (syn.: n.v.p.), D. Dietr. 1847. — Lenz 1831: 80 *pl. 10 f. 43*; Harzer 1842–5: 29 *pl. 13*.

Polyporus politus Fr. 1836 A.S.: 59; fide Bres. 1916 (Am 14): 226 (forma). — *Caloporus* Quél. 1886. — Fr. 1838: 429; 1882 & 1884 I. 79 *pl. 179 f. 2*; Lloyd 1911 (LMW 3, O.): 79.

Scutiger laeticolor Murrill 1903 (BTC 30): 428 (U.S.A., Alabama); fide Overh. 1953: 216, 217. — *Polyporus* Sacc. & D. Sacc. 1905, not ∼ Berk. 1843.

Scutiger whiteae Murrill 1903 (BTC 30): 432 (U.S.A., Maine); fide Murrill 1920 (M 12): 10, & Overh. 1953: 216. — *Polyporus* Sacc. & D. Sacc. 1905.

? *Polyporus preslianus* Velen. 1922: 668 *f. 103: 9* [see Pilát 1948: 254 for Latin translation] (Czechoslovakia).

M.—*Boletus acanthoides* Bull. sensu Velen. 1922: 662; fide Kotl. & P. 1966 (ČM 20): 99, 102.

cristatus (Schaeff. per Fr.) Kotl. & P. 1957. — *Boletus* Schaeff. 1774 (Germany) (d.n.), not ∼ Gouan 1765 (d.n.), not ∼ Baumg. 1790 (d.n.), J. F. Gmel. 1792 (d.n.); *Polyporus* (Schaeff.) per Fr. 1821, not ∼ Fr. 1838; *Grifola* S. F. Gray 1821; *Boletus* Pollini 1824; *Caloporus* Quél. 1888; *Scutiger* Bond. & S. 1941. — Schaeff. 1774: 93 [*pls. 316, 317*]; Pers. 1801: 522 (*Boletus*); Rostk. 1828 (StP 4): 35 *pl. 16*; Krombh. 1841 S. 7: 8 *pl. 48 fs. 15, 16* (*Polyporus*); Quél. 1888: 406 (*Caloporus*); Michael 1901 F.P. 2: *no. 33*; Lloyd 1911 (LMW 3, O.): 80 *f. 501*; Bourd. & G. 1928: 520; Bres. 1931 (BIm 20): *pl. 971*; Coker 1948 (JMS 64): 294 *pls. 47, 48, pl. 53 f. 113*; Overh. 1953: 221 *pl. 29 f. 178, pl. 31 f. 185, pl. 32 f. 194, pl. 129 fig.* (*Polyporus*); Pouz. 1966 (Fgp 1): 360 (*Albatrellus*); & cf. Donk 1969 (Pe 5): 246, notes. — Sensu Pers. → *Polyporus cristatus* Fr. 1838, see this sp., cited below.

Boletus flabelliformis Schaeff. 1774: 81 [*pl. 113*] (Germany) (d.n.), not ∼ Leyss. 1761 (d.n.) per Opiz 1823, not ∼ Scop. 1772 (d.n.); fide Pers. 1801:522 (*"β. B. floriformis"* [!]) (var.), & Fr. 1821: 356. — *Polyporus* Secr. 1833 (as a sp. of *Boletus*: n.v.p.), (Schaeff.) per Sacc. 1916, not ∼ Pers. 1825, not ∼ Kl. 1833; *Cladomeris* Big. & Guill. 1909 (syn.: n.v.p.). — Secr. 1833 M. 3: 54 (*Polyporus*).

Polyporus agilis Viv. 1834–8: 42 *pls. 37, 38* (Italy); fide Fr. 1874: 525 = *Polyporus virellus* (var.).

Polyporus virellus Fr. 1838: 429 (Germany, Saxonia or Switzerland). — *Cladomeris* Big. & Guill. 1909 (syn.: n.v.p.).

Polyporus cristatus Fr. 1838, not ∼ (Schaeff.) per Fr. 1821. — *Merisma* Gillet 1878; *Polypilus* P. Karst. 1882; *Cladomeris* Quél. 1886; [≡ *Boletus cristatus* Schaeff. sensu Pers. 1801 (Germany)]. — Pers. 1801: 522 (*Boletus*); Fr. 1838: 447; Konr. & M. 1929 I. 5: *pl. 423* (*Polyporus*); Pilát 1936 (ACE 3): 24 *f. 6* (*Caloporus*); & cf. Donk 1969 (Pe 5): 246, notes.

Polyporus flavovirens B. & Rav. apud B. & C. 1853 (U.S.A., South Carolina); fide Lloyd 1910 (LMW 3, O.): 81, 89, 1910 (LMW 3, L. 29): 7, & Overh. 1953: 221. — Berk. 1872 (G 1): 38; Overh. 1914 (AMo 1):

111. — [Rea apud] A. L. Sm. 1902 (TBS 1): 200 is a European record (wrong spores).

M.—*Boletus subsquamosus* L. sensu Secr. 1833 M. 3: 53 (*Polyporus*); fide Fr. 1838: 447 = *Polyporus cristatus* Fr.

? M.—*Boletus lobatus* J. F. Gmel. sensu Fr. 1838: 448 (*Polyporus*); fide Bres. 1897 (AAR III 3): 69 ("status vetustus, induratus"). — Cf. Donk 1969 (Pe 5): 251, notes.

M. ?—*Polyporus poripes* Fr. (**O**) sensu Ravenel 1855 F.c. 4: No. 4, no descr.; fide Lloyd 1911 (LMW 3, O.): 90. — Murrill 1904 (BTC 31): 335; 1907 (NAF 9): 68 (*Grifola*).

hirtus (Quél). Donk 1971. — *Polyporus* Quél. 1873 (France), not ∼ (P. Beauv.) per Fr. 1821; *Fomes* Cooke 1885; *Cerioporus* Quél. 1886; *Leucoporus* Pat. 1900; *Polyporellus* Pilát 1936; *Favolus* Imaz. 1943, not ∼ (P. Beauv. per Fr.) Fr. 1825; *Piptoporus* Bond. & Ljub. apud Ljub. 1962. — Quél. 1873 (MMb II 5): 356 *pl. 2 f. 7* (*Polyporus*); 1886: 408 (*Cerioporus*); Lloyd 1912 (LMW 3, S.P.): 130 *f. 426*; Shope 1931 (AMo 18): 357 *pl. 27 f. 3*; Povah 1935 (PMi 20): 145 *pl. 24*; Lowe 1942: 32; Overh. 1953: 270 *pl. 34 fs. 201–203, pl. 102 f. 574, pl. 130 fig.* (*Polyporus*); Ljub. 1962 (BMs 15): 123 *f. 10* (*Piptoporus*); Donk 1971 (PNA 74): 3.

Polyporus hispidellus Peck 1899 (BNS 25): 649 (U.S.A., New York); fide Lloyd 1912 (LMW 3, S.P.): 146. — *Scutiger* Murrill 1915.

ovinus (Schaeff. per Fr.) Kotl. & P. 1957. — *Boletus* Schaeff. 1774 (Germany) (d.n.); *Polyporus* Pers. 1818 (d.n.); *Polyporus* (Schaeff.) per Fr. 1821; *Boletus* Pollini 1824; *Caloporus* Quél. 1886; *Scutiger* Murrill 1920; ≡ *Boletus crispus* Batsch 1783 (d.n.), not ∼ Pers. 1799 (d.n.). — Schaeff. 1774: 83 [*pls. 121, 122*] (*Boletus*); Fr. 1860 S.S.: 10 *pl. 8*; Gillet 1874–90 P.: *pl. 565/452*; Lloyd 1911 (LMW 3, O.): 76 *f. 497*; Gramb. 1913 P.H. 2: 19 *pl. 19*; Bourd. & G. 1928: 519; Bres. 1931 (BIm 19): *pl. 948*; Shope 1931 (AMo 18): 350 *pl. 24 f. 2*; Konr. & M. 1935 I. 5: *pl. 421* (*Polyporus*); Pilát 1936 (ACE 3): 15 *f. 1, pls. 1, 2* (*Caloporus*); Nannf. & Du R. 1952: 251 *pl. 130*; Overh. 1953: 215 *pl. 28 f. 170, pl. 30 f. 179, pl. 131 fig.* (*Polyporus*); Poelt & Jahn 1963: *pl. 43* (*Scutiger*); Pouz. 1966 (Fgp 1): 359 (*Albatrellus*).

Boletus subsquamosus L. 1753: 1178 (Sweden) (d.n.), not ∼ Batsch 1783 (d.n.); fide Donk 1969 (Pe 5): 255, & cf. (**17–19**). — *Agaricus* Lam. 1783 (d.n.); *Polyporus* Fr. 1815 (d.n.), (L.) per Fr. 1821; *Boletus* Pollini 1824, not ∼ Batsch per Bergam. 1823; *Caloporus* Quél. 1886, misapplied; *Boletopsis* Kotl. & P. 1957, misapplied. — L. 1755: 453; Wulf. 1787 (CoJ 1): 342. — Sensu Batsch → *Polyporus squamosus*; sensu Chev. → *Boletopsis griseus*; sensu Wulf. → *Boletus carinthiacus*, see this sp., cited below.

Boletus fragilis Pers. 1796 O. 1: 84 (Germany) (d.n.); fide Pers. 1800: 47 (with doubt), & Fr. 1821: 346 (for *B. albidus*). — ≡ *Boletus albidus*

Pers. 1801 (d.n.), not ∼ Schaeff. 1774 (d.n.); *Albatrellus* (Pers.) per S. F. Gray 1821; *Boletus* Zant. 1822, not ∼ Roques 1832, not ∼ (Romagnoli) ex Maire & al. 1903.

Boletus carinthiacus Pers. 1801 (d.n.) (**19**). — *Polyporus* (Pers.) per Roques 1832; [≡ *Boletus subsquamosus* L. sensu Wulf. 1787 (Austria)]. — Wulf. 1787 (CoJ 1): 342 (*Boletus subsquamosus*). — = *Albatrellus similis* Pouz.?

? *Polyporus punctiporus* Britz. 1893 (BCb 54): 102 ("*punctisporus*") [pl. 634 *f. 140*], 1894 (BnS 31): 216, 217 ("*punctiformis*") (*Germany*). — = *Albatrellus confluens*?

Polyporus luteolus (G. Beck) Velen. 1922; cf. Pilát 1936 (ACE 3): 16. — *Polyporus subsquamosus* var. G. Beck 1886 (VW 36): 469 (Austria).

? *Polyporus limonius* Velen. 1922: 668 *f. 103: 2* [see Pilát 1948: 254 for Latin translation] (Czechoslovakia); fide Pilát 1936 (ACE 3): 15, 16 (with doubt).

? *Polyporus lutescens* Velen. 1922: 669 [see Pilát 1948: 254 for Latin translation] (Czechoslovakia), not ∼ (Pers. per Nocca & Balb.) Pers. 1825; fide Pilát 1936 (ACE 3): 15, 16 (with doubt).

pes-caprae (Pers. per Fr.) Pouz. 1966. — *Polyporus* Pers. 1818 (France) (d.n.) per Fr. 1821; *Boletus* Cordier 1826, Spreng. 1827; *Caloporus* Pilát 1931; *Scutiger* Bond. & S. 1941. — Pers. 1818: 241 *pl. 3*; Quél. 1872 (MMb II 5): 272/256 *pl. 17 f. 2*; Gillet 1874–90 P.: *pl. 566/450*; Lloyd 1910 (LMW 3): 467 *f. 332*; 1911 (LMW 3, O.): 83 *f. 504*; Bourd. & G. 1928: 519 (*Polyporus*); Pilát 1936 (ACE 3): 17 *f. 2* (*Caloporus*); Coker 1948 (JMS 64): 291 *pl. 43, pl. 44 fig., pl. 53 f. 16*; Overh. 1953: 229 *pl. 29 fs. 173–175, pl. 98 f. 562, pl. 131 fig.* (*Polyporus*); Poelt & Jahn 1964: *pl. 42* (*Scutiger*); Pouz. 1966 (Fgp 1): 360 (*Albatrellus*). — Sensu Payer → *Abortiporus biennis*.

Fungus tuber Paul. 1793 T. 2: 122 (descr.), Ind. [*pl. 31 fs. 1–3*] (France) (d.n.); fide Bres. 1899 F.m.: 104 (for *Polyporus asprellus*) = *Polyporus scobinaceus*. — *Boletus* Cordier 1826; ≡ *Scutiger tuberosus* Paul. 1812–35 (generic name n.v.p.); ≡ *Polyporus asprellus* Lév. 1855. — Paul. 1793 T. 2: 122 ("Savatelle-truffe"), Ind. (*Fungus tuber*), & 1812–35: *pl. 31 fs. 1–3* (*Scutiger tuberosus*).

Fungus sapatella Paul. 1793 T. 2: 123 (descr.), Ind. [*pl. 21 f. 4*] (France) (d.n.); fide Bres. 1920 (Am 18): 68 = *Polyporus scobinaceus*. — ≡ *Scutiger badius* Paul. 1812–35 (generic name n.v.p.); ≡ *Polyporus pauletii* Fr. 1838; *Merisma* Gillet 1878; *Cladomeris* Quél. 1886.

Boletus scobinaceus Cumino 1805: 220 *pl. 2 fig.* (Italy) (d.n.); fide Fr. 1874: 524; fide Bres. 1899 F.m.: 104 (*Polyporus pes-caprae* cited as syn.). — *Polyporus* (Cumino) per Pers. 1825; *Caloporus* Quél. 1886 ("Scop."); *Cerioporus* Quél. 1888. — Bres. 1899 F.m.: 104 *pl. 95*; 1931 (BIm 19): *pl. 950* (*Polyporus*).

Boletus laxiporus Chev. 1837: no. 35 *plate* (Germany).

Cerioporus inflexus (S. Schulz.) ex Quél. 1888: 408 (Yugoslavia, Slavonia; "Alpes"); fide Pilát 1936 (ACE 3): 17, & Donk. — [*Polyporus*] S. Schulz. (in manuscr.: n.v.p.).

? *Polyporus bulbipes* G. Beck 1889 (Austria), not ∼ Fr. 1846. — G. Beck 1889 (VW 39): 605 *pl. 15 f. 2.*

Polyporus retipes Underw. 1897 (BTC 24): 85 (U.S.A., Alabama); fide Lloyd 1911 (LMW 3, O.): 83, 91, & Overh. 1953: 229. — *Scutiger* Murrill 1903.

Scutiger oregonense Murrill 1912 (M 4): 93 (U.S.A., Oregon); fide Overh. 1941 (M 33): 95, 1953: 229. — *Polyporus* Murrill 1912 (nom. altern.), not ∼ (Murrill) Yas. 1917.

similis Pouz. 1966 (Czechoslovakia). — Pouz. 1966 (Fgp 1): 274 *pls. 5, 6*; Canf. & Gilb. 1971 (M 63): 969 *fs. 3, 5.*

syringae (Parm.) Pouz. 1966 (3). — *Scutiger* Parm. 1962 (Estonia). — Parm. 1962 (BMs 15): 132 *f. 7* (*Scutiger*); Niemelä 1970 (Abf 7): 53 *fs. 1, 2, 4* (*Albatrellus*).

M.—*Polyporus peckianus* Cooke (**O**) sensu Malenç. 1966; fide Donk & Niemelä in litt. ad H. Jahn. ——Malenç. 1966 (BOy 16–18): 35 *f. 1, plate.*

AMYLOCYSTIS Bond. & S. ex Sing.

1944, Bond. 1953 [1960 (Pe 1): 185]. — *Amylocystis* Bond. & S. 1941 (lacking Latin description: n.v.p.). — Holotype: *Polyporus lapponicus* Romell.

SPECIAL LITERATURE.—Domański, *1959b*; Kotlaba & Pouzar, *1963b*; Pilát, *1965a.*

lapponica (Romell) Sing. 1944. — *Polyporus* Romell 1911 (Sweden); *Fomes* Bres. 1926 (error of printing: n.v.p.); *Ungulina* Pilát 1934; *Leptoporus* Pilát 1937, Murašk. 1939; *Amylocystis* Bond. & S. 1941 (generic name n.v.p.). — Romell 1911 (ABS 11³): 17 *pl. 2 f. 24* (*Polyporus*); Pilát 1936–8 (ACE 3): 179 *f. 47, pls. 101–104*; Murašk. 1939: 81 *f. 5* (*Leptoporus*); Overh. 1953: 276 *pl. 23 fs. 139–141, pl. 130 fig.* (*Polyporus*); Domański 1959 (Mob 8): 171, 180 *fs. 1–5* (*Leptoporus*); Kotl. & P. 1963 (ČM 17): 179, 185 *fs. 4–8*; Pilát 1965 (ČM 19): 9 *pl. 56* (*Amylocystis*).

Polyporus ursinus Lloyd 1915 (Canada, Ontario), not ∼ (Link per Fr.) 1821; fide Bres. 1926: 80, & Overh. 1953: 276. — Lloyd 1915 (LMW 4, Ap.): 319 *fs. 659, 660*; Shope 1932 (AMo 18): 332 *pl. 19 fs. 5–8*; Lowe 1942: 75.

ANTRODIA P. Karst. (4, 5)

1879 [1960 (Pe 1): 186]. — Lectotype: *Trametes serpens* (Fr. per Fr.) Fr. — Sensu Murrill → *Datronia* Donk.

Coriolellus Murrill 1905 [1960 (Pe 1): 102]. — Holotype: *Trametes sepium* Berk. *Cartilosoma* Kotl. & P. 1958 [1960 (Pe 1): 194]. — Holotype: *Trametes subsinuosa* Bres.

SPECIAL LITERATURE.—General: Sarkar, *1959b*.
Antrodia heteromorpha: Ebert, *1967*; Příhoda, *1953b*.
Antrodia malicola: Domański, *1966c*.
Trametes sepium: Bakshi, Singh, & Gibson, *1958*.
Antrodia serialis: Nobles, *1943*.
Antrodia ramentacea: Domański, *1969a*; Kotlaba, *1955b*.

albida (Fr. per Fr.) Donk 1966 (**6, 7, 8**). — *Daedalea* Fr. 1815 (Sweden) (d.n.) per Fr. 1821, not ∼ Purt. 1821, not ∼ Schw. 1822; *Lenzites* Fr. 1838; *Cellularia* O.K. 1898; *Trametes* Bres. ex Killerm. 1928, not ∼ (Schaeff. per Trog) Fr. 1849, not ∼ Lév. ex Bres. 1907; *Coriolellus* Bond. 1953; ≡ *Agaricus albulus* E. Krause 1933. — Fr. 1828 E. 1: 70 (*Daedalea*); 1863 M. 2: 245 (*Lenzites*); Bres. 1908 (Am 6): 40 (*Daedalea*); Bourd. & G. 1928: 591 *f. 168* (*Trametes serpens* subsp.); Bres. 1932 (BIm 21): *pl. 1012*; Pilát 1939 (ACE 3): 300, in part, *f. 125, pl. 205 f. a* (*Trametes*); D. Reid 1958 (TBS 41): 437 *f. 21* (*Trametes serpens* subsp.); Domański 1965 (FpG 2): 176 *pl. 49 fs. 1, 2, pl. 50 f. 1* (*Coriolellus*). — Sensu Fr. 1882 → *Antrodia heteromorpha*.

heteromorpha (Fr. per Fr.) Donk 1966 (**6**). — *Daedalea* Fr. 1815 (Sweden) (d.n.) per Fr. 1821; *Lenzites* Fr. 1838; *Cellularia* O.K. 1898; *Trametes* Neuman 1914; *Polystictus* Lloyd 1916 (nom. altern.), not ∼ Torrend 1940; *Coriolellus* Bond. & S. 1941. — Fr. 1863 M. 2: 248; 1882 I. 2: 77 *pl. 177 f. 3* (*Lenzites*); Lloyd 1919 (LMW 5): 848 *fs. 1416–1419*; Shope 1931 (AMo 18): 366 *pl. 31 f. 2* (*Trametes*); Bourd. 1932 (BmF 48): 227 (*Lenzites*); Jørst. 1937 (KnS 1936¹⁰): 30 *fs. 7–9*; Imaz. 1939 (JJB 24): 305 *f. 3*; Pilát 1939 (ACE 3): 304 *f. 127, pls. 207, 208*, excl. var.; D. Baxt. 1942 (PMi 27): 152, 160 *f. 4, pls. 10, 11*; Overh. 1953: 141 *pl. 85 f. 488, pl. 88 fs. 501–503, pl. 111 f. 612, pl. 125 fig.* (*Trametes*); Sarkar 1959 (CJB 37): 1261 *fs. 44–57, pl. 1 f. 7*; Domański 1963 (Mob 15): 339 *fs. 3, 4*; 1965 (FpG 2): 183 *f. 63, pl. 53* (*Coriolellus*).

Coriolus hexagoniformis Murrill 1907 (NAF 9): 20 (U.S.A., Alabama); fide Lloyd 1919 (LMW 5): 848, & Overh. 1953: 141. — *Polystictus* Sacc. & Trott. 1912.

Trametes lacerata Lloyd 1916 (LMW 5): 604 *f. 854* (U.S.A., Montana); fide Lloyd 1919 (LMW 5): 850 ("with erroneous spore measurements"), & Overh. 1953: 141.

M.—*Daedalea albida* Fr. sensu Fr. 1882 (*Lenzites*); fide Lundell 1941 (LNF 21–22): 6 Nos. 1011a, b ("young specimens of typical *Tr*[*ametes*] *heteromorpha*"). — Fr. 1882 I. 2: 76 *pl. 177 f. 1* (*Lenzites*).

malicola (B. & C.) Donk 1966 (**4, 5**). — *Trametes* B. & C. 1856; *Coriolellus* Murrill 1920; *Daedalea* Aosh. 1967; [≡ *Polyporus populinus* (Schum.)

per Fr. sensu Schw. 1832 (U.S.A., Pennsylvania)]. — Overh. 1914 (AMo 1):
140; Bourd. & G. 1928: 595; Pilát 1939 (ACE 3): 295 *f. 120, pl. 201*;
D. Baxt. 1940 (PMi 25): 160; Lowe 1942: 89; Overh. 1953: 150 *pl. 55 f.
334, pl. 87 f. 498, pl. 125 fig.* (*Trametes*); Sarkar 1959 (CJB 37): 1268
fs. 103–116, pl. 1 fs. 11, 12; Domański apud Domański & al. 1963 (Mob 15):
45 *fs. 15–17*; Domański *1960c*, with cult. char. (*Coriolellus*). — V.s. (error):
'*marimalicola*'.

Polyporus mali Velen. 1922: 656 *f. 106* [see Pilát 1948: 251 for Latin
translation] (Czechoslovakia); fide Pilát 1939 (ACE 3): 295.

Trametes kuzyana Pilát 1953 (U.S.S.R., Ukraine); fide Domański 1966
(APo 35): 607. — *Trametes* Pilát 1939 (lacking Latin descr.: n.v.p.);
Funalia Bond. 1953 (invalid ref. & lacking Latin descr.: n.v.p.), Kotl.
& P. 1957. — Pilát 1939 (ACE 3): 285 *f. 144* (*Trametes*); Kotl. & P. 1957
(ČM 11): 218, 223 *fig.* (*Funalia*).

ramentacea (B. & Br.) Donk 1966 (**4, 5**). — *Polyporus* B. & Br. 1879
(Scotland); *Poria* Cooke 1886; *Daedalea* Aosh. 1967. — Reid & Austw.
1963 (GN 18): 310 (*Poria*); Domański *1969a*, with cult. char. (*Antrodia*).

Trametes subsinuosa Bres. 1903 (Poland); fide Reid & Austw. 1963
(GN 18): 310 (*Polyporus ramentaceus* cited as a syn.). — *Polyporus* Lind
1913; *Poria* Torrend 1913; *Coriolellus* Bond. & S. 1941; *Cartilosoma*
Kotl. & P. 1958. — Bourd. & G. 1928; 593; Bres. 1932 (BIm 21): *pl. 1021*;
Donk 1933: 191; Pilát 1939 (ACE 3): 307 *f. 129, pl. 206 f. b*; Kotl. 1955
(ČM 9): 83, 89 *3 figs.*; Kotl. & P. 1958 (ČM 12): 102 *3 figs.* (*Cartilosoma*).

Trametes **salicina** Bres. apud Egeland 1914 (Norway) (**6**), not Bres. 1920.
— Egeland 1914 (NMN 52): 166; cf. Donk 1966 (Pe 4): 340, note.

Trametes salicina Bres. 1920 (Italy), not ~ Egeland 1914; *Antrodia*
Parm. 1968 (incomplete ref.: n.v.p.). — Bres. 1920 (Am 18): 40 (*Trametes*);
Bourd. & G. 1928: 592 (*Trametes serpens* subsp.); Pilát 1939 (ACE 3):
302 *f. 126, pl. 206 f. a* (*Trametes albida* var.); M. P. Christ. 1960 (DbA 19):
370 *f. 371* (*Coriolellus*).

Trametes **sepium** Berk. 1847 (U.S.A., Ohio) (**6, 8**). — *Daedalea* Ravenel
1852; *Coriolellus* Murrill 1915; *Polyporus* G. Cunn. 1948; *Tyromyces*
G. Cunn. 1965; *Daedalea* Aosh. 1967 (preoccupied). — Overh. 1915
(WUS 3[1]): 67 *tpl. 7 f. 39*; Lloyd 1919 (LMW 5): 850 *f. 1420*; Lowe 1942:
86; Overh. 1953: 136 *pl. 89 fs. 506, 507, pl. 125 fig.*; Bakshi & al. 1958
(CJB 36): 603 *fs. 6–14, pl. 1* (*Trametes*); Sarkar 1959 (CJB 37): 1266
fs. 88–102, pl. 1 f. 10 (*Coriolellus*). — Cf. Bres. 1903 (Am 1): 81; Donk
1966 (Pe 4): 340, note.

? *Polyporus favescens* Schw. 1832: 158 (U.S.A., Pennsylvania); fide
Lloyd 1913 (LMW 4, L. 50): 9, & Overh. 1923 (M 15): 214. — *Poria*
Cooke 1886; *Trametes* Lloyd 1913 ("McGinty"; not accepted: n.v.p.).
— Overh. 1923 (M 15): 214 *fs. 10–11, pl. 22 fs. 5, 6* (*Poria*). — Correctly
identified by Overh.?

? *Polyporus rhododendri* Schw. 1832: 158 (U.S.A., Pennsylvania);

fide Overh. 1923 (M 15): 221. — *Poria* Cooke 1886. — Overh. 1923 (M 15): 221 *f. 18, pl. 22 f. 8* (*Poria*). — Correctly identified by Overh.?

serialis (Fr.) Donk 1966 (**4, 9**). — *Polyporus* Fr. 1821 (Sweden); *Boletus* Spreng. 1827; *Trametes* Fr. 1849 (n.v.p.), 1863 (**9**), 1874; *Fomitopsis* P. Karst. 1881 (generic name n.v.p.); *Polystictus* Cooke 1886; *Pycnoporus* P. Karst. 1888; *Coriolellus* Murrill 1907; *Coriolus* E. Komar. 1964 (n.v.p.); *Daedalea* Aosh. 1967. — Fr. 1884 I. 2: 90 *pl. 192 f. 2*; Bres. 1897 (AAR III 3): 92, spores; Lloyd 1923 (LMW 7): 1191 *pl. 235 f. 2393*; Bourd. & G. 1928: 596; Pilát 1939–40 (ACE 3): 314 *f. 133, pl. 211, pl. 212 f. a*; Robak 1942 (MVf 7[3]): 36, 149, cult. char.; D. Baxt. 1942 (PMi 27): 140 *f. 1, pl. 1 f. 1, pls. 3, 4, 9*; Overh. 1953: 138 *pl. 89 fs. 508–510, pl. 125 fig.* (*Trametes*); Sarkar 1959 (CJB 37): 1259 *fs. 28–43, pl. 1 f. 6, pl. 2 fs. 13, 14* (*Coriolellus*); Donk 1971 (PNA 74): 26, notes.

? *Poria turrita* Scop. 1772 (Hungary, now Czechoslovakia) (d.n.). — *Boletus* Humb. 1793 (d.n.); *Polyporus* (Scop.) per Steud. 1824. — Scop. 1772 P.s.: 106 *pl. 29 f. 1* (*Poria*). — Scopoli's figure agrees well with *Trametes* [*Antrodia*] *serialis f. corallopoda* Pilát 1927 (Sčz 2): 480 *pl. 10 f. 23*, 1940 (ACE 3): 316 *pl. 212 f. a.*

Poria rubella Pers. 1796 O. 1: 14 (Germany) (nom. conf.) (d.n.); cf. Donk 1971 (PNA 74): 26.

Poria echinata Hoffm. 1797–1811 V.s.: 12 ["21"] *pl. 8* (Germany) (d.n.). — *Polyporus* (Hoffm.) per Pers. 1825; *Polyporus vaporaria* var. Harz 1888 (BCb 36): 379.[4]

Polyporus callosus Fr. 1821 (**11**); fide P. Karst. 1882 (Mfe 9): 69, & Bres. 1897 (AAR III 3): 86, 92 (forma). — *Physisporus* P. Karst. 1881; *Poria* Cooke 1886; *Coriolellus* M. P. Christ. 1960. — Bres. 1897 (AAR III 3): 86 (*Poria*); M. P. Christ. 1960 (DbA 19): 369 *f. 369* (*Coriolellus*).

Polyporus scalaris Pers. 1825: 90 (Switzerland), not ∼ Berk. 1856; fide Fr. 1828 E. 1: 93 (forma), & Donk 1967 (Pe 5): 111.

Polyporus cruentus Pers. 1825: 92 *pl. 16 f. 4* (Switzerland) (nom. conf.); fide Donk 1967 (Pe 5): 88; 1971 (PNA 74): 26.

Trametes contigua Wettst. 1888 (VW 38): 180; fide Fr. 1828 E. 1: 93 (for *Boletus contiguus* var. *dimidiatus* A. & S.). — ≡ *Boletus contiguus* var. *dimidiatus* A. & S. 1805: 255 (Germany) (d.n.); *Microporus* (A. & S.) per O.K. 1898. — Cf. Donk 1971 (PNA 74): 26, note.

Polyporus favogineus (Hoffm. ex Harz) Wettst. 1888 (VW 38): 181. — *Poria favoginea* Hoffm. 1797–1811 V.s.: 12 *pl. 11 fs. 1, 2, pl. 12 f. 3* (Germany) (as a var. of *Poria echinata*); *Polyporus vaporarius* var. (Hoffm.) ex Harz 1888 (BCb 36): 379.[5]

? *Bjerkandera roseomaculata* P. Karst. 1891 (H 30): 247 (Finland)

---

[4] In this publication the name "8. *Polyporus vaporarius* (Pers.) Fr." was transferred by a printer's error from the top of page 379 to foreign context at the bottom of p. 374; cf. *in* 1889 (H 28): 85.

[5] See preceding foot-note.

(nom. conf.); fide Donk 1967 (Pe 5): 107; 1971 (PNA 74): 26. — *Polyporus* Sacc. 1895; *Poria* Lowe 1959 (syn.: n.v.p.). — Dr. J. L. Lowe (in litt., March 1968) informs me that the type has the aspect of *A. serialis* but the hyphal structure is distinctly different.

? *Polyporus pallidissimus* Velen. 1922: 639 [see Pilát 1948: 243 for Latin translation] (Czechoslovakia); fide Pilát 1942 (ACE 3): 611 (resupinate; with doubt).

Polyporus pseudoannosus Velen. 1922: 659 [see Pilát 1948: 253 for Latin translation] (Czechoslovakia); fide Pilát 1939 (ACE 3): 314.

Polyporus fechtneri Velen. 1922: 659 [see Pilát 1948: 253 for Latin translation] (Czechoslovakia); fide Pilát 1939 (ACE 3): 314. — *Trametes* Velen. 1922 (nom. altern.).

M.—*Poria incarnata* Pers. sensu Pers. 1825: 98 (*Polyporus*); cf. Donk 1971 (PNA 74): 26.

M.—*Poria vaporaria* Pers. sensu R. Hartig 1878 (*Polyporus*). — R. Hartig 1878: *45 pl. 8 fs.* (n.v.); Harz 1888 (BCb 36): 379[5] (*Polyporus*).

serpens (Fr. per Fr.) P. Karst. 1879 (**6, 7, 9**) . — *Polyporus* Fr. 1818 (Sweden) (d.n.); *Daedalea* (Fr.) per Fr. 1821; *Polyporus* Secr. 1833 (as a sp. of *Boletus*: n.v.p.), Lind 1913, not ∼ Pers. apud Gaud. 1827; *Trametes* Fr. 1874, not ∼ Fr. 1849; *Physisporus* P. Karst. 1881; *Agaricus* E. Krause 1932; *Coriolellus* Bond. 1953; ≡ *Daedalea proserpens* E. Krause 1928; *Agaricus* E. Krause 1932 (syn.: n.v.p.). — Fr. 1828 E. 1: 71 (*Daedalea*); 1884 I. 2: 90 *pl. 192 f. 3*; Egeland 1914 (NMN 52): 165; Bourd. & G. 1928: 591; Bres. 1932 (BIm 21): *pl. 1022 f. 2* (*Trametes*); Pilát 1939 (ACE 3): 301 *f. 124, pl. 205 f. b* (*Trametes albida* var.). — Sensu Murrill = *Poria subserpens* (Murrill) ex Weir, extra-European.

Polyporus stephensii B. & Br. 1848 (AM II 2): 264, Berk. 1860: 252 ("Fr.") (England); fide Fr. 1874: 586. — *Trametes* Cooke & Q. 1878.

sinuosa (Fr.) P. Karst. 1881 (**4, 10, 11**). — *Polyporus* Fr. 1821 (Sweden); *Physisporus* Gillet 1878, P. Karst. 1881; *Poria* Cooke 1886; *Trametes* Cooke & Q. 1878; *Coriolus* Bond. & S. 1941; *Coriolellus* Sarkar 1959. — Fr. 1884 I. 2: 88 *pl. 190 f. 1,* "exempl. original. sicca" (*Polyporus*); Bres. 1903 (Am 1): 78 (*Poria*); Romell 1926 (SbT 20): 17, notes (*Polyporus*); Bourd. & G. 1928: 672; Overh. 1942: 37; Malenç. 1956 (BmF 71): 310 (*Poria*); Sarkar 1959 (CJB 37): 1264 *fs. 72–87, pl. 1 f. 9* (*Coriolellus*); Lowe 1966: 103 *f. 87* (*Poria*), & cf. Donk 1966 (Pe 4): 340, note.

Polyporus incertus Pers. 1825, not ∼ Currey 1876; fide Bourd. & G. 1925 (BmF 41): 232 (subsp.) (for *Poria vaporaria* Bres.). — *Poria* Murrill 1920; [≡ *Polyporus vaporarius* (Pers.) per Fr. sensu Fr. 1821: 382 (Sweden)]; ≡ *Poria vaporaria* Bres. 1897, typonym, not ∼ Pers. 1784 & (Pers. per Fr.) Cooke 1886; *Coriolellus* Domański 1965. — Bres. 1903 (Am 1): 78 (*Poria*); Bourd. & G. 1928: 673 (*Poria sinuosa* subsp.); Malenç. 1956 (BmF 71): 310 (*Poria*); Domański 1965 (FpG 2): 187 *f. 60G, pl. 54, pl. 55 f. 1* (*Coriolellus*).

Poria sylvestris (Romell) ex D. Baxt. 1932 (PMi 15): 200. — *Poria* Romell 1926 (*"silvestris"*; nom. prov.: n.v.p.); [≡ *Polyporus vaporarius* (Pers.) per Fr. sensu Romell 1911 (ABS 11³): 25 (Sweden)].

M.—*Poria vaporaria* Pers. sensu Fr. 1818 O. 2: 260, 1821: 382 (*Polyporus*). — Romell 1911 (ABS 11³): 25 (*Polyporus*); D. Baxt. 1941 (PMi 26): 118 (*Poria*); M. P. Christ. 1960 (DbA 19): 363 *f. 362* (*Tyromyces*); Lomb. & Gilb. 1965 (M 57): 71 *f. 3A*, with cult. char. (*Poria*). → *Polyporus incertus* Pers., → *Poria vaporaria* Bres.; & cf. *Poria sylvestris* (Romell) ex D. Baxt.

Trametes **subalutacea** Bourd. & G. 1925 (France). — Bourd. & G. 1928: 590.

M.—*Polyporus ornatus* Peck sensu Pilát 1939 (ACE 3): 306 *f. 128, pl. 191 f. b* (*Trametes*); fide Pilát, l.c. (*Trametes subalutacea* cited as a syn.).

APOXONA Donk

1969 (Ta 18): 666. — Holotype: *Hexagonia nitida* Dur. & Mont.
M.—*Hexagonia* Pollini per Fr. sensu Bond. & S. 1941, in part: as to lectotype [1960 (Pe 1): 226, in obs.]. → *Hexagonia* Bond. 1953 (n.v.p.) (**O**).

SPECIAL LITERATURE.—des Abbayes, *1967*; Demoulin, *1967*; Donk, *1969c*.

nitida (Dur. & Mont.) Donk 1969. — *Hexagonia* Dur. & Mont. 1846–9 (Algeria); *Scenidium* O.K. 1898; *Trametes* Pilát 1939, not ∼ Pat. 1890. — Dur. & Mont. 1846–9: *pl. 33 f. 1*; apud Mont. 1856: 170; Lloyd 1910 (LMW 3, H.): 14 *f. 290*; Maire 1917 (BAN 8): 78 *f. 3*; Bourd. & G. 1928: 598 *f. 169*; Pilát 1937 (BmF 53): 88 *pl. 1* (*Hexagonia*); 1939 (ACE 3): 282 *f. 112, pl. 192* (*Trametes*); Farinha 1957 (PAb 6): 9 *pl. 1*; Demoulin 1967 (BmF 82): 517 *f. 1*, distribution; K. Fid. 1968 (MNY 17²): 96 (*Hexagonia*).

Trametes guyoniana Mont. 1856 (ASn IV 5): 332 (Algeria); fide Bres. 1920 (Am 18): 69 (f. trametoidea).
Hexagonia marcucciana Bagl. & De-Not. 1868 (Italy, Sardinia) [repr. 1868 (H 7): 121]; fide Lloyd 1910 (LMW 3, H.): 15, & Bres. 1920 (Am 18): 69. — *Scindalma* O.K. 1898; *Trametes* Big. & Guill. 1913. — Bagl. & De-Not. 1868 (ECi II): No. 90.

M.—*Hexagonia mori* Pollini sensu Marcucci ?1867 U.: No. 69, no descr.; fide Bagl. & De-Not. 1868 (ECi II): No. 90 = *Hexagona marcucciana*.

BJERKANDERA P. Karst. (12)

1879 [1960 (Pe 1): 190]. — Lectotype: *Polyporus adustus* (Willd.) per Fr.
Myriadoporus Peck 1884 (nom. monstr.) [1960 (Pe 1): 247]. — Monotype: *Myriadoporus adustus* Peck.

Special literature.—

Bjerkandera adusta: Brooks, *1925*; Kennedy & Larcade, *1971*; Prior, *1913*; Vandendries, *1936a*.

Bjerkandera fumosa: Bondarcev, *1924*.

adusta (Willd. per Fr.) P. Karst. 1879 (**13, 14**). — *Boletus* Willd. 1787 (Germany) (d.n.); *Polyporus* (Willd.) per Fr. 1821, Endl. 1830 (*"adnatus"*); *Boletus* Pollini 1824; *Leptoporus* Quél. 1886, Pat. apud Pat. & Har. 1903; *Polystictus* Gillot & Luc. 1890, not ∼ Lloyd 1917; *Gloeoporus* Pilát 1937; *Tyromyces* Maire apud Maire & Wern. 1938 ("Donk"), Pouz. 1966. — Schrad. 1794: 168; Pers. 1801: 529 (*Boletus*); Overh. 1915 (AMo 2): 693 *pl. 23 f. 8* (*Polyporus*); Bourd. & G. 1928: 551 (*Leptoporus*); Shope 1931 (AMo 18): 339 *pl. 21 fs. 1–4* (*Polyporus*); Konr. & M. 1932 I. 5: *pl. 432* (*Leptoporus*); Cartwr. 1932 (TBS 16): 305 *pl. 16 f. 2*, cult. char. (*Polyporus*); Donk 1933: 161 (*Bjerkandera*); Pilát 1937 (ACE 3): 157 *f. 41*, *pls. 77–79* (*Gloeoporus*); Overh. 1953: 364 *pl. 12 f. 72, pl. 13 fs. 73, 74, pl. 128 fig.*; Westh. 1971 (Bo. 10): 162 *fs. 1, 2*, with cult. char. (*Polyporus*).

Boletus fuscoporus Plan. 1788 I.F.: 26 (Germany) d.n.); fide Fr. 1821: 363, & Donk.

Boletus pelloporus Bull. 1790 (France) (d.n.); fide Fr. 1821: 363, & Bres. 1890 (BmF 6): xxxix. — *Boletus* Bull. per Mérat 1821; *Polyporus* Secr. 1833 (as a sp. of *Boletus*: n.v.p.), misapplied fide Fr. 1874: 550 "(n. 73"), E. Krause 1928; = *Boletus nigricans* Schum. 1803: 390 (d.n.), not ∼ (Fr.) Spreng. 1827. — Bull. 1790: *pl. 501 f. 2*; 1791 H.: 365. — Sensu Sow. = ?, perhaps *Abortiporus biennis*; fide Fr. 1818 O. 2: 256 ("Sowerb. t. 250") = *Polyporus pallescens* ("optime") (**16**).

Boletus candidus Roth 1797 C. 1: 244 (Germany) (d.n.), not ∼ Pers. 1801 (d.n.) per Steud. 1824; fide Bres. 1920 (Am 18): 60 (forma). — *Polyporus caesius* var. (Roth) per Pers. 1825; *Polyporus* Fr. 1838, not ∼ (Pers.) per Pers. 1825, not ∼ (Speg.) Lloyd (n.v.p.); *Meripilus* P. Karst. 1882; *Leptoporus* Quél. 1886.

Boletus crispus Pers. 1799 O. 2: 8 (Germany) (d.n.), not ∼ Batsch 1783 (d.n.); fide Quél. 1886: 177 (var.). — *Polyporus* Fr. 1815 (d.n.) & (Pers.) per Fr. 1821; *Boletus* Pollini 1824; *Leptoporus* Har. & Pat. 1914; *Gloeoporus* G. Cunn. 1965. — Secr. 1833 M. 3: 122; Lloyd 1915 (LMW 4, Ap.): 329 *f. 672*; Overh. 1915 (AMo 2): 694; Shope 1931 (AMo 18): 340 *pl. 23 f. 1* (*Polyporus*); Konr. & M. 1932 I. 5: *pl. 432 f. 1* (*Leptoporus adustus f.*).

Boletus carpineus Sow. 1799 (England) (d.n.); fide Pers. 1825: 64 (var.). — *Polyporus* Fr. 1818 (d.n.); *Boletus* Sow. per S. F. Gray 1821; *Polyporus* Secr. 1833 (as a sp. of *Boletus*: n.v.p.), Trog 1844, not ∼ S. Schulz. 1866 (n.v.p.), not ∼ Velen. 1922; *Polystictus* Konr. 1923. — Sow. 1799: *pl. 231* (*Boletus*); Secr. 1833 M. 3: 123 (*Polyporus*); Gillet 1890–6 S.P.: *pl. 554* (*Polyporus adustus* var.); Konr. 1923 (BmF 39): 40 (*Polystictus*); Konr. & M. 1932 I. 5: *pl. 432 f. 2* (*Leptoporus adustus* f.).

? *Agaricus nigrescens* Dubois 1803: 178 ("*nigescens*") (France) (d.n.), not ∼ Pers. (n.v.p.), not ∼ Lasch 1829. — *Agaricus* Dubois per Dubois 1833.

Boletus concentricus Schum. 1803: 387 (Denmark) (d.n.); fide Fr. 1821: 364.

? *Boletus plumbeus* Cumino 1805: 223 (Italy) (d.n.); fide Sacc. 1916: 985. — *Boletus* Cumino per Pollini 1824.

Poria argentea Ehrenb. 1818: 19, 31 (Germany) (d.n.); fide Pers. in herb., Fr. 1821: 364, & Donk 1967 (Pe 5): 79 (resupinate).

Boletus isabellinus Schw. 1822: 96 (U.S.A., North Carolina); fide Murrill 1906 (BTC 32): 634. — *Polyporus* Steud. 1824: Fr. 1828, not ∼ (Fr.) Romell 1917; *Bjerkandera* P. Karst. 1879.

Polyporus tristis Pers. 1825: 94 (Switzerland), not ∼ Lév. 1846; fide Lloyd 1915 (LMW 4, Ap.): 388. — *Polystictus* Cooke 1886; *Microporus* O.K. 1898.

? *Polyporus murinus* Rostk. 1838 (StP 4): 117 *pl. 57* (Germany), not ∼ Lév. 1844, not ∼ Kalchbr. apud Thüm. 1875; fide Killerm. 1927 (ZP 6): 137 (resupinate).

Polyporus subcinereus Berk. 1839 (AM 3): 391 ("North America"); fide Mont. apud Berk. 1841 (AM 7): 452.

Polyporus scanicus Fr. 1851, 1863 (Sweden); fide Lloyd 1915 (LMW 4, Ap.): 385. — *Bjerkandera* P. Karst. 1882. — Fr. 1863 M. 2: 269.

Polyporus halesiae B. & C. 1853 (AM II 12): 434, apud Berk. 1872 (U.S.A., Georgia); fide Murrill 1906 (BTC 32): 634 & Lloyd 1915 (LMW 4, Ap.): 379.

Polyporus lindheimeri B. & C. apud Berk. 1872 (G 1): 50 (U.S.A., Texas); fide Murrill 1907 (NAF 9): 40 & Lloyd 1915 (LMW 4, Ap.): 381.

Polyporus fumosogriseus Cooke & Ell. 1881 (G 9): 103 (U.S.A., New Jersey); fide Cooke 1886 (G 15): 55.

Trametes tristis Roum. 1883 F.g.: No. 2609 & 1883 (Rm 5): 225 (France); fide Lloyd 1915 (LMW 4, Ap.): 388.

Myriadoporus adustus Peck 1884 (BTC 11): 27 (U.S.A., Ohio) (nom. monstr.); fide Pat. 1889 (BmF 5): 84.

? *Poria separabilis* P. Karst. 1888 (Mfe 16): 21 (Finland); cf. Lowe 1956 (M 48): 121 (a resupinate condition?).

Polystictus gloeoporoides Speg. 1889 (BCó 11): 451 (Brazil, São Paulo); fide Bres. 1916 (Am 14); 224. — *Microporus* O.K. 1898.

Polyporus macrosporus Britz. 1894 (BnS 31): 174 [*pl. 642 f. 166*] (Germany), wrong spores.

Polyporus ochraceocinereus Britz. 1895 (BCb 62): 311 [*pl. 645 f. 175*], 1910 (BbC 26): 212 (Germany); fide Killerm. 1922 (Dba 15): 72. — Spores too small.

Polyporus burtii Peck 1897 (BTC 24): 146 (U.S.A., Vermont); fide Bres. 1920 (Am 18): 67. — Overh. 1915 (AMo 2): 695 *pl. 23 f. 4.*

Daedalea fennica (P. Karst.) P. Karst. 1906; fide Lowe 1956 (M 48): 105. — *Daedalea oudemansii* var. P. Karst. 1882 (Mfe 9): 69 (Finland).

Coriolus alabamensis Murrill 1907 (NAF 9): 19 (U.S.A., Alabama); fide Overh. 1953: 364, 366. — *Polystictus* Sacc. & Trott. 1912.

Poria luteogrisea Bond. 1912 (TlR 37): 37 (U.S.S.R., European Russia); fide Bond. 1953: 237 (resupinate).

Polyporus excavatus Velen. 1922: 641 [see Pilát 1948: 244 for Latin translation] (Czechoslovakia); fide Pilát 1937 (ACE 3): 159 (sterile form). — *Poria* Velen. 1922 (nom. altern.).

Polyporus cinerascens Velen. 1922: 642 [see Pilát 1948: 245 for Latin translation] (Czechoslovakia), not ∼ (Schw.) Steud. 1824, not ∼ Lév. 1844, not ∼ Bres. apud Strass. 1900; fide Pilát 1937 (ACE 3): 159 (resupinate form).

Polyporus emergens Velen. 1922: 657 [see Pilát 1948: 252 for Latin translation] (Czechoslovakia); fide Pilát 1937 (ACE 3): 158.

Polyporus aberrans Velen. 1925 (MP 2): 73, 74 (Czechoslovakia); fide Pilát 1937 (ACE 3): 158.

Polyporus atropileus Velen. 1925 (MP 2): 73, 74 (Czechoslovakia); fide Pilát 1937 (ACE 3): 160 (forma).

Polyporus tegumentosus Velen. 1925 (MP 2): 73, 74 *fig.* (Czechoslovakia); fide Pilát 1937 (ACE 3): 160 (forma).

Daedalea solubilis Velen. 1926 (MP 3): 102, 103 *f. 2* (Czechoslovakia); fide Pilát 1937 (ACE 3): 160 (resupinate form).

fumosa (Pers. per Fr.) P. Karst. 1879 (**13, 15**). — *Boletus* Pers. 1801 (Germany) (d.n.); *Polyporus* Fr. 1818 (d.n.) & (Pers.) per Fr. 1821; *Boletus* Spreng. 1827; *Leptoporus* Quél. 1886; *Polystictus* Gillot & Luc. 1890, not ∼ (Murrill) Murrill 1912; *Gloeoporus* Pilát 1937; *Tyromyces* Pouz. 1966. — Pers. 1801: 530 (*Boletus*); Rostk. 1837 (StP 4): 87 *pl. 42*; Fr. 1874: 549; Lloyd 1915 (LMW 4, Ap.): 308 *f. 647*; Overh. 1915 (AMo 2): 695 *pl. 23 f. 3*; Cartwr. 1932 (TBS 16): 304 *pl. 15 fs. 2, 4, pl. 16 f. 1*, cult. char. (*Polyporus*); Donk 1933: 163 (*Bjerkandera*); Pilát 1937 (ACE 3): 161 *f. 42, pls. 84, 85* (*Gloeoporus*); Overh. 1953: 366 *pl. 13 fs. 75–77, pl. 130 fig.* (*Polyporus*).

Boletus salicinus Bull. 1789 (France) (d.n.), not ∼ Pers. apud J. F. Gmel. 1792 (d.n.) (**15**); fide Fr. 1874: 549 = *Polyporus albus* [sensu Fr.]. — *Boletus* Bull. per Hook. 1821, not ∼ (Pers. per Fr.) Wahl. 1826; ? = *Boletus salignus* J. F. Gmel. 1792 (d.n.); = *Polyporus inodorus* Chev. 1826. — Bull. 1789: *pl. 433 f. 1*; 1791 H.: 340 (*Boletus*). — Sensu Bres. (syn.) = *Polyporus albus* sensu Bres. → *Tyromyces fissilis*; sensu Sow. 1799: *pl. 277*, cf. *Trametes suaveolens*.

Boletus imberbis Bull. 1789: *pl. 445 f. 1* & 1791 H.: 339 (France); fide Bres. 1892 F.t. 2: 29 = *Boletus fumosus* (cited as a syn.). — *Boletus* Bull. per Mérat 1821; *Daedalea* Chev. 1826; *Polyporus* Fr. 1838; *Merisma* Gillet 1877; *Cladomeris* Quél. 1886; *Leptoporus* Quél. 1888; *Agaricus* E. Krause 1932. — Sensu Quél. 1888: 388 (*Leptoporus*); Bres. 1890 (BmF 6): xxxix; 1892 F.t. 2: 29 *pl. 125*, pl. poor (*Polyporus*); Bourd. & G.

1928: 550 (*Leptoporus*). — In my opinion Bulliard's fungus is doubtfully *B. fumosa*; the same applies to Secr. 1833 M. 2: 485 (*Daedalea*).

? *Boletus caeruleus* Schum. 1803: 387 (Denmark) (d.n.); fide Fr. 1832[Ind.]: 57, & Bres. 1903 (Am 1): 74. — *Polyporus* (Schum.) per Pers. 1825; *Leptoporus* Quél. 1886; *Hemidiscia* Lázaro 1916 ("*caerulescens*"), misapplied. — Hornem. 1829 (Fd 11 / F. 33): 11 *pl. 1963 f. 2 below*, presumably Schum.'s original fig. (*Polyporus fumosus*). — Sensu Lázaro → *Tyromyces caesius.*

Daedalea saligna Fr. 1818 O. 2: 241 (Sweden) (d.n.); fide Bres. 1892 F.t. 2: 29 = *Polyporus imberbis* [sensu Bres.]. — *Daedalea* Fr. per Fr. 1821, Quél. 1892; *Polyporus* Fr. 1838; *Merisma* Gillet 1877; *Meripilus* P. Karst. 1882; *Cladomeris* Quél. 1886. — Fr. 1828 E. 1: 69 (*Daedalea*); cf. Donk 1933: 164, with notes (*Bjerkandera fumosa f.*).

? *Polyporus pallescens* Fr. 1818 O. 2: 256 (Sweden) (d.n.) (**16**). — *Polyporus* Fr. per Fr. 1821, not ~ Romell 1911; *Boletus* Wahl. 1826, not ~ Schrad. apud J. F. Gmel. 1792 (d.n.); *Tyromyces* P. Karst. 1881; *Bjerkandera* P. Karst. 1881; *Polystictus* Big. & Guill. 1913; *Polystictoides* Lázaro 1916; *Coriolus* Pilát 1932, misapplied. — Sensu P. Karst., in part → "*Trametes*" *semisupinus.*

Polyporus alligatus Fr. 1828 E. 1: 78 (Sweden); fide Bres. 1890 (Rm 12): 103 = *Polyporus imberbis* [sensu Bres.]. — *Polypilus* P. Karst. 1882; *Cladomeris* Big. & Guill. 1909 (syn.: n.v.p.); *Boletus* Pilát 1937 (syn.: n.v.p.). — Donk 1971 (PNA 74): 27, note.

? *Polyporus rosarum* Weinm. 1836: 319 (U.S.S.R., European Russia). — Referred by Fr. 1838: 449 to *Polyporus candidus* Roth.

Daedalea puberula B. & C. apud Berk. 1872 (G 1): 67 (U.S.A., Pennsylvania); fide Murrill 1920 (M 12): 8. — *Striglia* O.K. 1891; *Bjerkandera* Murrill 1907.

Polyporus holmiensis (Fr.) Cooke 1878 (**15**); fide Bres. 1892 F.t. 2: 29 = *Polyporus imberbis* [sensu Bres.]. — *Polyporus salignus* **P. holmiensis* Fr. 1874; *Bjerkandera* P. Karst. 1889; [≡ *Daedalea saligna* Fr. sensu Fr. 1851 (Sweden) (*Polyporus*)]. — Fr. 1884 I. 2: 80 *pl. 181 f. 2*; Lloyd 1915 (LMW 4, Ap.): 308 *f. 648.*

Polyporus fragrans Peck 1878 (RNS 30): 45 (U.S.A., New York); fide Overh. 1915 (AMo 2): 692. — Peck 1897 (RNS 49): 44 (*Polyporus*); Murrill 1906 (BTC 32): 636 (*Bjerkandera*); Overh. 1914 (AMo 1): 103 (*Polyporus*). — Recorded for Europe by Plowr. 1899 (TBS 1): 54 (*Polyporus*).

Polyporus hederae Ade 1911 (Mba 2): 371 (Germany). — Referred here on the basis of the descr. only.

Polyporus eminens Velen. 1922: 639 [see Pilát 1948: 243 for Latin translation] (Czechoslovakia); fide Pilát 1937 (ACE 3): 164 (forma). — Pilát 1937 (ACE 3): 164 (*Gloeoporus fumosus* f.).

Polyporus decurrens Velen. 1922: 657 [see Pilát 1948: 252 for Latin translation] (Czechoslovakia), not ~ Underw. 1897; fide Pilát 1937 (ACE 3): 162.

Polyporus robiniae Velen. 1922: 658 *f. 107* [see Pilát 1948: 253 for Latin translation] (Czechoslovakia); fide Pilát 1937 (ACE 3): 162.

Polyporus tyttlianus Velen. 1922: 686 [see Pilát 1948: 259 for Latin translation] (Czechoslovakia); fide Pilát 1937 (ACE 3): 162.

? M.—*Boletus albidus* Schaeff. sensu Wahl. 1812: 531 ("Sowerb. engl. fung. t. 226"); fide Fr. 1828 E. 1: 96 = *Polyporus pallescens.*

M.—*Boletus albus* Huds. sensu Fr. 1838: 456 (*Polyporus*); fide Donk 1933: 164 (forma). — Donk 1933: 162 (*Bjerkandera fumosa* f. *alba*).

M.—*Bjerkandera cinerata* P. Karst. sensu Bourd. & G. 1925 (BmF 41): 132, 1928: 551 (*Leptoporus imberbis* f. "*P*[*olyporus*] *cineratus* Karst.").

BOLETOPSIS Fay.

1889, not ∼ P. Henn. 1897 (Boletaceae) [1960 (Pe 1): 190]. — Monotype: *Polyporus leucomelas* (Pers. per Schw.) Pers. (as *Boletopsis "melaleuca"*).

SPECIAL LITERATURE.—Fayod, *1889.*

grisea (Peck) Bond. & S. 1941 (**17–19**). — *Polyporus* Peck 1873, 1874 (U.S.A., New York), not ∼ Bres. 1912, not ∼ (Wint.) Pilát (n.v.p.); *Scutiger* Murrill 1903. — Lloyd 1911 (LMW 3, O.): 78 *fs. 499, 500*; Lowe 1942: 35; Overh. 1953: 228 *pl. 168, pl. 130 fig.* (*Polyporus*); Malenç. 1956 (BmF 71): 272 (*Boletopsis*).

Polyporus involutus Britz. 1896 (Germany) (**17**). — Britz. 1896 (BCb 68): 140 [*pl. 648 f. 183*]; 1910 (BbC 26): 208.

Polyporus earlei Underw. 1897 (BTC 24): 84 (U.S.A., Alabama); fide Murrill 1903 (BTC 30): 431, & Overh. 1953: 228, 229; fide Bres. 1920 (Am 18): 67 = *Polyporus subsquamosus* [sensu Bres.].

Polyporus maximovici Velen. 1922 (Czechoslovakia); fide Pilát 1936 (ACE 3): 20, 21 = *Caloporus leucomelas* f. *subsquamosus* [sensu Pilát]. — Velen. 1922: 669 [see Pilát 1948: 254 for Latin translation]; ? Fremr 1929 (MP 6): 79 *fig.*

M.—*Boletus subsquamosus* L. sensu Chev. 1837: no. 38 *plate* (*Polyporus*); fide Donk, & Pegl. in litt.; & cf. (**18**). — Bres. 1931 (BIm 19): *pl. 947* (*Polyporus*).

leucomelaena (Pers. per Pers.) Fay. 1889 (**17**). — *Boletus* Pers. 1801 (Germany) (d.n.); *Polyporus* (Pers.) per Pers. 1825; *Caloporus* Pilát 1931. — Fr. 1838: 429; Gillet 1874–90 P.: *pl. 561/451*; Fr. 1882 I. 2: 78 *pl. 179 f. 1*; Britz. 1887 (BnS 29): 273 [*pl. 592 f. 6*] (*Polyporus*); Fay. 1889 (Mal 3): 72 *fs. 1–3* (*Boletopsis "melaleuca"*); Britz. 1894 (BnS 31): 174 [*pl. 641 f. 163*]; Michael 1905 F.P. 3: *no. 42*; Boud. 1904–11 I. 1: 77 *pl. 151*; Lloyd 1911 (LMW 3, O.): 77 *f. 498*; Vuyck 1921–4 (Fb 26): *pl. 2013*; Bourd. & G. 1928: 519; Bres. 1931 (BIm 19): *pl. 949* (*Polyporus*); Pilát 1931 (BbC 48): 423 *fs. 1, 3–6* (*Caloporus*); Donk 1933: 65 (*Boletopsis*); Konr. & M. 1935 I. 5: *pl. 422* (*Polyporus*).

Fungus porcinus Paul. 1793 (d.n.); fide Lév. 1855: 90. — [≡ *Polyporus esculentus, parvus* ... Fungo Corvo, o Carbonajo ... Mich. 1729: 131 *pl. 70 f. 2* (Italy)]; ≡ *Polyporus carbonarius* Paul. 1812–35 (n.v.p.?), not ∼ Fr. 1821; ≡ *Polyporus *compactus* Pers. 1825, typonym, not ∼ Overh. 1922; ≡ *Boletus fongo-corvo* E. Duchesne 1836, typonym. — Paul. 1793 T. 2: 361 ("Le Porcelet brun") [*pl. 164 fs. 3, 4 (Polyporus carbonarius)*] is a mere adaption from Micheli's account.

*Polyporus *scoparius* Pers. 1825. — [≡ *Polyporus esculentus, parvus, pileolo desuper obscuro* ... Scopetino ... Mich. 1729: 130 *pl. 70 f. 3* (Italy)]; ≡ *Boletus scopetino* E. Duchesne 1836, typonym.

? *Polyporus repandus* (Fr. ex Pers.) P. Karst. 1879 (**20**), not ∼ Pat. 1896. — [*Polyporus subsquamosus* var. "*β. P. repandus*" Fr. 1821: 346 (Sweden, by exclusion of Mich *pl. 70 f. 2*, cf. Fr. 1838: 429)]; *Polyporus subsquamosus* var. *repandus* Fr. ex Pers. 1825.

Polyporus formatus Britz. 1887 (BnS 29): 273 [*pl. 592 f. 5*] & 1910 (BbC 26): 208 (Germany) (**17**).

? *Polyporus conspicabilis* Britz. 1887 (BnS 29): 274 [*pl. 611 f. 69*] (Germany) (**17**); fide Ade 1923 (ZP 2): 41. — Britz., l.c., & 1910 (BbC 26): 208 [*pl. 621 f 106*].

Boletus violaceus C. E. Mart. 1894 (Switzerland); cf. C. E. Mart. apud Sacc. & Trott. 1912 (SF 21): 255 (var.). — *Polyporus* C. E. Mart. 1904. — C. E. Mart. 1894 (BGe 7): 192 (*Boletus*); [C. E. Mart. apud] Sacc. & Trott. 1912 (SF 21): 255 (*Polyporus*).

BONDARZEWIA Sing.

1940 [1960 (Pe 1): 191]. — Holotype: *Polyporus montanus* (Quél.) Cost. & Duf.

SPECIAL LITERATURE.—Buchs, *1930a*; Killermann, *1919*; Pilát, *1965b*; Kreisel, *1968*; Pegler & Young, *1972*; Singer, *1940*.

montana (Quél.) Sing. 1940. — *Cerioporus* Quél. 1888 (France); *Polyporus* Cost. & Duf. 1891; *Cladomeris* Big. & Guill. 1908; *Grifola* Pilát 1934. — Quél. 1888 (Crf 16²): 589 *pl. 21 f. 10*, 1888: 407 (*Cerioporus*); Bres. 1897 (AAR III 3): 69; Ferry 1897 (Rm 19): 144 *pl. 180 f. 27*; Lloyd 1910 (LMW 3, P.I.): 38 *f. 364*; Killerm. 1919 (H 61): 1 *pl. 1*; Konr. & M. 1929 I. 5: *pl. 424*; Buchs 1930 (ZP 9): 8; Sing. 1930 (BbC 46): 81; Bourd. 1932 (BmF 48): 224; Sing. 1936 (BbC 56): 173, spores; 1940 (RM 5): 3 *f. 1* (*Polyporus*); Pilát 1936 (ACE 3): 55 *fs. 14, 15 pls. 17–19 (Grifola)*; Overh. 1953: 237 *pl. 104 f. 581, pl. 108 f. 598, pl. 130 fig.* (*Polyporus*); Kotl. & P. 1957 (ČM 11): *figs.* on pp. 163, 165; Poelt & Jahn 1964: *pl. 44*; Tort. & Jel. 1969 (Abc 28): 380 *f. 2, pl. 1*; Pegl. & Young 1972 (TBS 58): 50 *fs. 1A, 2, pl. 6, pl. 7 fs. 9, 10, pl. 9 f. 23 (Bondarzewia)*.

Boletus mesentericus Schaeff. 1774: 91 [*pl. 267*] (Germany) (d.n.). — *Grifola* (Schaeff.) per Murrill 1920, misapplied. — Sensu **Murrill** → *Meripilus giganteus*.

M.—*Boletus acanthoides* Bull. sensu Quél. 1886: 168 (*Cladomeris*); fide Quél. 1888: 408, & Lloyd 1910 (LMW 3, P.I.): 38.

M.—*Boletus imbricatus* Bull. sensu Britz. 1887 (BnS 29): 276 [*pl. 597 f. 18*] (*Polyporus*); fide Killerm. 1919 (H 61): 3.

BUGLOSSOPORUS Kotl. & P.

1966 (ČM 20): 82, 88. — Holotype: *Polyporus quercinus* (Schrad.) per Fr.

SPECIAL LITERATURE.—Eliade, *1960*; Jahn, *1967b*; Kotlaba & Pouzar, *1966a*; Ritter, *1964*.

pulvinus (Pers. per Pers.) Donk 1970. — *Boletus* Pers. 1799 (Germany) (d.n.) per Pers. 1825. — Pers. 1799 O. 2: 7 (*Boletus*); Donk 1970 (PNA 74): 4, note.

Boletus quercinus Schrad. 1794 (Germany) (d.n.). — *Polyporus* (Schrad.) per Fr. 1838, not ∼ (Lázaro) Sacc. & Trott. apud Trott. 1925; *Piptoporus* P. Karst. 1881, Pilát 1937; *Placodes* Quél. 1886; *Fomes* Gillot & Luc. 1890; *Ungulina* Pat. 1900; *Placoderma* Ulbr. 1928. — Fr. 1838: 441; Hussey c. 1847 I. 1: *pl. 52*; Bres. 1897 (AAR III 3): 73, spores; Boud. 1904–11: 79 *pl. 154*; Lloyd 1915 (LMW 4, Ap.): 298, wrong spores (*Polyporus*); Bourd. & G. 1928: 606 (*Ungulina*); Pilát 1937 (ACE 3): 124 *f. 30, pls. 53, 54, 58* (*Piptoporus*); Kotl. & P. 1966 (ČM 20): 84, 89 *fs. 1, 2 pls. 9, 10, 61* (*Buglossoporus*); Tort. & Jel. 1969 (Abc 28): 382 *f. 3, pl. 2 f. a* (*Piptoporus*); Grög. 1970 (MMH 14): 94 (*Buglossoporus*). — Sensu Cartwr. → *Fomitopsis pinicola*.

Polyporus cadaverinus S. Schulz. apud Fr. 1874; fide Pilát 1937 (ACE 3): 124, & Tort. & Jel. 1969 (Abc 28): 382. — *Polystictus* J. Rick 1937. — Kalchbr. 1877: 55 *pl. 35 f. 1.* — Höhn. 1908 (SbW 117): 1027 suggested *Fistulina hepatica*.

Caloporus fuscopellis Quél. 1892 (Crf 20²): 469 *pl. 3 f. 33* [not 35]; fide Bres. apud Lloyd 1915 (LMW 4, Ap.): 379 = *Polyporus quercinus*. — *Polyporus* Sacc. 1896; *Polystictus* Big. & Guill. 1913; [*Polyporus helveolus* Rostk. (**O**) sensu Quél. 1890 (*Coriolus*) (France)].

Polyporus quercicola Velen. 1922: 646 [see Pilát 1948: 247 for Latin translation] (Czechoslovakia); fide Pilát 1937 (ACE 3): 124 = *Piptoporus quercinus*.

M.—*Polyporus suberosus* Fr. sensu Krombh. 1831; fide Fr. 1838: 441 = *Polyporus quercinus*. — Krombh. 1831 S. 1: 75 *pl. 5 fs. 3–5*; 1841 S. 7: 7 *pl. 48 fs. 11–15*.

M.—*Polyporus helveolus* Rostk. sensu Quél. 1890 (Crf 18²): 512 (*Coriolus*). → *Caloporus fuscopellis* Quél.

M.—*Daedalea rubescens* A. & S. sensu Velen. 1922: 658 (*Polyporus, Trametes, Daedalea*); fide Pilát 1937 (ACE 3): 124, 125 = *Piptoporus quercinus* (f. monstrosa).

CERIPORIA Donk (21)

1933 (*"Ceraporia"*) [1960 (Pe 1): 197]. — Holotype: *Poria viridans* (B. & Br.) Cooke.

Poria **blepharistoma** (B. & Br.) Cooke 1886. — *Polyporus* B. & Br. 1875 (Scotland). — Reid & Austw. 1963 (GN 18): 309 (*Poria*). — Referred by Lowe 1966: 30 to *Poria rhodella* [sensu Lowe (**22**)].

bresadolae (Bourd. & G.) Bond. & S. 1941 (**21**). — *Poria* Bourd. & G. 1925 (France); *Polyporus* Lowe apud Gilb. & Lowe 1962 (incomplete ref.: n.v.p.), not ∼ S. Schulz. 1885. — Bourd. & G. 1928: 662 (*Poria*).
M.—*Boletus sanguinolentus* A. & S. sensu Bres. 1903 (Am 1): 79, in part (*Poria*); fide Bourd. & G. 1928: 662.

camaresiana (Bourd. & G.) Bond. & S. 1941. — *Poria* Bourd. & G. 1925 (France). — Bourd. & G. 1928: 663; ? Lowe 1962 (PMi 47): 182 (*Poria*).

excelsa (Lundell) Parm. 1959. — *Poria* Lundell 1946; [≡ *Polyporus rhodellus* Fr. sensu Bres. 1897 (Hungary, now Czechoslovakia) (*Poria*)]. — Lundell 1946 (LNF 27–28): 14 No. 1328 (*Poria*); Domański 1965 (FpG 2): 66 *f. 17* (*Ceriporia*).
M.—*Polyporus rhodellus* Fr. sensu Bres. 1897 (**22**) (*Poria*). — Bres. 1897 (AAR III 3): 80 (*Poria*). → *Poria excelsa*.

mellita (Bourd.) Bond. & S. 1941. — *Poria* Bourd. 1916 (France). — Bourd. in Lloyd 1916 (LMW 4): 543 *f. 743*; Bourd. & G. 1928: 664 (*Poria*).

purpurea (Fr.) Donk 1971. — *Polyporus* Fr. 1821 (Sweden); *Poria* Cooke 1886, Quél. 1886; *Physisporus* Gillet 1878; *Meruliopora* Bond. & S. 1941 (generic name n.v.p.); *Meruliopsis* Bond. apud Parm. 1959, Bond. 1961; *Ceriporia* E. Komar. 1964 (incomplete ref.: n.v.p.). — Sensu Bres. 1897 (AAR III 3): 80; Egeland 1914 (NMN 52): 157 (*Poria*); Romell 1926 (SbT 20): 14, & cf. p. 17 in obs. (*Polyporus*); Bourd. & G. 1928: 660; D. Baxt. 1939 (PMi 24): 186; Pilát 1941 (ACE 3): 396 *f. 167, pl. 252 f. a*; Overh. 1942: 49 (*Poria*); M. P. Christ. 1960 (DbA 19): 336 *f. 331* (*Merulioporia*); Lowe 1966: 34 *f. 19* (*Poria*), Donk 1971 (PNA 74): 28, note.
? *Polyporus brunneus* Pers. 1825: 95 (Switzerland), not ∼ Schw. (n.v.p.), not ∼ Opiz (n.v.p.); fide Romell 1926 (SbT 20): 8 ("most probably the old stage").
M.?—*Polyporus rhodellus* Fr. sensu Fr. 1884 I. 2: 88 *pl. 189 f. 2*; fide Bres. 1897 (AAR III 3): 81, & Lundell 1946 (LNF 27–28): 14 No. 1328.

reticulata (Hoffm. per Fr.) Domański 1963. — *Mucilago* Hoffm. 1796 (Germany) (d.n.); *Boletus* Pers. 1801 (d.n.), not ∼ Schaeff. 1774 (d.n.); *Polyporus* Fr. 1815 (d.n.); *Polyporus* (Hoffm.) per Fr. 1821; *Boletus* Zant. 1822, not ∼ Hook. 1822, not ∼ Secr. 1833, not ∼ (Schaeff.) per Boud. 1877; *Polysticta* Pers. 1825 (syn.: n.v.p.); *Merulius* Kl. in herb., misapplied & Berk. 1836 (syn.: n.v.p.); *Physisporus* Gillet 1877, P. Karst.

1882; *Poria* Quél. 1886 ("Pers."), not ∼ (Fr.) Cooke 1886; *Fibuloporia* Bond. 1953. — Hoffm. 1796: *pl. 12 f. 2* with text [*fs.* copied by Nees 1816: *pl. 29a f. 225*] (*Boletus*); Fr. 1884 I. 2: 89 *pl. 190 f. 3* (*Polyporus*); sensu Bres. 1897 (AAR III 3); 89; 1903 (Am 1): 80 (*Poria*); Romell 1911 (ABS 11³): 21 *pl. 2 f. 10* (*Polyporus*); Bourd. & G. 1928: 665 *f. 183*; Overh. 1942: 35 (*Poria*); M. P. Christ. 1960 (DbA 19): 340 *f. 337* (*Fibuloporia*); Lomb. & Gilb. 1965 (M 57): 64 *fs. 2B, 6C*, with cult. char. (*Poria*); Domański 1965 (FpG 2): 68 *f. 18, pl. 13 f. 2* (*Ceriporia*); Lowe 1966: 33 (*Poria*); H. Jahn 1971 (WPb 8): 52 *f. 6* (*Ceriporia*); Donk 1971 (PNA 74): 28, notes. — Sensu Carm. → *Polyporus carmichaelianus* Grev. = *Rigidoporus sanguinolentus*; sensu Famintzin & Vor. 1872 (BZ 30): 616 = plasmodial stage of myxomycete.

? *Polyporus farinellus* Fr. 1821: 384 (Sweden); fide Romell 1911 (ABS 11³): 21 ("... the authentic specimen ... in the Kew herbarium belongs here"). — *Physisporus* Gillet 1878; *Poria* Cooke 1886, Quél. 1886. — Wak. & Pears. 1920 (TBS 6): 321 *fig.* (*Poria*).

Polyporus reticulatus Fr. 1874: 580 (Sweden), not ∼ (Hoffm.) per Fr. 1821. — *Poria* Cooke 1886, not ∼ (Hoffm. per Fr.) Quél. 1886 ("Pers."): [≡ *Polyporus reticulatus* (Hoffm.) per Fr. sensu Fr. 1821: 385, excl. of type (Sweden)]. — Donk 1971 (PNA 74): 28, note.

? *Poria chakasskensis* Pilát 1934 (U.S.S.R., Russia, Siberia); cf. Domański 1964 (APo 33): 170. — Pilát 1934 (BmF 49): 276 *pl. 14 fs. 3, 4*; Domański 1964 (APo 33): 170 (*Poria*). — Referred by Lowe 1962 (PMi 47): 185 to *Poria rhodella* [sensu Lowe (**22**)].

viridans (B. & Br.) Donk 1933 (**22**). — *Polyporus* B. & Br. 1861 (England); *Poria* Cooke 1886; *Physisporus* Cost. & Duf. 1895; *Leptoporus* Pat. 1900; *Tyromyces* Park.-Rh. 1956 (incomplete ref.: n.v.p.). — B. & Br. 1861 (AM III 7): 380; Bres. 1897 (AAR III 3): 83; Bourd. & G. 1928: 661 (*Poria*); Donk 1933: 171 (*Ceriporia*); D. Baxt. 1939 (PMi 24): 177 *pl. 2 f. 2, pl. 3* (*Poria*); M. P. Christ. 1960 (DbA 19): 346 *f. 342*; Domański 1965 (FpG 2): 63 *f. 15, pl. 13 f. 1* (*Ceriporia*).

? *Polyporus rhodellus* Fr. 1818 O. 2: 261 (Sweden) (d.n.) per Fr. 1821 (nomen ambiguum) (**22, 23**); *Physisporus* Gillet 1877; *Poria* Cooke 1886, Quél. 1886; *Ceriporia* Bond. & S. 1941; *Fibuloporia* M. P. Christ. 1960, misapplied. — Sensu Fr. 1884 → *Ceriporia purpurea*?; sensu Bres. → *Ceriporia excelsa*; sensu Lowe 1959 & afterwards [1966: 30 *f. 16*], ? Lomb. & Gilb. 1965 (M 57): 65 *fs. 2C, 6D* (*Poria*), in part, = *Ceriporia viridans*; with cult. char.; sensu M. P. Christ. = *Ceriporia* sp.

? *Polyporus griseo-albus* Peck 1885 (RNS 38): 91 [repr. Murrill 1920 (M 12): 302] (U.S.A., New York); fide Lowe 1966: 30 = *Poria rhodella* [sensu Lowe (**22**)]. — Overh. 1919 (BNS 205–206): 80 *pl. 5f. 8, pl. 6 fs. 1–5*; Lowe 1946: 44; R. L. Gilb. 1956 (Ll 19): 75 (*Poria*).

Physisporus inconstans P. Karst. 1887 (Rm 9): 10, 1887 (Mfe 14): 81 (Finland); fide Bres. 1897 (AAR III 3): 83. — *Poria* Sacc. 1888.

Poria aurantiocarnescens P. Henn. in Syd. 1898 M.m.: No. 4706 & 1898 (VBr 40): 125 (Germany); fide Bourd. & G. 1925 (BmF 41): 222 (forma).

Polyporus nuoljae Romell 1911 (Sweden); fide Bourd. & G. 1928: 661. — *Poria* Lloyd 1912 (nom. nud.: n.v.p.). — Romell 1911 (ABS 11³): 18 *pl. 2 f. 11.*

Poria turkestanica Pilát 1937 (U.S.S.R., Kazakstan); fide Domański 1964 (APo 33): 169. — *Amyloporia* Bond. 1953. — Pilát 1937 (BmF 52): 319 *f. 28, pl. 4 f. 4;* 1942 (ACE 3): 417 *f. 185, pl. 262 f. b;* Domański 1964 (APo 33): 169 *f. 3.*

Poria nathorst-windahlii Pilát apud Nath.-W. 1949 (BoN): 208 (Sweden); fide Lowe 1966: 31 = *Poria rhodella* [sensu Lowe (**22**), in part]. — "... sordide albida, sordide griseola vel subisabellina ...."

CERRENA S. F. Gray

1821 [1960 (Pe 1): 197]. — Monotype: *Sistotrema cinereum* Pers.

Sistotrema Pers. per Nocca & Balb. 1821, not ∼ Fr. 1821 Jan. 1 [1960 (Pe 1): 281]. — *Sistotrema* Pers. 1794 (d.n.). — Lectotype: *Sistotrema cinereum* Pers.

Phyllodontia P. Karst. 1883 [1960 (Pe 1): 255; 1962 (Pe 2): 208]. — Lectotype: *Phyllodontia magnusii* P. Karst. — Karsten introduced the genus with three species.

Bulliardia Lázaro 1916, not *Bulliarda* DC. 1801 (Crassulaceae), not *Bullardia* Jungh. 1830 (Gastromycetes) [1960 (Pe 1): 191]. — Lectotype: *Daedalea unicolor* (Bull.) per Fr.

SPECIAL LITERATURE.—Baccarini, *1911*; W. A. Campbell, *1939*; Tobler, *1954*; Van der Westhuizen, *1963*.

unicolor (Bull. per Fr.) Murrill 1905. — *Boletus* Bull. 1788 (France) (d.n.); *Daedalea* (Bull.) per Fr. 1821; *Boletus* St-Am. 1821, Laterr. 1821, not ∼ Schw. 1822, not ∼ Frost apud Peck 1889 (Boletales); *Sistotrema* Secr. 1833; *Striglia* O.K. 1891; *Trametes* Pilát 1939; *Coriolus* Pat. 1897; *Cerrena* Murrill 1903 (n.v.p.); *Bulliardia* Lázaro 1916; *Phyllodontia* Bond. & S. 1941; *Lenzites* G. Cunn. 1949. — Bull. 1790: *pl. 408, pl. 501 f. 3;* 1791 H.: 365 (*Boletus*); Fr. 1821: 336 (*Daedalea*); Secr. 1833 M. 2: 498 var. A (*Sistotrema*); Konr. 1923 (BmF 39): 42 (*Daedalea*); Bourd. & G. 1928: 563; Konr. & M. 1932 I. 5: *pl. 436;* Donk 1933: 184 (*Coriolus*); Pilát 1939 (ACE 3): 279 *f. 111* (*Trametes*); Overh. 1953: 125 *pl. 83 fs. 475–478, pl. 97 f. 554, pl. 125 fig.* (*Daedalea*); Westh. 1963 (CJB 41): 1487 *fs. 1–14 pl. 1* (*Cerrena*); H. Jahn 1963 (WPb 4): 68 *Abb. 50* (*Trametes*).

Sistotrema cinereum Pers. 1794 (NMB 1): 109 (Germany); fide Pers. in herb. (cf. Donk 1933: 185) & Fr. 1821: 337. — *Sistotrema* Pers. per S. F. Gray 1821.

Daedalea cinerea Pers. 1801 (d.n.); fide Fr. 1838: 494; fide Pers. 1825: 204 (sensu Fr.) = *Sistotrema cinereum* (var. "β?"). — *Daedalea* [Pers. per] Fr. 1821; *Sistotrema* Secr. 1833; *Lenzites* Quél. 1888; *Striglia* O.K. 1891; *Agaricus* E. Krause 1933; [≡ *Agaricus squamosus daedalaeis sinibus minoribus* Batt. 1755: 72 *pl. 38 f. G* (Italy)]. — Fr. 1815 O. 1: 105; 1884 I. 2: 91 *pl. 192 f. 2.*

? *Hericius strigiliformis* Dubois 1803: 183 (France) (d.n.); from descr. — *Hericius* Dubois per Dubois 1833.

Polyporus latissimus Fr. 1815 (Sweden) (d.n.); fide Bourd. & G. 1928: 564 (forma). — *Daedalea* Fr. per Fr. 1821; *Physisporus* P. Karst. 1882; *Trametes* Quél. 1888. — Fr. 1828 E. 1: 71 (*Daedalea*); Bourd. & G. 1928: 564 (*Coriolus unicolor* var. [unnamed]).

Polyporus argyraceus Pers. 1825: 73 (Switzerland); fide Donk 1933: 185. — Sensu Fr., Pilát → *Coriolus versicolor.*

Phyllodontia magnusii P. Karst. 1883 (H 22): 163 (Germany); fide P. Henn. 1899 (VBr 40): 140 (var.), & cf. Lowe 1956 (M 48): 109.

Daedaleopsis incana P. Karst. 1904 (ÖfF 46[11]): 4 (Finland); fide Lowe 1956 (M 48): 105. — *Daedalea* Sacc. & D. Sacc. 1905; *Antrodia* P. Karst. 1911.

? *Bulliardia nigrozonata* Lázaro 1916 (RMa 14): 843 / 1917: 155 (Spain); fide [Bres.? apud] Trott. 1925 (SF 23): 450 ("verisimiliter"). — *Daedalea* Sacc. & Trott. apud Trott. 1925.

Polyporus rohlenae Velen. 1922: 655 [see Pilát 1948: 251 for Latin translation] (Czechoslovakia); fide Pilát 1939 (ACE 3): 281 (forma).

M.—*Agaricus decipiens* Willd. sensu J. F. Gmel. 1792: 1437, & Schrad. 1794: 169 (*Boletus*), not *Hydnum decipiens* (Willd.) Schrad. 1794; fide J. F. Gmel., Schrad., ll.cc. (*Boletus unicolor* cited as syn.); fide Fr. 1821: 337 ("Schrad.").

CHAETOPORELLUS Bond. & S. ex Sing.

1944 [1960 (Pe 1): 198]. — Holotype: *Poria latitans* Bourd. & G. — Cf. Donk 1967 (Pe 5): 69.

latitans (Bourd. & G.) Sing. 1944. — *Poria* Bourd. & G. 1925 (France); *Chaetoporellus* Bond. & S. 1941 (generic name n.v.p.); *Chaetoporus* Parm. 1963. — Bourd. & G. 1928: 666; Gilb. & Lowe 1962 (PMi 47): 172 *pl. 2 fs. 5E, F*; Lowe 1966: 72 *f. 50* (*Poria*); Donk 1967 (Pe 5): 69, descr. generico-specifica (*Chaetoporellus*).

CLIMACOCYSTIS Kotl. & P.

1958 [1960 (Pe 1): 200]. — *Polyporus borealis* Fr.

Special literature.—O. Fidalgo, *1958b*; Robak, *1932*; Tikka, *1934*.

borealis (Fr.) Kotl. & P. 1958. — *Polyporus* Fr. 1821 (Sweden); *Boletus* Wahl. 1826; *Postia* P. Karst. 1879; *Bjerkandera* P. Karst. 1881; *Daedalea* Quél. 1886; *Polystictus* Gillot & Luc. 1890; *Spongipellis* Pat. 1900; *Leptoporus* Pilát 1937; *Heteroporus* Bond. & S. 1941; *Tyromyces* Imaz. 1943, Kotl. & P. 1956; *Abortiporus* Sing. 1944. — P. Karst. 1882 (BFi 37):

40 (*Bjerkandera*); R. Hartig 1878: 55 *pl. 10*; 1882: 84 *pl. 5*; J. Schroet. 1888: 471; Lloyd 1914 (LMW 4, Ap.): 326 *fs. 668–670*; Shope 1931 (AMo 18): 335 *pl. 20 (Polyporus)*; Pilát 1937–8 (ACE 3): 234 *f. 85, pls. 147–151, 153, 154 (Leptoporus)*; Jørst. & Juul 1939 (MnS 6³): 346, 476 *fs. 12–18*; Overh. 1953: 312 *pl. 22 fs. 132–135, pl. 97 f. 555, pl. 101 f. 572, pl. 108 f. 599, pl. 114 f. 626, pl. 128 fig. (Polyporus)*; Fid. 1958 (Myp 10): 2 *fs. 1–8 (Spongipellis)*; Siepm. & Zycha 1968 (NH 15): 566 *pl. 79 f. 4, pl. 81 f. 2*, cult. char. (*Polyporus*).

? *Boletus albus* Schaeff. 1774: 92 [*pl. 314*] (Germany) (d.n.), not ∼ Huds. 1762 (d.n.), &c.; cf. Fr. 1821: 366 ("forsan") & Romell 1926 (SbT 20): 6. — *Polyporus* (Schaeff.) per E. Krause 1928. — Rather a nomen dubium.

? *Trametes nivea* Otth 1871 (MiB 1870): 94 (Switzerland).

Bjerkandera irpicoides P. Karst. 1905 (Afe 27⁴): 5 (Finland); fide Lowe 1956 (M 48): 101. — *Polyporus* Sacc. & Trott. 1912.

Spongipellis ambiens P. Karst. 1906 (Ttk 8¹): 61 (U.S.S.R., 'Transbaical'); fide P. Karst. 1911 (Ttk 12¹,²): 110 ("verisimiliter" var.).

Polyporus piceus Velen. 1922: 645 [see Pilát 1948: 246 for Latin translation] (Czechoslovakia), not ∼ Ces. 1879; fide Pilát 1926 (MP 3): 78, 79 & 1926 (BmF 42): 100.

Polyporus pacificus C. H. Kauffm. 1930 (U.S.A., Washington); fide Overh. 1953: 312, 314. — *Grifola* C. H. Kauffm. 1930 (nom. prov. & altern.: n.v.p.). — C. H. Kauffm. 1930 (PMi 11): 178 *pls. 24–26*.

M.—*Boletus mollis* Pers. sensu A. & S. 1805: 247, in part: var. α; sensu Romell 1926 (SbT 20): 5, 8, 14 (as syn.).

M.—*Merulius gibbosus* Pers. sensu Wahl. 1820: 454 (*Daedalea*); fide Fr. 1828 E. 1: 85 (var.).

COLTRICIA S. F. Gray

1821 [1960 (Pe 1): 200]. — Lectotype: *Coltricia connata* S. F. Gray.

Strilia S. F. Gray 1821 [1960 (Pe 1): 285]. — Monotype: *Boletus cinnamomeus* Jacq.

Polystictus Fr. 1851 [1960 (Pe 1): 264; 1962 (Pe 2): 208]. — Lectotype: *Polyporus perennis* (L.) per Fr. — Sensu Rea 1922 → *Coriolus*; sensu Sing. 1944 → *Onnia*.

Pelloporus Quél. 1886, Torrend 1920 [1960 (Pe 1): 250]. — Lectotype: *Polyporus perennis* (L.) per Fr. — Sensu Sing., Bond. → *Ischnoderma*.

Xanthochrous Pat. 1896 (n.v.p.), 1897 [1960 (Pe 1): 293; 1962 (Pe 2): 209]. — Lectotype: *Polyporus perennis* (L.) per Fr.

Coltriciella Murrill 1904 [1960 (Pe 1): 201]. — Holotype: *Polyporus dependens* B. & C.

Cycloporus Murrill 1904 [1960 (Pe 1): 203]. — Holotype: *Cyclomyces greenei* Berk.

? *Volvopolyporus* Lloyd ex Sacc. & Trott. 1912 [1960 (Pe 1): 292]. — *Volvopolyporus* Lloyd 1909 (n.v.p.). — Monotype: *Polyporus peronatus* S. Schulz. — A very doubtful syn.

SPECIAL LITERATURE.—Coker, *1946*; Gilbertson, *1954*; Imazeki & Kobayasi, *1966*; Lloyd, *1908, 1909c*; Van Bambeke, *1909*.

cinnamomea (Jacq. per S. F. Gray) Murrill 1904. — *Boletus* Jacq. 1787 (Austria) (d.n.), not ~ Schum. 1803 (d.n.); *Strilia* (Jacq.) per S. F. Gray 1821; *Polyporus* Pers. 1825, Sacc. 1878, not ~ Trog 1832; *Polystictus* Sacc. 1888; *Pelloporus* Quél. 1888; *Microporus* O. K. 1898; *Xanthochrous* Pat. 1900. — Jacq. 1787 (CoJ 1): 116 *pl. 2 (Boletus)*; Sacc. 1878 (Mi 1): 362; Bres. 1887 F.t. 1: 88 *pl. 99 (Polyporus)*; Murrill 1904 (BTC 31): 343 *(Coltricia)*; Lloyd 1908 (LMW 3, P.I.): 6 *f. 200*; Bamb. 1909 (BBB 46): 25 *pl. 1 fs. 1–7*; Shope 1931 (AMo 18): 345 *pl. 24 f. 3 (Polyporus)*; Donk 1933: 239 *(Polystictus)*; Coker 1946 (JMS 62): 96 *pl. 18 fs. 1–3, pl. 22 fs. 1, 2 (Coltricia)*; Overh. 1953: 386 *pl. 41 fs. 244–246, pl. 128 fig. (Polyporus)*. — Sensu G. Cunn. 1948 = *Coltricia oblectans* (Berk.) G. Cunn. (extra-European), fide G. Cunn. 1965: 267.

Polyporus parvulus Kl. 1833 (North America), not ~ Schw. 1832, not ~ Lázaro 1916; not ~ (Lázaro) Trott. 1925 (n.v.p.); fide Murrill 1908 (NAF 9): 91 & Lloyd 1908 (LMW 3, P.I.): 7. — *Polystictus* Fr. 1851; *Microporus* O.K. 1898; *Xanthochrous* Pat. 1900; *Coltricia* Murrill 1904, misapplied. — Fr. 1838: 435; Berk. 1839 (AnH 3): 384 *(Polyporus)*. — Sensu Murrill 1904 → *Coltricia focicola*.

Polyporus splendens Peck 1873 (BBf 1): 61 & 1874 (RNS 26): 68 (U.S.A., New York), not ~ Lév. 1844; fide Bres. 1890 (BmF 6): xlii; Murrill 1904 (BTC 31): 343, 344, & Overh. 1953: 386, 387. — *Xanthochrous* Pat. 1900; ≡ Polyporus *subsericeus* Peck 1880; *Polystictus* Peck 1907. — Lowe 1934 (PMi 19): 143 *f. 5H*, spores.

Stereum unicum Lloyd 1913 (LMW 4, S.S.): 35 *f. 555* (U.S.A., New York) (nom. monstr.); fide Donk 1933: 239–240 & cf. D. Reid 1962 (Pe 2): 141.

? *Polyporus baudyšii* Kavina apud Velen. 1922: 681 *f. 114: 2* [see Pilát 1948: 258 for Latin translation] (Czechoslovakia) (26).

? *Polyporus casimiri* Velen. 1922: 681 *f. 114: 1* [see Pilát 1948: 258 for Latin translation] (Czechoslovakia) (26). — Pilát 1942 (ACE 3): 581 *pl. 371 f. b (Polystictus perennis* f.).

M.—*Polyporus montagnei* Fr. ex Mont. sensu Dur. & Mont. 1846–9: 15 *pl. 33 f. 2.*

focicola (B. & C.) Murrill 1908 (24). — *Polyporus* B. & C. 1868; *Xanthochrous* Pat. 1900; *Polystictus* Lloyd 1908, Sacc. & Trott. 1912; ≡ *Polyporus connatus* Schw. 1832 (U.S.A., North Carolina), not ~ Weinm. 1826, not *Coltricia connata* S. F. Gray 1821; *Xanthochrous* Pat. 1900. — Lloyd 1908 (LMW 3, P.I.): 8 *f. 203, f. 204 at right (Polystictus)*; Coker 1946 (JMS 62): 97 *pl. 17, pl. 22 fs. 3, 4 (Coltricia)*; Overh. 1953: 389 *pl. 41 fs. 250, 251, pl. 44 f. 265, pl. 129 fig. (Polyporus)*.

M.—*Polyporus parvulus* Kl. of "American mycology"; fide Lloyd 1908 (LMW 3, P.I.): 7, 10 & Murrill 1908 (NAF 9): 92. — Murrill 1904 (BTC 31): 345.

montagnei (Fr. ex Mont.) Murrill 1920 (25). — *Polyporus* Fr. ex Mont.

1836 (France); *Polystictus* Fr. 1851; *Pelloporus* Quél. 1886; *Microporus* O.K. 1898; *Xanthochrous* Pat. 1900; *Coltricia* Kotl. & P. 1964, preoccupied. — Quél. 1872 (MMb II 5): 269 *pl. 17 f. 4* [→ *Polyporus montagnei* Bres.] (*Polyporus*); Coker 1962 (JMS 62): 100 *pls. 19, 20, pl. 22 fs. 7–11* (*Coltricia*); Overh. 1953: 393 *pl. 45 fs. 270, 271, pl. 107 f. 594, pl. 130 fig.*; R. L. Gilb. 1954 (M 46): 229 *fs. 1–9* (*Polyporus*; & cf. Donk 1969 (Pe 5): 251, notes. — Sensu Dur. & Mont. → *Coltricia cinnamomea*.

Cyclomyces greenei Berk. 1845 (U.S.A., Massachusetts) (**25**); fide R. L. Gilb. 1954 (M 46): 232 & cf. *fs. 7–9* (var.). — *Xanthochrous* Pat. 1900; *Cycloporus* Murrill 1904; *Polystictus* A. Ames 1913; *Coltricia* Imaz. 1943. — Berk. 1845 (LJB 4): 306 *pl. 11*; Lloyd 1910 (LMW 3): 488 *fs. 380, 382*; 1917 (LMW 5): 633 *f. 902*; Farl. & Burt 1929: 103 *pl. 95*; Lowe 1942: 100, Overh. 1953: 116 *pl. 81 f. 465, pl. 125 fig.*; R. L. Gilb. 1954 (M 46): *fs. 7–9* (*Cyclomyces*); Imaz. & Hongo 1965 C.J. 2: 153 *pl. 51 f. 298* (*Coltricia*).

Polyporus saxatilis Britz. 1896 (Germany); fide Bres. 1916 (Am 14): 225, 240 = *Polyporus montagnei* Bres. (*P. "lignatilis"*). — ≡ *Polyporus lignatilis* Bres., l.c. (error, syn.: n.v.p.). — Britz. 1896 (BCb 68): 141 [*pl. 649 f. 184*]; 1910 (BbC 26): 209.

Polystictus obesus Ell. & Ev. 1897 (U.S.A., New Jersey); fide Bres. 1916 (Am 14): 240 & Coker 1946 (JMS 62): 101. — *Coltricia* Murrill 1904; *Polyporus* Overh. 1914, not ∼ (Pat.) Sacc. & Trott. 1912. — Murrill 1904 (BTC 31): 346 (*Coltricia*); Overh. 1914 (AMo 1): 121 (*Polyporus*).

Coltricia memmingeri Murrill 1904 (U.S.A., North Carolina); fide Coker 1946 (JMS 62): 100, 102. — Coker 1939 (JMS 55): 381 *pl. 40, pl. 44 fs. 9–11* [*Coltricia*; & cf. Coker 1946 (JMS 62): 103, in obs.].

Polystictus cuticularis Lloyd 1908 (LMW 3, P.I.): 12 *f. 205* (U.S.A., Massachusetts), not ∼ (Bull. per Fr.) Gillot & Luc. 1890; fide Coker 1946 (JMS 62): 100 & Overh. 1953: 393.

Polyporus montagnei Bres. 1916 (Am 14): 240, not ∼ Fr. ex Mont. 1836. — [≡ *Polyporus montagnei* Fr. ex Mont. sensu Quél. 1872 (MMb II 5): 269 *pl. 17 f. 4*, excl. of type, (France)]. — Bres. 1931 (BIm 20): *pl. 961 f. 1*.

Polyporus greenei Yas. 1919 June (BMT 33): 141, (140) *f. 3* (Japan); fide Imaz. apud Imaz. & Toki 1954 (BFJ No. 67): 25 = *Coltricia greenei* — ≡ *Polyporus greenei* Lloyd 1919 Apr. (LMW 5): 843 *f. 1410* (as a "form" of *Cyclomyces greenei*: n.v.p.).

perennis (L. per Fr.) Murrill 1903. — *Boletus* L. 1753 (Sweden) (d.n.), not ∼ Batsch 1783 (d.n.); *Polyporus* (L.) per Fr. 1821; *Boletus* Mérat 1821; *Trametes* Fr. 1848 (nom. nud.: n.v.p.), 1849; *Polystictus* P. Karst. 1879; *Pelloporus* Quél. 1886; *Ochroporus* J. Schroet. 1888; *Microporus* O.K. 1898; *Xanthochrous* Pat. 1900; *Placodes* L. Maire 1910 (error); *Fomes* Maubl. 1927; ≡ *Coltricia connata* S. F. Gray 1821, not *Polyporus connatus* Schw. 1832. — Sow. 1799: *pl. 192* (*Boletus*); J. Schroet. 1888: 488 (*Ochroporus*): Murrill 1904 (BTC 31): 344 (*Coltricia*); Lloyd 1908

(LMW 3, P.I.): 7 *f. 201* (*Polystictus*); Gramb. 1913 P.H. 2: 27 *pl. 27* (*Polyporus*); Bourd. & G. 1928: 630 (*Xanthochrous*); Donk 1933: 238 (*Polystictus*); Konr. & M. 1935 I. 5: *pl. 456* (*Fomes*); Coker 1946 (JMS 62): 99 *pl. 18 fs. 4–6, pl. 22 fs. 5, 6* (*Coltricia*); Overh. 1953: 387 *pl. 41 fs. 247, 248, pl. 44 f. 266, pl. 131 fig.* (*Polyporus*).

Boletus coriaceus Scop. 1772: 465 (Yugoslavia, Carniola) (d.n.), not ~ Huds. 1778 (d.n.), not ~ Batsch 1783 (d.n.), not ~ Batsch 1786 (d.n.); fide Fr. 1821: 350. — *Agaricus* Lam. 1783 (d.n.), not ~ Scop. per St-Am. 1821; *Boletus* Scop. per Bergam. 1823; *Polyporus* Endl. 1830, not ~ (Batsch) per D. Dietr. 1847. — Emend. Schaeff. 1774: 84 [*pl. 125*], restriction to var. 1; Bull. 1780: *pl. 28*; 1789: *pl. 449 f. 2*; 1791 H.: 334 (*Boletus*).

? *Boletus infundibuliformis* Batsch 1783: 97 (d.n.), not ~ Pers. 1794 (d.n.); fide Fr. 1874: 531 as to lectotype. — [≡ (by lecto-typification) *Polyporus lignosus, fulvus, infundibuli forma* . . . Mich. 1729: 130 *pl. 70 f. 8* ≡ *Fungus porosus, & lignosus, infundibuli forma* Breyne apud. Mich. (presumably northern Europe) (syn.)].

Boletus zonatus Batsch 1783: 105 (Germany) (d.n.), not ~ Nees 1816 fide Fr. 1832[Ind.]: 64. — Lectotype: Schaeff. 1763: *pl. 125* [in Schaeff. 1774 called *Boletus coriaceus* Scop.].

Boletus fimbriatus Bull. 1785 (France) (d.n.), not ~ (Pers.) Pers. 1801; fide Fr. 1821: 330. — *Boletus* Bull. per St-Am. 1821; *Polyporus* Secr. 1833 (as a sp. of *Boletus*: n.v.p.), Torrend 1902, not ~ (Pers.) per Fr. 1821, not ~ Fr. 1830; *Pelloporus* Quél. 1886; *Polystictus* Sacc. 1916, not ~ (Fr.) Cooke 1886. — Bull. 1785: *pl. 254*; 1791 H.: 332 (*Boletus*); Quél. 1888: 402 (*Pelloporus perennis* var.); Bourd. & G. 1928: 630 (*Xanthochrous perennis* var.).

Boletus cyathiformis Vill. 1789: 1040 ("*cyatiformis*") (France) (d.n.); fide Fr. 1821: 350.

Boletus leucoporus Holmskj. 1790 F.d. 1: 57 *pl. 30* (Denmark); fide Schum. 1803: 379 = *Boletus confluens* Schum.; fide Fr. 1821: 350 ("junior, poris albidis").

Boletus infundibulum Roth 1797 C. 1: 244 (Germany) (d.n.); fide Fr. 1821: 350.

Boletus confluens Schum. 1803: 378 (Denmark) (d.n.), not ~ A. & S. 1805 (d.n.); fide Fr. 1821: 350 ("junior, poris albidis").

Boletus pictus K. F. Schultz 1806: 485 (Germany) (d.n.), not ~ Ehrenb. 1818 (d.n.), not ~ Peck 1872 (Boletales); fide Fr. 1861 (ÖVS 18): 30 = *Polyporus fimbriatus* Bull. (cited as syn.). — *Polyporus* (K. F. Schultz) per Fr. 1838; *Polystictus* P. Karst. 1882; *Microporus* O.K. 1898; *Xanthochrous* Pat. 1900. — Sensu Fr. 1874: 531 (*Polyporus*).

Agaricus cyathiformis Paul. 1812–35: *pl. 4 f. 3* (d.n.?); fide Donk 1971 (PNA 74): 4. — [≡ *Fungus Campanulus lignosus* Sterb. 1675 & 1712: 258 *pl. 27 f. I* (Belgium)].

Boletus lejeunii L. March. 1826 (BnW 1): 413 (Luxemburg). — Unpublished plate (L).

Boletus perfossus L. March. 1826 (BnW 1): 414 (Luxemburg). — Unpublished plate (L).

? *Polyporus scutellatus* I. Boršč. 1856: 144 (U.S.S.R., Russia, Siberia), not ∼ Schw. 1832. — *Polystictus* P. Karst. 1882.

? *Polyporus peronatus* S. Schulz. apud Fr. 1874 (Hungary, now Romania); fide Pilát 1942 (ACE 3): 580. — *Polystictus* Cooke 1886; *Microporus* O.K. 1898; *Xanthochrous* Pat. 1900; *Volvopolyporus* Lloyd 1909 (not accepted: n.v.p.), Sacc. & Trott. 1912. — Kalchbr. 1877: 53 *pl. 32 f. 3* (*Polyporus*). — A very doubtful syn.

Polyporus simillimus Peck 1880 (RNS 32): 34 (U.S.A., New York); fide Lowe 1934 (PMi 19): 143 & Overh. 1953: 387, 389. — *Polystictus* Peck 1907. — Fide Coker 1946 (JMS 62): 97, 98 = *Coltricia focicola*.

Polystictus decurrens Lloyd 1908 (LMW 3, P.I.): 12 *f. 206* (U.S.A., Massachusetts); fide Overh. 1953: 387, 389.

Pelloporus parvulus Lázaro 1916 (RMa 15): 118 / 1917: 210; fide Bres. apud Trott. 1925 (SF 23): 371. — *Polyporus* Trott. 1925 (syn.: n.v.p.), not ∼ Schw. 1832, not ∼ Kl. 1833, not ∼ Lázaro 1916; ≡ *Polyporus lazaroanus* Trott. 1925.

M.—*Boletus subtomentosus* L. (**O**) sensu Bolt. 1788: 87 *pl. 87*; fide Fr. 1821: 350.

CORIOLUS Quél.

1886 [1960 (Pe 1): 201]. — Lectotype: *Polyporus versicolor* (L.) per Fr.

Cellularia Bull. per Corda 1842 [1960 (Pe 1): 194]; fide Donk 1971 (PNA 74): 5. — *Cellularia* Bull. 1788 (nom. monstros.) (d.n.). — Monotype: *Cellularia cyathiformis* Bull. — Sensu O.K. → *Lenzites*.

Hansenia P. Karst. 1879, not ∼ Turč. 1844 (Umbelliferae), not ∼ Zopf 1883 (Ascomycetes; generic rank dubious), not ∼ P. Lindner 1904 (Ascomycetes), not ∼ Zikes 1911 (Deuteromycetes) [1960 (Pe 1): 221]. — Lectotype. *Polystictus versicolor* (L. per Fr.) Fr.

Poronidulus Murrill 1904 [1960 (Pe 1): 273]. — Holotype: *Boletus conchifer* Schw. *Polystictus* Rea 1922: 608, not ∼ Fr. 1851, not ∼ Bond. 1953 (n.v.p.). — [≡ *Polystictus* Fr. sensu Rea, l.c., excl. of type]. — Lectotype: *Polystictus versicolor* (L. per Fr.) Fr.

M.—*Polystictus* Fr. sensu Rea → *Polystictus* Rea.

SPECIAL LITERATURE.—*Coriolus hirsutus*: Bose, *1934, 1943a*; Oehm, *1937c*; Pilát, *1930b*.

Coriolus versicolor: Bayliss, *1908*; Boulter & Burges, *1955*; W. G. Campbell, *1930*; Day & al., *1953*; Dion, *1952*; Fåhraeus, *1954*; Elliott, *1918*; Girbardt, *1955, 1956, 1957, 1958, 1960, 1961a, 1961b, 1962*; Hirt & Lowe, *1945b*; Jay, *1934*; La Fuze, *1937*; Lange, *1966*; Lindeberg & Fåhraeus, *1952*; Lyr, *1963*; Pelczar & al., *1950*; Pešek, *1969*; von Schrenck, *1914*; Shimano & al., *1953*; Smith, *1924*; Stevens, *1912*; Strunk, *1963, 1968*; Takemaru & Sano, *1970*; Ulbrich, *1939b*.

Coriolus zonatus: Jahn, *1961*; Lindeberg & Fåhraeus, *1952*.

hirsutus (Wulf. per Fr.) Quél. 1886. — *Boletus* Wulf. 1788 (CoJ 2): 149 (Austria) (d.n.), not ∼ Scop. 1772 (d.n.), not ∼ Batsch 1783 (d.n.), not ∼ Latourr. 1785 (d.n.); *Polyporus* (Wulf.) per Fr. 1821; *Boletus* Wahl. 1826, not ∼ Scop. per Pollini 1824; *Polystictus* Fr. 1851; *Hansenia* P. Karst. 1881; *Microporus* O.K. 1898; *Polystictoides* Lázaro 1916; *Trametes* Pilát 1939, not ∼ Lloyd 1924 (n.v.p.); ≡ *Boletus hirsutulus* J. F. Gmel. 1792 (d.n.); ≡ *Boletus wulfenii* Humb. 1793 (d.n.). — Schrad. 1794: 169 (*Boletus*); Fr. 1821: 367; Secr. M. 3: 127 (*Polyporus*); Konr. 1923 (BmF 39): 41 *pl. 3 fs. 5, 6* (*Polystictus*); Bourd. & G. 1928: 361; Konr. & M. 1932 I. 5: *pl. 435* (*Coriolus*); Pilát 1939 (ACE 3): 265 *f. 104, pl. 179 f. a, pl. 180 f. b* (*Trametes*); Overh. 1953; 345 *pl. 3 f. 14, pl. 5 f. 26, pl. 6 fs. 31, 32, pl. 8 f. 49, pl. 10 fs. 60, 61, pl. 130 fig.* (*Polyporus*); Grög. 1970 (MMH 14): 99, substrata (*Trametes*). — Cf. *Polyporus pininus* E. Krause (**O**).

Boletus velutinus Plan. 1788 I.F.: 26 (Germany) (d.n.), not ∼ Vahl 1794 (d.n.), not ∼ With. 1796 (d.n.); fide Fr. 1832[Ind.]: 64. — *Polyporus* (Plan.) per Fr. 1821 ("P[ers]."), misapplied, not ∼ Fr. 1832; *Boletus* Zant. 1822, not ∼ With. per Purt. 1821; *Hansenia* P. Karst. 1879 ["(Fr.)"], misapplied; *Bjerkandera* P. Karst. 1881 ["(Fr.)"], 1882 ["(Pers.)"], misapplied; *Polystictus* Cooke 1886, misapplied; *Coriolus* Quél. 1886, misapplied; *Microporus* O.K. 1898; *Fomes* Bres. 1926 (error: n.v.p.); *Trametes* G. Cunn. 1965, misapplied. — Sensu Pers. 1794 (NMB 1): 29; 1801: 539; Donk 1973 (PNA 76): 221, notes. — Sensu Sw., Fr. 1821 → *Coriolus pubescens*.

Boletus lutescens Pers. 1794 (ABU 11): 29 (Germany); fide Donk 1973 (PNA 76): 217. — *Polyporus* Fr. 1818 (nom. nud.); *Boletus* Pers. per Nocca & Balb. 1821; *Polyporus* Pers. 1825, not ∼ Velen. 1922; *Polystictus* Cooke 1886; *Trametes* Bres. 1896, misapplied, not ∼ Lázaro 1916; *Microporus* O.K. 1898. — Pers. 1801: 539 (*Polyporus*); Lloyd 1910 (LMW 3): 468 *f. 334* (*Polystictus*). — Sensu Sw., Fr. 1821, Bres. 1916 → *Coriolus pubescens*; sensu Bres. 1896, in part → *Funalia gallica & F. trogii*; sensu Michael → *Coriolus versicolor* (forma).

Boletus nigromarginatus Schw. 1822: 98 (U.S.A., North Carolina); fide B. & C. 1856 (JAP II 3): 209 & Overh. 1953: 345. — *Polyporus* Steud. 1824, Schw. 1832; *Polystictus* Cooke 1886; *Microporus* O.K. 1898; *Coriolus* Murrill 1906. — Murrill 1906 (BTC 32): 649 (*Coriolus*).

Polyporus cinereus Lév. 1846 (ASn III 5): 140 (Brazil), not ∼ Schw. 1832; fide Bres. 1916 (Am 14): 223. — ≡ *Polyporus cinerellus* Cooke 1878.

Coriolus sulcatus (P. Karst.) P. Karst. 1911; fide Bond. 1953: 489 & Lowe 1956 (M 48): 104 (forma). — *Coriolus velutinus* subsp. *C. sulcatus* P. Karst. 1906 (Ttk 8[1]): 61 (U.S.S.R., 'Transbaikal').

Polyporus reisneri Velen. 1922: 654 *f 103: 8* [see Pilát 1948: 250 for Latin translation] (Czechoslovakia); fide Pilát 1939 (ACE 3): 267 (forma).

Polyporus fagicola Velen. 1922: 654 [see Pilát 1948: 250 for Latin description] (Czechoslovakia), not ∼ Murrill 1906; fide Pilát 1939 (ACE 3): 266 (forma).

M.—*Boletus fibula* Sow. (**O**) sensu Fr. 1838 (*Polyporus*); fide Bourd. & G. 1928:562 (subsp.).—Fr. 1838:475; 1874:567 (*Polyporus*): Bres. 1897 (AAR III 3): 77 (*Polystictus*); Bourd. & G. 1928: 562 (*Coriolus hirsutus* subsp.).

pubescens (Schum. per Fr.) Quél. 1888. — *Boletus* Schum. 1803 (Denmark) (d.n.); *Polyporus* Fr. 1815 (d.n.); *Polyporus* (Schum.) per Fr. 1821; *Boletus* Spreng. 1827; *Bjerkandera* P. Karst. 1882; *Hansenia* P. Karst. 1888; *Polystictus* Gillot & Luc. 1890; *Leptoporus* Pat. 1900; *Coriolus* Murrill 1906, preoccupied; *Agaricus* E. Krause 1932, not ∼ Scop. 1780 (d.n.), not ∼ (Schw.) Fr. 1838, &c.; *Trametes* Pilát 1939; *Tyromyces* Imaz. 1943. — Fr. 1815 O. 1: 126; Hornem. 1823 (Fd 10 / F. 50): 12 *pl. 1790 f. 1*; Bres. 1897 (AAR III 3): 72 (*Polyporus*); Bourd. & G. 1928: 560 (*Coriolus*); Pilát 1939 (ACE 3): 268 *f. 106, pls. 183, 184* (*Trametes*); Overh. 1953: 346 *pl. 7 fs. 41–43, pl. 8 f. 44, 45, pl. 101 f. 571, pl. 131 fig.* (*Polyporus*); H. Jahn 1963 (WPb 4): 70 (*Trametes*); Westh. 1971 (Bo 10): 237 *fs. 27, 28*, with cult. char. (*Polyporus*); Donk 1973 (PNA 76): 219, notes.

Polyporus velutinus Fr. 1832[Ind.]: 149, not ∼ (Plan.) per Fr. 1821: 368. — Donk 1973 (PNA 76): 220, notes.

Polyporus sullivantii Mont. 1842 (ASn II 18): 243 [repr. Mont. 1856: 165] (U.S.A., Ohio); fide Murrill 1907 (NAF 9): 19. — *Polystictus* Cooke 1886; *Microporus* O.K. 1898; *Coriolus* Murrill 1906.

Hansenia imitata P. Karst. 1886 (Mfe 13): 161 (Finland); fide P. Karst. 1889 (BFi 47): 30.

Polyporus subluteus Ell. & Ev. 1897 (AN 31): 339 (Canada, Ontario); fide Overh. 1953: 346, 347. — *Coriolus* Murrill 1906.

Coriolus applanatus P. Karst. 1904 (ÖfF. 46[11]): 3 ("Karelia"); fide Lowe 1956 (M 48): 104. — *Polystictus* Sacc. & D. Sacc. 1905.

Coriolus lloydii Murrill 1907 (NAF 9): 23 (U.S.A., Kentucky); fide Overh. 1953: 348 = *Polyporus pubescens* var. *grayi*. — *Polyporus* Overh. 1914. — Overh. 1914 (AMo 1): 95 (*Polyporus*).

Coriolus concentricus Murrill 1907 (NAF 9): 23 (Canada, Ontario); fide Overh. 1953: 348 = *Polyporus pubescens* var. *grayii*. — *Polystictus* Sacc. & Trott. 1912.

Trametes merisma Peck 1910 (BNS 139): 31 (U.S.A., New York); fide Overh. 1953: 346, 347.

Polystictus grayii (Cooke & Ell. ex Macbr.) Lloyd 1915; fide Murrill 1920 (M 12): 7. — *Polyporus pubescens* var. Cooke & Ell. apud Ell. & Ev. 1888 N.A.F. II: No. 1933 (nom. nud.) ex Macbr. 1895 (U.S.A., Ohio); *Polystictus* Lloyd 1911 (nom. nud.: n.v.p.). — Lloyd 1915 (LMW 4, L. 60): 12, in obs. (*Polystictus*); Overh. 1953: 348 *pl. 8 fs. 46, 47* (*Polyporus pubescens* var.).

? *Trametes quercina* Lloyd 1922 (LMW 7): 1114 *pl. 195 f. 2088* (U.S.A., Michigan), not ∼ (L. per Fr.) Pilát 1939; fide Bres. 1926: 81 = *Polyporus velutinus*; fide Overh. 1953: 346, 347.

M.—*Boletus lutescens* Pers. sensu Sw. 1810: 89; fide Fr. 1821: 368 = *Polyporus velutinus* [sensu Fr. 1821] (forma). — Bres. 1916 (Am 14): 239, notes.

M.—*Boletus velutinus* Plan. sensu Sw. 1810 (SVH 31): 90. — Fr. 1821: 368 (*Polyporus*); Quél. 1888: 389 (*Coriolus*): Bres. 1932 (BIm 21): *pl. 1019* (*Polystictus*); Overh. 1953: 354 *pl. 6 fs. 29, 30, pl. 132 fig.* ? (*Polyporus*); Donk 1973 (PNA 76): 220, notes. — Sensu Overh. 1953, cf. *Coriolus zonatus.*

M.—*Boletus zonatus* Nees "of American authors"; fide Lowe & Gilb. 1962 (M 53): 489 = *Polyporus velutinus* "Fries". — Sensu Overh. 1915 → *Coriolus versicolor.*

versicolor (L. per Fr.) Quél. 1886. — *Boletus* L. 1753 (Sweden) (d.n.); *Poria* Scop. 1772 (d.n.); *Agaricus* Lam. 1783 (d.n.), Paul. 1812–35 (d.n.?), not ∼ Plan. 1788 (d.n.); *Agarico-suber* Paul. 1793 (d.n.); *Sistotrema* Tratt. 1804 (d.n.), misapplied; *Polyporus* (L.) per Fr. 1821; *Boletus* St-Am. 1821, Laterr. 1821; *Polystictus* Fr. 1851; *Hansenia* P. Karst. 1881; *Bjerkandera* P. Karst. 1881; *Microporus* O.K. 1898; *Agaricus* E. Krause 1932, not ∼ Plan. 1788 (d.n.); *Trametes* Pilát 1939, not ∼ Lloyd 1921 (n.v.p. ?); ≡ *Polyporus variegatus* Endl. 1830 (error ?), not ∼ (Sow.) per Loud. 1829. — L. 1755: 453; Bull. 1781: *pl. 86*; 1791 H.: 367; Sow. 1799: *pl. 229* (*Boletus*); Bourd. & G. 1928: 562 (*Coriolus*); Konr. & M. 1932 I. 5: *pl. 438* (*Coriolus*; f. *flavo-aureus*); Pilát 1939 (ACE 3): 261 *f. 101, pl. 177, pl. 178 f. a, pl. 180 f. a* (*Trametes*); R. W. Davids. & al. 1942 (TUS 785): 39 *f. 5J, pl. 3 f. B*, cult. char. (*Polyporus*); Overh. 1953: 342 *pl. 3 fs. 11–13, pl. 132 fig.* (*Polyporus*); H. Jahn 1961 (WPb 3): 11 *f. 1b*; 1963 (WPb 4): 72; Poelt & Jahn 1965: *pl. 36 fig.* (*Trametes*); Westh. 1971 (Bo 10): 219 *fs. 21, 22*, with cult. char. (*Polyporus*).

Boletus imbricatus Scop. 1772: 467 (Yugoslavia, Carniola) (d.n.), not ∼ Bull. 1787 (d.n.); fide Fr. 1821: 369 (as to var. 1).

Boletus atrofuscus Schaeff. 1774: 91 [*pl. 268*] (Germany) (d.n.); fide Fr. 1821: 369 & cf. Pers. 1800: 107 ("*atrorufus*"). — *Polyporus* Secr. 1833 (as a sp. of *Boletus*: n.v.p.), not ∼ Velen. (n.v.p.); ≡ *Boletus atrorufus* Pers. 1800 (d.n.) & Fr. 1821 (syn.: n.v.p.; error?); *Polyporus* (Pers.) ex E. Krause 1925 ("Sch. t. 268"); *Polystictus* E. Krause 1934. — Secr. 1833 M. 3: 151 (*Polyporus atrofuscus*).

Cellularia cyathiformis Bull. 1788: *pl. 414* (France) (nom. monstr.) (d.n.); fide Donk 1971 (PNA 74): 5. — *Cellularia* Bull. per Corda 1842. — Sensu O.K. → *Lenzites betulina.*

Boletus cyaneus Pers. 1794 (ABU 11): 30 (Germany) (d.n.).

? *Boletus plicatus* Schum. 1803: 389 (Denmark) (d.n.); fide Fr. 1821: 369. — Rather an indeterminable sp. of *Coriolus.*

Polyporus fuscatus Fr. 1818 (Sweden) (d.n.); fide Bres. 1932 (BIm 21): *pl. 1017* (var.). — *Polyporus versicolor* [subsp.?] *P. *fuscatus* Fr. 1838; *Polyporus* Cooke 1878, not ∼ Lloyd 1920; *Polystictus* Cooke 1886;

Microporus O.K. 1898. — Bres. 1932 (BIm 21): *pl. 1017 f. 1* (*Polystictus versicolor* var.).

Polyporus humboldtii Pers. 1825. — ≡ *Boletus versicolor* var. *stipitatus* Humb. 1793: 94–95 *pl. 2 f. 5* (Germany) (d.n.).

Polyporus hirsutulus Schw. 1832 (U.S.A., Pennsylvania); fide Murrill 1920 (M 12): 7 & Overh. 1953: 342, 344 (forma). — *Polystictus* Cooke 1886; *Microporus* O.K. 1898; *Coriolus* Murrill 1906. — Overh. 1914 (AMo 1): 92; 1915 (WUS 3¹): 30 *pl. 3 f. 12*; Lowe 1942: 60 (*Polyporus*).

Polyporus apophysatus Rostk. 1848 (StP Fs. 27–28): 7 *pl. 4* (Germany/ Poland); fide Bres. 1916 (Am 14): 223. — *Microporus* O.K. 1898.

Polystictus azureus Fr. 1851 (Mexico); fide Murrill 1906 (BTC 32): 643 & 1907 (NAF 9): 18. — *Polyporus* Cooke 1878; *Microporus* O.K. 1898; *Coriolus* G. Cunn. 1948; *Trametes* G. Cunn. 1965. — Fr. 1851 (NAu III 1): 93/77 (*Polystictus*); G. Cunn. 1965: 165 *pl. 2 f. c* (*Trametes*).

Polyporus nigricans Lasch in Rab. 1859 F.e.: No. 15 (Germany); fide Fr. 1874: 568 (var.). — *Polystictus* Sacc. 1895.

? *Polyporus cerebrinus* B. & Br. 1879 (AM V 3): 209 (Scotland); cf. Reid & Austw. 1963 (GN 18): 308. — Fide Bres. 1916 (Am 14): 223 = *Polyporus amorphus* (myriadoporous).

Polystictus aequus Lloyd 1920 (LMW 6): 933 *pl. 149 f. 1696* (Australia, Tasmania); fide G. Cunn. 1965: 166, 167.

? *Trametes versicolor* Lloyd 1921 (LMW 6): 1045 *pl. 177 f. 1927* (Chile); fide Lloyd, l.c. ("impresses us as being only a . . . form").

Polyporus aculeatus Velen. 1922: 646 [see Pilát 1948: 246 for Latin translation] (Czechoslovakia), not ∼ Lév. 1846, not ∼ Sacc. & Trav. 1911; fide Pilát 1939 (ACE 3): 261.

Polyporus vitellinus Velen. 1922: 652 [see Pilát 1948: 250 for Latin translation] (Czechoslovakia), not ∼ (Schw.) Fr. 1828; fide Pilát 1939 (ACE 3): 262 (forma).

Polyporus irpiciformis Velen. 1922: 655 [see Pilát 1948: 251 for Latin translation] (Czechoslovakia); fide Pilát 1939 (ACE 3): 261, 262 (forma).

Polyporus picicola Velen. 1922: 655 [see Pilát 1948: 251 for Latin translation] (Czechoslovakia); fide Pilát 1939 (ACE 3): 262 (forma).

Daedalea lobata Velen. 1922: 692 [see Pilát 1948: 262 for Latin translation] (Czechoslovakia); fide Pilát 1939 (ACE 3): 262 (forma).

Ceriomyces versicolor Seyn. 1890 (BbF 37): 112 (nom. anam.). — [≡ *Polyporus versicolor* (L.) per Fr. "les formes vernales" Pat. 1883 T.a. 1: 62 *f. 143* (France)].

? **M.**—*Hydnum tomentosum* L. (**O**) sensu Oed. 1770 (Fd 3 / F. 9): 7 *pl. 534 f. 3* (lacking descr.).) — Referred to *Polyporus adustus* by Fr. 1821: 406 ("potissimum"), to *Polyporus populinus* by Hornem. 1827: 23.

M.—*Boletus lutescens* Pers. sensu Michael 1905 F.P. 3: no. 41 & 1917 F.P., B. 3: no. 244 (*Polyporus*); fide Konr. & M. 1935 I. 5: *pl. 438* (forma).

? **M.**—*Boletus zonatus* Nees sensu Overh. 1915 (WUS 3¹): 31; fide Overh. 1953: 344 (forma).

M.—*Polyporus argyraceus* Pers. sensu Fr. 1828 E. 1: 94 (syn.); Pilát 1939 (ACE 3): 262 (*Trametes versicolor* f.).

zonatus (Nees per Fr.) Quél. 1886 (**27**). — *Boletus* Nees 1816 (Germany) (d.n.), not ∼ Batsch 1783 (d.n.); *Polyporus* (Nees) per Fr. 1821; *Polystictus* Fr. 1851; *Hansenia* P. Karst. 1879; *Bjerkandera* P. Karst. 1881; *Trametes* Pilát 1939, not ∼ Wettst. 1885. — Nees 1816: 221 *pl. 28 f. 221* & 1817: 57 (*Boletus*); Fr. 1821: 368 (*Polyporus*); Bres. 1897 (AAR III 3): 77 (*Polystictus*); Bourd. & G. 1928: 562; Donk 1933: 182 (*Coriolus*); Pilát 1939 (ACE 3): 263 *f. 102, pl. 181* (*Trametes*); Nannf. & Du R. 1952: 244 *pl. 126* (*Polyporus*); H. Jahn 1961 (WPb 3): 10 *fig., Bildbeilage,* 1963 (WPb 4): 73 *Abb. 28*; Balabán & Kotl. 1970: 68 t*plate* (*Trametes*). — Sensu Overh. 1915 → *Coriolus versicolor*; "of [other] American authors" → *Coriolus pubescens*.

Boletus multicolor Schaeff. 1774: 91 [*pl. 269*] (Germany) (d.n.); fide Fr. 1821: 368 (forma). — *Polyporus zonatus* var. (Schaeff.) per Opiz 1855; *Microporus* (Schaeff.) per O.K. 1898, typonym; *Polystictus* Sacc. 1916; *Polyporus* E. Krause 1925.

Boletus ochraceus Pers. 1794 (ABU 11): 29 (Germany) (d.n.) (**28**), not ∼ Cumino 1805 (d.n.); fide Fr. 1821: 368 = *Polyporus zonatus* (forma) & cf. Fr. 1828 E. 1: 94. — *Polyporus* Fr. 1815 (d.n.); *Polyporus* Pers. per Sommerf. 1826, Wettst. 1885; *Boletus* Schw. 1822 (n.v.p.), G. F. Re 1827, Spreng. 1827; *Polystictus* Killerm. 1928. — Pers. 1801: 539.

Boletus placenta Schum. 1803: 387 (Denmark) (d.n.); fide Fr. 1821: 368 (forma). — *Polyporus angulatus* var. (Schum.) per Pers. 1825; *Polyporus* Fr. 1832 (syn.: n.v.p.), Secr. 1833 (as a sp. of *Boletus*: n.v.p.), not Fr. 1861.

Boletus angulatus Schum. 1803: 388 (Denmark) (d.n.); fide Fr. 1821: 368 (forma). — *Polyporus* (Schum.) per Pers. 1825.

CRISTELLA Pat., in part

Cristella Pat. 1887 [1957 (Ta 6): 68; & cf. Donk 1952 (Re 1): 485]. — Lectotype: "*Crist. cristata*".

Trechispora P. Karst. 1890 [1960 (Pe 1): 288], not *Trachyspora* Fuck. 1861 (Uredinales). — Monotype: *Trechispora onusta* P. Karst.

Sulphurina Pilát 1942 (n.v.p.), 1953 [1960 (Pe 1): 286]. — Lectotype: *Sistotrema sulfureum* (Quél.) Bourd. & G.

Special literature.—Donk, *1968*; Liberta, *1966*; Weresub, *1967*.

Nota bene.—Most species of *Cristella* are not 'polypores'.

Daedalea **sulphurea** Quél. 1894 ("*sulfurea*"); *Sistotrema* Bourd. & G. 1914; *Sulphurina* Pilát 1942 (generic name n.v.p.), 1953 (indirect ref.: n.v.p.). — Quél. 1894 (Crf 22²): 487 *pl. 3 f. 10* (*Daedalea*); Bourd. & G. 1928: 438 (*Sistotrema*).

mollusca (Pers. per Fr.) Donk 1967. — *Boletus* Pers. 1801 (Germany) (d.n.); *Polyporus* Fr. 1815 (d.n.); *Polyporus* (Pers.) per Fr. 1821; *Boletus* Pollini 1824; *Physisporus* Chev. 1826; *Poria* Cooke 1886, Quél. 1886; *Leptoporus* Pat. 1900; *Fibuloporia* Bond. ex S. 1941 (generic name n.v.p.), Sing. 1944, Bond. 1953, misapplied. — Cf. Donk 1967 (Pe 5): 95–98. — Sensu Berk. 1860 = *Merulius taxicola* (**O**); sensu Bres. → *Poria mucida*.

Boletus subtilis Schrad. 1794 (Germany) (d.n.); fide Romell 1911 (ABS 10³): 14 = *Polyporus hymenocystis*. — *Polyporus* Fr. 1815 (d.n.); *Polyporus* (Schrad.) per Fr. 1821, misapplied; *Boletus* Zant. 1822; *Porotheleum* Fr. 1832^Ind.: 150, misapplied; *Poria* Bres. 1897. — Sensu Bres. 1897 (AAR III 3): 88; Bourd. & G. 1928: 656 (*Poria*). — Cf. Donk 1967 (Pe 5): 111.

Polyporus candidissimus Schw. 1832 (U.S.A., Pennsylvania), not ~ Velen. 1922; fide Donk 1967 (Pe 5): 95. — *Poria* Cooke 1886; *Trechispora* Bond. & S. 1941; *Cristella* Donk apud W. Cooke 1943; *Phlebiella* W. Cooke 1952, Bond. 1953 (generic name n.v.p.). — Overh. 1923 (M 15): 208 *fs. 3*: 1–5, *pl. 21 fs. 1–5*; Pilát 1942 (ACE 3): 389 *f. 161, pl. 249, pl. 250 f. a*; Lowe 1946: 71 *f. 17* (*Poria*); M. P. Christ. 1960 (DbA 19): 100 *f. 87*; Domański 1965 (FpG 2): 21 *pl. 1 f. 1* (*Cristella*); Lowe 1966: 57 *f. 33* (*Poria*).

Polyporus gordoniensis B. & Br. 1865 (AM III 15): 319 (Scotland); fide Lowe 1962 (PMi 47): 182 & Reid & Austw. 1963 (GN 18): 309 = *Poria candidissima*. — *Poria* Cooke 1886.

Polyporus hymenocystis B. & Br. 1879 (AM V 3): 210 (Scotland); fide Bres. 1926: 80 = *Poria subtilis* [sensu Bres.]; fide Reid & Austw. 1963 (GN 18): 309 = *Poria candidissima*. — *Poria* Cooke 1886. — Romell 1911 (ABS 11³): 13 *pl. 2 f. 9* (*Polyporus*); Egeland 1914 (NMN 52): 141 (*Poria*).

Trechispora onusta P. Karst. 1890 (H 29): 147; fide Lowe 1956 (M 48): 123 = *Poria candidissima*. — *Poria* Sacc. 1895. — Sensu Bres. → *Sistotrema eluctor*.

Physisporinus fragillimus P. Karst. 1898 T. 3: 15 / 1903 (BFi 62): 79 (Finland); fide Lowe 1956 (M 48): 112 = *Poria candidissima*. — *Poria* Romell; D. Baxt. 1932 (*"fragillina"*; syn.: n.v.p.).

M.—*Boletus byssinus* Schrad. (**O**) sensu Quél. 1888: 383 (*Poria*); cf. Bourd. & G. 1928: 691, "est vraisemblablement *Poria subtilis*". — Sensu originario = *Stromatoscypha fimbriatum* (**O**), cf. Donk 1959 (Pe 1): 81, 82.

DAEDALEA Pers. per Fr. (4)

1821 [1960 (Pe 1): 204; 1962 (Pe 2): 203]. — *Daedalea* Pers. 1801 (d.n.). — Lectotype: *Agaricus quercinus* L. — Sensu Quél. → *Daedalea* Quél.

Agrico-suber Paul. 1791 (d.n.) [1960 (Pe 1): 176, 178; 1962 (Pe 2): 202]. — ≡ *Agaricus* Paul. 1808 (d.n.), not ~ L. 1753 (d.n.) per Fr. 1821 (Agaricaceae), not ~ [Tourn.] Adans. 1763 ("*Agaricon*"), Paul. 1808 ("*Agaricum*"), (d.n.) & *Agaricus* [Tourn.] Rafin. 1830, not ~ Haller 1768 (d.n.) [1960 (Pe 1): 181, in obs.]. — Lectotype: *Agarico-suber daedaleum* Paul.

Merulius Pers. 1794 (NMB 1): 106 / 1797 T.: 26 (d.n.), not ∼ [Haller] Boehm. 1760 (d.n.) per St-Am. 1821, not ∼ Fr. 1821 [1958 (Fu 28): 10–11, in obs.]. — Lectotype: "*M. quercinus*" ≡ *Agaricus quercinus* L.

Striglia Adans. per O.K. 1891, not *Strilia* [!] S. F. Gray 1821 ("Polyporaceae") [1960 (Pe 1): 285]. — *Striglia* Adans. 1763 (d.n.). — Lectotype: "*Agaricus daedalaeis sinibus excavatus* Tou. J.R.H. 562" sensu Batt.

Agaricus Murrill 1903, 1905, not ∼ L. per Fr. 1821 (Agaricaceae), not ∼ [Tourn.] Rafin 1830 (Polyporaceae) [1960 (Pe 1): 182; 1962 (BnH 5): 11, in obs.]. — Holotype: *Agaricus quercinus* L.

? *Xylostroma* Tode per Fr. 1821 (nom. anam.) [1962 (Ta 11): 102]. — *Xylostroma* Tode 1790 (d.n.). — Monotype: *Xylostroma giganteum* Tode (29).

SPECIAL LITERATURE.—Aoshima, *1967*; Buchwald, *1941a*; O. Fidalgo, *1958a*.

quercina (L.) per Fr. 1821 (66). — *Agaricus* L. 1753 (Sweden) (d.n.); *Merulius* Pers. apud J. F. Gmel. 1792 (d.n.); *Daedalea* Pers. 1801 (d.n.); *Agaricus* Mérat 1821, Laterr. 1821; *Lenzites* P. Karst. 1882; *Striglia* O.K. 1891; *Trametes* Pilát 1939, not ∼ Lloyd 1922; ≡ *Agaricus labyrinthiformis* Bull. 1787 (d.n.), not ∼ Hoffm. 1789 (d.n.); *Merulius* Lam. 1797 (d.n.); *Agaricus* Bull. per St-Am. 1821; ≡ *Agaricus querneus* Schrank 1789 (d.n.); ≡ *Merulius labyrinthiformis* L. March. 1828, not ∼ (Hoffm.) J. F. Gmel. 1792 (d.n.). — Bull. 1787: *pl. 352*; 1789: *pl. 442 f. 1*, in part; 1809 H.: 377 (*Agaricus labyrinthiformis*); Sow. 1799: *pl. 181* (*Agaricus quercinus*); Fr. 1821: 333; Grev. 1826 S. 4: *pl. 238*; Gramb. 1913 P.H. 2: 2 *pl. 2* (*Daedalea*); Bourd. & G. 1928: 578 *f. 166* (*Lenzites*); Donk 1933: 195 (*Daedalea*); Pilát 1940 (ACE 3): 329 *f. 143, pl. 221, pl. 227 f. b* (*Trametes*); R. W. Davids. & al. 1942 (TUS 785): 16 *f. 3D, pl. 1 f. C*, cult. char.; Overh. 1953: 122 *pl. 84 fs. 479, 480, pl. 125 fig.* (*Daedalea*); Birkf. & Hersch. 1966: *pl. 154*, coalescence of fruitbodies (*Trametes*); Westh. 1971 (Bo 10): 191 *fs. 11, 12*, with cult. char. (*Daedalea*). — Sensu Schaeff. (pl. 57) → *Lenzites betulina*.

Agaricus dubius Schaeff. 1774: 56 [*pl. 231*] (Germany) (d.n.); fide Pers. 1800: 95 & Fr. 1821: 333 (var.).

Agaricus antiquus Willd. 1787: 376 (Germany) (d.n.); Willd. cited as syn. "Battara fung. *t. 38, f. A, B* (optima)" [= *Daedalea quercina*]. — Referred by Fr. 1821: 340 (as to descr.) to *Daedalea* [*Gloeophyllum*] *sepiaria*.

Agaricus labyrinthiformis Hoffm. 1789: 256 & 1790: 55 (d.n.), not ∼ Bull. 1787 (d.n.); fide Hoffm. 1790: 55 (*Agaricus quercinus* cited as syn.). — *Merulius* J. F. Gmel. 1792 (d.n.) ?, not ∼ Lam. 1797 (d.n.) & ∼ L. March. 1828; [≡ (by lectotypification) "*Agaricus daedalaeis, sinibus excavatus* Tou. J.R.H. 562" sensu Batt. 1755: 72 *pl. 38 f. A* (Italy)]. — J. F. Gmel. 1792: 1431 cited *Agaricus labyrinthiformis* Bull. as syn. of *Merulius* [*Daedalea*] *quercinus*; his phrase is a copy of that of *Agaricus antiquus*, cited above; Schaeffer's *pl. 231* [of *Agaricus dubius*] is cited as syn.

Agarico-suber daedaleum Paul. 1793 T. 2: 75 (descr.), Ind. (d.n.); fide

Paul. 1812–35: *pl. 1* (as *Agaricus quercinus* L.). — [≡ (by lectotypification) *Agaricus Daedalaeis sinibus excavatus* Tourn. 1700 (presumably Italy)].

Daedalea nigricans Pers. 1801 (d.n.). — [≡ "*Agaricus daedaleis sinibus excavatus, nigricans* Tou. J.R.H. 562" sensu Batt. 1755: 72 (Italy)].

Trametes hexagonoides Fr. apud Quél. 1872 (France); fide Bourd. & G. 1928: 579. — *Antrodia* P. Karst. 1879; *Polyporus* Feltg. 1907. — Quél. 1872 (MMb II 5): 287/282 *pl. 22 f. 2* (*Trametes*); Bourd. & G. 1928: 579 (*Lenzites quercina* f. 3). — Sensu Mez 1908: 129 *f. 54* = ?

Daedalea inzengae Fr. apud Inz. 1869 (GSP 5): 203 *pl. 2*; Fr. 1874 (Italy, Sicilia). — *Striglia* O.K. 1891. — Inz. 1879 F.s. 2: 8 *pl. 2*.

Hexagona minor Lázaro 1916 (RMa 14): 514 / 1917: 65 *pl. 10 f. 22* (Spain).

[*Fungus coriaceus Quercinus haematodes* Breyne "in Ephemer. German. An. 4. & 5. Obser. 150" (n.v.) cited by Ray 1686: 110 (nom. anam.). — Ray 1724: 25 (". . . Oak-Leather Hibernis") (**29**).

Xylostroma giganteum Tode 1790 (nom. anam.) (d.n.) (**29**). — *Byssus* DC. 1805 (d.n.) per St-Am. 1821; *Xylostroma* Purt. 1821; *Dematium* Chev. 1826; ≡ *Racodium xylostroma* Pers. 1801 (d.n.); ≡ *Xylostroma corium* Pers. 1822. — Tode 1790 F.m. 1: 36 *pl. 6 f. 51*; Sow. 1802: *pl. 358*.

M.—*Merulius gibbosus* Pers. sensu Purt. 1821: 248 *pl. 14* (*Daedalea*); fide Berk. 1836: 131.

DAEDALEOPSIS J. Schroet.

1888 [1960 (Pe 1): 205]. — Monotype: *Daedalea confragosa* (Bolt.) per Fr.

SPECIAL LITERATURE.—Allen, *1907*; Schmitz & Zeller, *1919*.

confragosa (Bolt. per Fr.) J. Schroet. 1888 (**30, 31**). — *Boletus* Bolt. 1791 (England) (d.n.); *Daedalea* Pers. 1801 (d.n.); *Daedalea* (Bolt.) per Fr. 1821; *Trametes* Rab. 1844; *Striglia* O.K. 1891; *Lenzites* Pat. 1900; *Agaricus* Murrill 1905. — Bolt. 1791: 160 *pl. 160* (*Boletus*); Fr. 1821: 336 (*Daedalea*); Bourd. & G. 1928: 580 (*Lenzites tricolor* f. *daedalea*); Donk 1933: 197 (*Daedaleopsis*); R. W. Davids. & al. 1942 (TUS 785): 16 *f. 3C*, cult. char.; Overh. 1935: 4 *tpls. 1, 2*; 1953: 120 *pl. 82 fs. 467–472, pl. 125 fig.* (*Daedalea*); H. Jahn 1963 (WPb 4): 67 *Abb. 26a* (*Trametes*); Westh. 1971 (Bo 10): 276 *fs. 39, 40*, with cult. char. of Canadian material (*Daedalea*). — Sensu Imaz. (cf. Imaz. & Hongo 1957 C.J. 1: 113 *pl. 53 f. 295*), may be a different sp.?; sensu G. Cunn. 1965 is not this sp., fide D. Reid 1967 (TBS 50): 163.

Boletus labyrinthiformis Bull. 1790 (France) (d.n.); fide Pers. 1801: 501 & Fr. 1821: 336. — *Boletus* Bull. per St-Am. 1821. — Bull. 1790: *pl. 491 f. 1*; 1791 H.: 357.

Boletus angustatus Sow. 1799: *pl. 193* (England) (d.n.); fide Berk. 1860: 254. — *Daedalea* Pers. 1801 (d.n.); *Daedalea* (Sow.) per Fr. 1821, Lapl. 1894 ("*angusta*").

Daedalea suaveolens Pers. 1801: 502 (d.n.); fide Quél. 1888: 373 (for Bull. *pl. 310 fs. B, C*) = *Trametes rubescens*; fide Bres. 1897 (AAR III 3); 92 (for *Trametes bulliardii*). — *Daedalea* Pers. per Fr. 1821; *Boletus* Bergam. 1823; [≡ *Boletus suaveolens* L. sensu Bull. 1786: *pl. 310*, in part. excl. of type (France)]; ≡ *Trametes bulliardii* Fr. 1838; *Polyporus* Kumm. 1871, not ∼ (Fr.) Pers. 1825.

? *Boletus horizontalis* Thore 1803: 489 (*"Horisontalis"*) (France) (d.n.). — *Boletus* Thore per Duby 1830. — The identity is suggested for the fruitbodies described from willows.

Daedalea rubescens A. & S. 1805 (Germany) (d.n.); fide Donk 1933: 197, 198. — *Daedalea* A. & S. per Fr. 1821; *Trametes* Fr. 1838; *Polyporus* Kumm. 1871; *Daedaleopsis* Imaz. 1943. — A. & S. 1805: 238 *pl. 11 f. 2*; Fr. 1821: 336 (*Daedalea*); Allen 1907 (TBS 2): 161; Joach. & Dumée apud Dumée 1925 (BmF 41): 77, notes; Bourd. & G. 1928: 590; Konr. & M. 1930 I. 5: *pl. 446* (*Trametes*). — Sensu Velen. → *Buglossoporus pulvinus*. — *Polyporus* "*rubescens* Fr." of Bres. is an error for *P. erubescens* Fr. (= *Tyromyces mollis*].

Daedalea pruinata Secr. 1833 M. 2: 481 (Switzerland).

Lenzites septentrionalis P. Karst. 1866: 199 & 1876 (BFi 25): 239 & 1882 (Finland); fide Lowe 1956 (M 48): 109 ("a lamellate form"). — *Gloeophyllum* P. Karst. 1882; *Lenzitina* P. Karst. 1889; *Cellularia* O.K. 1898.

Trametes erubescens S. Schulz. 1882 (RjA 64): 179 [repr. 1883 (BCb 15): 4] (Yugoslavia, Slavonia). — If correctly interpreted from the descr. then the hymenium and spores were wrongly described.

Trametes zonata Wettst. 1885 (VW 35): 561 [repr. 1886 (BCb 27): 86] (Austria), not ∼ (Nees per Fr.) Pilát 1939; from descr.

? *Polyporus capreae* Britz. 1896 (BCb 6): 142 [*pl. 652 f. 194*] (Germany) & 1910 (BbC 26): 211; fide Ade 1923 (ZP 2): 60 = *Trametes rubescens*. — Identification doubtful.

Daedalea ochracea Velen. 1922: 693 *f. 118* [see Pilát 1948: 262 for Latin translation] (Czechoslovakia); fide Pilát 1939 (ACE 3): 286.

M.—*Boletus suaveolens* L. sensu Bull. 1786 → *Daedalea suaveolens* Pers.; → *Trametes bulliardii* Fr.

tricolor (Bull. per Mérat) Bond. & S. 1941 (**30, 32**). — *Agaricus* Bull. 1791 (France) (d.n.), not ∼ A. & S. 1805 (d.n.), not ∼ Tratt. 1806 (d.n.); *Agaricus* Bull. per Mérat 1821, not ∼ A. & S. per Fr. Jan. 1, 1821; *Daedalea* Pers. 1828: Fr. 1832; *Merulius* L. March. 1828; *Lenzites* Fr. 1838; *Cellularia* O.K. 1898. — Bull. 1791: *pl. 541 f. 2*; 1809 H.: 380 (*Agaricus*); Quél. 1873 (MMb II 5): 355; Kallenb. 1925 (ZP 4): 65; Konr. & M. 1927 I. 5: *pl. 440 f. 1*; Bourd. & G. 1928: 580, excl. of forms (*Lenzites*); Pilát 1939 (ACE 3): 188 (*Trametes confragosa* var. *Lenzites tricolor*).

Lenzites atropurpurea Sacc. 1873 (ASv 2): 93/45 *pl. 6 fs. 15–19* (Italy); fide Bres. 1890 (Rm 12): 102. — *Cellularia* O.K. 1898.

DATRONIA Donk

1966 (Pe 4): 337. — Holotype: *Trametes mollis* (Sommerf.) Fr.
M.—*Antrodia* P. Karst. sensu Murrill 1905 (BTC 32): 354.

SPECIAL LITERATURE.—Pilát, *1926b*; Ryvarden, *1968b*.

epilobii (P. Karst.) Donk 1966; fide Romell 1911 (ABS 11[3]): 24 & Lowe 1956 (M 48): 122 = *Polyporus/Trametes stereoides* [sensu Romell]. — *Trametes* P. Karst. 1868 (NfF 9): 361 [repr. 1869 (H 8): 76] (Finland); *Polyporus* P. Karst. 1876; *Antrodia* P. Karst. 1879; *Pycnoporus* P. Karst. 1889. — Ryv. 1968 (SbT 62): 508 *f. 4* (*Datronia stereoides* var.); Donk 1966 (Pe 4): 338, 1971 (PNA 74): 5, notes.

Polyporus planus Peck 1879 (U.S.A., New York), not ∼ Wallr. 1833; fide Romell 1911 (ABS 11[3]): 24 = *Polyporus stereoides* [sensu Romell]; fide Bres. 1920 (Am 18): 68 = *Trametes kmetii.* — *Polystictus* Cooke 1886; *Microporus* O.K. 1898; ≡ *Coriolus planellus* Murrill 1906; *Polyporus* Overh. 1915. — Murrill 1907 (NAF 9): 21 (*Coriolus planellus*); Shope 1931 (AMo 18): 344 *pl. 22 f. 1*; Lowe 1942: 63; Overh. 1953: 377 *pl. 45 f. 273, pl. 49 f. 296, pl. 132 fig.* (*Polyporus planellus*).

Trametes kmetii (Bres.) Bres. 1920; fide Romell 1911 (ABS 11[3]): 24 = *Polyporus stereoides* [sensu Romell]. — *Trametes stereoides* var. Bres. 1897 (AAR III 3): 92 (Hungary, now Czechoslovakia).

M.—*Polyporus stereoides* Fr. (**O**) sensu Romell 1911 & apud Lloyd 1916; fide Romell 1911 (ABS 11[3]): 23 (*Trametes stereoides* var. *kmetii* cited as a syn.). — Romell 1911 (ABS 11[3]): 23 *pl. 2 f. 2* (*Polyporus*); Bourd. & G. 1928: 596 (*Trametes*); D. Baxt. 1939 (PMi 24): 184 *pl. 7* (*Polyporus*); Ryv. 1968 (SbT 62): 504 *fs. 3–5* (*Datronia*).

mollis (Sommerf.) Donk 1966. — *Daedalea* Sommerf. 1826 (Norway), not ∼ (Pers.) Fr. 1815 (d.n.), not ∼ Velen. 1922; *Trametes* Fr. 1874; *Polyporus* P. Karst. 1876, not ∼ (Pers.) per Fr. 1821; *Antrodia* P. Karst. 1879; *Daedaleopsis* P. Karst. 1899. — Bourd. & L. Maire 1920 (BmF 36): 82; Bourd. & G. 1928: 595; D. Baxt. 1951 (PMi 35): 45; Overh. 1953: 146 *pl. 86 fs. 491–495, pl. 125 fig.* (*Trametes*); Domański 1965 (FpG 2): 195 *pl. 56 f. 2* (*Antrodia*); Al. David 1967 (Nca 94): 563 *tpl. 1 f. 5, tpl. 2 f. d,* cult. char. (*Trametes*); Ryv. 1968 (SbT 62): 501 *fs. 1, 2* (*Datronia*). — Sensu P. Karst. → *Polyporus sommerfeldtii* P. Karst., see below.

Polyporus cervinus Pers. 1825 (Switzerland); not ∼ (Schw.) Steud. 1824, not ∼ (Quél.) Sacc. 1895; fide Fr 1863 M. 2: 256, Romell 1911 (ABS 11[3]): 23, & Donk 1933: 193, 194. — *Trametes* Lloyd 1913, not ∼ (Schw.) Bres. 1903. — Lloyd 1913 (LMW 3): 470 (*Trametes*).

Trametes serpens Fr. 1849: 324 (Sweden), not ∼ (Fr. per Fr.) Fr. 1874; fide Fr. 1863 M. 2: 256 (9).

Polyporus sommerfeldtii P. Karst. 1878; fide P. Karst. 1882 (BFi 37): 53. — [≡ *Polyporus mollis* (Sommerf.) P. Karst. sensu P. Karst. 1876 (BFi 25): 280 (Finland)].

Daedalea lassbergii Allesch. 1889 (BLa 11): 23 (Germany); fide Bres. 1920 (Am 18): 69 = *Trametes stereoides* [sensu Bres.].

'M.'—*Daedalea mollis* Sommerf. sensu P. Karst. 1876 (BFi 25): 280 (*Polyporus*) → *Polyporus sommerfeldtii* P. Karst., this sp., above.

M.—*Polyporus stereoides* Fr. (O) sensu Fr. 1884; fide Donk 1933: 194, in obs. — Fr. 1884 I. 2: 86 *pl. 187 f. 2* (*Polyporus*); Bres. 1897 (AAR III 3): 92, excl. of var.; Shope 1931 (AMo 18): 367 *pl. 31 f. 1* (*Trametes*).

Antrodia **sajanensis** Parm. 1962 (U.S.S.R., Russia, Siberia). — Parm. 1962 (BMs 15): 134 *fs. 8, 9*; 1967 (MF 1): 283, European record.

DICHOMITUS D. Reid

1965 (RBL 5): 149. — Holotype: *Trametes squalens* P. Karst.

SPECIAL LITERATURE.—*Dichomitus campestris*: Domański & Orlicz, *1966*.

Dichomitus squalens: Andrews, *1971*; Baxter & Manis, *1939*; Kauffman, *1926*; Long, *1916, 1917*; Perlman, *1949*.

campestris (Quél.) Dom. & Orl. 1966. — *Trametes* Quél. 1872 (France); *Antrodia* P. Karst. 1879; *Coriolellus* Bond. 1953. — Quél. 1872 (MMb II 5): 286/271 *pl. 2 f. 6*; Bourd. & G. 1928: 597; Pilát 1939–40 (ACE 3): 311 *f. 131, pl. 210 f. a*; Teston 1953 (BOy 7): 94 *tpl. 5 f. 5*; R. L. Gilb. 1961 (NwS 35): 16 *f. 13* (*Trametes*); Dom. & Orl. *1966*, with cult. char. (*Dichomitus*); Al. David 1967 (Nca 94): 561 *tpl. 1 f. 4, tpl. 2 f. c*, cult. char. (*Trametes*).

squalens (P. Karst.) D. Reid 1965. — *Trametes* P. Karst. 1886 (Finland); *Bjerkandera* P. Karst. 1887; *Polyporus* Sacc. 1888; *Coriolellus* Bond. & S. 1941; *Tyromyces* Imaz. 1943, Kotl. & P. 1957, misapplied; *Poria* Lowe 1956, misapplied, not ∼ Lloyd 1908. — P. Karst. 1887 (Mfe 4): 79 (*Bjerkandera*); Romell 1926 (SbT 20): 7, in obs. (*Polyporus*); Bourd. & G. 1928: 592; Pilát 1939–40 (ACE 3): 312 *f. 132, pl. 176, pl. 210 f. b*; Nikol. 1940 (TSR 4): 403 *fs 15, 16* (*Trametes*); Bond. 1953: 46, 507 *f. 131, pl. 135 f. 2, pl. 144* (*Coriolellus*); Lundell 1953 (LNF 43–44): 2 No. 2102 (*Polyporus*); Jo. Erikss. 1958 (Sbu 16¹): 146 *f. 46* (*Trametes*); D. Reid 1965 (RBL 5): 150 *fs. 7, 13, 14* (*Dichomitus*); Donk 1962 (Pe 2): 235, notes. — Sensu Lowe 1956 → *Poria albobrunnea*.

Polyporus anceps Peck 1895 (U.S.A., Massachusetts); fide Romell 1926 (SbT 20): 7. — *Tyromyces* Murrill 1907; *Coriolellus* Parm. 1959. — C. H. Kauffm. 1926 (M 18): 27 *pl. 5*; Shope 1931 (AMo 18): 343 *pl. 23 f. 2*; Baxt. & Manis 1939 (PMi 24): 190 *fs. 1, 2, pls. 1–3*; Overh. 1953: 279 *pl. 15 f. 87, pl. 121 f. 662, pl. 128 fig.* (*Polyporus*).

Tyromyces ellisianus Murrill 1907 (NAF 9): 34 (U.S.A., New Jersey); fide Baxt. & Manis 1939 (PMi 24): 195 & Overh. 1953: 279, 280 = *Polyporus anceps*. — *Polyporus* Sacc. & Trott. 1912.

Leptoporus bulgaricus Pilát 1937 (BmF 53): 84 (Bulgaria); fide Domański 1964 (APo 33): 174. — *Tyromyces* Bond. 1953 (incomplete ref.: n.v.p.). — Pilát 1937–8 (ACE 3): 224 *f. 75, pl. 139 f. b*; Domański 1964 (APo 33): 174 *f. 6* (*Leptoporus*).

Leptoporus dalmaticus Pilát 1937 (Yugoslavia); fide Domański 1964 (APo 33): 174. — Domański 1964 (APo 33): 175.

M.—*Boletus stypticus* Pers. sensu Bres. in herb. & "Bresad. et von Post (teste Bres. in litt.)" apud Höhn 1909 (ÖbZ 59): 108, as to bigger-spored material, & Bres. 1931 (BIm 20): *pl. 977 f. 2* (*Polyporus*); fide Bres., l.c. (*Trametes squalens* cited as syn.) & Romell 1926 (SbT 20): 7. — Killerm. 1943 (Am 41): 243 (*Trametes*).

ECHINOTREMA Park.-Rh.

1955 [1960 (Pe 1): 207]. — Holotype: *Echinotrema clanculare* Park.-Rh.

SPECIAL LITERATURE.—Parker-Rhodes, *1955*.

clanculare Park.-Rh. 1955 (England). — Park.-Rh. 1955 (TBS 38): 368 *f. 1*.

FISTULINA Bull. per Fr.

1821 [1960 (Pe 1): 214]. — *Fistulina* Bull. 1791 (d.n.). — Monotype: *Boletus hepaticus* Schaeff. — V.s.: *Tubulina* Paul. 1808 (error); *Fistulinia*; *Fistularia*; *Fistulinus* E. Krause 1925.
Agarico-carnis Paul. 1791 [1960 (Pe 1): 176, 178; 1962 (Pe 2): 201, 202] (d.n.). — Lectotype: *Agarico-carnis lingua-bovis* Paul.
Hypodrys [Solenander] Pers. 1818 (d.n.) [1960 (Pe 1): 228]. — [*Hypodrys* Solenander 1596 (n.v.) (pre-Linnean name)]. — *Hypodrys* Pers. per Pers. 1825. — Monotype: *Boletus hepaticus* Schaeff.
Buglossus Wahl. 1820 [1960 (Pe 1): 191] (d.n.). — *Buglossus* Wahl. per Wahl. 1826. — Monotype: *Buglosus quercinus* Wahl.

SPECIAL LITERATURE.—Arcangeli, *1878*; Braid, *1924*; Cartwright, *1937*; H. Lohwag & Follner, *1936*; Molliard, *1909*; Rothberg, *1937*; Runge, *1966*; Sartory & Maire, *1924*; de Seynes, *1863, 1867, 1874a, 1874b*; dal Vesco, *1963*.

hepatica (Schaeff.) per Fr. 1821. — *Boletus* Schaeff. 1774 (Germany) (d.n.); *Fistulina* Sibth. 1794 (d.n.); *Dendrosarcus* Paul. 1812–35 (gen. name n.v.p.); *Boletus* (Schaeff. per Fr.) Hook. 1821, Laterr. 1821; *Hypodrys* Pers. 1825; = *Fistulina buglossoides* Bull. 1789 (d.n.) per St-Am. 1821; *Hypodrys* Krombh. 1841 (syn.: n.v.p.); = *Boletus globatus* J. F. Gmel. 1792 (d.n.; error for '[sub]lobatus'; = *Boletus hepaticus* Schaeff. sensu Huds. 1778, Willd. 1787); = (by lecto-typification) *Agarico-carnis lingua-bovis* Paul. 1793 (d.n.); = *Fistulina hypodris* Rafin. 1830 (= *Fistulina buglossoides*). — Schaeff. 1774: 82 [*pls. 116–120*] (*Boletus h.*); Bull. 1789: *pl. 464*; 1791 H.: 314 (*Fistulina buglossoides*); Pers. 1800: 46 (*Boletus h.*);

1818: 245 (*Fistulina buglossoides*); Vitt. 1835: 280 *pl. 36* (*Fistulina h.*); Krombh. 1841 S. 7: 5 *pl. 47* (*Boletus h.*); Seyn. 1874 F.: 7, 68 *pls. 1–6, 7 fs. 8–11*; Arcangeli *1878*; Boud. 1904–11: 84 *pl. 164*, chlamydosporous state only; Lloyd 1908 (LMW 3, O.S. 1): 6 *fs. 219–221*; Gramb. 1913 P.H. 2: 3 *pl. 3*; Bourd. & G. 1928: 686; Lohw. & Follner *1936*; Lohw. 1941: 79, 111, 305 *fs. 64, 89, 213*; Pilát 1942 (ACE 3): 588 *f. 272, pl. 344 f. b, pls. 373, 374*; R. W. Davids. & al. 1942 (TUS 785): 17 *f. 3E, pl. 1 f. D*, cult. char.; Poelt & Jahn 1963: *pl. 60* (*Fistulina h.*).

Boletus buglossum Retz. 1769 (SVH 30): 253 (Sweden) (d.n.); fide Fr. 1821: 396. — *Fistulina* Pers. 1794 (d.n.); *Fistulina* (Retz.) per E. Krause 1934; ≡ (by lecto-typification) *Buglossus quercinus* Wahl. 1820 (d.n.) per Wahl. 1826.

Boletus esculentus With. 1776: 769 (d.n.). — [≡ *Agaricus porosus, rubens, carnosus, hepatis facie* Dill. 1719: 192; Ray 1724: 23. — ≡ *Fungus arboreus rubens carnosus, hepatis facie* Ray 1696: 340 (England)].

Boletus bulliardii J. F. Gmel. 1792: 1434 (d.n.). — [≡ *Boletus hepaticus* Schaeff. sensu Bull. 1781: *pl. 74*, excl. of type (France)]. — J. F. Gmel. treated *Boletus hepaticus* as a distinct species—inclusive of Bulliard's plate 74!

Thelephora bucreas Spreng. 1806: 380 (d.n.) (Germany), citing *Boletus buglossum* Retz. sensu Vahl 1794 (Fd 7 / F. 19): 8 *pl. 1136*, depicting a young stage.

Fistulina sarcoides St-Am. 1821: 547 (France); fide Seyn. 1874 F.: 16, 69 (var.).

Ceriomyces hepaticus Sacc. 1888 (SF 6): 54, 388 (nom. anam.). — *Ptychogaster* Lloyd 1909; [≡ *Fistulina hepatica*, "status omnimo gasterosporus *Ceriomycetem* sistens" Sacc. 1878 (Mi 1): 362 (Italy)]. — Lloyd 1909 (LMW 3, P.I. 2): 32 *f. 266* (*Ptychogaster*).

M.—*Boletus sanguineus* L. (**O**) sensu Plan. 1788 I.F.: 25; fide Pers. 1800: 47 & Fr. 1832[Ind.] 62; & sensu Vill. 1789: 1041 → *Boletus sanguineus* Fr. 1821 (incidental mention) (**O**).

FLAVIPORUS Murrill

1905 [1960 (Pe 1): 215]. — Holotype: *Polyporus rufoflavus* B. & C.

Baeostratoporus Bond. & S. 1941 ex Sing. 1944 [1960 (Pe 1): 189]. — Holotype: *Polyporus braunii* Rab.

brownii (Humb. per Steud.) Donk 1960; fide Bres. 1897 (AAR III 3): 76 = *Fomes braunii* ("juxta diagnosin omnino"). — *Boletus* Humb. 1793: 101 (Germany) (d.n.) per Steud. 1824 ("*Brownei*"); *Polyporus* Pers. 1825.

Boletus paradoxus Humb. 1793: 103 *pl. 3 f. 15* (Germany) (d.n.); fide Bres. 1897 (AAR III 3): 76 = *Fomes braunii*. — *Boletus* Humb. per Steud. 1824; ≡ *Polyporus clavus* Pers. 1825.

? *Polystictus cyphelloides* Fr. 1851 (NAu III 1): 88/72 (Mexico). —

Polyporus Fr. apud Kalchbr. 1868, typonym; *Hansenia* P. Karst. 1879; *Microporus* O.K. 1898 (*"cyphellodes"*); *Coriolus* Murrill 1907. — Kalchbr. 1868 (VW 18): 431 [repr. 1868 (H 7): 183] (*Polyporus*); Murrill 1907 (NAF 9): 26 (*Coriolus*).

Polyporus rufoflavus B. & C. 1868 (Cuba); fide Pat. 1900: 85 = *Leptoporus braunii*. — *Fomes* Cooke 1885, C. & D. Over. 1922 (*"rufo-flavens"*, error); *Scindalma* O.K. 1898; *Flaviporus* Murrill 1905; *Leptoporus* Mang. & Pat. 1922; Pilát 1937. — Murrill 1908 (NAF 9): 84 (*Flaviporus*); Lloyd 1915 (LMW 4, F.): 220; Pilát 1937–8 (ACE 3): 220 *fs. 73, 96, pl. 120*; Wakef. 1952 (TBS 35): 35; D. Reid 1955: 12 *f. 2* (*Leptoporus*).

Polyporus braunii Rab. 1876 (Germany); fide Lloyd 1915 (LMW 4, F.): 220 & Bres. 1920 (Am 18): 68 = *Fomes rufoflavus*; fide Donk 1960 (Pe 1): 189. — *Polystictus* Cooke 1886; *Ochroporus* J. Schroet. 1888; *Fomes* Bres. 1896; *Microporus* O.K. 1898; *Leptoporus* Pat. 1900; *Baeostratoporus* Bond. & S. 1941. — Pat. & Lag. 1893 (BmF 9): 128 (*Polyporus*); Bres. 1897 (AAR III 3): 76 (*Fomes*); Bataille 1907 (BSD No. 14): reprint p. 2; Lloyd 1910 (LMW 3): 458 (*Polyporus*); Pilát 1927 (Sčz 2): 477 *pl. 10 fs. 20–22*; Bourd. & G. 1928: 553 (*Leptoporus*).

Polyporus lucens Wettst. 1885 (ÖbZ 35): 151 (Austria); fide Pilát 1927 (Sčz 2): 477, 478, 479 = *Leptoporus braunii* (resupinate).

Polyporus silaceus Wettst. 1885 (ÖbZ 35): 152 (Austria); fide Pilát 1937 (Sčz 2): 477, 479 = *Leptoporus braunii*. — *Polystictus* Sacc. 1895; *Microporus* O.K. 1898.

Polyporus engelii Harz 1888 (Germany); fide Bres. 1897 (AAR III 3): 76 = *Fomes braunii*. — *Polystictus* Sacc. 1891; *Microporus* O.K. 1898. — Harz 1888 (BCb 36): 379.

M.—*Boletus cryptarum* Bull. sensu Mont. 1858 (ASn VIII 9): 158 (*Polyporus*); fide Mang. & Pat. 1922 (CrP 175): 393 = *Leptoporus rufoflavus*.

FOMES (Fr.) Fr.

1849 (nom. altern.) [1960 (Pe 1): 215; 1962 (Pe 2): 204]. — *Polyporus* subgen. *Fomes* Fr. 1836. — *Polyporus* [sect.] *Fomes* Fr. 1849 (nom. altern.); ≡ *Ungulina* Pat. 1897 (nom. prov.), 1900 [1960 (Pe 1): 291]. — Lectotype: *Polyporus fomentarius* (L.) per Fr. — Sensu Murrill → *Fomitopsis*; *Ungulina* sensu Kotl. & P. 1957 → *Truncospora*.

Agarico-igniarium Paul. 1791 (d.n.) [1960 (Pe 1): 176, 178; 1962 (Pe 2): 202]; ≡ *Pyreium* Paul. 1808 (d.n.), 1812–35 (n.v.p.) [1960 (Pe 1): 176, 179]. — Lectotype: *Agarico-igniarium foliaceum* Paul. [in part = *Pyreium fomentarium* (L.) Paul.].

Placodes Quél. 1886 [1960 (Pe 1): 259]. — Lectotype: *Polyporus fomentarius* (L.) per Fr.

Elfvingiella Murrill 1914 [1960 (Pe 1): 207]. — Holotype: *Fomes fomentarius* (L. per Fr.) Fr.

SPECIAL LITERATURE.—Benzoni, *1942*; Biers, *1922*; Bondarcev, *1934*; Buchwald, *1930, 1934, 1938*; Buchwald & Hansen, *1934*; Buchwald & Hellmers, *1946*; Campbell, *1938*; Guinier & Maire, *1908*; Heim & Lami, *1950*; Herrmann, *1962*; Herschel & Huth, *1970* (galls)?; Hilborn, *1942*;

Hilborn & Linder, *1939*; Ingold, *1965*; Jahn, *1965b*; Kreisel, *1957*; Krull, *1892, 1894*; Leuchs, *1832*; Lloyd, *1915a*; H. Lohwag, *1931c*; Lowe, *1957a*; Macdonald, *1938*; Mangin, *1907*; Meyer, *1936*; Nordin, *1967*; Piccone, *1876*; Priehäusser, *1931*; Runge, *1959*; Sargent, *1901*; Suvorov, *1966, 1967*; Teixeira, *1958a, 1962a*; Wulff, *1909*; Zier, *1832*.

fomentarius (L. per Fr.) Fr. 1849. — *Boletus* L. 1783 (Sweden) (d.n.); *Agaricus* Lam. 1783 (d.n.); *Pyreium* Paul. 1812–35 (generic name n.v.p.); *Polyporus* G. Meyer 1818 (d.n.); *Polyporus* (L.) per Fr. 1821; *Boletus* Hook. 1821; *Placodes* Quél. 1886; *Ochroporus* J. Schroet. 1888; *Scindalma* O.K. 1898; *Ungulina* Pat. 1900; *Elfvingia* Murrill 1903; *Elfvingiella* Murrill 1914; *Ganoderma* A. Ames 1913; *Pyropolyporus* Teng 1964. — Pers. 1799 O. 2: 1; 1801: 536, in part: var. [α] only; Sow. 1798: *pl. 133* (*Boletus*); Fr. 1864–5 S.S.: 37 *pl. 62* (*Polyporus*); Bres. 1897 (AAR III 3): 74, spores (*Fomes*); T. Wulff 1909 (SkT 7): 11 *f. 6 pl. 2*; Gramb. 1913 P.H. 2: 23 *pl. 23* (*Polyporus*); Lloyd 1915 (LMW 4, F.): 225 *f. 584* (*Fomes*); Bourd. & G. 1928: 601 (*Ungulina*): Pilát 1940 (ACE 3): 346 *f. 149, pl. 229–231*; Overh. 1953: 91 *pl. 59 f. 351, pl. 70 fs. 407, 408, pl. 104 f. 582, pl. 126 fig.*; Kawam. 1954 I. 1: 157 *f. 139*; Lowe 1957 F.: 43 *f. 27*; A. Teix. 1962 (Ri 1): 68 *tpl. 2*; H. Jahn 1963 (WPb 4): 50 *f. 1m, Abb. 8–10*; Poelt & Jahn 1964: *pl. 38*; H. Jahn 1965 (WPb 5): 117 *fs. 1–9*; Nordin, *1967* (*Fomes*).

? *Boletus fagineus* With. 1776 (d.n.), not ∼ Schrad. apud J. F. Gmel. 1792 (d.n.). — [≡ *Agaricus porosus igniarius Fagi, superne candicans, inferne fuscus* Dill. 1719: 193 (Germany), Ray 1724: 24].

Agarico-igniarium foliaceum Paul. 1793 T. 2: 87 (descr.), Ind. (France?) (d.n.); fide Paul. 1812–35: *pl. 7 fs. 2, 3* (as *Pyreium fomentarium*) and Donk 1960 (Pe 1): 178, 1971 (Pe 6): 201.

? *Agarico-igniarium lignoso-tomentosum* Paul. 1793 T. 2: 92–93 (descr., restricted to the form on oak), Ind. (France) (d.n.); fide Lév. 1855: 5. — ≡ *Pyreium lignosum* Paul. 1812–35: *pl. 8 fs. 2–4* (d.n.?).

Polyporus inzengae Ces. & De-Not. 1861 (Italy, Sicilia); fide Sacc. 1881 (Mi 2): 377 & Bres. 1890 (Rm 12): 104. — *Fomes* Cooke 1885, Bizzoz. 1885; *Ungulina* Biers 1922 (nom. altern.?), M. Bon 1970 (n.v.p.). — Ces. & De-Not. 1861 (ECi I): No. 636, with descr.; De-Not. apud Inz. 1869 F.s. 1: 17 *pl. 2 f. 1*; Ripart 1876 (BbF 23): 213 (*Polyporus*); Sacc. 1916: 998 (*Fomes*).

Polyporus mirus Kalchbr. 1877 (BSM 52): 145 & 1878 (EtK 8¹⁶): 13 *pl. 1 f. 3* (U.S.S.R., Russia, Siberia); fide Lloyd 1915 (LMW 4, F.): 283. — *Fomes* P. Karst. 1882; *Scindalma* O.K. 1898.

Polyporus introstuppeus B. & Cooke ("in Herb. Berk.") apud Cooke 1884 (G 13): 2 (NW. India or Malay Peninsula); fide Bres. 1916 (Am 14): 225 = *Polyporus inzengae*. — Sensu P. Henn. ≡ "*Trametes*" *floccosus* Bres. (extra-European).

Fomes excavatus (Berk.) Cooke 1885; fide Lowe 1955 (M 47): 223 &

1957 F.: 43. — *Polyporus fomentarius* var. Berk. 1839 (AM 3): 387 (Canada).
— Lloyd 1915 (LMW 4, F.): 280 *f. 584*, photograph of type (*Fomes fomentarius*).

Fomes griseus Lázaro 1916 (RMa 14): 658 / 1917: 97 (Spain); fide [Bres.? apud] Trott. 1925 (SF 23): 390 ("proximus").

Ungularia tuberosa Lázaro 1916–7 (RMa 14, 15): 672 / 1917: 111 *pl. 9 fs. 20, 21*; fide [Bres.? apud] Trott. 1925 (SF 23): 388 ("verisimiliter ... f. crassior").

Ungulina nivea Lázaro 1917 (RMa 15): 373 / 1917: 287 *pl. 5* (Spain); fide Bres. apud Trott. 1925 (SF 23): 388 = *Fomes inzengae*. — *Fomes* Sacc. & Trott. apud Trott. 1925.

Ungularia albescens Lázaro 1917 (RMa 15): 374 / 1917: 288 (Spain); *Fomes* Sacc. & Trott. apud Trott. 1925.

Ungularia subzonata Lázaro 1917 (RMa 15): 375 / 1917: 289 *pl. 6* (Spain); fide Bres. apud Trott. 1925 (SF 23): 388. — *Fomes* Sacc. & Trott. apud Trott. 1925.

Ungulina nigricans (Bourd. & G.) Pilát 1932. — *Ungulina fomentaria* subsp. Bourd. & G. 1925 ["(Fr., p.p.)"]. — [≡ *Polyporus nigricans* Fr. sensu Fr. 1838: 466, in part: as to specimen from Klotzsch (Scotland) excl. of type (cf. Donk 1971 (PNA 74): 30)]. — *Fomes nigrescens* Lloyd 1915 (LMW 4, F.): 237, 288, typonym (as a form of *Fomes fomentarius*: n.v.p.). — Bourd. & G. 1928: 601 (*Ungulina fomentaria* subsp.); Donk 1971 (PNA 74): 30, notes.

M.—*Boletus ungulatus* Schaeff. sensu Bull. 1788; fide Fr. 1821: 374 (citing Bull. pl. 491 [f. 2]) & Donk. — Bull. 1788: *pl. 401*; 1790: *pl. 491 f. 2*; 1791: H.: 357; Pers. 1799 O. 2: 4, at least in part (*Boletus*); 1801: 537 (*Boletus fomentarius* var. "δ. *B. ungulatus*").

M.—*Boletus igniarius* L. sensu Pers. 1818: 91; fide Pers., l.c.

M.—*Polyporus nigricans* Fr. sensu Fr. 1838: 466, in part; cf. Donk 1971 (PNA 74): 30. — Fr. 1884 I. 2: 83 *pl. 184 f. 2*, in part: as to depicted form (*Polyporus*); Bres. 1897 (AAR III 3): 74; Lloyd 1908 (LMW 2): 373 *f. 194* & (LMW 3, P.I.): 15 *f. 210*, in part: as to "second" form (*Fomes*); Pilát 1940 (ACE 3): 348 *pl. 231* (*Fomes fomentarius* subsp.). → *Ungulina nigricans* (Bourd. & G.) Pilát; → *Fomes nigrescens* Lloyd (n.v.p.) (**O**).

M.—*Polyporus roburneus* Fr. (**O**) sensu Bres. 1897 (AAR III 3): 74 (as a syn. of *Fomes nigricans* [sensu Bres.]).

M.—*Boletus fulvus* Scop. sensu Velen. 1922: 679 *f. 113* (*Polyporus*); fide Pilát 1941 (ACE 3): 346.

FOMITOPSIS P. Karst. (33).

1881 [1960 (Pe 1): 217]. — Lectotype: *Polyporus pinicola* (Sw.) per Fr.

Placodes Pat. 1887: 139, not ~ Quél. 1886 [cf. 1960 (Pe 1): 259, in obs.]. — [≡ *Placodes* Quél. sensu Pat., l.c., excl. of type.] — Lectotype: *Placodes marginatus* (Pers. per Fr.) Quél.

Fomes Murrill 1903 (BTC 30): 225, not ~ (Fr.) Fr. 1849 [1960 (Pe 1): 216, in

obs.]. — [= *Fomes* (Fr.) Fr. sensu Murrill, l.c., excl. of type, (as *Fomes* "Gillet")].
— Holotype: *Fomes marginatus* (Pers. per Fr.) Fr.
  M.—*Placodes* Quél. sensu Pat. 1887 → *Placodes* Pat.
  M.—*Fomes* (Fr.) Fr. sensu Murrill 1903 → *Fomes* Murrill.

SPECIAL LITERATURE.—*Fomitopsis cytisinus* (*Fomes fraxineus* pl. auctt.): Aoshima, *1963*; Baxter, *1925*; Dumée, *1917*; Igmándy, *1961*, *1963*; Montgomery, *1936*; Müller & Jahn, *1966*; Nicolas, *1926*; Pilát, *1947*.

  Fomitopsis pinicola: Cartwright, *1951b*; Guinier & Maire, *1908*; Jahn, *1968*; LaFuze, *1937*; K. Lohwag, *1955b*; Mounce, *1929*; Mounce & Macrae, *1938*; Pennington, *1927*; Příhoda, *1957*; Schmitz, *1925*; Veselý, *1957*.

  Fomitopsis cajanderi ('ca') & *F. rosea* ('ro'): Adams & Roth, *1967* (ca); *Göpfert*, 1970 (ro); *Madhosingh*, 1964 (ca & ro), Mounce & Macrae, *1937* (ca & ro); *Neuhauser & Gilbertson*, 1971 (ca); *von Schrenck*, 1900 (ca); *Snell, Hutchinson, & Newton*, 1928 (ca & ro); *Svrček*, 1960 (ro); *Weir*, 1923 (ca & ro); *Zeller*, 1926 (ca).

cajanderi (P. Karst.) Kotl. & P. 1957 (33). — *Fomes* P. Karst. 1904 (U.S.S.R., Russia, Siberia). — Lowe 1956 (M 48): 106, note; 1957 F.: 70 *f. 53*; Westh. 1971 (Bo 10): 208 *fs. 17, 18*, with cult. char. (*Fomes*); Donk 1971 (PNA 74): 31, notes.

  Pycnoporus mimicus P. Karst. 1906 (Ttk 8[1]): 62 (U.S.S.R., 'Transbaikal'); fide Lowe 1956 (M 48): 121. — *Polystictus* Sacc. & Trott. 1912.

  Polyporus palliseri Berk. ex Lloyd 1911 (LMW 3, L. 34): 2 ("*Palliser*") (Canada, Saskatchewan), not ∼ Cooke 1882; fide Lloyd 1911 (LMW 3, L. 32): 4 = *Polyporus carneus* [sensu Lloyd] (forma). — *Polyporus palliseri* Berk. in herb., Cooke 1885 (syn.: n.v.p.), not ∼ Cooke 1882; Lloyd 1911 (LMW 3, L. 32): 4 ("*Palliser*"; "certainly only a form ... of *Polyporus carneus*": n.v.p.); *Fomes* Cooke 1885 (nom. nud.: n.v.p.), not ∼ (Cooke) Sacc. 1888; *Fomes* (Berk. ex Lloyd) Burt 1931 ("Berk."), not ∼ (Cooke) Sacc. 1888; = *Trametes subrosea* Weir 1923; *Fomes* Humphr. & Sigg. 1933, Bond. apud Ljub. 1934, Overh. 1935, Bond. 1935; *Ungulina* Murašk. 1939; *Fomitopsis* Bond. & S. 1941. — Bres. 1916 (Am 14): 237 (*Polyporus palliseri*); Weir 1923 (Rh 25): 219; Shope 1931 (AMo 18): 372 *pl. 32 f. 1* (*Trametes subrosea*); Mounce & Macrae 1937 (CJR 15): 154 *fs. 7–12, pl. 1 fs. 3, 4* (*Fomes subroseus*); Murašk. 1939: 86 *f. 11A* (*Ungulina subrosea*); D. Baxt. 1951 (PMi 35): 49 *pl. 4*; Overh. 1953: 57 *pl. 64 fs. 376–379, pl. 126 fig.* (*Fomes subroseus*); Bond. 1953: 41, 291 *f. 66, pl. 67 f. 3, pl. 88 f. 1* (*Fomitopsis subrosea*); Lowe 1955 (M 47): 221 (*Fomes subroseus*); Westh. 1971 (Bo 10): 208 *fs. 17, 18*, with cult. char. (*Fomes*).

  Trametes carnea Lloyd 1915 (LMW 4, F.): 224 *f. 577*. — [= *Trametes carnea* (Bl. & Nees) Lloyd sensu Lloyd, l.c., excl. of type (U.S.A.)].

  M.—*Polyporus carneus* Bl. & Nees (**0**) sensu Ravenel 1860 F.c. 5: No. 14, no descr., & Berk. 1872 (G 1): 51; fide Bres. 1916 (Am 14): 228 ("ex America boreali") = *Polyporus palliseri* Berk. ex Lloyd, non Cooke;

fide Lowe 1957 F.: 70. — Overh. 1914 (AMo 1): 131 (*Fomes*); 1915 (WUS 3[1]): 68 *pl. 8 f. 41*; Pilát 1940 (ACE 3): 320 *f. 136, pl. 124* (*Trametes*).

cytisina (Berk.) Bond. & S. 1941 (33). — *Polyporus* Berk. 1836 (England); *Fomes* Gillet 1878; *Scindalma* O.K. 1898; *Ungulina* Murašk. 1940. — Sensu Lloyd 1915 (LMW 4, F.): 230, 279 (as syn.); Donk 1933: 210; Pilát 1940 (ACE 3): 357 *f. 154, pl. 239 f. b, pl. 240* (*Fomes*); Murašk. 1940: 9 (*Ungulina*); Müll. & Jahn 1966 (WPb 6): 13 *fig.* (*Fomitopsis*); Donk 1971 (PNA 74): 33, notes.

   Polyporus gibbosus Pers. 1825, not ~ Bl. & Nees 1826: Fr. 1828; not ~ (Pers. per Fr.) Kumm. 1871; fide Berk. 1836: 142 (as to Sow. pl. 288). — [≡ *Boletus suberosus* L. sensu Sow. 1800: *pl. 288* (England)].

   Polyporus sublinguaeformis S. Schulz. in Linhart 1882 F.h.: No. 54 [repr. 1883 (H 22): 22] (Hungary); fide Lloyd 1914 (LMW 4, L. 52): 26 = *Fomes fraxineus* [sensu Lloyd] & Donk 1971 (PNA 74): 37.

   Placodes incanus Quél. 1886: 172 (France); fide Bourd. & G. 1928: 602 = *Ungulina fraxina* [sensu Bourd. & G.] & cf. Donk 1971 (PNA 74): 37. — *Polyporus* Pat. apud Rolland 1890, not ~ (Lév.) Lév. 1846.

   Polyporus induratus Lloyd 1918 (LMW 5, L. 68): 11 (U.S.A., Illinois), not ~ Berk. 1875, not ~ Peck 1879; fide Overh. 1953: 50, 51 & Lowe 1955 (M 47): 218 = *Fomes fraxineus* [sensu Lloyd].

   Fomes fraxineus Lloyd 1915, not ~ (Bull. per Fr.) Cooke 1885; fide Lloyd 1915 (LMW 4, F.): 230 = *Fomes cytisinus* (cited as syn.). — [≡ *Boletus fraxineus* Bull. sensu Lloyd, l.c., excl. of type (Europe)]. — Lloyd 1915 (LMW 4, F.): 230 (*Fomes*).

   M.—*Boletus suberosus* L. (**O**) sensu Sow. 1800: *pl. 288* → *Polyporus gibbosus* Pers.

   M.—*Boletus ulmarius* Sow. sensu Gillet 1890–6 S.P.: *pl. 292* (*Fomes*).

   M.—*Boletus fraxineus* Bull. sensu auctt. plur.; Lloyd 1915 (LMW 4, F.): 230, excl. of type (*Fomes*). — Dumée 1917 (BmF 33): 31 (*Polyporus*); Bourd. & G. 1928: 602 (*Ungulina*); Vuyck 1924–30 (Fb 27): *pl. 2162*, plate only; Overh. 1953: 50 *pl. 75 f. 433, pl. 98 f. 559, pl. 126 fig.*; Lowe 1957 F.: 66 *f. 48* (*Fomes*). → *Fomes fraxineus* Lloyd.

pinicola (Sw. per Fr.) P. Karst. 1881. — *Boletus* Sw. 1810 (Sweden) (d.n.); *Polyporus* (Sw.) per Fr. 1821; *Boletus* Wahl. 1826, not ~ Vitt. 1834, Rea 1913; *Fomes* Fr. 1849 (nom. altern.); *Fomitopsis* P. Karst. 1881; *Trametes* P. Karst. 1881; *Placodes* Pat. 1887 (nom. nud.: n.v.p.); *Pseudofomes* Lázaro 1916; *Ungulina* Sing. 1930; ≡ *Trametes pini* Thüm. 1876 ("Fckl."), not ~ (Brot. per Fr.) Fr. 1838; not ~ (Thore per Pers.) Britz. 1887. — Sw. 1810 (SVH 31): 88 (*Boletus*); Secr. 1833 M. 3: 75; Fr. 1874: 561 (*Polyporus*); Gillet 1874–90 P.: *pl. 291/464*; Lloyd 1915 (LMW 4, F.): 219; 1923 (LMW 7): 1207, a chemical test; Farl. & Burt 1929: 101 *pl. 94 fig.*; Mounce *1929*; Donk 1933: 207 (*Fomes*); Jørst. & Juul 1939 (MnS 6[3]): 327, 475 *fs. 6–11*; Nannf. & Du R. 1952: 248 *pl. 128* (*Polypors*); Overh. 1953: 42 *pl. 59 f. 352, pl. 65 f. 383, pl. 67 f. 393, pl. 126 fig.*; Lowe 1957 F.:

85 *f. 66*; Westh. 1971 (Bo 10); 186 *fs. 9, 10*, with cult. char. (*Fomes*); Donk 1973 (PNA 76): 222, notes.

Boletus ungulatus Schaeff. 1774 (Germany) (d.n.); fide Bres. 1890 (Rm 12): 105 = *Fomes marginatus*. — *Pyreium* Paul. 1812–35 (generic name n.v.p.), misapplied; *Boletus* Schaeff. per St-Am. 1821, Nocca & Balb. 1821, Laterr. 1821, misapplied, *Polyporus* Balbis 1828, misapplied, Sacc. 1879, not ∼ (Berk.) Cooke 1885; *Fomes* Bizzoz. 1885, Sacc. 1888, not ∼ Lázaro 1916; *Scindalma* O.K. 1898; *Fomitopsis* P. Karst. 1898; *Ungulina* Pat. 1914; *Placodes* Rick. 1918. — Schaeff. 1774: 88 [*pls. 137, 138*] (*Boletus*); Bres. 1897 (AAR III 3): 76; 1931 (BIm 20): *pl. 990.* — Sensu Bull. → *Fomes fomentarius*; sensu Secr., in part, → *Phellinus pomaceus*.

Boletus fulvus Schaeff. 1774: 89 [*pl. 262*] (Germany) (d.n.), not ∼ Scop. 1772 (d.n.), not ∼ Willd. 1787 (d.n.); fide Pers. 1800: 106 & Fr. 1821: 372 = *Boletus/Polyporus marginatus*; fide Fr. 1838: 468. — *Fomitopsis* (Schaeff.) per P. Karst. 1891.

Boletus semiovatus Schaeff. 1774: 92 [*pl. 270*] (Germany) (d.n.), Pers. 1800: 108 ("*semiovoideus*"; syn.: n.v.p.); fide Fr. 1874: 561 (var.); fide Bres. 1897 (AAR III 3): 76 = *Fomes ungulatus*. — *Polyporus* (Schaeff.) per Britz. 1889, misapplied; *Scindalma* O.K. 1898; ≡ (by lecto-typification) *Boletus ellipticus* Pers. 1800 (d.n.) per Spreng. 1827. — Sensu Britz. → *Heterobasidion annosum*.

? *Boletus hippocrepis* Schrank 1789: 616 (Germany) (d.n.); fide Schrank, l.c. = Schaeff. *pl. 137* [=*Boletus ungulatus*] (cited as syn.). — *Agaricus* Humb. 1793 (syn.; error). — Fr. 1832[Ind.]: 59 referred this to *Polyporus igniarius* [sensu lato].

Boletus resinosus Schrad. 1794 (Germany) (d.n.), not ∼ Rubel 1778 (d.n.); fide Bres. 1897 (AAR III 3): 73, 76 = *Polyporus marginatus* & *Fomes ungulatus*. — *Polyporus* (Schrad.) per Fr. 1821, misapplied; *Boletus* Spreng. 1827, misapplied; *Ischnoderma* P. Karst. 1879, Litsch. & Lohw. 1939 ("*Inoderma resinaceus*"), misapplied; *Placodes* Quél. 1886, misapplied; *Ochroporus* J. Schroet. 1888, misapplied; *Ganoderma* Hariot 1902 ("Pat."), presumably misapplied; *Fomes* Big. & Guill. 1913, misapplied; *Ungulina* Pat. fide Maire & al. 1903; *Placodes* Ulbr. 1928, misapplied. — Schrad. 1794: 171 (*Boletus*). — Sensu Fr. → *Ischnoderma benzoinum*; sensu Rostk. → *Heterobasidion annosum*; sensu Quél. → *Ganoderma pfeifferi* (**43**).

Boletus marginatus Pers. 1794 (Germany) (d.n.); fide Fries 1828 E. 1: 105–106 (*Polyporus pinicola* cited as a syn.). — *Polyporus* (Pers.) per Fr. 1821, not ∼ Fr. 1838, not ∼ Krause 1925; *Boletus* Spreng. 1827; *Placodes* Quél. 1886; *Mensularia* Lázaro 1916. — Pers. 1799 O. 2: 6; 1801: 534 (*Boletus*); Fr. 1828 E. 1: 105 (*Polyporus*); Bres. 1890 (Rm 12): 105, notes (*Fomes*); Bourd. & G. 1928: 601 (*Ungulina*); Bres. 1931 (BIm 20): *pl. 991?*; Konr. & M. 1932 I. 5: *pl. 449* (Fomes); Donk 1973 (PNA 76): 222, notes. — Some refs. to descrs. and notes that should have been cited under *Polyporus marginatus* Fr. (see this sp. below) are included here. — Sensu Rostk. → *Polyporus marginatus* E. Krause, see this sp., below.

Polyporus cinnamomeus Trog 1832 (Fl 15): 556 (Switzerland), not ~
(Jacq. per S. F. Gray) Pers. 1825. — *Boletus* Lenz 1840, not ~ Jacq. 1787
(d.n.), not ~ Schum. 1803 (d.n.); *Fomes* Fr. 1849 (nom. altern.), misapplied fide Fr. 1863 M. 2: 253; *Scindalma* O.K. 1898; *Placodes* Rick.
1918, misapplied. — This has been referred to *Phellinus pomaceus*, certainly
in error; cf. Quél. 1888: 399 (=*Placodes igniarius* var. *pomaceus*), Bres.
1897 (AAR III 3): 75 (= *Fomes fulvus* sensu Bres.).

Polyporus helveolus Rostk. 1837 (StP 4): 73 *pl. 35* (Germany/Poland),
not ~ Fr. 1828; fide Bres. 1920 (Am 18): 67 = *Fomes ungulatus*. —
Ischnoderma P. Karst. 1879; *Piptoporus* P. Karst. 1882; *Placodes* Quél.
1886; *Coriolus* Quél. 1890, misapplied. — Sensu Quél. 1890 → *Buglossoporus pulvinus*. — The original descr. contains, "Die Substanz ist . . .
weich, saftig, getrocknet sehr leicht." I have seen specimens of *Fomitopsis
pinicola* with such a context, which was at the same time quite tough.

Polyporus marginatus Fr. 1838: 468 (Sweden), not ~ (Pers.) per Fr.
1821, not ~ E. Krause 1925; *Fomes* Fr. 1849 (nom. altern.) ["Epicr.
(non P.)"]; *Fomitopsis* P. Karst. 1881 ("Fr."; and also admitting *F.
pinicola* as a distinct sp.); *Trametes* P. Karst. 1882 (same remarks);
Scindalma O.K. 1898 ["Fries"; following Sacc. 1886 (SF 6): 168 who
cites "Fr. Epicr. p. 468 . . ."]; *Ungulina* Pat. 1900 ("Fr."). — For descrs.
and notes, see under *Boletus marginatus*, above, and (34).

Antrodia tuber (P. Karst.) P. Karst. 1898; fide Lowe 1956 (M 48): 100
(young pulvinate specimen). — *Antrodia serpens* var. P. Karst. 1889
(BFi 48): 324 (Finland).

Polyporus ponderosus Schrenk 1903, not ~ Kalchbr. 1882; fide Overh.
1953: 42 & Lowe 1957 F.: 85. — *Fomes* Murrill 1908 (syn.: n.v.p.), not ~
(Kalchbr.) Cooke 1885. — Schrenk 1903 (BPI 36): 30 *pl. 7 fs. 10, 11,
pl. 11 f. 3, pl. 13 f. 2.*

Friesia rubra Lázaro 1916 (RMa 14): 589 & 1917 (RMa 15): *pl. 2* /
1917: 92 *pl. 2* (Spain); fide Bres. apud Trott. 1925 (SF 23): 402, "potius" =
Fomes ungulatus. — *Ganoderma* Sacc. & Trott. apud Trott. 1925.

Fomes ganodermicus Lázaro 1916 (RMa 14): 664 / 1917: 103 (Spain);
fide Bres. apud Trott. 1925 (SF 23): 398 = *Fomes ungulatus*.

Fomes fuscatus Lázaro 1916 (RMa 14): 666 / 1917: 105 (Spain); fide
[Bres. in litt.? apud] Trott. 1925 (SF 23): 391, "An *Fomes ungulatus* Sacc.?"

Polyporus marginatus E. Krause 1925 (ANM II 1): 129 ["Sturm 16 t. 43
(vix Fries)"], not ~ (Pers.) per Fr. 1821, not ~ Fr. 1838. — [≡ *Polyporus
marginat(s* (Pers.) per Fr. sensu Rostk. 1837 (StP 4): 89 *pl. 43* ("Fries")
(Germany/Poland)]; ≡ *Polyporus marginatoides* E. Krause 1928 B.r.: 54.

M.—*Boletus igniarius* L. sensu Vahl 1787 (Fd 6 / F. 16): 8 *pl. 953*; cf. Pers.
1818: 94 & fide Fr. 1838: 468 = *Polyporus pinicola* ("eleganter"). — Pers.
1799 O. 2: 5; 1801: 534; Donk 1971 (PNA 74): 408, note.

M.—*Boletus sanguineus* L. (**O**) sensu Liljebl. 1798: 453; fide Sw. 1810
(SVH 31): 89.

M.—*Boletus pini* Brot. sensu Fuck. 1871 (Jna 25–26): 290 ("*T*[*rametes*]

Pini Fr. Epicr. p. 489"; lacking descr.); fide Thüm. 1876 M.u.: No. 7 corrected label & 1876 (Fl 54): 204 (*Polyporus pinicola* cited as syn.). → *Trametes pini* Thüm., see this species, cited above.

M.—*Boletus quercinus* Schrad. sensu Cartwr. 1951 (TBS 34): 604 *pls. 29, 30* (*Polyporus*).

M.—*Polyporus erubescens* Fr. sensu Kotl. & P. 1959 (ČM 13): 36 (as a syn.); fide Kotl. & P., l.c., = *Fomitopsis pinicola.*

rosea (A. & S. per Fr.) P. Karst. 1881 (33). — *Boletus* A. & S. 1805 (Germany) (d.n.); *Polyporus* Fr. 1818 (d.n.); *Polyporus* (A. & S.) per Fr. 1821, not ∼ (Mont.) Speg. 1889; *Boletus* Spreng. 1827; *Fomes* Fr. 1849 (nom. altern.); *Trametes* P. Karst. 1881, not ∼ Lloyd 1915; *Placodes* Quél. 1886; *Scindalma* O.K. 1898; *Ungulina* Pat. 1900. — Lloyd 1915 (LMW 4, F.): 223 *f. 576*; Bres. 1931 (BIm 20): *pl. 1000*; Shope 1931 (AMo 18): 387 *pl. 29 fs. 4–6* (*Fomes*); Bourd. 1932 (BmF 48): 228 (*Ungulina*); Donk 1933: 208; Overh. 1953: 56 *pl. 63 f. 375, pl. 64 fs. 380, 381, pl. 126 fig.*; Lowe 1957 F.: 71 *f. 54*; Pilát 1940 (ACE 3): 352 *f. 151, pl. 235 f. b, pl. 236* (*Fomes*); Pegl. & Wat. 1968 (CDp): no. 191 *figs.* (*Fomitopsis*). — V.s.: "*Ungulina noxa* [!] ..." R. Heim 1942: 23.

Polyporus rufopallidus Trog 1832 (Switzerland); fide Quél. 1888: 397. —*Fomitopsis* P. Karst. 1881; *Fomes* Cooke 1885; *Scindalma* O.K. 1898. — Trog 1832 (Fl 15): 556; 1844 (MiB[15–23]): 54; Fr. 1884 I. 2: 84 *pl. 186 f. 1*, poor.

Trametes přibramensis Pilát 1927 (Sčz 2): 484 *pl. 11 f. 26* (Czecho-slovakia); fide Pilát 1941 (ACE 3): 353 (forma).

FUNALIA Pat. (35)

1900 [1960 (Pe 1): 218]. — Lectotype: *Polyporus funalis* Fr.
Trametella Pinto-L. 1952 [1960 (Pe 1): 288]. — Holotype: *Trametes hispida* Bagl.

SPECIAL LITERATURE.—Anonymus, *1892*; Birkfeld, *1965*; Cartwright, *1951a*; Fritzsche & Herschel, *1969*; Guillemot, *1894*; Hariot, *1891* (cf. Anonymus, *1892*); Kreisel, *1962*; Schulzer von Müggenburg, *1883*; Smith, *1930*.

gallica (Fr.) Bond. & S. 1941 (**36, 37**). — *Polyporus* Fr. 1821 Jan. 1; *Trametes* Fr. 1838; *Hexagona* Quél. 1886; [≡ *Boletus favus* L. sensu Bull., excl. of type 1788, (France); Lloyd 1912 (LMW 4): 521 ("*flavus*"); error & incidental mention)]; ≡ *Poria vulgaris* S. F. Gray 1821, typonym, not ∼ (Fr.) Cooke 1886; ≡ *Hexagona favus* Quél. 1888, typonym, not ∼ (L. per Spreng.) Hariot 1891 (nom. prov.: n.v.p.); ≡ *Trametes favus* Bres. 1908, typonym. — Quél. 1881 (Crf 9[2]): 670, in part; sensu Boud. apud Lloyd 1912 (LMW 4): 520 *f. 517* (*Trametes*); Bond. 1953: 529 *f. 141, pl. 147 fs. 1, 2, pl. 155* (*Funalia*).

Polyporus extenuatus Dur. & Mont. ex Mont. 1856 (Algeria), not ∼

J. Rick ex Lloyd 1924; cf. Pat. 1897: 45 & fide Bres. 1920 (Am 18): 67.
— *Polystictus* Cooke 1886; *Trametes* Pat. 1897; *Microporus* O.K. 1898;
Funalia Domański apud Domański & al. 1967 (incomplete ref.: n.v.p.);
Trametella Domański 1968. — Pat. 1897: 45, note on nomenclature;
Malenç. 1956 (BmF 71): 293; Kreisel 1962 (Bba 35): 55 *f. 2* (*Trametes*);
Domański *1968*, with cult. char. (*Trametella*); Mal. & Bert. 1971 (Apb 8):
30, notes (*Trametes*).

Trametes hispida Bagl. 1865 (Italy); fide Bres. 1908 (Am 6): 39 = *Trametes favus* [sensu Bull., Bres.]. — *Polyporus* J. Rick 1940, misapplied,
not ∼ (Bull.) Fr. 1821, which is the name Rick had in mind; *Daedalea*
E. Krause 1928; ≡ *Agaricus hispidissimus* E. Krause 1932. — Fr. 1874:
583; Bres. 1890 (BmF 6): xlvi, in part; Gillet 1893 S.P.: *pl. 657* & 1898
L.H.: 22; Guillemot 1894 (BmF 10): 74; Bourd. & G. 1928: 587; Al.
David 1967 (Nca 94): 565 *tpl. 1 f. 6, tpl. 2 f. e*, cult. char. (*Trametes*).
— Sensu Gillot & Luc. → *Funalia trogii*. — References to descrs. and
illustrations based on North American material are not cited (**36**).

Polyporus schulzeri Kalchbr. 1868 (VW 18): 431 [repr. 1868 (H 7): 182]
(Yugoslavia, Slavonia), not ∼ Fr. 1874; fide Bres. 1890 (BmF 6): xlvi (as
to *P. vulpinus* Fr. sensu Kalchbr.) = *Trametes hispida* [sensu Bres.,
conceived incl. of *Funalia trogii*]; fide Donk 1972 (PNA 75): 166–168.
— ≡ *Polyporus tyrolensis* Sacc. & Cub. apud Sacc. 1888. — Kalchbr.
1877: 58 *pl. 37 f. 1a* (*P. schulzeri*, as part & syn. of *P. vulpinus* Fr. sensu
Kalchbr.); S. Schulz. 1880 (ÖbZ 30): 109; Donk, l.c., notes.

Polyporus sarrazinii S. Schulz. 1883 (Yugoslavia, Slavonia); fide Bres.
apud Bourd. & G. 1928: 535 = *Trametes hispida* [sensu Bres., conceived
incl. of *Funalia trogii*]. — S. Schulz. 1883 (Rm 5): 257; Donk 1972 (PNA
75): 168, note.

M.—*Boletus favus* L. (**0**) sensu Bull. 1788. — Bull. 1788: *pl. 421* &
1791 H.: 363, rather an aberrant form (*Boletus*). → *Polyporus gallicus*
Fr.; → *Poria vulgaris* S. F. Gray; → *Hexagona favus* Quél.; → *Trametes
favus* Bres.

M.—*Polyporus vulpinus* Fr. 1852 sensu Kalchbr. 1877: 56 *pl. 37*, in
part, viz. *f. 1a*, not "fig. b" as stated in the text; fide Kalchbr., op cit.,
p. 57 = *Polyporus schulzeri* Kalchbr. (cited as a syn.) *q.v.*

M.—*Boletus lutescens* Pers. sensu Bres. 1896 (H 35): 284, in part (*Trametes*). — Bres. 1897 (AAR III 3): 89, in obs. (*Trametes lutescens* f. umbrina).

trogii (Berk. apud Trog) Bond. & S. 1941 (**36**). — *Trametes* Berk. apud
Trog 1850 (Switzerland). — Trog 1850 (MiB): 52; Cooke 1892 (G 21):
46; Gillet 1893 S.P.: *pl. 659* & 1898 L.H.: 23; Guillemot 1894 (BmF 10):
74; Bourd. & L. Maire 1920 (BmF 36): 82 (*Trametes*); Bourd. & G. 1928:
588 (*Trametes hispida* subsp.); Konr. & M. 1930 I. 5: *pl. 444*; Donk 1933:
189; Kreisel 1962 (Bba 35): 55 *f. 2*; Poelt & Jahn 1965: *pl. 37 fig.*; Al.
David 1967 (NCa 94): 566 *f. 7, tpl. 2 f. f.*, cult. char. (*Trametes*); Wojew.
1967 (Ffg 13): 155, 161 *fs. 1, 2* (*Funalia*); Domański 1968 (APo 37):

126, 137, 143 *fs. 3, 4, 5 at left, 8, 10, 12* (*Trametella*). — References to descr. and illustrations based on North American material are not cited on purpose (36).

Polyporus hausmannii Fr. ex Kalchbr. 1868 (VW 18): 430 [repr. 1868 (H 7): 182] & Fr. 1874: 552 (Italy, Tirol); fide Bres. 1897 (AAR III 3): 90 = *Trametes lutescens* [sensu Bres.] ("neque . . . diversus videtur"). — Donk 1972 (PNA 75): 169, notes.

Inodermus maritimus Quél. 1887 (Crf 15²): 487 *pl. 9 f. 8* (France); cf. Bourd. & G. 1928: 561, in obs., & fide Bourd. apud Konr. & M. 1930 I. 5: *pl. 444*, in obs. — *Polyporus* Rolland 1890, Sacc. 1891; *Coriolus* Boud. apud Niel 1894 (incidental mention: n.v.p.); *Trametes* Pat. 1900; *Polystictus* Cost. & Duf. 1895; *Polystictoides* Lázaro 1916.

Trametes rhodostoma (Forq. ex Quél.) Lloyd 1908 (n.v.p.?), Big. & Guill. 1913; fide Bourd. & G. 1928: 589 (var.). — *Trametes hispida* var. Forq. ex Quél. 1888: 372 (France).

M.—*Boletus populinus* Schum. sensu Passerini 1872 (NGi 4): 155 (*Trametes*).

M.—*Polyporus vulpinus* Fr. (1852) sensu Kalchbr. 1877: 56 *pl. 37*, in part: as to *f. 1a* (not "fig. b" as stated in text); fide Kalchbr., op. cit., p. 57 = *Polyporus schulzeri* Kalchbr. (cited as a syn.).

M.—*Trametes hispida* Bagl. sensu Gillot & Luc. 1889. — Gillot & Luc. 1889 (BAt 2): *pl. 4 f. 1* & 1890 (BAt 3): 193.

M.—*Boletus lutescens* Pers. sensu Bres. 1896 (H 35): 284 (*Trametes*), in part. — Bres. 1897 (AAR III 3): 89 [*Trametes lutescens* f. *trogii* (Berk. apud Trog) Bres.].

GANODERMA P. Karst. (38).

1881 [1960 (Pe 1): 219]. — Monotype: *Polyporus lucidus* (Curt.) per Fr.

Elfvingia P. Karst. 1889 [1960 (Pe 1): 207]. — Monotype: *Polyporus applanatus* (Pers. per S. F. Gray) Wallr.

Dendrophagus Murrill 1905, not ~ Toumey 1900 (n.v.; Loranthaceae) [1960 (Pe 1): 206]. — ≡ *Tomophagus* Murrill 1905 [1960 (Pe 1): 287]. — Holotype: *Polyporus colossus* Fr.

Friesia Lázaro 1916, not ~ Spreng. 1818 (Euphorbiaceae); not ~ DC. 1824 (Elaeocarpaceae); not ~ P. Wieselgren 1846 (Orchidaceae), not ~ Frič ex Kreuzinger 1929 (Cactaceae) [1960 (Pe 1): 218]. — Lectotype: *Polyporus applanatus* (Pers. per S. F. Gray) Wallr.

Trachyderma (Imaz.) Imaz. 1952, not ~ Norm. 1853 (Lichenes) [1960 (Pe 1): 287]. — *Ganoderma* subgen. *Trachyderma* Imaz. 1939. — Monotype: *Ganoderma tsunodae* (Yas.) Trott.

SPECIAL LITERATURE.—General & miscellaneous: Bondarcev, *1934, 1960*; Bose, *1933, 1935*; Coleman, *1927*; Džafarov, *1957*; Furtado, *1962, 1965*; Haddow, *1931*; Hansen, *1958*; Heim, *1963*; Kreisel, *1963b*; Patouillard, *1889a*; Schulzer von Müggenburg, *1878*; Steyaert, *1961a, 1967*; Walch, *1967, 1968*; Walch & Kühlwein, *1968*.

Ganoderma adspersum: Donk, *1969b*; Jahn, *1960, 1964*; Kotlaba &

Pouzar, *1971*; Kreisel, *1960*; Schulzer von Müggenburg, *1880b*; Tortić, *1971*.

Ganoderma applanatum (for galls, see next paragraph): Aoshima, *1953a*, *1954a*, *1954b*; Atkinson, *1908b*; Banerjee & Sarkar, *1958*; Berlese, *1889*; Bondarcev, *1936b*; Buller, *1922c*; Eyerdam, *1953*; Humphrey & Leus/ Leus-Palo, *1931, 1932*; Josserand, *1935*; Kramer & Long, *1971*; Lazebniček, *1962*; Romell *1916*; Schulzer von Müggenburg, *1880b*; Sreeramulu, *1963*; White, *1919*; Yamano, *1931*.

Ganoderma applanatum (galls): Eisfelder & Herschel, *1966*; Haase, *1961*; Hennig, *1935*; Jahn, *1959*; Kallenbach, *1933–4a*; Koppe, *1956*; Kreh, *1939*; *1954*; Moens, *1971*; Ricek, *1968*; Ulbrich, *1939a*; Weidner & Schremmer, *1962*.

Ganoderma lucidum: Atkinson, *1908a*; Banerjee & Sarkar, *1958, 1959*; Bose, *1929, 1940*; Boudier, *1899*; Butler, *1909*; Chow & Chen, *1935*; Demelius, *1917*; Edgerton, *1907*; Guéguen, *1901*; Høeg & Jørstad, *1938*; Krebs, *1961*; Reichert & Avizohar, *1939*; Sarkar, *1959a, c, d*; Schmitz & Zeller, *1919*; Steyaert, *1961b*; Ulbrich, *1940*; Van Bambeke, *1895*; West, *1919* (*G. tsugae*).

Ganoderma pfeifferi: Leontovyč, *1956*; Schröder, *1961*; von Wettstein, *1885*.

Ganoderma resinaceum: Džafarov, *1955, 1959*; Jahn, *1966b*; Kallenbach, *1932*; H. Lohwag, *1938b*.

adspersum (S. Schulz.) Donk 1969 (**42**). — *Polyporus* S. Schulz. 1878 (Fl 61): 11 (Yugoslavia, Slavonia); *Ganoderma* Bres. (in herb.) apud Bourd. & G. 1925 (syn.: n.v.p.). — Kotl. & P. 1971 (ČM 25): 88 *fs. 1–5*; Tortić 1971 (Abc 30): 113 *ill.*; Donk 1969 (PNA 72): 278, notes.

Polyporus linhartii Kalchbr. in Linhart 1884 F.h.: No. 252 (with descr.) (Hungary); fide Donk 1969 (PNA 72): 280. — *Ganoderma* Igmándy 1968.

Fomes undatus Lázaro 1916 (RMa 14): 661 & 1917 (RMa 15): *pl. 3 /* 1917: 100 *pl. 3*; fide Bres. apud Trott. 1925 (SF 23): 395 = *Ganoderma australe* [sensu Bres.]; & cf. Donk 1969 (PNA 72): 281.

Ganoderma europaeum Steyaert 1961 (Belgium); fide Donk 1969 (PNA 72): 278. — Steyaert 1961 (BBr 31): 70; H. Jahn 1963 (WPb 4): 85 *Abb. 3, 4, 5 at right, 47 in part*; Steyaert 1967 (BBB 100): 205 *fs. 21, 22, pl. 3*, on pl. colour of context poor.

M.—*Polyporus australis* Fr. (**O**) sensu Fr. in herb. K, & Fr. 1874: 556, at least in part: as to certain European collections; fide Steyaert 1961 (BBr 31): 71 = *Ganoderma europaeum*. — Humphr. & Leus 1932 (PJS 42): 160, 176 *pl. 5 f. 1, pl. 8 f. 1, pl. 9 f. 4, pl. 12 f. 1*, descr. of Fries's specimen in K; & cf. Donk 1969 (PNA 72): 276, notes (*Ganoderma*).

M.—*Boletus applanatus* Pers. sensu Pat. 1889 (BmF 5): 67 (*Ganoderma*), at least in part: as to European collections. — Bres. 1890 (Rm 12): 104, in obs.; Ingold 1953: 87 *f. 51*, fide H. Jahn 1963 (WPb 4): 90 = *G. europaeum* (*Ganoderma*); & cf. Donk 1969 (PNA 72): 274, notes.

M.—*Ganoderma pfeifferi* Bres. apud Pat. sensu Kreisel 1960; fide H. Jahn 1963 (WPb 4): 86 = *Ganoderma europaeum.* — Kreisel 1960. (WPb 2): 85; H. Jahn 1960 (WPb 2): 90 *photogr.* [*& lower fig. on p. 86*].

applanatum (Pers. per S. F. Gray) Pat. 1889, not ∼ Bres. 1890 (**39–41**) — *Boletus* Pers. 1799 (Germany) (d.n.) per S. F. Gray 1821; *Polyporus* Secr. 1833 (as a sp. of *Boletus*: n.v.p.), Wallr. 1833; *Fomes* Fr. 1849 (nom. altern.); *Placodes* Quél. 1886; *Ganoderma* Pat. 1887 (nom. nud.: n.v.p.); *Phaeoporus* J. Schroet. 1888; *Elfvingia* P. Karst. 1889; *Friesia* Lázaro 1916. — Gillet 1874–90 P.: *pl. 466/287* (*Fomes*); Gramb. 1913 P.H. 2: 26 *pl. 26* (*Polyporus*); Konr. & M. 1932 I. 5: *pl. 448*, colour of context abnormal; Pilát 1939 & 1942 (ACE 3): only *pl. 248, pl. 308 f. a, pls. 309, 310, pl. 311 f. a* (*Ganoderma*); R. W. Davids. & al. 1942 (TUS 785): 18 *f. 3D, pl. 1 f. C*, cult. char. (*Fomes*); Nannf. & Du R. 1952: 249 *f. 194, pl. 129* (*Polyporus*); Overh. 1953: 98 *pl. 72 f. 419, pl. 79 fs. 449–452, pl. 105 f. 587, pl. 106 f. 590, pl. 125 fig.* (*Fomes*); L. Hansen 1958 (BT 54): 336 *fs. 1–4*, hyphal analysis; A. Teix. 1961 (M 52): 33 *f. 3*, generative hyphae; H. Jahn 1963 (WPb 4): 89 *Abb. 1, Abb. 5 in part*; Poelt & Jahn 1963: *pl. 58*; Birkf. & Hersch. 1964: *pl. 97*, galls; Steyaert 1967 (BBB 100): 207 *fs. 23, 24, pl. 4*; Donk 1969 (PNA 72): 273, notes (*Ganoderma*). — Sensu Pat., at least in part, → *Ganoderma adspersum.* — V.s.: *Polyporus "adplanatus"*: Saut. 1878.

Boletus lipsiensis Batsch 1786 (Germany); fide Pers. 1801: 537 (with doubt), Atk. 1908 (Am 6): 189 (*Boletus applanatus* Pers. cited as a syn.). — *Scindalma* (Batsch) per O.K. 1898; *Elfvingia* Murrill 1903; *Ganoderma* Atk. 1908; *Polyporus* E. Krause 1928; *Agaricus* E. Krause 1932. — Atk. 1908 (Am 6): 179 (*Polyporus applanatus*), 189 *pls. 2–4* (*Ganoderma lipsiensis*).

? *Boletus rubiginosus* Schrad. 1794: 168 (Germany) (d.n.), not ∼ Retz. 1769 (d.n.) per Steud. 1824, not ∼ Fr. 1874 (Boletales). — *Placodes* (Schrad.) per Quél. 1892, misapplied?; *Ganoderma* Bres. 1897, misapplied? — Sensu Quél. 1892 (Crf 20²): 469 (*Placodes*); Bres. 1897 (AAR III 3): 74 & cf. p. 73 (*Ganoderma*); fide Quél., l.c. (*Polyporus applanatus* Pers. cited as syn.). — Donk 1969 (PNA 72): 276, notes.

Polyporus merismoides Corda 1837 (StP 3): 139 *pl. 63* (Czechoslovakia); fide Fr. 1874: 558 ("forma monstrosa"). — *Fomes* Bizzoz. 1885.

Polyporus stevenii Lév. 1842 D.: 91 [*pl. 2*] (U.S.S.R., Ukraine) (**39**); fide Lloyd 1915 (LMW 4, F.): 269, 285 = *Fomes leucophaeus.*[6] — *Fomes* P. Karst. 1882, Lloyd 1915 ("*stevensii*", error), J. Rick 1940 ("*stephenii*", syn.); *Scindalma* O.K. 1898.

Polyporus megaloma Lév. 1846 (ASn III 5): 128 (U.S.A., New York)

---

[6] Kotl. & P. 1971 (ČM 25): 88, 99 thought that the possibly oldest name given to *Ganoderma adspersum* could be *Polyporus stevenii*. I am convinced that this is not the case. The colour of the cap as rendered on the plate is a diluted café-au-lait and comes near to that of the form of *G. applanatum* that has been called *Polyporus leucophaeus* Mont.

(39); fide Atk. 1908 (Am 6): 189 = *Ganoderma lipsiense*; fide Bres. 1912 (H 53): 54 (*Ganoderma leucophaeum* cited as a syn.); fide Overh. 1953: 99. — *Fomes* Cooke 1885; *Scindalma* O.K. 1898; *Elfvingia* Murrill 1903; *Ganoderma* Bres. 1912. — Murrill 1903 (BTC 30): 300 (*Fomes*).

Polyporus leucophaeus Mont. 1856 (U.S.A., Ohio) (39); fide Atk. 1908 (Am 6): 180–189 = *Ganoderma lipsiense*; fide Bres. 1912 (H 53): 54 = *Ganoderma megaloma*. — *Fomes* Cooke 1885; *Ganoderma* Pat. 1889, J. Rick 1940 ("*leucotephrum*", syn., error); *Scindalma* O.K. 1898; *Placodes* L. Maire 1910 (nom. nud.: n.v.p.). — Bres. 1897 (AAR III 3): 72 (*Ganoderma*); Lloyd 1909 (LMW 3): 437; 1915 (LMW 4, F): 264 (*Fomes*); Humphr. & Leus 1932 (PJS 49): 168 *pl. 1 fs. 12–15, pl. 5 f. 2, pl. 7 f. 5, pl. 8 f. 8, pl. 10 f. 3, pl. 11 f. 2, pl. 12 f. 8*; Kawam. 1954 I. 2: 220 *f. 210* (*Ganoderma*).

Fomes gelsicola Berlese 1889 (Italy); Lloyd 1915 ("*gelsicolor*"); fide Lloyd 1915 (LMW 4, F.): 280. — *Scindalma* O.K. 1898; *Ganoderma* Sacc. 1916. — Berlese 1889 (Mal 3): 373 *pl. 12* (*Fomes*); Sacc. 1916: 1010 (*Ganoderma*).

Ganoderma applanatum Bres. 1890 (excl. of *Boletus applanatus* Pers.), not ∿ (Pers. per S. F. Gray) Pat. 1889; fide Donk 1969 (PNA 72): 275. — [≡ *Polyporus applanatus* (Pers. per S. F. Gray) Wallr. sensu Fr. 1838: 465, excl. of type, (Sweden)]. — Bres. 1890 (Rm 10): 104.

M.—*Polyporus australis* Fr. (**O**) sensu Pat. 1889 (BmF 5): 71 (*Ganoderma*), at least in part: as to certain European collections. — Pilát 1942 (ACE 3): *pl. 313 f. a*, not described; Domański & al. 1960 (Mob 10): 196, 198 *f. 7* (*Ganoderma applanatum* f.).

carnosum Pat. 1889 (France). — *Fomes* Sacc. 1891; *Polyporus* Cost. & Duf. 1895, not ∿ Rostr. 1902; *Scindalma* O.K. 1898. — Pat. 1889 (BmF 5): 66.

lucidum (Curt. per Fr.) P. Karst. 1881 (**44**). — *Boletus* Curt. 1781 (England) (d.n.); *Polyporus* Fr. 1815 (nom. nud.); *Polyporus* (Curt.) per Fr. 1821; *Boletus* Hook. 1821, not ∿ Velen. 1922; *Grifola* S. F. Gray 1821; *Fomes* Fr. 1849 (nom. altern.); *Placodes* Quél. 1886; *Phaeoporus* J. Schroet. 1888; *Trametes* Hendr. 1948 (syn.: n.v.p.). — Curt 1781 F.l. 2/F. 4/no. 37: pl. 72/216/224; Leyss. 1783: 300 [cf. Steyaert 1961 (Ta 10): 251]; Schrad. 1794; 163; Sow. 1798: *pl. 134* (*Boletus*); Grev. 1826 S. 5: *pl. 245*; Gillet 1874–90 P.: *pl. 457/562* (*Polyporus*); Bamb. 1895 (BJG 7): 93 *pls. 1, 2*, monstrosities (*Ganoderma*); Rolland 1910: 84 *pl. 95*; Dumée 1911 N.A. 2: 50 *pl. 50* (*Polyporus*); Bourd. & G. 1928: 609; Haddow 1931 (JAA 12): 30, 39 *pl. 29 fs. 1–3, pl. 30 fs. 8–11*, inclusive of *G. tsugae*; Donk 1933: 230; Imaz. 1939 (BTS 1): 34 *fs. 1a, b, 2a, 3, 4*; Murašk. 1939: 87 *fs. 12–14*; inclusive sense; Pilát 1941–2 (ACE 3): only *pls. 304, 307, pl. 308 f. b*; L. Hansen 1958 (BT 54): 344 *f. 8*, hyphal analysis; H. Jahn 1963 (WPb 4): 81 *Abb. 1 in part*; Poelt & Jahn 1963: *pl. 59*; Kreisel 1963: 51, 52; Steyaert 1967 (BBB 100): 197 *fs. 11–14, 16–18, pl. 1*, too inclusive (**46**) (*Ganoderma*).

Boletus flabelliformis Leyss. 1761: 219 (Germany) (d.n.), not ~ Scop. 1772 (d.n.), not ~ Schaeff. 1774. — *Boletus* Leyss. per Opiz 1823.

Boletus flabelliformis Scop. 1772: 466, in part: var. 1 (Yugoslavia, Carniola), not ~ Leyss. 1761 (d.n.) per Opiz 1823, not ~ Schaeff. 1774; fide Fr. 1821: 353. — *Ganoderma* (Scop.) per Murrill 1903, 1905, in part.

Agaricus pseudoboletus Jacq. 1773 (Austria) (d.n.); fide Fr. 1821: 533. — *Ganoderma* (Jacq.) per Murrill 1902; *Agaricus* E. Krause 1934. — Jacq. 1773 F.a. 1: 26 *pl. 41*, young, incompletely developed fruitbody with initial growth of cap and pores (*Agaricus*); Atk. 1908 (BG 46): 334 *fs. 1, 2 pl. 9 fs. 1, 2, 6*, inclusive of *G. tsuga* (*Ganoderma*).

Boletus rugosus Jacq. 1774 (Austria) (d.n.), not ~ Sow. 1805 (d.n.), not ~ Fr. 1835 ("Sow.") (Boletales); fide Curt. 1781 F.l. 2; pl. 244 & Fr. 1821: 533. — *Polyporus* (Jacq.) per E. Krause 1928, not ~ (Sow.) per. Pers. 1825, not ~ Bl. & Nees 1826: Fr. 1828, not ~ Trog 1844. — Jacq. 1774 F.a. 2: 44 *pl. 169* [text, repr. Murrill 1902 (BTC 29): 603].

Boletus obliquatus Bull. 1780 (France) (d.n.); fide Fr. 1821: 353. — *Boletus* Bull. per St-Am. 1821, Laterr. 1821; *Ganoderma* Big. & Guill. 1913 (syn.: n.v.p.). — Bull. 1780 *pl. 7*; 1789: *pl. 459*; 1791 H.: 335.

Boletus resupinus O. F. Müll. 1782 (Denmark) (d.n.); Schrad. 1794 ("*resupinatus*", incidental mention); Fr. 1821: 353 (syn.: n.v.p.); fide Fr. 1821: 353 ("*resupin[atus]*"; "monstros."). — O. F. Müll. 1782 (Fd 5 / F. 15): 6 *pl. 894*.

Agaricus lignosus Lam. 1783 (EmB 1): 51 (France?) (d.n.). — Batt. 1755: 70 *pl. 36 f. C* (*Agaricus Dactyloides*, a non-binomial name) cited as syn. by Lam.

Boletus vernicosus Bergeret ?1783 P.: 99 *plate* ("*vernigosus*") (France) (d.n.); fide Fr. 1821: 533. — *Pyreium* Paul. 1812–35 (generic name n.v.p.); *Polyporus* (Bergeret) ex Chev. 1826, not ~ Berk. 1856. — Paul. 1812–35: *pl. 10 | pl. 8 fs. 1, 2* (*Pyreium*).

Boletus crustatus Plan. 1788 I.F.: 23 (Germany) (d.n.); fide Fr. 1821: 533.

Boletus laccatus Timm 1788: 269 (Germany) (d.n.); fide Fr. 1821: 353; fide Pers. 1825: 54 (citing *Boletus lucidus* as a syn.). — *Polyporus* (Timm) per Pers. 1825, not ~ Kalchbr. apud Wettst. 1885, not ~ Velen. 1922; *Ganoderma* Pat. 1897, misapplied, not ~ Bres. 1912, not ~ (Kalchbr. apud Wettst.) Bourd. & G. 1925. — Pers. 1818: 96 (*Boletus*); Bres. 1932 (BIm 21): *pl. 1004* (*Ganoderma*). — Sensu Lév. apud Zoll. 1854 = *Ganoderma laccatum* Bres. (tropical).

Boletus nitens var. *crocatus* Batsch 1789 (Germany) (d.n.); fide Fr. 1821: 353 ("*B. nitens*, Batsch f. 225"). — Batsch 1789: 113 *pl. 41 f. 225* → *Ganoderma nitens* Lázaro, see this sp., below.

Agarico-igniarium trulla Paul. 1793 T. 2: 95 (descr.), Ind. (name) (France) (d.n.). — Corresponding fig. (Paul. 1812–35: *pl. 10 f. 1 | pl. 8 fs. 1, 2*) published as *Pyreium vernicosum* q.v.

Boletus verniceus Brot. 1804: 468 (Portugal) (d.n.); fide Fr. 1821: 533.

Polyporus semipatera Pers. 1825: 55; fide Fr. 1838: 442. — [≡ *Agaricus fulvus semipateram referens* Batt. 1755: 69 *pl. 35 f. D* (Italy)]. — The picture suggests *Ganoderma lucidum*, but Battarra's text states, "interne fuscum pulpam gerebat".

? *Ganoderma tsugae* Murrill 1902 (U.S.A., New York) (**44, 45**); fide Atk. 1908 (BG 46): 335 = *Ganoderma pseudoboletus* (forma). — *Fomes* Sacc. & D. Sacc. 1905; *Polyporus* Overh. 1915. — Murrill 1902 (BTC 29): 601; 1908 (NAF 9): 118 (*Ganoderma*); Atk. 1908 (BG 46): 335 *pl. 19 f. 2* (*Ganoderma pseudoboletus* f.); Overh. 1915 (AMo 2): 709, 714 *f. 6*; 1915 (WUS 3[1]): 45 *tpl. 5 f. 23* (*Polyporus*); Murrill 1919 (M. 11): 101 *pl. 6 f. 1*; Imaz. 1939 (BTS 1): 41 *fs. 9, 10* (*Ganoderma*); Overh. 1953: 210 *pl. 57 fs. 341–345, pl. 85 f. 489, pl. 105 f. 585, pl. 132 fig.* (*Polyporus*).

Ganoderma nitens Lázaro 1916 (RMa 15): 104 / 1917 :196 ["(Batsch)"], not ∼ (Fr.) Pat. 1889; fide Fr. 1821: 353 (for "*B. nitens*, Batsch f. 225"). — ≡ *Boletus nitens* var. *crocatus* Batsch 1789: 113 *pl. 41 f. 225* (d.n.), not *B. nitens* Batsch, the type of the specific name being excluded when the epithet 'nitens' was restricted to 'var. crocatus'. — Donk 1971 (PNA 74): 10, note.

? *Ganoderma ostreatum* Lázaro 1916 (RMa 15): 110 & 1917 (RMa 15): 380 / 1917: 202, 294 *pl. 8* (Spain).

M.—*Boletus variegatus* Schaeff. (**O**) sensu O.K. 1898: 517 (*Scindalma*); name used to replace *Fomes lucidus*.

pfeifferi Bres. apud Pat. 1889 (Germany) (**42**). — *Fomes* Sacc. 1891; *Scindalma* O.K. 1898. — Pat. 1889 (BmF 5): 70; Donk 1933: 233; L. Hansen 1958 (BT 54): 341 *fs. 5–7*, hyphal analysis; A. Schröder 1961 (WPb 3): 44 *fig.*; H. Jahn 1963 (WPb 4): 82 *Abb. 2, 54*; Steyaert 1967 (BBB 100); 204 *fs. 19, 20, pl. 2*, on pl. colour of context poor. — Sensu Kreisel 1960 → *Ganoderma adspersum*.

"*Fomes Advena*" Quél. 1872 (MMb II 5): *pl. 19 f. 1 & Polyporus* "*Resinosus* v[ar.] *Advena*" Quél. 1872: 318 (in index of reprint only), both, alternative names (validly published?) to *Polyporus resinosus* (Schrad.) per Fr. sensu Quél. 1872 (MMb II 5): 278/262 (France); fide Quél. 1892 (Crf 20[2]): 469 = *Ganoderma pfeifferi* (cited as a syn.) & cf. (**43**).

Polyporus laccatus Kalchbr. apud Wettst. 1885 (Austria), not ∼ (Timm) per Pers. 1825, not ∼ Velen. 1922; fide Lloyd 1915 (LMW 4, F.); 267 = *Fomes pfeifferi* (cited as a syn.). — *Fomes* Sacc. 1891; *Scindalma* O.K. 1898; *Ganoderma* Bourd. & G. 1925, not ∼ (Timm per Pers.) Pat. 1897, not ∼ Bres. 1912. — Lloyd 1915 (LMW 4, F.): 267 *f. 603* (*Fomes*); Bourd. & G. 1928: 612 (*Ganoderma*).

Ganoderma soniense Steyaert 1961 (Belgium); fide Steyaert 1967 (BBB 100): 204. — Steyaert 1961 (BBr 31): 71.

M.—*Boletus resinosus* Schrad. sensu Quél. 1872 (MMb II 5): 278/262 (*Polyporus*) & 1888: 400 (*Placodes*) at least in part (**43**); fide Bres. 1890 (Rm 12): 103; & cf. Quél. 1892 (Crf 20[2]): 469. → *Polyporus resinosus* var. *advena* & *Fomes advena* Quél.

M.—*Polyporus vegetus* Fr. (**O**) sensu Rick. 1918 (*Placodes*); fide H. Jahn 1963 (WPb 4): 82. — Rick. 1918: 223; 1920: 232 (*Placodes*).

resinaceum Boud. apud Pat. 1889 (France) (**46**). — *Fomes* Sacc. 1891, Rea 1911; *Polyporus* Rolland 1890; *Scindalma* O.K. 1898 ("*retinaceum*"); *Friesia* Lázaro 1916. — Pat. 1889 (BmF 4): 72; 1897: 51 (*Ganoderma*); Bourd. & G. 1928: 610 (*Ganoderma lucidum* subsp.); Kallenb. 1932 (ZP 11): 18 *pl. 3, pl. 4 below*; L. Hansen 1958 (BT 54): 347 *f. 9*, hyphal analysis; H. Jahn 1963 (WPb 3): 82; Steyaert 1972 (Pe 7): 95 (*Ganoderma*).

Ganoderma chaffangeonii Pat. 1889 (BmF 5): 74 (Venezuela); fide Steyaert 1972 (Pe 7): 95.

Ganoderma resinaceum var. *martelli* Bres. 1892 (Italy). — Bres. 1892 F.t. 2: 31 *pl. 137*; 1932 (BIm 21): *pl. 1013*.

Ganoderma sessile Murrill 1902 (U.S.A., New York) (**46**). — *Fomes* Sacc. & D. Sacc. 1905, Poll. & Kauffm. 1905; *Polyporus* Lloyd 1914 (nom. nud.: n.v.p.), 1915. — Murrill 1902 (BTC 29): 604; 1919 (M 11): 102 *pl. 3*; Haddow 1931 (JAA 12): 40 *pl. 29 fs. 4, 5, pl. 30 fs. 12–15*; Imaz. 1939 (BTS 1): 43 *fs. 1d, 11* & Kawam. 1954 I. 2: 216 *f. 208*, pictures of same specimen (*Ganoderma*).

Polyporus polychromus Copel. 1904 (Am 2): 507 (U.S.A., California); fide Steyaert 1972 (Pe 7): 95. — *Ganoderma* Murrill 1908.

Ganoderma subperforatum Atk. 1908 (U.S.A., Ohio); fide Haddow 1931 (JAA 12): 32, 40 = *Ganoderma sessile*; fide Steyaert 1972 (Pe 7): 95. — Atk. 1908 (BG 46): 337 *f. 5, pl. 19 f. 5*.

Ganoderma praelongum Murrill 1908 (NAF 9): 121 (Cuba); fide Steyaert 1972 (Pe 7): 95.

Ganoderma argillaceum Murrill 1908 (NAF 9): 122 (Cuba); fide Steyaert 1972 (Pe 7): 95.

? *Mensularia vernicosa* Lázaro 1916 (RMa 14): 740 & 1917 (RMa 15): *pl. 4* / 1917: 125 *pl. 4* (Spain); fide Bres. apud Trott. 1925 (SF 23): 398, "Verisimiliter . . . est *Ganoderma resinaceum*." — *Fomes* Sacc. & Trott. apud Trott. 1925.

valesiacum Boud. 1895 (Switzerland) (**47**). — & *Ganoderma* Boud. apud Boud. & Fisch. 1895; *Polyporus* Sacc. 1896; *Fomes* Sacc. & Syd. 1899; *Scindalma* O.K. 1898. — Boud. 1895 (BmF 11): 28; apud Boud. & Fisch. 1895 (BbF 41): ccxxv; Sacc. 1916: 1008 (*Ganoderma*); Bourd. & G. 1928, 609 (*Ganoderma lucidum* subsp.); Bres. 1932 (BIm 21): *pl. 1016*, poor (*Ganoderma*).

GLOEOPHYLLUM (P. Karst.) P. Karst. (49)

1882, not ~ Koršikov 1953 (n.v.p.; Chlorophyta) [1960 (Pe 1): 220; 1962 (Pe 2): 204] *Lenzites* subgen. *Gloeophyllum* P. Karst. 1881. — ≡ *Lenzitina* P. Karst. 1889 [1960 (Pe 1): 235]. — Lectotype: *Lenzites sepiaria* (Wulf. per Fr.) Fr.

Serda Adans. 1763 (d.n.) [1960 (Pe 1): 279]. — Monotype: *Agaricus de St. Clou, nigerrimus* Vaill.

Sesia Adans. per O.K. 1891, misapplied [1960 (Pe 1): 280]. — *Sesia* Adans. 1763 (d.n.). — Monotype: *Agaricus de St. Clou* Vaill. — Sensu O.K. & emend. P. Karst. = *Serpula* (Pers.) per S. F. Gray (Coniophoraceae).

Lenzites Falck 1909 [1960 (Pe 1): 237, foot-note], not Fr. 1835. — [*Lenzites* Fr. sensu Falck 1909 (HF 3): 36, excl. of type.] — Lectotype: *Lenzites abietina* (Bull. per Fr.) Fr.

Reisneria Velen. 1922 [see Pilát 1948: 271 for Latin translation] [1960 (Pe 1): 277]. — Monotype: *Reisneria papyracea* Velen.

Osmoporus Sing. 1944 [1960 (Pe 1): 249]. — Holotype: *Trametes odorata* (Wulf. per Fr.) Fr.

Phaeocoriolellus Kotl. & P. 1957 [1960 (Pe 1): 252]. — Holotype: *Daedalea trabea* (Pers.) per Fr.

Ceratophora Humb. per Corda 1842 (nom. anam. & monstr.), not ∼ Pant. 1889 (Bacillariophyta) [1960 (Pe 1): 195; 1962 (Ta 11): 82]. — *Ceratophora* Humb. 1793 (d.n.). — Monotype: *Ceratophora fribergensis* Humb. — Sensu Bond. & S. 1941, incorrect application, emendation to include perfect and normal fruitbody → *Ceratophora* Bond. & S. (n.v.p.) (**O**).

M.—*Lenzites* Fr. sensu Falck 1909 (HF 3): 36, excl. of type → *Lenzites* Falck.

SPECIAL LITERATURE.—General: David, *1968*; R. Falck, *1909*; Mizumoto, *1955?–7*; Mounce & Macrae, *1936*.

Gloeophyllum odoratum & al. (*"Osmoporus"*): M. E. P. K. Fidalgo, *1962*.

Gloeophyllum saepiarium: Anderson, *1931*; Cartwright, *1929*; Hyde & Walkinshaw, *1966*; Němec, *1925b*; Scheld & Perry, *1970a, 1970b*; Spaulding, *1911*; Ulbrich, *1942*; Zeller, *1916*.

Gloeophyllum trabeum (*"Phaeocoriolellus"*): Amburgey, *1967*; Barnett & Lilly, *1949*; Da Costa & Kerruish, *1965*; Domański, *1960b*; Lilly & Barnett, *1948*; Kerruish & Da Costa, *1963*; K. Lohwag, *1964*; Madhosingh, *1967*; Příhoda, *1953a*.

abietinum (Bull. per Fr.) P. Karst. 1882. — *Agaricus* Bull. 1789 (France) (d.n.); not ∼ Schrad. 1794 (d.n.), not ∼ Otto 1816 (d.n.); *Daedalea* (Bull.) per Fr. 1821; *Agaricus* Mérat 1821, G. F. Re 1821, not ∼ Otto per Steud. 1824; *Merulius* L. March. 1828; *Lenzites* Fr. 1838; *Lenzitina* P. Karst. 1889; *Cellularia* O.K. 1898. — Bull. 1789: *pl. 442 f. 2*; 1791: *pl. 541 f. 1*; 1809 H.: 379; Pers. 1801: 486 (*Agaricus*); Fr. 1828 E. 1: 66 (*Daedalea*); 1863 M. 2: 247; Falck 1909 (HF 3): 38 *pl. 1 fs. 1, 2, pl. 2*; *pl. 3 fs. 1, 3, pl. 5 f. 1, pl. 7 fs. 1–3, 5–6*; Konr. & M. 1927 I. 5: *pl. 440 f. 2*; Bourd. & G. 1928: 581 *f. 167* (*Lenzites*); Donk 1933: 213; Pilát 1940 (ACE 3): 335 *f. 146, pls. 225, 228*; Macrae & Aosh. 1967 (M 58): 922 *f. 21*; Al. David 1968 (BmF 84): 121 *fs. 3, 4*, cult char. (*Gloeophyllum*).

Agaricus asserculorum Schrad. 1794: 134 (Germany) (d.n.), not ∼ Batsch 1783 (d.n.); fide Pers. 1801: 486 & Fr. 1821: 334. — *Daedalea* (Schrad.) Pers. 1815 (d.n.) & (Schrad.) per Schleich. 1821, Secr. 1833.

Irpex umbrinus Weinm. 1836: 372 (U.S.S.R., "Russia minor"); fide Bourd. & G. 1928: 582 (var.). — *Xylodon* O.K. 1898; *Sistotrema* Quél. apud Moug. & Ferry 1887.

Reisneria papyracea Velen. 1922: 739 *fs. 130, 131* [see Pilát 1948: 271 for Latin translation] (Czechoslovakia); fide Pilát 1940 (ACE 3): 335.

odoratum (Wulf. per Fr.) Imaz. 1943. — *Boletus* Wulf. 1788 (Austria) (d.n.); *Polyporus* (Wulf.) per Fr. 1821; *Boletus* Wahl. 1826; *Trametes* Fr. 1838, P. Karst. 1889; *Fomes* P. Karst. 1881; *Ochroporus* J. Schroet. 1888; *Daedaleopsis* P. Karst. 1899; *Anisomyces* Pilát 1940 (generic name n.v.p.); *Ceratophora* Bond. & S. 1941; *Osmoporus* Sing. 1944. — Wulf. 1788 (CoJ 2): 150 (*Boletus*); Secr. 1833 M. 3: 96 (*Polyporus*); Bourd. & G. 1928: 586; Konr. & M. 1930 I. 5: *pl. 443*; Bres. 1932 (BIm 21): *pl. 1025* (*Trametes*); Pilát 1940 (ACE 3): 331 *f. 144, pl. 217 f. b, pls. 222, 223* (*Anisomyces*); Robak 1942 (MVf 7³): 32, 150, cult. char. (*Trametes*); K. Fid. 1962 (Ri 1): 102 *fs. 1–9*; Poelt & Jahn 1963: *pl. 40* (*Osmoporus*); Al. David 1968 (BmF 84): 124, cult. char. (*Gloeophyllum*); D. Reid 1969 (RM 33): 240 (*Osmoporus*). — Sensu Shope → *Gloeophyllum protractum*.

    Boletus annulatus Schaeff. 1774 (Germany) (d.n.), not ∼ Pers. 1801 (d.n.) per Krombh. 1821; fide Pers. 1800: 42 & Fr. 1821: 373. — *Polyporus* S. Schulz. 1866 (nom. nud.: n.v.p.), not ∼ Jungh. 1838; *Trametes* Sacc. 1916 ("*anulata*"). — Schaeff. 1774: 77 [*pl. 106*].

    ? *Helvella subterranea* Houtt. 1783: 649 *pl. 105 f. 3* (Germany) (nom. anam.) (d.n.); fide Schrad. 1794: 170 & Fr. 1821: 373. — Rather a nomen dubium.

    Ceratophora fribergensis Humb. 1793 (Germany) (nom. anam.) (d.n.); fide Schrad. 1794: 170 (var.) & Fr. 1821: 373 (monstrosity). — ≡ *Boletus ceratophora* Hoffm. 1794 (d.n.). — Humb. 1793: 112 *pl. 1* (*Ceratophora fribergensis*); Hoffm. 1794 (GöA³⁹): 378; 1796 (CSg 12): 33 *pl. 6*; 1797–1811 V.s.: 1 *pls. 1, 2, 5* (*Boletus ceratophora*); Pilát 1927 (Sčz 2): 483 *pl. 11 fs. 24, 25* (*Trametes odorata* f. *ceratophora*); 1940 (ACE 3): 333 *pl. 223* (*Anisomyces odoratus* f. *ceratophora*); K. Fid. 1962 (Ri 1): 112 (*Osmoporus odoratus* f. *ceratophora*).

    Boletus polymorphus Hoffm. 1797–1811 V.s.: 3 *pl. 2* (Germany) (d.n.), not ∼ Bull. 1782 (d.n.); fide Fr. 1821: 373 (monstrosity).

    Polyporus vajsii Velen. 1922: 686 [see Pilát 1948: 259 for Latin translation] (Czechoslovakia); fide Pilát 1940 (ACE 3): 331. — *Trametes* Velen. 1922 (nom. altern.)

    M.—*Boletus pini* Thore sensu Britz. 1887 (BAg 29): 280 [*pl. 613 f. 76*] (*Trametes*).

protractum (Fr.) Imaz. 1943. — *Trametes* Fr. 1851 (Sweden); *Fomes* P. Karst. 1881; *Mucronoporus* Sacc. 1891; *Polyporus* Mez 1908; *Lenzites* Lloyd 1910, misapplied, not ∼ Fr. 1851; *Osmoporus* Bond. 1959. — Fr. 1863 M. 2: 272; 1884 I. 2: 89 *pl. 191 f. 3*; Romell 1911 (ABS 11³): 26 *pl. 2 f. 21*; Lundell 1952 (LNF 43–44): 6 No. 2108 (*Trametes*); Bond. 1959 (BŽ 44): 452 *fs. 3–5* (*Osmoporus*). — Sensu Lloyd 1910 → *Gloeophyllum trabeum*.

    Fomes thelephoroides P. Karst. 1887 (Finland); fide Lowe 1956 (M 48): 107. — *Trametes* P. Karst. 1889; *Scindalma* O.K. 1898. — P. Karst. 1887 I. 2: 13 / 1888 (ASf 16): 527 *pl. 11 f. 57* (*Fomes*).

Inonotus ufensis P. Karst. 1904 (ÖfF 46[11]): 3 (U.S.S.R., Baškiria); fide Lowe 1956 (M 48): 108 ("labyrinthiform condition"). — *Polyporus* Sacc. & D. Sacc. 1905.

Trametes americana Overh. 1935 (U.S.A., Pennsylvania). — Lowe 1942: 91; Overh. 1953: 151 *pl. 90 fs. 513–516, pl. 125 fig. (Trametes)*; K. Fid. 1962 (Ri 1): 124 *fs. 8, 9 (Osmoporus odoratus* subsp.).

Ceratophora caucasica (Bres. apud Voronich.) Bond. & S. 1941; fide Bond. 1959 (BŽ 44): 452. — *Trametes odorata* var. Bres. apud. Voronich. 1927 (U.S.S.R., Georgia); *Osmoporus* Sing. 1944; *Anisomyces* Bond. 1953 (generic name n.v.p.). — Nikol. 1940 (TSR 4): 397 *fs. 6, 7 (Trametes odorata* var.); Bond. 1953: 40, 282 *pl. 44 f. 1 (Anisomyces)*; K.Fid. 1962 (Ri 1): 120 *fs. 4–6 (Osmoporus odoratus* var.).

M.—*Boletus odoratus* Wulf. sensu Shope 1931 *(Trametes)*; fide Overh. 1935; 15 (of "American authors") = *Trametes americana*. — Shope 1931 (AMo 18): 370 *pl. 31 f. 4*; R. L. Gilb. 1961 (NwS 35): 8 *f. 3 (Trametes)*.

sepiarium (Wulf. per Fr.) P. Karst. 1882. — *Agaricus* Wulf. 1787 (Austria) (d.n.); *Merulius* Schrank 1789 (d.n.); *Daedalea* P. Gärtn. & al. 1802, Sw. 1810 (d.n.); *Daedalea* (Wulf.) per Fr. 1821; *Agaricus* Zant. 1821; *Lenzites* Fr. 1838; *Lenzitina* P. Karst. 1889. — Wulf. 1787 (CoJ 1): 339 *(Agaricus)*; Secr. 1833 M. 2: 492 *(Daedalea)*; Fr. 1863 M. 2: 247; Falck 1909 (HF 3): 39 *pl. 1 fs. 4, 5* [= *pl. 1a fs. 1, 2*], *pl. 5 f. 2, pl. 7 f. 4*; Michael 1905 F.P. 3: no. 44; Bourd. & G. 1928: 581 *(Lenzites)*; Donk 1933: 214; Pilát 1940 (ACE 3): 337 *f. 147, pl. 226 (Gloeophyllum)*; Robak 1942 (MVf 7[3]): 29, 147, cult. char.; Nannf. & Du R. 1952: 252 *f. 195, pl. 131*; D. Baxt. 1952 (PMi 36): 71 *pls. 2, 4*; Overh. 1953: 111 *pl. 94 fs. 533–536, pl. 100 f. 567, pl. 127 fig. (Lenzites)*; S. Ito 1955: 237 *f. 184*; Al. David 1968 (BmF 84): 120 *f. 1*, cult. char. *(Gloeophyllum)*; Westh. 1970 (Bo 10): 175 *fs. 5, 6*, with cult. char. *(Lenzites)*.

Agaricus hirsutus Schaeff. 1774: 33 [*pl. 76*] (Germany) (d.n.); fide Fr. 1821: 334. — *Cellularia* (Schaeff.) per O.K. 1898; *Sesia* Murrill 1903 (n.v.p.), 1904; *Gloeophyllum* Murrill 1903; *Lenzites* Sacc. 1915; *Daedalea* E. Krause 1925, misapplied; *Agaricus* E. Krause 1934.

Agaricus asserculorum Batsch 1783: 95 (Germany) (d.n.), not ∼ Schrad. 1794 (d.n.); fide Batsch, l.c. = "Sch. LXXVI" [= *Agaricus hirsutus* Schaeff.]; fide Pers. 1801: 487 & Fr. 1821: 334.

Agaricus fasciatus Hoffm. 1789: 255 (Germany?) (d.n.), not ∼ Scop. 1772, not ∼ Schaeff. 1774. — Hoffmann's specific phrase seems to support the view that this was not merely a new name for the three non-binomial synonyms cited, but that he himself had also seen fruitbodies.

Merulius squamosus Schrad. apud J. F. Gmel. 1792: 1431 ("2431") & Schrad. 1794: 140 (Germany) (d.n.); fide Fr. 1821: 324. — Schrader cited several earlier synonyms, but his specific phrase and description strongly suggest that they are drawn up from specimens of *G. sepiarium*.

Merulius umbrinus Pers. 1794 (NMB 1): 106 / 1797 T.: 26 (Germany) (d.n.), not ∼ Pers. 1797; fide Pers. 1799 O. 2: 93 ("Disp. method. fung. p. 26", citing no name).

Agaricus undulatus Hoffm. 1797–1811 V.s.: 7 *pl. 4* (Germany) (d.n.), not ∼ Hoffm. 1789 (d.n.), not ∼ Bull. 1791 (d.n.) per Pers. 1828, not ∼ Jungh. 1830: Fr. 1832; fide Pers. apud Fr. 1821: 334 ("monstr.") & Donk. — *Daedalea sepiaria* var. (Hoffm.) per Pers. 1828; *Lenzites* Sacc. & Trav. apud Sacc. & Trott. 1912.

? *Daedalea aurea* Pers. 1801 (d.n.). — [≡ *Agaricus aureus daedalaeis sinibus* Batt. 1755: 72 *pl. 35 f. F.* (Italy)]; ≡ *Daedalea aurea* Fr. 1821. typonym; *Striglia* O.K. 1891.

Agaricus boletiformis Sow. 1814: *pl. 418* (England) (d.n.); fide Fr. 1832[Ind.]: 9.

Lenzites rhabarbarina B. & C. 1853 (AM II 12): 428 & apud Berk 1872 (G 1): 35 (U.S.A., South Carolina), not ∼ (B. & Cooke) Lloyd 1914; fide Murrill 1904 (BTC 31): 603 = *Sesia hirsuta*. — *Cellularia* O.K. 1898. — Not to be confused with *Daedalea rhabarbarina* B. & Cooke apud Cooke 1878.

Lenzites crocata (Sacc.) Sacc. 1882; fide Lloyd 1923 (LMW 7): 1216. — *Lenzites sepiaria* subsp. Sacc. 1879 (Mi 1): 539 (Italy).

Gloeophyllum abietinellum Murrill 1908 (North America, Rocky Mts.); fide Lowe & Gilb. 1962 (M 53): 506. — *Lenzites* Sacc. & Trott. 1912. — Lloyd 1924 (LMW 7): 1273 *pl. 287 f. 2806*; Overh. 1953: 114 *pl. 127 fig.* (*Lenzites*).

M.—*Agarico-suber applicatus* (Batsch) Paul. (O) sensu Paul. 1793 T.2: 77 (descr.), Ind. (name), in part; fide Donk 1971 (PNA 74): 6.

M.—*Agaricus resupinus* Paul. (O) sensu Paul. 1812–35: *pl. 2 f. 5*, in part; fide Donk 1971 (PNA 74): 6.

trabeum (Pers. per Fr.) Murrill 1908. — *Agaricus* Pers. 1801 (Germany) (d.n.); *Daedalea* (Pers.) per Fr. 1821; *Agaricus* Schw. 1822; *Lenzites* Fr. 1838; *Trametes* Bres. 1897; not ∼ Otth 1871; *Cellularia* O.K. 1898; *Coriolopsis* Bond. & S. 1941; *Phaeocoriolellus* Kotl. & P. 1957. — Bres. 1897 (AAR III 3): 91; Bourd. & G. 1928: 586 (*Trametes*); Shope 1931 (AMo 18): 392 *pl. 38 f. 1* (*Lenzites*); Donk 1933: 215 (*Gloeophyllum*); Konr. & M. 1935 I. 5: *pl. 442 f. 1* (*Trametes*); Pilát 1940 (ACE 3): 339 *f. 145, pl. 224* (*Gloeophyllum*); Overh. 1953: 110 *pl. 61 f. 366, pl. 94 fs. 537, 538, pl. 127 fig.* (*Lenzites*); Al. David 1968 (BmF 84): 124, cult. char. (*Gloeophyllum*); Westh. 1971 (Bo 10): 180 *fs. 7, 8*, with cult. char. (*Lenzites*).

Lenzites mollis Heufl. ex Kalchbr. 1868 (VW 18): 431 [repr. 1868 (H 7): 183] (Austria); fide Pilát 1942 (ACE 3): 608. — *Cellularia* O.K. 1898.

Trametes trabea Otth 1871 (MiB 1870): 92 (Switzerland), not ∼ (Pers. per Fr.) Bres. 1879; fide Bres. 1918 (Am 16): 39.

Lenzites vialis Peck 1873 (U.S.A., New York); fide Lloyd 1916 (LMW 5, L. 62): 5 & Overh. 1953: 110. — Overh. 1914 (AMo 1): 146; 1915 (WUS 3[1]): 74 *pl. 8 f. 46*.

Daedalea eatonii Berk. 1876 (JBL 14): 174; Berk. ex Cooke 1891 (G 19): 93, typonym (South Africa); fide Bres. 1916 (Am 14): 230 & Wakef. apud Lloyd 1918 (LMW 5, L. 68): 9. — Sensu Lloyd 1915, 1916 = *"Polystictus" obstinatus* (Cooke) Cooke [= *"Polyporus" meyenii* Kl., an extra-European sp.], fide Wakef. apud Lloyd 1918 (LMW 5, L. 68): 9.

Daedalea poetschii S. Schulz. apud Poetsch 1879 (ÖbZ 29): 290 (Austria); fide Bres. 1908 (Am 18): 69. — *Striglia* O.K. 1891.

Daedalea mutabilis Quél. 1896 (Crf 24^2): 620 *pl. 6 f. 12* (France); fide Bres. 1908 (Am 6): 39 & Bourd. & G. 1928: 587.

Trametes sordida Speg. 1899 (ABA 6): 173 (Argentina); fide Bres. 1916 (Am 14): 229.

Lenzites thermophila Falck 1909 (Germany); fide Cartwr. 1932 (TBS 16): 307. — Falck 1909 (HF 3): 40 *f. 13: 3, pl. 1 f. 3.*

Daedalea reisneri Velen. 1922: 693 [see Pilát 1948: 263 for Latin translation] (Czechoslovakia); fide Pilát 1940 (ACE 3): 339.

Lenzites roburnea Velen. 1930 (MP 7): 18, 19 (Czechoslovakia); from descr.

M.—*Daedalea pallidofulva* Berk (**O**) sensu Bres. 1897 (AAR III 3): 91 (cited as a syn. of *Trametes trabea*). — Murrill 1904 (BTC 31): 605 (*Sesia*).

M.—*Trametes protracta* Fr. sensu Lloyd 1910 (LMW 3, L. 29): 7 (*Lenzites*); fide Lloyd 1912 (LMW 4, L. 39 [bis]): 4. — Sensu Overh. 1915 (WUS 3^1): 69 = *Gloeophyllum protractum.*

GLOEOPORUS Mont. (**50**)

1842 [1960 (Pe 1): 220]. — Monotype: *Gloeoporus conchoides* Mont.

SPECIAL LITERATURE,—Domański, *1966b, 1970d*; Hansen, *1956 (G. dichrous)*; Kotlaba & Pouzar, *1964a (G. pannocinctus).*

dichrous (Fr. per Fr.) Bres. 1912 (**51**). — *Polyporus* Fr. 1815 (Sweden) (d.n.) per Fr. 1821; *Boletus* Spreng. 1827 (*"dichrus"*); *Bjerkandera* P. Karst. 1879; *Leptoporus* Quél. 1886; *Polystictus* Gillot & Luc. 1890, not ∼ Lloyd 1924; *Tyromyces* Maire apud Maire & Wern. 1938. — Bres. 1912 (H 53): 75 (*Gloeoporus*); Bourd. & G. 1928: 550 (*Leptoporus*); Pilát 1937 (ACE 3): 151 *f. 43, pls. 86, 87*, in part (*Gloeoporus*); L. Hansen 1956 (Fr 5): 253 *fig*; Domański 1966 (AmW 2): 151 *fs. 1A, 2, 4A, 5, 6* (*Polyporus*); 1970 (APo 39): 532, 538 *fs. 1a, 2, 3*, with cult. char. (*Gloeoporus*).

Polyporus erythroporus Otth 1869 (MiB 1868): 38 (Switzerland); from descr.

pannocinctus (Romell) Jo. Erikss. 1958. — *Polyporus* Romell 1911 (Sweden); *Poria* Lloyd 1912 (nom. nud.: n.v.p.), Lowe 1946; *Tyromyces* Kotl. & P. 1964. — Romell 1911 (ABS 11^3): 20 *pl. 2 f. 8* (*Polyporus*); Lowe 1946: 55 *f. 13* (*Poria*); Jo. Erikss. 1958 (Sbu 16^1): 136 *f. 43* (*Gloeo-*

porus); Kotl. & P. 1964 (ČM 18): 65, 72 (*Tyromyces*); Domański 1965 (FpG 2): 159 *f. 57, pl. 45* (*Gloeoporus*); Lowe 1966: 71 (*Poria*); Domański 1966 (AmW 2): 151 *fs. 1A, 3, 4A* (*Polyporus*); 1970 (APo 39): 532, 538 *fs. 1b, 4, 5*, with cult. char. (*Gloeoporus*).

Leptoporus bourdotii Pilát 1932 (U.S.S.R., Ukraine); fide Lowe 1962 (PMi 47): 184. — *Poria* Pilát apud Pil. & Lindtn. 1938, restricted; *Gloeoporus* Bond. & S. 1941. — Pilát 1936 (BmF 51): 357, in part: excl. of "b)"; 1941 (ACE 3): 404 *pl. 255 f. b* (*Poria*).

Leptoporus zameriensis Pilát 1936 (U.S.S.R., Ukraine); fide Lowe 1946: 55. — *Poria* Overh. 1942; *Polyporus* Killerm. 1943; *Tyromyces* Bond. 1953. — Pilát 1936 (BmF 51): 356 ["256"] *pl. 6 f. 2* (*Leptoporus*); Overh. 1942: 43 (*Poria*); Pilát 1937–8 (ACE 3): 191 *f. 52, pl. 112* (*Leptoporus*).

Poria tacamahacae D. Baxt. 1939 (Canada, Northwest Territories); fide Lowe 1966: 71. — D. Baxt. 1939 (PMi 24): 169 *pl. 1.*

GRIFOLA S. F. Gray (52)

1821 [1960 (Pe 1): 221]. — Lectotype: *Boletus frondosus* Dicks.

Merisma (Fr.) Gillet 1878, not ∼ Pers. per S. F. Gray 1821 [1960 (Pe 1): 241, 242]. — *Polyporus* trib. *Merisma* Fr. 1821. — ≡ *Polypilus* P. Karst. 1881 [1960 (Pe 1): 260], synisonym. — ≡ *Cladomeris* Quél. 1886 [1960 (Pe 1): 199], synisonym. — ≡ ["Stipitate Polyporoids" sect. *Merismus* Lloyd 1912]; *Merismus* (Lloyd) ex Torrend 1920 [1960 (Pe 1): 242], synisonym & later homonym. — Lectotype: *Polyporus frondosus* (Dicks.) per Fr.

Cladodendron Lázaro 1916 [1960 (Pe 1): 199]. — Lectotype: *Polyporus frondosus* (Dicks.) per Fr.

SPECIAL LITERATURE.—General: Pilát, *1934a.*

Grifola frondosa: Buchs, *1930b*; Cartwright, *1940*; Gillot, *1909*; Hennings, *1901*; Kirchmayr, *1914*; Wiepken, *1937.*

Grifola umbellata: Herink, *1955*; Pilát, *1950*; Shirai, *1905.*

frondosa (Dicks. per Fr.) S. F. Gray 1821. — *Boletus* Dicks. 1785 (d.n.); *Polyporus* Fr. 1815 (d.n.); *Polyporus* (Dicks.) per Fr. 1821, not ∼ Secr. 1833 (n.v.p.), not ∼ Fr. 1838; *Boletus* Purt. 1821, misapplied, Nocca & Balb. 1821; *Merisma* Gillet 1877; *Polypilus* P. Karst. 1881; *Cladomeris* Quél. 1886; *Caloporus* Quél. 1888; *Cladodendron* Lázaro 1916; [≡ (by lectotypification) *Polyporus frondosus, cespitosus, imbricatus, spadiceus, poris albidis* Haller 1768 H. 3: 139 no. 2276, in part]. — Vahl 1789 (Fd 6 / F. 16): 8 *pl. 952*; Schrad. 1794: 159; Pers. 1801: 520 (*Boletus*); Viv. 1834–8: 41 *pl. 36*; Bres. 1899 F.m.: 105 *pls. 97, 97bis* & 1931 (BIm 20): *pl. 968*; Bourd. & G. 1928: 521; Buchs 1930 (ZP 9): 74 *pl. 4* (*Polyporus*); Donk 1971 (Pe 6): 202, notes. — No North American authors cited. — Sensu With., Fr., in part, Rostk. → *Meripilus giganteus.*

[? *Fungus Intybaceus* Bauh. & Cherl. 1651 H. 3: 839 *fig.*, in part (non-binomial) name); cf. Donk 1971 (Pe 6): 206–207. — ≡ *Agaricus intybaceus* Tourn. 1700: 562 (non-binomial name), no descr.; fide Fr. 1838: 447 = *Polyporus intybaceus.*]

[*Polyporus frondosus* ... Haller 1768 → *Boletus frondosus* Dicks.]

Boletus ramosissimus var. *cristatus* Schaeff. 1774: 80, in "Nota", 85, & Ind. secundus ("*christatus*") [*pls. 127–129*] (Germany); fide Pers. 1800: 49. — Lectotype: represented by *pls. 128, 129.* — Fide Fr. 1838: 446 in obs. under *Polyporus frondosus*, "meo sensu" = *P. intybaceus.*

Boletus frondosus Schrank 1789: 616 (d.n.), not ∼ Dicks. 1785 (d.n.) & (Dicks. per Fr.) Purt. or Nocca & Balb. 1821; fide Schrad. 1794: 159–160 (for *B. frondosus* & *B. intybaceus*); and Pers. 1801: 520 (for *B. intybaceus*). — [≡ "*Polyporus (Boletus) imbricatus, squamosus* ... Gleditsch Meth. fung." [then unpublished] apud Boehm. 1750: 325 ≡ *IX. Boletus*; *imbricatus, squamosus* ... Gled. 1753: 75 (Germany)]; ≡ *Boletus intybaceus* Baumg. 1790: 631 (d.n.), not *Polyporus intybaceus* Fr. 1838. — Donk 1971 (Pe 6): 207, note.

Fungus squamatim-incumbens Paul. 1793 (France) (d.n.); fide Fr. 1838: 446 ("Paul. t. 30") = *Polyporus intybaceus* Fr., & Donk. — ≡ *Polyporus multiconcha* Paul. 1812–35 (n.v.p.?). — Paul. 1793 T. 2: 121 (descr.; "Coquillier en plateau") & 1812–35: *pl. 30 (Polyporus multi-concha).*

Polyporus barrelieri Viv. 1834–8 (Italy); fide Bres. 1899 F.m.: 105. — Viv. 1834–8: 28 *pl. 28* (on plate in error as "*Polyporus umbellat(s* Pers.").

Polyporus intybaceus Fr. 1838: 446 (Sweden), not ∼ Berk. 1842, not *Boletus intybaceus* Baumg. 1790 (d.n.); fide Donk 1971 (Pe 6): 202. — *Merisma* Gillet 1878; *Polypilus* P. Karst. 1882; *Cladomeris* Quél. 1886; *Caloporus* Quél. 1888; *Grifola* Imaz. 1943, misapplied. — Hussey c. 1847 I. 1: *pl. 6 (Polyporus)*; Quél. 1888: 406 (*Caloporus*); Dumée 1905 N.A. [1]: 44 *pl. 44* & 1917 N.A. 1: 47 *pl. 47*; Wak. & Denn. 1950: 228 *pl. 93 f. 2* (*Polyporus*). — Sensu Rolland 1910: *pl. 92 f. 202?*, Kawam. 1929 → *Meripilus giganteus*; sensu Hussey, l.c., = typical *Grifola frondosa.*

M.—*Agaricus ramosus* Lam. sensu Lam. 1783 (EmB 1): 51.

umbellata (Pers. per Fr.) Pilát 1934 (53). — *Boletus* Pers. 1801 (Germany) (d.n.); *Polyporus* (Pers.) per Fr. 1821; *Boletus* Krombh. 1821; *Merisma* Gillet 1878; *Polypilus* P. Karst. 1882, Bond. 1953; *Cladomeris* Quél. 1886; *Cerioporus* Quél. 1888; *Cladodendron* Lázaro 1916; *Grifola* Imaz. 1943, preoccupied. — Krombh. 1841 S. 7: 16 *pl. 52 fs. 3–9*; Quél. 1872 (MMb II 5): 272/255 *pl. 18 f. 1 (Polyporus)*; Quél. 1888: 409 (*Cerioporus*); Atk. 1900: 189 *f. 178* / 1901: 189 *f. 183*; Rostr. 1902: 370 *fs. 150, 151* [*f. 150* = Lind 1913: *f. 30*]; Hard 1908: 390 *f. 320/tpl. 144*; Lloyd 1912 (LMW 3, S.P.): 150 *f. 450*; Overh. 1915 (WUS 3[1]): 24 *tpl. 2 f. 7*; Bourd. & G. 1928: 250; Bres. 1931 (BIm 20): *pl. 969 (Polyporus)*; Pilát 1936 (ACE 3): 29 *f. 7, pls. 8–10 (Grifola)*; Overh. 1953: 247 *pl. 30 f. 180, pl. 132 fig.*; Kawam. 1954 I. 1: 130 *fs. 119, 120 (Polyporus)*; S. Ito 1955: 335 *f. 243*; Poelt & Jahn 1963: *pl. 34*; H. Jahn 1963 (WPb 4): 38 *Abb. 7 (Grifola)*; Pouz. 1966 (Fgp 1): 361 (*Polyporus*); Birkf. & Hersch. 1967: pl. 162, sclerotium (*Grifola*).

Boletus ramosissimus Scop. 1772 (Austria) (d.n.); fide Pers. 1801: 519 (as to "Schaeff. fung. *t. 111*" cited by Scop.) & Fr. 1821: 355. — *Fungus* Paul. 1793 (d.n.), misapplied; *Polyporus* (Scop.) Secr. 1833 (as a sp. of *Boletus*: n.v.p.), misapplied; *Polyporus* (Scop.) per J. Schroet. 1888; *Cladomeris* Murrill 1903 (n.v.p.); *Grifola* Murrill 1904; *Polypilus* Bond. & S. 1941; ≡ *Agaricus ramosus* Lam. 1783 (d.n.), misapplied, not ∼ Bull. 1782 (d.n.) per Fr. 1821, not ∼ Lour. 1790 (d.n.). — Schaeff. 1774: 80, Ind. secundus, in part: [*pls. 111, 265, 266*] (*Boletus ramosissimus* var. *pileatus*); Jacq. 1774 F.a. 2: 45 *pl. 172* (*Boletus*); J. Schroet. 1888: 481; Bres. 1897 (AAR III 3): 69; Gramb. 1913 P.H. 2: 22 *pl. 22* (*Polyporus*). — Sensu Lam. (*Agaricus ramosus*) & Schaeff. in part (var. *cristatus*), → *Grifola frondosa*.

Boletus ramosissimus var. *pileatus* Schaeff. 1774: 80 & Ind. secundus [*pls. 111, 265, 266*] (Germany).

Boletus ramosus Vahl 1797 (Denmark) (d.n.), not ∼ Bull. 1788 (d.n.) per Mérat 1821; fide Fr. 1821: 355 (forma). — Vahl 1797 (Fd 7 / F. 20): 8 *pl. 1197*.

Boletus polycephalus Pers. 1801 (Germany) (d.n.); fide Fr. 1821: 355. — *Boletus* Pers. per Krombh. 1821. — Pers. 1801: 520; Krombh. 1821: 26.

Hydnum ramosissimum March. & Court. apud L. March. 1828 (BnW 3): 268 (Luxemburg): Fr. 1832. — ≡ *Hydnum umbellatum* Fr. 1838; *Dryodon* Quél. 1886.

Polyporus chuling Shirai 1905 (Japan); fide Lloyd 1916 (LMW 5, L. 63): 14 & Imaz. 1943 (JJB 19): 384. — Shirai 1905 (BMT 19): 92 *pl. 4*.

Sclerotium giganteum Rostr. 1889 (PT 17): 231 (Denmark) (nom. anam.), & cf. Rostr. 1892 (BT 18): 73. — Rostr. 1899 (BT 22): 259–262 *fig.*; 1902: 370 *fs. 150, 151* [Lind 1913: *f. 30*, copy of *f. 150*] (*Polyporus umbellatus*).

HAPALOPILUS P. Karst.

1881 [1960 (Pe 1): 222]. — Monotype: *Polyporus nidulans* Fr.
Inonotus Pat. 1887: 140 ("Karst."), not ∼ P. Karst. 1879. — [*Inonotus* P. Karst. sensu Pat., l.c., excl. of type]. — Monotype: *Inonotus nidulans* (Fr.) P. Karst.
M.—*Inonotus* P. Karst. sensu Pat. 1887 [1962 (Pe 2): 206] → *Inonotus* Pat.

SPECIAL LITERATURE.—Bamberger & Landsiedl, *1909*; Kögl, *1926*; Klingemann, *1893*; Příhoda, *1952*; Stahlschmidt, *1877, 1879*.

rutilans (Pers. per Fr.) P. Karst. 1899. — *Boletus* Pers. 1798 (Germany) (d.n.); *Polyporus* Fr. 1818 (d.n.); *Polyporus* (Pers.) per Fr. 1821; *Inonotus* P. Karst. 1882; *Leptoporus* Quél. 1886; *Inodermus* Quél. 1888; *Hapalopilus* Murrill 1904, preoccupied; *Polystictus* Big. & Guill. 1913; *Hemidiscia* Lázaro 1916; *Hydnum* Hruby 1932 (error?); *Phaeolus* Sart. & M. 1921. — Pers. 1798 I.D.: 18 *pl. f. 17* (*Boletus*); Secr. 1833 M. 3: 97; Lloyd 1915 (LMW 4, Ap.): 331 *f. 674* (*Polyporus*); Bourd. & G. 1928: 554;

Pilát 1937 (ACE 3): 139 *f. 35, pls. 65–68*; Al. David 1969 (Nca 96): 214 *f. 2*, cult. char. (*Phaeolus*).

Boletus suberosus Batsch 1783 (d.n.), non/an ∼ L. 1753 (d.n.), not ∼ Bull. 1790 (d.n.); fide Fr. 1821: 362 (as to Schaeff. *pl. 136*) = *Polyporus nidulans.* — [≡ (by lecto-typification) *Boletus versicolor* L. sensu Schaeff. 1774 (Germany)]. — Schaeff. 1774: 88 [*pl. 136*] (*Boletus versicolor*).

Boletus suberosus Bull. 1790 (France) (d.n.), not ∼ L. 1753 (d.n.), not ∼ Batsch 1783 (d.n.); fide Fr. 1838: 455 = *Polyporus nidulans.* — *Boletus* Bull. per St-Am. 1821; *Polyporus* Chev. 1826, Sacc. 1916, not ∼ Fr. 1821; *Daedalea* Duby 1830; ≡ *Daedalea bulliardii* Fr. 1821; *Polyporus* Pers. 1825, Lind 1913, not ∼ (Fr.) Kumm. 1871. — Bull. 1790: *pl. 482*; 1791 H.: 354.

Polyporus nidulans Fr. 1821 (Sweden); fide Quél. 1886: 177 & Bres. 1897 (AAR III 3): 71. — *Boletus* Spreng. 1827; *Hapalopilus* P. Karst. 1881; *Inonotus* P. Karst. 1881; *Polystictus* Gillot & Luc. 1890; *Phaeolus* Pat. 1900 ("Pers."); *Ochroporus* Feltg. 1907; *Agaricus* E. Krause 1933, not ∼ Pers. 1798 (d.n.) per Fr. 1821. — Gillet 1874–90 P.: *pl. 460/563*, too pale (*Polyporus*); Donk 1933: 172 (*Hapalopilus*); D. Baxt. 1943 (PMi 28): 228; Overh. 1953: 398 *pl. 46 fs. 277, 278 pl. 131 fig.* (*Polyporus*).

Polyporus pallidocervinus Schw. 1832 (U.S.A., Pennsylvania); fide Lloyd 1913 (LMW 4, L. 50): 7, 1915 (LMW 4, Ap.): 383; fide Overh. 1953: 398 = *Polyporus nidulans.* — *Leptoporus* Pat. apud Duss 1903. — Sensu Mont. 1842 C.: 397 ≡ "*Coriolus*" *pallidofulvellus* Murrill (Cuba) (extra-European).

Trametes lignicola Lasch in Rab. 1854 Kl.: No. 1809, with descr., as part of the name *T. lignicola* var. *populina* Lasch [repr. 1854 (BZ 12): 185; 1854 (Fl 37): 203] (Germany); fide Pilát 1937 (ACE 3): 139.

? *Polyporus schaefferi* Heufl. 1867 (ÖbZ 17): 315 (Germany).

Polyporus purpurascens Stahlschmidt 1877 (LAC 187): 179 (Germany), not ∼ (DC. per Steud.) Pers. 1825, not ∼ (Hook.) Fr. 1838; fide Bamberger & Landsiedl *1909*; fide Falck apud Kögl 1926 (ZP 5): 260 = *Polyporus nidulans.* — Donk 1971 (PNA 74): 7, note.

Trametes ribicola P. Karst. 1881 (H 20): 178 (Finland); fide Lowe 1956 (M 48): 122 = *Polyporus nidulans.*

Polyporus ramicola Velen. 1922: 647 [see Pilát 1948: 247 for Latin translation] (Czechoslovakia); fide Pilát 1937 (ACE 3): 139.

Polyporus conicus Velen. 1922: 683 *f. 103: 7* [see Pilát 1948: 259 for Latin translation] (Czechoslovakia); fide Pilát (ACE 3): 596.

? *Polyporus salmonicolor* Lloyd 1925 (LMW 7): 1360 *pl. 341 f. 3234* (U.S.A., Maine), not ∼ B. & C. 1849.

? *Hapalopilus* **taxi** Bond. 1940 (BMa 5): 17 (U.S.S.R., Caucasia). — *Polyporus* Bond. 1953 (syn.: n.v.p.). — Bond. 1953: 267.

HAPLOPORUS Bond. & S. ex Sing.

1944, not ∼ Bond. 1953 [1962 (Pe 2): 205]. — *Haploporus* Bond. & S. 1941 (lacking Latin descr.; n.v.p.). — Holotype: *Trametes odora* (Sommerf.) Fr. [sensu Nikol.; Bond. & S.].

Sᴘᴇᴄ𝀵ᴀʟ ʟɪᴛᴇʀᴀᴛᴜʀᴇ.—Bondarcev, *1950*; Niemelä, *1971*.

suaveolens (L. per Fr.) Donk 1971. — *Boletus* L. 1753 (d.n.); *Agaricopulpa* Paul. 1793 (d.n.), misapplied; *Agaricum* Paul. 1812–35 (generic name n.v.p. at the time of publication?), misapplied; *Polyporus* (L.) per Fr. 1821, misapplied, not ∼ Fr. 1828; *Boletus* St-Am. 1821, misapplied (= *Boletus suaveolens* L. sensu Bull.); *Pycnoporus* P. Karst. 1911, misapplied?; [≡ *Boletus acaulis, superne laevis, salicini insidens* L. 1737: 368 (Sweden, Lapland)]; ≡ *Agaricus odoratus* Lam. 1783 (d.n.). — L. 1753: 1177; 1755: 453 (*Boletus*); Donk 1971 (PNA 74): 19, notes. — Sensu Enslin, Fr. 1821 → *Trametes suaveolens*; sensu Bull. → *Daedaleopsis confragosa*.

Polyporus odorus Sommerf. 1826 (Norway): Fr. 1828, not ∼ Peck 1885; fide Jo. Erikss. 1958 (Sbu 16¹): 142. — *Trametes* Fr. 1838; *Fomitopsis* P. Karst. 1881 (generic name n.v.p.), M. Bond. 1964 (indirect ref.: n.v.p.); *Pycnoporus* P. Karst. 1889; *Ochroporus* Feltg. 1907 ("Linn."; error through confusion with *Ochroporus* [*Gloeophyllum*] *odoratus*); *Haploporus* Sing. 1944. — Fr. 1828 E. 1: 90 (*Polyporus*; Romell apud Vleugel 1908 (SbT 2): 308; Nikol. 1940 (TSR 4): 411; Jo. Erikss. 1958 (Sbu 16¹): 140 *f. 44*; R. L. Gilb. 1961 (NwS 35): 12 *f. 7* (*Trametes*); Pouz. 1966 (Fgp 1): 363; Niemelä 1971 (Abf 8): 237 *fs. 1–6* (*Haploporus*).

Fomitopsis odoratissima Bond. 1950 (BŽ 35): 76 *fs. 1, 2* (U.S.S.R.): fide Bond., l.c., = *Trametes odora* sensu Nikol.; fide M. Bond. 1964 (NSn): 191. — Bond. 1953: 300 *f. 96, pl. 91, pl. 143 f. 2*; Parm. 1963 (IDV 1): 259.

HETEROBASIDION Bref.

1888, not *Heterobasidium* Mass. 1889 (nom. conf.) [1960 (Pe 1): 223]. — Monotype: *Polyporus annosus* Fr. — Donk 1971 (PNA 74): 7, note.

Sᴘᴇᴄ𝀵ᴀʟ ʟɪᴛᴇʀᴀᴛᴜʀᴇ (see p. 404).—Bakshi, *1952*; Bratus' & Černick, *1966*; Brefeld, *1888b*; Conference, *1962, 1970*; Etheridge, *1955*; Griffith & Wilson, *1967*; Koenigs, *1960*; Laine & Nuorteva, *1970*; Liese, *1936*; K. Lohwag, *1956*; Low & Gladman, *1960*; Molin, *1957*; Negruckij, *1966*; Olson, *1941*; Orlos & Dominik, *1960*; Parmasto, *1956b*; Rennerfeldt, *1946*; Rishbeth, *1950, 1951a, 1951b, 1957, 1959, 1961, 1963*; Roll-Hansen, *1940*; Sinclair, 1964; Tikka, *1934*; M. Wilson, *1927*; C. L. Wilson & al., *1967*; Yde-Anderson, *1961b*; Zycha, *1964*.

annosum (Fr.) Bref. 1888. — *Polyporus* Fr. 1821 (Sweden); *Boletus* Spreng. 1827; *Trametes* Fr. 1849 (nom. nud.: n.v.p.), Otth 1863, P. Karst. 1882;

Fomes P. Karst. 1879; *Fomitopsis* P. Karst. 1881 (gen. name n.v.p.); *Placodes* Quél. 1886; *Scindalma* O.K. 1898; *Pycnoporus* P. Karst. 1898; *Ungulina* Pat. 1900; *Friesia* Lázaro 1916; *Fomitopsis* Bond. & S. 1941; ≡ *Polyporus serpentarius* Pers. 1825. — Fr. 1838: 471; 1884 I. 2: 85 *pl. 186 f. 2* (*Polyporus*); Bref. 1888 U. 8: 154 *pls. 9–11*; Warm. 1890: 113 *f. 148* (*Heterobasidion*); Bres. 1903 (Am. 1): 75 (*Fomes*); Bourd. & G. 1928: 604 (*Ungulina*); Konr. & M. 1934 I. 5: *pl. 451* (*Fomes*); Roll-H. 1940 (MnS 7[1]): 79 *f. 22*, spores, 79 *fs. 17–21*, clamps (*Polyporus*); Pilát 1940–1 (ACE 3): 362 *f. 157 pl. 243 f. a, pl. 244, f. b, pls. 245–247*; D. Baxt. 1941 (PMi 26): 113 *pl. 7* (*Fomes*); Rennerf. 1946 (MSS 35[8]): 2 *fs. 1–4* (*Polyporus*); Overh. 1953: 40 *pl. 59 fs. 353, 354, pl. 74 f. 426, pl. 80 f. 460, pl. 125 fig.*; Lowe 1957 F.: 85 *f. 67* (*Fomes*); Pouz. 1966 (Fgp 1): 362; Pegl. & Wat. 1968 (CDp): no. 192 *figs.* (*Heterobasidion*). — Sensu Fr. 1828 → *Polyporus roburneus* Fr. (**O**).

Poria perspicillum Scop. 1772 P.s.: 103 *pl. 21 f. 1* (Hungary, now Czechoslovakia) (d.n.); fide Hoffm. 1797–1811 V.s.: 14 = *Poria scutata*.

Poria plicata Scop. 1772 P.s.: 103 *pl. 21 f. 2* (Hungary, now Czechoslovakia) (d.n.).

Poria scutata Scop. 1772 (Hungary, now Czechoslovakia) (d.n.); fide Harz 1888 (BCb 36): 378 = *Trametes radiciperda* (cited as a syn.). — *Polyporus* (Scop.) per Pers. 1825, Mez 1908; *Trametes* Harz 1888. — Scop. 1772 P.s.: 106 *pl. 27*; Hoffm. 1797–1811 V.s.: 13 *pl. 9, pl. 10 f. 1* [text, repr. Pers. 1825: 86] (*Poria*); Secr. 1833 M. 3: 93; Harz 1888 (BCb 36): 378 (*Polyporus*).

Boletus cryptarum Bull. 1789: *pl. 478* & 1791 H.: 350 (France) (d.n.); fide Harz 1888 (BCb 36): 378 = *Trametes scutata*; fide Bres. 1903 (Am 1): 75, & cf. Bourd. & G. 1928: 686 in obs. under *Poria spongiosa*. — *Polyporus* (Bull.) per Fr. 1821; *Boletus* St-Am. 1821, Laterr. 1821; *Poria* S. F. Gray 1821; *Trametes* Fr. 1848 (nom. nud.: n.v.p.); *Fomes* Cooke 1885; *Inodermus* Quél. 1886; *Phellinus* Quél. 1888, P. Karst. 1889; *Trametes* Harz 1889; *Scindalma* O.K. 1898; *Polystictus* Gillot & Luc. 1890; *Spongioides* Lázaro 1916, misapplied (**107**). — Sensu Schum. → *Polyporus epiphegum* (O); sensu Mont. → *Flaviporus brownii*; sensu Quél., in part, Mang. & Pat. → *Poria expansa*.

Poria encephalum Hoffm. 1797–1811 V.s.: 18 *pl. 12 fs. 1, 2, 4* (Germany) (d.n.); fide Pilát 1941 (ACE 3): 362. — *Polyporus* (Hoffm.) per Pers. 1825.

Polyporus subpileatus Weinm. 1826 (SPR 2): 102 (U.S.S.R., European Russia); fide Fr. 1849: 323 ("status ... vulgatioris").

Polyporus scoticus Kl. ex Berk. 1836: 142 (Scotland); fide Berk. 1860: 248.

Polyporus makraulos Rostk. 1838 (StP 4): 113 *pl. 55* (Germany/Poland); fide Bres. 1890 (Rm 12): 106. — *Physisporus* P. Karst. 1882; *Poria* Quél. 1886.

Trametes radiciperda R. Hartig 1874 (Germany); fide R. Hartig 1878: 14, 19 & Bref. 1888 U. 8: 150. — *Polyporus* Rostr. 1902. — R. Hartig

1874: 62 *pl. 3 fs. 20–29*; 1878: 14 *pls. 1–4* (*Trametes*); Rostr. 1902: 354 *fs. 145–148* (*Polyporus*).

Polyporus gillotii Roum. 1882 F.g.: No. 2207 & 1882 (Rm 4): 215 (lacking descr.: n.v.p.) apud Gillot 1882 (Rm 4): 234 *pl. 32 fs. A, B* (France); fide Bres. 1903 (Am 1): 75 & Donk 1933: 209. — *Ungulina* Maheu 1906.

Polyporus irregularis Underw. 1897 (BTC 24): 85 (U.S.A., Alabama), not ∼ (Sow.) per Pers. 1825; fide Murrill 1903 (BTC 30): 227 & Overh. 1953: 40, 41.

Polyporus atrannosus E. Krause 1928 B.r.: 54 (Germany); fide E. Krause, l.c. (*Polyporus annosus* Fr. 1884 I. 2: *pl. 186 f. 2* ("media") cited as syn.); fide E. Krause 1929 B.r.: 79 = *Polyporus medulla-panis* [sensu E. Krause p. p. maj.].

Polyporus marginatoides E. Krause 1928 B.r.: 54 (Germany); fide E. Krause 1929 B.r.: 79 = *Polyporus medulla-panis* [sensu E. Krause p. p. maj.]. — Lectotype: Rostock, Alter Friedhof.

Cunninghamella meineckella A. J. Olson 1941 (Ph 31): 1076 *fs. 2F, G* (U.S.A., California) (nom. anam.); fide S. Hugh. 1953 (CJB 31): 593 (conidial state). — *Oedocephalum* Donk 1971 (PNA 74): 7.

Oedocephalum lineatum Bakshi 1950 (Scotland) (nom. anam.); fide Bakshi *1952*. — Bakshi 1950 (TBS 33): 114 *f. 1, pl. 9 fs. 4, 5, 8, 9*.

M.—*Boletus resinosus* Schrad. sensu Rostk. 1830 (StP 4): 61 *pl. 29* (*Polyporus*); fide Fr. 1874: 564.

M.—*Boletus semiovatus* Schaeff. sensu Britz. 1889 (BnS 29): 279 [*pl. 609 f. 57*] (*Polyporus*).

M.—*Boletus medulla-panis* Jacq. sensu Fr. 1884 I. 2: *pl. 190 f. 2* (*Polyporus*)?; cf. Lundell 1953 (LNF 43–44): 3 No. 2103; E. Krause 1929 B.r.: 79 (*Polyporus*), in part: excl. of "Resupinate Formen auf Rinde . . .".

M.—*Boletus igniarius* L. sensu E. Krause 1931 B.r.: 123; fide E. Krause 1931 B.r.: 123 = *Polyporus medulla-panis* [sensu E. Krause p. p. maj.] (cited as a syn.).

HIRSCHIOPORUS Donk

1933 [1960 (Pe 1): 227]. — Holotype: *Polyporus abietinus* (Pers.) per Fr.

SPECIAL LITERATURE.—General & miscellaneous: Bakshi & Chroudhury, *1961*; Nikolajeva, *1953*.

Hirschioporus abietinus complex: Fries & Aschan, *1952*; Fries & Jonassen, *1941*; Garren, *1938*; K. Lohwag, *1965*; Macrae, *1967*; Macrae & Aoshima, *1967*; Morot, *1888*; Raestadt, *1941*; Takemaru & Fujioka, *1968*.

Hirschioporus pargamenus: Elliott, *1918*; Lambinon & Mathot, *1965*; Rhoads, *1918*, Webster, *1899*.

abietinus (Pers. per Fr.) Donk 1933. — *Boletus* Pers. apud J. F. Gmel. 1792 (Germany) (d.n.), not ∼ Anon. 1790 (d.n.), not ∼ Cumino 1805 (d.n.): *Polyporus* (Pers.) per Fr. 1821; *Boletus* S. F. Gray 1821, not ∼ (Anon.) per Purt. 1821; *Polystictus* Cooke 1886; *Coriolus* Quél. 1886; *Microporus* O.K. 1898; = *Boletus purpurascens* Pers. 1796, not ∼ DC. 1815 (d.n.) per Steud. 1824, not ∼ Hook. 1822. — Dicks. 1793 P.c. 3: 21 *pl. 9 f. 9*; Pers. 1801: 541 (*Boletus*); Fr. 1828 E. 1: 97 (*Polyporus*); Bourd. & G. 1928: 567 (*Coriolus*); Donk 1933: 168 (*Hirschioporus*); Konr. & M. 1935 I. 5: *pl. 439 f. 1* (*Coriolus*); Pilát 1939 (ACE 3): 273 *f. 110, pls. 187–190, pl. 191 f. a*, in part (*Trametes*); Robak 1942 (MVf 7[3]): 27, 145 *fs. 42c-m*; D. Baxt. 1948 (PMi 32): 194 *pl. 4*, in part; Overh. 1953: 333 *pl. 1 fs. 1–4, pl. 8 f. 48, pl. 128 fig.*, in part (*Polyporus*); Macrae *1967* (*Hirschioporus*); Donk 1971 (Pe 6): 208, note.

Boletus abietinus Anon. 1790 (MB 4 / Stück 12): 19 (Germany) (d.n.), not ∼ Pers. apud J. F. Gmel. 1792 (d.n.), not ∼ Cumino 1805 (d.n.); fide Pers. 1796 O. 2: 112 (as to "Dicks.") = *Boletus purpurascens* & Pers. 1801: 541. — *Boletus* Anon. per Purt. 1821, not ∼ (Pers. per Fr.) S. F. Gray 1821; *Bjerkandera* P. Karst. 1881; *Hansenia* P. Karst. 1889; *Polystictoides* Lázaro 1916; *Trametes* Pilát 1939, not ∼ (P. Karst.) Sacc. 1888. — For descrs., illustrations, & note see under correct name. — Sensu DC. → *Skeletocutis amorphus*.

? *Boletus incarnatus* Schum. 1803: 391 (Denmark) d.n.); non/an ∼ (Pers.) Pers. 1801 (d.n.); fide Fr. 1821: 371. — Hornem. 1806 (Fd 8 / F. 22): 7 *pl. 1298*.

Polyporus dolosus Pers. 1825: 77 (Switzerland); fide Fr. 1828 E. 1: 97 & Donk 1933: 168, 169. — Lectotype: Switzerland, "Neuchâtel (misit Chaillet)" in herb. Pers., L 910.263–1016.

Polyporus dentiporus Pers. 1825: 104 (Switzerland); fide Romell 1911 (ABS 11[3]): 10 & Donk 1967 (Pe 5): 88. — *Poria* Cooke 1886. — Sensu Bres. → *Poria dentipora* Pilát & *Coriolus dentiporus* Bond. & S., listed under *Poria*.

Polyporus parvulus Schw. 1832: 157 (U.S.A., Pennsylvania), not ∼ Kl. 1833; fide Murrill 1907 (NAF 9): 27 & Lloyd 1913 (LMW 4, L. 50): 8. — *Polystictus* Cooke 1886, not ∼ (Kl.) Fr. 1851; = *Polystictus pusio* Sacc. & Cub. apud Sacc. 1888; *Microporus* O.K. 1898.

Physisporus caesio-albus P. Karst. 1883 (H 22): 177 (Finland); fide Lowe 1956 (M 48): 110 ("the pale resupinate form"). — *Poria* Sacc. 1888; *Polyporus* All. & Schn. 1891. — P. Karst. 1885 I. 1: 14 / 1888 (ASf 15): 194 *pl. [9] f. 30*.

fuscoviolaceus (Ehrenb. per Fr.) Donk 1933. — *Sistotrema* Ehrenb. 1818 (Germany) (d.n.); *Hydnum* (Ehrenb.) per Fr. 1821; *Daedalea* Fic. & Sch. 1823; *Irpex* Fr. 1828; *Lenzites* Pat. 1885; *Sistotrema* J. Schroet. 1888; *Xylodon* O.K. 1898. — Fr. 1828 E. 1: 144 (*Irpex*); Donk 1933: 169; Pilát 1939 (ACE 3): 275 *pl. 188* (*Trametes abietina* var.); Nikol. 1953 (TSR 8): 191 *fs. 11, 13: 5*; Macrae *1967* (*Hirschioporus*).

Agaricus decipiens Willd. 1788 (MB 2 / St. 4): 12 *pl. 2 f. 5* (Germany) (d.n.). — *Boletus* J. F. Gmel. 1792, misapplied; *Hydnum* Schrad. 1794; *Hydnum* (Willd.) per St-Am. 1821, in part; *Daedalea* Sommerf. 1826. — DC. 1805: 112 (*Hydnum*); Sommerf. 1826: 271 (*Daedalea*). — Fide Fr. 1821: 370 = *Polyporus abietinus*. — Sensu J. F. Gmel. (phrase) → *Cerrena unicolor*.

Sistotrema violaceum Pers. 1801: 551 (Germany) (d.n.); fide Pers. l.c. = *Agaricus decipiens* (cited as syn.); fide Bres. 1897 (AAR III 3): 100 = *Irpex fuscoviolaceus* (cited as a syn.); fide Donk 1933: 169, 170. — *Sistotrema* Pers. per Schleich. 1821, not ∼ Secr. 1833; *Irpex* Quél. 1888. — Quél. 1888: 376; Bres., l.c.; Bourd. & G. 1928: 572. — *Sistotrema violaceum* Secr. 1833 (see below) is this species, but technically a different taxon ("excl. syn. Pers."); Secr. also cited *Hydnum decipiens* (Willd.) Schrad. as syn.

Sistotrema hollii J. C. Schmidt 1817 (MH 1): 87 (Germany) (d.n.); fide Donk 1933: 169, 170. — *Hydnum* (J. C. Schmidt) per Fr. 1821; *Acia* P. Karst. 1879; *Odontia* Quél. 1886, Rea 1922.

Sistotrema candidum Ehrenb. 1818: 19, 30 (Germany) (d.n.); fide Bres. 1897 (AAR III 3): 100 = *Irpex violaceus* ("status resupinatus in prima evolutione"). — *Hydnum* (Ehrenb.) per Schlechtend. 1824, not ∼ Willd. 1788 (d.n.), not ∼ J. C. Schmidt 1818 (d.n.) per Fr. 1821; *Sistotrema* Pers. 1825; *Irpex* Weinm. 1836; *Xylodon* P. Karst. 1882.

Sistotrema carneum Ehrenb. 1818: 19, 30 (Germany) (d.n.), not ∼ Fr. 1818 (d.n.), not ∼ Bon. 1857; fide Fr. 1821: 420 = *Hydnum hollii* ("ipse inventor suspicatus").

Sistotrema violaceum Secr. 1833 M. 3: 501 ("excl. syn. Pers.") (Switzerland), not ∼ Pers. 1801 (d.n.) per Schleich. 1821.

M.?—*Hydnum parasiticum* L. (**O**) sensu Willd. 1787: 396 & Timm 1788: 273; fide Schrad. 1794: 180 = *Hydnum decipiens*.

laricinus (P. Karst.) Teramoto 1951. — *Lenzites* P. Karst. 1905 (U.S.S.R., Russia, Siberia); *Hirschioporus* Nikol. 1953 (preoccupied). — Nikol. 1953 (TSR 8): 193 *f. 12, f. 13: 3* (*Hirschioporus*); Macrae & Aosh. 1967 (M 58): 915, 924 *fs. 1–4, 17* (*Lenzites* & *Hirschioporus*), Macrae *1967* (*Hirschioporus*). — Descriptions based on Japanese material not included.

Lenzites ambigua P. Karst. 1905 (Finland); fide Macrae & Aosh. 1967 (M 58): 924. — Macrae & Aosh. 1967 (M 58): 917 *fs. 15, 16, 20*.

Lenzites pinicola P. Karst. 1906 (TtK 8¹): 62. (U.S.S.R., 'Transbaikal'); fide Macrae & Aosh. 1967 (M 58): 924. — Macrae & Aosh. 1967 (M 58): 917 *fs. 7–9, 19*.

Lenzites abietis Lloyd 1920 (U.S.A., Colorado) ("lenzitoid form of *Polystictus abietinus*": n.v.p.); *Polyporus abietinus* var. *abietis* (Lloyd) ex Overh. 1933; *Hirschioporus* Imaz. 1933 (n.v.p.). — Lloyd 1920 (LMW 6): 909 *pl. 141 f. 1607* (*Lenzites*); Overh. 1953: 334 (*Polyporus abietinus* var.); Macrae & Aosh. 1967 (M 58): 915 *fs. 5, 6, 18* (*Lenzites*). — Descrs. based on Japanese specimens not included.

Coriolus abietinus f. *lenzitoideus* Murašk. apud Pilát 1932 (U.S.S.R., Russia, Siberia); fide Pilát 1936 (BmF 51): 367 (*Lenzites laricinus* cited as a syn.) & Macrae & Aosh. 1967 (M 58): 924. — Pilát 1937 (ACE 3): 276 (*Trametes abietina* f. *abietis*) *pl. 189 f. a* (*Trametes abietina* f. *lenzitoidea*).

pargamenus (Fr.) Bond. & S. 1941. — *Polyporus* Fr. 1838 (Canada); *Polystictus* Fr. 1851; *Microporus* O.K. 1898; *Coriolus* Pat. 1900, G. Cunn. 1950; *Trametes* Kotl. & P. 1957; *Trichaptum* G. Cunn. 1965. — Fr. 1851 (NAu III 1): 85/69 (*Polystictus*); Peck 1883 (RNS 33): 36 (*Polyporus*); Hard 1908: 417 *f. 345* (*Polystictus*); Bourd. & G. 1928: 567 (*Coriolus*); Shope 1931 (AMo 18): 329 *pl. 16 f. 4*; R. W. Davids & al. 1942 (TUS 785): 37 *f. 5 F*, cult. char.; Overh. 1953: 336 *pl. 2 fs. 6–9, pl. 7 fs. 39, 40, pl. 131 fig.* (*Polyporus*); Bond. 1953: 49, 561 *f. 151: 2, 4, pl. 143 f. 1, pl. 179 f. 3* (*Hirschioporus*).

Polyporus prolificans Fr. 1838: 443 (Mexico); fide Bres. 1916 (Am 14): 226 = *Polyporus pargamenus* (cited as a syn.). — *Polystictus* Fr. 1851; *Microporus* O.K. 1898; *Coriolus* Murrill 1907.

Polyporus laceratus Berk. 1839 (AM 3): 392 (U.S.A., Louisiana); fide Murrill 1906 (BTC 32): 654, 655; & cf. Fr. 1851 (NAu III 1): 85/69. — *Polystictus* Fr. 1851; *Microporus* O.K. 1898; *Coriolus* Pat. 1900.

Polyporus menandianus Mont. 1843 (ASn II 20): 362 (U.S.A., New York); fide Berk. apud Mont. 1856: 165–166, B. & C. 1856 (JAP II 3): 209; fide Bres. 1916 (Am 14): 226 = *Polyporus prolificans*.

Hydnum schizodon Lév. 1844 (ASn III 2): 204 (Mexico); fide Bres. 1916 (Am 14): 231 = *Polyporus prolificans* Fr. ("vetusta"). — *Polystictus* Lloyd 1924.

Polyporus xalappensis Berk. apud B. & C. 1849 (HJB 1): 103 (Mexico); fide Fr. 1851 (NAu III 1): 78/62 & Bres. 1916 (Am 14): 228 = *Polystictus/ Polyporus prolificans*. — *Polystictus* Cooke 1886; *Microporus* O.K. 1898; *Leucoporus* Pat. apud Duss 1903.

Polyporus sartwellii B. & C. apud Berk. 1872 (G 1): 51 (U.S.A., New York); fide Murrill 1907 (NAF 9): 27 = *Coriolus prolificans*. — *Polystictus* Cooke 1886; *Microporus* O.K. 1898; *Coriolus* Murrill 1906.

Polyporus ilicincola B. & C. apud Berk. 1872 (G 1): 52 (U.S.A., Alabama), not *P. illicicola* [!] P. Henn. 1903; fide Murrill 1907 (NAF 9): 27 = *Coriolus prolificans*. — *Polyporus* B. & C. in Ravenel 1860 (nom. nud.); *Polystictus* Cooke 1886; *Microporus* O.K. 1898; *Coriolus* Murrill 1906.

Polyporus pseudopargamenus Thüm. 1878 M.u.: No. 1102 [repr. 1879 (Fl 62): 95] (U.S.A., New York); fide Bres. 1897 (AAR III 3): 77 ("*subpergamenus*") = *Polystictus biformis* [sensu Bres.] & 1903 (Am 1): 76 ("*subpergamenus*"). — *Polystictus* Peck 1901.

Polyporus simulans Błoński 1889 (H 28): 280 ("Lithuania" = Poland), not ~ Berk. ex Sacc. 1888, not ~ P. Karst. 1888, not ~ Wakef. ex Lloyd 1924; fide Bres. 1903 (Am): 76, 78 & 1920 (Am 18): 68. — *Polystictus* Sacc. 1891; *Microporus* O.K. 1898; *Coriolus* P. Karst. 1904.

M.—*Polyporus biformis* Fr. apud Kl. sensu Berk. 1839 (AM 3): 392, in part; fide Berk. 1841 (AM 7): 452. — Bres. 1897 (AAR III 3): 77 (*Polystictus*); Pat. 1897 T.: 48 (*Coriolus*); Boud. 1904–11 I. 81 *pl. 159* (*Polyporus*); Murrill 1907 (NAF 9): 26; Konr. & M. 1935 I. 5: *pl. 439 f. 2* (*Coriolus*); Pilát 1939 (ACE 3): 277 *f. 109, pl. 179 f. b* (*Trametes*); Donk 1973 (PNA 76): 229.

INCRUSTOPORIA Domański (5)

1963 (APo 32): 737. — Holotype: *Poria stellae* Pilát. — Donk 1971 (PNA 74): 37, notes.

Leptotrimitus Pouz. 1966 (ČM 20): 175. — Holotype: *Polyporus semipileatus* Peck.

M.—*Poria* "(Fr.)" sensu Bond. 1953: 156. — *Polyporus* "series" [subgen.] *Poria* Fr. 1851 (NAu III 1): 70/54 ≡ *Poria* Pers. 1794 (d.n.) per S. F. Gray 1821, in part: excl. of type. → *Poria* Bond. 1953 (n.v.p.) (O).

SPECIAL LITERATURE.—David, *1971* (*Incrustoporia percandida*); Domański, *1969e* (*I. tschulymica*).

alutacea (Lowe apud Overh. & L.) D. Reid 1969. — *Poria* Lowe apud Overh. & L. 1946 (U.S.A., New York); *Fibuloporia* M. P. Christ. 1960 — M. P. Christ. 1960 (DbA 19): 339 *f. 335* (*Fibuloporia*); Lowe 1966: 92 *f. 73* (*Poria*); D. Reid 1969 (RM 33): 237 *tpl. 2 f. 6* (*Incrustoporia*); H. Jahn 1971 (WPb 8): 60 *f. 8* (*Poria*).

Poria calcea f. radicata Bourd. & G. 1925 (France); fide Lowe 1962 (PMi 47): 182. — Bourd. & G. 1928: 674.

M.—*Boletus vaillantii* DC. sensu Lib. 1837 P.A.: No. 321, no descr. (*Polyporus*); fide Lowe 1962 (PMi 47): 182.

percandida (Mal. & Bert.) Donk 1972 (53bis). — *Poria* Mal. & Bert. 1971 (Morocco); *Incrustoporia* Al. David *1971* (incompl. ref.: n.v.p.). — Mal. & Bert. 1971 (Apb 8): 31, 35 *f. 6* (*Poria*); David 1971, cult. char. (*Incrustoporia*).

semipileata (Peck) Donk 1971. — *Polyporus* Peck 1883 (U.S.A., New York); *Tyromyces* Murrill 1907; *Leptoporus* Pilát 1937; *Leptotrimitus* Pouz. 1966; ≡ *Polyporus subpileatus* Lloyd 1920 (error; cf. Lowe 1942: 82), not ∼ Weinm. 1826. — Overh. 1914 (AMo 1): 1953: 295 *pl. 12 fs. 70, 71, pl. 14 f. 79, pl. 132 fig.* (*Polyporus*); M. P. Christ. 1960 (DbA 19): 365 *f. 364* (*Tyromyces*); Lowe 1966: 93 *f. 74* (*Polyporus*); Pouz. 1966 (ČM 20): 175 *pl. 14* (*Leptotrimitus*); Donk 1971 (PNA 74): 38, notes.

? *Leptoporus cervinus* Quél. 1892 (Crf 20²): 468 *pl. 3 f. 32*, wrong spores ? (France); cf. Bourd. & G. 1928: 553 who suggested the identity with *Leptoporus chioneus* [sensu Bourd. & G.]. — *Polyporus* Sacc. 1895, Cost. & Duf. 1895, not ∼ (Schw.) Steud. 1824: Fr. 1828; not ∼ Pers. 1825; *Polystictus* Big. & Guill. 1913, not ∼ (Schw.) Cooke 1886.

Poria hymenicola Murrill 1920 (M 12): 305 (U.S.A., Maine); fide Lowe 1947 (Ll 10): 52 (resupinate).

Poria coprosmae G. Cunn. 1947 (BPZ 72): 18, 38 *f. 13* (New Zealand); fide G. Cunn. 1965: 130 = *Tyromyces chioneus* [sensu G. Cunn.].

Polyporus atromaculatus Lloyd ex G. Cunn. 1950 (PNW 75): 270 ("*atro-maculus*") (Tasmania); fide G. Cunn. 1965: 130 = *Tyromyces chioneus* [sensu G. Cunn.]. — *Polyporus* Lloyd 1922 (nom. nud.: n.v.p.); Stev. & Cash 1936: 98, descr. (incidental mention: n.v.p.; "*atro-maculus*"); = *Polyporus alboniger* Lloyd in herb. (n.v.p.), cf. G. Cunn. 1965: 130, 264.

M.—*Polyporus nivosus* Berk. (**O**) sensu Morg. 1885 (JCi 8): 101; fide Lloyd 1915 (LMW 4, Ap.): 316 = *Polyporus semisupinus* [sensu Lloyd].

M.—*Polyporus chioneus* Fr. sensu Quél. 1888 (*Leptoporus*); cf. Bourd. & G. 1928: 54 ("praeter sporam"). — Quél. 1888: 385, wrong spores (*Leptoporus*); Bres. 1897 (AAR III 3): 70, wrong spores; 1908 (Am 6): 37 (*Polyporus*); Bourd. & L. Maire 1920 (BmF 36): 83; Bourd. & G. 1928: 543 *f. 154*; Teston 1953 (AUL III C 7): 17 *tpl. 1 f. 11* & 1953 (BOy 7): 84 *tpl. 1 f. 4*, hyphal analysis (*Leptoporus*).

M.—*Polyporus semisupinus* B. & C. apud Berk. sensu Lloyd 1915 (LMW 4, Ap.): 316 *fs. 654, 655*; fide Bourd. & G. 1928: 544 = *Leptoporus chioneus* [sensu Bourd. & G.].

stellae (Pilát) Domański 1963. — *Poria* Pilát 1941 (lacking Latin descr.: n.v.p.), 1953 (U.S.S.R., Ukraine); *Fomitopsis* Bond. 1953 (lacking descr. & valid ref.: n.v.p.); *Fomes* E. Komar. 1964. — Pilát 1941 (ACE 3): 464 *f. 226*; Jo. Erikss. 1958 (Sbu 16[1]): 151 *f. 49, pl. 23* (*Poria*); Domański 1965 (FpG 2): 96 *f. 30, pl. 21 f. 2, pl. 22* (*Incrustoporia*); Lowe 1966: 91 *f. 71* (*Poria*).

subincarnata (Peck) Domański 1963. — *Poria attenuata* var. Peck 1896 (U.S.A., New York); *Poria* Murrill 1921. — Overh. 1919 (BNS 205–206): 73 *pl. 2 fs. 3–6*; 1942; D. Baxt. 1943 (PMi 28): 229; Jo. Erikss. 1949 (SbT 43): 7 *f. 2, pl. 1*; M. P. Christ. 1960 (DbA 19): 350 *f. 347* (*Poria*); Domański 1965 (FpG 2): 98 *f. 31, pl. 23* (*Incrustoporia*); Lowe 1966: 92 *f. 72* (*Poria*).

Poria calcea var. *fragilis* Bourd. & G. 1925 (France); fide Bourd. & G. 1928: 675 (*Polyporus vulgaris* sensu Romell cited as a syn.). — Bourd. & G. 1928: 675.

Poria biguttulata Romell ex Pilát 1932. — *Poria* Romell 1926 (Sweden) (nom. prov.: n.v.p.). — Bourd. & G. 1928: 675 (*Poria calcea* f. *Poria biguttulata*); Pilát 1932 (BmF 48): 44 (*Poria*).

Poria krawtzewii Pilát 1932 (U.S.S.R., Russia, Siberia); fide Jo. Erikss. 1949 (SbT 43): 7. — *Chaetoporellus* Bond. & S. 1941 (generic name n.v.p.). — Pilát 1932 (BmF 48): 34.

M.—*Polyporus vulgaris* Fr. sensu Romell 1926 (SbT 20): 21; fide Jo. Erikss. 1949 (SbT 43): 2, 7. — Bond. 1953: 156 (*Poria*).

M.—*Polyporus vulgaris* var. "*β. P. calceus*" Fr. [= *Polyporus calceus*

(Fr. ex Pers.) Schw. (**O**)] sensu M. P. Christ. 1960 (DbA 19): 347 *f. 343* (*Amyloporia*); from descr.

tschulymica (Pilát) Domański 1963. — *Poria* Pilát 1932 (U.S.S.R., Russia, Siberia); *Gloeoporus* Bond. & S. 1941. — Pilát 1932 (BmF 48): 35 *f. 5, pl. 6 fs. 3, 4*; Jo. Erikss. 1958 (Sbu 16[1]): 154 *f. 50* (*Poria*); E. Komar. 1959 (BMs 12): 252 *f. 3* (*Gloeoporus*); Domański 1964 (APo 33): 171 (*Poria*); 1965 (FpG 2): 100 *f. 32, pls. 24, 25* (*Incrustoporia*); Kartav. 1967: 96. (*Gloeoporus*); Domański *1969e*, cult. char. (*Incrustoporia*).

Poria gilvella Pilát 1942 (lacking Latin descr.: n.v.p.), 1953 (U.S.S.R., Ukraine); fide Domański 1963 (APo 32): 734, 739 & 1964 (APo 33): 171. —*Ceriporia* Bond. 1953 (lacking Latin descr. & valid ref.: n.v.p.). — Pilát 1941 (ACE 3): 405 *f. 174, pl. 256.*

INONOTUS P. Karst. (54)

1879 [1960 (Pe 1): 230; 1962 (Pe 2): 206]. — Lectotype: *Polyporus hispidus* (Bull.) per Fr. — Sensu Pat. → *Hapalopilus*.

Inoderma P. Karst. 1879, not ∼ (Ach.) S. F. Gray 1821 (Lichenes), not ∼ Kütz. 1833 (Chlorophyta or diatoms), not ∼ Berk. 1881 (Ascomycetes) [1960 (Pe 1): 229]. — Lectotype: *Polyporus radiatus* (Sow.) per Fr. — Cf. (**55**).

Inodermus Quél. 1886 [1960 (Pe 1): 229; 1962 (Pe 2): 206]. — Lectotype: *Polyporus radiatus* (Sow.) per Fr. — Cf. (**55**).

Phaeoporus J. Schroet. 1888 [1960 (Pe 1): 252; 1962 (Pe 2): 208]. — Lectotype: *Polyporus cuticularis* (Bull.) per Fr.

Mensularia Lázaro 1916 [1960 (Pe 1): 240]. — Lectotype: *Polyporus radiatus* (Sow.) per Fr. — Sensu Pinto-L. → *Rigidoporus*.

Polystictoides Lázaro 1916 [1960 (Pe 1): 264]. — Lectotype: *Polyporus cuticularis* (Bull.) per Fr.

Xanthoporia Murrill 1916 [1960 (Pe 1): 294]. — Holotype: *Mucronoporus andersonii* Ell. & Ev.

S pecial literature. — General: Nikolajeva, *1955*; Pegler, *1964*.

Inonotus andersonii: Campbell & Davidson, *1939*; Černý, *1963a*; Murrill, *1946*.

Inonotus cuticularis: Brown, *1930*; Igmándy, *1964a*; H. Lohwag, *1929, 1930*; Pilát, *1926c*.

Inonotus dryophilus: Bailey, *1941*.

Inonotus dryadeus: Buchwald & Jørgersen, *1949*; Fergus, *1956*; Lloyd, *1910*; Long, *1913a, 1913b, 1930*; Weir, *1921*.

Inonotus glomeratus: Campbell & Davidson, *1939a, 1939b*; Good & Nelson, *1951*; Hirt, *1949b*; Overholts, *1917b*; Shigo, *1969*.

Inonotus hispidus: Baxter, *1924*; Berlese, *1889a, 1889b*; W. G. Campbell, *1931*; Hazslinszky, *1870*; H. Lohwag, *1930*; Martelli, *1889*; McCracken & Toole, *1969*; Messikommer, *1950*; Nannfeldt, *1956*; Nutman, *1929*; Pilát, *1928, 1930a*; Prilleux, *1893*; Prutenskaja, *1965*; Sleeth & Bidwell, *1937*; Toole, *1955*.

Inonotus nidus-pici: Benkert, *1971*; Černý, *1959, 1964, 1965, 1970*; Haracsi, *1941*; Haracsi & Igmándy, *1956*; K. Lohwag, *1966*; Skorič, *1937*.

Inonotus obliquus: Aoshima, *1951*; Batko, *1950*; Bulatov & al. (ed.), *1959*; Campbell & Davidson, *1938*; Černý, *1963b, 1964, 1965*; Findlay, *1939*; Hirt, *1949b*; von Höhnel, *1907*; Kartajevskaja, *1928*; Ljubarskij, *1960*; H. Lohwag, *1935, 1936*; K. Lohwag, *1960*; Manka & Stube, *1952*; Nizkowskaja, *1963* (n.v.); Parmasto, *1957*; Winters & al., *1961*.

Inonotus radiatus complex: Jahn, *1965c* (*I. "polymorphus"*).

Inonotus rheades complex: Bailey, *1941*; Hedgcock & Long, *1914*; Kotlaba & Pouzar, *1969*; Sinadskij & Bondarceva, *1956*.

Inonotus weirii: Bier & Buchanan, *1947*; Buchanan, *1948*; Buckland & al., *1954*; Childs, *1963*; Childs & Nelson, *1960*; Mounce, Bier, & Nobles, *1940*; Nelson, *1967*; Wallis & Reynolds, *1962, 1965*.

andersonii (Ell. & Ev.) Černý 1963. — *Mucronoporus* Ell. & Ev. 1890 (U.S.A., New Jersey); *Scindalma* O.K. 1898; *Xanthochrous* Pat. 1900; *Poria* Lloyd 1912; *Inonotus* Nikol. 1955 (incomplete ref.: n.v.p.). — Campb. & Dav. 1939 (M 31): 161, 163 *fs. 1A–E, fs. 2A, B*, with cult. char.; R. W. Davids. & al. 1942 (TUS 785): 41 *f. 5L, pl. 3 f. D*, cult. char.; Overh. 1942: 54 (*Poria*); Nikol. 1955 (BŽ 40): 263 *f. 3*; Černý 1963 (ČM 17): 1 *fs. 1–7*; Pegl. 1964 (TBS 47): 186 *fs. 1: 1, 2: 1* (*Inonotus*); Lowe 1966: 140 *f. 125* (*Poria*).

Poria xanthospora Underw. 1894 (PIA 1893): 61 (U.S.A., Indiana; fide Overh. 1942: 55 & Pegl. 1964 (TBS 47): 186.

Inonotus leei Murrill 1915 W.P.: 21 (U.S.A., California); fide Campb. & Dav. 1939 (M 31): 162 & Overh. 1942: 55. — *Poria* Sacc. & Trott. apud Trott. 1925.

Xanthochrous krawtzewii Pilát 1934 (U.S.S.R., Russia, Siberia); fide Černý 1963 (ČM 17): 2. — *Xanthochrous* Pilát 1932 (nom. nud.: n.v.p.); *Inonotus* Pilát 1940. — Pilát 1934 (BmF 49): 273 *pl. 22* (*Xanthochrous*); 1940 (Am 38): 81; 1942 (ACE 3): 571 *f. 267, pl. 362*; Bond. 1961 (TSR 14): 201 (*Inonotus*).

Corticium pactolinum Cooke & Harkn. 1881 (G 9): 81 (U.S.A., California) (nom. anam.); fide Ell. & Ev. 1890 (JM 6): 79, based on a spore print on bark. — *Chromosporium* Cooke 1888, not ∼ (Cooke) Cooke 1904. — Donk 1971 (PNA 74): 8, note.

Chromosporium vitellinum Sacc. & Ell. apud Sacc. & Berl. 1885 (AIv VI 3): 728 (U.S.A., New Jersey) (nom. anam.); fide Ell. & Ev. 1890 (JM 6): 79, based on a spore print on bark. — Donk 1972 (PNA 74): 8, note.

M.—*Boletus obliquus* Ach. ex Pers. sensu Bourd. 1932 (BmF 48): 229 (*Xanthochrous*); fide Pilát 1942 (ACE 3): 571 = *Inonotus krawtzewii*.

M.—*Boletus hispidus* Bull. sensu Burt 1931 (AMo 18): 472, "resupinate" (*Polyporus*); fide Pilát 1942 (ACE 3): 571 = *Inonotus krawtzewii*.

cuticularis (Bull. per Fr.) P. Karst. 1879. — *Boletus* Bull. 1789 (France) (d.n.) not ∼ Thore 1803 (d.n.); *Polyporus* (Bull.) per Fr. 1821; *Boletus* Mérat 1821; *Inodermus* Quél. 1886; *Phaeoporus* J. Schroet. 1888; *Polystictus*

Gillot & Luc. 1890, not ∼ Lloyd 1908; *Xanthochrous* Pat. 1900; *Polystic-toides* Lázaro 1916; *Agaricus* E. Krause 1933; *Fomes* Konr. & M. 1937. — Bull. 1789: *pl. 462*; 1791 H.: 350 (*Boletus*); Lloyd 1915 (LMW 4, Ap.): 359 *fs. 693, 694* (*Polyporus*); Bourd. & G. 1928: 635 *f. 179*; Lohw. 1929 (ArP 65): 321 *fs. 1–5*; 1930 (ArP 72): 420 *fs. 1–3* (*Xanthochrous*); Donk, 1933: 243 (*Inonotus*); Overh. 1953: 412 *pl. 49 f. 295, pl. 99 f. 564, pl. 103 f. 579, pl. 129 fig.* (*Polyporus*); A. Teix. 1961 (M 52); 14 *f. 14*, hyphae; H. Jahn 1963 (WPb 4): 110 *Abb. 21*; Pegl. 1964 (TBS 47): 185 *fs. 1: 4, 9, f. 2: 4* (*Inonotus*). — Sensu Wahl., Rostk. → *Inonotus rheades*; sensu Imaz. & Hongo = *Inonotus mikadoi* (Lloyd) Bond. (cited as syn.; extra-European) (**O**).

Boletus triqueter Pers. 1796 O. 1: 86 (Germany) (d.n.); fide Fr. 1838: 458 (var.) & Lloyd 1910 (LMW 3): 466, 1915 (LMW 4, Ap.): 360. — *Polyporus* (Pers.) per Pers. 1825, not ∼ Fr. 1838. — Sensu A. & S., Secr. → *Onnia triqueter*; sensu Quél. → *Ischnoderma trogii*.

? *Xanthochrous fuscovelutinus* Pat. 1908 (BmF 24): 6 (U.S.A., Louisiana); fide Lloyd 1915 (LMW 4, Ap.): 379. — *Polyporus* Sacc. & Trott. 1912. — Cf. Pat., l.c., "Proche . . . de *X. cuticularis*, mais presque glabre, plus dur et sans [!] cystides."

M.—*Polyporus perplexus* Peck (**O**) sensu Murrill 1904 (BTC 31): 596 & 1908 (NAF 9): 88 (*Inonotus*); fide Lloyd 1908 (LMW 2): 378 & Murrill 1920 (M 12): 12.

dryadeus (Pers. per Fr.) Murrill 1908. — *Boletus* Pers. 1799 (Germany) (d.n.); *Polyporus* (Pers.) per Fr. 1821; *Boletus* S. F. Gray 1821; *Ischnoderma* P. Karst. 1879; *Placodes* Quél. 1886; *Fomes* Gillot & Luc. 1890; *Phellinus* Pat. 1900; *Ungularia* Lázaro 1916; *Placoderma* Ulbr. 1928; *Inonotus* G. Cunn. 1958 (preoccupied); *Xanthochrous* Igmándy 1965 (incomplete ref.: n.v.p.). — Fr. 1874: 553; Hussey c. 1847 I. 1: *pl. 21*; Long 1913 (JaR 1): 239 *pl. 21 f. 4, pl. 22 fs. 2, 4–6*; Lloyd 1910 (LMW 3): 490 *f. 383*; 1915 (LMW 4, Ap.): 352 (*Polyporus*); Bourd. & G. 1928: 615 *f. 171* (*Phellinus*); Konr. & M. 1930 I. 5: *pl. 464* (*Fomes*); Pilát 1942 (ACE 3): 554 *f. 261, pls. 355, 356* (*Inonotus*); R. W. Davids & al. 1942 (TUS 785): 29 *f. 4H, pl. 2 f. D*, cult. char.; Overh. 1953: 408 *pl. 50 f. 300, pl. 112 f. 616, pl. 128 fig., pl. 129 fig.* (*Polyporus*); Kreisel 1961: 122 *f. 76*; H. Jahn 1963 (WPb 4): 105 *f. 2, Abb. 18*; Pegl. 1964 (TBS 47): 168 in obs., 180 *fs. 1: 3, 2: 5* (*Inonotus*). — Sensu R. Hartig → *Inonotus dryophilus*; sensu Clel., G. *Cunn.* = *Inonotus chondromyelus* Pegl. (extra-European).

Boletus pseudo-igniarius Bull. 1789 (France); fide Fr. 1821: 374. — *Boletus* Bull. per St-Am. 1821; *Polyporus* Chev. 1826, Balbis 1828, Sacc. 1916; *Ochroporus* J. Schroet. 1888; *Fomes* Big. & Guill. 1913 (syn.: n.v.p.). — Bull. 1789: *pl. 458*; 1791 H.: 356 (*Boletus*).

dryophilus (Berk.) Murrill 1904 (**59**). — *Polyporus* Berk. 1847 (U.S.A., Ohio); *Phellinus* A. Ames 1913; *Inonotus* Bond. & S. 1941 (preoccupied). — Hedgc. & Long 1914 (JaR 3): 72 *pls. 8–10*, excl. of specimens on

Populus; R. W. Davids. & al. 1942 (TUS 785): 29 *f. 4I, pl. 2 f E*, cult. char.; Overh. 1953: 417 *pl. 50 fs. 302, 303, pl. 52 f. 312, pl. 110 fs. 608, 609, pl. 129 fig.*, excl. of var. (*Polyporus*); H. Jahn 1963 (PWb 4): 106 *Abb. 12, 13*; Pegl. 1964 (TBS 47): 187 *f. 2: 6* (*Inonotus*); Igmándy 1965 (EFE): 216 *fs. 5–7* (*Xanthochrous*). — Sensu auctt. nonn. → *Inonotus rheades.*

Polyporus corruscans Fr. 1851 (ÖVS 8): 52 (Sweden); fide Pegl. 1964 (TBS 47): 187. — *Inonotus* P. Karst. 1882. — Fr. 1863 M. 2: 269; sensu Lloyd 1909 (LMW 2, L. 24): 1 (*Polyporus*); Bourd. & G. 1928: 637 (*Xanthochrous rheades* subsp.). — Sensu Speg. = *Inonotus rickii* (Pat.) D. Reid (extra-European); sensu Bres. 1931 → *Inonotus tamaricis.* — V.s.: "*coruscans*".

M.—*Boletus dryadeus* Pers. sensu R. Hartig 1878 (*Polyporus*); fide Hedgc. & Long 1914 (JaR 3): 68. — R. Hartig 1878: 124 *pl. 17* (*Polyporus*).

hispidus (Bull. per Fr.) P. Karst. 1879. — *Boletus* Bull. 1784 (France) (d.n.); *Polyporus* Fr. 1818 (d.n.); *Polyporus* (Bull.) per Fr. 1821, not ∼ (Bagl.) J. Rick 1938 (error); *Boletus* St-Am. 1821; *Inodermus* Quél. 1886; *Phaeoporus* J. Schroet. 1888; *Polystictus* Gillot & Luc. 1890; *Xanthochrous* Pat. 1897; *Hemidiscia* Lázaro 1916; *Fomes* Maubl. 1927. — Bull. 1784: *pl. 210*; 1790: *pl. 493*; 1791 H.: 351; Bolt. 1791; 161 *pl. 161* (*Boletus*); Krombh. 1841 S. 7: 7 *pl. 48 fs. 7–10*; Gillet 1874–90 P.: *pl. 560/461*; Boud. 1904–11: 81 *pl. 158*; Lloyd 1915 (LMW 4, Ap.): 359 (*Polyporus*); Bourd. & G. 1928: 638; Lohw. 1930 (ArP 72): 422 *pls. 25–28* (*Xantho-chrous*); Donk 1933: 242; Pilát 1942 (ACE 3): 567 *f. 264, pls. 365–367* (*Inonotus*); R. W. Davids & al. 1942 (TUS 785): 34 *f. 5B, pl. 2 f. J*, cult. char.; Overh. 1953: 423 *pl. 52 f. 314, pl. 53 f. 319, pl. 105 f. 586, pl. 114 f. 623, pl. 117 f. 637, pl. 130 fig.* (*Polyporus*); A. Teix. 1961 (M 52): 37 *f. 13*, hyphae; H. Jahn 1963 (WPb 4): 109 *Abb. 19*; Pegl. 1964 (TBS 47): 184 *f. 1: 7, f. 2: 9*; Pegl. & Wat. 1968 (CDp): no. 193 *figs.* (*Inonotus*). — Sensu Rostk. 1830 (StP 4): 65 *pl. 31* = ? (referred by Fr. 1863 M. 2: 270 to *Polyporus* [*Inonotus*] *vulpinus*); sensu Burt 1931 → *Inonotus andersonii.*

? *Boletus hirsutus* Scop. 1772: 468 (Yugoslavia, Carniola) (d.n.), not ∼ Batsch 1783 (d.n.), not ∼ Latourr. 1785 (d.n.), not ∼ Wulf. 1788 (d.n.); fide Fr. 1821: 362. — *Boletus* Scop. per Pollini 1824, not ∼ (Wulf. per Fr.) Wahl. 1826; *Inonotus* Murrill 1904. — Murrill 1904 (BTC 31): 594 (*Inonotus*).

Boletus spongiosus Lightf. 1778: 1033 (Scotland) (d.n.), not ∼ Pers. 1801; fide Bolt. 1791: 161 & Fr. 1821: 362. — *Boletus* Lightf. per Hook. 1821.

Boletus villosus Huds. 1778: 626 (England) (d.n.), not ∼ With. 1776 (d.n.), not ∼ Sw. 1788 (d.n.), not ∼ Gaut. 1884; fide Fr. 1821: 562.

Boletus velutinus With. 1796: 331 (England) (d.n.), not ∼ Plan. 1788 (d.n.), not ∼ Vahl 1794 (d.n.); fide Berk. 1836: 138 (as to "Sow.", "Purt.").

— *Boletus* With. per Purt. 1821; *Polyporus* E. Krause 1931; *Agaricus* E. Krause 1933. — Sow. 1802: *pl. 345*.

Polyporus endocrinus Berk. 1847 (LJB 6): 320 (U.S.A., Ohio); fide Murrill 1904 (BTC 31): 595 = *Inonotus hirsutus*; fide Pegl. 1964 (TBS 47): 184. — *Phaeolus* Pat. 1900. — Bres. 1916 (Am 14): 235 (*Polyporus*), "Cum *Pol. hispido* non identicus, probabiliter ejus varietas."

Polyporus pollinii Heufl. 1871 (VW 21): 289–290; fide Sacc. 1873 (ASv 2): 99; Bres. 1891 (BSb 9): 30 (for *Boletus flavus*). — [≡ *Boletus flavus* Pollini 1824: 607 ("*Boleti igniarii* varietas memorabilis, vel distincta forte species est: *Boletus flavus*" = nom. prov.: n.v.p.) (Italy), not ∼ With. per Purt. 1821, not ∼ Jungh. 1838].

M.—*Polyporus gelsorum* Fr. (O) sensu Martelli 1889 (NGi 21): 293; fide Berlese 1889 (NGi 21): 531 & Bres. 1891 (BSb 9): 30.

nidus-pici Pilát 1953 (Yugoslavia, Croatia) — *Inonotus* Pilát 1942 (lacking Latin descr.: n.v.p.); *Xanthochrous* Igmándy 1965 (incomplete ref.: n.v.p.). — Pilát 1942 (ACE 3): 574 *fs. 269, 276: 2, pl. 360 f. b, pl. 361*; Černý 1965 (SnP 21): 160 *fs. 1–56* (*Inonotus*); Igmándy 1965 (EFE): 219 *fs. 14–17, 21* (*Xanthochrous*).

Poria obliqua (Pers. per Fr.) P. Karst. f. "S u r   c h ê n e" Bourd. & G. 1925 (France); fide Pilát 1953 (SnP 9²): 108. — Bourd. & G. 1928: 643.

M.—*Boletus obliquus* Ach. ex Pers. sensu Škorič 1937 (*Poria*); fide Černý 1959: 85. — Škorič 1937 (GšP 5): 275 *fs. 1, 2, pls. 1–4* (*Poria*); Haracsi *1941*; Haracsi & Igmándy *1956*;—all fide Černý *1959*.

nodulosus (Fr.) P. Karst. 1882 (**56**). — *Polyporus* Fr. 1838 (Sweden); *Trametes* Fr. 1848, 1849 (nom. nud.: n.v.p.); *Inoderma* P. Karst. 1879; *Inonotus* Pilát 1937, preoccupied; *Polystictus* Cooke 1886; *Inodermus* Keissl. 1933 (syn.: n.v.p.); *Microporus* O.K. 1898; *Xanthochrous* Pat. 1900; *Poria* Jaap apud Lloyd 1908 (nom. nud.: n.v.p.). — Fr. 1884 I. 2: 86 *pl. f. 2* (*Polyporus*); Bres. 1897 (AAR III 3): 72 (*Polyporus radiatus* var.); A. L. Sm. & Rea 1907 (TBS 2): 171 *pl. 16 fig.*; Lloyd 1915 (LMW 4, Ap.): 352 *f. 689* (*Polyporus*); Bourd. & G. 1928: 634 (*Xanthochrous radiatus* var.); H. Jahn 1963 (WPb 4): 104 *Abb. 15* (*Inonotus*); Pegl. 1964 (TBS 47): 181 *f. 1: 15* (*Inonotus radiatus* var.); H. Jahn 1967 (WPb 6): 98 (*Inonotus*).

Polyporus polymorphus Rostk. 1838 (Germany/Poland) (**56**), not ∼ Holterm. 1898; fide Bres. 1897 (AAR III 3): 73 & Lloyd 1915 (LMW 4, Ap.): 352. — *Polystictus* Cooke 1886; *Inodermus* Quél. 1886; *Ochroporus* J. Schroet. 1888; *Microporus* O.K. 1898: *Xanthochrous* Bourd. & L. Maire 1920, misapplied; *Inonotus* Pilát 1940, Bond. & S. 1941. — Rostk. 1838 (StP 4): 115 *pl. 56*. — Sensu Bourd. & L. Maire → *Inonotus polymorphus* sensu Bourd. & L. Maire; sensu Bourd. 1932 ("forme") → *Inonotus weirii*.

Trametes fagi Otth 1871 (MiB 1870): 93 (Switzerland); from descr.

? *Polyporus dentifer* Velen. 1922: 673 [see Pilát 1948: 256 for Latin translation] (Czechoslovakia); fide Pilát 1942 (ACE 3): 557, 558. — (**57**).

Polyporus armatus Velen. 1922: 680 [see Pilát 1948: 258 for Latin translation] (Czechoslovakia), not ∼ (Pat.) Sacc. & Trott. apud Trott. 1925; fide Pilát 1942 (ACE 3): 557, 558, 559.

obliquus (Pers. per Fr.) Pilát 1942. — *Boletus* Ach. ex Pers. 1801 (Sweden) (d.n.); not ∼ Bolt. 1788 (d.n.); *Polyporus* (Pers.) per Fr. 1821; *Physisporus* Chev. 1826; *Boletus* Lenz 1840; *Poria* P. Karst. 1881; *Fomes* Cooke 1885; *Phaeoporus* J. Schroet. 1888; *Mucronoporus* Ell. & Ev. 1889; *Scindalma* O.K. 1898; *Phellinus* Pat. 1900, Bond. & S. 1941; *Xanthochrous* Bourd. & G. 1928; *Fuscoporia* Aosh. 1951; ≡ *Polyporus incrustans* Pers. 1825; *Poria* Lloyd 1910 (incidental mention: n.v.p.); ≡ *Polyporus obliquiporus* E. Krause 1934. — Fr. 1821: 378; 1884 I. 2: 86 *pl. 188 f. 1*, poor (*Polyporus*); Höhn. 1907 (ÖbZ 57): 177 (*Poria*); Bourd. & G. 1928: 642 *f. 180*, in part: "Sur Orme" (*Xanthochrous*); Lohw. 1936 (ÖbZ 85): 270 *fs. 1–6*; Lundell 1950 (LNF 37–38): 5 No. 1809 (*Polyporus*); H. Jahn 1963 (WPb 4): 111 *f. 7, Abb. 20*; Pegl. 1964 (TBS 47): 186 *fs. 1: 12, 2: 15*; Černý 1965 (SnP 21): 217 *fs. 1 [bis]-25 [bis]* (*Inonotus*); Lowe 1966: 139 *f. 124* (*Poria*). — Sensu Bourd. & G. in part: f. "Sur chêne" → *Inonotus nidus-pici*; sensu Bourd. 1932 → *Inonotus andersonii*; sensu Škorič → *Xanthoporia nidus-pici*.

M.—*Polyporus nigricans* Fr. sensu Lindr. 1904 (NZL 2): 393, in part: as to "abnorme Fruchtkörper" p. 395 *f. 1*.

polymorphus (Rostk.) Pilát sensu Bourd. & L. Maire 1920 (*Xanthochrous*) (**57**). — Bourd. & L. Maire 1920 (BmF 36): 85 (*Xanthochrous*); Bourd & G. 1928: 634 (*Xanthochrous radiatus* subsp.); Domański 1965 (FpG 2): 208 *f. 68, pl. 59 f. 1* (*Inonotus radiatus* subsp.); H. Jahn 1965 (WPb 5): 131 *fs. 1, 2* (*Inonotus*). — Sensu typi → *Inonotus nodulosus*. — Cf. *Polyporus salebrosus* Lasch (**O**).

radiatus (Sow. per Fr.) P. Karst. 1881 (**58**). — *Boletus* Sow. 1799 (England) (d.n.); *Polyporus* (Sow.) per Fr. 1821; *Boletus* Spreng. 1827; *Trametes* Fr. 1849; *Inoderma* P. Karst. 1879; *Inonotus* P. Karst. 1881; *Polystictus* Cooke 1886; *Inodermus* Quél. 1886; *Ochroporus* J. Schroet. 1888; *Microporus* O.K. 1898; *Xanthochrous* Pat. 1900; *Mensularia* Lázaro 1916; *Fomes* Konr. & M. 1935. — Fr. 1874: 565; Bres. 1897 (AAR III 3): 72; Lloyd 1915 (LMW 4, Ap.): 351 *f. 688* (*Polyporus*); Bourd. & G. 1928: 633 (*Xanthochrous*); Donk 1933: 245 (*Inonotus*); Konr. & M. 1935 I. 5: *pl. 454* (*Fomes*); Overh. 1953: 406 *pl. 49 f. 294, pl. 121 f. 658, pl. 131 fig.* (*Polyporus*); H. Jahn 1963 (WPb 4): 103 *Abb. 14*; Pegl. 1964 (TBS 47): 180 *fs. 1, 4, 2: 23* (*Inonotus*);—mostly 'sensu lato'. — Sensu G. Cunn. 1927 = *Inonotus nothofagi* G. Cunn. (extra-European), fide G. Cunn. 1965: 205.

Boletus alneus Pers. 1796 O. 1: 86 (Germany) (d.n.); fide Fr. 1832[Ind.]: 35 & Lloyd 1910 (LMW 3): 467, 1915 (LMW 4, Ap.): 352. — *Boletus* Pers. per S. F. Gray 1821; *Polyporus* Secr. 1833 (as a sp. of *Boletus*: n.v.p.), E. Krause 1928. — Secr. 1833 M. 3: 98 (*Polyporus*).

Polyporus plicatus Pers. 1825: 212 (France), not ∼ Bl. ex Lév. 1844; fide Donk 1933: 245, 246.

Polyporus cucullatus B. & C. apud Berk. 1872 (G 1): 51 (U.S.A., New England) (**58**); fide Bres. 1916 (Am 14): 224 (forma); fide Pegl. 1964 (TBS 47): 181 (var.). — Pegl., l.c. [*Inonotus radiatus* var. *cucullatus* (B. & C. apud Berk.) Pegl.], extra-European.

Polyporus aureonitens Pat. apud Peck 1889 (RNS 42): 121 (U.S.A., New York); fide Overh. 1953: 406, 407. — *Polystictus* Sacc. 1891; *Microporus* O.K. 1898; *Xanthochrous* Pat. 1900. — Cf. Murrill 1904 (TBC 31): 601.

Polyporus coffeaceus Velen. 1922: 684 *f. 111: 1* [see Pilát 1948: 259 for Latin translation] (Czechoslovakia); fide Kotl. & P. 1969 (ČM 23): 169.

Polyporus minutus Vanin 1923 (BMs 2): 15 (U.S.S.R., European Russia); fide Bond. 1953: 324.

Polyporus illinoisensis Baxt. 1939 (U.S.A., Illinois) (**58**); fide Overh. 1953: 407 = *Polyporus radiatus* var. *cephalanti* Overh., extra-European. — D. Baxt. 1939 (PMi 24): 180 *pl. 4 f. 2.*

rheades (Pers.) P. Karst. 1882 (**59**). — *Polyporus* Pers. 1825 (France); *Inodermus* Quél. 1886; *Xanthochrous* Pat. 1897, misapplied; *Polystictus* Big. & Guill. 1913; *Hemidiscia* Lázaro 1916; *Inonotus* Bond. & S. 1941 & Pilát 1942 (preoccupied). — Pat. 1904 (BmF 20): 51, in obs.; Bres. 1908 (Am 6): 38; Lloyd 1910 (LMW 3): 467 *f. 333* (*Polyporus*); Bourd. & G. 1928: 635 (*Xanthochrous*); Bres. 1921 (BIm 20): *pl. 980 f. 1* (*Polyporus*); Pegl. 1964 (TBS 47): 188; Kotl. & P. *1969*; 1970 (ČM 24): 150 (*Inonotus*). — Sensu Bres. 1892 → *Inonotus tamaricis*; sensu G. Cunn. 1965 = *Inonotus pirisporus* Pegl. ? (extra-European), cf. D. Reid 1967 (TBS 50): 164. — V.s.: *"rhaeades"*, *"rhoeades"*.

Polyporus rubiginosus Fr. 1838: 460, not ∼ Wallr. 1833, not ∼ Berk. 1839; fide Donk 1971 (Pe 6): 209. — *Boletus* Lenz 1840 ("Rostkov."); *Trametes* Fr. 1849; [≡ *Polyporus cuticularis* Bull. sensu Rostk. 1830 (StP 4): 67 *pl. 32* (Germany/Poland)]. — Sensu Bres. → *Tyromyces fissilis.*

Polyporus vulpinus Fr. 1852 (Sweden) (**59**), not ∼ (Link) per Fr. 1821; fide Bres. 1908 (Am 6): 38 & Lloyd 1910 (LMW 3): 467. — *Polystictus* Fr. 1852 (nom. altern.); *Inoderma* P. Karst. 1879; *Inonotus* P. Karst. 1882; *Inodermus* Quél. 1886; *Ochroporus* J. Schroet. 1888; *Microporus* O.K. 1898; *Xanthochrous* Pat. 1900. — Fr. 1863 M. 2: 270 (*Polyporus*); Bourd. & G. 1928: 639 (*Xanthochrous rheades* subsp.); Donk 1933: 244 (*Inonotus*); Overh. 1953: 418 *pl. 50 f. 304, pl. 51 f. 310; pl. 129 fig.* (*Polyporus dryophilus* var.); H. Jahn 1963 (WPb 4); 108 *Abb. 11* (*Inonotus*). — Sensu Kalchbr. 1877, in part, → *Funalia trogii*; sensu Kalchbr. 1877, in part, → *"Trametes" cervina.*

Polyporus fulvus Fr. 1863 M. 2: 253 ("Fr. excl. syn."), 270, not ∼ (Scop.) per Fr. 1838; fide Lloyd 1915 (LMW 4, Ap.): 362 = *Polyporus rheades* [s. str.]. — [≡ *Polyporus fulvus* (Scop.) per Fr. sensu Fr. 1838: 466 (Sweden)]. — Donk 1971 (Pe 6): 209, 1971 (PNA 74): 405, notes.

Inonotus hisingeri (P. Karst.) P. Karst. 1882; fide P. Karst. 1890

(Mfe 16): 103 = *Inonotus vulpinus* (forma); fide Bres. 1908 (Am 6): 38. — *Inonotus vulpinus* var. P. Karst. 1882 (BFi 37): 238 (Finland); ≡ *Polyporus inonotus* Sacc. 1888.

Inonotus levis P. Karst. 1887 (H 26): 112; fide P. Karst. 1898 T. 3: 18 / 1903 (BFi 62): 82 = *Inonotus fulvus* (Fr.) P. Karst. [sensu P. Karst.]; fide Lowe 1956 (M 48): 107 = *Polyporus dryophilus* [broad sense]. — P. Karst. in Radde & Walt. 1889 (AHp 10^2): 550 *pl. 1 f. 2.*

Polyporus friesii Bres. 1905 (Am 3): 163 (Germany) (**60**), not ∼ Kl. 1833; fide Bres., l.c. (*Polyporus fulvus* Fr. cited as a syn.) — *Polystictus* Rick. 1918; *Fomes* Killerm. 1922.

M.—*Boletus cuticularis* Bull. sensu Wahl. 1826; fide Fr. 1852 (ÖVS 9): 130, 1874: 565 = *Polyporus vulpinus.* — Sensu Rostk. → *Polyporus rubiginosus* Fr., cited above.

M.—*Boletus fulvus* Scop. (**O**) sensu Fr. 1838: 466 (*Polyporus*); fide Bres. 1905 (Am 3): 163 = *Polyporus friesii*; fide Lloyd 1908 (LMW 2, L. 24): 1 = *Polyporus corruscans* [sensu Lloyd] (old fruitbodies), 1915 (LMW 4, Ap.): 362 → *Polyporus fulvus* Fr., see this sp. cited above.

M.—*Polyporus castaneus* Fr. (**O**) sensu Britz. 1887 (BnS 29): 279 [*pl. 610 f. 62*] ("Rostk."); fide Bres. apud Killerm. 1940 (Bba 21): 70 ("alt").

M.—*Polyporus dryophilus* Berk. sensu auctt. nonn. — Hedgc. & Long 1914 (JaR 3): 68, 71 in obs., *pl. 10 fs. 1–3, 5, 6, 8.*

subiculosus (Peck) Erikss. & Strid 1969 (**61**). — *Polyporus* Peck 1879 (U.S.A., New York); *Fuscoporia* Murrill 1907. — Overh. 1919 (BNS 205–206): 115 *pl. 21 fs. 1–5*; D. Baxt. 1937 (PMi 24^I): 178; Overh. 1942: 63; Lowe 1966: 164 *f. 153* (*Poria*); Erikss. & Strid 1969: 135 *f. 2* (*Inonotus*).

tamaricis (Pat.) Maire apud Maire & Wern. 1938. — *Xanthochrous* Pat. 1904; *Polyporus* Sacc. & D. Sacc. 1905; *Inonotus* Bond. & S. 1941 (preoccupied); [≡ *Xanthochrous rheades* (Pers.) Pat. sensu Pat. 1897 (BmF 13): 199 (Algeria)]. — Pat. 1904 (BmF 20): 51 (*Xanthochrous*); Bourd. & G. 1928: 636 (*Xanthochrous rheades* subsp.); Pegl. 1964 (TBS 47): 187 *f. 2: 28* (*Inonotus*).

M.—*Polyporus rheades* Pers. sensu Bres. 1892; fide Pat. 1904 (BmF 20): 51. — Bres. 1892 F.t. 2: 30 *pl. 136.* → *Polyporus corruscans* sensu Bres. 1931.

M.—*Polyporus corruscans* Fr. sensu Bres. 1931. — Bres. 1931 (BIm 20): *pl. 981.*

weirii (Murrill) Kotl. & P. 1970. — *Fomitoporia* Murrill 1914 (U.S.A., Idaho); *Poria* Murrill 1914 (nom. altern.); *Fuscoporia* Aosh. 1953. — Murrill 1914 (M 6): 93 *pl. 122*; Overh. 1931 (M. 23): 122 *pl. 12 fs. 1, 2, pl. 13 fs. 9, 11, 13*; D. Baxt. 1934 (PMi 9): 329 *pl. 65*; Mounce & al. 1940 (CJR 18): 257 *fs. 4–14, pl. 2*; D. Baxt. 1953 (PMi 38): 116 *pls. 1, 4*; 1954 (PMi 39): 128 *pl. 2* (*Poria*); Imaz. & Aosh. 1955; 241, 268 *pl. 27* (*Fuscoporia*);

Lowe 1966: 145 *f. 130* (*Poria*); Kotl. & P. 1970 (ČM 24): 148; Pegl. & Gibs. 1972 (CDp): no. 323 *figs.* (*Inonotus*).

Xanthochrous heinrichii (Pilát) Pilát 1934; fide Kotl. & P. 1970 (ČM 24): 148, 150. — *Xanthochrous glomeratus* subsp. Pilát 1932 (U.S.S.R., Russia, Siberia); *Xanthochrous* Pilát 1932 (nom. prov.: n.v.p.); *Inonotus* Bond. & S. 1941, Pilát 1942. — Pilát 1932 (BmF 48): 28 *f. 4, pl. 6 fs. 1, 2* (*Xanthochrous glomeratus* subsp.); Murašk. 1939: 93 *f. 15A* (*Xanthochrous*); Pilát 1942 (ACE 3): 575 *pl. 359*; Parm. 1959 (EAT 8[4]): 266, 278 *fs. 1–3, pl. (1) fs. 1, 2*; Kartav. 1961 (TSR 14): 192; Pegl. 1964 (TBS 47): 179; Kartav. 1967: 119 (*Inonotus*); Kotl. & P. 1970 (ČM 24): 147 (*Xanthochrous glomeratus* var.).

Phellinus sulphurascens Pilát 1936 (U.S.S.R., Russia, Siberia); fide Kotl. & P. 1970 (ČM 24): 148, 150. — *Poria* Killerm. 1943. — Pilát 1936 (BmF 51): 372 *f. 5*; 1942 (ACE 3): 550 *pl. 331 f. a.*

Xanthochrous heinrichii f. *nodulosus* Pilát 1936 (BmF 51): 376 (U.S.S.R., Russia, Siberia); fide Kotl. & P. 1970 (ČM 24): 148. — *Inonotus heinrichii* forma Pilát 1942 (ACE 3): 576.

M.—*Polyporus polymorphus* Rostk. sensu Bourd. 1932 (BmF 48): 229 (*Xanthochrous*; "forme"); fide Pilát 1942 (ACE 3): 575 = *Inonotus heinrichii.*

IRPEX Fr.

1825 [1956 (Ta 5): 100; 1963 (Ta 12): 154]. — Lectotype: *Hydnum lacteum* (Fr.) per Fr.

M.—*Xylodon* (Pers.) ex S. F. Gray sensu O.K. 1898: 540 [1956 (Ta 5): 115, in obs.].

Special literature.—*Irpex* s.lat.: Pilát, *1925.*

Irpex lacteus complex: Darley & Christensen, *1945*; Domański & Orlicz, *1969*; Nisizawa, *1955.*

lacteus (Fr. per Fr.) Fr. 1828 (62). — *Sistotrema* Fr. 1818 (Sweden) (d.n.); *Hydnum* (Fr.) per Fr. 1821; *Dryodon* Pat. 1887 (nom. nud.: n.v.p.), misapplied; *Xylodon* O.K. 1898; *Coriolus* Pat. 1900; *Irpiciporus* Murrill 1907; *Trametes* Pilát 1939, not ∼ Fr. 1851; *Hirschioporus* Teng 1964. — Bres. 1897 (AAR III 3): 101; Bourd. & G. 1928: 573 (*Irpex*); Pilát 1940 (ACE 3): 322 *f. 137, pls. 215, 216* (*Trametes*); Nannf. & Du R. 1952: 238 *f. 185, pl. 119*; Nikol. 1953 (TSR 8): 183 *fs. 1–7*; Angerer 1958 (Dba 32): 141 *plate f. 2*; Dom. & Orl. *1969* (*Irpex*). — Sensu Pat. → *Spongipellis pachyodon*; sensu Maas. G. 1963 (PNA 66): 453 fs. 11–13 (*Irpex*) = *Steccherinum ochraceum* (Pers. per Fr.) S. F. Gray (personal communication).

[*Agaricum squamosum album, superne subhirsutum, inferne pectinatum* Mich. 1729 → *Hydnum pectiniforme* Batsch, → *Agarico-suber dentatum* Paul., → *Xylometron spinosum* Paul., → *Hydnum pectinatum* Fr.]

Hydnum pectiniforme Batsch 1783 (d.n.); [≡ (by lectotypification) *Agaricum squamosum album, superne subhirsutum, inferne pectinatum* Mich. 1729: 122 *pl. 64 f. 4* (Italy)]; ≡ (by lectotypification) *Agaricosuber dentatum* Paul. 1793 (d.n.) ≡ *Xylometron spinosum* Paul. 1812–35 (generic name n.v.p.); ≡ *Hydnum pectinatum* Fr. 1821, typonym.

Boletus tulipiferae Schw. 1822 (U.S.A., North Carolina) (**62**); fide Bres. 1897 (AAR III 3): 101. — *Polyporus corticola* var. Fr. 1828; *Irpex* Schw. 1832; *Polystictus* Cooke 1886; *Coriolus* Pat. 1900; *Irpiciporus* Murrill 1905; *Poria* Lloyd 1906 (syn.: n.v.p.); *Polyporus* Lloyd 1906 (syn.: n.v.p.), Overh. 1915. — Morg. 1887 (JCi 10): 15; Hard 1908: 448 *f. 376* (*Irpex*); Overh. 1953: 329 *pl. 4 fs. 15–18, pl. 5 f. 28, pl. 132 fig.*; Bakshi & al. 1956 (IF 82): 449 *f. 1, pl. 1 fs. 1, 2* (*Polyporus*); Maas G. 1963 (PNA 66): 454, in obs. (*Irpex*); Lowe 1966: 22 *f. 6* (*Polyporus*); Al. David 1969 (BlL 38): 199, cult. char. (*Irpex*). — V.s.: "*tulipifera*", an error, the correct epithet refers to *Liriodendron tulipifera*, the substratum.

Irpex sinuosus Fr. 1828 E. 1: 145 (Sweden) (**62**); fide Bres. 1897 (AAR III 3): 101 ("forma valde effusa"). — *Xylodon* O.K. 1898. — Fr. 1828 E. 1: 145 (*Irpex*); Bourd. & G. 1928: 573 (*Irpex lacteus* subsp.).

Irpex canescens Fr. 1828 E. 1: 145, in obs. ("Forsan in Europa australi lectus") (**62**); fide Bres. 1897 (AAR III 3): 101 ("forma cyclomycetoidea"). — *Xylodon* O.K. 1898; *Coriolus* Pat. 1900; *Agaricus* E. Krause 1932, misapplied.

Irpex pallescens Fr. 1838: 522 (U.S.A.) (**62**); fide Murrill 1907 (NAF 9): 15.

Polyporus chartaceus B. & C. 1849 (HJB 1): 103 & apud Berk. 1872 (G 1): 53 (U.S.A., North Carolina); fide Bres. 1926: 79 = *Irpex tulipiferae*. — *Polystictus* Cooke 1886; *Microporus* O.K. 1898. — Referred to *Polystictus* [*Hirschioporus*] *pargamenus* by Fr. 1851 (NAu III 1): 85/69 and to *Coriolus biformis* [sensu Murrill = "*Trametes*" *cervina*] by Murrill 1906 (BTC 23): 653.

? *Irpex johnstonii* Berk. 1860: 262 (England); cf. Fr. 1874: 621 ("v.s."), "*I. lacteo* . . . proximus." — *Xylodon* O.K. 1898. — At first Berk. 1836: 161 referred this taxon to *Irpex lacteus.*

Irpex hirsutus Kalchbr. 1878 (EtK 8[16]): 17 *pl. 2 f. 1* (U.S.S.R., Russia, Siberia); fide P. Karst. 1882 (BFi 37): 56 (subsp.). — *Xylodon* O.K. 1898.

Irpex bresadolae S. Schulz. 1885 (H 24): 146 (Yugoslavia, Slavonia); fide Bres. 1897 (AAR III 3): 101. — *Xylodon* O.K. 1898.

Poria cincinnati (Berk.) ex Cooke 1886 (G 15): 27 (U.S.A., Ohio); fide Bres. 1926: 30 & Lowe 1959 (Ll 11): 111 = *Irpex*/*Polyporus tulipiferae*. — *Polyporus* Berk. in herb. & Cooke 1886 (G 15): 27 (syn.: n.v.p.); *Poria* Cooke 1886 (G 14): 114 (nom. nud.: n.v.p.).

Irpex rimosus Peck 1890 (RNS 43): 68/22 (U.S.A., New York); fide R. L. Gilb. 1963 (M 54): 668 = *Polyporus tulipiferae.* — R. L. Gilb., l.c.

Irpex raduloides Pilát 1937 (BmF 52): 308 *f. 9, pl. 7 f. 2* (U.S.S.R., Russia, Siberia); fide Domański 1964 (APo 33): 174.

IRPICODON Pouz. (63)

1966 (Fgp 1): 371. — Holotype: *Hydnum pendulum* (A. & S.) per Fr.

pendulus (A. & S. per Fr.) Pouz. 1966. — *Sistotrema* A. & S. 1805 (Germany) (d.n.); *Hydnum* (A. & S.) per Fr. 1821; *Sistotrema* Pers. 1825; *Irpex* Fr. 1828; *Xylodon* O.K. 1898; *Trametes* Pilát 1939; = *Radulum pendulinum* Nikol. 1961. — A. & S. 1805: 261 *pl. 6 f. 2* (*Sistotrema*); Fr. 1821: 413 (*Hydnum*); Bourd. & G. 1928: 574 (*Irpex*); Nikol. 1861 (FsR 6): 94 *fs. 38–40, pl. 7 fs. 3–5* (*Radulum pendulinum*); Reid & Austw. 1963 (GN 18): 314 (*Irpex*); Pouz. 1966 (Fgp 1): 371 (*Irpicodon*).

Sistotrema conchatum Ehrenb. 1818: 19, 30 (Germany) (d.n.) per Pers. 1825; fide Fr. 1821: 413.

ISCHNODERMA P. Karst.

1879 [1960 (Pe 1): 231]. — Lectotype: *Polyporus resinosus* (Schrad.) per Fr. sensu Fr.
Podofomes Pouz. 1966 (ČM 20): 174; fide Dom. & Orl. 1967 (Ffg 12): 549. —
Holotype: *Polyporus corrugis* Fr.
M.—*Pelloporus* Quél. sensu Bond. & S. 1941, excl. of type [cf. 1960 (Pe 1): 250, in obs.]. → *Pelloporus* Bond. & S. (n.v.p.) (**O**).

SPECIAL LITERATURE.—General: Domański & Orlicz, *1967c*.
Ischnoderma benzoinum: Birkinshaw, Morgan, & Findlay, *1952*: Imazeki, *1957*.
Ischnoderma trogii: Jelić & Tortić, *1968*; Wojewoda, *1966*.

benzoinum (Wahl.) P. Karst. 1881 (**64**). — *Boletus* Wahl. 1826 (Sweden); *Polyporus* Fr. 1828; *Trametes* Fr. 1838; *Ochroporus* Błoński 1890; *Ungulina* Pat. 1900, Sing. 1930; *Polystictus* Big. & Guill. 1913 ("*Polyporus*"); *Placoderma* Ulbr. 1928; *Fomes* Konr. & M. 1935; *Ischnoderma* M. Bon 1970, preoccupied (n.v.p.). — Fr. 1828 E. 1: 100; Kalchbr. 1877: 56 *pl. 36 f. 1*; Fr. 1884 I. 2: 82 *pl. 183 f. 2*; A. L. Sm. & Rea 1906 (TBS 2): 130 *pl. 12 fig.*; Lloyd 1915 (LMW 4, Ap.): 333 (*Polyporus*); Konr. & M. 1935 I. 5: *pl. 452* (*Fomes*); Pouz. 1971 (ČM 25): 16 (*Ischnoderma*); Donk 1971 (PNA 74): 8, notes.

Boletus fuliginosus Scop. 1772 (Yugoslavia, Carniola); fide Quél. 1888: 393 & Bres. 1890 (Rm 12): 103 = *Polyporus benzoinum* (cited as a syn.). — *Boletus* Fr. 1821 (incidental mention: n.v.p.); *Boletus* Scop. per Schleich. 1821 ("Schr."), Zant. 1822; *Polyporus* Fr. 1838; *Fomes* Cooke 1885; *Cladomeris* Quél. 1886; *Polystictus* Pat. 1887 (nom. nud.: n.v.p.); *Inodermus* Quél. 1888; *Scindalma* O.K. 1898; *Ungulina* Pat. 1900; *Ischnoderma* Murrill 1904; *Polystictus* Big. & Guill. 1913. — Quél. 1888: 393 (*Inodermus*); Bres. 1890 (Rm 12): 103; 1897 (AAR III 3): 73 (*Polyporus*); Bourd. & G. 1928: 605 (*Ungulina*); Bres. 1931 (BIm 20): *pl. 984*; Kawam. 1954 I. 1: 119 *f. 111* (*Polyporus*).

Boletus velutinus Vahl 1794 (Fd 7 / F. 19): 8 *pl. 1138* (Denmark) (d.n.), not ∼ Plan. 1788 (d.n.), &c.; fide Fr. 1821: 361 = *Polyporus resinosus* [sensu Fr.].

Boletus fuscus Pers. 1794 (NMB 1): 108 / 1797 T.: 28 (Germany) (d.n.), not ∼ With. (mentioned by Fr. 1821: 395 as a syn.: n.v.p.), not ∼ Rostk. 1844; fide Fr. 1821: 361 = *Polyporus resinosus* [sensu Fr.]. — *Boletus* Pers. per Wahl. 1826; *Polyporus* Lloyd 1915, not ∼ Lév. 1846, not ∼ (Lázaro) Sacc. & Trott. apud Trott. 1925. — Lloyd 1915 (LMW 4, Ap.): 334 (*Polyporus*). — Cf. (**64**).

Polyporus guttatus Weinm. 1826 (SPR 2): 101 (U.S.S.R., European Russia); fide Fr. 1828 E. 1: 100 = *Polyporus resinosus* [sensu Fr.].

Polyporus morosus Kalchbr. 1869 (BZ 27): 496 (Hungary, now Czechoslovakia); fide Fr. 1874: 554 & Kalchbr. 1877: 56.

Polyporus pini-silvestris Allesch. 1889 (BLa 11): 28 (Germany); fide Lloyd 1905 (LMW 4, Ap.): 384.

Ischnoderma resinosum Donk 1933: 175, not ∼ (Fr.) Karst. 1879. — [≡ *Polyporus resinosus* (Schrad.) per Fr. sensu Fr. 1821: 361 (Sweden), excl. of type, mentioned as "*Boletus resinosus* Schrad. . . ."].

M.—*Boletus resinosus* Schrad. sensu Fr. 1821: 361 (*Polyporus*); fide Bres. 1890 (Rm 12): 103 = *Polyporus fuliginosus*. — Fr. 1828 E. 1: 100 (*Polyporus*); J. Schroet. 1888: 484 (*Ochroporus*); Overh. 1915 (WUS 3[1]): 47; Shope 1931 (AMo 18): 338 *pl. 22 f. 2* (*Polyporus*); Pilát 1937 (ACE 3): 129 *f. 33, pls. 59–64* (*Ischnoderma*); Overh. 1953: 301 *pl. 46 f. 279, pl. 53 f. 318, pl. 131 fig.* (*Polyporus*); Poelt & Jahn 1964: *pl. 39*; Dom. & Orl. 1967 (Ffg 12): 537, 548 *fs. 1b, 3, 5*; Pouz. 1971 (ČM 25): 16 (*Ischnoderma*). → *Ischnoderma resinosum* Donk.

M.—*Boletus confluens* A. & S. sensu Rostk. 1837 (StP 4): 71 *pl. 34* (*Polyporus*); fide Fr. 1874: 554 = *Polyporus resinosus* [sensu Fr.] status "sitaneus".

trogii (Fr.) Donk 1971. — *Polyporus* Fr. 1851; *Pelloporus* Kotl. & P. 1957; *Podofomes* Pouz. 1971; ≡ *Polyporus rugosus* Trog 1844 (Switzerland), not ∼ (Sow.) per Pers. 1825, not ∼ Bl. & Nees 1826: Fr. 1828, not ∼ (Jacq.) per E. Krause 1928; ≡ *Polyporus corrugis* Fr. 1874, not ∼ (Fr.) Cooke 1898; *Fomes* Cooke 1885; *Scindalma* O.K. 1898; *Lignosus* Torrend 1924; *Ungulina* Bourd. & G. 1925; *Pelloporus* Bond. & S. 1941; *Podofomes* Pouz. 1966; *Ischnoderma* Dom. & Orl. apud Domański & al. 1967. — Lloyd 1912 (LMW 3, S.P.): 122 *f. 423* (*Polyporus? corrugis*); Bourd. & G. 1928: 606 (*Ungulina corrugis*); Konr. & M. 1934 I. 5: *pl. 450*; Pilát 1940–1 (ACE 3): 361 *f. 156, pl. 241 f. b, pl. 242, pl. 244 f. b*; 1953 (SnP 9[2]): 82 *f. 82* (*Fomes corrugis*); Pouz. 1966 (ČM 20): 174 (*Podofomes corrugis*); Wojew. 1966 (Ffg 12): 513, 517 *2 plates* (*Ungulina corrugis*); Dom. & Orl. 1967 (Ffg 12): 536, 547 *fs. 1a, 2, 4*; Jel. & Tort. 1968 (GbB II 3): 233 *f. 1, pls. 1–3* (*Ischnoderma corrugis*); Donk 1971 (PNA 74): 9, note.

Trametes butignotii Boud. ex Lloyd 1910 (LMW 3, L. 28): 1 (Switzer-

land); fide Lloyd 1912 (LMW 3, S.P.): 123 & Bres. 1920 (Am 18): 69
= *Polyporus corrugis.*

M.—*Boletus triqueter* Pers. sensu Quél. 1888 (*Pelloporus*); fide Bourd.
& G. 1928: 606 = *Ungulina corrugis.* — Quél. 1888: 401 (*Pelloporus*).

JUNGHUHNIA Corda

1842, not ∼ Miq. 1859 (Euphorbiaceae) [1960 (Pe 1): 231]. — ≡ *Laschia* Jungh.
1838, not ∼ Fr. 1830 (Auriculariaceae) [1960 (Pe 1): 232]. — ≡ *Aschersonia* Endl.
1842, nomen rejiciendum not ∼ Mont. 1948, nomen conservandum (Deuteromycetes,
Necterioidaceae) [1960 (Pe 1): 187], — Lectotype: *Laschia crustacea* Jungh.
Chaetoporus P. Karst. 1890 [1960 (Pe 1): 198]. — Monotype: *Chaetoporus tenuis*
P. Karst. — Cf. Donk 1967 (Pe 5): 71.

SPECIAL LITERATURE.—Lombard & Gilbertson, *1967* (various species):
Parmasto, *1959a* (*J. pseudozillingiana*); Ryvarden, *1972.*

collabens (Fr.) Ryv. 1972. — *Polyporus* Fr. 1874: 572 (Sweden); *Poria*
P. Karst. 1882; *Physisporus* Gillet 1884; *Chaetoporus* Pouz. 1967. — Cf.
Donk 1967 (Pe 5): 107, in obs.
Polyporus emollitus (Fr.) Cooke 1878; fide Bres. 1897 (AAR III 3):
82 = *Poria blyttii* [sensu Bres.]. — *Polyporus laevigatus* [subsp.?] **P.
emollitus* Fr. 1874: 571 (Sweden); *Poria* Cooke 1886.
Polyporus rixosa (P. Karst.) P. Karst. 1879 (Finland); fide Bres. 1897
(AAR III 3): 82 = *Poria blyttii* [sensu Bres.]; fide Lowe 1956 (M 48):
118. — *Polyporus contiguus* subsp. P. Karst. 1876 (BFi 25): 272 (Finland);
Poria P. Karst. 1881: *Chaetoporus* P. Karst. 1899. — Egeland 1914
(NMN 52): 158; Pilát 1936 (BmF 51): 385 *f. 7*; Overh. 1942: 50 (*Poria*);
Domański 1963 (Mob 15); 304 (*Chaetoporus*); Lowe 1966: 95 *f. 77*; Lomb.
& Gilb. 1967 (M 58): 840 *fs. 3, 4, 10*, with cult. char. (*Poria*).
Poria dodgei Murrill 1921 (M 13): 87 (U.S.A., Wisconsin); fide Overh.
1942: 50 = *Poria rixosa.*
? M.—*Poria incarnata* Pers. sensu Fr. 1884 I. 2: 87 *pl. 189 f. 1* (*Poly-
porus*); fide Romell 1911 (ABS 11³): 13, 1912 (SbT 6): 638 = *Polyporus
rixosus?* — Fr., l.c.: "Nominis auctor A.S. (vel potius forte Pers.) nec Fr."
M.—*Polyporus blyttii* Fr. sensu Bres. 1897 (AAR III 3): 82, in obs.,
in part ("forma typica"), fide Bres., l.c. (*Poria collabens* cited as a syn.);
fide Donk 1967 (Pe 5): 81 = *Poria rixosa.*

luteo-alba (P. Karst.) Ryv. 1972. — *Physisporus* P. Karst. 1887 (Finland);
Poria Sacc. 1888; *Chaetoporellus* Bond. 1953 (incomplete ref.: n.v.p.);
Chaetoporus M. P. Christ. 1960. — Jo. Erikss. 1949 (SbT 43): 15 *f. 4,
pl. 3, pl. 5 f. 2* (*Poria*); Lowe 1956 (M 48): 113, notes (*Physisporus*):
M. P. Christ. 1960 (Dba 19): 353 *f. 351* (*Chaetoporus*); Lowe 1966: 106
f. 90; Lomb. & Gilb. 1967 (M. 58): 828 *fs. 1, 5, 8*, with cult. char. (*Poria*).
— Sensu Bres. 1897 → *Poria xantha.*
M.—*Physisporus flavicans* P. Karst. (**106**) sensu Lowe 1956 (M 48):
111 (syn.); fide Lowe, l.c.

M.—*Physisporus variecolor* P. Karst. sensu Lowe 1956 (M 48): 115;
cf. Lowe, l.c., "If correctly interpreted *P. variecolor* is the prior and valid
name for [*Poria luteoalba*]", but cf. Donk 1967 (Pe 5): 119. — Parm.
1959 (TSR 12): 224 *f. 4* (*Poria*); Domański 1963 (Mob 15): 303 *fs. 5, 6*;
1965 (FpG 2): 109 *f. 37, pl. 28 f. 2* (*Chaetoporus*).

nitida (Pers. per Fr.) Ryv. 1972. — *Poria* Pers. 1799 (Germany) (d.n.);
Boletus Pers. 1801 (d.n.); *Polyporus* Fr. 1818 (d.n.); *Polyporus* (Pers.)
per Fr. 1821, not ∼ I. Boršč. 1856, not ∼ (Murrill) Overh. 1936;
Physisporus Chev. 1826; *Boletus* Spreng. 1827; *Poria* Cooke 1886, Quél.
1886; *Chaetoporus* Donk 1967. — Lloyd 1910 (LMW 3): 472; Egeland
1914 (NMN 52): 151 (*Poria*), & cf. Donk 1967 (Pe 5): 99; *Chaetoporus*
Donk 1967. — Sensu Bres. 1903 → *Poria salmonicolor*; sensu Boud.
→ *Oxyporus obducens* ?

Poria micans Ehrenb. 1818: 19, 30 (Germany) (d.n.); fide Romell
1926 (SbT 20): 22. — *Polyporus* (Ehrenb.) per Fr. 1821; *Physisporus*
P. Karst. 1882 ("Chev."): *Poria* Cooke 1886, Quél. 1886; *Trametes* Bres.
1897, misapplied. — Cf. Donk 1967 (Pe 5): 95. — Sensu Rostk. → *Poria
fulgens* (**O**); sensu Bres. → *Pachykytospora tuberculosa*; sensu Romell
1926 (SbT 20): 13 (specimen from near Stockholm) = ?

Polyporus euporus P. Karst. 1867 & 1868 (Finland); fide Donk 1933:
217, 226. — *Physisporus* P. Karst. 1881; *Poria* Cooke 1886; *Chaetoporus*
P. Karst. 1898, Bond. & S. 1941. — Bres. 1897 (AAR III 3): 82 (*Poria*);
Romell 1911 (ABS 11³): 12 *pl. 2 f. 20* (*Polyporus*); Overh. 1942: 45;
Malenç. 1956 (BmF 71): 306 *tpl. 8* (*Poria*); M. P. Christ. 1960 (DbA 19):
355 *f. 353*; Domański 1965 (FpG 2): 104 *f. 34, pls. 26, 27* (*Chaetoporus*);
Lowe 1966: 122 *f. 109*; Lomb. & Gilb. 1967 (M 58): 834 *fs. 2, 6, 9*, with
cult. char. (*Poria*).

Polyporus attenuatus Peck 1873 (BBf 1): 61, 1874 (RNS 26): 70; fide
Overh. 1919 (BNS 205–206): 72 = *Poria eupora*; fide Bres. 1920 (Am 18):
68. — Overh. 1919 (BNS 205–206): 71 *pl. 1, pl. 2 fs. 1, 2* (*Poria*).

Polyporus blyttii Fr. 1874: 571 (Norway); fide Donk 1967 (Pe 5): 81.
— *Poria* P. Karst. 1882; *Phellinus* Pat. 1900. — Sensu Bres., in part,
→ *Chaetoporus collabens*.

Chaetoporus tenuis P. Karst. 1890 (H 29): 148 (Finland), not *Physisporus
tener* Har. & Karst. 1890; fide Donk 1960 (Pe 1): 198 = *Poria eupora*
— *Mucronoporus* Sacc. 1891, not ∼ O.K. 1898 (error: n.v.p.); ≡ *Scindalma
chaetoporus* O.K. 1898.

pseudozillingiana (Parm.) Ryv. 1972. — *Chaetoporus* Parm. 1959 (Estonia).
— Parm. 1959 (EAT 8²): 113 *fs. 1–3, 2 plates*, 1959 M.e. 2: 11 No. 38.

separabilima (Pouz.) Ryv. 1972. — *Chaetoporus* Pouz. (Czechoslovakia).
— Pouz. 1967 (ČM 21): 210.

M.—*Poria radula* Pers. sensu Bres. 1897; fide Pouz. 1967 (ČM 21): 210.
— Bres. 1897 (AAR II 3): 87, spore measurements incorrect, cf. Bres. 1903

(Am 1): 80; Egeland 1914 (NMN 52): 143; Romell 1926 (SbT 20): 15, notes; Bourd. & G. 1928: 678; Pilát 1932 (BmF 48): 47 *f. 8, pl. 4 f. 3*; Lowe 1946: 30 *f. 5* (*Poria*); Domański 1965 (FpG 2): 106 *f. 35, pl. 38 f. 1* (*Chaetoporus*); Lowe 1966: 114 *f. 100* (*Poria*), & cf. Donk 1967 (Pe 5): 104, notes. — Sensu originario → *Schizopora paradoxa*.

LAETIPORUS Murrill

1904 [1960 (Pe 1): 232]. — Holotype: *Agaricus speciosus* Batt.
    Polyporus (Pers.) per S. F. Gray 1821 (nom. monstr.), not ∼ [Mich.] Fr. 1821, not ∼ P. Karst. 1881 [1960 (Pe 1): 263]. — *Boletus* sect. *Polyporus* Pers. 1801 (d.n.). — Monotype: *Boletus ramosus* Bull.; → *Cladoporus* Chev., typonym.
    Cladoporus Chev. 1826 ("*Cladosporus*") (nom. monstr.) [1960 (Pe 1): 199]. — *Cladoporus* Pers. 1818 (nom. nud.), Brongn. 1823 (nom. nud.) & *Polyporus* sect. *Cladoporus* Pers. 1825, typonym. — Monotype: *Cladoporus fulvus* Chev. ≡ *Boletus ramosus* Bull.; → *Polyporus* (Pers.) per S. F. Gray, typonym.

SPECIAL LITERATURE.—Bourquelot & Hérissey, *1895*; Chifflot, *1921*; Corner, *1953*; Heufler, *1870*; Jahn, *1970b*; List & Menssen, *1959a, 1959b*; K. Lohwag, *1951*; van Overeem, *1925*; Patouillard, *1885*; Pilát, *1930a*; Rahm, *1956*; Rosen, *1927*; Roumeguère, *1881*; de Seynes, *1878, 1879, 1888a*; Suvorov, *1963*; Valadon & Mummury, *1969*; Van Bambeke, *1902*; Van der Westhuizen, *1959*; Wolf, *1941*.

sulphureus (Bull. per Fr.) Murrill 1920 (**65**). — *Boletus* Bull. 1788 (France (d.n.); *Sistotrema* Reb. 1804 (d.n.); *Polyporus* (Bull.) per Fr. 1821, not ∼ Chev. 1837 (n.v.p.); *Boletus* Hook. 1821, not ∼ Fr. 1838; *Agaricus* Boisd. 1828 (nom. nud. & error: n.v.p.); *Merisma* Gillet 1878; *Polypilus* P. Karst. 1881; *Cladomeris* Quél. 1886; *Leptoporus* Quél. 1888; *Tyromyces* Donk 1933; *Grifola* Pilát 1934, G. Cunn. 1965; *Laetiporus* Bond. & S. 1941 (preoccupied). — Bull. 1788: *pl. 429*: 1791 H.: 347: Sow. 1798: *pl. 135* (*Boletus*); Grev. 1824 S. 2: *pl. 113*; Secr. 1833 M. 3: 71; Hussey *c.* 1847 I. 1: *pl. 46*, abnormal: Fr. 1866 S.S.: 51 *pl. 88*, poor; R. Hartig 1878: 109 *pl. 14*; Seyn. 1888 P.: *1 pls. 1–3, pl. 4 fs. 1, 5–11*; Peck 1895 (RNS 48): 301 *pl. 37 fs. 1–4*; Atk. 1900: 190 *pls. 62, 64*; Bamb. 1902 (BmF 18): 54 *pls. 2–4*, monstr.; Bourd. & G. 1928: 524 (*Polyporus*); Pilát 1936 (ACE 3): 40 *f. 10, pls. 20–24* (*Grifola*); Imaz. 1940 (JJB 16): 264, 267 *fs. 1, 2*; R. W. Davids, 1942 (TUS 785): 38 *f. 5J, pl. 3 f. A*, cult. char; Overh. 1953: 234 *pl. 27 f. 166, pl. 30 f. 181, pl. 40 f. 238, pl. 132 fig.*; Corner 1953 (Phm 3): 157 *fs. 4–7*, hyphal analysis (*Polyporus*): Poelt & Jahn 1963: *pl. 33* (*Laetiporus*).
    [*Agaricum laricibus innascuntur . . . aureo colore . . . Mattioli* 1554: 485 (n.v.) [repr. Heufl. 1870 (ÖbZ 20): 193]. — *Fungus lariceus aurei coloris* C. Bauh. 1623: 371.]
    [*Agaricus speciosus* Batt. 1755: 68 *pl. 34 f. B* (Italy) (non-binomial name) → *Laetiporus speciosus* Murrill.]
    Boletus caudicinus Scop. 1772, in part: viz. var. 2 (Yugoslavia, Carniola);

fide Fr. 1821; 357 & Donk 1971 (PNA 74): 10 (as to var. 2). — *Boletus*
Scop. per Pollini 1824; *Polyporus* J. Schroet. 1888, not ∼ Murrill 1903;
Polypilus P. Karst. 1889; *Leptoporus* Quél. 1894 (combination implied,
not actually made: n.v.p.); *Cladomeris* Big. & Guill. 1909 (syn.: n.v.p.);
≡ (by lectotypification) *Boletus nitens* Batsch 1783, in part: as to var. α
(d.n.) [≡ *Boletus caudicinus* Scop. emend. Schaeff. 1774 per Pollini 1824
[see Donk 1971 (PNA 74): 10]. — Emend. Schaeff. 1774: 86 [*pls, 131,
132*], restricted to "Scop. . . . var. 2 ?"; Vahl 1790 (Fd / F. 17): 10 *pl. 1019*
(*Boletus*); J. Schroet. 1888: 471; Anon. 1908 (GwB 12): 634 *fig.*, monstrosity
(*Polyporus*). — Sensu Murrill → *Polyporus caudicinus* Murrill [= *Polyporus squamosus*].

    Boletus citrinus With. 1776, not ∼ Plan. 1788 (d.n.) per S. F. Gray
1821, not ∼ Lumn. 1791 (d.n.). — [≡ *Fungus arboreus major aureus,
nulla membrana superne tectus* Ray 1696: 336 & 1724: 22 (England)].

    Boletus tenax Lightf. 1778: 1031 (Scotland) (d.n.), citing Schaeff.
pls. 131, 132 [≡ *Boletus caudicinus* Scop. emend. Schaeff.] as "opt."
— Bolt. 1797: 75 *pl. 75*.

    Boletus coriaceus Huds. 1778: 625 (England) (d.n.), not ∼ Scop. 1772
(d.n.) per Bergam. 1823, not ∼ Batsch 1783 (d.n.), not ∼ Batsch 1786
(d.n.); fide Fr. 1821: 357. — ≡ *Boletus lobatus* J. F. Gmel. 1792 (d.n.);
Polyporus (J. F. Gmel.) per Fr., misapplied, not ∼ Schw. 1832; *Boletus*
Lenz 1840, misapplied; *Meripilus* P. Karst. 1882, misapplied; *Cladomeris*
Quél. 1886, misapplied. — Cf. Donk 1969 (Pe 5): 251, note. — Sensu Fr.
(*Polyporus lobatus*) → *Albatrellus cristatus* ?

    Boletus imbricatus Bull. 1787 (France) (d.n.), not ∼ Scop. 1772 (d.n.);
fide Quél. 1894 (Crf 22²): 488 ("un état, à peine une forme"); fide Bres.
1897 (AAR III 3): 70, "videtur tantum statum adultum *Pol. sulphurei*
sistere." — *Polyporus* (Bull.) per Fr. 1821; *Boletus* Mérat 1821; *Daedalea*
Purt. 1821, misapplied; *Merisma* Gillet 1877; *Polypilus* P. Karst. 1882;
Cladomeris Quél. 1886; *Leptoporus* Quél. 1888; ≡ *Boletus amaricans*
Pers. 1801 (d.n.): *Cladomeris* Big. & Guill. 1909 (syn.: n.v.p.); *Polyporus*
Pilát 1934 (syn.: n.v.p.). — Bull. 1787: *pl. 366*; 1791 H.: 349 (*Boletus*);
Quél. 1894 (Crf 22²): 488 (*Leptoporus*). — Sensu Sow. → *Meripilus
giganteus*; sensu Britz. → *Bondarzewia montana*.

    Boletus ramosus Bull. 1788 (France) (d.n.) (nom. monstr.), not ∼ Vahl
1797; fide Sow. 1798: text to pl. 135; fide Fr. 1821: 357 = *Polyporus
imbricatus* (forma). — *Boletus* Bull. per Mérat 1821; *Polyporus* S. F. Gray
1821; ≡ *Cladoporus fulvus* Chev. 1826: 261. — Bull. 1788: *pl. 418*; 1791 H.:
349 (*Boletus*).

    ? *Boletus citrinus* Plan. 1788 J.F.: 26 (Germany) (d.n.), not ∼ With.
1776 (d.n.), non/an ∼ Lumn. 1791 (d.n.); fide Fr. 1821: 357. — *Boletus*
Plan. per S. F. Gray 1821, not ∼ Venturi 1863; *Polyporus* Pers. 1825,
not ∼ Chev. 1837 (n.v.p.). — Sensu Pers. 1801: 524 (*Boletus*).

    Boletus lingua-cervina Schrank 1789: 618 (Germany) (d.n.); fide Fr.
1821: 357.

Boletus citrinus Lumn. 1791: 525 (Czechoslovakia), not ∼ With. 1776 (d.n.), non/an ∼ Plan. 1788 (d.n.) per S. F. Gray 1821, not ∼ Venturi 1863.

Agarico-carnis flammula Paul. 1793 T. 2: 100 (descr.), Ind. (France); fide Fr. 1874: 542 ("Paulet t. 14"). — ≡ *Dendrosarcos imbricatus* Paul. 1812–35: *pl. 14.*

Agarico-pulpa styptica Paul. 1793 T. 2: 101 (descr.), Ind. (France) (d.n.); fide Lév. 1855: 7 & Donk. — *Agaricum* Paul. 1812–35: *pl. 15* (generic name n.v.p.).

Polyporus casearius Fr. 1838: 449 (Sweden); cf. Quél. 1888: 387. — *Polypilus* P. Karst. 1882; *Cladomeris* Quél. 1886; *Leptoporus* Quél. 1888.

Polyporus ceratoniae Risso ex Barla 1859 (France); fide Comes 1879 (G 7): 112–113 & Bres. 1890 (Rm 12): 103. — *Cladomeris* Big. & Guill. 1909 (syn.: n.v.p.). — Barla 1859: 60 *pl. 30 fs. 1–3.*

Polyporus todari Inz. 1866 (Italy, Sicily); fide Comes 1879 (G 7): 112–113. — Inz. 1866 (GSP 2): 98 *pl. 7 f. 2*; 1869 F.s. 1: 38 *pl. 3 f. 2.*

Stereum speciosum Fr. apud Inz. 1871 & Fr. 1874 (Italy, Sicily); fide Donk 1971 (PNA 74): 11. — Inz. 1871 (GSP 7): 158 *pl. 23 f. 1*; 1879 F.s. 2: 20 *pl. 4 f. 1.* — Fruitbodies lacking tubes, margins desiccated.

Polyporus cincinnatus Morg. 1885 (JCi 8): 97 (U.S.A., Ohio); fide Lloyd 1912 (LMW 3, S.P.): 156 (forma). — Overh. 1953: 245 *pl. 27 f. 165, pl. 37 f. 226, pl. 111 f. 613* (*Polyporus sulphureus* var.).

Polyporus rostafinskii Błoński 1888 (H 28): 280 ("Lithuania" = Poland); fide Pilát 1936 (ACE 3): 41.

Laetiporus speciosus Murrill 1904; fide Murrill 1904 (BTC 31): 607 (*Boletus sulphureus* cited as a syn.). — *Polypilus* Murrill 1903 (generic name not accepted: n.v.p.); [≡ *Agaricus speciosus* Batt. 1755: 68 *pl. 34 f. B* (non-binomial name) (Italy)].

Ptychogaster aurantiacus Pat. 1885 (nom. anam.); cf. Boud. 1887 (JBM 1): 12 & fide Seyn. 1888 P.: 35. — *Ceriomyces* Sacc. 1888; ≡ *Ceriomyces sulfureus* Seyn. 1890. — Pat. 1885 (Rm 7): 28 *pl. 50 f. 10*; 1886 T.a. 1: 201 *f. 458*; Seyn. 1888 P.: 35 *pl. 4 fs. 17, 18* (*Ptychogaster*); H. Jahn 1970 (NaH 30): 85 *fs. 1, 2* (*Ceriomyces*).

LENZITES Fr.

1835 [1960 (Pe 1): 235]. — Lectotype: *Daedalea betulina* (L.) per Fr.

Leucolenzites Falck 1909 [1960 (Pe 1): 237]. — Monotype: *Lenzites betulina* (L. per Fr.) Fr.

M.—*Cellularia* Bull. per Corda sensu O.K. 1898, but not excl. of type. — Cf. Donk 1971 (PNA 74): 5.

SPECIAL LITERATURE.—*Lenzites betulina*: Brodie, *1936*; M. E. P. K. Fidalgo, *1959*; Kotlaba, *1956b*; K. Lohwag *1955a*; Vandendries & Brodie, *1933*.

Lenzites warnieri: David, *1967b*; Igmándy, *1962*; Montagne & Durieu de Maisonneuve, *1876*.

betulina (L. per Fr.) Fr. 1838. — *Agaricus* L. 1753 (Sweden) (d.n.); *Merulius* J. F. Gmel. 1792 (d.n.); *Daedalea* Reb. 1804 (d.n.); *Daedalea* (L.) per Fr. 1821; *Agaricus* Purt. 1821; *Cellularia* O.K. 1898; *Trametes* Pilát 1939. — Sow. 1799: *pl. 182* (*Agaricus*); Fr. 1863 M. 2: 246; Cooke 1881–91 I.: *pl. 1100/1145 f. A*; Michael 1905 F.P. 3: *no 43*; Lloyd 1925 (LMW 7): 1339 *pl. 325 f. 3114*; Bourd. & G. 1928: 579; Bres. 1929 (BIm 11): *pl. 523*; Donk 1933: 199 (*Lenzites*); Pilát 1940 (ACE 3): 327 *f. 142, pl. 220* (*Trametes*); Overh. 1953: 108 *pl. 91 fs. 517, 518, pl. 92 f. 525, pl. 127 fig.*; Kotl. 1965 (ČM 19): 79 *pl. 57*; Westh. 1971 (Bo 10): 231 *fs. 25, 26*, with cult. char. (*Lenzites*). — Sensu Wulf. = *Merulius tremellosus* Fr. ('Corticiaceae').

? *Agaricus flabelliformis* Scop. 1772: 460 (Yugoslavia, Carniola) (d.n.), not ∼ Schaeff. 1774 (d.n.), &c., not ∼ Fr. 1821, &c.; fide Fr. 1821: 333.

Agaricus tomentosus Lam. 1778 F.f. 1: (118) (France) (d.n.), not ∼ Bull. 1782 (d.n.), &c., not ∼ Bull. per St-Am. 1821, &c.

Agaricus coriaceus Bull. 1788 (France) (d.n.), not ∼ (Scop.) Lam. 1783 (d.n.); fide Fr. 1821: 333 (as to Bull. pl. 394) = *Lenzites flaccida*. — *Agaricus* Bull. per St-Am. 1821, Laterr. 1821; *Daedalea* P. Gärtn. & al. 1802 (d.n.); *Apus* S. F. Gray 1821; *Merulius* L. March. 1828; *Daedalea* Rox. Clem. 1864 ("Pers.") ?; ≡ *Merulius alutaceus* J. F. Gmel. 1792 (d.n.). — Bull. 1788: *pl. 394*; Bolt. 1791: 158 *pl. 158*.

Agaricus versicolor Plan. 1788 J.F.: 4 (Germany) (d.n.), not ∼ (L.) Lam. 1783 (d.n.), not ∼ With. 1796 (d.n.); fide 1821: 333 (forma). — *Agaricus* Plan. per Zant. 1821, not ∼ With. per Fr. 1821; *Cellularia* O.K. 1898. — *Daedalea versicolor* Secr. 1833 may be a recombination, *A. versicolor* not being definitely included/excluded.

? *Agaricus fagineus* Schrad. apud J. F. Gmel. 1792 (Germany) (d.n.), not ∼ Schum. 1803 (d.n.) per Pers. 1828. — Schrad. 1794: 134.

Daedalea variegata Fr. 1818 O. 2: 240 (Sweden) (d.n.); fide Quél. 1888: 367 = *Lenzites flaccida* (var.); fide Donk 1933: 200 (forma). — *Daedalea* Fr. per Fr. 1821, not ∼ P. Karst. 1911 (n.v.); *Lenzites* Fr. 1838; ≡ *Agaricus taeniatimpictus* E. Krause 1933. — Secr. 1833 M. 2: 498 (*Daedalea*); Fr. 1863 M. 2: 246 (*Lenzites*); Quél. 1888: 367 (*Lenzites flaccida* var.); Bourd. & G. 1928: 580 (*Lenzites betulina* subsp.); Bres. 1929 (BIm 11): *pl. 525* (*Lenzites*).

Daedalea furcata Link ex Fr. 1830 (Li 5): 513 (Brazil); 1832; cf. Lloyd 1912 (LMW 4, L. 39): 6. — *Lenzites* Fr. 1838. — Theiss. 1911 (DAW 83): 222 *pl. 1 f 32*; Lloyd 1922 (LMW 7): 1106 *pl. 187 f. 2029*; 1925 (LMW 7): 1339 *pl. 325 f. 3117* (*Lenzites*).

? *Daedalea cinnabarina* Secr. 1833 M. 2: 482 (Switzerland). — *Lenzites* Quél. 1888; *Striglia* O.K. 1891. — The descr. suggests a rather daedaleoid condition of the form that has been called *L. variegata*.

Lenzites flaccida Fr. 1838 (Sweden); fide Bres. 1912 (H 53): 50, & Donk 1933: 200 (forma). — *Cellularia* O.K. 1898; *Daedalea* E. Krause 1928. — Fr. 1863 M.: 246; Quél. 1888: 366 (*Lenzites*); Bourd. & G. 1928:

580 (*Lenzites betulina* subsp.); Bres. 1929 (BIm 11): *pl. 324* (*Lenzites*).

Lenzites berkeleyi Lév. 1846 (ASn III 5): 122 (U.S.A., New York); fide Bres. 1920 (Am 18): 66 = *Lenzites flaccida.*

Lenzites guineensis (Afz. ex Fr.) Fr. 1851; fide Lloyd 1917 (LMW 5): 626 (forma). — *Lenzites betulina* var. Afz. ex Fr. 1838: 405 (Guinea); *Cellularia* O.K. 1898. — Fr. 1860 R.A.: [4] *pl. 12 f. 26*; Lloyd 1917 (LMW 5): 626 *f. 890* (*Lenzites*).

Lenzites cinnamomea Fr. 1851 (Sweden); fide K. Fid. 1959 (M 50): 756. — *Gloeophyllum* P. Karst. 1882; *Cellularia* O.K. 1898; *Daedalea* E. Krause 1928. — Fr. 1863 M. 2: 247; 1874: 494; 1882 I. 2: 77 *pl. 177 f. 2*; K. Fid. 1959 (M 50): 754 *fs. 1, 2* (*Lenzites*).

Lenzites pinastri Kalchbr. apud Fr. 1874 (Hungary); fide Pilát 1942 (ACE 3): 612. — *Cellularia* O.K. 1898. — Fr. 1874: 495; Kalchbr. 1875: 49 *pl. 30 f. 3.*

? *Lenzites sorbina* P. Karst. 1881 (Afe 2[1]): 15; cf. Lowe 1956 (M 48): 109, probably *Lenzites betulina.*

Cellularia velutina (Berk.) O.K. 1898. — *Daedalea betulina* var. Berk. 1839 (AM 3): 381 (North America); fide Lév. 1846 (ASn III 5): 122 = *Lenzites berkeleyi.*

Lenzites betuliniformis Murrill 1908 (NAF 9): 128 (Mexico); fide Lloyd 1922 (LMW 7): 1107 = *Lenzites furcata.*

? *Lenzites connata* Lázaro 1916 (RMa 14): 850 / 1917: 162 (Spain).

Lenzites ochracea Lloyd 1922 (LMW 7): 1130 (U.S.A., New Hampshire), non/an ∼ Lloyd 1922 (n.v.p.), simultaneously published; fide Overh. 1953: 108.

M.—*Agaricus quercinus* L. sensu Schaeff. 1774: 25 [*pl. 57*]; fide Fr. 1874: 493 ("Speciosa forma").

M.—*Cellularia cyathiformis* Bull. sensu O.K. 1898, without descr. — Cf. Donk 1971 (PNA 74): 5.

warnieri Dur. & Mont. apud Mont. 1860 (Algeria) (**66**). — Mont. 1860 (ASn IV 4): 182; Mont. & Dur. 1876; Donk 1971 (PNA 74): 11, notes.

? *Lenzites gigantea* Černj. 1845 (BSM 18[2]): 140 (U.S.S.R., Ukraine) (nomen subnudum); cf. Donk 1971 (PNA 74): 11. — *Cellularia* O.K. 1898.

Lenzites faventina Cald. 1869 (NGi 1): 133; fide Donk 1971 (PNA 74): 11. — *Lenzites* Cald. 1868 (nom. nud.: n.v.p.); *Cellularia* O.K. 1898.

Lenzites reichardtii S. Schulz. in Thüm. 1880 M.u.: No. 1501 (Yugoslavia, Slavonia) [repr. 1881 (Fl 64): 237]; fide Höhn. 1906 (SbW 115): 689 = *Lenzites faventina.* — Nikol. 1940 (TSR 4): 427 *fs. 39–41*; Igmándy *1962*; Švarcm. 1964: 619 *fs. 275–277*; Al. David 1967 (BIL 36): 155 *fs. 1–7, tpls. 1, 2,* with cult. char.

LINDTNERIA Pilát

1938 [1960 (Pe 1): 239]. — Monotype: *Poria trachyspora* Bourd. & G.

SPECIAL LITERATURE.—Hansen, *1960*; Pilát, *1958b*.

trachyspora (Bourd. & G.) Pilát 1938. — *Poria* Bourd. & G. 1925 (France);
Trechispora Bond. & S. 1941; *Phlebiella* Bond. 1953 (indirect ref.; generic
name n.v.p.). — Bourd. & G. 1928: 659 *f. 182*; Lowe 1946: 43 *f. 9 (Poria)*;
Litsch. 1939 (Gsk 20): 22; Pilát 1941 (ACE 3): 472 *fs. 230, 231, pl. 303 f. b*;
Wakef. 1952 (TBS 35): 38 *f. 1*; L. Hansen *1960*; Svrček 1960 (Sy 14):
243; Pilát 1960 (ČM 14: 39 *4 figs.*; Poelt & Oberw. 1962 (Bba 35): 94
f. 23, spores; Lowe 1966: 46 *f. 31 (Lindtneria)*.

Sistotrema sulphureum var. *retirugum* Bourd. & G. 1914 (BmF 30):
274 (France); fide Bourd. & G. 1925 (BmF 41): 219 ("état jeune").

MERIPILUS P. Karst. (52)

1882 [1960 (Pe 1): 241]. — Lectotype: *Polyporus giganteus* (Pers.) per Fr.
Flabellopilus Kotl. & P. 1957 [1960 (Pe 1): 215]. — Holotype: *Polyporus giganteus*
(Pers.) per Fr.

SPECIAL LITERATURE.—Kallenbach, *1934c*; van der Lek, *1921*; K.
Lohwag, *1940c*; Niel, *1896*.

giganteus (Pers. per Fr.) P. Karst. 1882 (67). — *Boletus* Pers. 1794
(Germany) (d.n.); *Polyporus* Fr. 1815 (d.n.); *Polyporus* (Pers.) per Fr.
1821; *Boletus* Opiz. 1823; *Merisma* Gillet 1877; *Cladomeris* Quél. 1886;
Fomes Bres. 1926 (error: n.v.p.); *Polypilus* Donk 1933; *Grifola* Pilát
1934; *Flabellopilus* Kotl. & P. 1957. — Schum. 1803: 385 [cf. Hornem.
1823 (Fd 10 / F. 30): *pl. 1793* for plate, & cf. Donk 1971 (Pe 6): 206]
(*Boletus*); Fr. 1821: 356; Hussey *c.* 1847 I. 1: *pl. 82*; Bres. 1892 F.t. 2:
28 *pl. 134* & 1931 (BIm 20): *pl. 972*; Boud. 1904–11: 78 *pl. 153*; Bourd.
& G. 1928: 523; Kallenb. 1930 (ZP 9): *pl. 5 (Polyporus)*; Pilát 1936
(ACE 3): 37 *f. 9, pls. 12, 13 (Grifola)*: Overh. 1953: 242 *pl. 52 f. 313, pl. 101
f. 570, pl. 108 f. 600, pl. 130 fig. (Polyporus)*. — Sensu Harzer → *Polyporus
squamosus*.

[*Fungi esculenti, Genus XXI* Clus. 1601: cclxxv [repr. Istv. 1900:
79, Codex *pl. 67*] (Hungary); fide Donk 1971 (Pe 6): 203. — ≡ *Florum
fasciculatus* Sterb. 1675 & 1712: 269 *pl. 28 f. A* ("*faciculatus*").]

[*Florum fasciculatus* Sterb. 1675 → *Fungi esculenti, Genus XXI* Clus.]

Boletus proliferus With. 1776 (d.n.). — [≡ *Agaricus multiplex porosus*
Dill. sensu Ray 1724: 23 (England)]].

Boletus elegans Bolt. 1788: 76 *pl. 76* (England) (d.n.), not ∼ Bull.
1780 (d.n.), not ∼ Schum. 1803 (d.n.) per Fr. 1838; fide Fr. 1821: 356.
— ≡ *Boletus cornutus* J. F. Gmel. 1792 (d.n.).

Boletus acanthoides Bull. 1790 (France) (d.n.); fide Fr. 1821: 356;
fide Quél. 1888: 407 (*Boletus giganteus* cited as a syn.). — *Boletus* Bull.
per Mérat 1821; *Polyporus* Fr. 1838, misapplied, not ∼ Rostk. 1848;
Merisma Gillet 1877, misapplied; *Cladomeris* Quél. 1886, misapplied;
Caloporus Quél. 1888; *Grifola* Pilát 1934, misapplied; *Polypilus* Bond. 1953
(lacking ref.: n.v.p.), misapplied. — Bull. 1790: *pl. 486*, very poor; 1791

H.: 337 (*Boletus*); Quél. 1888: 406 (*Caloporus*). — Sensu Fr. → *Aborti-porus biennis*; sensu Quél. 1886 → *Bondarzewia montana*; sensu Velen. → *Albatrellus confluens*.

Clavaria aequivoca Holmskj. 1790 F.d. 1: 32 pl. [13] (Denmark) (d.n.); fide Schum. 1803: 383; Fr. 1874: 540 (var.). — *Agaricus* (Holmskj.) per E. Krause 1932; *Polyporus* E. Krause 1934.

Polyporus frondosus Fr. 1838: 446, not ∼ (Dicks.) per Fr. 1821, not ∼ Secr. 1833 (n.v.p.). — Fr. 1863 S.S.: 28 *pl. 44*; Donk 1971 (Pe 6): 204–205, notes. — Recombinations referring to "Fr". are listed in connection with *P. frondosus* (Dicks.) per Fr. 1821 even when the descrs. point to *Meripilus giganteus*.

Grifola sumstinei Murrill 1904 (BTC 31): 335 (U.S.A., Pennsylvania); fide Lloyd 1912 (LMW 3, S.P.): 157 & Overh. 1914 (AMo 1): 242, 1953: 242. — *Polyporus* Sacc. & D. Sacc. 1905; *Polypilus* Bond. & S. 1941.

M.—*Boletus imbricatus* Bull. sensu Sow. 1797: *pl. 86*; fide Fr. 1821: 356; fide Donk, recognizable from descr.

M.—*Boletus frondosus* Dicks. sensu With. 1801: 315, in obs.; sensu Rostk. 1830 (*Polyporus*). — Rostk. 1830 (StP 4): 39 *pl. 18*; Krombh. S. 7: 8 *pl. 48 fs. 17–20*; Fr. 1863 S.S.: 28 *pl. 44*; Kallenb. 1934 (ZP 13): 22 *pls. 2, 3*, fide K. Lohw. *1940*.

M.—*Polyporus acanthoides* Rostk. sensu Rostk. 1848. — Rostk. 1848 (StP F. 27–28): 37 *pl. 19*. This has been variously interpreted, cf. Donk 1971 (PNA 74): 2.

M.—*Boletus mesentericus* Schaeff. sensu Murrill 1920 (M 12): 10 (*Grifola*); fide Murrill, l.c. (*Polyporus giganteus* cited as a syn.).

M.—*Polyporus intybaceus* Fr. sensu Kawam. 1929; fide Imaz. 1943 (JJB 19): 389. — Kawam. 1929: no. 161, text with *plate f. 3*.

OLIGOPORUS Bref. (68)

1888 [1960 (Pe 1): 248]. — Lectotype: *Oligoporus farinosus* Bref. — Cf. Donk 1971 (Pe 6): 210.

SPECIAL LITERATURE.—Boudier, *1887*; Brefeld, *1888a*; Jahn, *1970a*; Kallenbach, *1934*.

rennyi (B. & Br.) Donk 1971. — *Polyporus* B. & Br. 1875 (England); *Poria* Cooke 1886; *Strangulidium* Pouz. 1967. — Reid & Austw. 1963 (GN 18): 310 (*Poria*); Pouz. 1967 (ČM 21): 208 (*Polyporus*); H. Jahn 1971 (WPb 8): 13 *fig.*; 1971 (WPb 8): 58 *f. 9* (*Strangulidium*).

Oligoporus farinosus Bref. 1888 (Germany); fide Donk 1971 (Pe 6): 212. — *Polyporus* Sacc. 1891, not ∼ J. Rick. 1907. — Bref. 1888 U. 8: 118 *pl. 7 fs. 12–22*.

Ptychogaster citrinus Boud. 1887 (France) [nom. anam., cf. Donk 1971 (Pe 6): 212]; fide Bref. 1888 U. 8: 117 = *Oligoporus farinosus*.

— Boud. 1887 (JBM 1): 8 *pl. 1 f. 1.* — Cf. *Oligoporus citrinus* R. & O. Falck **(O)**.

M.—*Polyporus sericeo-mollis* Romell sensu Romell 1911 (ABS 11[3]): 22, in part: as to sulphureous chlamydospores producing specimens **(111)**; fide Pouz. 1967 (ČM 21): 206, 208. — Wak. & Pears. 1918 (TBS 6): 75 *fig.*, in part: chlamydosporous state producing collections only (*Poria*); Bourd. & G. 1925 (BmF 41): 128; 1928: 548 (*Leptoporus destructor* subsp.).

M.—*Polyporus apalus* Lév. **(O)** sensu Kallenb. 1934 (ZP 13): 66 *pl. 10.*

ONNIA P. Karst. (69)

1889 [1960 (Pe 1): 248]. — Lectotype: *Polyporus circinatus* Fr. — Cf. Donk 1971 (PNA 74): 12.

M.—*Polystictus* Fr. sensu Sing. 1944 (provisional emendation), Bond. 1953 → *Polystictus* Bond. (n.v.p.) **(O)**.

M.—*Mucronoporus* Ell. & Ev. sensu Domański apud Domański & al. 1967, excl. of type, → *Mucronoporus* Domański **(O)**.

SPECIAL LITERATURE.—Christensen, *1940*; Domański, *1960a*; Gosselin, *1944*; Haddow, *1941*; Hubert, *1929b*; Myron & Patton, *1970, 1971*; Sartory & Maire, *1923*; Solov'ev, *1927*; Whitney, *1962, 1963, 1965, 1966a, 1966b*; Whitney & Bohaychuk, *1969, 1971*.

tomentosa (Fr.) P. Karst. 1889 **(70)**. — *Polyporus* Fr. 1821 (Sweden); *Boletus* Spreng. 1827; *Trametes* Fr. 1848, 1849 (nom. nud.: n.v.p.), Strauss 1850, not ∼ Bijl 1922; *Polystictus* P. Karst. 1881; *Pelloporus* Quél. 1886; *Mucronoporus* Ell. & Ev. 1889; *Microporus* O.K. 1898; *Xanthochrous* Pat. 1900; *Coltricia* Murrill 1904; *Fomes* Dambl. & Moureau 1937; *Inonotus* Teng 1964. — Ell. & Ev. 1889 (JM 5): 28 *pl. 8* (*Mucronoporus*); Lloyd 1908 (LMW 3, P.I.): 2 *f. 197* (*Polystictus*); Bres. 1931 (BIm 20): *pl. 960*; Haddow 1941 (TBS 25): 187 *pl. 6 f. 1*, excl. of var.; Coker 1946 (JMS 62): 104 *pl. 21, pl. 22 fs. 14–16*; Overh. 1953: 390 *pl. 43 fs. 257, 260, pl. 128 fig.*, excl. of var. (*Polyporus*). — Sensu Rostk. → *Polyporus ciliatus*.

Polyporus scutiger Kalchbr. 1868 (MtK 5): 259 *pl. 2 f. 3* (Hungary, now U.S.S.R., Ukraine), not Fr. 1828; fide Bres. 1931 (BIm 20): text to pl. 960. — ≡ *Polyporus kalchbrenneri* Fr. 1874; *Polystictus* Cooke 1886; *Microporus* O.K. 1898. — Kalchbr. 1877: 57 *pl. 38 f. 1* (*P. kalchbrenneri*).

? *Polyporus peakensis* Lloyd 1920 (LMW 6): 933 *pl. 149 f. 3009* (U.S.A., Colorado); fide Overh. 1953: 395.

M.—*Trametes circinata* Fr. sensu Pat. 1900: 100 *f. 4*; cf. Haddow *1941.* — Hubert 1929 (Ph 19): 745; 1931: 351 *f. 92*; Shope 1931 (AMo 18): 349 *pl. 26 fs. 1, 2*; C. M. Christ. 1940 (Ph 30): 957 *f. 1* (*Polyporus*).

triqueter (Fr.) Imaz. apud S. Ito 1955 **(70)**. — *Polyporus* Fr. 1838, not ∼ (Pers.) per Pers. 1825; fide Bres. 1903 (Am 1): 75 (as a var. of *Polyporus circinatus*). — *Boletus* Lenz 1840; *Trametes* Fr. 1849; *Inoderma* P. Karst. 1879; *Inonotus* P. Karst. 1882; *Polystictus* Cooke 1886; *Pelloporus* Quél.

1886; *Ochroporus* J. Schroet. 1888; *Microporus* O.K. 1898; *Xanthochrous* Siem. 1931 (nom. nud.: n.v.p.), Maire 1933; [≡ *Polyporus triqueter* Pers. sensu Secr. 1833 M. 3: 101 (Switzerland)]. — Bres. 1903 (Am 1): 75 (*Polyporus circinatus* var.); Lloyd 1915 (LMW 4, Ap.): 353 (*Polyporus*); Bourd. & G. 1928: 632 (*Xanthochrous circinatus* var.); Jørst. 1937 (KnS 1936[10]): 27 (*Polyporus circinatus* var.); Donk 1971 (PNA 74): 14, note. — Sensu Fr. 1884 I. 2: *pl. 187 f. 1* = ?

Trametes circinata Fr. 1848 & 1849 (Sweden) (**70**); *Polyporus* Fr. 1851; *Polystictus* Fr. 1851, P. Karst. 1882; *Pelloporus* Quél. 1886; *Mucronoporus* Ell. & Ev. 1889: *Onnia* P. Karst. 1889; *Microporus* O.K. 1898; *Xantho-chrous* Pat. 1900; *Fomes* Konr. & M. 1926; *Inonotus* Teng 1964 (indirect ref.: n.v.p.). — Fr. 1863 M. 2: 268; 1882 & 1884 I. 2: 79 *pl. 180 f. 1* (*Polyporus*); Ell. & Ev. 1889 (JM 5): 28 (*Mucronoporus*); Bres. 1903 (Am 1): 75 (*Polystictus*); Konr. & M. 1926 I. 5: *pl. 457* (*Fomes*); Bourd. & G. 1928: 631 (*Xanthochrous*; Jørst. & Juul 1939 (MnS 6): 426, 478 *fs. 28–32*; Overh. 1953: 392 *pl. 42 f. 252, pl. 43 fs. 256, 259, pl. 44 f. 261, pl. 128 fig.*; Domański 1960 (Mob 10): 130 *fs. 1–9* (*Polyporus tomentosus* var.) — Sensu Pat. → *Onnia tomentosa.*

Polyporus leporinus Fr. 1852 (Sweden); fide Pegl. 1964 (TBS 47): 191. — *Polystictus* Fr. 1852 (nom. altern.); *Inoderma* P. Karst. 1879; *Inonotus* P. Karst. 1882; *Inodermus* Quél. 1886; *Microporus* O.K. 1898. — Fr. 1863 M. 2: 271; 1884 I. 2: 85 *pl. 186 f. 3*, poor (*Polyporus*).

Polyporus dualis Peck 1878 (RNS 30): 44 (U.S.A., New York); fide Bres. 1920 (Am 18): 67; fide Haddow 1941 (TBS 25): 186 = *Polyporus circinatus*. — *Mucronoporus* Ell. & Ev. 1889; *Polystictus* Lloyd 1908 — Ell. & Ev. 1889 (JM 5): 28 (*Mucronoporus*); Lloyd 1908 (LMW 3, P.I.): 4 *f. 199* (*Polystictus*); Lowe 1942: 45 (*Polyporus*).

Polyporus aduncus Lloyd 1915 (LMW 4, L. 56): 5 & 1915 (LMW 4, Ap.): 354 (U.S.A., California). — Referred by Pegl. 1964 (TBS 47): 191 to *Phaeolus schweinitzii*, but Lloyd described large setae "with peculiar, hooked points". Overholts (*1953*: 390) listed the name as a syn. of *Poly-porus tomentosus* sensu stricto (that is, exclusive of var. *circinnatus*).

Daedalea pinacea Velen. 1926 (MP 3): 101, 102 *f. 1* (Czechoslovakia); fide Pilát 1942 (ACE 3): 583, 612 = *Polystictus tomentosus* f. *leporinus*.

M.—*Boletus triqueter* Pers. sensu A. & S. 1805: 248, var. α; & sensu Secr. 1833 M. 3: 101 (*Polyporus*) → *Polyporus triqueter* Fr.

M.—*Polyporus perplexus* Peck (**O**) sensu Lloyd 1912 (LMW 4, L. 39): 7.

OSTEINA Donk

1966 (SZP 44): 86. — Holotype: *Polyporus osseus* Kalchbr.

SPECIAL LITERATURE.—Aoshima & Furukawa, *1963*; Donk, *1966a*; Igmándi, *1956.*

obducta (Berk.) Donk 1966. — *Polyporus* Berk. 1845 (Canada); *Tyromyces*

Murrill 1907; *Grifola* Aosh. & Furuk. 1963. — Aosh. & Furuk. 1963 (TmJ 4): 91 *fs. 1, 2*.

Polyporus osseus Kalchbr. 1865 (Hungary, now Czechoslovakia); fide [Bres. apud] Lloyd 1915 (LMW 4, Ap.): 383 & Bres. 1920 (Am 18): 67 = *Polyporus obductus* (cited as a syn.). — *Leptoporus* Quél. 1886; *Leucoporus* Quél. 1888; *Grifola* Pilát 1934; *Polypilus* Parm. 1963; *Scutiger* Kartav. 1967 (indirect ref.: n.v.p.). — Kalchbr. 1865 (MtK 3): 217 *pl. 1 f. 2*; Fr. 1874: 541; Kalchbr. 1877: 54 *pl. 34 f. 2*; Lloyd 1912 (LMW 3, S.P.): 191 *f. 496*; Bres. 1931 (BIm 20): *pl. 974*; Shope 1931 (AMo 18): 360 *pl. 29 f. 3* (*Polyporus*); Pilát 1936 (ACE 3): 53 *f. 13, pls. 14–16* (*Grifola*); Murašk. 1939: 76 *f. 16*; Overh. 1953: 226 *pl. 37 f. 225, pl. 44 f. 263, pl. 100 f. 566, pl. 131 fig.*; Pilát 1960 (ČM 14): 33 *4 figs.* (*Polyporus*).

Polyporus zelleri Murrill 1915 W.P.; 13 (U.S.A., Washington); fide Overh. 1953: 226, 227.

OXYPORUS (Bourd. & G.) Donk (71)

1933 [1960 (Pe 1): 249]. — *Coriolus* sect. *Oxyporus* Bourd. & G. 1925. — Monotype: *Polyporus connatus* Weinm. [sensu Bourd. & G.].

SPECIAL LITERATURE.—W. A. Campbell, *1937* (*O. populinus*); Juel, *1914* (*O. corticola*); Le Breton, *1887* (cf. *O. obducens*); Pilát, *1935b* (*Poria pearsonii*), *1939* (*Leptoporus werneri*); Robbins & Hervey, *1961* (*O. late-marginatus*); DeVay & al., *1968* (*O. late-marginatus*).

T y p i c a l   s p e c i e s

obducens (Pers.) Donk 1933. — *Polyporus* Pers. 1825 (Germany); *Physi-sporus* Gillet 1878 ("*obducans*"); *Poria* Cooke 1886, Quél. 1886; *Fomitopsis* P. Karst. 1899; *Trametes* Pat. 1900; *Coriolus* Pilát 1932; *Physisporinus* Pilát 1941 (syn.: n.v.p.); *Polystictus* Killerm. 1943; *Rigidoporus* Pouz. 1966. —Bres. 1897 (AAR III 3): 85? (*Poria*); Romell 1911 (ABS 11³): 19 (*Polyporus*); Bourd. & G. 1928: 570 *f. 165* (*Coriolus connatus* subsp.); Donk 1933: 202; M. P. Christ. 1960 (DbA 19): 373 *f. 374?*; Domański 1965 (FpG 2): 198 *f. 66, pl. 57 f. 1* (*Oxyporus*); & cf. Donk 1967 (Pe 5): 100, notes.

Polyporus inhalatus Velen. 1922: 636 [see Pilát 1948: 242 for Latin translation] (Czechoslovakia): fide Bourd. apud Pilát 1942 (ACE 3): 604 = *Oxyporus populinus* var. *obducens* (conidiferous).

? M.—*Poria nitida* Pers. sensu Boud. 1904–11: 82 *pl. 160* (*Polyporus*); cf. Donk 1967 (Pe 5): 100, note.

populinus (Schum. per Fr.) Donk 1933 (72). — *Boletus* Schum. 1803 (Denmark) (d.n.); *Polyporus* (Schum.) per Fr. 1821, not ∼ Fr. 1863; not ∼ (Lázaro) Sacc. & Trott. apud Trott. 1925; *Boletus* Spreng. 1827; *Trametes* Fr. 1848 (n.v.p.), 1849, not ∼ Bres. apud Sacc. 1896; *Fomes* Cooke 1885; *Scindalma* O.K. 1898; *Polystictus* Rick. 1918; *Rigidoporus* Pouz.

1966. — Hornem. 1823 (Fd 10 / F. 30): 12 *pl. 1791*; sensu Secr. 1833 M. 3:
113 (*Polyporus*); Bres. 1897 (AAR III 3): 76 (*Fomes*); Donk 1933: 204;
Pilát 1940 (ACE 3): 341 *f. 148, pl. 227 f. a*; W. Cooke 1949 (M 11): 452;
H. Jahn 1963 (WPb 4): 60 *f. 2d, Abb. 46* (*Oxyporus*). — Sensu Schw. →
Antrodia malicola; sensu Passerini → *Funalia trogii*; sensu S. Schulz.
→ *"Trametes" cervina*.

Polyporus connatus Weinm. 1826 (U.S.S.R., European Russia) **(72)**,
not ∼ Schw. 1832; fide Bres. 1897 (AAR III 3): 76. — *Trametes* Fr. 1849,
P. Karst. 1882; *Fomes* Gillet 1878; *Fomitopsis* P. Karst. 1881, G. Cunn.
1950; *Leptoporus* Quél. 1886; *Coriolus* Quél. 1888; *Scindalma* O.K. 1898:
Boudiera Lázaro 1916; *Lazaroa* Gonz. 1917; *Placodes* Rick. 1918; *Oxyporus*
Kühner 1950; *Flaviporus* G. Cunn. 1965. — Sensu Fr. 1838: 472 (*Polyporus*); Gillet 1874–90 P.: *pl. 465/288* (*Fomes*); Fr. 1884 I. 2: 84 *pl. 185 f. 2*
(*Polyporus*); Quél. 1888: 391 (*Coriolus*); Boud. 1904–11 I. 1: 80 *pl. 157*
(*Polyporus*); Lloyd 1915 (LMW 4, F.): 216 *f. 572* (*Fomes*); Bourd. & G.
1928: 569 (*Coriolus*); D. Baxt. 1948 (PMi 32): 205 *pl. 8*; Overh. 1953:
52 *pl. 72 f. 420, pl. 28 fs. 446–448, pl. 84 fs. 483, 484, pl. 126 fig.*; Lowe
1957 F.: 74 *f. 55* (*Fomes*).

Polyporus populinus Fr. 1863 M. 2: 254 (Sweden) **(72)**, not ∼ (Schum.)
per Fr. 1821, not ∼ (Lázaro) Sacc. & Trott. apud Trott. 1925.

Trametes secretanii Otth 1866 (MiB 1865): 157 (Switzerland); fide
Lloyd 1915 (LMW 4, F.): 285 = *Fomes connatus*. — [≡ *Polyporus*
populinus (Schum. per Fr.) sensu Secr. 1833 M. 3: 113].

Polyporus cremeus Bres. ex Lloyd 1915 (LMW 4, Ap.): 311, 390 (Brazil),
not *Polystictus cremeus* Bres. 1920; fide Bres. 1920 (Am 18): 67 = *Fomes*
populinus.

ravidus (Fr.) Bond. & S. 1941 **(73)**. — *Polyporus* Fr. 1838 (presumably
France); *Polystictus* Cooke 1886; *Cladomeris* Big. & Guill. 1909 (syn.;
n.v.p.); *Coriolus* Bourd. & G. 1928; *Trametes* Pilát 1939; *Rigidoporus*
Pouz. 1966. — Sensu Bres. 1903 (Am 1): 75 (*Polystictus*); Höhn. 1909
(ÖbZ 59): 63 (*Polyporus*); Bourd. & G. 1928: 564 (*Coriolus*); Pilát 1939
(ACE 3): 272 *f. 108, pl. 193 f. a* (*Trametes*); Domański & al. 1963 (Mob 15):
49 *fs. 18–20*; 1965 (FpG 2): 200 *pl. 57 f. 2, pl. 58 fs. 1, 2* (*Oxyporus*).

? *Sistotrema* **lutescens* Pers. 1825: 205 (France); fide Fr. 1874: 566
("*Sistotrema cinereo-lutescens*"). — [≡ *Sistotrema cinereo-lutescens* Fr.
1874: 566 (error; as a syn.)]; *Microporus cinereo-lutescens* (Fr.) ex O.K.
1898. — Cf. **(73)**.

Daedalea rugosa Allesch. 1886 (BLa 9): 61 (Germany); fide Bres. 1903
(Am 1): 75.

F u r t h e r   s p e c i e s :
always strictly resupinate (*Poria* p.p.)

corticola (Fr.) Ryv. 1972. — *Polyporus* Fr. 1821 (Sweden); *Trametes*
Fr. 1849 (nom. nud.: n.v.p.); *Physisporus* Gillet 1877, P. Karst. 1882;

Poria Cooke 1886, Quél. 1886; *Muciporus* Juel 1897, mixtum compositum; *Coriolus* Pat. 1900; *Chaetoporus* Bond. & S. 1941; *Rigidoporus* Pouz. 1966; *Oxyporus* E. Komar. 1964 ("Parm."; indirect ref.: n.v.p.); ≡ *Polyporus polystictus* Pers. 1825. — Juel 1914 (ABS 14[1]): 3 *pl. 1* (*Polyporus*); Egeland 1914 (NMN 52): 143; Overh. 1923 (BTC 50): 245 *f. 1, pl. 13*; 1929 (M 21): 285 *pl. 25 fs. 17, 18*; D. Baxt. 1936 (PMi 21): 247; Overh. 1942: 25; Lowe 1946: 50 *f. 11* (*Poria*); Domański 1965 (FpG 2): 111 *f. 38, pls. 30, 31* (*Chaetoporus*); Lowe 1966: 19 *f. 2* (*Poria*). — Illustrative specimen ('neotype' distribution): Lundell 1935 (LNF 3–4): 23 No. 154 as *Poria corticola*, on decaying trunks of *Populus tremula*.

? *Polyporus salviae* B. & C. apud Berk. 1872 (G 1): 54 (U.S.A., South Carolina) (nom. monstr.); cf. Lowe 1959 (Ll 21): 105 ("type sterile and myriadoporous") & 1966: 19 (with "?"). — According to Pilát 1941 (ACE 3): 427 referred to *Poria ambigua* by Lloyd. — Type on *Salvia*, surrounding the branches.

? *Polyporus vesiculosus* B. & C. apud Berk. 1872 (G 1): 65 (U.S A., Alabama) (nom. monstr.); fide Lowe 1947 (Ll 10): 59 (sterile), 1959 (Ll 21): 105 ("type sterile and myriadoporous"), & 1966: 19. — *Poria* Cooke 1886. — Cf. Bres. 1916 (Am 14): 241, "Est status myriadoporus alicuius *Poriae* sp. sterilis, indeterminabilis." Referred by Peck 1885 (RNS 38): 93 to *Polyporus* [*Perenniporia*] *subacida*, apparently in error. — Type on pine planks. — Sensu Peck → *Perenniporia subacida*.

Polyporus rostafinskii P. Karst. 1876 (BFi 25): 274 (Finland), not ∼ Błoński 1888; fide P. Karst. 1892 (H 31): 293 (var.), Romell 1926 (SbT 20): 15 (forma or var.), & Lowe 1956 (M 48): 119. — *Physisporus* P. Karst 1881: *Poria* Cooke 1886. — Type on bark of *Alnus*.

Physisporus tener Har. & Karst. 1890 (Rm 12): 128 (Finland), not *Chaetoporus tenuis* P. Karst. 1890; fide Lowe 1956 (M 48): 115 (sterile). — *Poria* Sacc. 1891. — Cf. Donk 1960 (Pe 1): 198, note. — Type on rotten wood of *Populus tremula*.

? *Physisporus confusus* P. Karst. 1899: 132. — [≡ *Physisporus corticola* (Fr.) Gillet sensu P. Karst. 1889 (BFi 48): 320 (Finland), indirectly cited by P. Karst. 1899: 131 in key]. — Type on "Aspbark" (= bark of *Populus*).

Poria separans Murrill 1920 (M 12): 305 (U.S.A., New Hampshire); fide Lowe 1947 (Ll 10): 56 & 1966: 19. — Type on a dead log.

Poria vicina Bres. 1925 (M 17): 76 (U.S.A., Washington); fide R. L. Gilb. 1956 (Ll 19): 66, 84. — (**74**).

M.—*Polyporus aneirinus* Sommerf. sensu Fr. in herb., in part; & Bres. 1897 (AAR III 3): 85 (*Poria*); fide Romell 1911 (ABS 11[3]): 21. — Cf. also Litsch. apud Pilát 1932 (BmF 48): 40 & Donk 1967 (Pe 5): 79, notes.

late-marginatus (Dur. & Mont. ex Mont.) Donk 1966 (**71**). — *Polyporus* Dur. & Mont. ex Mont. 1856: 163 (Algeria); *Poria* Cooke 1886; *Rigidoporus* Pouz. 1966. — Mont. 1856: 163 (*Polyporus*); Lowe 1966: 17 *f. 1* (*Poria*); Pouz. 1966 (Fgp 1): 368 (*Rigidoporus*).

Poria ambigua Bres. 1897 (Hungary, now Czechoslovakia; or, rather, Italy); fide Lowe 1963 (M 55): 455 & 1966: 17. — *Chaetoporus* Bond. & S. 1941; *Oxyporus* E. Komar. 1964 (incomplete ref.: n.v.p.). — Bres. 1897 (AAR III 3): 84; Overh. 1922 (M 14): 1 *fs. 1, 2, pl. 1 fs. 5, 6*; Bourd. & G. 1928: 669; Pilát 1932 (BmF 48): 37 *pl. 4 f. 4, pl. 7 f. 2*; Overh. 1942: 21; Lowe 1946: 49; D. Reid 1958 (TBS 41): 436 *fs. 20, 24*; Lombard & al. 1961 (M 52): 287 *fs. 2B, 3B*, cult. char., decay (*Poria*): Domański 1965 (FpG 2): 113 *f. 39, pl. 32 f. 1* (*Chaetoporus*). — Speg. 1925 (BCó 28): 385 confused the name *Poria ambigua* with that of *Trametes ambigua* (Berk.) Fr. (extra-European).

Poria geoderma Speg. 1898 (ABA 6): 171 (Argentina); fide Lowe 1963 (PMi 48): 168 = *Poria ambigua*. — Lowe, l.c. — Lowe found no cystidia in the type.

Irpex concrescens Lloyd 1915 (LMW 4, L. 60): 9 (U.S.A., Alabama); fide R. L. Gilb. 1963 (PMi 48): 147 = *Poria ambigua*. — R. L. Gilb., l.c.

Poria lacerata Murrill 1920 (M 12): 92 (U.S.A., Arkansas); fide Bres. 1926: 30, Overh. 1942: 21, & Lowe 1947 (Ll 10): 46 = *Poria ambigua*; fide Lowe 1966: 18.

Poria salicina Murrill 1920 (M 12): 304 (U.S.A., Pennsylvania), not ~ (Pers. apud J. F. Gmel.) Pers. 1794 (d.n.); fide Overh. 1942: 21 & Lowe 1947 (Ll 10): 46 = *Poria ambigua*. — ≡ *Poria sumstinei* Seav. 1934 (M, Gen. Ind.): 226.

Poria roseitingens Murrill 1920 (M 12): 305 (Jamaica); fide Lowe 1963 (M 55): 455. — Lowe 1947 (Ll 10): 56. — According to Lowe the type lacks cystidia.

Poria cokeri Murrill 1920 (M 12): 306 (U.S.A., North Carolina); fide Lowe 1947 (Ll 10): 46 = *Poria ambigua*.

Leptoporus zilingianus Pilát 1934 (U.S.S.R., Russia, Siberia): fide Domański 1964 (APo 33): 171 = *Chaetoporus ambiguus* (forma). — *Polyporus* Killerm. 1943; *Tyromyces* Bond. 1953. — Pilát 1934 (BmF 49): 258 *pl. 12 f. 3*; 1937–8 (ACE 3): 223 *f. 74, pl. 134 f. 2*; Domański 1964 (APo 33): 171 *f. 4.*

Poria consobrinoides Pilát 1936 (U.S.S.R., Russia, Siberia); fide Pilát 1941 (ACE 3): 427 = *Poria ambigua* ("forma acystidiata"). — Pilát 1936 (BmF 51): 380 *pl. 10 f. 1.*

Poria reticulato-marginata Pilát 1936 (U.S.S.R., Russia, Siberia); fide Pilát 1941 (ACE 3): 427 = *Poria ambigua*. — *Fibuloporia* Bond. & S. 1941 (nom. prov., & generic name n.v.p). — Pilát 1936 (BmF 51): 380 *pl. 10 f. 4.*

? *Leptoporus werneri* Pilát 1939 (Sbč 2): 61 *pl. 2* (Morocco); from descr.

Poria herbicola D. Baxt. 1948 (PMi 32): 190 *pl. 1* (U.S.A., Michigan); fide Lowe 1966: 18.

pearsonii (Pilát) E. Komar. 1964 (n.v.p.) (74). — *Poria* Pilát 1935 (Czechoslovakia, now U.S.S.R., Ukraine); *Chaetoporus* Bond. 1953; *Oxyporus*

E. Komar. 1964 (indirect ref.: n.v.p.; comb. cited by author, not made in 1959). — Pilát 1935 (TBS 19): 195 *pl. 3*; 1941 (ACE 3): 450 *f. 212, pl. 288, pl. 289 f. b* (*Poria*); Domański 1965 (FgP 2): 116 *f. 40, pl. 33* (*Chaetoporus*). — Referred to *Poria* [*Oxyporus*] *corticola* by Lowe 1962 (PMi 47): 183. — Type on *Abies alba*.

? M.—*Boletus medulla-panis* Jacq. sensu A. Pears. apud Bourd. 1932 (*Poria*): fide Pilát 1935 (TBS 19): 195 — Bourd. 1932 (BmF 48): 231 (*Poria*). — "Sur pin."

philadelphi (Parm.) Ryv. 1972. — *Chaetoporus* Parm. 1959 (Estonia); *Rigidoporus* Pouz. 1966. — Parm. 1959 (BMs 12): 237 *fs. 1, 4, plate f. 2.*

PACHYKYTOSPORA Kotl. & P.

1963 (ČM 17): 27. — Holotype: *Polyporus tuberculosus* Fr.

SPECIAL LITERATURE.—Jahn, *1965a*, Kotlaba & Pouzar, *1963a*.

tuberculosa (Fr.) Kotl. & P. 1963. — *Polyporus* Fr. 1821, not ∼ (Pers.) per Pers. 1825: *Strigilia* O.K. 1891, typonym; [≡ *Boletus tuberculosus* (Pers.) Pers. sensu DC. 1815: 40, Fr. 1821, excl. of type, (Switzerland)]; ≡ *Polyporus colliculosus* Pers. 1825, typonym; *Trametes* Lundell 1937, not ∼ Berk. 1847; *Coriolellus* Bond. 1953. — R. L. Gilb 1961 (PMi 46): 211 *tpl. 1 fs. 3, 4, tpl. 1 cont'd fs. C, D* (*Trametes col.*): Kotl. & P. 1963 (ČM 17): 28 *f. 1, pls. 1, 2* (*Pachykytospora tub.*). — Cf. Donk 1967 (Pe 5): 86 (*Polyporus col.*), 114 (*Polyporus tub.*).

Polyporus albo-carneo-gilvidus Romell 1890 (Sweden); fide Bres. 1897 (AAR III 3): 93 = *Trametes micans* [sensu Bres.]; fide Romell 1926 (SbT 20): 6, 14 = *Polyporus colliculosus.* — *Poria* Sacc. 1891; *Trametes* Lundell 1946: ≡ "*Poria* (or *Trametes*)" *carneogilvida* Romell 1926 (nom. prov.: n.v.p.). — Romell 1890 F.s.: No. 17 [repr. 1890 (BoN): 151; 1890 (H 29): 363]; 1926 (SbT 20): 6, notes.

Trametes morganii Lloyd 1919 (LMW 5, L. 69): 15 (U.S.A., Ohio + Sweden), in part: as to spores; cf. Donk 1971 (Pe 6): 216.

Polyporus weinzettlii Velen. 1922: 673 *f 112* [see Pilát 1948: 673 for Latin translation] (Czechoslovakia): fide Pilát 1940 (ACE 3): 309 = *Trametes colliculosa.*

Poria subtrametea Pilát 1942 (lacking Latin descr.: n.v.p.), 1953; fide Domański 1964 (APo 33): 177 (forma). — Pilát 1942 (ACE 3): 453 *f. 215, pl. 291*: Domański 1964 (APo 33): 177 *f. 8* (*Poria*); 1965 (FpG 2): 172 *pl. 48 f. 2* (*Pachykytospora tuberculosa* f.).

M.—*Poria tuberculosa* Pers. (**O**) sensu DC. 1815: 40 (*Boletus*). → *Polyporus tuberculosus* Fr. & *Striglia tuberculosa* O.K.; → *Polyporus colliculosus* Pers.

M.—*Poria incarnata* Pers. sensu Gillet 1874–90; fide Bres. 1897 (AAR III 3): 93 = *Trametes micans* [sensu Bres.]. — Gillet 1874–90 P.: *pl. 532/471 fig.*

M.—*Poria micans* Ehrenb. sensu Bres. 1897 (AAR III 3): 93 (*Trametes*); fide Romell 1926 (SbT 20): 13–14 = *Polyporus albo-carneo-gilvidus* = *Polyporus colliculosus*. — Egeland 1919 (NMN 52): 167 (*Trametes*).

PARMASTOMYCES Kotl. & P.

1964 (FR 69): 138. — Holotype: *Tyromyces kravtzevianus* Bond. & Parm. apud Parm.

kravtzevianus (Bond. & Parm. apud Parm.) Kotl. & P. 1964 (**75**). — *Tyromyces* Bond. & Parm. apud 1957: [≡ *Leptoporus bourdotii* Pilát 1936, in part: sub "b" [u n n a m e d] (U.S.S.R., Russia, Siberia). — Parm. 1959 (TSR 12): 236 *fs. 12–16* (*Tyromyces*): Kotl. & P. 1964 (FR 69): 139 (*Parmastomyces*).

Polyporus subcartilagineus Overh. 1941 (Canada, Québec) (lacking Latin descr.: n.v.p.); fide Pouz. 1966 (ČM 20): 176. — *Tyromyces* Domański 1965 (lacking Latin descr.: n.v.p.). — Overh. 1941 (M. 33): 90 *f. 1*; Davids. & Campb. 1943 (Ph 33): 976, 982 *fs. 3D, 6F, 7C*, cult. char.: Overh. 1953: 358; Lowe 1966: 71 *f. 49*.

PERENNIPORIA Murrill (104)

1942 [1960 (Pe 1): 251]. — Lectotype: *Polyporus unitus* Pers. [sensu **Murrill**]. — Cf. Donk 1967 (Pe 5): 74.

Special literature.—Baxter, *1935*; Domański, *1964c*.

Nota bene.—See remark under *Poria*.

Poria **fulviseda** Bres. 1920 (Italy). — *Poria* Bres. apud Lloyd 1912 (nom. nud.: n.v.p.); = *Poria medulla-panis* var. Bres. apud Bres. & Cav. 1901, presumably typonym, but not cited by Bres. in 1920. — Bourd. & G. 1928: 685; Thind & Ratt. 1971 (JPh 24): 5.5 *fs. 19–22, 27* (*Poria*).

medulla-panis (Jacq. per Fr.) Donk 1967. — *Boletus* Jacq. 1778 (Austria) (d.n.); *Poria* Pers. 1794 (d.n.); *Polyporus* (Jacq.) per Fr. 1821, misapplied?; *Boletus* Mérat 1821; *Physisporus* Chev. 1826; *Poria* Cooke 1886, Quél. 1886; *Fomes* Bres. apud Lloyd 1899 (incidental mention: n.v.p.): *Trametes* Pat. 1900; *Fomitopsis* Bond. & S. 1941; ≡ *Poria medullaris* S. F. Gray 1821; *Perenniporia* Domański 1966. — Bres. 1897 (AAR III 3): 84; 1903 (Am 1): 79; Murrill 1920 (M 12): 48; Overh. 1923 (BTC 50): 249 *f. 3, pl. 14 fs. 1–4*: Bourd. & G. 1928: 684; Lowe 1966: 107 *f. 92* (*Poria*); & cf. Donk 1967 (Pe 5): 76, 92, notes. — Sensu Sow. → *Polyporus rangiferinus* (**O**); sensu Fr. 1884?, E. Krause 1929, in part, = *Heterobasidion annosum*; sensu A. Pears. apud Bourd. → *Oxyporus pearsonii*.

Polyporus limitatus B. & C. apud Berk. 1872 (G 1): 54 (U.S.A., South Carolina): fide Lowe 1966: 107. — *Poria* Cooke 1886.

Polyporus xylostromatis Fuck. 1873 (Jna 27–28): 86 (Germany): fide

Lowe 1947 (Ll 10): 59 = *Poria unita* [sensu Lowe]. — *Poria* Cooke 1886.

? *Polyporus albo-incarnatus* Pat. & Gaill. 1888 (BmF 4): 35 (Venezuela); fide Lowe 1947 (Ll 10): 46 = *Poria unita* [sensu Lowe] — *Poria* Sacc. 1891. — Lowe 1947 (Ll 10): 46 (*Poria*).

M.—*Polyporus unitus* Pers. sensu Donk 1933: 234, at least in part; fide Donk, l.c. (*Poria medulla-panis* cited as a syn.); 1967 (Pe 5): 116. — Overh. 1942: 40; Lowe 1946: 27; Malenç. 1952 (BbF 99^{10}): 45; Lowe 1955 (M 47): 222 (*Poria*): 1957 F.: 83 *f. 64* (*Fomes*): D. Reid 1963 (KB 17): 295 (*Poria*).

Poria **pulchella** (Schw.) Cooke 1886 (**76**). — *Polyporus* Schw. 1832 (U.S.A., Pennsylvania), not ∼ Sacc. 1873. — Overh. 1923 (M. 15): 219 *fs. 16, 17, pl. 23 fs. 7–9* (*Poria*): Bourd. & G. 1928: 684 (*Poria medulla-panis* var.): Overh. 1942: 49 (*Poria*): Lowe 1946: 28; 1966: 110 (*Poria tenuis* var.).

? *Polyporus tenuis* Schw. 1832 (U.S.A., Pennsylvania) (**76**), not ∼ (Hook.) Kl. 1833; cf. Overh. 1923 (M 15): 225, who compared it, *inter alia*, with *Poria pulchella*. — *Poria* Cooke 1886. — Overh. 1923 (M 15): 225 *fs. 23, 24, pl. 24 f. 4; 1942: 40*; R. L. Gilb 1956 (Ll 19): 73; Lowe 1966: 110 *f. 24* (*Poria*).

Polyporus xantholoma Schw. 1832 (U.S.A., North Carolina); fide Overh. 1942: 49. — *Poria* Cooke 1886. — Overh. 1923 (M. 15): 230 *fs. 29, 30, pl. 24 f. 9* (*Poria*).

Polyporus cremor B. & C. 1849 (HJB 1): 104 (U.S.A., South Carolina); fide Bres. 1926: 80. — *Poria* Cooke 1886. — Lowe 1947 (Ll 10): 48 (*Poria*).

Polyporus dryinus B. & Cooke ex Cooke 1878 (G 6): 130 (U.S.A., South Carolina); fide Bres. 1926: 80. — *Polyporus* B. & Cooke 1878 (nom. nud.: n.v.p.): *Poria* Cooke 1886.

Poria holoxantha (B. & Cooke) ex Cooke 1886 (G 15): 26 (U.S.A., South Carolina): fide Bres. 1926: 80. — *Polyporus* B. & Cooke 1879 (nom. nud.: n.v.p.).

Poria tomentocincta (B. & Rav.) ex Cooke 1886 (G 15): 26 (U.S.A., South Carolina); fide Bres. 1926: 81 (forma). — *Polyporus* B. & Rav. "in Herb." (n.v.p.).

Poria leonidis Manc. & Sacc. apud Sacc. 1888. — ≡ *Physisporus vitellinus* P. Karst. 1881, protonym (nom. nud.: n.v.p.): ≡ *Physisporus nitidus* subsp. *vitellinulus* P. Karst. 1882 (BFi 37): 59 (Finland): *Physisporus* P. Karst. 1889; *Poria* Egeland 1914. — Egeland 1914 (NMN 52): 150 (*Poria vitellinulus*).

Poria chrysella Egeland 1913 (NMN 51): 77 (Norway); fide Bres. apud Egeland 1914 (NMN 51): 383 = *Poria pulchella*.

subacida (Peck) Donk 1967. — *Polyporus* Peck 1885 (U.S.A., New York); *Poria* Sacc. 1888: *Chaetoporus* Bond. & S. 1941: *Oxyporus* E. Komar. apud Parm. 1961; *Perenniporia* Domański 1967 (incomplete ref.: n.v.p.). — Overh. 1919 (BNS 205–206): 111 *pl. 19 f. 1, pl. 20 fs. 1, 2, 4–7, pl. 21* *fs. 1–5*; D. Baxt. 1935 (PMi 20): 273 *pls. 55–58*; Overh. 1942: 38; Domański 1964 (APo 33): 661, 676 *fs. 1, 2, 4*, with cult. char.; 1965 (FpG 2): 128 *f. 45, pl. 35*; Lowe 1966: 109 *f. 63 (Poria)*.

? *Physisporus variecolor* P. Karst. in Thüm. 1881 M.u.: No. 1803 & 1881 (Mfe 6): 10 (Finland); fide [Bres. apud] J. Rick 1898 (ÖbZ 48): 137 & Bres. 1920 (Am 18): 69. — *Poria* Cooke 1886; *Chaetoporellus* Parm. 1959, presumably misapplied; *Chaetoporus* Parm. 1961, Domański 1963, M. Bond. 1964, presumably misapplied. — Sensu Lowe → *Chaetoporus luteo-albus*. — As yet rather a nomen dubium but it may appear an earlier name for *Perenniporia subacida* rather than a syn. of *Poria pulchella*: cf. Donk 1967 (Pe 5): 119.

Polyporus ornatus Peck 1885 (RNS 38): 92 (U.S.A., New York): fide Lowe 1946: 22. — *Poria* Sacc. 1888: *Trametes* Pilát 1939, misapplied: *Coriolellus* Bond. 1953 (incomplete ref.: n.v.p.), Kotl. & P. 1957. — Overh. 1919 (BNS 205–206): 99 *pl. 14 (Poria)*. — Sensu Pilát → *Trametes* [*Antrodia*] *subalutacea*.

Poria fuscomarginata (Berk.) ex Cooke 1886 (G 15): 24 (U.S.A., Rhode Island): fide Bres. 1926: 80. — *Polyporus Berk*. "in Herb." (n.v.p.).

Poria omaena (Berk.) ex Cooke 1886 (G 15): 26 (U.S.A., South Carolina); fide Bres. 1926: 80. — *Polyporus* Berk. "in Herb." (n.v.p.). — V.s.: "homaema", "homema"; & (Overh. 1942: 41; syn.) "amoema".

Poria subaurantia (Berk.) ex Cooke 1886 (G 15): 27 (U.S.A., "Carolina"); fide Bres. 1926: 81. — *Polyporus* Berk. "in Herb." (n.v.p.).

Poria trametopora Pilát 1936 (U.S.S.R., Russia, Siberia): fide Domański 1964 (APo 33): 174. — Pilát 1936 (BmF 51): 381 *pl. 6 f. 3*.

Poria colorea Overh. & Engl. apud Englerth 1942 (BFY 50): 20 *pl. 4* (U.S.A., Oregon); fide Lowe 1966: 109.

M.—*Polyporus vesiculosus* B. & C. apud Berk. sensu Peck 1885 (RNS 38): 93 (*Polyporus subacidus* var.): fide Overh. 1919 (BNS 205–206): 114. — Overh. 1919 (BNS 205–206): 114 *pl. 19 f. 3 (Poria subacida* var.).

PHAEOLUS Pat.

1897 ("subgen." in error for "gen."?), 1900 [1960 (Pe 1): 252]. — Lectotype: *Polyporus schweinitzii* Fr.
Romellia Murrill 1904, not ∼ Berlese 1900 (Pyrenomycetes) [1960 (Pe 1): 277]. — Holotype: *Boletus sistotremoides* A. & S.
Choriphyllum Velen. 1922 [1960 (Pe 1): 198]. — Monotype: *Daedalea fusca* Velen.

SPECIAL LITERATURE.—Boyce, *1924*; Childs, *1937*; David, *1969b*; Gray, *1947*; Hedgcock & al., *1925*; Nagatomo, *1931*; Pilát, *1934a*; Wean, *1937*; Yde-Andersen, *1961a*.

schweinitzii (Fr.) Pat. 1900. — *Polyporus* Fr. 1821 ("*Schweinizii*"), 1832 ("*Schweinitzii*"); *Trametes* Fr. 1849 (nom. nud.: n.v.p.), Kalchbr. 1865; *Polystictus* P. Karst. 1879: *Cladomeris* Quél. 1886; *Inodermus* Quél. 1888; *Phaeolus* Pat. 1897 (n.v.p.?); *Pelloporus* Sart. & M. 1921; *Ovinus*

Sart. & M. 1924; *Spongiosus* Torrend 1924: *Hapalopilus* Donk 1933; *Coltricia* G. Cunn. 1948; ≡ *Boletus sistotremoides* A. & S. 1805 (Germany) (d.n.), not ~ Fr. 1815 (d.n.); *Boletus* A. & S. per Schw. 1822; *Polyporus* Duby 1830; *Daedalea* Fr. 1832; not ~ Velen. 1926, not ~ Beeli 1930; *Ochroporus* J. Schroet. 1888; *Romellia* Murrill 1904; *Phaeolus* Murrill 1905; ≡ *Sistotrema ferrugineum* Pers. 1825; ≡ *Boletus sistotrema* Fr. 1874 (error, syn.: n.v.p.); *Cladomeris* Big. & Guill. 1909 (syn.: n.v.p.); *Polyporus* Velen. 1922. — Fr. 1874: 529; 1882 & 1884 I. 2: 79 *pl. 179 f. 3*; Schrenk 1900 (BVP 25): 18 *f. 1, pl. 1 f. 1*; Lloyd 1908 (LMW 3, P.I.): 13 *f. 208* (*Polyporus*); Konr. & M. 1925 I. 5: *pl. 433*; Bourd. & G. 1928: 554 (*Phaeolus*): Shope 1931 (AMo 18): 347 *pl. 25 fs. 1, 2* (*Polyporus*); Donk 1933: 173 (*Hapalopilus*); Jørst. & Juul 1939 (MnS 6): 310, 474 *fs. 3–5*; Overh. 1953: *395 pl. 42 fs. 253, 254, pl. 45 f. 272, pl. 98 f. 557, pl. 103 f. 580, pl. 132 fig.* (*Polyporus*), Al. David 1969 (Nca 96): 213 *f. 1*, cult. char. (*Phaeolus*).

Hydnum spadiceum Pers. 1800 (Germany) (d.n.), not ~ Desv. (n.v.). — *Sistotrema* Sw. 1812 (d.n.); *Daedalea* (Pers.) per Fr. 1821; *Sistotrema* Pers. 1825; *Hydnellum* P. Karst. 1879; *Calodon* Quél. 1886; *Phaeodon* J. Schroet. 1888. — Pers. 1800 I.D.: 34 *pl. 9 fs. 1–3*.

Boletus maximus Brot. 1804 (Portugal) (d.n.), not ~ Schum. 1803 (d.n.); fide Torrend 1913 (Bro 11): 61. — *Daedalea* (Brot.) per Fr. 1821; *Polyporus* Fr. 1838; not ~ Overh. 1926; *Cladomeris* Big. & Guill. 1909 (syn.: n.v.p.); *Heteroporus* Lázaro 1916; *Polystictus* J. Rick 1936 (citing wrong basionym): ≡ *Sistotrema lusitanicum* Pers. 1825. — Brot. 1804: 468.

Daedalea epigaea Lenz 1831 (Germany); fide Fr. 1838: 433 ("bene!"). — Lenz 1831: 62.

Polyporus holophaeus Mont. 1843 (ASn II 20): 36 (France): fide Lloyd 1910 (LMW 3, M.): 68. — *Polystictus* Cooke 1886; *Phaeolus* Pat. ex R. Heim 1931.

Polyporus tabulaeformis Berk. 1845 (LJB 4): 302 (U.S.A., Georgia), Sacc. 1888 ("*tubulaeformis*"); fide Berk. 1872 (G 1): 37. — *Phaeolus* Pat. 1900.

Polyporus herbergii Rostk. 1848 (Germany/Poland); fide Lloyd 1915 (LMW 4, Ap.): 380. — *Polystictus* P. Karst. 1887, apparently misapplied; *Inonotus* P. Karst. 1887; *Boletus* Mussat 1901 (error, syn.: n.v.p.), Pilát 1934 (syn.: n.v.p.). — Rostk. 1848 (StP Fs. 27–28): 35 *pl. 18*.

Polyporus spectabilis Fr. 1851 (NAu III 1): 48/32 (U.S.A., North Carolina); fide Berk. 1854 (HJB 6): 136 = *Polyporus tabulaeformis*; fide Murrill 1904 (BTC 31): 339.

Polyporus spongia Fr. 1852 (Sweden); fide Quél. 1888: 393 (var.) & Lloyd 1912 (LMW 3, S.P.): 162. — *Inonotus* P. Karst. 1882; *Merisma* Gillet 1884; *Cladomeris* Quél. 1886; *Mucronoporus* Ell. & Ev. 1889: *Ochroporus* Błoński 1890; *Inodermus* Bourd. 1898; *Phaeolus* Pat. 1900. — Fr. 1863 M. 2: 268; Quél. 1880 (BbF 26): 230; Fr. 1884 I. 2: 80 *pl. 180 f. 2* (*Polyporus*); Konr. & M. 1935 I. 5: *pl. 434* (*Phaeolus schweinitzii* f.).

? *Merulius giganteus* Saut. 1877 (H 16): 73 (Austria); cf. Höhn. 1906 (SbW 115): 688 ("vielleicht"). — *Sesia* O.K. 1891. — P. Henn. 1903 (H 42): 181 thought of *Merulius [Serpula] lacrimans* (Wulf.) per Fr. (**O**).

Polyporus hispidoides Peck 1880 (RNS 33): 21 *pl. 1 fs. 18–20* (U.S.A., New York); fide Peck 1887 (RNS 39): 42 (forma) & Overh. 1953: 395, 397. — Reported for Bohemia (Czechoslovakia) by Velen. 1922: 684, but Pilát 1942 (ACE 3): 56 listed this ref., with doubt, under *Inonotus rheades* [sensu lato].

Inonotus sulphureo-pulverulentus P. Karst. 1904 (ÖfF 46[11]): 8 (U.S.S.R., 'Baikal'); fide P. Karst. 1905 (Afe 27[4]): 16 = *Inonotus herbergii*; fide Lowe 1956 (M 48): 108.

Daedalea fusca Velen. 1922: 689 [see Pilát 1948: 261 for Latin translation] (Czechoslovakia), not ∼ Link per Fr. 1821; fide Pilát 1942 (ACE 3): 601.

M.—*Boletus mollis* Pers. sensu R. Hartig 1878 (*Polyporus*). — R. Hartig 1878: 49 *pl. 9* (*Polyporus*).

PHELLINUS Quél. (77)

1886 [1960 (Pe 1): 253]. — ≡ *Pyropolyporus* Murrill 1903 [1960 (Pe 1): 276]. — Lectotype: *Polyporus rubriporus* Quél.

Mison Adans. 1763 (d.n.), not ∼ Fr. 1835 ("Tuberacei") [1960 (Pe 1): 245]. — ≡ *Scindalma* Hill 1751 (pre-Linnaean name) ≡ *Scindalma* [Hill] O.K. 1898, typonym. — Lectotype: *Agaricum durum, crassum . . .* Mich.

Boletus S. F. Gray 1821, not ∼ [Dill.] L. 1753 (d.n.) per St-Am. 1821 (Boletaceae), not ∼ Fr. Jan. 1, 1821 (nom. cons.; Boletaceae), not ∼ [Mich.] Maratti 1822 (Morchellaceae) [1960 (Pe 1): 190; 1962 (Pe 2): 202]. — Lectotype: *Boletus igniarius* L.

Poria P. Karst. 1881, not ∼ [Hill] Adans. 1763 (d.n.), not ∼ Pers. per S. F. Gray 1821 [1960 (Pe 1): 268, in obs.]. — [≡ *Poria* Pers. sensu P. Karst. 1881, in part: excl. of type]. — Lectotype: *Poria salicina* Pers.

Ochroporus J. Schroet. 1888 [1960 (Pe 1): 247; 1962 (Pe 2): 207]. — Lectotype: *Polyporus igniarius* (L.?) per Fr.

Mucronoporus Ell. & Ev. 1889 [1960 (Re 1): 246; 1962 (Pe. 2): 207; but cf. Donk 1971 (PNA 74): 12]. — Lectotype: *Polyporus gilvus* (Schw.) Steud. — Sensu Domański → *Onnia*.

Porodaedalea Murrill 1905 [1960 (Pe 1): 270]. — Holotype: *Boletus pini* Thore.

Fomitiporia Murrill 1907 [1960 (Pe 1): 217]. — Holotype: *Fomitiporia langloisii* Murrill.

Fuscoporia Murrill 1907 [1960 (Pe 1): 219]. — Holotype: "*Boletus ferruginosus* Schrad." [sensu Murrill].

Fulvifomes Murrill 1914 [1960 (Pe 1): 218]. — Holotype: *Pyropolyporus robiniae* Murrill.

? *Pseudofomes* Lázaro 1916 [1960 (Pe 1): 275]. — Lectotype: "*Pseudofomes nigricans* (Bull.) Láz."

Daedaloides Lázaro 1916 [1960 (Pe 1): 206; 1962 (Pe 2): 203]. — Monotype: *Daedalea pinicola* Lázaro.

Boudiera Lázaro 1916, not ∼ Cooke 1877 (Ascomycetes) [1960 (Pe 1): 191; 1962 (Pe 2): 203]. — ≡ *Lazaroa* Gonz. *in* Boln R. Soc. esp. Hist. nat. 17: 459. 1917. — Lectotype (revised): *Polyporus rubriporus* Quél., cf. Donk 1971 (PNA 74): 406.

Cryptoderma Imaz. 1943 [1960 (Pe 1): 201]. — Holotype: *Fomes ribis* (Schum. per Fr.) Gillet. — Cf. Donk 1963 (Ta 12): 327.

M.—*Poria* Pers. sensu P. Karst. 1881 → *Poria* P. Karst.

SPECIAL LITERATURE.—General & miscellaneous: Ellis & Everhart, *1889a, 1889b*; Ferry, *1894*; Gricius, *1967a*; Hartig, *1893*; Igmándy, *1970*; Jahn, *1967a*; Kirschstein, *1936*; Martirosjan, *1965a*; Mikelaikevičius, *1962*.

Phellinus conchatus: Bondarcev, *1955*; Ulbrich, *1939a*.

Phellinus ferrugineo-fuscus: Göpfert, 1971; Kotlaba, *1965*.

Phellinus igniarius complex (see p. 127): Borisov, *1940*; Bratus' & Cyljuryk, *1964*; Domański, *1954, 1960d*; Göpfert, *1968*; Good & Spanis, *1958*; Hiorth, *1965*; Hirt, *1949b*; Hirt & Hopp, *1936, 1942*; Hopp, *1936*; Jahn, *1962, 1966a*; Klingemann, *1893*; Kotlaba & Pouzar, *1968b*; Manion & French, *1968, 1969*; Nannfeldt, *1967*; Patouillard, *1889b*; Riley, *1952*; Riley & Bier, *1936*; Roll-Hansen, *1967*; Surovov, *1967a, 1970*; Verrall, *1937*; Wall & Kuntz, *1964*.

Phellinus laevigatus: Bergstädt, 1967; Campbell & Davidson, *1941*; Mayr, *1884*?; Niemelä, *1972*.

Phellinus lundellii: Niemelä, *1972*.

Phellinus nigrolimitatus: K. Lohwag, *1950*; Kotlaba, *1972*; Weir, *1914b*.

Phellinus pilatii: Černý, *1968*; Igmándy & Pagony, *1965*; Kotlaba, *1968b*.

Phellinus pini complex: Bratus', *1959*; Cool, *1920*; De Groot, *1964, 1965*; Durand, *1924*; Guinier, *1919*; Haddow, *1938a, 1938b*; Hole, *1915*; Igmándy, *1955a*; Imazeki, *1951*; Kalandra, *1968*; Khan, *1910*; Möller, *1904, 1910, 1914*; Ohlmann, *1960*; Owens, *1936*; Percival, *1933*; Pilát, *1926a*; Singh, *1924*; Suri, *1926*; Tubeuf, *1906*; Weir, *1914a*.

Phellinus pomaceus: E. Fisher, *1934*; Kotlaba, *1966*; Pešek, *1969*.

Phellinus pouzari: Kotlaba, *1968a*.

Phellinus punctatus: Domański, *1956*; Gard, *1922*; Rezende-Pinto, *1940, 1942*.

Phellinus ribis: Farinha & Rosado, *1952*; Kallenbach, *1936b–7*; Saccas, *1946*; Voronichin, *1925*.

Phellinus robiniae: Van der Bijl, *1917*; von Schrenk, *1901*.

Phellinus robustus complex: Bondarcev, *1949c*; Jacquiot, *1960*; Jahn, *1965d*; H. Lohwag, *1943*; K. Lohwag, *1937, 1940b*.

Phellinus torulosus: Bondarceva, *1959*; Constantin & Dufour, *1923*; Voronichin, *1925*.

Phellinus viticola: Weir, *1914b*.

ampelinus Bond. 1953 (U.S.S.R., Uzbekistan or Georgia). — *Fomes* Bond. (in sched.: n.v.p.): *Phellinus* Bond. & S. 1941 (nom. nud.: n.v.p.). — Bond. 1953: 400 *pl. 74 f. 3* (Latin descr. on p. 401]. — Recorded from Europe for U.S.S.R., Georgia (Caucasia).

chrysoloma (Fr.) Donk 1969 (**84**). — *Polyporus* Fr. 1861 (Sweden): *Daedalea* Cooke & Q. 1878; *Physisporus* P. Karst. 1882; *Poria* Cooke

1886. — Fr. 1861 (ÖVS 18): 30; 1863 M. 2: 271; 1884 I. 2: 88 *pl. 189 f. 3* (*Polyporus*); Donk 1971 (PNA 74): 39, note.

Fomes abietis P. Karst. 1882 (Finland) (**84**); cf. Romell 1911 (ABS 11^3): 26. — *Trametes* Sacc. 1888; *Mucronoporus* Ell. & Ev. 1892; *Polyporus* Vleugel 1908; *Xanthochrous* Pilát 1932; *Inonotus* Pilát 1940; *Phellinus* Pilát 1950. — Lloyd 1915 (LMW 4, F.): 277 (*Trametes*); Bourd. & G. 1928: 633 (*Xanthochrous pini* subsp.); Jørst. 1937 (KnS 1926^{10}): 38 (*Polyporus pini* var.); Pilát 1942 (ACE 3): 520 *pls. 351–353, pl. 354 f. a* (*Phellinus pini* var.); Jo. Erikss. 1958 (Sbu 16^1): 157 (*Phellinus pini* var.); H. Jahn 1967 (WPb 6): 83 *Abb. 14–17, 34* (*Phellinus*); & cf. Haddow 1938 (TBS 22): 186, 187. — Descriptions based on extra-European material not cited.

Daedalea indurata Velen. 1922: 894 [see Pilát 1948: 263 for Latin translation] (Czechoslovakia), not ∼ Berk. 1877; fide Pilát 1942 (ACE 3): 518, 520, at least in part; and from descr.

conchatus (Pers. per Fr.) Quél. 1886. — *Boletus* Pers. 1795 (Germany) (d.n); *Polyporus* (Pers.) per Fr. 1821, not ∼ Lloyd 1917; *Boletus* Mérat 1821; *Trametes* Fr 1848 (nom. nud.: n.v.p.), 1849 not ∼ Berk 1877 not ∼ (Bres.) Pat. 1900; *Fomes* Gillet 1877; *Ochroporus* J. Schroet. 1888; *Mucronoporus* Ell. & Ev. 1889; *Xanthochrous* Pat. 1897; *Scindalma* O.K. 1898; *Pyropolyporus* Murrill 1903; *Placodes* Rick. 1918 (**78**). — Lloyd 1915 (LMW 4 F.): 244 (*Fomes*); Donk 1933: 254 (*Ochroporus*); Pilát 1942 (ACE 3): 533 *f. 252 pls. 337, 338* (*Phellinus*); D. Baxt. 1950 (PMi 34): 46; Overh. 1953; 69 *pl. 77 fs. 440–442*; Lowe 1957 F.: 49 *f. 33* (*Fomes*); H. Jahn 1967 (WPb 6): 96 *f. 9, Abb. 44, 50* (*Phellinus*). — Sensu Quél. → *Phellinus ribis*.

Polyporus loricatus Pers. 1825: 86, var. α (Switzerland): fide Bres. 1910 (Am 8): 586 = *Fomes salicinus* [sensu Bres.]; fide Donk 1933: 254, 255. — *Fomes* Cooke 1885; *Scindalma* O.K. 1898; *Ganoderma* Speg. 1919 misapplied: = *Polyporus loricatus* var. α *glaucoporus* Pers. 1825: 86. — Sensu Speg. = *Ganoderma* sp. (extra-European).

Polyporus fuscolutescens Fuck. 1864 F.r.: No. 1371 & 1870 (Jna 23–24): 18 (Germany); fide Bres. 1916 (Am 14): 224 = *Fomes salicinus* (sensu Bres.] "juvenilis resupinatus"). — *Poria* Cooke 1886, Quél. 1886.

Fomes densus Lloyd ex Overh. 1931 (M 23): 127, misapplied; fide Lowe 1949 (Ll 11): 166 & 1957 F.: 37. — *Fomes* Lloyd 1915 (LMW 4, F.): 245 (lectotype from France) (as a "form" of *F. salicinus*: n.v.p.); = *Phellinus densus* (Lloyd) Teng 1964 (lacking Latin descr. & valid ref.: n.v.p.), typonym. — Sensu Overh. = *Fomes* [*Phellinus*] *johnsonianus* (Murrill) Lowe (extra-European), fide Lowe 1957 F.: 37.

M.—*Boletus salicinus* Pers. sensu Fr. 1884 (*Polyporus*). — Fr. 1884 I. 2: 84 *pl. 185 f. 1* (*Polyporus*); P. Karst. 1885 I. 1: 4 / 1888 (ASf 15): 184 *plate f. 5* (*Fomes*); Quél. 1888: 395 (*Phellinus conchatus* var.); Bres. 1897 (AAR III 3): 75; Konr. & M. 1925 I. 5: *pl. 460*, spores too dark

(*Fomes*); Bourd. & G. 1928: 621 *f. 174* (*Phellinus*): cf. Donk 1967 (Pe 5): 109, in obs., notes.

 M.—*Polyporus lonicerae* Weinm. sensu Bond. 1934 (TSR 2): 499 (*Fomes conchatus* f.); fide Bond. 1953: 371 (forma). — Bond. 1953: 371 *pl. 121 f. 3*; 1955 (BMs 10): 191 (*Phellinus conchatus* f.).

contiguus (Pers. per Fr.) Pat. 1900. — *Boletus* Pers. 1801 (Germany) (d.n.); *Polyporus* Fr. 1815 (d.n.); *Polyporus* (Pers.) per Fr. 1821; *Boletus* G. F. Re 1821; *Physisporus* Gillet 1877; *Poria* P. Karst. 1881: *Ochroporus* J. Schroet. 1888; [for *Trametes* Wettst. 1888 see Donk 1971 (PNA 74): 27]; *Mucronoporus* Ell. & Ev. 1889; *Fuscoporia* G. Cunn. 1948. — Pers. 1818: 90 (*Boletus*); Bres. 1897 (AAR III 3): 79 (*Poria*); Bourd. & G. 1928: 624; Donk 1933: 257, 258; Domański 1965 (FpG 2): 222 *pl. 60 f. 1, pl. 63 f. 2* (*Phellinus*); Lowe 1966: 154 *f. 140* (*Poria*); H. Jahn 1967 (WPb 6): 68 *fs. 2a, 3i Abb. 6, 7, 28, 31, 37, 53* (*Phellinus*); Donk 1967 (Pe 5): 87, notes.

 Polyporus cribosus Pers. 1825: 96 (nom. monstr.); fide Donk 1933: 257, 258 & 1967 (Pe 5): 87. — *Poria* Lloyd 1910 (incidental mention: n.v.p.); ≡ *Boletus fuliginosus* Schleich. 1821: 56 (Switzerland) (nom. nud.: n.v.p.), not ∼ (Scop.) per Fr. 1838.

 Polyporus racodioides Pers. 1825: 113 (France); fide Bourd. & G. 1928: 625 & Donk 1967 (Pe 5): 104. — *Poria* Bres. 1897. — Bres. 1897 (AAR III 3): 80 (*Poria*).

 Polyporus cellaris Desm. 1826 P.c.: No. 72 (France); fide Donk 1967 (Pe 5): 85. — Sensu Lib. 1834 P.A.: No. 223 (n.v.) = *Poria versipora* [=*Schizopora paradoxa*], fide Pilát (ACE 3): 458.

 Polyporus suberis Dur. & Mont. apud Mont. 1856: 162 (Algeria): fide Lowe 1963 (M 55): 470. — *Poria* Cooke 1886. — Lowe 1947 (Ll 10): 57 (*Poria*).

 Polyporus floccosus Fr. 1874: 572 (Sweden), not ∼ Jungh. 1838; fide Bres. 1897 (AAR III 3): 80 = *Poria racodioides*; fide Donk 1933: 258 & Lundell 1950 (LNF 37–38): 6 No. 1812. — *Poria* Cooke 1886, Quél. 1886; *Mucronoporus* Sacc. 1891; *Physisporus* Cost. & Duf. 1891; *Phellinus* Bond. & S. 1941. — Sensu Quél. 1892 → *Phellinus ferruginosus*.

 Polyporus holubyanus Velen. 1922: 673 [see Pilát 1948: 256 for Latin translation]: fide Pilát 1942 (ACE 3): 536, 537 & H. Jahn 1967 (WPb 6): 68, 69 (forma).

ferreus (Pers.) Bourd. & G. 1928. — *Polyporus* Pers. 1825 (France): Fr. 1832, not ∼ Berk. 1847: *Poria* Lloyd 1910 (incidental mention: n.v.p.), Bourd. & G. 1925; *Ochroporus* Donk 1933; *Fuscoporia* G. Cunn. 1948. — Romell 1926 (SbT 20): 10 (*Polyporus*); Bourd. & G. 1928: 627 *f. 178* (*Phellinus*); Overh. 1931 (M 23): 117 *pl. 12 f. 5, pl. 13 fs. 10, 12* (*Poria*); Donk 1933: 255 (*Ochroporus*); D. Reid 1958 (TBS 41): 434 *f. 16*; M. P. Christ. 1960 (DbA 19): 310 *f. 307* (*Phellinus*); Lowe 1966: 149 *f. 135* (*Poria*); H. Jahn 1967 (WPb 6): 63 *fs. 2b, 3e, 6, Abb. 2, 29 46, 51* (*Phellinus*); Donk 1967 (Pe 5): 90, notes.

Mucronoporus fulvidus Ell. & Ev. 1894 (PAP): 323 (U.S.A., California); fide Overh. 1931 (M 23): 117. — *Scindalma* O.K. 1898; *Fuscoporia* Murrill 1907; *Poria* Lloyd 1915 (nom. nud.: n.v.p.).

Poria "(or *Fomes*)" *cylindrospora* Lloyd 1917 (LMW 5, L. 65): 8 (U.S.A., Oregon); fide Lowe 1949 (Ll 11): 164. — *Fomitiporia* Murrill 1920.

ferrugineo-fuscus (P. Karst.) Bourd. 1932. — *Poria* P. Karst. 1887 (Finland); *Polyporus* Romell 1898; *Ochroporus* Litsch. 1935. — P. Karst. 1887 I. 2: 13 / 1888 (ASf 16): 527 *pl. 10 f. 56*, "poris nimis magnis"; Egeland 1914 (NMN 52): 162 (*Poria*); Bourd. 1932 (BmF 48): 228; Pilát 1942 (ACE 3): 548 *f. 260, pls. 335, 336*; Kotl. 1965 (ČM 19): 21 *pls. 1, 2*; Domański 1965 (FpG 2): 227 *pl. 63 f. 1* (*Phellinus*); Lowe 1966: 147 *f. 132* (*Poria*); H. Jahn 1967 (WPb 6): 81 *fs. 2h, 3c, 7, Abb. 33, 47* (*Phellinus*).

Polyporus marginellus Peck 1889 (U.S.A., New York); fide Overh. 1942: 56. — *Poria* Sacc. 1891; *Fuscoporia* Murrill 1907. — Overh. 1919 (BNS 205–206): 88 *pl. 9* (*Poria*).

Poria labyrinthica P. Karst. 1891 (H 30): 298 (Finland); fide Bres. 1920 (Am 18): 69 & Lowe 1956 (M 48): 120. — Egeland 1914 (NMN 52): 164.

ferruginosus (Schrad. per Fr.) Pat. 1900. — *Boletus* Schrad. apud J. F. Gmel. 1792, Schrad. 1794 (Germany) (d.n.); *Polyporus* Fr. 1818 (d.n.); *Polyporus* (Schrad.) per Fr. 1821; *Boletus* Spreng. 1827; *Physisporus* Gillet 1878; *Poria* P. Karst. 1881; *Ochroporus* J. Schroet. 1888; *Mucronoporus* Ell. & Ev. 1889; *Fomes* Mass. 1892; *Fuscoporia* Murrill 1907; *Agaricus* E. Krause 1933. — Sensu Fr. in herb., cf. Lundell 1941 (LNF 19–20): 44 No. 61 corrected label; Bres. 1897 (AAR III 3): 78; Overh. 1922 (M 14): 5 *fs. 3, 4, pl. 1 fs. 3, 4* (*Poria*); Konr. & M. 1927 I. 5: *pl. 548* (*Fomes*); Bourd. & G. 1928: 625 *f. 177* (*Phellinus*); Lundell, l.c. (*Polyporus*); Domański 1965 (FpG 2): 223 *pl. 60 f. 1* (*Phellinus*); Lowe 1966: 160 *f. 149* (*Poria*); H. Jahn 1967 (WPb 6): 60 *fs. 2d, e, 3f, Abb. 3–5, 39* (*Phellinus*); Donk 1967 (Pe 5): 90, notes.

Boletus salicinus Pers. apud J. F. Gmel. 1792 (Germany) (d.n.), not ∼ Bull. 1789 (d.n.); fide Donk 1933: 254–256 & 1967 (Pe 5): 108. — *Poria* Pers. 1794 (d.n.); *Polyporus* Fr. 1815 (d.n.); *Polyporus* (Pers.) per Fr. 1821; *Boletus* Nocca & Balb. 1821, not ∼ Bull. per Hook. 1821; *Physisporus* Chev. 1826; *Fomes* Kickx f. 1867; *Phellinus* Quél. 1886; *Ochroporus* J. Schroet. 1888; *Mucronoporus* Ell. & Ev. 1889, misapplied; *Scindalma* O.K. 1898; *Chaetoporus* Romell 1901 (nom. nud.: n.v.p.); *Poria* Lloyd 1910 (incidental mention: n.v.p.); *Placodes* Rick. 1918 (78); *Poria* D. Baxt. 1950 (syn.: n.v.p.); ≡ *Ochroporus confusus* Donk 1933; *Polyporus* Lundell 1934. — Pers. 1801: 543 (*Poria salicina*); Donk 1933: 256 (*Ochroporus confusus*). — Sensu Fr. 1838 → *Phellinus 'trivialis'* = *P. igniarius* ?; sensu Fr. 1884 → *Phellinus conchatus*.

Polyporus umbrinus Fr. 1863 (nom. nud.: n.v.p.), 1874 (Sweden),

not ~ Pers. 1825, not ~ Beeli 1929; fide Bres. 1897 (AAR III 3): 78.
— *Physisporus* Gillet 1877; *Poria* Cooke 1886, Quél. 1886; *Phellinus*
Corb. 1929. — Sensu Bres. in litt. apud Bourd. & G. 1928: 627 (*Phellinus
ferruginosus* subsp.); Donk 1971 (PNA 74): 418, note.

Polyporus macounii Peck 1879 (Canada, Ontario); fide Overh. 1931
(M 23): 127 & 1942: 57. — *Poria* Overh. 1919. — Overh. 1919 (BNS
205–206): 86 *pl. 8 fs. 3–6* (*Poria*) — Cf. also under *Polyporus macowanii*
Kalchbr. (**O**).

M.—*Polyporus floccosus* Fr. sensu Quél. 1892 (*Poria*). — Quél. 1892
Crf 20²): 468 *pl. 2 f. 18* (*Poria*); Bourd. & G. 1928: 626 (*Phellinus ferrugi-
(nosus* subsp.).

hartigii (All. & Schn.) Pat. 1903 (**89**). — *Polyporus* All. & Schn. 1890
(Germany); *Fomes* Bres. 1897; *Phellinus* Imaz. 1943 & Bond. 1953
(preoccupied); *Placodes* H. Huber 1927. — Bres. 1897 (AAR III 3):
74 (*Fomes*); Bourd. & G. 1928: 617 (*Phellinus robustus* f.); Konr. & M.
1932 I. 5: *pl. 463* (*Fomes robustus* f.); K. Lohw. 1937 (Am 35): 339 *f. 1*
(*Fomes*); Pilát 1942 (ACE 3): 506 *pl. 322 f. b, pl. 323* (*Phellinus robustus* f.);
Bond. 1953: 365 *pl. 105 f. 1*; S. Ito 1955: 373 *f. 267*; Jacquiot *1960*,
cult. char.; Domański 1961 (Ffg 7): 209 *f. 6*; H. Jahn 1968 (WPb 7):
24, 36 *fs. 1, 2, 8* (*Phellinus*).

Fomes igniarius var. *pinuum* Bres. 1890 (Rm 12): 105 (Hungary);
fide Bres. 1897 (AAR III 3): 74. — Bres. 1931 (BIm 20): *pl. 993* (*Fomes
robustus* var.); Kravc. 1933: 14 *f. 5* (*Fomes robustus* f.).

M.—*Boletus fulvus* Scop. (**O**) sensu Quél. 1872 (*Polyporus*). — Quél.
1872 (MMb II 5): 280/264; R. Hartig 1878: 40 *pl. 7*; 1882: 83 *pl. 4*
(*Polyporus*; J. Schroet. 1888: 487 (*Ochroporus*).

M.—*Fomes robustus* P. Karst. sensu Lloyd 1924 (LMW 7): 1317 *pl. 312
f. 3032*, in part: "on *Abies*"; fide Lloyd, l.c. (*Fomes hartigii* cited as a
syn.).

igniarius (L. ? per Fr.) Quél. 1886 (**81**). — *Boletus* L. 1753 (d.n.); *Agaricus*
Lam. 1783 (d.n.); *Pyreium* Paul. 1812–35 (generic name n.v.p.), misapplied;
Polyporus (L.?) per Fr. 1821, Opiz 1855; *Boletus* Hook. 1821, not ~
Mérat 1821; *Fomes* Fr. 1849 (nom. altern.); *Placodes* Quél. 1888; *Ochroporus*
J. Schroet. 1888; *Mucronoporus* Ell. & Ev. 1889; *Scindalma* O.K. 1898;
Pyropolyporus Murrill 1903; *Agaricus* E. Krause 1932; [≡ *Boletus acaulis
pulvinatus laevis, poris tenuissimis* L. 1745: 382?]. — Sow. 1798: *pl. 132*
as to plate and corresponding descr. only (*Boletus*); Bourd. & G. 1928:
617 (*Phellinus*); Bres. 1931 (BIm 20): *pl. 994*, poor; Konr. & M. 1934 I. 5:
pl. 465, in part; Overh. 1953: only as to *pl. 76 f. 436*, "typical form of
species" (*Fomes*); H. Jahn 1963 (WPb 4): 96 in part: *f. 6a–c Abb. 61* ?*62*;
Poelt & Jahn 1965: *pl. 56* (*Phellinus*); Donk 1971 (PNA 74): 407. notes.
— Sensu Vahl. Pers. 1799 → *Fomitopsis pinicola*; sensu Bull., in part,
→ *Boletus igniarius* Mérat & *Phellinus robustus*, & → *Phellinus ribis*, &
→ *Phellinus torulosus*; sensu Pers. 1818 → *Fomes fomentarius*; sensu Fr.

1821 & auctt. plur., in part, = *Phellinus pomaceus*; sensu Berk. "in early American work", fide Lloyd 1915 (LMW 4, F.): 246, = *Fomes rimosus* **(0)** [sensu Lloyd = *Phellinus robiniae*]; sensu E. Krause → *Heterobasidion annosum*.

Some very inclusive conceptions are: Romell 1911 (ABS 11³): 14 *f. 23* (*Polyporus*): Lloyd 1915 (LMW 4 F.): 245 (*Fomes*); Pilát 1942 (ACE 3): 508 *fs. 240, 274 pl. 324*; Bond. 1953: 349 *f. 82, pl. 1, pl. 9 f. 2, pl. 83 f. 9, pl. 100 fs. 2, 3, 5, 7, pl. 121 f. 2* (*Phellinus*): Overh. 1953: 60 *pl. 60 f. 359 pl. 73 f. 424 pl. 74 f. 430 pl. 76 fs. 435–438 pl. 104 f. 583 pl. 126 fig.*; Lowe 1957 F.: 56 *f. 40* (*Fomes*).

? [*Agaricum durum, crassum* . . . Mich. 1729: 121 *pl. 62* (Italy). — = *Scindalma laminis tenuioribus* Hill 1751;] = *Fomes igniarius* var. *resupinatus* Sacc. apud All. & Schn. 1874. — Presumably cushion-shaped, resupinate fruitbodies of *Phellinus igniarius* or *P. robustus*, rather than *P. laevigatus*.

Boletus obtusus Pers. 1795 (ABU 15): 24 / 1796 O. 1: 24 ["*Boletus fulvus* (potius *obtusus*)", altern. name] (d.n.), in part: var. *salignus* (Germany). — *Boletus* Pers. per St-Am. 1821; *Polyporus* E. Krause 1931; *Agaricus* E. Krause 1932; *Phellinus* E. Krause 1934. — Pers. 1799 O. 2: 4; Donk 1971 (PNA 74): 416, notes.

Fomes trivialis Bres. 1931; fide H. Jahn in litt. — *Fomes* Bres. apud P. Magn. 1905 (syn.: n.v.p.); *Phellinus* Kreisel 1961 (incomplete ref.: n.v.p.), 1964; [= *Polyporus nigricans* "forma . . . trivialis" Fr. 1874: 558, at least in part (Sweden) (unnamed form)]; *Fomes igniarius* var. *trivialis* Bres. ex Killerm. 1928. — Bres. 1931 (BIm 20): *pl. 995* (*Fomes*); Pilát 1942 (ACE 3): 510 *f. 241, pl. 327, pl. 328 f. a*, in part (*Phellinus igniarius* subsp.); Kreisel 1961: 138 *fs. 87, 88*; H. Jahn 1963 (WPb 4): 98 *f. 6k–o, Abb. 63* (*Phellinus trivialis* f. *salicum*); Donk 1971 (PNA 74): 416, notes.

Polyporus laccatus Velen. 1922: 678 [see Pilát 1948: 257 for Latin translation] (Czechoslovakia), not ~ (Timm) per Pers. 1825, not ~ Kalchbr. apud Wettst. 1885; fide Pilát 1942 (ACE 3): 510 = *Phellinus igniarius* subsp. *trivialis*.

? *Polyporus foeniculaceus* Velen. 1922: 678 [see Pilát 1948: 257 for Latin translation] (Czechoslovakia). — Pilát 1942 (ACE 3): 508 referred this to *Phellinus igniarius* subsp. *igniarius*. — Type on *Alnus*.

Phellinus trivialis var. *salicum* H. Jahn 1963 (Europe) (lacking Latin description: n.v.p.): fide H. Jahn in litt. — H. Jahn 1963 (WPb 4): 98 *f. 6k–o, Abb. 63*.

? M.—*Boletus salicinus* Pers. sensu Fr. 1838: 467 (*Polyporus*); cf. Donk 1967 (Pe 5): 109. — Quél. 1872 (MMb II 5): 180 *pl. 17 f. 6* (*Polyporus*)?

M.—*Polyporus nigricans* Fr. — (i) Sensu Fr. 1874: 558–559, in part: as to refs. cited under "forma . . . trivialis" but restricted to specimens from *Salix*, — Rostk. 1838 (StP 4): 105 *pl. 51*; Quél. 1872 (MMb II 5):

279 *pl. 19 f. 3*, cf. Donk 1971 (PNA 74): 417. — (ii) Sensu auctt. nonn. — Boud. 1904–11: 79 *pl. 155* fide Donk 1971 (PNA 74): 414.

laevigatus (Fr.) Bourd. & G. 1928 (82). — *Polyporus* Fr. 1874 (Sweden), not ∼ (Pers.) Duby 1830; *Poria* P. Karst. 1881; *Physisporus* Cost. & Duf. 1895; *Agaricus* E. Krause 1932; *Fuscoporia* G. Cunn. 1948. – Bres. 1897 (AAR III 3): 79 (*Poria*); Romell 1911 (ABS 11[3]): 16 (*Polyporus*); Pilát 1942 (ACE 3): 538 *f. 254, pl. 329* (*Phellinus*); Lundell 1953 (LNF 43–44): 1 No. 2101, in obs. (*Polyporus*); Domański 1956 (FpG 2): 225 *pl. 63* (*Phellinus*); Lowe 1966: 156 *f. 143* (*Poria*); H. Jahn 1967 (WPb 6): 86 *fs. 2f, 3g, 8, Abb. 26 27 36, 38, 48*; Niemelä 1972 (Abf 9): 42 *fs. 1–7* (*Phellinus*). — Sensu Bourd. & G. → *Phellinus rhamni*.

Some inclusive conceptions are: Overh. 1935: 7, 12 (*Fomes igniarius* var.); 1942: 58; Lowe 1946: 82 (*Poria*); D. Baxt. 1952 (PMi 37): 97 *pl. 4 f. 1*; Overh. 1953: 63 *pl. 70 f. 409* (*Fomes igniarius* var.).

Fuscoporella ludoviciana Murrill 1907 (NAF 9): 6 (U S A , Louisiana); fide Niemelä 1972 (Abf 9): 42, & Donk (type studied). — Niemelä, op. cit., p. 51.

Fomitiporella betulina Murrill 1907 (NAF 9): 12 (U.S.A., Maine); fide Niemelä 1972 (Abf 9): 42 and Donk (type studied). — *Poria* Sacc. & Trott. 1912. — D. Baxt. 1934 (PMi 19): *pl. 49 f. 1*, photograph of type; Niemelä, op. cit., p. 50.

lonicerinus (Bond.) Bond. & S. 1941. — *Fomes* Bond. 1935 (U.S.S.R., European Russia): *Cryptoderma* Imaz. 1943; *Phellinus* Pilát 1942 (preoccupied). — Bond. 1935 (TSR 2): 501 *f. 4* [repr. Pilát 1942 (ACE 3): 535] (*Fomes*); 1953: 43, 372 *fs. 86, 87* (*Phellinus*); S. Ito 1955: 366 *f. 260?* (*Cryptoderma*); Švarcm. 1964: 431 *f. 187*; Domański & al. 1967 (FpG 3): 291 (*Phellinus*).

lundellii Niemelä 1972 (Finland). — Niemelä 1972 (Abf 9): 51 *fs. 8–11, 13*.

Phellinus nigricans var. *subresupinatus* (Lundell) H. Jahn 1967 (WPb 6): 92 *Abb. 25* (lacking Latin descr.: n.v.p.); fide Niemelä 1972 (Abf 9): 51, "pro p. max." — *Polyporus igniarius* f. *subresupinatus* Lundell in herb. [& *P. igniarius* "subresupinate form" Lundell 1953 (LNF 43–44): 1 No. 2101 (Sweden)]

nigricans (Fr) P. Karst. 1899 (81). — *Polyporus* Fr. 1821: 375 (Sweden), not ∼ Lasch 1859; *Boletus* Spreng. 1827, not ∼ Schum. 1803 (d.n.); *Fomes* Fr. 1849 (nom. altern.); *Placodes* Quél. 1886; *Mucronoporus* Ell. & Ev. 1889; *Scindalma* O.K. 1898; *Pseudofomes* Lázaro 1916; *Ganoderma* C. & O. Over. 1922, misapplied. — Sensu lato: Lloyd 1908 (LMW 2): 372 *f. 193*, in part: "First, . . ." (*Fomes*); Bourd. & G. 1928: 618 *f. 172* (*Phellinus igniarius* subsp.). — H. Jahn 1967 (WPb 6): 93, in obs. (*Phellinus*); Donk 1971 (PNA 74): 412, notes. — Sensu Fr. 1838, in part, Bres. → *Fomes fomentarius*; sensu Fr. 1874, in part, Rostk., Quél. (specimens on *Salix*) → *Phellinus 'trivialis' = P. igniarius?*; sensu auctt.

nonn. → *Phellinus igniarius*; sensu Lindr. "abnorme Fruchtkörper" → *Inonotus obliquus.*

Fomes igniarius f. *betulae* Bond. 1912 (U.S.S.R., European Russia). — *Phellinus igniarius* f. Bond. 1953. — Bond. 1912: 21 *fs. 4, 5, pl. 1 f. 8*; 1935 (TSR 2): 493.

Phellinus trivialis f. *betularum* H. Jahn 1963 (Sweden) (lacking Latin descr.: n.v.p.); fide H. Jahn 1967 (WPb 6): 92. — H. Jahn 1963: 100 *f. 6p, Abb. 64, 65.*

nigrolimitatus (Romell) Bourd. & G. 1925. — *Polyporus* Romell 1911 (Sweden); *Fomes* Egeland 1913; *Cryptoderma* Imaz. & Aosh. 1952 (incomplete ref.: n.v.p.); *Poria* Killerm. 1943 (syn.: n.v.p.). — Romell 1912 (SbT 6): 641 (*Polyporus*); Bourd. & G. 1928: 622; Murašk. 1939: 91 (*Phellinus*); Overh. 1953: 74 *pl. 68 fs. 399–401, pl. 113 f. 620, pl. 126 fig.*; Lowe 1957 F.: 53 *f. 38* (*Fomes*); Parm. 1959 (TSR 12): 249 *f. 22*; Domański 1965 (FpG 2): 218 *pl. 62 fs. 1, 2* (*Phellinus*); Lowe 1966: 152 *f. 138* (*Fomes*); Kotlaba *1972* (*Phellinus*). — Sensu G. Cunn. = "*Fuscoporia*" *kamahi* G. Cunn. (extra-European), fide G. Cunn. 1965: 214.

Fomes putearius Weir 1914 (U.S.A., Idaho); fide Hubert 1924 (JaR 29): 528 & Overh. 1953: 74, 76. — Weir 1914 (JaR 2): 163 *pl. 9*, wrong spores, fide Overh. 1931 (M 23): 127.

Xanthochrous suberosomollis Pilát 1937 (U.S.S.R., Russia, Siberia); fide Pilát 1942 (ACE 3): 544, 545 (forma). — Pilát 1937 (BmF 52): 317 *f. 24, pl. 6 f. 4.*

M.—*Boletus spongiosus* Pers. (**O**) sensu Romell 1907 (*Polyporus*); fide Romell 1911 (ABS 11³): 18 = *Polyporus nigrolimitatus.* — Romell 1907: 266; Vleugel 1908 (SbT 2): 309 (*Polyporus*).

pilátii Černý 1968 (Czechoslovakia). — Černý 1968 (ČM 22): 2 *fs. 1–11*; Kotl. 1968 (ČM 22): 179, in obs., *fs. 3, 4.*

pini (Brot. per Fr.) A. Ames 1913, not ∼ (Thore per Pers.) Pilát 1942 (*83, 84*). — *Boletus* Brot. 1804 (Portugal) (d.n.), not ∼ Thore 1803; *Daedalea* (Brot.) per Fr. 1821; *Trametes* Fr. 1838, not ∼ Thüm. 1876, not ∼ (Thore per Pers.) Britz. 1887; *Polyporus* Kumm. 1871, not ∼ (Thore) per Pers. 1825, not ∼ Rostk. 1838; *Fomes* P. Karst. 1881; *Xanthochrous* Pat. 1897; *Inonotus* Maire apud Maire & Wern. 1938; *Phellinus* Bond. & S. 1941 (preoccupied); *Cryptoderma* Imaz. 1943, 1951. — In the following no distinction is made between 'pini Brot.' and 'pini Thore'. Most of the authors cited include *Phellinus abietinus* [= *P. chrysoloma*] in their conception; American authors prefer an even wider circumscription of this sp. — Fr. 1830 (Li 5): 514 (*Daedalea*); R. Hartig 1882: 8 *f. 35*; Bres. 1891 (BSb 9): 31; Boud. 1904–11: 82 *pl. 161* (*Trametes*); Lloyd 1915 (LMW 4, F.): 275 *fs. 608, 609*; Overh. 1915 (AMo 2): 723 (*Fomes*); Bourd. & G. 1928: 632 (*Xanthochrous*); Shope 1931 (AMo 18): 379 *pl. 33 fs. 2, 3* (*Fomes*); Jørst. & Juul 1939 (MnS 6): 362, 477 *fs. 19–27*

(*Polyporus*); Imaz. 1951 (FPJ 4): 175?, always on *Larix* (*Cryptoderma*); Overh. 1953: 76 *pl. 66 f. 392, pl. 77 fs. 393–395, pl. 126 fig.*; Kawam. I. 2: 222 *f. 211?* (*Trametes*); Lowe 1959 F. : 47 *f.31* (*Fomes*). Sensu Fuc → *Fomitopsis pinicola*.

Boletus pini Thore 1803 [repr. 1938 (TBS 22): 183] (France) (d.n.), not ∼ Brot. 1804; fide Fr. 1828 E. 1: 68. — *Polyporus* (Thore) per Pers. 1825, not ∼ Rostk. 1838, not ∼ Kumm. 1871; *Boletus* Duby 1830; *Trametes* Britz. 1887, misapplied, not ∼ (Brot. per Fr.) Fr. 1838, not ∼ Thüm. 1876; *Ochroporus* J. Schroet. 1888; *Porodaedalea* Murrill 1905; *Phellinus* Pilát 1942, not ∼ (Brot. per Fr.) A. Ames 1913. — For references to descrs. and illustrations, see above, under 'pini Brot.' — Sensu Britz. 1887 → *Gloeophyllum odoratum*.

Polyporus pinicola var. b [u n n a m e d] Kl. ex Berk. 1836: 43; fide Haddow 1938 (TBS 22): 183 = *Trametes pini* "(Thore) Fries".

Polyporus pini Rostk. 1838 (StP 4): 103 *pl. 50* (Germany/Poland), not ∼ (Thore) per Pers. 1825, not ∼ (Brot. per Fr.) Kumm. 1871: fide Fr. 1874: 582.

Daedaloides pinicola Lázaro 1916 (RMa 14): 675 / 1917: 114 (Spain); fide Bres. apud Trott. 1925 (SF 23): 448: "est e diagn. *Trametes Pini*, sed sporis si genuina (?) diversa." — *Daedalea* Sacc. & Trott. apud **Trott.** 1925.

pomaceus (Pers. per S. F. Gray) Maire 1933 (85). — *Boletus* Pers. 1799 (Germany) (d.n.) per S. F. Gray 1821; *Polyporus* Pers. 1825, not ∼ Velen. 1922; *Placodes* Bourd. 1894; *Fomes* Lloyd 1908; *Ochroporus* Donk 1933. — Pers. 1799 O. 2: 5 (*Boletus*); Lloyd 1908 (LMW 2, L. 23): 8; 1910 (LMW 3): 469; 1915 (LMW 4, F.): 241 *f. 588* (*Fomes*); Konr. & M. 1927 I. 5: *pl. 459* (*Fomes pomaceus* f. *prunastri*); Pilát 1942 (ACE 3): 512 *f. 242, pls. 325, 326* (*Phellinus igniarius* subsp.); D. Baxt. 1952 (PMi 37): 97 *pl. 4, pl. 5 f. 2*; Overh. 1953: 64 *pl. 74 fs. 427–429, pl. 126 fig.*; Lowe 1957 F.: 57 *f. 41* (*Fomes*); Pegl. & Wat. 1968 (CDp): no. 196 *figs.* (*Phellinus*).

[*Agaricum tuberosum, durum, Prunis, aut Malis adnascens* ... Mich. 1729: 119 *pl. 61 f.* "Ordo II" (Italy).]

Boletus tuberculosus Baumg. 1790: 635 (d.n.), not ∼ (Pers.) Pers. 1801 (d.n.). — [≡ XII. *Boletus*; ... e. *Boletus*; *tuberculosus, durus et sessilis* ... Gled. 1753: 80 (Germany)].

Boletus scutiformis Tratt. 1804: 49 *pl. 5 f. 9* (Austria) (d.n.); fide Pers. 1825: 85 (var.).

Hemidiscia prunorum Lázaro 1916 (RMa 14): 581 & 1917 (RMa 15): 377 / 1917: 82, 291 (Spain); fide Bres. apud Trott. 1925 (SF 23): 390, "est *Fomes fulvus* [sensu Bres.] f. annua." — *Fomes* Sacc. & Trott. apud Trott. 1925.

Pseudofomes prunicola Lázaro 1916 (RMa 14): 585 & 1917 (RMa 15): 378 / 1917: 87, 292 (Spain); fide Bres. apud Trott. 1925 (SF 23): 389 =

"synon. *Fom. fulvi* [sensu Bres.] vetusti." — *Fomes* Sacc. & Trott. apud Trott. 1925, not ∼ Lázaro 1916.

Fomes prunicola Lázaro 1916 (RMa 14): 663 / 1917: 102 (Spain), not ∼ (Lázaro) Sacc. & Trott. apud Trott. 1925. — = *Fomes lazaroi* Sacc. & Trott. apud Trott. 1925.

Boudiera scalaria Lázaro 1916 (Spain). — *Fomes* Sacc. & Trott. apud Trott. 1925; *Lazaroa* Gonz. 1917. — Lázaro 1916 (RMa 14): 838 & 1917 (RMa 15): *pl. 7* / 1917: 150 *pl. 7*.

Polyporus sorbi Velen. 1922: 687 [see Pilát 1948: 260 for Latin translation] (Czechoslovakia); fide Kotl. 1966 (ČM 20): 185, 187. — *Phellinus* Pilát 1942. — Pilát 1942 (ACE 3): 516 *f. 244, pl. 328 f. b*, type.

M.—*Boletus fulvus* Scop. (O) sensu Pers. 1796 O. 1: 24, excl. of var. *saligna*; Bres. 1891 (*Fomes*). — Bres. 1891 (BSb 9): 30; 1897 (AAR III 3): 75, in part; 1911 (Am 9): 426; 1931 (BIm 20): *pl. 977* (*Fomes*): Murrill 1903 (BTC 30): 112 (*Pyropolyporus*); Overh. 1915 (WUS 3¹): 61 *pl. 7 f. 35* (*Fomes*); Bourd. & G. 1928: 619 (*Phellinus igniarius* subsp.); Shope 1931 (AMo 18): 383 *pl. 34 fs. 4–6* (*Fomes*); Donk 1971 (PNA 74): 419, notes.

M.—*Boletus ungulatus* Schaeff. sensu Secr. 1833, in part (*Polyporus*). — Secr. 1833 M. 3: 79, as to vars. C, D only.

pouzari Kotl. 1968 (Czechoslovakia). — Kotl. 1968 (ČM 22): 24 *fig. at left, fs. 1, 2*.

punctatus (Fr.) Pilát 1942 (86, 87). — *Polyporus* Fr. 1874 (Norway), not ∼ Jungh. 1838; *Poria* P. Karst. 1882; *Fuscoporia* G. Cunn. 1948; *Fomitiporia* Murrill 1948. — Egeland 1914 (NMN 52): 160; Overh. 1931 (M. 23): 119 *pl. 12 f. 3, pl. 14 fs. 15, 17* (*Poria*); Lundell 1936 (LNF 5–6): 21 No. 244 (*Polyporus*); Overh. 1942: 63 (*Poria*); Pilát 1942 (ACE 3): 530 *f. 251, pl. 330* (*Phellinus*); D. Baxt. 1952 (PMi 37): 99 *pl. 7 f. 3* (*Poria*); M. P. Christ. 1960 (DbA 19): 312 *f. 310*; Domański 1965 (FpG 2): 221 *pl. 61 fs. 2, 3*; H. Jahn 1966 (WPb 6): 56 *fs. 3k, 4, 5, Abb. 1, 41–43* (*Phellinus*); Lowe 1966: 163 *f. 152*, in part (*Poria*).

Fomitiporia obliquiformis Murrill 1907 (NAF 9): 9 (U.S.A., Ohio); fide Overh. 1931 (M. 23): 119, 121. — *Poria* Sacc. & Trott. 1912.

Fomitiporia laminata Murrill 1907 (NAF 9): 11 (U.S.A., Maine); fide Bres. 1926: 80 = *Poria friesiana*. — *Poria* Sacc. & Trott. 1912.

Poria friesiana Bres. 1908 (France); fide Egeland 1914 (NMN 52): 160, 161. — *Phellinus* Bourd. & G. 1928. — Bres. 1908 (Am 6): 40 (*Poria*); Bourd. & G. 1928: 623 *f. 175* (*Phellinus*).

Poria viticola Lázaro 1917 (RMa 15): 370 / 1917: 284 (Spain) (87), not ∼ (Schw.) Cooke 1886.

Fomes igniarius var. *viticidus* Gard 1922 (BPv 9): 25 *fs. 1, 2* (France) (87).

rhamni (M. Bond.) H. Jahn 1967. — *Phellinus laevigatus* f. M. Bond. apud Sin. & M. Bond. 1960 (U.S.S.R.). — Sin. & M. Bond. 1960 (BMs 13): 230 *f. 6* (*Phellinus laevigatus* f.); H. Jahn 1967 (WPb 6): 89 *fs. 2g, 3d, Abb. 20–22, 24*; Niemelä 1972 (Abf 9): 57 (*Phellinus*).

M.—*Polyporus laevigatus* Fr. sensu Bourd. & G. 1925 (*Poria*); fide H. Jahn 1967 (WPb 6): 89. — Bourd. & G. 1928: 624 *f. 176*.

ribis (Schum. per Fr.) P. Karst. 1889. — *Boletus* Schum. 1803 ("*ribi*") (Denmark) (d.n.), not ∼ Pers. ex DC. 1815 (d.n.); *Polyporus* Fr. 1815 (d.n.); *Polyporus* (Schum.) per Fr. 1821; *Boletus* Spreng. 1827, not ∼ Pers. ex DC. per Mérat 1821; *Trametes* Fr. 1848 (nom. nud.: n.v.p.), 1849; *Fomes* Gillet 1878; P. Karst. 1882; *Ochroporus* J. Schroet. 1888; *Scindalma* O.K. 1898; *Chaetoporus* Romell 1901; *Pyropolyporus* Murrill 1903; *Fulvifomes* Murrill 1914; *Placodes* Rick. 1918; *Xanthochrous* Bourd. & G. 1925, not ∼ (Pers. ex DC. per Mérat) Pat. 1900; *Inonotus* Maire & Wern. 1938; *Phellinus* Bond. & S. 1941 (preoccupied); *Cryptoderma* Imaz. 1943. — Hornem. 1823 (Fd 10 / F. 30): 12 *pl. 1790 f. 2*, leg. Schum.!, perhaps type; Secr. 1833 M. 3: 96 (*Polyporus*); Bres. 1897 (AAR III 3): 75); Lloyd 1915 (LMW 4, F.): 252 *f. 594* (*Fomes*); Bourd. & G. 1928: 638 (*Xanthochrous*); Konr. & M. 1928 I. 5: *pl. 453* (*Fomes*); Pilát 1942 (ACE 3): 527 *f. 250, pl. 345, pl. 346 f. a, pls. 347, 348* (*Phellinus*): Overh. 1953: 95 *pl. 66 fs. 388–390, pl. 126 fig.*; Lowe 1957 F.: 19 *f. 3* (*Fomes*).

Agaricus ribis Dubois 1803: 178 (France), not *Boletus ribis* Schum. 1803 (d.n.), not *B. ribis* Pers. ex DC. 1815 (d.n.); fide Fr. 1821: 375. — *Agaricus* Dubois per Dubois 1833.

Boletus ribis Pers. ("ined.") ex DC. 1815: 41 (France) (d.n.), not ∼ Schum. 1803 (d.n.); fide Fr. 1828 E. 1: 110. — *Boletus* Pers. ex DC. per Mérat 1821, not ∼ (Schum. per Fr.) Spreng. 1827; *Xanthochrous* Pat. 1900, not ∼ (Schum. per Fr.) Bourd. & G. 1925; ≡ (by lectotypification) *Polyporus ribesius* Pers. 1825, fide Lloyd 1910 (LMW 3): 470 & Donk 1933: 253; *Boletus* Konr. & M. 1928 (syn.: n.v.p.).

Polyporus lonicerae Weinm. 1826: Fr. 1828; fide Quél. 1876 (BbF 23): 149 (var.), Bourd. & G. 1925 (BmF 41): 205 (forma). — *Trametes* Fr. 1849; *Fomes* Gillet 1877; *Scindalma* O.K. 1898. — Weinm. 1826 (SPR 2): 102; Fr. 1828 E. 1: 110; Mont. 1836 (ASn II 5): 341 *pl. 12 f. 6* ? (*Polyporus*). — Sensu Bond. → *Phellinus conchatus*.

Polyporus evonymi Kalchbr. 1868 [repr. 1869 (H 8): 116], not *Trametes evonymi* Fuck. 1870; fide Quél. 1876 (BbF 23): 149 (var.), Bres. 1897 (AAR III 3): 75. — *Fomes* Gillet 1877; *Phellinus* Bourd. 1894; *Xantho-chrous* Pat. 1900; *Polystictus* Lloyd 1907 ("*euonymus*"; nom. nud.: n.v.p.). — Kalchbr. 1868 (MtK 5): 261 *pl. 2 f. 4*; 1877: 55 *pl. 35 f. 3* (*Polyporus*). — V.s.: "*euonymus*".

Trametes evonymi Fuck. 1870 (Jna 23–24): 21 (Germany), not *Polyporus evonymi* Kalchbr. 1868.

Polyporus cytisi Britz. 1887 (BnS 29): 278 [*pl. 606 f. 51*] & 1909 (BbC II 26): 212 (Germany); fide Killerm. 1922 (Dba 15): 81. — *Fomes* Sacc. & Trav. 1910.

Phellinus versatilis Quél. 1890 (Crf 18²): 512 (78); fide Bres. apud Quél., l.c., & 1897 (AAR III 3): 75. — [≡ (by lectotypification) *Polyporus conchatus* (Pers.) per Fr. sensu Quél. 1872 (MMb II 5): 280/264 *pl. 17 f. 5* (France) & Cooke & Q. 1878: 181; ≡ *Polyporus pectinatus* Kl. sensu Fr. 1874: 559, in part: European specimens].

Pyropolyporus langloisii Murrill 1903 (BTC 30): 118 (U.S.A., Louisiana); fide Lloyd 1914 (LMW 4, F.): 282 & Lowe 1957 F.: 19. — *Fomes* Sacc. & D. Sacc. 1905.

Fomes jasmini (Quél.) Lloyd 1915; fide Bres. 1908 (Am 6): 38 (forma). — *Phellinus pectinatus* var. Quél. 1892 (France): *Fomes* Lloyd 1908 (nom. nud.: n.v.p.); *Phellinus* Voronich. 1925 (incidental mention: n.v.p.). — Quél. 1892 (Crf 20²): 469 *pl. 3 f.* "*33*" [= *34*] (*Phellinus pectinatus* var.); Bres. 1908 (Am 6): 38 (*Fomes pectinatus* f.); Lloyd 1915 (LMW 4, F.): 254 *f. 596* (*Fomes*); Bourd. & G. 1928: 640 (*Xanthochrous ribis* f.).

Polyporus ulicis Boud. 1917 (France): fide Bourd. & G. 1925 (BmF 41): 205 (forma). — Boud. 1917 (BmF 33): 10 *pl. 3 f. 1*; Farinha & Rosado 1952 (Bsb II 26): 193 *fs. 1–43, pls. 1, 2* (*Polyporus*).

Fomes ephedrae Voronich. 1925 (Am 23): 300 (U.S.S.R., Georgia); fide Bres., determination of type (forma).

Polyporus pedatus Velen. 1925 (MP 2): 98 (Czechoslovakia); fide Pilát 1942 (ACE 3): 528.

M.—*Boletus igniarius* L. sensu Bull., in part: 1789: *pl. 454 f. E*: fide Pers. 1818: 94 = *Boletus ribis* Pers. ex DC. [= *Phellinus ribis* (Schum. per Fr.) P. Karst.].

M.—*Boletus conchatus* Pers. sensu Quél. 1872 (*Polyporus*). — Quél. 1872 (MMb II 5): 280/264 *pl. 17 f. 5* (*Polyporus*): Bres. 1890 (Rm 12): 105 (*Fomes*), fide Bres. 1897 (AAR III 3): 75 = *Fomes ribis* (Schum. per Fr.) Gillet.

M.—*Polyporus pectinatus* Kl. (**O**) sensu Fr. 1874: 559, in part: European specimens. — Quél. 1888: 395 (*Phellinus*). → *Phellinus versatilis* Quél.

robiniae (Murrill) A. Ames 1913 (**88**). — *Pyropolyporus* Murrill 1903 (U.S.A., Virginia): *Fomes* Sacc. & D. Sacc. 1905; *Fulvifomes* Murrill 1914. — Murrill 1908 (NAF 9): 105 (*Pyropolyporus*): Lowe 1957 F.: 22 *f. 6* (*Fomes*).

Xanthochrous tuniseus Pat. 1897 (Tunisia) (**88**); fide Lloyd 1915 (LMW 4, F.): 286 = *Fomes rimosus* [sensu Lloyd]. — *Fomes* Sacc. & Syd. 1899. — Pat. 1897 (BmF 13): 200 *pl. 13 f. 1*.

Pyropolyporus cedrelae Murrill 1908 (Jamaica); fide Lowe 1957 F.: 22. — Murrill 1908 (NAF 9): 105.

M.—*Polyporus rimosus* Berk. (**O**) sensu Cooke apud Morg. 1885 (JCi 8): 104, in obs. (**88**); fide Lowe 1957 F.: 22 ("of many authors"). — Schrenk 1901 (RMo 12): 27 *pls. 1, 2* (*Polyporus*); Hard 1908: 418 *f. 347*; Overh. 1914 (AMo 1): 133; Lloyd 1915 (LMW 4, F.): 248, in part (*Fomes*): Pilát 1942 (ACE 3): 526 *f. 249* (*Phellinus*); Overh. 1953: 96 *pl. 95 fs. 384–386* (*Fomes*); Bond. 1953: 397 *fs. 94, 95, pl. 183 f. 3*; Malenç. 1956 (BmF 71): 289 *f. 7, tpl. 5* (*Phellinus*).

robustus (P. Karst.) Bourd. & G. 1925 (**89**). — *Fomes* P. Karst. 1889 (Finland): *Ochroporus* J. Schroet. 1888: *Scindalma* O.K. 1898; *Phellinus* P. Karst. apud Sacc. 1891 (syn.: n.v.p.); *Polyporus* Lundell 1936. — Lloyd 1915 (LMW 4. F.): 242 *f. 589* (*Fomes*); Bourd. & G. 616 (*Phellinus*); Konr. & M. 1932 I. 5: *pl. 462* (*Fomes*; f. *fuscescens* Bourd. apud Konr. & M.); Donk 1933: 248 (*Ochroporus*); R. W. Davids. 1942 (TUS 785): 20 *f. 3J, pl. 1 f. I*, cult. char.; D. Baxt. 1952 (PMi 37): 102, 103 *pl. 14, pl. 15 f. 1*; Overh. 1953: 87 *pl. 61 f. 362, pl. 63 fs. 371–373, pl. 123 f. 669, pl. 126 fig.*; Lowe 1954 (M 46): 494 *f. 1a–c*, 1957 F.: 54 *f. 39*, in part (*Fomes*); H. Jahn 1967 (WPb 6): 94 *f. 10*, resupinate forms (*Phellinus*). — Some of the cited descriptions are rather inclusive. — Sensu Lloyd, in part, → *Phellinus hartigii*; sensu G. Cunn., to be excluded.

Boletus igniarius Mérat 1821: 41, in part, not ~ L. 1753 (d.n.) & (L. per Fr.) Hook. 1821. — [≡ *Boletus igniarius* L. sensu Bull. 1791 H.: 361 [*pls. 82, 454*] (France), in part]. — Bull. 1781: *pl. 82*, at least as to depicted fruitbodies; 1789: *pl. 454 fs. B* ?, *D, F* (*Boletus igniarius*): Donk 1971 (PNA 74): 410, notes.

Trametes inaequalis P. Karst. 1890 (H 29): 177 (Finland); fide Lowe 1956: 122.

Polyporus buxi Branke 1894 (LŽS 24): 473 (n.v.) (U.S.S.R., Caucasia); fide Bond. 1934 (TSR 2): 497 (with "?") = *Fomes robustus* var. *buxi* (Bourd. & G.) Bond. ≡ *Phellinus robustus* var. *buxi* Bourd. & G. 1928: 617. — *Fomes* Bond. 1935 (syn.: n.v.p.).

Pyropolyporus calkinsii Murrill 1903 (U.S.A., Florida); fide Bres. 1926: 79 & Lowe 1957 F.: 55. — *Fomes* Sacc. & D. Sacc. 1905. — Murrill 1903 (BTC 30): 113; 1908 (NAF 9): 105 (*Pyropolyporus*).

Pyropolyporus crustosus Murrill 1903 (BTC 30): 113 (Jamaica): fide Lowe 1954 (M 46): 491.

Pyropolyporus bakeri Murrill 1908 (U.S.A., Wisconsin); fide Bres. 1916 (Am 14): 233, Overh. 1953: 87, & Lowe 1957 F.: 55. — *Fomes* Sacc. & Trott. 1912; *Phellinus* A. Ames 1913. — Neuman 1914: 79 *pl. 6 f. 27a, pl. 7 f. 27 b–d*; Overh. 1915 (WUS 3^1): 60 *pl. 6 f. 31*; 1920 (M 12): 136 *pl. 9*; 1922 (BTC 49): 172 (*Fomes*). — On *Betula* spp.

Pyropolyporus abramsianus Murrill 1915 W.P.: 26 (U.S.A., California); fide Overh. 1953: 87, 88 & Lowe 1957 F.: 55. — *Fomes* Sacc. & Trott. apud Trott. 1925. — Type on willow stump.

Polyporus aestriplex E. Krause 1931 B.r.: 122 (Germany); fide E. Krause 1933 B.r.: 164 (f. *fuscescens*) & Pilát 1942 (ACE 3): 504. — *Agaricus* E. Krause 1932; *Fomes* Pilát 1933 (syn.: n.v.p.); *Phellinus* E. Krause 1934.

M.—*Boletus igniarius* L. sensu Bull. 1781: *pl. 82*, at least as to depicted fruitbodies; 1789: *pl. 454 fs. B?, D, F*. — Gillet 1874–90 P.: *pl. 290/468* (*Fomes*); Quél. 1888: 399 (*Placodes*). → *Boletus igniarius* Mérat.

M.—*Polyporus roburneus* Fr. (**0**) sensu Quél. as to determinations; fide Bourd. & G. 1928: 618. — Quél. 1888: 400? (*Placodes*). — Cf. (**48**).

torulosus (Pers. per Pers.) Bourd. & G. 1925. — *Boletus* Pers. 1818 (France) (d.n.); *Polyporus* (Pers.) per Pers. 1825; *Boletus* Rox. Clem. 1864 (syn.: n.v.p.); *Fomes* Lloyd 1910. — Lloyd 1910 (LMW 3): 470; 1915 (LMW 4, F.): 243 (*Fomes*); Bourd. & G. 1928: 619 (*Phellinus*); Konr. & M. 1930 I. 5: *pl. 461* (*Fomes*); Wakef. 1952 (TBS 35): 35 (*Phellinus*). — American descrs. doubtful: Overh. 1953: 71 *pl. 120 f. 657, pl. 127 fig.*; Lowe 1954 (M 46): 496; 1957 F.: 90 (*Fomes*).

Agarico-igniarium tegularium Paul. 1793 T. 2: 87 (descr.), Ind. (France) (d.n.): fide Lév. 1855: 5. — I would consider the citation in the original descr. of Paul. 1812–35 *pl. 7 f. 1 | pl. 6 f. 1* (as *Pyreium igniarium*) an error.

Polyporus rubriporus Quél. 1881 (France): fide Lloyd 1910 (LMW 3): 470. — *Phellinus* Quél. 1886; *Fomes* Bres. & Cav. 1900, Sacc. 1916; *Boudiera* Lázaro 1916; *Lazaroa* Gonz. 1917. — Quél. 1881 (Crf. 9^2): 669 (*Polyporus*); 1888: 394 (*Phellinus*).

Polyporus fuscopurpureus Boud. 1881 (France), not ∼ Pers. 1827; fide Quél. 1888: 394 = *Phellinus rubriporus*; fide Lloyd 1910 (LMW 3): 470. — *Fomes* Cooke 1885; *Scindalma* O.K. 1898. — Boud. 1881 (BbF 28): 92 *pl. 2 f. 3*; 1904–11: 80 *pl. 156*.

Fomes castaneae Voronich. 1925 (Am 23): 298 (U.S.S.R., Georgia): fide Bond. 1953: 377. — Described with wrong spores?

? *Polyporus assimilis* Velen. 1922: 677 [see Pilát 1948: 257 for Latin translation] (Czechoslovakia); fide Pilát 1942 (ACE 3): 501 (with "?").

M.—*Boletus igniarius* L. sensu Bull., in part, 1789: *pl. 454 f. C*; fide Pers. 1818: 94 & Lév. 1855: 5. — Paul. 1812–35: *pl. 7 f. 1* (*Pyreium*; for corresponding descr., see Paul. 1793 T. 2: 87, "Le Roux-plat en toit").

M.—*Polyporus vegetus* Fr. sensu Velen. 1922: 685; fide Pilát 1942 (ACE 3): 501.

tremulae (Bond.) Bond. & Boris. apud Bond. 1953. — *Fomes igniarius* f. Bond. 1912; *Fomes* Borisov 1940 ["(Fr.)"]. — Bond. 1912 (TlR 37): 20 *fs. 7, 8* (*Fomes igniarius* f.); 1953: 43, 356, 358 *pl. 87 f. 2, pl. 100 fs. 1, 4, 6, pl. 106 f. 1, pl. 121 f. 1*; H. Jahn 1962 (WPb 3): 94 *fs. 1–4, Bild-beilage*; 1966 (ZP 32): 30 *fig.*; Domański & al. 1967 (FpG 3): 278 *pl. 25 f. 4*; Roll-Hansen *1967*; Kotl. & P. 1968; Balabán & Kotl. 1970: 98 *tplate* (*Phellinus*).

viticola (Schw. apud Fr.) Donk 1966; fide Overh. 1931 (M 23): 128 = *Trametes tenuis*. — *Polyporus* Schw. apud Fr. 1828 (U.S.A., North Carolina): *Poria* Cooke 1886, not ∼ Lázaro 1917; *Fuscoporia* Murrill 1907; *Fomes* Lowe 1957. — Overh. 1923 (M 15): 228 *fs. 27, 28 pl. 24 fs. 5–8* (*Poria*); Lowe 1957 F.: 45 *f 30*; 1966: 149 *f. 134* (*Fomes*); H. Jahn 1967 (WPb 6): 73 *fs. 1a, 2c, 3a, Abb. 18, 19, 32, 40 45, 49* (*Phellinus*). — Sensu G. Cunn. = *Fuscoporia* [*Phellinus*] *cryptacantha* (Mont.) G. Cunn. (extra-European: correctly interpreted?), fide G. Cunn. 1965: 212, 283.

Boletus superficialis Schw. 1822: 99 (U.S.A., North Carolina): fide Overh. 1923 (M 15): 225, 229. — *Polyporus* Steud. 1824, Schw. 1832;

Poria Cooke 1886. — Overh. 1923 (M 15): 224 *f. 22, pl. 24 fs. 2, 3 (Poria)*.

Trametes isabellina Fr. 1874 (Norway); fide Lowe 1957 F.: 45. —
Antrodia P. Karst. 1879; *Physisporus* P. Karst. 1882; *Polyporus* Romell
1917, not ∼ (Schw.) Fr. 1828; *Phellinus* Bourd. & G. 1925; *Poria* Overh.
1942, not ∼ (Pat.) Sacc. 1891; *Bjerkandera* Jørst. 1937 (syn.: n.v.p.;
"Donk", error with respect to *B. isabella* (Fr.) P. Karst., a different sp.).
— Egeland 1914 (NMN 52): 167: Bres. 1920 (Am 18): 62 (*Trametes*);
Bourd. & G. 1928: 622 (*Phellinus*); Shope 1931 (AMo 18): 363 *pl. 30
fs. 1, 2*; D. Baxt. 1937 (PMi 22): 292 *pl. 37 f. 3 (Trametes)*; Overh. 1942:
57 (*Poria*); Bond. 1953: 387 *fs. 91, 92, pl. 116 f. 2, pl. 169 f. 2*; Domański
1963 (Mob 15): 347 *fs. 8, 9*; 1965 (FpG 2): 220 *pl. 61 f. 1 (Phellinus)*.

Fomes tenuis P. Karst. 1887 ₍Finland₎; fide Bres. 1920 (Am 18): 62,
68 = *Trametes isabellinus.* — *Scindalma* O.K. 1898; *Mucronoporus* O.K.
1898: 518 (error with respect to *Scindalma tenuis* on p. 519: n.v.p.);
Polyporus Romell 1907 not ∼ (Hook.) Kl. 1833, not ∼ (Link ex Sacc.)
Overh. 1926; *Trametes* Overh. 1923; *Mucronoporus* Cash 1953 (incidental
mention: n.v.p.). — P. Karst. in Roum. 1887 F.g.: No. 4019 & 1887
(Rm 9): 102; P. Karst. 1887 (Mfe 14): 81; 1887 I. 2: 14 *pl. 11 f. 58* / 1888
(Afe 16): 528 (*Fomes*; Romell 1911 (ABS 11³): 24 (*Polyporus*); Lowe
1942: 91 (*Trametes*), Overh. 1953: 67 *pl. 61 fs. 361, 363, pl. 64 f. 382,
p . 127 fig. (Fomes)*.

Trametes setosa Weir 1914 (U.S.A., Idaho); fide Overh. 1931 (M 23):
128 = *Trametes tenuis*; fide Lowe 1957 F.: 45. — *Polyporus* Lloyd 1915.
— Weir 1914 (JaR 2): 164 *pl. 10 (Trametes)*; Lloyd 1915 (LMW 4, Ap):
350, 392 *fs. 686, 687 (Polyporus)*.

PIPTOPORUS P. Karst.

1881 [1960 (Pe 1): 257]. — Lectotype: *Polyporus betulinus* (Bull.) per Fr.
  Ungularia Lázaro 1916 [1960 (Pe 1): 291]. — Lectotype: *Polyporus betulinus*
(Bull.) per Fr.
  Placoderma (Rick.) Ulbr. 1928 [1960 (Pe 1): 258]. — *Placodes* [sect. ?] *Placoderma*
Rick. 1918. — [≡ *Polyporus* B. *Placodermei* sect. *Suberosi* Fr. 1838]. — Lectotype:
Polyporus betulinus (Bull.) per Fr.

SPECIAL LITERATURE.— Corner, *1935*; Dörfelt, *1970*; LaFuze, *1937*;
Hemmi & Kurata, *1931*; K. Lohwag *1957, 1965a*; Macdonald, *1937*;
Mayr, *1884*; Wulff, *1909*.—All on *Piptoporus betulinus*.

betulinus (Bull. per Fr.) P. Karst. 1881. — *Boletus* Bull. 1786 (France)
(d.n.); *Polyporus* Fr. 1815 (d.n.); *Polyporus* (Bull.) per Fr. 1821; *Boletus*
Mérat 1821; *Placodes* Quél. 1886; *Fomes* Gillot & Luc. 1890; *Ungulina*
Pat. 1900; *Ungularia* Lázaro 1916: *Placoderma* B. Henn. 1927 (generic
name n.v.p.), Ulbr. 1928. — Bull. 1786: *pl. 312*; 1791 H.: 348; Sow.
1799: *pl. 212 (Boletus)*; Gillet 1874–90 P.: *pl. 556/462*; Gramb. 1913
P.H. 2: *pl. 25*; Lloyd 1915 (LMW 4, Ap): 293 *fs. 631, 632 (Polyporus)*;
Bourd. & G. 1928: 606 (*Ungulina*); Pilát 1937 (ACE 3): 121 *f. 26, pl. 55*

fs. 4, 5, pls. 56, 57 (*Piptoporus*); Overh. 1953: 269 *pl. 26 fs. 161, 162, pl. 27 fs. 163, 164, pl. 103 f. 577, pl. 128 fig.*; Corner 1953 (Phm 3): 162 *fs. 10, 11*, hyphal analysis (*Polyporus*); Poelt & Jahn 1963: *pl. 35* (*Piptoporus*).

Boletus sutorius Scop. 1770: 149 (Hungary, now Czechoslovakia) (d.n.); fide Fr. 1828 E. 1: 89 ("totus albus hic aut saltim magis affinis") & Donk.

Agarico-pulpa pseudo-agaricon Paul. 1793 T. 2: 105 [*pl. 18*] (descr.) Ind. (France); = *Agaricum conchatum* Paul. 1812–35 (n.v.p.?). — Said to grow on *Quercus* hence referred reluctantly to *Polyporus* [*Buglossoporus*] *quercinus* by Lév. 1855: 7. The plate strongly suggests *Piptoporus betulinus*.

M.—*Boletus suberosus* L. (**O**) sensu L. 1755: 453: fide Fr. 1821: 358 (Wulf.) & 1828 E. 1: 89 (' Linn. Suec. 2 n. 1253"). — Wulf. 1787 (CoJ 1): 345 (*Boletus*); Murrill 1903 (BTC 30): 425 (*Piptoporus*).

pseudobetulinus (Murašk. ex Pilát) Pilát 1937. — *Polyporus* Murašk 'in sched." [1929] (n.v.p.); *Ungulina* (Murašk.) ex Pilát 1932 (U.S.S.R. Russia Siberia). — Pilát 1937 (ACE 3): 123 *f. 29 pl. 55 f. 3*; Murašk 1940: 8; Bond. 1953: 273 *f. 61: 2 pl. 79*; E. Komar. 1964: 122.

? soloniensis (Dubois per Fr.) Pilát 1937. — *Agaricus* Dubois 1803 (France) (d.n.); *Boletus* DC. 1815 (d.n.); *Polyporus* (Dubois) ex Fr. 1821; *Boletus* Cordier 1826; *Agaricus* Dubois 1833; *Ischnoderma* P. Karst. 1879; *Fomes* Big. & Guill. 1913 (syn.: n.v.p.); *Ungulina* Bourd. & G. 1928. — DC. 1815: 41 (*Boletus*); Lloyd 1912 (LMW 4, L. 42): 3 (*Polyporus*); Bourd. & G. 1928: 607 (*Ungulina*); Pilát 1937 (ACE 3): 126 *f. 31 pl. 55 fs. 1, 2* (*Piptoporus*).

? *Polyporus paradoxus* Fr. 1873 (ÖVS[5]): 8 (Sweden); cf. Bourd. & G. 1928: 608. — *Piptoporus* P. Karst. 1881.

POLYPORUS [Mich.] Fr.

1821, not ~ (Pers.) per S. F. Gray 1821 (Polyporaceae), not ~ Murrill 1904 (monadelphous homonym) [1960 (Pe 1): 261]. — [*Polyporus* Mich. 1729 (pre-Linnean name).] — *Polyporus* [Mich.] Adans. 1763, typonym (d.n.). — *Polyporus* [Mich.] Fr. 1815 (d.n.). — Lectotype: *Polyporus esculentus* . . . Mich. *pl. 71 f. 1* [= *Boletus tuberaster* Pers.]. — Sensu P. Karst. → *Albatrellus*.

Hexagonia Pollini per Fr. 1835 ("*Hexagona*") [1960 (Pe 1): 224]. — *Hexagonia* Pollini 1816 (d.n.). — Monotype: *Hexagonia mori* Pollini. — Sensu auctt. plur. = *Scenidium* O.K. (extra-European); sensu Bond. & S. → *Apoxona*.

Mycelithe Gasparrini 1842 (nom. anam. & conf.), in part [1962 (Ta 11): 90]. — Monotype: *Mycelithe fungifera* Gasparrini.

Polyporellus P. Karst. 1879 [1960 (Pe 1): 260]. — Lectotype: *Polyporus brumalis* (Pers.) per Fr.

Bresadolia Speg. 1883 ("*Bredasolia*") [1960 (Pe 1): 191]. — Monotype: *Bresadolia paradoxa* Speg.

Cerioporus Quél. 1886 [1960 (Pe 1): 196]. — Lectotype: *Polyporus squamosus* (Huds.) per Fr.

Leucoporus Quél. 1886 [1960 (Pe 1): 238]. — Lectotype: *Polyporus brumalis* (Pers.) per Fr.

Melanopus Pat. 1887 [1960 (Pe 1): 240]. — Lectotype: *Polyporus melanopus* (Pers.) per Fr.

Polyporus Murrill 1904, not ∽ [Mich.] Fr. 1821, monadelphous homonym, not ∽ (Pers.) per S. F. Gray 1821 [1960 (Pe 1): 263, in obs.]. — Holotype: *Polyporus ulmi* Paul.

Lentus (Lloyd) ex Torrend 1920 [1960 (Pe 1): 233]. — ["Stipitate Polyporoids" sect. *Lentus* Lloyd 1912 (n.v.p.).] — Lectotype: *Polyporus brumalis* (Pers.) per Fr.

Petaloides (Lloyd) ex Torrend 1920 & 1924 [1960 (Pe 1): 251]. — ["Stipitate Polyporoids" sect. *Petaloides* Lloyd 1912 (n.v.p.).] — Lectotype: *Polyporus petaloides* Fr.

SPECIAL LITERATURE.—General & miscellaneous: *Briganti, 1840*; Codina Viñas *1925*; Jahn *1969*; Klotzsch *1833*; Kreisel *1961*; Patouillard *1890*; Pilát *1936a*; Pouzar, *1972*; Ryvarden, *1969*.

Polyporus arcularius: Gibson & Trapnell, *1957*; Vandendries, *1936b*.

Polyporus brumalis complex: Kreisel, *1963a*; Plunkett, *1958, 1961*; Vandendries, *1936c*.

Polyporus floccipes: Hansen, *1955*; Malençon, *1929*; Sumstine, *1907*.

Polyporus mori: Donk, *1969c*; M. E. P. K. Fidalgo, *1968*; Paravicini, *1919*.

Polyporus rhizophilus: Bondarcev & Kravcev, *1952*; Favre & Ruhlé *1947*; Hruby *1931*; Moesz, *1913*; Pilát, *1952*; Rauschert, *1962*; Sebek, *1962*; Suza, *1933*, Zerova, *1957*.

Polyporus squamosus: Bakshi, *1956*, Brooks, *1909*, Brunnthaler, *1896*; Buller, *1906a, 1906b*; Campbell & Munson, *1936*; Corner, *1953*; Constantin, *1894*; Graff, *1936*; Jačevskij, *1911*; von Keissler, *1907*; Magnus, *1910*; Martyn, *1744*; Oehm, *1933a, 1933b, 1937b*; Pilát, *1934b–1935a*; Price, *1913*; Reichardt, *1866*; Sadebeck, *1886*; Ulbrich, *1940*; Van Bambeke, *1906*; Vandendries, *1937*.

Polyporus tuberaster: Brunner, *1842*; Constantin, *1895*; Decker, *1970*; Gasparrini, *1842*; Güssow, *1919*; van der Lek, *1921, 1926*; K. Lohwag, *1935*; Mattirolo, *1914*; Schmid, *1934*; de Secondat *1785*; Shirai, *1911*; Trotter, *1946*; Vanterpool & Macrae, *1951*; Zaneveld, *1940*.

alveolarius (Bosc) per Fr. 1821, not *P. alveolaris* (DC. per Fr.) Bond. & S. 1941 (*"alveloarius"*) (**90**). — *Boletus* Bosc 1811 (U.S.A., South Carolina) (d.n.); *Favolus* (Bosc per Fr.) Fr. 1825, not *F. alveolaris* (DC. per Fr.) Quél. 1886; *Boletus* Spreng. 1827; *Polyporellus* P. Karst. 1879, not *P. alveolaris* (DC. per Fr.) Pilát 1936 (*"alveolarius"*). — Bosc 1811 (MBe 5): 84 *pl. 4 f. 1* (*Boletus*); Rostk. 1848 (StP Fs. 27–28): 29 *pl. 15?* (*Polyporus*); Donk 1971 (PNA 74): 15, notes.

Polyporus calaber F. Brig. 1840 (AII 6): 146 151 *plate fs. 1–4* (Italy); fide Donk 1971 (PNA 74): 15.

Polyporus anisoporus Del. & Mont. apud Mont. 1845 (France). — Mont. 1845 (ASn III 4): 367; Donk 1969 (Pe 5): 239, note.

Favolus curtisii Berk. 1872 (G 1): 68 (U.S.A., North Carolina); fide
Lloyd 1912 (LMW 3, S.P.): 176 ,"The late summer form of [*Polyporus*]
arcularius" [sensu Lloyd in part: as to North American specimens].
— ≡ *Polyporus arculariellus* Murrill 1904 (BTC 31): 36.

Polyporus arculariformis Murrill 1904 (To 4): 151 *fs. 1–4* (U.S.A.,
Tennessee).

M.—*Boletus arcularius* Batsch sensu Fr. ex Berk. 1847 (LJB 6): 497,
lacking descr.; indirect inference. — Murrill 1907 (NAF 9): 59 & Lloyd
1912 (LMW 3, S.P.): 175 & Overh. 1953: 271 *pl. 35 f. 206 pl. 36 fs. 215,
216, pl. 97 f. 553, pl. 128 fig.* —these refs. are based on North American
collections, at least for the largest part: Kreisel 1963 (FR 68): 136 *pl. 2 f. 1*,
fide Donk 1969 (Pe 5): 240 in obs. = *Polyporus anisoporus*; H. Jahn
1963 (WPb 4): 32 *Abb. 41c* (*Polyporus*).

M.—*Polyporus floccipes* Rostk. sensu Bres. 1903. — Bres. 1903 (Am 1):
72, spores; Bres. ("MS") apud Sacc. 1916: 950; & cf. Bres. 1915 (H 65):
291, in obs.

M.?—*Polyporus agariceus* (Kön.) ex Berk. (**O**) sensu Bres. 1915 (H 56):
291, at least in part: as to European collections; fide Bres., l.c. = *Polyporus
anisoporus* (cited as a syn.). — Bres., l.c. (*Polyporus*); Bourd. & G. 1928:
531 (*Leucoporus*); Donk 1969 (Pe 5): 238, notes.

M.—*Favolus boucheanus* Kl. (**O**) sensu Bres. 1915 (H 56): 291, in obs.
(*Polyporus*); used by Bres. from 1916 onward to replace the name *Polyporus agariceus* [sensu Bres.]. — Bres. 1920 (BIm 31): *pl. 955* (*Polyporus*)?

arcularius (Batsch) per Fr. 1821 (**90, 91**). — *Boletus* Batsch 1783 (d.n.);
Boletus (Batsch per Fr.) Schw. 1822, misapplied, Pollini 1824; *Polyporellus*
P. Karst. 1879, Pilát 1936; *Leucoporus* Quél. 1886, Pat. apud Har. & Pat.
1903; *Favolus* Lév. apud Moritzi 1846 (nom. nud.: n.v.p.), Lév. 1855;
Heteroporus Lázaro 1916; [≡*Polyporus exiguus, pileo hemisphaerico* . . .
Mich. 1729 (Italy)]; ≡ *Polyporus arculatus* Kumm. 1871 (error?). — Mich.
1729: 130 *pl. 70 f. 5* (*Polyporus exiguus, pileo hemisphaerico* . . .); Sacc.
1916: 958 (*Polyporus*); Bourd. & G. 1928: 532, in part (*Leucoporus arcula-
rius* var. *strigosus*); Konr. & M. 1935 I. 5: *pl. 428 f. 2* ? (*Leucoporus*);
Bres. 1931 (BIm 20): *pl. 956* (*Polyporus*); Donk 1969 (Pe 5): 239, notes.
— Sensu Gillet → *Polyporus floccipes*; sensu auctt. nonn., Lundell →
Polyporus brumalis; sensu auctt. Amer. & auctt. nonn. Europ., Kreisel
→ *Polyporus alveolaris*.

? *Boletus exasperatus* Schrad. apud J. F. Gmel. 1792: 1433; Schrad.
1794: 155 (Germany) (d.n.); fide Schrad., l.c. (*B. arcularius* cited as a
syn.). — *Polyporus* J. Rick 1938 (syn.: n.v.p.).

Polyporus rhombiporus Pers. 1825: 211 (France); fide Fr. 1832[Ind.]:
148 (**91**).

Polyporus intermedius Rostk. 1837 (StP 4): 69 *pl. 33* (Germany/Poland),
not ∼ Sing. 1922; cf. Fr. 1874: 526 in obs.; fide Konr. & M. 1935 I. 5:
text to pl. 428 f. 2 (**91**).

Leucoporus arcularius var. *strigosus* Bourd. & G. 1925 (France) (**91**). — Bourd. & G. 1928: 532.

badius (Pers. per S. F. Gray) Schw. 1832, not ∼ Berk. 1841, not ∼ (Berk.) Lév. 1846. not ∼ Jungh. ex Bres. 1912. — *Boletus* Pers. 1801 (Germany) (d.n.), not ∼ Fr. 1821; *Polyporus* Fr. 1815 (nom. nud.); *Grifola* (Pers.) per S. F. Gray 1821. — Pers. 1801: 523, excl. of var. (*Boletus*); Donk 1969 (Pe 5): 241, notes; H. Jahn 1969 (SZP 47): 219 (*Polyporus*).

Boletus perennis Batsch 1783 (Germany) (d.n.), not ∼ L. 1783 (d.n.); fide Pers. 1801: 523. — ≡ *Boletus durus* Timm 1788 (d.n.); ≡ *Boletus batschii* J. F. Gmel. 1792 (d.n.). — Batsch 1783: 103 & 1786: 181 *pl. 25 f. 129*; Donk 1969 (Pe 5): 254 note. — Sensu Vahl → *Polyporus varius*.

Polyporus nigripes Wallr. 1833: 598 (Germany) not ∼ Fr. 1830; fide Fr. 1838: 440 = *Polyporus picipes.*

Polyporus fissus Berk. 1847 (LJB 6): 318 (U.S.A., Ohio); fide Lloyd 1912 (LMW 3 S.P.): 187 ("probably depauperate *picipes*"). — *Polyporellus* P. Karst. 1879; *Favolus* A. Ames 1913 — Murrill 1904 (BTC 31): 42 (*Polyporus*).

Polyporus trachypus B. & Mont. apud Mont. 1856: 154 (U.S A., Ohio), not ∼ Rostk. 1848; fide Murrill 1904 (BTC 31): 42 = *Polyporus fissus*; fide Lloyd 1912 (LMW 3, S.P.): 188 ("abortive *picipes*"). — *Polyporus aculeatus* Sacc. & Trav. 1911 not ∼ Mont. 1840, not ∼ Lév. 1846, not ∼ Velen. 1922.

Polyporellus tubaeformis (P. Karst.) P. Karst. 1887; fide Pilát 1936 (BbC 56): 65 & 1937 (ACE 3): 103 = *Polyporus picipes* (forma). — *Polyporellus varius* subsp. P. Karst. 1882 (Finland). — P. Karst. 1887 I. 2: 12 / 1888 (ASf 16): 526 *pl. 10 f. 53* (*Polyporellus*).

M.—*Boletus varius* Pers. sensu Grev. 1825 (*Polyporus*); fide Donk. — Grev. 1825 S. 4: *pl. 202* excl. of descr ; Trapp. 1846–9 (Fb 10): *pl. 755*; Bres. 1931 (BIm 20): *pl. 966*, excl. of descr., fide Pilát 1937 (ACE 3): 99 = *Polyporellus picipes.*

M.?—*Polyporus picipes* Fr. sensu auctt. nonn. — Bres. 1931 (BIm 20): *pl. 965* (*Polyporus*); Konr. & M. 1935 I. 5: *pl. 427* (*Melanopus varius* var.); Pilát 1936–7 (ACE 3) 99 *f. 24 f. B on p. 105*, *pl. 44 fs. 1–3*, *pl. 46 f. b*, *pl. 47* (*Polyporellus*); Overh 1953: 262 *pl. 33 fs. 196, 197*, *pl. 109 f. 601*, *pl. 120 fs. 650–655*, *pl. 131 fig.* (*Polyporus*); Donk 1969 (Pe 5): 254, 1973 (PNA 76): 227, notes.

brumalis (Pers.) per Fr. 1821 (**90, 93**). — *Boletus* Pers. 1794 (Germany) (d.n.); *Polyporus* Fr. 1818 (d.n.); *Boletus* Wahl. 1826; *Polyporellus* P. Karst. 1879; *Leucoporus* Quél. 1886; *Favolus* A. Ames 1913 — Overh. 1953: 273 *pl. 35 fs. 213, 214*, *pl. 123 f. 168*, *pl. 128 fig.*; Kreisel 1963 (FR 68): 130 *pl. 2 f. 2*; H. Jahn 1963 (WPb 4): 30 *f. 3, Abb. 38, 40, 41c*; Poelt & Jahn 1964: in obs. *pl. 62 fig.*; Donk 1969 (Pe 5): 242, notes; Ryv. 1969 (NyM 16): 151 *f. 1*, notes (*Polyporus*). — Sensu Fr. 1821, in part, Bres. → *Polyporus ciliatus.*

? *Boletus pusillus* Schrad. apud J. F. Gmel. 1792 & Schrad. 1794: 152 (Germany) (d.n.), not ∼ Rafin. 1815 (n.v.p.), not ∼ Berk. 1854; fide Fr. 1838: 430 ("sitaneus").

Boletus fasciculatus Schrad. apud J. F. Gmel. 1792 & Schrad. 1794: 154 (Germany) (d.n.); fide Fr. 1874: 526 ("hornotinum")

Boletus trichocephalus Ehrenb. 1818: 19, 31 (Germany) (d.n.); fide Fr 1821: 348.

Polyporus nanus F. Brig. 1840 (AII 6): 148, 151 *plate fs. 5–7* (Italy), not ∼ Dur. & Mont. ex Mont. 1856; fide Donk 1972 (PNA 75): 170.

Polyporus trachypus Rostk. 1848 (StP Fs. 27–28): 27 *pl. 14* (Germany/ Polen), not ∼ B. & Mont. apud Mont. 1856.

Polyporus dibaphus B. & C. apud Berk. 1872 (G 1): 36 (U.S.A., Alabama); fide Lloyd 1912 (LMW 3, S.P.): 178.

Polyporus luridus B. & C. apud Berk. 1872 (G 1): 37 (U.S.A.), not ∼ Kalchbr. apud Cooke 1883; fide Lloyd 1912 (LMW 3, S.P.): 178.

Polyporus subcularius (Donk) Bond. 1953; fide Kreisel 1963 (FR 68): 131 & Donk. — *Polyporus brumalis* f. *subcularius* Donk 1933 (Netherlands); *Polyporus* Bond. apud Bond. & S. 1941 (nom. nud.: n.v.p.). — Donk 1933: 133, 134 (*Polyporus brumalis* f.); 1969 (Pe 5): 255, note.

M.—*Boletus arcularius* Batsch sensu auctt. nonn. — Bourd. & G. 1925 (BmF 41): 115 & 1928: 331 (*Leucoporus*), in part: as to var. *scabellus* Bourd & G.: Lundell 1932 (KSN 22): 16 & 1937 (LNF 9–10): 14 No. 438 (*Polyporus*); 1939 (LNF 15–16): 12 No. 728 (*P. arcularius* f. *griseus*); 1941 (LNF 21–22): 1 No. 1001 (*Polyporus*).

ciliatus (Fr.) per Fr. 1821 (**93, 94**). — *Polyporus* Fr. 1815 (Sweden) (d.n.), not *Boletus ciliatus* Hornem. 1806 (d.n.); *Polyporellus* (Fr. per Fr.) P. Karst. 1879. — Fr. 1815 O. 1: 123 (*Polyporus*); Kreisel 1963 (FR 68): 135 *pl. 3*; Poelt & Jahn 1964: *pl. 62 fig.*; Donk 1969 (Pe 5): 245, notes; H. Jahn 1969 (SZP 47): 224 (*Polyporus*).

Boletus substrictus Bolt. 1791: 170 *pl. 170* (England) (d.n.). — *Polystictus* (Bolt.) per Sacc. 1916. — Referable to '*P. lepideus*'. Referred by Fr. 1838: 431 to *Polyporus fuscidulus*.

Polyporus lepideus Fr. 1818 (Sweden), 1821 (incidental mention: n.v.p.); fide Kreisel 1963 (FR 68): 135 (forma). — *Polyporus* Fr. per Steud. 1824: Fr. 1832; *Polyporellus* P. Karst. 1879; *Leucoporus* Quél 1886. — Fr. 1818 O. 2: 253; 1838: 430 (*Polyporus*); Pilát 1936 (ACE 3): 68 (*Polyporellus brumalis* f.) *pl. 28* (*Polyporellus brumalis*); Kreisel 1963 (FR 68): 135 *pl. 3*; H. Jahn 1963 (WPb 4): 29 *Abb. 39, 41c* (*Polyporus ciliatus* f.); Donk 1969 (Pe 5): 245, notes.

Polyporus substriatus Rostk. 1828 (StP 4): 21 *pl. 9* (Germany/Poland): Fr. 1832; fide Bres. 1916 (Am 14): 227 = *Polyporus brumalis* [sensu Bres.]. — *Polystictus* Cooke 1886; *Microporus* O.K. 1898. — Referable to '*P. lepideus*'.

Boletus nummulariformis L. March. 1828. — [= *Boletus nummularius*

Bull. sensu L. March. 1826 (BnW 1): 145 (Luxemburg)]. — Unpublished plate (as *Boletus ambiguus* L. March.; L).

Polyporus vernalis Quél. ex Fr. 1874: 527, Quél. 1880 ("in litt. ad E. Fries, 1873"). — *Leucoporus* Bourd. 1898; [≡ *Polyporus cyathoides* (Sw. per Fr.) Quél. sensu Quél. 1872 (France)]; = *Polyporus queletianus* Sacc. & Trav. 1911, Sacc. & Trav. apud Sacc. & Trott. 1912, typonym. — Quél. 1872 (MMb II 5): 270/253 (*Polyporus cyathoides*); 1880 (BRo II 15): 195/47 *pl. 3 f. 13*; 1888: 403 (*Polyporus vernalis*); Bourd. & G. 1928: 530 (*Leucoporus brumalis* f.); Donk 1969 (Pe 5): 259, notes. — Bres. 1931 (BIm 20): *pl. 952* (*Polyporus vernalis*) looks quite different from what Quél. described.

Polyporus vossii Kalchbr. apud W. Voss 1879 (VW 29): 689 (Yugoslavia, Carniola); fide Bres. 1897 (AAR III 3): 68 = *Polyporus brumalis* [sensu Bres.]. — Referable to '*P. lepideus*'.

Polyporus esculentus Britz. 1895 (BCb 62): 311 [*pl. 644 f. 172*] & 1910 (BbC 26): 209 (Germany). — Referable to '*P. lepideus*'.

? *Polyporus caerulescens* Velen. 1922: 672 [see Pilát 1948: 255 for Latin translation] (Czechoslovakia); cf. Pilát 1936 (ACE 3): 70.

? *Polyporus formosus* Laubert 1923 (ZP 2): 67 (Germany). — Referable to '*P. lepideus*'?

M.—*Polyporus tomentosus* Fr. sensu Rostk. 1828 (StP 4): 19 *pl. 8*. — Referable to '*P. lepideus*'.

M.—*Boletus melanopus* var. *cyathoides* Sw. sensu Quél. 1872 (MMb II 5): 270/253 (*Polyporus cyathoides*) → *Polyporus vernalis* Quél. ex Fr., listed above.

M.—*Boletus brumalis* Pers. sensu Fr. 1821, in part, Bres. 1897 (*Polyporus*); fide Bres. apud Sacc. 1916: 959 & Bres. 1931 (BIm 20): *pl. 951* = *Polyporus lepideus* (cited as a syn.); fide Kreisel 1963 (FR 68): 131, 134 = *Polyporus ciliatus* emend. Kreisel. — Bres. 1897 (AAR III 3): 68 (*Polyporus*); Konr. 1925 (BmF 41): 67; Bourd. & G. 1928: 530 (*Leucoporus*); Bres. 1931 (BIm 20): *pl. 951* (*Polyporus*); Konr. & M. 1935 I. 5: *pl. 429* (*Leucoporus*).

corylinus Mauri 1832 (Italy). — Mauri 1832 (GaS 54): 65 *plate f. 2*: Viv. 1834–8: *pl. 1*. — (**96**).

floccipes Rostk. 1848 (Germany/Poland). — Rostk. 1848 (StP Fs. 27-28): 25 ("*floccopes*") *pl. 13* ("*floccopus*"); sensu Donk 1969 (Pe 5): 249 & 1971 (PNA 74): 17, notes. — Sensu Bres. → *Polyporus alveolarius* (Bosc) per Fr. — Pouz. 1972 (ČM 26): 86, in obs., refers *P. floccipes* to *P. anisoporus* [= *P. alveolarius*]; I cannot agree with this conclusion.

Polyporus lentus Berk. 1860 (England); fide Donk 1969 (Pe 5): 249. — *Leucoporus* Pat. 1900, Maire 1907 misapplied; *Melanopus* Pilát 1936 (syn.: n.v.p.). — Berk. 1860: 237 *pl. 16 f. 1* (*Polyporus*); Bourd. & G. 1928: 526 (*Melanopus squamosus* subsp.); Malenç. 1952 (BbF 99[10]): 43, in obs.; L. Hansen 1955 (BT 52): 170 (*Polyporus*); Donk 1969 (Pe 5):

250, notes. — Sensu Bres. 1896 = *"Favolus" apiahynus* Speg. (Brazil), extra-European, fide Bres. 1920 (Am 18): 40; sensu Bres. 1903 (Am 1): 72 = a sp. of *Polyporus* with 'small' spores.

Polyporus tiliae S. Schulz. apud Fr. 1874 (Yugoslavia, Slavonia); fide Donk 1970 (PNA 74): 18. — *Polyporus* S. Schulz. 1866 (nom. nud.: n.v.p.). — Fr. 1874: 528; Kalchbr. 1877: 58 *pl. 38 f. 3*; Donk 1971 (PNA 74: 17, 18, notes.

Polyporus forquignonii Quél. apud Guillaud & al. 1884 (France); fide Bres. 1915 (H 56): 291 & 1920 (Am 18): 67 = *Polyporus lentus.* — *Cerioporus* Quél. 1886; *Leucoporus* Pat. 1900; *Melanopus* Pilát 1936 (syn.: n.v.p.), Pinto-L. 1952. — Quél. 1885 (Crf 13[2]): 281 *pl. 8 f. 12*; Boud. 1886 (BmF 2): 34 (*Polyporus*); Quél. 1888: 408 (*Cerioporus*); Bourd. & G. 1928: 526 (*Melanopus squamosus* subsp.); Malenç. 1929 (BmF 45, Atl.): *pl. 34* (*Leucoporus*); Pilát 1936 (ACE 3): 93 *pl. 42 f. b* (*Polyporellus squamosus* f.); Bond. 1953: 444 *f. 111* (*Polyporus*).

? *Polyporus latiporus* Britz. 1893 (BCb 54): 103 [*pl. 627 f. 124*] & 1910 (BbC 26): 210 (*"latisporus"*) (Germany) (**95**).

Polyporus fagicola Murrill 1906 (U.S.A., Maine), not ∼ Velen. 1922; fide Donk 1970 (PNA 74): 17, 18. — *Favolus* A. Ames 1913. — Murrill 1906 (To 6): 35 & 1907 (NAF 9): 55, spores wrongly described; Lowe 1934 (PMi 19): 143 *f. 5E, pl. 15 fs. 3, 4*; Overh. 1953: 258 *pl. 54 f. 325, pl. 98 f. 560, pl. 129 fig.*

Polyporus pennsylvanicus Sumstine 1907 (U.S.A., Pennsylvania); fide Overh. 1953: 258 = *Polyporus fagicola*; fide Donk 1970 (PNA 74): 17, 18. — Sumstine 1907 (JM 13): 137; Overh. 1914 (AMo 1): 108; 1915 (WUS 3[1]): 19; Lowe 1934: 29.

Polyporus mcmurphyi Murrill 1915 (U.S.A., California); fide Donk 1970 (PNA 74): 17, 18. — Murrill 1915 W.P. 12; Overh. 1953: 257 *pl. 53 f. 317, pl. 130 fig.*

M.—*Boletus arcularius* Batsch sensu Gillet 1874–90 P.: *pl. 553/453* (*Polyporus*).

M.?—*Favolus boucheanus* Kl. (**O**) sensu Lloyd 1911 (*Polyporus*): fide Bres. 1915 (H. 56): 291 = *Polyporus lentus.* — Lloyd 1911 (LMW 3, O.): 86 *f. 506* (*Polyporus*).

M.—*Polyporus coronatus* Rostk. sensu Malenç. 1952 (*Melanopus*). — Malenç. 1952 (BbF 99[10]): 41 (*Melanopus*); H. Jahn 1969 (SZP 47): 221 *fs. B–E*; Pouz. 1972 (ČM 26): 86, notes (*Polyporus*).

incendiarius (Bong. apud Weinm.) Fr. 1838 (**96**). — *Polyporus brumalis* var. Bong. apud Weinm. 1836 (U.S.S.R., "Rossia minor" = Ukraine in part); *Polyporellus* P. Karst. 1881. — Weinm. 1836: 309 (*Polyporus brumalis* var.); Fr. 1838: 431; Pilát 1937 (ACE 3): 104, in obs. (*Polyporus*); Murašk. 1940: 2 (*Polyporellus varius* var.).

melanopus (Pers.) per Fr. 1821, not ∼ (Mont.) Lloyd 1912. — *Boletus* Pers. 1797 (Germany) (d.n.); *Boletus* Mérat 1821; *Polyporellus* P. Karst.

1879, Pilát 1936, 1937; *Leucoporus* Quél. 1886; *Pelloporus* Lázaro 1916; *Melanopus* Sart. & M. 1921. — Fr. 1838: 439; Romell 1911 (ABS 11³): 17; Rea 1922: 576 (*Polyporus*); Bourd. & G. 1928: 529 (*Melanopus*); Bres. 1931 (BIm 20): *pl. 964* (*Polyporus*); Pilát 1937 (ACE 3): 111 *f. 27, f. C on p. 105, pl. 45, pl. 46 f. a* (*Polyporellus*): Overh. 1953: 232 *pl. 32 fs. 188, 189, pl. 130 fig.*; H. Jahn 1969 (SZP 47): 220; Hagström *1971* (*Polyporus*); Donk 1973 (PNA 76): 225, notes.

   Boletus infundibuliformis Pers. 1794 (NMB 1): 107 / 1797 T.: 27 (Germany) (d.n.), not ∼ Batsch 1783 (d.n.); fide Fr. 1838: 439, 440 ("Pers. syn."). — *Polyporus melanopus* var. (Pers.) per Pers. 1825; *Polyporus* Chev. 1826, not ∼ Rostk. 1830, not ∼ (Wakef.) Lloyd 1924; *Polyporellus* P. Karst. 1881; ≡ *Polyporus melanopus* var. *infundibulum* Fr. 1821: 347. — Donk 1973 (PNA 76): 225, notes.

   ? *Polyporus picipes* Fr. 1838; fide Hagström, *1971*. — *Polyporellus* P. Karst. 1879: *Leucoporus* Quél. 1886; *Melanopus* Pat. 1887 (nom. nud. n.v.p.), 1895; [≡ *Polyporus* sp., u n n a m e d, Fr. 1821: 353 (Sweden)]; ≡ *Polyporus picipes* Rostk. 1848, misapplied. — Fr. 1838: 440; cf. Donk 1973 (PNA 76): 227, note. — Sensu auctt. nonn → *Polyporus badius*; *Polyporus picipes* Rostk. sensu Rostk. (plate) → *Polyporus varius*.

   Polyporus flavescens Rostk. 1848 (StP Fs. 27–28): 45 *pl. 23* (Germany/Poland): fide Fr. 1863 M. 2: 339 (var.). — *Polyporus badius*?

   Polyporus cyathoides (Sw. per Fr.) Quél. 1872, misapplied; fide Fr. 1821: 348. — *Boletus melanopus* var. Sw. 1810 (SVH 31): 10 (Sweden) (d.n.); *Polyporus melanopus* var. (Sw.) per Fr. 1821; *Boletus* Mussat 1901 (syn.: n.v.p.). — Donk 1969 (Pe 5): 248; 1972 (PNA 76): 225, notes. — Sensu Quél. → *Polyporus ciliatus*.

   Scutiger subradicatus Murrill 1903 (BTC 30): 430 (U.S.A., New York); fide Overh. 1953: 232, 233. — *Polyporus* Sacc. & D. Sacc. 1905, Sing. 1951.

   M.?—*Boletus umbilicatus* Scop. (**O**) sensu Fr. 1821: 348 (as syn. of *Polyporus melanopus* var. *cyathoides* [= *P. melanopus* s. str.]), Spreng. 1827: 472.

michelii Fr. 1821 (**97**). — *Boletus* Pollini 1824; *Cerioporus* Quél. 1886; *Melanopus* Maire 1933; [≡ *Agaricum esculentum, candidum, flabelliforme* . . . Mich. 1729 (Italy)]. — Mich. 1729: 120 *pl. 61 f. 2* (*Agaricum* . . .); Secr. 1833 M. 3: 51, in part: var. A? (*Polyporus favolus albus*).

mori (Pollini) per Fr. 1821 (**98**). — *Hexagonia* Pollini 1816 (Italy) (d.n.); *Favolus* Pollini 1816 [repr. K. Fid. 1968 (Ta 17): 40]; *Boletus* Pollini 1824, not ∼ Pollini in herb. (n.v.p.); *Favolus* Fr. 1825; *Hexagonia* Fr. 1838. — Pollini 1916 H.: 35 *pl. 1 fs. 2, 3* (*Hexagonia*): Kotl. & P. 1957 (ČM 11): 214, 215, notes; H. Jahn 1969 (SZP 47): 227 (*Polyporus*). — Sensu Marcucci → *Apoxonia nitida*.

   Merulius alveolaris DC. 1815 (France) (d.n.); fide Sacc. 1873 (ASv 2): 104 & Bres. 1882 F.t. 1: 22, 23 (for *Favolus europaeus*) = *Hexagonia mori* (cited as a syn.). — *Cantharellus* (DC.) per Fr. 1821: *Merulius* Pers.

1825; *Favolus* Quél. 1866, Fairm. 1892, not *F. alveolarius* (Bosc per Fr.)
Fr. 1825; *Hexagonia* Hariot 1891 (not accepted: n.v.p.), Murrill 1904:
Polyporellus Pilát 1936 (*"alveolarius"*), not *P. alveolarius* (Bosc) per Fr.
1821; = *Favolus extratropicus* Fr. 1825; = *Favolus europaeus* Fr. 1838.
— Bres. 1881 F.t. 1: 22 *pl. 27*; Pat. 1883 T.a. 1: 56 *f. 131* (*F. europaeus*);
Quél. 1888: 369 (*F. alveolaris*); Bataille 1911 (BmF 27): 379 (*F. europaeus*);
Paravicini 1919 (SZF 70): 16; Bourd. & G. 1928: 533; May 1929 (ZP 8):
46 (*F. europaeus*); Pilát 1936 (ACE 3): 83 *f. 20, pl. 33, pl. 34 f. a* (*F.
alveolaris*): Imbach 1939 (SZP 17): 145 *2 figs.* (*F. europaeus*). — Sensu
Velen. (*Favolus europaeus*) = *Plicatura faginea* (Schrad. per J. Schroet.)
P. Karst. [= *Plicatura crispa* (Pers. per Fr.) Rea (**O**)], fide Pilát 1936
(ACE 3): 83; sensu auctt. Amer. → *Favolus [Polyporus] canadensis* (**O**).

   Daedalea broussonetiae Cappelli 1821: 64 [repr. G. F. Re 1827: 309]
(Italy), fide Sacc. apud Trav. 1905 (Fic 1): 27 = *Favolus europaeus*.

   Polyporus favoloides Doass. & Pat. 1880 & apud Roum. 1881 (France),
not ∼ P. Henn. 1897: fide Bres. 1882 F.t. 1: 22, 23 & 1882 (Rm 4): 20,
88 = *Favolus europaeus*. — Doass. & Pat. apud Roum. 1881 (Rm 3 /
No. 11): 21 *pl. 18 f. 5*.

podlachicus Bres. 1903 (Poland). — *Melanopus* Sart. & M. 1921. — Bres.
1903 (Am 1): 73; Lloyd 1922 (LMW 7): 111 *pl. 192 f. 2062* (*Polyporus*);
Bourd. & G. 1928: 529 (*Melanopus varius* subsp.).

rhizophilus Pat. 1894 (Tunesia). — *Leucoporus* Pat. 1897; *Melanopus*
Hruby 1931; *Polyporellus* Pilát apud Suza 1933 (n.v.). — Pat. 1897:
46 *pl. 5 f. 2* (*Leucoporus*): Moesz 1913 (BK 12): 231, (63) (*Polyporus*):
Pilát 1936 (BbC 56): 78 *fs. 10, 11*: 1936–7 (ACE 3): 97 *f. 23, pl. 34 f. b,
pls. 36–38*; Favre & Ruhlé 1947 (SZP 25): 57 *f. 1*; Pilát 1952 (ČnM 121):
136 *2 figs.* (*Polyporellus*): Zerova 1957 (UbŽ 14²): 69 *f. 1*; Švarcm. 1964:
512 *f. 227* (*Polyporus*).

squamosus (Huds.) per Fr. 1821 (**99, 100**). — *Boletus* Huds. 1778 (England)
(d.n.), non/an ∼ Timm 1788 (d.n.); *Boletus* (Huds. per Fr.) Hook. 1821,
not ∼ Roques 1841 (Boletales): *Favolus* Kl. 1833: *Polyporellus* P. Karst.
1879; *Cerioporus* Quél. 1886; *Melanopus* Pat. 1887. — Vahl 1797 (Fd 7 /
F. 22): 8 *pl. 1196*; Sow. 1800: *pl. 266* (*Boletus*); Fr. 1821: 343; Grev.
1825 S. 4: *pl. 207*; Gillet 1874–90 P.: *pl. 568/456*; Bres. 1897 (AAR III 3):
68, spores (*Polyporus*); Bourd. & G. 1928: 525 (*Melanopus*); Bres. 1931
(BIm 20): *pls. 962, 963*; Shope 1931 (AMo 18): 356 *pl. 29 fs. 1, 2*; Donk
1933: 130; Overh. 1953: 256 *pl. 44 fs. 204, 205, pl. 132 fig.*; Corner 1953
(Phm 3): 157 *fs. 8, 9*, hyphal analysis (*Polyporus*); Pilát 1936 (ACE 3):
86 *f. 21 pls. 39–41*, in part (*Polyporellus*): Malenç. 1956 (BmF 71): 275
tpl. 2 (*Melanopus*): Poelt & Jahn 1964: *pl. 61* (*Polyporus*).

   [*Fungi esculenti. Genus V*. Clus. 1601 → *Boletus testaceus*.]
   [*Fungus angulosus pediculo exiguo* C. Bauh. 1623 → *Boletus testaceus*.]
   [*Auricula flammea Machi* Sterb. 1675 → *Boletus testaceus*.]

[*Fungus maximus arboreus porosus pediculo limbo affixo* Ray 1696:
336 (England): fide Ray 1724: 11 = *Fungus angulosus pediculo exiguo.*]

[*Agaricus ramosus cornu reniferi (potius rangiferi) referendi* Blackstone
1746: 3 *frontisp.* (England); fide Bolt. 1783: 138 *pl. 138* = *Boletus
rangiferinus.*

Boletus juglandis Schaeff. 1774 (Germany) (d.n.); fide Pers. 1800:
40 & Fr. 1821: 343. — *Agaricus* Paul. 1793 (d.n.): *Boletus* Schaeff. per
St-Am. 1821; *Polyporus* Pers. 1825; ⩵ *Boletus platyporus* Pers. 1794
(d.n.) **(101)**; *Polyporus* Fr. 1815 (d.n.), not ∼ Secr. 1833 (n.v.p.) **(O)**,
not ∼ Berk. 1851; *Grifola* (Pers.) per S. F. Gray 1821. — Bull. 1780:
pl. 19; 1791 H.: 344 (*Boletus*). — V.s.: "*iuglandis*".

Boletus testaceus With. 1776 (d.n.). — [⩵ (by lecto-typification) *Fungus
angulosus pediculo exiguo* C. Bauh. 1623 (non-binomial name); ⩵ *Auricula
flammea Machi* Sterb. 1675 (non-binomial name): ⩵ *Fungi esculenti.
Genus V*. Clus. 1601 (Hungary)]. — Clus. 1601: cclxv *fig.* & Codex apud
Istv. 1900: *pls. 19, 19 [bis], 20* (*Fungi esculenti. Genus V*.); Ray 1724: 11
(*Fungus maximus* ... Ray & *Fungus angulosus* ... C. Bauh.).

Boletus cellulosus Lightf. 1778: 1032 (Scotland) (d.n.), not ∼ O. F. Müll.
1777 (d.n.): fide Pers. 1800: 40 & Fr. 1821: 343.

Boletus polymorphus Bull. 1782 (France) (d.n.), not ∼ Hoffm. 1791–1811
(d.n.): fide Bull. 1791 H.: 345 = *Boletus juglandis*; fide Fr. 1821: 343.
— *Boletus* Bull. per Dubois 1833. — Bull. 1782: *pl. 114*.

Boletus subsquamosus Batsch 1783 (Germany) (d.n.), not ∼ L. 1753
(d.n.); fide Pers. 1800: 40. — *Boletus* Batsch per Bergam. 1823; ⩵ *Boletus
squamosus* Timm 1788 (d.n.), non/an ∼ Huds. 1778 (d.n.). — Batsch
1783: 97, 177 *pl. 10 f. 41*.

Boletus rangiferinus Bolt. 1789: 138 *pl. 138* ("*rengiferinus*") (England)
(nom. monstr.) (d.n.); fide Fr. 1821: 344 (monstrosity). — *Sistotrema*
C. & T. Nees 1820 (d.n.). — Bolt. 1789: 138 *pl. 138*. — V.s.: "*rhangi-
ferinus*".

Agarico-pulpa ulmi Paul. 1793 T. 2: 102 (descr.), Ind. (France?) (d.n.);
fide Lév. 1855: 8. — *Polyporus* Paul. 1812–35: *pl. 16* (d.n.?); *Polyporus*
(Paul.) per Over. 1924.

Boletus maximus Schum. 1803: 381 (Denmark) (d.n.), not ∼ Brot.
1804 (d.n.); fide Fr. 1821: 344.

Polyporus flabelliformis Pers. 1825, not ∼ Kl. 1833, not ∼ (Schaeff.)
per Sacc. 1916; fide Fr. 1828 E. 1: 73. — [⩵ *Agaricus aureus flabelli
effigie* Batt. 1755: 68 *pl. 37 f. A* (Italy)].

*Polyporus *tigrinus* Pers. 1825: 54 (Europe): fide Fr. 1828 E. 1: 73.
— Descr. recalls the form called *P. coronatus* Rostk.

? *Polyporus dissectus* Letell. 1826: 48 *pl. 2 f. 22* (France), not ∼ Lév.
1846.

Polyporus infundibuliformis Rostk. 1830 (StP 4): 37 *pl. 17*: Fr. 1832
(Germany/Poland), not ∼ (Pers. per Pers.) Chev. 1826, not ∼ (Wakef.)
Lloyd 1924; fide Donk 1933: 130, 131 (forma). — ⩵ (by lecto-typification)

Polyporus rostkowii Fr. 1838; *Polyporellus* P. Karst. 1879; *Cerioporus* Quél. 1886. — Bamb. 1906 (BBB 43): 256 *pls. 1, 2*; Keissl. 1907 (AW 22): 143 *pl. 2* (*Polyporus rostkowii*).

Polyporus coronatus Rostk. 1848 (Germany/Poland): fide Donk 1968 (Pe 5): 245. — *Melanopus* Pilát 1936 (syn.: n.v.p.), Malenç. 1952, misapplied. — Rostk. 1848 (StP Ts. 27–28): 33 *pl. 17* (*Polyporus*): Bourd. & G. 1928: 523 (*Melanopus squamosus subsp.*); Pilát 1936 (ACE 3): 91 *pl. 42 f. a* (*Polyporus squamosus* f.); Maas G. 1955 (Fu 25): 50 (*Polyporus*). — Sensu Malenç. → *Polyporus floccipes*.

Polyporus pallidus S. Schulz. apud Fr. 1874 (Yugoslavia, Slavonia), not ∼ Berk. 1856; fide Bres. 1892 F.t. 2: 27, 28 (f. *erectus* Bres.). — Kalchbr. 1877: 57 *pl. 38 f. 2*.

Polyporus squamatus Schwalb 1891: 167, 218 (Austria), not ∼ Lloyd 1911.

Polyporus clusianus Britz. 1894 (BnS 31): 174 [*pl. 639 f. 158*] (Germany). — Britz. l.c. & 1910 (BbC 26): 209.

Polyporus caudicinus Murrill 1903 (JM 9): 89, 100 & 1904 (BTC 31): 40, not ∼ (Scop. per Pollini) J. Schroet. 1888. — [≡ *Boletus caudicinus* Scop. 1772: 469, in part: viz. var. 1 (Yugoslavia, Carniola)]. — Murrill 1907 (NAF 9): 60; Donk 1971 (PNA 74): 10, note.

Polyporus helopus Har. & Pat. 1904 (France): fide Bres. 1916 (Am 14): 225 ("f. caespitosa"). — Har. & Pat. 1904 (BmF 20): 63 *fig. at right*.

Bresadolia caucasica Šestunov apud P. Magn. 1910 (H 50): 101 *pl. 2* (U.S.S.R., Russia, Caucasia): fide P. Magn. 1910 (H 50): 102 ("monströse Form"); fide Jač. 1911 (H 50): 253.

M.—*Boletus favus* L. (**O**) sensu O. F. Müll. 1782 (Fd 5 / F. 15): 6 *pl. 893* ("?"); fide Fr. 1832[Ind.].: 56.

M.—*Boletus subsquamosus* L. sensu Batsch 1783 → *Boletus squamosus* Timm.

M.—*Boletus giganteus* Pers. sensu Harzer 1842–5 (*Polyporus*); fide Fr. 1863 M. 2: 326. — Harzer 1842–5: 59 *pl. 32* (*Polyporus*).

M.—*Boletus caudicinus* Scop. sensu Murrill 1903 (JM 9): 89, 100 (*Polyporus*) → *Polyporus caudicinus* Murrill.

tubarius Quél. 1879 (France) (**102**). — *Leucoporus* Quél. 1886; *Pelloporus* Lázaro 1916; *Polystictus* Nentien 1924. — Quél. 1879 (BbF 25): 289 & 1880 (BRo II 15): *pl. 3 f. 12*; Bres. 1897 (AAR III 3): 68?; Boud. 1904–11: 78 *pl. 152*. — Referred to *Polyporus agariceus* [sensu Bres. = *P. anisoporus*] by Bres. 1915 (H 56): 291.

tuberaster (Pers.) per Fr. 1821 (**103**), non/an ∼ Paul. 1812–35 (d.n.?). — *Boletus* Pers. 1801; *Boletus* (Pers. per Fr.) Pollini 1824; *Polyporellus* Pilát 1937; [≡ *Boleti* nova forte species Jacq. 1796 (Italy)]. — Jacq. 1796 (CoJ 5): 160 *pl. 8, pl. 9, f. 1* (*Boletus* sp.); Lloyd 1911 (LMW 3, O.): 92 *f. 309*; Mattirolo *1914*, also for previous literature; Lek 1921 (MmV 11): 85 *pl. 1*; 1926 (MmV 15): 131 *plate*; Bres. 1931 (BIm 19): *pl. 946*; Zaneveld

1941 (Fu 13): 1 *fs. 1, 2*; Trotter *1946*; Vanterpool & Macrae *1951*; Overh. 1953: 240 *pl. 121 f. 659, pl. 132 fig.*; Kawam. 1954 I. 1: 132 *f. 121*; Imaz. & Hongo 1957 C.I. [1]: 110 *pl. 51 f. 285*; Bergstädt & al. 1970 (MMH 13): 83 *fs. 5, 7* (*Polyporus*).

[*Tuberaster Fungos ferens Ital.* occone 1697 M.F.: 300 *fig.* B & cf. pp. 292–293 (Italy)].

[*Polyporus esculentus* ... Mich. 1729 (non-binomial name) (Italy). — Mich. 1729: 131 *pl. 71 f. 1*].

[*Ceriomyces* Batt. 1755: 61 *pl. 24 f. A* (Italy) (non-binomial, specific name). → *Fungus tuberaster* Paul.]

["Champignon de Velletri" Secondat 1785: 37 *pls. 11, 12*.]

Boletus calabricus Latourr. 1785: 39 (Italy) (d.n.).

Fungus tuberaster Paul. 1793 (d.n.). — *Polyporus* Paul. 1812–35 (d.n.?), non/an ∼ (Pers.) per Fr. 1821; [≡ (by lecto-typification) *Ceriomyces* Batt. 1755 (Italy)]. — Batt. 1755: 61 *pl. 24 f. A* (*Ceriomyces*): Paul. 1793 T. 2: 361 ("La Truffe ou Pièrre a champignons"): 1812–35: *pls. 165, 166* [*pl. 165 f. 1* after Batt.] (*Polyporus tuberaster*).

Mycelithe fungifera Gasparrini 1842 (AAp 2⁴): 221 *pl. 1, pl. 2 fs. 1–4, 6, 7* (Italy); nom. anam. & conf., fide Donk 1962 (Ta 11): 90.

Grifola tuckahoe Güssow 1919 (Canada); fide Vanterpool & Macrae *1951*. — *Polyporus* Lloyd 1920, Sacc. & Trott. apud Trott. 1925. — Güssow 1919 (M 11): 109 *pls. 7–9*.

varius (Pers.) per Fr. 1821. — *Boletus* Pers. 1796 (Germany) (d.n.): *Polyporus* Fr. 1815 (nom. nud.) (d.n.); *Grifola* S. F. Gray 1821: *Boletus* Pollini 1824; *Polyporellus* P. Karst. 1879: *Melanopus* Pat. 1887 (nom. nud.: n.v.p.), 1893; *Favolus* A. Ames 1913; *Leucoporus* R. Heim 1957 (n.v.p.). — Fr. 1838: 440 (*Polyporus*); Bourd. & G. 1928: 527, in part (*Melanopus*): Shope 1931 (AMo 18): 358 *pl. 27 fs. 4–7*; Donk 1933: 136, excl. of some refs.; Lowe 1942: 25; Overh. 1953: 265 *pl. 33 fs. 199, 200, pl. 36 f. 222, pl. 132 fig.* (*Polyporus*); Donk 1969 (Pe 5): 259, notes. — Sensu Grev. → *Polyporus badius.*

Boletus leptocephalus Jacq. 1778 (Austria) (d.n.). — *Polyporus* (Jacq.) per Fr. 1821: *Coltricia* S. F. Gray 1821; *Boletus* Zant. 1822, not ∼ Peck 1898; *Polyporellus* P. Karst. 1879; *Leucoporus* Quél. 1886. — Jacq. 1778 (MaJ 1): 142 *pl. 12* (*Boletus*); Secr. 1833 M. 3: 63 (*Polyporus*) may represent *P. varius* sensu stricto; Donk 1969 (Pe 5): 250, note.

Boletus elegans Bull. 1780: *pl. 46* (France) (d.n.), not ∼ Bolt. 1788 (d.n.), not ∼ Schum. 1803 (d.n.) per Fr. 1838; fide Bull. 1791 H.: 338 = *Boletus calceolus*; fide Donk. — *Polyporus* (Bull.) per Trog 1832, misapplied; *Polyporellus* P. Karst. 1879; *Melanopus* Pat. 1887 (nom. nud.: n.v.p.), 1900, not ∼ Konr. & M. 1935; *Favolus* A. Ames 1913. — Donk 1969 (Pe 5): 248, 1971 (Pe 6): 215, notes. — Sensu Fr. → *Polyporus varius* sensu stricto.

Boletus nummularius [!] Bull. 1782 (France) (d.n.); fide Donk 1969

(Pe 5): 253. — *Polyporus* Fr. 1815 (*"numularius"*) (d.n.); *Polyporus varius* var. (Bull.) per Fr. 1821; *Boletus* Mérat 1821; *Coltricia* S. F. Gray 1821; *Polyporus* Pers. 1825, Wallr. 1833; *Leucoporus* Bourd. 1894; *Melanopus* Pat. 1887 (nom. nud.: n.v.p.), Maire & al. 1903 ("Pat."); ≡ *Boletus ramulorum* J. F. Gmel. 1792 (d.n.). — Bull. 1782: *pl. 124*; 1791 H.: 335, Schrad. 1794: 152, excl. of var. β (*Boletus*); Trog 1832 (Fl 15): 553; Secr. 1833 M. 3: 64 (*Polyporus*); Quél. 1888: 403 (*Leucoporus leptocephalus* var.); Bourd. & G. 1928: 528 (*Melanopus varius* subsp.); Donk 1933: 138 (*Polyporus varius* subsp.); Konr. & M. 1935 I. 5: *pl. 428 f. 1* (*Melanopus*).

Boletus calceolus Bull. *1787*, at least in part (France) (d.n.); fide Fr. 1838: 440. — *Boletus* Bull. per St-Am. 1821; *Polyporus* Balbis 1828, Murrill 1904; *Agaricus* Boisd. 1828 (nom. nud., error: n.v.p.): *Leucoporus* Quél. 1886; *Melanopus* Riel 1919. — Bull. 1789: *pl. 445 f. 2* & 1791 H.: 338, in part (*Boletus*); Quél. 1888: 404 (*Leucoporus*); Donk 1968 (Pe 5): 243, notes.

Boletus lateralis Bolt. 1788: 83 *pl. 83* (England) (d.n.); fide Pers. 1796 O. 1: 85 & Donk 1969 (Pe 5): 250; fide Fr. 1838: 441 = *Polyporus elegans* [sensu Fr.] ("h.l. e descript."). — *Boletus* Bolt. per Hook. 1821, not ∼ Bundy 1883.

? *Boletus pallescens* Schrad. apud J. F. Gmel. 1792 & Schrad. 1794: 154 (Germany) (d.n.), not ∼ (Konr.) Sing. 1936; fide Fr. 1838: 432 = *Polyporus leptocephalus*.

Boletus nigripes With. 1796: 316 (England) (d.n.).

? *Polyporus globularis* Pers. 1825; cf. Donk 1969 (Pe 5): 249. — [≡ *Polyporus exiguus, coriaceus albus, lignis adnescens* Mich. 1729: 130 *pl. 70 f. 7* (presumably northern Germany/Poland)].

Polyporus leprodes Rostk. 1828 (StP 4): 33 *pl. 15* (Germany/Poland); fide Fr. 1863 M. 2: 251 (subsp.).

Polyporus petalodes Fr. 1838 (Germany), not ∼ Berk. 1856; fide Bourd. & G. 1925 (BmF 41): 111 = *Melanopus varius* subsp. *nummularius* (forma). — *Polyporellus* P. Karst. 1879; *Leucoporus* Quél. 1892. — Fr. 1838: 444 (*Polyporus*); Bourd. & G. 1928: 528 (*Melanopus varius* subsp. *nummularius* f.). — Sensu Quél. 1892 (Crf 20²): 469 *pl. 3 f. 25* (*Leucoporus*) = ?

? *Polyporus boltonii* Rostk. 1848 (StP Fs. 27–28): 47 *pl. 24* (Germany/ Poland); fide Fr. 1863 M. 2: 340. — Perhaps rather *Polyporus badius*?

Polyporus minimus (Fr. per Fr.) Cooke 1878. — *Polyporus nummularius* var. Fr. 1815 O. 1: 123 (Sweden) (d.n.) per Fr. 1821.

Polyporus gintlianus Velen. 1922: 687 *f. 103: 4* [see Pilát 1948: 260 for Latin translation] (Czechoslovakia); fide Pilát 1937 (ACE 3): 106. — If correctly identified by Pilát, the spores described must have been foreign.

Melanopus elegans Konr. & M. 1935, not ∼ (Bull. per Trog) Pat. 1900. — [≡ *Polyporus elegans* (Bull.) per Trog sensu Fr. 1838: 440, but excl.

of type, (Sweden)]; ≡ *Polyporus varius* subsp. *elegans* Donk 1933, typonym, in part. — Donk 1933: 139, excl. of cited specimen (*Polyporus varius* subsp.): Konr. & M. 1935 I. 5: *pl. 426 f. 2* (*Melanopus*): Donk 1969 (Pe 5): 248, notes.

M.—*Boletus perennis* Batsch sensu Vahl 1792 (Fd 6 / F. 18): 8 *pl. 1075 f. 1*; fide Fr. 1838: 440 = *Polyporus elegans* [sensu Fr.] ("optima").

M.—*Boletus elegans* Bull. sensu Fr. 1838 (*Polyporus*); fide Donk 1969 (Pe 5): 248. — Fr. 1838: 440; Pat. 1883 T.a. 1: 59 *f. 137* (*Polyporus*); Bourd. & G. 1928: 527 (*Melanopus varius* subsp.): Bres. 1931 (BIm 20): *pl. 967* (*Polyporus*). → *Polyporus varius* subsp. *elegans* Donk, → *Melanopus elegans* Konr. & M.

M.—*Polyporus picipes* Rost. [≡ Fr.] sensu Rostk. 1848 (StP Fs. 27–28): *pl. 20*, excl. of descr.; fide Donk 1933: 136, & cf. Donk 1968 (Pe 5): 254.

PORIA (104)

Poria Pers. per S. F. Gray 1821, in part, excl. of lectotype, not ∼ P. Karst. 1881 [1960 (Pe 1): 266]. — *Poria* Pers. 1794 (d.n.), Roussell 1806 (d.n.), not ∼ [Hill] Adans. 1763 (d.n.); ≡ *Physisporus* Chev. 1826 [1960 (Pe 1): 256]. — Lectotype: *Boletus medulla-panis* Jacq. sensu Pers., listed under *Perenniporia*. — Sensu P. Karst. → *Poria* P. Karst. → *Phellinus*; sensu Bond. → *Incrustoporia*.

Theleporus Fr. 1847 ("*Thelepora*"), 1848 [1960 (Pe 1): 286]. — Monotype: *Theleporus cretaceus* Fr. (extra-European).

Poroptyche G. Beck 1888 [1960 (Pe 1): 273] (nom. monstr.). — Monotype: *Poroptyche candida* G. Beck.

Sarcoporia P. Karst. 1894 [1960 (Pe 1): 277]. — Monotype: *Sarcoporia polyspora* P. Karst.

Melanoporella Murrill 1907 [1960 (Pe 1): 239]. — Holotype: *Polyporus carbonaceus* B. & C. (extra-European).

Melanoporia Murrill 1907 [1960 (Pe 1): 239]. — Holotype: *Polyporus niger* Berk. (extra-European).

Amyloporia Bond. & S. ex Sing. 1944, Bond. 1953 [1960 (Pe 1): 185]. — *Amyloporia* Bond. & S. 1941 (lacking Latin descr.: n.v.p.). — Holotype: *Poria calcea* (**Fr.** ex Pers.) Cooke [sensu Bond. & S. = *Poria xantha*]. — Cf. Donk 1967 (Pe 5): 67.

Fibuloporia Bond. & S. ex Sing. 1944 [1960 (Pe 1): 214]. — *Fibuloporia* Bond. & S. 1941 (lacking Latin descr.: n.v.p.). — *Holotype: Poria mollusca* Pers. [sensu Bres. = *Poria mucida* (Pers. per Fr.) Cooke].

Ceriporiopsis Domański 1963 (APo 32): 731. — Holotype: *Poria gilvescens* Bres.

Anomoporia Pouz. 1966 (ČM 20): 172. — Holotype: *Poria bombycina* (Fr.) Cooke.

Strangulidium Pouz. 1967 (ČM 21): 206. — Holotype: *Polyporus sericeo-mollis* Romell, in part.

Fibroporia Parm. 1968: 176. — Holotype: *Poria vaillantii* (DC. per Fr.) Cooke.

Riopa D. Reid 1969 (RM 33): 244. — Holotype: *Riopa davidii* D. Reid. — (71).

Sporotrichum Link per Fr. 1821: xliv [1962 (Ta 11): 100] (nom. anam.). — *Sporotrichum* Link 1809 (d.n.), — Lectotype: *Sporotrichum aureum* Link.

SPECIAL LITERATURE.—General & various species: Dománski, *1959a, 1963, 1964b*; Donk, *1967*; Egeland, *1914*; Eriksson, *1949*; Gricius, *1967*; Guillemin, *1923*; von Höhnel, *1907*; Jahn, *1971*; Lowe, *1966*; Sartory & Maire, *1920*; Wright, *1964a, 1964b*.

"Poroptyche": Beck von Mannagetta, *1888*.

Poria albidofusca: Domański, *1966a*.

Poria aurea sensu Pilát: Pilát, *1933a*.

Poria expansa: Campbell & Bryant, *1940*: Domański & Orlicz, *1967a*, *1967b*; Jahn, *1967a*; Mangin & Patouillard, *1922*.

Poria gilvescens: Domański, *1971*.

Poria placenta sensu lato: Bagchee & Bakshi, *1951* (*P. monticola*): Brushhaber & Jenkins, *1971* (*P. monticola*): Cartwright, *1930* (*Trametes serialis*): Domański, *1965a*, *1970a*; Hirt & Lowe, *1945* (*P. microspora*): Overholts, *1946*.

Poria rivulosa: David, *1972*.

Poria šimanii: Parmasto, *1961*.

Poria subvermispora: Domański, *1969d*.

Poria terrestris: Zak, *1969*.

Poria vaillantii: Da Costa & Kerruish, *1965*; Sison, Schubert, & Nord, *1958*.

Poria xantha: Hirt, *1949a*; Razzore, *1911* (*P. suaveolens*).

Poria sp.: Schorstein, *1907*.

NOTA BENE.—The name *Poria* is applied here as if its type is as yet of uncertain taxonomic position and not referable to any of the segregates from the conventional genus *Poria*, although on this Check list the lectotype is considered to belong to *Perenniporia*.

albidofusca Domański 1966 (Poland). — Domański *1966a*, with cult. char.

albobrunnea (Romell) Lloyd 1912. — *Polyporus* Romell 1911 (Sweden); *Leptoporus* Pilát 1937; *Poria* D. Baxt. 1939 (preoccupied); *Tyromyces* Bond. 1953. — Romell 1911 (ABS 11³): 10 *pl. 1 f. 6*; 1912 (SbT 6): 641, foot-note (*Polyporus*); D. Baxt. 1939 (PMi 24): 172; Overh. 1940 (M 32): 262 *fs. 4–6*; Lomb. & Gilb. 1965 (M 57): 46 *fs. 1A, 5A*, with cult. char. (*Poria*); Domański 1965 (FpG 2): 144 *f. 49, pl. 33 f. 2* (*Tyromyces*); Lowe 1966: 104 *f. 89*; 144 *f. 49, pl. 33 f. 2* (*Tyromyces*); Lowe 1966: 104 *f. 89*; Donk 1967 (Pe 5): 78, notes (*Poria*). — Sensu Kotl. & P. 1956 → *Tyromyces leucomallelus*.

Poria dichroa Bres. 1925 (M 17): 75 (U.S.A., Idaho); fide D. Baxt. 1939 (PMi 24): 172, 175 & Lowe 1966: 104, 106.

M.—*Trametes squalens* P. Karst. sensu Lowe 1956 (M 48): 123 (*Poria*); fide Lowe, l.c. (*Poria albobrunnea* cited as a syn.) & cf. Donk 1962 (Pe 2): 235–237.

albogrisea (Britz.) Sacc. & Syd. 1899, not ∼ [Ell. & Ev. apud] Burt 1923 (n.v.p.). — *Polyporus* Britz. 1897 (BCb 71): 58 [*pl. 657 fs. 217, 218*] (Germany). — Insufficiently known.

albolutescens (Romell) Bourd. & G., June 1914. — *Polyporus* Romell 1911 (Sweden); *Poria* Egeland, Sept. 1914 (preoccupied); *Anomoporia* Pouz.

1966. — Romell 1911 (ABS 11³): 11, excl. of specimen from Rydbo (*Polyporus*); Lomb. & Gilb. 1965 (M 57): 52 *fs. 1B, 5B*, with cult. char.; Lowe 1966: 59 *f. 35*; Donk 1967 (Pe 5): 79, notes (*Poria*).

Poria grandis Overh. 1943 (U.S.A., Tennessee/North Carolina): fide Lowe 1946: 54 & 1966: 59. — Overh. 1943 (M 35): 249 *f. 2*.

Tyromyces **allantoideus** M. P. Christ. 1960 (Denmark). — M. P. Christ. 1960 (DbA 19): 364 *f. 363*. — H. Jahn 1971 (WPb 8): 59 suggested = *Ceriporiopsis* [*Poria*] *gilvescens*.

alpina Litsch. 1939 (Austria). — Litsch. 1939 (ÖbZ 88): 143 *f. 6*. — Sensu Lomb. & Gilb. 1965 (M 57): 53 *fs. 1C, 5C*, with cult. char., & Lowe 1966: 89 *f. 69*, same species?

aneirina (Sommerf.) Cooke 1886. — *Polyporus* Sommerf. 1826: Fr. 1828 (Norway); *Physisporus* Gillet 1878; P. Karst. 1882; *Trametes* Cooke & Q. 1878, Cooke 1878; *Caloporia* P. Karst. 1898; *Tyromyces* Bond. & S. 1941; *Ceriporiopsis* Domański 1963. — Bourd. & G. 1928: 667 *f. 185A*; Overh. 1942: 21; Lowe 1946: 63; Malenç. 1952 (BbF 99¹⁰): 44 (*Poria*); Domański 1963 (APo 32): 732 *f. 3*; 1965 (FpG 2): 81 *f. 23, pl. 15* (*Ceriporiopsis*); Lomb. & Gilb. 1965 (M 57): 55 *fs. 1D, 5D*, with cult. char.; Lowe 1966: 68 *f. 45*; Donk 1967 (Pe 5): 79, note (*Poria*). — Sensu Fr. in herb., in part, & Bres. 1897. → *Oxyporus corticola*.

Polyporus macer Sommerf. 1826: 279 (Norway), not ∼ Berk. 1856; fide Lowe 1966: 68.

? *Polyporus pyrrhoporus* Dur. & Mont. apud Mont. 1856: 162 (Algeria); fide Lowe 1963 (M 55): 463. — *Poria* Cooke 1886. — Lowe 1947 (Ll 10): 55 (*Poria*). — The specific epithet and the colours mentioned in the protologue seem to clash with Lowe's identification; & cf. Pat. 1897: 50.

Antrodia serena P. Karst. 1881 (Mfe 6): 10 (Finland); fide Romell 1911 (ABS 11³): 21, Bres. 1920 (Am 18): 69, & Lowe 1956 (M 48): 100. — *Physisporus* P. Karst. 1881; *Trametes* Sacc. 1888; *Poria* Bres. 1897.

Poria fulvescens Bres. 1897 (AAR III 3): 81 (Hungary, now Czechoslovakia); fide Bres. 1920 (Am 18): 69.

? *Poria wasjuganica* Pilát 1936 (U.S.S.R., Russia, Siberia); fide Domański 1964 (APo 33): 171, but cf. Lowe 1966: 69. — *Gloeoporus* Bond. 1953 (incomplete ref.: n.v.p.). — Pilát 1936 (BmF 51): 382 *f. 6, pl. 6 f. 4, pl. 11 f. 2*; 1941 (ACE 3): 437 *f. 204, pl. 277 f. b, pl. 228 f. 8*; Domański, l.c.

Poria pulvinascens Pilát 1942 (lacking Latin descr.: n.v.p.), 1953 (Sweden); fide Ryv. 1971 (NMN 17): 166. — Pilát 1942 (ACE 3): 451 *f. 213, pl. 289*; Domański 1964 (APo 33): 176. — Domański, l.c., referred this to *Coriolellus* [*Poria*] *crustulina*.

aurea Peck (O) sensu Pilát 1933. — Pilát 1933 (H 73): 31 *f. 1, pl. 1*; 1941 (ACE 3): 401 *f. 171, pl. 254* (*Poria*); Domański 1965 (FpG 2): 120 *f. 42* (*Chaetoporellus*).

bibula (Pers.) Pilát 1932. — *Polyporus* Pers. 1825; *Poria* Lloyd 1910 (incidental mention: n.v.p.); *Tyromyces* Bond. & S. 1941. — Sensu Bourd. & G. 1928: 668 (*Poria aneirina* subsp.).

bombycina (Fr.) Cooke 1886. — *Polyporus* Fr. 1827: 1828 (Sweden); *Physisporus* Gillet 1878, P. Karst. 1882; *Trametes* Cooke & Q. 1878, not ~ Pat. 1889; *Poria* Cooke 1886; *Fibuloporia* Bond. & S. 1941 (generic name n.v.p.), Bond. 1953; *Anomoporia* Pouz. 1966. — Bres. 1897 (AAR III 3): 81; Egeland 1914 (NMN 52): 152; Bourd. & G. 1928: 670; Pilát 1941 (ACE 3): 431 *f. 199, pls. 273, 274*; Overh. & L. 1946 (M 38): 206 *f. 2E*; D. Baxt. 1948 (PMi 32): 204 *pl. 7*; Jo. Erikss. 1958 (Sbu 16[1]): 148; Lomb. & Gilb. 1965 (M 57): 57 *fs. 1E, 6A*, with cult. char. (*Poria*); Domański 1965 (FpG 2): 40 *f. 6, pl. 5* (*Fibuloporia*); Lowe 1966: 65 *f. 42* (*Poria*). — Sensu Wirtgen → *Polyporus wirtgenii* Fr. (**0**).

Polyporus hians P. Karst. 1867, 1868 (Finland); fide Bres. 1897 (AAR III 3): 81 & Lowe 1956 (M 48): 119. — *Physisporus* P. Karst. 1881; *Antrodia* P. Karst. 1882; *Poria* Cooke 1886. — P. Karst. 1868 (NfF 9): 360 [repr. 1869 (H 8): 75].

Poria fulvella Bres. 1925 (M 17): 76 (U.S.A., Idaho); fide R. L. Gilb. 1956 (Ll 19): 66, 82.

? *Poria coniferarum* D. Baxt. 1938 (PMi 23): 300 *pl. 6 f. 1* (U.S.A., Idaho); fide Lowe 1966: 66 (paratype studied). — The spores were described "with slightly roughened or echinulate walls".

buxi Bond. 1940 (U.S.S.R., Caucasia). — *Aporpium* Bond. & S. 1941 (generic name n.v.p.), Bond. 1953; *Fibuloporia* Bond. & M. Bond. 1960 (incomplete ref.: n.v.p.). — Bond. 1940 (BMs 5): 20.

carneo-lilacea (Britz.) Sacc. & Syd. 1899. — *Polyporus* Britz. 1897 (BCb 71): 58 [*pl. 656 f. 211*] (Germany). — Britz. 1910 (BbC 26): 212 (*Polyporus*). — Insufficiently known.

Multiporus **chlamydoformans** R. & O. Falck 1937 (presumably Germany) (lacking Latin descr.: n.v.p.). — R. &. O. Falck 1937 (HF 12): 32, 58, 59 *fs. 26–39, pl. 3 fs. 2–5, pl. 3b fs. 1–4*.

confusa Bres. 1897 (Hungary, now Czechoslovakia). — *Aporpium* Bond. 1953. — Bres. 1897 (AAR III 3): 87; Lowe 1962 (PMi 47): 183. — Cf. Lowe 1966: 133. — Sensu Malenç. 1952 (Bbf 99[10]): 45 = *Incrustoporia percandida*, fide Mal. & Bert. 1971 (Apb 8): 31, 33.

consobrina Bres. apud Bourd. & G. 1925 (France) (**105**). — *Fibuloporia* Bond. & S. 1941 (generic name n.v.p.). — Bourd. & G. 1928: 671.

crassa (P. Karst.) Sacc. 1891. — *Physisporus* P. Karst. 1889 (Finland); *Amyloporia* Bond. & S. 1941 (generic name n.v.p.); *Fomitopsis* Bond. 1953; *Fomes* E. Komar. 1964 (incomplete ref.: n.v.p.), not ~ (Fr.) Cooke 1885. — Bourd. & G. 1928: 691, excl. of information on spores taken from Höhn.; Litsch. 1939 (ÖbZ 88): 145, in obs.; Overh. 1942: 26; Jo. Erikss. 1949 (SbT 16[1]): 22 *f. 6* (*Poria*); Domański 1965 (FpG 2): 92 *fs. 28, 29, pls. 19, 20, pl. 21 f. 1* (*Amyloporia*); Lowe 1966: 70 *f. 46*, "monomitic" [?] (*Poria*). — Sensu Höhn., D. Baxt. → *Poria xantha.*

crustulina Bres. 1925 (U.S.A., Idaho; lectotype, Weir 15,025). — *Tyromyces* Parm. 1959; *Coriolellus* Domański 1964; *Antrodia* Parm. 1968; *Diplomitoporus* Domański 1970. — D. Baxt. 1937 (PMi 22): 290; Overh. 1942: 44; Jo. Erikss. 1958 (Sbu 16[1]): 148 *f. 47* (*Poria*); Parm. 1959 (EAT 8[4]): 271 *fs. 4–6, pl. (2) fs. 3, 4* (*Tyromyces*); Domański 1965 (FpG 2): 191 *f. 65, pl. 41 f. 2* (*Coriolellus*); Lowe 1966: 101 *f. 84* (*Poria*); Domański 1970 (APo 39): 193 *fs. 2a, 3c, d, 7–9*, with cult. char. (*Diplomitoporus*).

    Poria chromatica Overh. 1939 (lacking Latin descr.: n.v.p.) (U.S.A., New Hampshire), not ∼ (B. & Cooke) Cooke 1886; fide Overh. 1942: 45. — Overh. 1939 (PPA 13): 123 *f. 2.*

    Poria conwayana Pilát 1940 (U.S.A., New Hampshire); cf. Domański 1964 (APo 33): 176 & fide Lowe 1966: 101, Domański 1970 (APo 39): 192, 199. — Pilát 1940 (Sbč 3): 3 *pl. 1 f. 1*; Domański 1964 (APo 33): 176 *f. 7* (*Poria*). — V.s.: "*convayana*".

Riopa **davidii** D. Reid 1969 (France, Corsica) (**71**). — D. Reid 1969 (RM 33): 247 *tpl. 2 f. 7.*

dentipora Pilát 1941, not ∼ (Pers.) Cooke 1886. — [≡ *Poria dentipora* (Pers.) Cooke sensu Bres. 1897, excl. of type, (Hungary, now Czechoslovakia)]; ≡ *Coriolus dentiporus* Bond. & S. 1941, typonym. — Bres. 1897 (AAR III 3): 82; Pilát 1941 (ACE 3): 440 *f. 206, pl. 281 f. 1* (*Poria*), & cf. Bourd. & G. 1928: 673, in obs.; Donk 1967 (Pe 5): 88, note.

eupatorii (P. Karst.) Sacc. 1891. — *Physisporus* P. Karst. 1884 (Rm 6): 214 (France). — Lowe 1956 (M 48): 111 (*Physisporus*).

expansa (Desm.) H. Jahn 1967 (**107**). — *Boletus* Desm. 1823 (France); *Polyporus* Desm. 1825; *Phellinus* Pat. & Har. 1900; *Fomes* Dom. & Orl. 1967. — Desm. 1823: 19 (*Boletus*); Donk 1933: 228 (*Polyporus*); H. Jahn 1967 (WPb 6): 100 *fs. 11, 12, Abb. 59–61*; Siepm. 1969 (NH 18): 187 *tpl. 2 f. 1, tpl. 4 f. 1*, cult. char. (*Poria*); Donk 1967 (Pe 5): 89, notes.

    Polyporus megaloporus Pers. 1825 (France), not ∼ Mont. 1854; fide Donk 1933: 228. — *Poria* Cooke 1886; *Fomes* Bres. apud Lloyd 1899 (incidental mention: n.v.p.); *Phellinus* R. Heim 1942, Bond. 1953; *Fomes* Dom. & Orl. *1967a* (incomplete ref.: n.v.p.), *1967b* (not accepted: n.v.p.). — Pers. 1825: 88 (*Polyporus*); Bres. 1897 (AAR III 3): 78; Bourd. & G. 1928: 686; R. Heim 1942: 18 *fs. 1, 16–18* (*Phellinus*); Cartwr. & Findl. 1958: 219 *pl. 39, pl. 41 fs. c, d*; D. Baxt. 1950 (PMi 34): 43; Lowe 1966: 168 *f. 157* (*Poria*); Dom. & Orl. 1967 (AmW 3): 51 *fs. 1, 3* (*Polyporus*); Donk 1967 (Pe 5): 94, notes.

    ? *Polyporus unitus* Pers. 1825: 93: Fr. 1828 (France), not ∼ Lloyd

1917; fide Bres. 1897 (AAR III 3): 78 & 1920 (Am 18): 69 = *Poria megalopora*; & cf. Lowe 1966: 108, in obs. **(107)**. — *Physisporus* Gillet 1878; *Poria* P. Karst. 1881; *Perenniporia* Murrill 1942, misapplied; *Fomitopsis* Bond. 1953, misapplied; *Fomes* Lowe 1955, misapplied, E. Komar. 1964 (indirect ref.: n.v.p.), misapplied. — Donk 1967 (Pe 5): 116, notes. Sensu Donk 1933 → *Perenniporia medulla-panis*.

Fomitoporia ohiensis Murrill 1907 (NAF 9): 11 (U.S.A., Ohio); fide Lowe 1949 (Ll 11): 167 & D. Baxt. 1950 (PMi 34): 43 = *Poria megalopora*; fide H. Jahn 1967 (WPb 6): 100, 103.

Phellinus cryptarum Cartwr. & Findl. 1946: 203 *pl. 35* ("Karst.") (lacking Latin descr. or valid ref.: n.v.p.); fide Bourd. apud Cartwr. & Findl. 1946: 203 = *Poria megalopora*. — [≡ *Phellinus cryptarum* (Bull. per Fr.) P. Karst. sensu Cartwr. & Findl. excl. of type ("The plant . . . is in no way related to the *Boletus cryptarum* Bull. . . ..")].

M.—*Polyporus undatus* Pers. sensu Quél. 1888: 380, at least in part (*Poria*); fide Bourd. & G. 1928: 685 = *Poria megalopora* (forma), with descr. & note on p. 686.

M.—*Boletus cryptarum* Bull. sensu Quél. 1888: 395, in part (*Phellinus*). — Mang. & Pat. 1922 (CrP 175): 389 *4 fs.* (*Phellinus*).

M.—*Boletus spongiosus* Pers. (**O**) sensu Quél. 1892 (Crf 20²): 468 *pl. 2 f. 19* (*Poria*); fide Bourd. & G. 1928: 685 = *Poria megalopora*, with descr. in obs. on p. 686. — Mentioned by D. Baxt. 1950 (PMi 34): 43 as "*Poria spongia* Quél. . . .".

flavicans (P. Karst.) Sacc. & Syd. 1899 **(106)**. — *Physisporus* P. Karst. 1896 & 1898 (Finland); *Polyporus* Romell 1911. — Romell 1911 (ABS 11³): 25, in obs. (*Polyporus*), Egeland 1914 (NMN 52): 148 (*Poria*); Romell 1926 (SbT 20): 11; D. Baxt. 1936 (PMi 21): 253 *pl. 32 f. 2* (*Polyporus*). Sensu Lowe → *Chaetoporus luteo-albus*.

Poria johnstonii Murrill 1920 (U.S.A., California); fide Ryv. 1970 (NyM 17): 166 ("*Poria flavicans* Karst. sensu Romell" cited as a syn.). — D. Baxt. 1938 (PMi 23): 298 *pl. 8*; Lowe 1947 (Ll 10): 52; 1966: 76 *f. 55*; Ryv. l.c.

Oligoporus **friesii** R. & O. Falck 1937 (lacking Latin descr.: n.v.p.) (presumably Germany) **(109)**. — R. & O. Falck 1937 (HF 12): 41, 59 *fs. 40–46*.

fusco-carnea (Pers.) Cooke 1886. — *Polyporus* Pers. 1825: 97 (France). — Insufficiently known.

Irpex **galzini** Bres. 1908 (France). — *Trametes* Pilát 1939–40; *Coriolus* Bond. & S. 1941. — Bourd. & G. 1928: 575 (*Irpex*); Pilát 1940 (ACE 3): 325 *f. 139 pl. 218 f. b* (*Trametes*); D. Baxt. 1941 (PMi 26): 112, in obs., *pl. 6* (*Irpex*). — This is supposed to be closely related to *Poria* [*Antrodia*] *sinuosa*.

gelatinoso-tubulosa Pilát 1936 (U.S.S.R., Russia, Siberia). — *Gloeoporus* Bond. 1953 (incomplete ref.: n.v.p.) & others (n.v.p.). — Pilát 1936

(BmF 51): 383 *pl. 11 f. 3*; 1941 (ACE 3): 426 *f. 196, pl. 269 f. b* (*Poria*); Kartav. 1961 (BMs 14): 190 *fig.*; Parm. 1963 (TÜT 136): 114, 128 *f. 5* (*Gloeoporus*).

gilvescens Bres. 1908 (Hungary, now Czechoslovakia). — *Ceriporia* Bond. & S. 1941; *Ceriporiopsis* Domański 1963. — Wak. & Pears. 1920 (TBS 6): 321 *fig.*; Bourd. & L. Maire 1920 (BmF 36): 85; Bourd. & G. 1928: 663; Malençç. 1956 (BmF 71): 308 *f. 9* (*Poria*); Domański 1963 (APo 32): 731, 733 *f. 1*; 1965 (FpG 2): 73 *f. 19*; *1971*, with cult. char. (*Ceriporiopsis*); Lowe 1966: 76 *f. 54* (*Poria*). — Sensu Overh. = *Aporpium caryae* (**O**).

    M.—*Boletus sanguinolentus* A. & S. sensu Bres. 1897 (AAR III 3): 83 (*Poria*); fide Bres. 1908 (Am 6): 40 & Bourd. & G. 1928: 663.

    M.—*Leptoporus micantiformis* Pilát sensu Lowe 1946: 56 (*Poria*); fide Lowe 1966: 76. — Lowe 1946: 56 (*Poria*).

greschikii Bres. 1920 (Hungary). — Bourd. & G. 1928: 666. — Referred by Lowe 1959 (Ll 21): 101, 104 to *Poria xantha* "sense of Lind . . . a weathered specimen".

hibernica (B. & Br.) Cooke 1886. — *Polyporus* B. & Br. 1871 (AM IV 7): 428 (Ireland). — Lowe 1962 (PMi 47): 184; 1966: 92, in obs. (*Poria*). — Sensu Bres. 1903 (Am 1): 79, insufficiently described, viz. spores only.

hippophaës Bres. 1926: 57 (Austria). — Insufficiently known.

kazakstanica Pilát 1937 (U.S.S.R., Kazachstan). — *Hapalopilus* Bond. & S. 1941; *Fibuloporia* Bond. 1953. — Pilát 1937 (BmF 52): 318 *f. 25, pl. 4 fs. 1, 2*; 1942 (ACE 3): 434 *f. 201, pl. 276 f. b, pl. 277 f. a* (*Poria*); Domański 1965 (FpG 2): 43 *f. 8, pl. 6 f. 1* (*Fibuloporia*). — Referred by Lowe 1966: 59 to *Poria albolutescens*.

lacera (P. Karst.) Sacc. 1888. — *Physisporus* P. Karst. 1882 (Mfe 9): 69 (Finland). — Cf. Lowe 1956 (M 48): 113.

lenis (P. Karst.) Sacc. 1888. — *Physisporus* P. Karst. 1886 (Finland); *Polyporus* Strass. 1900, not ∼ Lév. 1848; *Amyloporia* Bond. & S. 1941 (generic name n.v.p.), Bond. 1953. — Romell 1911 (ABS 11³): 17; 1926 (SbT 20): 12 (*Polyporus*); D. Baxt. 1936 (PMi 21): 248 *pl. 31*; Jo. Erikss. 1949 (SbT 43): 11 *f. 3, pl. 2* (*Poria*); Domański 1965 (FpG 2): 86 *fs. 25, 26, pl. 16, pl. 17 f. 1* (*Amyloporia*); Lowe 1966: 98 *f. 80* (*Poria*). — Jo. Erikss., l.c., mentions many herb. names by Romell; these are entered on the "List of omitted names" under *Poria jodiolens* (**O**).

    Polyporus vulgaris Fr. 1821: 381 (Sweden) (nomen ambiguum). — *Physisporus* Chev. 1826; *Poria* Cooke 1886, Quél. 1886, not ∼ S. F. Gray 1821; *Trametes* Pat. & Har. 1900; *Aporpium* Bond. & S. 1941 (generic name n.v.p.), Bond. 1953, misapplied; *Tyromyces* Bond. & M. Bond. 1960 (incomplete ref.: n.v.p.). — Cf. Romell 1926 (SbT 20): 20; and Donk 1967 (Pe 5): 123, typification and notes. — Sensu Bres. 1897, in part,

Romell 1911 → *Poria xantha*; sensu Bres. 1897, in part → *Poria romellii*; sensu Velen. → *Schizopora paradoxa*; sensu Romell 1926 → *Incrustoporia subincarnata*.

Poria tenuipora Murrill 1920 (M 12): 85 (Jamaica); fide Lowe 1966: 98. — Lowe 1947 (Ll 10): 58; 1963 (M 55): 468.

? *Poria earlei* Murrill 1920 (M 12): 86 (Jamaica); fide Lowe 1963 (M 55): 469 ("type sterile") = *Poria montana*; fide Lowe 1966: 98. — Lowe 1947 (Ll 10): 49.

Poria montana Murrill 1920 (M 12): 307 (Jamaica); fide Lowe 1966: 98. — Lowe 1947 (Ll 10): 53; 1963 (M 55): 469.

Poria calcea var. *coriacea* Bourd. & G. 1925 (France); fide Bourd. & G. 1925 (BmF 41): 233 (*Poria lenis* cited as a syn.). — Bourd. & G. 1928: 674.

Poria calcea f. *micropora* Bourd. & G. 1928 (France); fide Bourd. & G. 1928: 674 (*Poria lunata* Romell cited as a syn.). — Bourd. & G., l.c.

Poria lunulispora Pilát 1936 (U.S.S.R., Russia, Siberia); fide Pilát 1942 (ACE 3): 442 = *Poria lenis* (forma). — Pilát 1936 (BmF 51): 381 *pl. 10 f. 3* (*Poria*); 1941 (ACE 3): 442 *pl. 283 f. a* (*Poria lenis* f.).

M.—*Polyporus vulgaris* var. *"β. P. calceus"* Fr. [≡ *Polyporus calceus* (Fr. ex Pers.) Schw. (**O**)] sensu Bres. 1897 (AAR III 3): 86 (*Poria vulgaris* f. *calcea*) & 1908 (Am 6): 41 (*Poria*); cf. Donk 1967 (Pe 5): 85.

lindbladii (Berk.) Cooke 1886. — *Polyporus* B. & Br. 1865 (nom. prov.: n.v.p.), Berk. 1872 (G 1): 54 (Sweden); *Polystictus* Cooke 1886; *Coriolus* Pat. 1900. — Donk 1971 (Pe 6): 215, note. — Sensu Berk., l.c., in part: M. A. Curt. 1623 from South Carolina = *Polyporus* [*Hirschioporus*] *versatilis* (Berk.) Romell (extra-European), fide Lowe 1966: 134.

Polyporus cinerascens Bres. apud Strass. 1900 (Austria), not ∼ (Schw.) Steud. 1824, not ∼ Lév. 1844, not ∼ Velen. 1922. — *Poria* Sacc. & Syd. 1902 ("*cinerescens*"); *Tyromyces* Bond. & S. 1941; *Coriolus* E. Komar. 1964 (incomplete ref.: n.v.p.). — Bourd. & G. 1928: 667 *f. 184*; Overh. 1942: 24 (*Poria*); Domański 1959 (Mob 8): 159, 168 *fs. 6, 8*; 1965 (FpG 2): 147 *f. 51, pl. 29 f. 2* (*Tyromyces*); Lowe 1966: 104 *f. 88* (*Poria*).

Poria subavellanea Murrill 1920 (M 12): 88 (U.S.A., Alabama); fide Lowe 1947 (Ll 10): 57 = *Poria cinerascens*.

M.—*Polyporus subfuscoflavidus* Rostk. (**O**) sensu Romell 1926 (SbT 20): 19, interpretation erroneously attributed to Fr., & "of most North American and European determinations"; fide Romell, l.c., & Lowe 1966: 104 = *Poria cinerascens*.

mappa Overh. & L. 1946 (Canada, British Columbia). — *Fibuloporia* M. P. Christ. 1960. — Lowe 1946: 66 *f. 16* (*Poria*); M. P. Christ. 1960 (DbA 19): 340 *f. 336*; M. Bond. 1964 (NSn): 187 (*Fibuloporia*); Lowe 1966: 77 *f. 56* (*Poria*).

mentschulensis Pilát 1953 (U.S.S.R., Ukraine). — *Poria* Pilát 1942 (lacking Latin descr.: n.v.p.); *Tyromyces* Bond. 1953 (lacking Latin descr.: n.v.p.), Domański 1965. — Pilát 1941 (ACE 3): 414 *f. 182 pl. 261 f. b* (*Poria*); Domański 1965 (FpG 2): 149 *f. 52* (*Tyromyces*).

metamorphosa (Fuck.) Cooke 1886. — *Polyporus* Fuck. 1873 (Germany). — Fuck. 1873 (Jna 27–28): 87 (*Polyporus*); Bourd. & G. 1928: 668 (*Poria aneirina* f. 3); Wakef. 1952 (TBS 35): 38 (*Poria*); & cf. Lowe 1966: 69, in obs.

Sporotrichum aureum Link 1809 (Germany) (d.n.) per S. F. Gray 1821: 551, not ∼ (Pers. per Pers.) Fr. 1832; fide Donk — ≡ *Sporotrichum aurantiacum* Fr. 1832: 423 (nom. anam.). — J. W. Carm. 1962 (CJB 40): 1150 *pl. 8 fs. 59, 60*; Arx 1970: 164 *f. 101d*; 1971 (Pe 6): 179 *f. 1*; Donk.

Trichoderma aureum Pers. 1796, 1801 (Germany) (nom. anam.) (d.n.), not ∼ (Tode) Pers. 1794 (d.n.); fide S. Hugh. 1958 (CJB 36): 811 = *Sporotrichum aureum* Link 1809. — *Oidium* Link 1809 (d.n.), misapplied, not ∼ (Pers.) Link 1824: Fr. 1832; *Trichoderma* Pers. per Pers. 1822, in obs., Steud. 1824; *Sporotrichum* Fr. 1832, misapplied, not ∼ Link per S. F. Gray 1821; *Trichosporium* Fr. 1859 ("Lk."), misapplied. — Sensu Link 1809 → *Oidium aureum* (Pers.) Link (**O**); sensu Link 1824, Fr. 1832 → *Botrytis aurantiaca* Link per Pers. = *Verticillium* state of *Nectria inventa* Pethybr., fide S. Hugh. 1958 (CJB 36): 742, 823.

millavensis (Bourd. & G.) Overh. 1939, misapplied (**71**). — *Poria mucida* subsp. Bourd. & G. 1925 (France); *Xylodon* Bond. 1935 (indirect ref.: n.v.p.). — Bourd. & G. 1928: 681 (*Poria mucida* subsp.). — Sensu Overh. 1939 (PPA 13): 124 *f. 4,* "appears better refered to *P*[*oria*] *radiculosa* [(Peck) Sacc., extra-European]", fide Lowe 1966: 38, in obs.

mollicula Bourd. 1916 (presumably France). — Bourd. 1916 (LMW 4): 543 *f. 744*; cf. Donk 1967 (Pe 5): 113, in obs., notes.

Poria parksii Murrill 1921 (M 13): 175 (U.S.A., California); fide Lowe 1966: 38 = *Poria terrestris* Bourd. & G. — Lowe 1947 (Ll 10): 54.

Poria terrestris Bourd. & G. 1925 (BmF 41): 215, not ∼ (Pers. per Fr.) Cooke 1886. — *Byssocorticium* Bond. & S. 1941 (generic name n.v.p.), Bond. 1953; *Corticium* Park.-Rh. 1954 (incomplete ref.: n.v.p.). — Bourd. & G. 1928: 655; Lowe 1966: 38 *f. 22* (*Poria*); Donk 1967 (Pe 5): 113, in obs., notes.

Poria mycorrhiza Killerm. 1927 (H 67): 130 *f. B* & 1928 (Dba 17): 74 *pl. 11 f. 4* (Germany); fide Killerm. 1928: 176 & Lowe 1966: 38 = *Poria terrestris* Bourd. & G.

M.—*Poria terrestris* Pers. sensu DC. 1815: 39 (*Boletus*) → *Poria terrestris* Bourd. & G.

mucida (Pers. per Fr.) Cooke 1886. — *Poria* Pers. 1796 (Germany) (d.n.); *Boletus* Pers. 1801 (d.n.), not ∼ Scop. 1770 (d.n.); *Polyporus* (Pers.) per Fr. 1821; *Boletus* Nocca & Balb. 1821 ("Spreng." = Pers.); *Physisporus* Gillet 1877, P. Karst. 1881; *Leptoporus* Pat. 1900; *Agaricus* E. Krause 1932; *Multiporus* R. & O. Falck 1937 (generic name n.v.p.), misapplied;

≡ *Fibuloporia donkii* Domański 1969 (APo 38): 454. — Romell 1926 (SbT 20): 14 (*Polyporus*), Donk 1933: 227, in obs., & 1967 (Pe 1): 98 (*Poria*), notes. — Sensu Bres. → *Schizopora paradoxa*.

Merulius lichenicola Burt 1917 (AMo 4): 329 *f. 12, pl. 21 f. 11* (U.S.A., New York); fide Ginns 1969 (M 60): 1217 = *Poria mollusca* [sensu Ginns = *Poria mucida*].

M.—*Boletus molluscus* Pers. sensu Bres. 1897 (*Poria*); fide Donk 1933: 227. — Bres. 1897 (AAR III 3): 86; 1903 (Am 1): 79, some microscopical data; Bourd. & G. 1928: 671; D. Baxt. 1939 (PMi 24): 183 *pls. 5, 6*; Overh. 1942: 31 (*Poria*); M. P. Christ. 1960 (DbA 19): 337 *f. 333*; Domański 1965 (FpG 2): 36 *f. 3, pl. 2* (*Fibuloporia*); Lowe 1966: 60 *f. 36* (*Poria*); Romell 1926 (SbT 20): 14, 22–23, notes.

myceliosa Peck 1901 (U.S.A., New York). — *Fibuloporia* Domański 1965; *Anomoporia* Pouz. 1966. — Overh. 1919 (BNS 205–206): 94 *pl. 10 f. 5, pl. 11*; D. Baxt. 1939 (PMi 24): 171; Overh. 1942: 32; J. E. Wright 1964 (M 56): 785 (*Poria*); Domański 1965 (FpG 2): 46 *f. 10* (*Fibuloporia*); Lowe 1966: 61 *f. 37* (*Poria*).

Poria perextensa Murrill 1920 (M 12): 304 (U.S.A., Maine); fide Overh. 1942: 32.

nordmannii (Lév.) Cooke 1886. — *Polyporus* Lév. 1842 D. 2: 95 [*pl. 1 f. 1*] (U.S.S.R., Ukraine). — Insufficiently known.

ochraceo-lateritia Bond. 1940 (U.S.S.R., European Russia). — *Hapalopilus* Bond. & S. 1941. — Bond. 1940 (BMs 5): 21 (*Poria*); 1953.: 270 *f. 62, pl. 74 fs. 1, 2*; Domański 1963 (Mob 15): 307 *f. 8* (*Hapalopilus*). — Distinct from *Poria aurantiaca* sensu Bres. [= *P. salmonicolor*]?

placenta (Fr.) Cooke 1886. — *Polyporus* Fr. 1861 (Sweden), not ∼ (Schum.) Secr. 1833 (n.v.p.). — *Physisporus* P. Karst. 1882; *Poria* Quél. 1886; *Leptoporus* Pat. 1900; *Ceriporiopsis* Domański 1963. — Fr. 1863 M. 2: 272; 1884 I. 2: 87 *pl. 188 f. 3* (*Polyporus*); sensu Lundell apud Jo. Erikss. 1958 (Sbu 16¹): 150 *f. 48* (*Poria*); Domański 1963 (APo 32): 732, 736. *f. 4*; 1965 (FpG 2): 83 *f. 24, pl. 10 f. 1; 1965a, 1970a*, with cult. char. (*Ceriporiopsis*); Lowe 1966: 81 *f. 62* (*Poria*). — Sensu Bres. → *Poria salmonicolor*.

Poria incarnata Pers. 1794 (Germany) (d.n.) (**110**); cf. Lloyd 1910 (LMW 3): 472, "the type ... may be [Fries's] *Poria placenta*"; fide Lowe & Gilb. 1962 (M 53): 495 (as named by Parmasto) = *Poria carnicolor*; fide Domański 1965 (APo 34): 515, 528 = *Ceriporiopsis placenta* sensu Jo. Erikss. — *Boletus* Pers. 1801 (d.n.), non/an ∼ Schum. 1803 (d.n.); *Polyporus* Fr. 1818 (d.n.), perhaps misapplied; *Polyporus* (Pers.) per Fr. 1821, presumably misapplied [→ *Polyporus incarnatus* Fr. 1832^Ind.: 149, not ∼ (Pers.) per Fr. 1821, = sp. dubia (**110**) (**O**)]; *Boletus* Schw. 1822 (nom. nud.: n.v.p.), Lenz 1840; *Physisporus* Gillet 1878; *Caloporus* P. Karst. 1881; *Poria* Quél. 1886; *Caloporia* P. Karst. 1898; *Ceriporia*

Bond. 1953, not ∼ Parm. 1963 (n.v.p.); *Ceriporiopsis* Domański 1963. — Pers. 1794 (ABU 11): 30 & 1797 T.: 70; sensu Bres. apud Bourd. & G. 1925 (BmF 41): 253, 1928: 689 (*Poria*); Parm. 1959 (TSR 12): 222 *fs. 2, 3* (*Ceriporia*); Domański 1965 (FpG 2): 75 *f. 20* (*Ceriporiopsis*). → *Ceriporia incarnata* Parm. (n.v.p.). — Sensu A. & S. → *Polyporus niskiensis* Pers. (**O**); sensu Fr. 1818, 1821 = sp. dubia; sensu Pers. 1825 → *Antrodia serialis*, fide Donk 1971 (PNA 74): 26; sensu P. Karst. 1870 F. F.: No. 904 (*Polyporus*) = *Merulius ravenelii* (**O**), fide Pat. 1900: 106, 107, = *Merulius taxicola* (**O**) fide Donk 1962 (Pe 1): 229; sensu Gillet → *Pachykytospora tuberculosa*; sensu Fr. 1884 → *Chaetoporus collabens?*; sensu P. Karst. 1896 = sp. dubia; sensu R. Fr. & sensu Romell, cf. Bourd. & G. 1928: 690, in obs.

Physisporus albolilacinus P. Karst. 1892 (H 31): 220 (Finland); fide Lowe 1966: 81. — *Poria* Sacc. 1895. — Cf. Lowe 1956 (M 48): 110 (*Physisporus*).

Poria monticola Murrill 1920 (M 12): 20 (U.S.A., Idaho), in part; fide Overh. 1946 (M 38): 676 (type) = *Poria microspora* (cited as a syn.) + *Trametes* [*Antrodia*] *serialis* (excluded); fide Domański 1965 (FpG 2): 252 & 1970 (APo 39): 61 = *Ceriporiopsis placenta* (forma). — Overh. 1946 (M 38): 671 *f. 1*; Nobles 1948 (CJR 26): 389 *pl. 15 f. 7, pl. 16 fs. 19–24*, cult. char.; Lowe 1966: 82 *f. 63*.

Poria carnicolor D. Baxt. 1941 (PMi 26): 109 *pl. 2* (U.S.A., Idaho); fide Lowe apud Domański 1965 (FpG 2): 252 & Lowe 1966: 81. — R. L. Gilb. 1956 (Ll 19): 76.

Poria microspora Overh. apud Nobles 1943 (Canada, British Columbia), not ∼ [correctly, *micropora*] (P. Karst.) Sacc. & D. Sacc. 1905; fide Overh. 1946 (M 38): 674 = *Poria monticola*. — Nobles 1943 (CJR 21): 220, 224 *fs. 46–53, pl. 1 f. 4, pl. 2 fs. 8–17, pl. 4 f. 38*, with cult. char.; Lowe 1946: 67; R. W. Davids. & al. 1947 (M 39): 314 *fs. 1G–H, 4A*.

Ceriporia incarnata Parm. 1963 (TÜT 136): 112 *f. 2* (lacking Latin descr.: n.v.p.), not ∼ (Pers. per Fr.) Bond. 1953 (110). — [≡ *Poria incarnata* Pers. sensu Bres. apud Bourd. & G. 1925 (BmF 41): 253, excl. of type, (Italy)].

M.—*Poria vaporaria* Pers. (**O**) "sensu Liese", "standard", "Eberswalde strain", British Standard No. 838 [cf. Cartwr. & Findl. 1946: 287, 1958: 324]; fide Cartwr. & Findl. 1958: 324 = *Poria monticola*; fide Domański 1965 (APo 34): 515, 527, 1970 (APo 39): 61 = *Ceriporiopsis placenta*.

rancida Bres. 1900 (Italy). — *Aporpium* Bond. & S. 1941 (generic name n.v.p.), Bond. 1953; *Coriolus* Park.-Rh. 1954 & Bond. & M. Bond. 1960 (both, incomplete ref.: n.v.p.). — Bres. 1900 F.t. 2: 96 *pl. 208 f. 1*; 1932 (BIm 21): *pl. 1020 f. 2*; D. Baxt. 1937 (PMi 22): 294 *pl. 37 f. 1*; Lowe 1966: 85 *f. 67*.

Poria cognata Overh. 1943 (U.S.A., Tennessee); fide Lowe 1954 (PMi 39): 33. — Overh. 1943 (M 35): 248.

reichertii (Velen.) Velen. 1922 (nom. altern.). — *Polyporus* Velen. 1922: 641 [see Pilát 1948: 245 for Latin translation] (Czechoslovakia). — Insufficiently known.

rensii Bres. 1926: 56 (Italy). — Insufficiently known.

resinascens (Romell) Lloyd 1912. — *Polyporus* Romell 1911 (Sweden); *Tyromyces* Bond. & S. 1941; *Ceriporiopsis* Domański 1963. — Romell 1911 (ABS 11³): 20 *pl. 2 f. 14*; Bourd. & G. 1928: 669, in obs. *f. 185R* (*Polyporus*); Pilát 1941 (ACE 3): 424 *f. 192, pl. 267 f. b, pl. 268* (*Poria*); D. Baxt. 1943 (PMi 28): 227 (*Polyporus*); M. P. Christ. 1960 (DbA 19): 362 *f. 361* (*Tyromyces*); Domański 1963 (APo 32): 732, 735 *f. 2*; 1965 (FpG 2): 77 *f. 21, pl. 14 f. 2* (*Ceriporiopsis*). — Cf. Lowe 1966: 87, "doubtfully distinct from *P*[*oria*] *aneirina*".

Leptoporus micantiformis Pilát 1936 (U.S.S.R., Russia, Siberia); fide Parm. 1963 (TÜT 136): 120 (forma). — *Poria* Lowe 1946, misapplied; *Tyromyces* Bond. 1953 (incomplete ref.: n.v.p.), Parm. 1959. — Pilát 1936 (BmF 51): 358 *pl. 7 f. 2*; 1937–8 (ACE 3): 201 *f. 58 pl. 119 [bis]* (*Leptoporus*); Parm. 1959 (TSR 12): 238 *fs. 17, 18, plate f. 19* (*Tyromyces*). — Sensu Lowe 1946 → *Poria gilvescens*.

Poria pseudo-gilvescens Pilát 1936 (U.S.S.R., Russia, Siberia); fide Domański 1963 (APo 32): 735, 738 (var.); & cf. Lowe 1962 (PMi 47): 184. — Pilát 1936 (BmF 51): 378 *pl. 11 f. 1*; 1941 (ACE 3): 407 *f. 176, pls. 258, 259*; Domański 1964 (APo 33): 170 (*Poria*); 1965 (FpG 2): 69 *f. 22, pl. 13 f. 3* (*Ceriporiopsis resinascens* var.).

Poria subpudorina Pilát 1941–2 (U.S.S.R., Ukraine) lacking Latin descr.: n.v.p.); fide Domański 1963 (APo 32): 735, 738 & 1964 (APo 33): 170 (var.). — *Ceriporia* Bond. 1953 (lacking Latin descr.: n.v.p.). — Pilát 1941 (ACE 3): 426 *pl. 269 f. a*; Domański 1964 (APo 33): 170 (*Poria*).

Tyromyces polyetes Parm. 1959 (Estonia); fide Parm. 1963 (TÜT 136): 120 (var.). — Parm. 1959 (TSR 12): 239 *fs. 5, 6*.

Tyromyces **resupinatus** (Bourd. & G.) Bond. & S. 1941. — *Leptoporus destructor* var. Bourd. & G. 1928 (France); *Leptoporus* Pilát 1932. — Bourd. & G. 1928: 547 *f. 156* (*Leptoporus destructor* var.); Pilát 1932 (BmF 48): 167 *pl. 14 fs. 1, 2*; 1937–8 (ACE 3): 195 *f. 55, pl. 115* (*Leptoporus*); Domański 1963 (Mob 15): 311; 1965 (FpG 2): 137 *f. 47, pl. 38* (*Tyromyces*). — Cf. *Poria gossypium* Speg. (**O**).

M.—*Boletus destructor* Schrad. (**O**) sensu Bres. 1908 (Am 6): 37, in part: resupinate element from France (*Polyporus*). — Bourd. & G. 1925 (BmF 41): 127, excl. of vars. (*Leptoporus*), not *Leptoporus destructor* sensu Bourd. & G. 1928. → *Leptoporus destructor* var. *resupinatus* Bourd. & G.

rivulosus (B. & C.) Cooke 1886. — *Polyporus* B. & C. 1868 (Cuba), not ∼ (Pat. & Har.) Lloyd 1912; *Rigidoporus* Al. David 1972 (lacking valid ref.: n.v.p.). — Lowe 1966: 70 *f. 47*; Al. David 1972 (BmF 87) 415 *fs. 1–5* (*Poria*) *pl. 1* (*Rigidoporus*).

Poria albipellucida D. Baxt. 1938 (Canada, British Columbia); fide Lowe 1963 (M 55): 465. — D. Baxt. 1938 (PMi 23): 291 *pl. 4.*

romellii Donk 1967 (**118, 119**). — [≡ *byssinus* Schrad. sensu Pers. in herb., in part (*Poria*) (Germany);] ≡ *Poria byssina* Romell 1926. — Romell 1926 (SbT 20): 8, 20 (*Poria/Polyporus byssina*); Jo. Erikss. 1949 (SbT 43): 3 *f. 1* (*Poria byssina*); Domański 1965 (FpG 2): 154 *f. 55, pl. 43* (*Tyromyces byssinus*); Donk 1967 (Pe 5): 82–85, notes. — Jo. Erikss., l.c., mentioned several herb. names by Romell; these are mentioned on the "List of omitted names" under *Polyporus byssinoides* (**O**).

M.—*Boletus byssinus* Schrad. (**O**) sensu Pers. in herb., in part. → *Poria byssina* Romell → *Poria romellii.* — (**119**).

M.—*Poria vitrea* Pers. sensu Bres. 1897; fide Bres. 1903 (Am 1): 78 = *Poria vulgaris* (Fr.) Cooke [sensu Bres.]. — Bres. 1897 (AAR III 3): 85 (*Poria*); Bourd. & G. 1928: 679 (*Poria vulgaris* f.).

M.—*Polyporus vulgaris* Fr. sensu Bres. 1897 (AAR III 3): 86 (*Poria*), in part: f. *typica* Bres.; fide Jo. Erikss. 1949 (SbT 43): 4, 5 = *Poria byssina* "(Pers.) Romell". — Egeland 1914 (NMN 52): 137; Bourd. & G. 1928: 679 (*Poria*); Bond. 1953: 163; M. P. Christ. 1960 (DbA 19): 349 *f. 345* (*Aporpium*).

salmonicolor (B. & C.) Cooke 1886. — *Polyporus* B. & C. 1849 (U.S.A., South Carolina), not ∼ Lloyd 1925; *Leptoporus* Pat. 1900; *Hapalopilus* Pouz. 1967. — Lowe 1961 (PMi 46): 206; 1966: 79 *f. 61* (*Poria*); Pouz. 1967 (ČM 21): 205, notes (*Hapalopilus*).

Polyporus aurantiacus Lasch in Rab. 1853 Kl.: No. 1714 [repr. 1853 (BZ 11): 235], (Germany), not ∼ Rostk. 1838, not ∼ Peck 1874; fide Donk 1967 (Pe 5): 81, 125.

? *Polyporus oxydatus* B. & C. 1868 (JLS 10): 317 (U.S.A., South Carolina); fide Lowe 1962 (PMi 47): 185 (sterile). — *Poria* Cooke 1886.

Sarcoporia polyspora P. Karst. 1894 (H 33): 15 (Finland); fide Lowe 1956 (M 48): 122 = *Physisporus aurantiacus* var. *saloisensis.* — *Poria* Sacc. 1895.

Poria rubens Overh. & L. 1946 (Canada, British Columbia); fide Lowe 1961 (PMi 46): 206. — Overh. & L. 1946 (M 38): 211 *f. 2F*; Lowe 1946: 60.

M.—*Polyporus xanthus* Fr. sensu Quél. 1888: 381 (*Poria*); fide Bres. 1903 (Am 1): 77 = *Poria nitida* [sensu Bres.]; fide Bourd. & G. 1928: 665, 676 = *Poria aurantiaca* [sensu Bres.].

M.—*Polyporus placenta* Fr. sensu Bres. 1903 (*Poria*); fide Domański 1965 (APo 34): 515, 518 & 1965 (FpG 2): 163 = *Hapalopilius aurantiacus* [sensu Bres.]; cf. Lowe 1966: 81. — Bres. 1903 (Am 1): 77; Bourd. & G. 1928: 664 (*Poria*).

M.—*Poria nitida* Pers. sensu Bres. 1903; fide Bres., l.c. = *Poria aurantiaca* [sensu Bres.] (cited as a syn.). — Bres. 1903 (Am 1): 77.

M.—*Polyporus aurantiacus* Rostk. (**O**) sensu Bres. apud Egeland 1914 (*Poria*); fide Lowe 1966: 79, 86. — Egeland 1914 (NMN 52): 155; Bourd. &

G. 1928: 665; Al. David 1969 (Nca 96): 219, cult. char. (*Poria*); Donk 1967 (Pe 5): 80, notes.

sartoryi Bourd. & L. Maire apud Sart. & M. 1922 (France). — *Byssocorticium* Bond. & S. 1941 (generic name n.v.p.), Bond. 1953. — Bourd. & G. 1928: 655 (*Poria terrestris* subsp.).

sericeo-mollis (Romell) Lloyd 1912 (**68, 111**). — *Polyporus* Romell 1911 (Sweden); *Leptoporus* Pilát 1936; *Tyromyces* Bond. & S. 1941; *Strangulidium* Pouz. 1967. — Romell 1911 (ABS 11³): 22 *pl. 2 f. 1*, in part, excl. of chlamydospores-producing specimens (*Polyporus*); Lomb. & Gilb. 1965 (M 57): 67 *f. 2D*, with cult. char. (*Poria*); Lowe 1966: 84 *f. 66* (*Polyporus*); Pouz. 1967 (ČM 21): 206 (*Strangulidium*). — Sensu Romell 1911, in part, → *Oligiporus rennyi*. — According to Lowe 1966: 84, "rarely pileate", hence listing of this sp. under *Tyromyces* would perhaps have been preferable.

  Leptoporus litschaueri Pilát 1932 (U.S.S.R., Russia, Siberia), not ∼ (Lohw.) Pilát 1939, not *Poria litschaueri* Pilát 1932; fide Kotl. & P. 1965 (ČM 19): 76. — *Chaetoporellus* Bond. 1953; ≡ *Leptoporus asiaticus* Pilát 1938; *Poria* Overh. 1939, 1942; *Polyporus* Overh. 1939 (syn.: n.v.p.), Killerm. 1943; *Chaetoporellus* M. P. Christ. 1960. — Pilát 1932 (BmF 48): 9 *f. 2, pl. 8 f. 1*; 1934 (BmF 49): 258; 1936 (BmF 51): 355; 1937 (BmF 52): 306 *f. 3, pl. 4 f. 3* (*Leptoporus l.*); 1937–8 (ACE 3): 194 *fs. 54, 92, pl. 114* (*Leptoporus a.*); Overh. 1939 (PPA 13): 121 *f. 6*; 1942: 22; Lowe 1946: 61 *f. 14* (*Poria a.*); M. P. Christ. 1960 (DbA 19): 352 *f. 349* (*Chaetoporellus a.*); Domański 1965 (FpG 2): 118 *f. 41, pl. 34 f. 1* (*Chaetoporellus l.*).
  Tyromyces sublacteus M. P. Christ. 1960 (Denmark); fide Lowe 1966: 84. — M. P. Christ. 1960 (DbA 19): 359 *f. 357*, no cystidia mentioned.

šimanii (Pilát) Gilb. & Lowe 1962. — *Leptoporus* Pilát 1938, 1953; *Chaetoporellus* Bond. 1953; *Tyromyces* Parm. 1961; ≡ *Leptoporus caesius* f. *porioides* Bourd. 1932 (Czechoslovakia). — Pilát 1937–8 (ACE 3): 181 *fs. 48, 91, pls. 105, 106* (*Leptoporus*); Parm. 1961 (EAT 10²): 118 *f. 1, plate f. 2* (*Tyromyces*); Gilb. & Lowe 1962 (PMi 47): 173 *pl. 2 fs 5A, B* (*Poria*); Domański 1965 (FpG 2): 121 *f. 43* (*Chaetoporellus*); Lowe 1966: 73 *f. 51* (*Poria*).
  M.—*Polyporus subsericeo-mollis* Romell "of most Scandinavian authors (Lundell, John Eriksson etc.)"; fide Kotl. & P. 1964 (ČM 18): 217 ("in most cases").

sinuascens Pilát 1940 (U.S.A., Pennsylvania). — *Poria* Pilát ex Overh. 1942 (lacking Latin descr.: n.v.p.); *Coriolellus* Domański 1965 (lacking Latin descr. & valid ref.: n.v.p.). — Overh. 1942: 37 (*Poria*); Domański 1965 (FpG 2): 190 *f. 64, pl. 50 f. 2* (*Coriolellus*). — Referred by Lowe 1954 (PMi 39): 32 to *Poria* [*Chaetoporus*] *luteo-alba*.

subvermispora Pilát 1940 (U.S.A., Missouri). — Pilát 1940 (Sbč 3): 2; Lowe 1966: 74 *f. 53* (*Poria*); Domański *1969d*, with cult. char. (*Fibuloporia*).

Poria notata Overh. 1942: 33 (lacking Latin descr.: n.v.p.) (U.S.A., Pennsylvania); fide Lowe 1959 (Ll 10): 101, 109. — Sensu Lowe 1946, in part, → *Tyromyces leucomallelus.*

Poria quercuum D. Baxt. 1949 (PMi 33): 10 *pl. 1* (U.S.A., Missouri); fide Lowe 1966: 74.

Polyporus **urineus** Velen. 1922: 637 [see Pilát 1948: 243 for Latin translation] (Czechoslovakia). — Insufficiently known. — Cf. Pilát 1942 (ACE 3): 622, "? *L[eptoporus] resupinatus.*"

vaillantii (DC. per Fr.) Cooke 1886. — *Boletus* DC. 1815 (Switzerland) (d.n.); *Polyporus* (DC.) per Fr. 1821; *Boletus* Mérat 1821; *Physisporus* Chev. 1826; *Porotheleum* Quél. 1886; *Leptoporus* Pat. 1900; *Coriolus* Maheu 1906 ("*Vallantii*"); *Fibuloporia* Bond. & S. 1941 (generic name n.v.p.), Bond. 1953; *Fibroporia* Parm. 1968. — Bres. 1897 (AAR III 3): 88; Egeland 1914 (NMN 52): 142; Bourd. & G. 1928: 677; Overh. 1942: 41; D. Baxt. 1955 (PMi 40): 104 *pl. 6* (*Poria*); M. P. Christ. 1960 (DbA 19): 339 *f. 334* (*Fibuloporia*); Reid & Austw. 1963 (GN 18): 309, in obs.; Lomb. & Gilb. 1965 (M 57): 70 *f. 2F*, with cult. char. (*Poria*); Domański 1965 (FpG 2): 37 *f. 5, pl. 3, pl. 4 f. 4* (*Fibuloporia*); Lowe 1966: 117 *f. 103* (*Poria*); Donk 1967 (Pe 5): 117, note. Sensu Lib. → *Incrustoporia alutacea*; sensu P. Henn. → *Polyporus [Tyromyces] henningsii*; sensu Pilát 1927 → *Polyporus [Tyromyces] fodinarum.* — V.s.: *Boletus* "*vaillantis*", L. Morel 1865; "*Vallantii*".

[? *Corallo-Fungus argenteus, Ormenti forma* Vaill. 1727: 41 *pl. 8 f. 1* (France); cf. DC. 1815: 38.]

Physisporus bombacinoides (P. Karst.) P. Karst. 1889; fide Lowe 1956 (M 48): 114. — *Physisporus molluscus* subsp. P. Karst. 1884 (Mfe 11): 21 (Finland); *Poria* Killerm. 1943.

Poria bergii Speg. 1899 (ABA 6): 171 (Argentina); fide Bres. 1916 (Am 14): 228 & Lowe 1966: 17.

M.—*Boletus hybridus* Sow. sensu Berk. in herb.; Lowe apud Reid & Austw. 1963 (GN 18): 309 & 1966: 107 (syn.); fide Reid & Austw., l.c., & Lowe, l.c.

xantha (Fr. per Fr.) Cooke 1886. — *Polyporus* Fr. 1815 (Sweden) (d.n.); *Polyporus* Fr. per Fr. 1821; *Physisporus* P. Karst. 1882; *Poria* Quél. 1886, misapplied; *Amyloporia* Bond. & S. 1941 (generic name n.v.p.), Bond. 1953. — Lind 1913: 389 (*Polyporus*); Romell 1926 (SbT 20): 21, 22; Bourd. & G. 1928: 675 (*Poria calcea* var.); D. Baxt. 1936 (PMi 21): 257; Jo. Erikss. 1949 (SbT 43): 18 *f. 5, pl. 4* (*Poria*); M. P. Christ. 1960 (DbA 19): 347 *f. 344* (*Amyloporia*); Lomb. & Gilb. 1965 (M 57): 72 *f. 3B*, with cult. char. (*Poria*); Domański 1965 (FpG 2): 90 *f. 27, pl. 17 f. 3, pl. 18* (*Amyloporia*); Lowe 1966: 88 *f. 68* (*Poria*); Donk 1967 (Pe 5): 124, notes. — Sensu Quél. → *Poria salmonicolor*; sensu Lind = *Poria xantha* f. *pachymeris* Jo. Erikss. (n.v.p.). — Jo Erikss., l.c., mentions several herb. names by

Romell; these are included on the "List of omitted names" under *Poria aniseus* (**O**).

Polyporus flavus P. Karst. 1859: 40 (Finland), not ∼ Jungh. 1838, not *Polyporus vulgaris* var. *flavus* Fr. 1874: 578; fide Jo. Erikss. 1949 (SbT 43): 18 & Lowe 1956 (M 48): 119. — ≡ *Polyporus selectus* P. Karst. 1868.

? *Polyporus laestadii* Fr. & Berk. apud B. & Br. 1883 (AM V 12): 373; cf. Cartwr. & Findl. 1946: 201 ("probably identical"). — *Poria* Cooke 1886; (≡ 'forma' mentioned by Fr. 1874: 575 in obs. under *P[olyporus] fulgens* (Sweden))].

Polyporus sulphurellus Peck 1889 (RNS 42): 123 (U.S.A., New York); fide Lowe 1959 (Ll 21): 104 (*Poria xantha* "sense of Lind"). — *Poria* Sacc. 1891. — Overh. 1919 (BNS 205–206): 117 *pl. 22 fs. 1–4*; Lowe 1946: 33 (*Poria*).

Poria suaveolens Bagl. & Razz. apud Razzore 1911 (ASl 22): 17 (Italy); fide Donk (from descr.).

Poria xantha f. *pachymeris* Jo. Erikss. 1949 ("nomen novum"; lacking Latin descr.: n.v.p.). — ≡ *Trametes cinereosulfurea* Ferd. & W. 1943: 85, in obs. (Denmark) (lacking Latin descr.: n.v.p.). — Jo. Erikss. 1949 (SbT 43): 3, 22 *pl. 5 f. 1* (*Poria xantha* f.); Domański 1965 (FpG 2): 92 *pl. 14 f. 1* (*Amyloporia xantha* f.).

M.—*Polyporus vulgaris* Fr. sensu Bres. 1897 (AAR III 3): 86 (as a syn. of *Poria vulgaris* f. *luteo-alba*), Romell 1911 (ABS 11³): 25. — Fide Romell 1926 (SbT 20): 20 = "*Pol. xanthus* sensu Lind" = *Poria xantha* (f. *pachymeris*).

M.—*Polyporus luteo-albus* P. Karst. sensu Bres. 1897 (AAR III 3): 86 (*Poria vulgaris* f.); fide Bourd. & G. 1928: 675.

M.—*Physisporus crassus* P. Karst. sensu Höhn. 1909 (ÖbZ 59): 108; fide Jo. Erikss. 1949 (SbT 43): 22 (*f. pachymeris*); & Donk [from descr.; excl. of specimen from Val di Sole = *Poria alpina* fide Litsch. 1939 (ÖbZ 88): 144–145]; & sensu D. Baxt. 1936 (PMi 21): 255 *pls. 34, 35*, fide Jo. Erikss. l.c. = *Poria xantha* (f. *pachymeris*).

M.—*Polyporus bullosus* Fr. apud Weinm. (**O**) sensu Pilát 1932 (BmF 48): 44 (*Poria callosa* var.); fide Domański 1964 (APo 33): 171 (f. *pachymeris*).

M.—*Polyporus vulgaris* var. "*β. P. calceus*" Fr. [≡ *Polyporus calceus* (Fr. ex Pers.) Schw. (**O**) sensu Bond. & S. 1941 (Am 39): 50 & Sing. 1944 (M 36): 67 (*Amyloporia*; lacking descr.); referred here by assuming a positive correlation between original descr. and type of the name *Amyloporia*.

xylostromatoides (Berk.) Cooke 1886. — *Polyporus* Berk. 1843 (Brazil); *Merulius* J. Rick 1938, W. Cooke 1957 ("*xylostromaticus*"; syn.). — Lowe 1966: 39 *f. 24* (*Poria*). — Recorded for Europe (Portugal) by Lowe 1966: 40.

Poria subambigua Bres. 1911 (Am 9): 268 (Congo-Kinshasa = Zaire); fide Lowe 1963 (M 55): 460. — Lowe 1954 (PMi 39): 34.

? *Poria aquosa* Petch 1916 (APe 6): 139 (Ceylon); fide Lowe 1963 (M 55): 460. — Petch 1919 (APe 7): 6, in obs.; Lowe 1963 (PMi 48): 165. — Referred by Petch 1919 (APe 7): 5 to *Poria interrupta* (**O**).

Poria subcollapsa Murrill 1920 (M 12): 20 (Jamaica); fide Lowe 1966: 39. — Lowe 1947 (Ll 10): 57.

Poria cremeicolor Murrill 1920 (M 12): 85 (Jamaica); fide Lowe 1963 (M 55): 460. — Lowe 1947 (Ll 10): 48.

Poria corioliformis Murrill 1920 (M 12): 86 (Cuba); fide Lowe 1963 (M 55): 460.

Poria subcorticola Murrill 1920 (M 12): 88 (Mexico); fide Lowe 1963 (M 55): 460 ("type sterile"), 1966: 39.

Poria byssopora J. Rick 1937 (Bro 6): 143 (Brazil); fide Lowe 1966: 39. — Referred by J. Rick 1960 (Ih 7): 271 as a var. to *Poria hyalina* (**O**).

Poria flaccida Overh. 1938 (U.S.A., Louisiana); fide Lowe 1963 (M 55): 460. — Overh. 1938 (BTC 65): 180.

Poria taxodium D. Baxt. 1941 (U.S.A., Illinois); fide Lowe 1966: 39. — D. Baxt. 1941 (PMi 26): 107 *pl. 1*.

PYCNOPORELLUS Murrill

1905 [1960 (Pe 1) 275]. — *Polyporus fibrillosus* P. Karst.

Aurantioporellus Murrill 1905 [1960 (Pe 1): 189]. — Holotype: *Polyporus alboluteus* (Ell. & Ev.) Ell. & Ev.

SPECIAL LITERATURE.—Barbas, *1957* (*P. fulgens*); Domański, *1959b* (*P. alboluteus*); Hintikka, *1970* (*P. alboluteus*); Kotlaba & Pouzar, *1963b* (*P. alboluteus, P. fulgens*).

alboluteus (Ell. & Ev.) Kotl. & P. 1963. — *Fomes* Ell. & Ev. 1895 (U.S.A., Colorado); *Polyporus* Ell. & Ev. 1898, not ~ Rostr. 1902; *Scindalma* O.K. 1898; *Aurantioporellus* Murrill 1905; *Phaeolus* A. Ames 1913, Pilát 1937, Murašk. 1939; *Hapalopilus* Bond. & S. 1941. — Ell. & Ev. 1898 (BTC 25): 518; Lloyd 1915 (LMW 4, Ap.): 340 *f. 678*; Shope 1931 (AMo 18): 333 *pl. 19 fs. 1–11* (*Polyporus*); Murašk. 1939: 82 *fs. 7, 8B* (*Phaeolus*); Overh. 1953: 382 *pl. 39 f. 287, pl. 40 f. 241, pl. 44 f. 262, pl. 112 f. 614, pl. 218 fig.* (*Polyporus*); Kotl. & P. 1963 (ČM 17): 174, 184 *fs. 1, 2*; Domański 1965 (FpG 2): 166 *f. 58, pls. 46, 47* (*Pycnoporellus*); Lowe 1966: 23 *f. 7* (*Polyporus*); Al. David 1969 (Nca 96): 221, cult. char. (*Phaeolus*).

Irpex woronowii Bres. 1920 (Am 18): 42 (U.S.S.R., Caucasus); fide D. Baxt. 1938 (PMi 23): 295.

fulgens (Fr.) Donk 1971. — *Hydnum* Fr. 1852 (Sweden); *Creolophus* P. Karst. 1879; *Dryodon* Quél. 1886; *Pleurodon* Rick. 1918. — Fr. 1852 (ÖVS 9): 130; 1863 M. 2: 278; 1867 I. 1: 10 *pl. 10 f. 2*; Donk 1971 (Pe 6): 216, note.

Polyporus fibrillosus P. Karst. 1859 (Finland); fide Lundell apud

Maas G. 1967 (Pe 5): 5. — *Trametes* P. Karst. 1866; *Inonotus* P. Karst. 1881; *Inoderma* P. Karst. 1881; *Pycnoporellus* Murrill 1905; *Phaeolus* A. Ames 1963 ("Pat."); *Hapalopilus* Bond. & S. 1941. — Bourd. & G. 1928: 558 (*Phaeolus*); Shope 1931 (AMo 18): 334 *pl. 17 fs. 4, 5* (*Polyporus*); Pilát 1937 (ACE 3): 142 *f. 36, pl. 70 f. b, pls. 71–75*; Muraŝk. 1939: 83 *fs. 8A, 9* (*Phaeolus*); Overh. 1953: 381 *pl. 47 fs. 283, 284, pl. 96 f. 548, pl. 129 fig.*; D. Baxt. 1954 (PMi 39): 134 *pl. 5* (*Polyporus*); S. Ito 1955: 291 *f. 216* (*Hapalopilus*); Lowe 1956 (M 48): 118, note (*Polyporus*); Kotl. & P. 1963 (ČM 17): 177, 185 *f. 3* (*Pycnoporellus*); Al. David 1969 (Nca 96): 221, cult. char. (*Phaeolus*).

Polyporus aurantiacus Peck 1874 (RNS 26): 69 (U.S.A., New York), not ∼ Rostk. 1838, not ∼ Lasch 1853; fide Bres. 1897 (AAR III 3): 72 = *Polyporus fibrillosus*. — *Polystictus* Cooke 1886; *Microporus* O.K. 1898; *Phaeolus* Pat. 1900. — Lloyd 1915 (LMW 4, Ap.): 341.

Ochroporus lithuanicus Błoński 1889 (H 28): 280 ('Poland'); fide P. Karst. 1889 (H 28): 366 & Bres. 1897 (AAR III 3): 72 = *Polyporus fibrillosus*. — *Polystictus* Sacc. 1891; *Microporus* O.K. 1898.

PYCNOPORUS P. Karst. (112, 114)

1881 [1960 (Pe 1): 276]. — Monotype: *Trametes cinnabarina* (Jacq. per Fr.) Fr. *Xylometron* Paul. 1808 (d.n.) [1960 (Pe 1): 176, 179]. — Monotype: *Boletus cinnabarinus* Jacq.

SPECIAL LITERATURE.—Bose, *1952*; Cavill & al., *1953, 1957*; W. Fischer, *1967*; Gripenberg, *1951, 1958*; Kříž, *1964*; Lemberg, *1952*; Matters, *1955*; McKay, *1959*; Nobles & Frew, *1962*; Noulet, *1871*; Routien, *1948*; E. Wolf, *1891*.

cinnabarinus (Jacq. per Fr.) P. Karst. 1881 (112). — *Boletus* Jacq. 1776 (Austria) (d.n.); *Polyporus* (Jacq.) per Fr. 1821 ("*cinnabarrinus*"); *Boletus* Schw. 1822 (n.v.p.?), Pollini 1824; *Trametes* Fr. 1849; *Polystictus* Cooke 1886; *Leptoporus* Quél. 1886; *Phellinus* Quél. 1888; *Hapalopilus* P. Karst. 1899; *Coriolus* G. Cunn. 1948. — Jacq. 1776 F.a. 4: 2 *pl. 304* (*Boletus*); Secr. 1833 M. 3: 99 (*Polyporus*); Murrill 1918 (M 10): 107 *pl. 6 f. 1* (*Pycnoporus*); Bourd. & G. 1928: 585; Konr. & M. 1935 I. 5: *pl. 442 f. 2*; Pilát 1940 (ACE 3): 318 *f. 135, pl. 213*, excl. of var. (*Trametes*); Overh. 1953: 379 *pl. 40 fs. 239, 240, pl. 47 f. 282, pl. 128 fig.*; McKay 1959 (M 51): 465 *f. 1* (*Polyporus*); Nobles & Frew 1962 (CJB 40): 1001 *fs. 31–53, pl. 2 fs. 16–18* (*Pycnoporus*). — Sensu Clel. & Cheel = *Pycnoporus coccineus* (Fr.) Bond. & S. (O).

Boletus coccineus Bull. 1790 (France) (d.n.) per St-Am. 1821, not ∼ Fr. 1838; fide Fr. 1821: 371. — Bull. 1790: *pl. 501 f. 1*; 1791 H.: 364.

Boletus miniatus Liboŝ. 1817 (MNM 5): 83 ("Ruthenia") (d.n.); *Polyporus* (Liboŝ.) per Steud. 1824: Fr. 1832, not ∼ Jungh. 1838.

RIGIDOPORUS Murrill (71)

1905 [1960 (Pe 1): 277]. — Holotype: *Polyporus micromegas* Mont.

Mensularia Pinto-L. 1952, not ∼ Lázaro 1916 [1960 (Pe 1): 241, in obs.]. — [≡ *Mensularia* Lázaro sensu Pinto-L. 1952 (MSb 8): 166, excl. of type). — Monotype: *Ungularia ulmaria* (Sow. per Fr.) Pat.

Leucofomes Kotl. & P. 1957 [1960 (Pe 1): 237]. — Holotype: *Polyporus ulmarius* (Sow.) per Fr.

Flabellophora G. Cunn. 1965: 88, 261. — Holotype: *Polyporus superpositus* Berk.

M.—*Podoporia* P. Karst. (O) sensu Donk 1933: 158. — Cf. Donk 1966 (Pe 4): 341, in obs.

M.—*Physisporinus* P. Karst. (O) sensu Pilát 1939 (ACE 3): 247. — 1966 (Pe 4): 341, in obs.

M.—*Mensularia* Lázaro sensu Pinto-L. 1952 → *Mensularia* Pinto-L.

SPECIAL LITERATURE.—Dumée, *1917* (*R. ulmarius*); Lombard, Davidson, & Lowe, *1961* (*R. ulmarius*); Hermann, *1971* (*R. vitreus*); Pilát, 1933b (*"Leptoporus undatus"*).

moeszii (Pilát) Pouz. 1966. — *Leptoporus* Pilát 1937 (lacking Latin descr.: n.v.p.), 1953 (Hungary). — Pilát 1937–8 (ACE 3): 217 *pl. 135* (*Leptoporus*).

nigrescens (Bres.) Donk 1966. — *Poria* Bres. 1897 (Hungary, now Czechoslovakia); *Physisporinus* Pilát 1941 (nom. prov.: n.v.p.); *Perenniporia* Murrill 1942; *Podoporia* Bond. 1953. — Bres. 1897 (AAR III 3): 83; Overh. 1922 (M 14): 7 *fs. 5, 6, pl. 1 fs. 1, 2*; D. Baxt. 1938 (PMi 23): 304; Overh. 1942: 48; Pilát 1942 (ACE 3): 412 *f. 181, pl. 261 f. a* (*Poria*); Domański 1965 (FpG 2): 60 *f. 14, pls. 11, 12* (*Podoporia*); Lowe 1966: 43 *f. 28* (*Poria*).

Physisporus microporus P. Karst. 1904 (ÖfF 46[11]): 8 (*"microsporus"*) (Finland); fide Lowe 1956 (M 48): 114. — *Poria* Sacc. & D. Sacc. 1905 (*"microspora"*), not *P. microspora* Overh. apud Nobles 1943.

sanguinolentus (A. & S. per Fr.) Donk 1966. — *Boletus* A. & S. 1805 (Germany) (d.n.); *Polyporus* Fr. 1815 (d.n.); *Polyporus* (A. & S.) per Fr. 1821; *Sistotrema* Secr. 1833; *Physisporus* Gillet 1877; *Poria* Cooke 1886, Quél. 1886; *Leptoporus* Pat. 1900; *Podoporia* Höhn. 1909 (nom. prov.: n.v.p.); *Daedalea* E. Krause 1928; *Podoporia* Donk 1933; *Xylomyzon* E. Krause 1934; *Physisporinus* Pilát 1939. — Höhn. 1907 (SbW 116): 93; Bourd. & G. 1928: 682 (*Poria*); Donk 1933: 158 (*Podoporia*); Pilát 1938–9 (ACE 3): 248 *f. 98, pls. 166, 167* (*Physisporinus*); Domański 1965 (FpG 2): 54 *f. 12, pl. 9 fs. 1, 2, pl. 10 f. 2* (*Podoporia*); Lowe 1966: 44 *f. 29* (*Poria*); & cf. Donk 1967 (Pe 5): 110. — Sensu Bres. 1897 → *Poria gilvescens*; sensu Bres. 1903, in part, → *Ceriporia bresadolae*.

Polyporus carmichaelianus Grev. 1826 S. 4: *pl. 224*: Fr. 1832 (*"Carmichaelinus"*); fide Reid & Austw. 1963 (GN 18): 324 (specimen at K). — *Merulius* Berk. 1836; *Hexagonia* Corda 1842 (*"Carmichelii"* & *"Carmicheliana"*); [≡ *Mucilago reticulata* Hoffm. sensu Carm. in litt. (*"[Polyporus] reticulatus* of Fries"*) (Scotland)]; ≡ *Boletus carmichaelii*

Spreng. 1827; ≡ *Hexagonia carmichelii* Corda 1842 (n.v.p.), Opiz 1855, typonym.

? *Polyporus decolorans* Schw. 1832 (U.S.A., Pennsylvania). — *Poria* Cooke 1886. — Overh. 1923 (M 15): 213 *fs. 8, 9, pl. 22 f. 4*; 1942: 27; Lowe 1946: 53 *f. 12* (*Poria*).

Polyporus nebulosus B. & C. 1868 (JLS 10): 317 (Cuba); fide Lowe 1947 (Ll 10): 53. — *Poria* Cooke 1886; *Fuscoporia* Murrill 1921.

? *Polyporus subgelatinosa* B. & Br. 1876 (AM IV 17): 136 [repr. 1876 (H 15): 46] (Scotland); cf. Lowe 1966: 44. — *Poria* Cooke 1886.

Caloporus expallescens P. Karst. 1883 (Mfe 9): 110 (Finland); fide Lowe 1956 (M 48): 104. — *Poria* Sacc. 1888; *Physisporus* P. Karst. 1889; *Caloporia* P. Karst. 1896; *Polyporus* Romell 1911 (syn.: n.v.p.); *Physisporinus* Pilát 1939; *Podoporia* Bond. & S. 1941 ("Donk"). — Bourd. & G. 1928: 683 (*Poria sanguinolenta* subsp.). — Compare with *Rigidoporus vitreus*.

Physisporus albo-ater P. Karst. 1892 (H 31): 293 (Finland; now U.S.S.R., European Russia); fide Lowe 1956 (M 48): 110. — *Poria* Sacc. 1895.

Merulius dubius Burt 1917 (AMo 4): 330 *f. 14, pl. 21 f. 13* (U.S.A., New York); fide Ginns 1969 (M 60): 1218.

M.—*Poria terrestris* Pers. sensu Fr. 1828 E. 1: 122 (syn.) & Bres. 1897 (AAR III 3): 83; fide Bres. 1908 (Am 6): 41.

M.—*Mucilago reticulatus* Hoffm. sensu Carm. in litt. [*Polyporus*] → *Polyporus carmichaelianus* Grev.

ulmarius (Sow. per Fr.) Imaz. apud S. Ito 1955. — *Boletus* Sow. 1797 (England) (d.n.); *Polyporus* (Sow.) per Fr. 1821; *Boletus* Purt. 1821; *Fomes* Gillet 1878; *Placodes* Quél. 1886; *Scindalma* O.K. 1898; *Ungulina* Pat. 1900; *Mensularia* Lázaro 1916; *Fomitopsis* Bond. & S. 1941, Bond. 1953; *Leucoporus* Kotl. & P. 1957. — Hussey c. 1847 I. 1: *pl. 64*; Berk. 1860: 246 *pl. 16 f. 5*; Quél. 1881 (Crf 9²): 669 (*Polyporus*); Bres. 1908 (Am 6): 38 (*Fomes*); Dumée 1917 (BmF 33): 29 (*Polyporus*); Pilát 1940 (ACE 3): 357 *f. 154, pl. 239 f. b, pl. 240*; Wak. & Denn. 1950: 235 *pl. 98 f. 2*; Lowe 1957 F.: 77 *f. 60*; Lombard & al. 1961 (M 52): 283 *fs. 3A, 4A*, cult. char. (*Fomes*); Pegl. & Wat. 1968 (CDp): no. 199 *figs.* (*Rigidoporus*); Thind & Ratt. 1971 (IPh 24): 53 *fs. 12–15, 25, 26* (*Fomes*); Donk 1971 (PNA 74): 33, notes. — Sensu Gillet → *Fomitopsis cytisina.*

Boletus fraxineus Bull. 1789 (France) (d.n.); fide Lloyd 1915 (LMW 4, F.): 228. — *Polyporus* Fr. 1815 (d.n.) & (Bull.) per Fr. 1821; *Boletus* Mérat 1821; *Agaricus* Boisd. 1828 (nom. nud. & error: n.v.p.); *Trametes* P. Karst. 1882; *Fomes* Cooke 1885, not ~ Lloyd 1915; *Placodes* Quél. 1886; *Ischnoderma* P. Karst. 1889; *Scindalma* O.K. 1898; *Ungulina* Sart. & M. 1921, misapplied. — Bull. 1789: *pl. 433 f. 2*; 1791 H.: 341 (*Boletus*); Berk. 1836: 142; Quél. 1881 (Crf 9²): 669 (*Polyporus*); Donk 1971 (PNA 74): 33, notes. — Sensu auctt. plur., Lloyd → *Fomitopsis cytisina.*

M.—*Polyporus geotropus* Cooke (**O**) of North American authors (*Fomes*); fide Lowe 1957 F.: 80. — Overh. 1953: 53 *pl. 71 f. 416, pl. 72 f. 418, pl. 126 fig.*

undatus (Pers.) Donk 1967. — *Polyporus* Pers. 1825 (Switzerland); *Polystictus* Cooke 1886; *Poria* Quél. 1888; *Microporus* O.K. 1898; *Leptoporus* Pilát 1932, 1933, not *L. undosus* (Peck) Pat. 1900 (in error as "*undatus*"); *Physisporinus* Pilát 1938. — Bres. 1897 (AAR III 3): 82, syns.; 1903 (Am 1): 78, note (*Poria*); H. Jahn 1971 (WPb 8): 51 (*Rigidoporus*); & cf. Donk 1967 (Pe 5): 114. — Formerly often incorrectly cited as a syn. of *Polyporus* [*Heterobasidion*] *annosus*. — Sensu Quél., at least in part, → *Poria expansa*; sensu Bourd. & G. → *Rigidoporus vitreus*.

    Polyporus frustulatus Pers. 1825: 91 (Switzerland); fide Bres. 1920 (Am 18): 67 & Donk 1967 (Pe 5): 91.

    Polyporus cereus Pers. 1825: 91, not ∼ Berk. 1854. — [≡ *Poria cerea* Scop. sensu Hoffm. 1797–1811 V.s.: 16 *pl. 11 f. 3* (Germany)]. — Compared by Hoffm. with *Poria* [*Rigidoporus*] *vitrea* Pers.

    Polyporus cinctus Berk. 1837 (MZB 1): 43 *pl. 2 f. 3* (England); fide Bres. 1897 (AAR III 3): 82 = *Poria undata* & cf. Lowe 1966: 42–43 who reports "coarsely incrusted cystidia in the context and in the adjacent tramal tissue."

    ? *Polyporus adiposus* B. & Br. 1854 (AM II 13): 404 (England); fide Bres. 1916 (Am 14): 222. — *Poria* G. Cunn. 1947, misapplied. — Sensu G. Cunn. = *Poria albolutescens*, fide G. Cunn. 1965: 264 [doubtful determination]. — The original descr. recalls rather *Rigidoporus vitreus*, but I have assumed that Bres. saw encrusted cystidia in the type when referring it to *Poria undata*.

    Corticium subterraneum Rab. 1866 F.e.: No. 1006 [repr. 1867 (H 6): 45] (Germany).

    Polyporus broomei Rab. 1876 F.e.: No. 2004 [cf. 1876 (H 15): 103]; fide Bres. 1897 (AAR III 3): 82 = *Poria undata*. — *Polystictus* Cooke 1886; *Microporus* O.K. 1898; *Leptoporus* Pat. 1900 ("*Bromei*").

vitreus (Pers. per Fr.) Donk 1966. — *Poria* Pers. 1795 (Germany) (d.n.); *Boletus* Pers. 1801 (d.n.); *Polyporus* Fr. 1818 (d.n.); *Polyporus* (Pers.) per Fr. 1821; *Boletus* Schleich. 1821; *Physisporus* Gillet 1877; *Poria* Cooke 1886, Quél. 1886, misapplied; *Physisporinus* P. Karst. 1889, misapplied; *Podoporia* Donk 1933 ["(Fr., non Pers.!)"]. — Donk 1933: 159 (*Podoporia*); Pilát 1939 (ACE 3): 250 *f. 100, pls. 168–175* may be mainly *R. undatus* (*Physisporus*); Lowe 1966: 41 *f. 27* (*Poria*); H. Jahn 1971 (WPb 8): 51 *f. 5* (*Rigidoporus*); Donk 1967 (Pe 5): 121, notes. — Sensu Fr. 1828 = ?, cf. Donk 1967 (Pe 5): 122; sensu Quél. apud Bourd. & G. → *Skeletocutis amorphus*; sensu Bres. 1897 → *Poria romellii*; sensu P. Karst. = ?, cf. *Podoporia* (**O**).

    ? *Poria alborosea* Murrill 1921 (M 13): 85 ("*albiroseus*") (U.S.A., Pennsylvania); fide Lowe 1966: 41.

? *Poria holoseparans* Murrill 1939 (BTC 65): 660 (U.S.A., Florida); fide Lowe 1966: 41.

M.—*Polyporus undatus* Pers. sensu Bourd. & G. 1925 (*Poria*). — Bourd. & G. 1928: 682; Lowe 1966: 41 *f. 27* (*Poria*); &c.;—at least no thick-walled, encrusted cystidia mentioned.

SCHIZOPORA Velen.

1922 [1960 (Pe 1): 278]. — Monotype: *Polyporus laciniatus* Velen. — Cf. Donk 1967 (Pe 5): 76.

Xylodon P. Karst. 1881 ["(Ehrenb.)"], O. K. 1891, not ∼ (Pers.) ex S. F. Gray ('Corticiaceae'; nomen dubium) [1956 (Ta 5): 114, in obs.]. — *Sistotrema* sect. ... (*Xylodon*) Pers. 1801 (d.n.) sensu P. Karst. 1881, excl. of type, viz. *Odontia quercina* Pers. — Lectotype: *Irpex paradoxus* (Schrad. per Fr.) Fr.

M.—*Xylodon* (Pers.) ex S. F. Gray sensu P. Karst. 1881 ["(Ehrenb.) Karst."] → *Xylodon* P. Karst.

SPECIAL LITERATURE.—Domański, *1969c* (*Schizopora phellinoides*); Neger, *1905* (*S. paradoxa*).

paradoxa (Schrad. per Fr.) Donk 1967 (**113**). — *Hydnum* Schrad. 1794 (Germany) (d.n.), not ∼ K. F. Schultz 1806 (d.n.); *Hydnum* Schrad. per Fr. 1821; *Xylodon* Chev. 1826, P. Karst. 1881, O.K. 1898; *Irpex* Fr. 1838; ≡ *Sistotrema digitatum* Pers. 1801 (d.n.); *Hydnum* Poir. 1808 (d.n.); *Xylodon* (Pers.) per S. F. Gray 1821; *Sistotrema* Secr. 1833; *Lenzites* Pat. 1885. — Schrad. 1794: 179 *pl. 4 f. 1* (*Hydnum*); Fr. 1838: 522; Bres. 1897 (AAR III 3): 101 (*Irpex*); Bourd. & G. 1928: 681 (*Poria mucida* f. *Irpex paradoxus*); Domański *1969b* with cult. char. H. Jahn 1971 (WPb 8): 62 *fs. 10–12* (*Schizopora*); Donk 1967 (Pe 5): 102, notes. — Sensu J. Schroet. 1888: 462 = *Hyphoderma radula* (Fr. per Fr.) Donk (Corticiaceae).

Hydnum obliquum Schrad. 1794 (Germany) (d.n.); fide Bourd. & G. 1925 (BmF 41): 237 = *Poria mucida* [sensu Bourd. & G.] (forma). — *Sistotrema* A. & S. 1805 (d.n.); *Hydnum* Schrad. per Fr. 1821; *Sistotrema* Pers. 1825; *Irpex* Fr. 1828; *Xylodon* P. Karst. 1881, O.K. 1898; *Coriolus* Pat. 1900; *Polyporus* E. Krause 1934, not ∼ (Pers.) per Fr. 1821. — Schrad. 1794: 179 (*Hydnum*); Bres. 1897 (AAR III 3): 102 (*Irpex*); Bourd. & G. 1928: 681 (*Poria mucida* f. *Irpex obliquus*).

Poria radula Pers. 1799 (Germany) (d.n.); fide Donk 1933: 226 & 1967 (Pe 5): 104 = *Polyporus versiporus*; & cf. Romell 1926 (SbT 20): 15. — *Boletus* Pers. 1801 (d.n.); *Polyporus* (Pers.) per Fr. 1821; *Physisporus* Chev. 1826; *Poria* Cooke 1886, Quél. 1886, not ∼ Romell 1926 (n.v.p.); *Chaetoporus* Bond. & S. 1941, misapplied. — Pers. 1799 O. 2: 14 (*Poria*); Bourd. & G. 1928: 681 (*Poria mucida* var.). — Sensu Bres. → *Chaetoporus separabilimus*.

Polyporus versiporus Pers. 1825 (France), not ∼ Lloyd 1915 (error for 'versisporus'). — *Poria* Lloyd 1910; *Agaricus* E. Krause 1932; *Xylodon*

Bond. 1953; ≡ (by lecto-typification) *Polyporus versiporus* var. *immutatus*
Pers. 1825. — Romell 1926 (SbT 20): 19; Donk 1933: 224 (*Polyporus*);
D. Baxt. 1940 (PMi 25): 150; Overh. 1942: 42 (*Poria*); M. P. Christ.
1960 (DbA 19): 342 *f. 339*; Domański 1965 (FpG 2): 48 *pl. 4 fs. 1–3,
pl. 7, pl. 8 f. 2* (*Xylodon*); Lowe 1966: 63 *f. 40* (*Poria*); D. Reid 1969
(RM 33): 248 (*Xylodon*); Donk 1967 (Pe 5): 120, notes.

Irpex deformis Fr. 1828 (Sweden); fide Bourd. & G. 1925 (BmF 41):
237 = *Poria mucida* [sensu Bourd. & G.] (var.); fide Romell 1926 (SbT 20):
20 = *Polyporus versiporus* (forma). — *Xylodon* P. Karst. 1882; *Polyporus*
Romell apud Lind 1913, not ~ (Schaeff. per Steud.) Fr. 1838. — Fr.
1828 E. 1: 147; Bres. 1897 (AAR III 3): 102 (*Irpex*); Bourd. & G. 1928:
681 (*Poria mucida* f. *Irpex deformis*).

Poria platensis Speg. 1902 (ABA 8): 53 (Argentina); fide Lowe 1963
(PMi 48): 169 = *Poria versipora*.

? *Poria eyrei* Bres. 1910 (TBS 3): 264 *pl. 14 fig.* (England); fide Lowe
1966: 65 = *Poria hypolateritia* [sensu Lowe in part = *Schizopora paradoxa*]. — Lowe 1966: *f. 41*, hymenium and spores. — Cf. (**105**).

Poria lignicola Murrill 1920 (M 12): 307 (Cuba); fide Lowe 1947 (L. 10):
58 & 1962 (PMi 47): 186 = *Poria versipora* ("typical material").

Poria ochracea Murill 1921 (M 13): 174 (U.S.A., Virginia); fide Lowe 1947
(Ll 10): 58 & 1962 (PMi 47): 186 = *Poria versipora* ("typical material").

Poria jalapensis Murrill 1921 (M 13): 177 (Mexico); fide Lowe 1947
(Ll 10): 58 & 1962 (PMi 47): 186 = *Poria versipora* ("typical material").

Polyporus laciniatus Velen. 1922: 638 [see Pilát 1948: 243 for Latin
translation] (Czechoslovakia), not ~ Pers. 1825; fide Pilát 1942 (ACE 3):
458 = *Poria versipora*. — *Poria* Velen. 1922 (nom. altern.); *Schizopora*
Velen. 1922 (nom. altern.).

? *Polyporus carpineus* Velen. 1922: 639 [see Pilát 1948: 243 for Latin
translation] (Czechoslovakia), not ~ (Sow. per S. F. Gray) Trog 1844,
not ~ S. Schulz. 1866 (n.v.p.); cf. Pilát 1941 (ACE 3): 458.

Irpex daedaleaeformis Velen. 1922: 743 [see Pilát 1948: 272 for Latin
translation] (*Czechoslovakia*); fide Pilát 1941 (ACE 3): 458 = *Poria
versipora*. — *Polyporus* Pilát 1941 (syn.: n.v.p.).

Daedalea mollis Velen. 1922: 690 [see Pilát 1948: 262 for Latin trans-
lation] (Czechoslovakia), not ~ (Pers.) Fr. 1815 (d.n.), not ~ Sommerf.
1826; fide Pilát 1942 (ACE 3): 608 = *Poria versipora*.

Irpex tiliaceus Pilát 1926 (Am 23): 306 (Czechoslovakia); fide Pilát
1941 (ACE 3): 458 = *Poria versipora*.

M.—*Odontia cerasi* Pers. (**O**) sensu Fr. 1821: 382 (*Polyporus*); fide
Fr. 1838: 523. — Cf. Donk 1967 (Pe 5): 86.

M.—*Polyporus cellaris* Desm. sensu Lib. 1834 P.A.: No. 223 (n.v.);
fide Pilát 1941 (ACE 3): 458 = *Poria versipora*.

M.—*Poria vaporaria* Pers. (**O**) sensu Berk. 1860: 252; fide Lloyd 1910
(LMW 3): 473 = *Poria versipora*. — Quél. 1892 (Crf 20^2): 471 *pl. 3 f. 25*
[cf. Bourd. & G. 1928: 673].

M.—*Poria mucida* Pers. sensu Bres. 1897 (AAR III 3): 84; fide Romell 1926 (SbT 20): 14 & Donk 1933: 227 = *Polyporus versiporus.* — Bourd. & G. 1928: 680.

M.—*Polyporus vulgaris* Fr. sensu Velen. 1922: 638; fide Pilát 1941 (ACE 3): 458 = *Poria versipora.*

phellinoides (Pilát) Domański 1969. — *Poria* Pilát 1936 (U.S.S.R., Russia, Siberia); *Fibuloporia* Bond. & S. 1941 (generic name n.v.p.); *Phellinus* Bond. 1953 (incomplete ref.: n.v.p.). — Pilát 1936 (BmF 41): 383 *pl. 4 f. 1*; 1941 (ACE 3): 461 *f. 222, pl. 295 f. b*; Domański 1964 (APo 33): 168 *f. 2* (*Poria*); 1965 (FpG 2): 52 *pl. 4 f. 3* (*Xylodon versiporus* var.); *1969c*, with cult. char.; H. Jahn 1971 (WPb 8): 63 *fs. 12, 13* (*Schizopora*).

Poria pseudo-obducens Pilát 1953 (U.S.S.R., Ukraine); fide Domański 1969 (APo 38): 267. — *Poria* Pilát 1942 (lacking Latin descr.: n.v.p.); *Oxyporus* Bond. 1953 (lacking Latin descr. & valid ref.: n.v.p.). — Pilát 1941 (ACE 3): 462 *f. 224, pl. 296 f. b, pl. 297 f. a*; Domański 1964 (APo 33): 167 *f. 1* (*Poria*); 1965 (FpG 2): 52 *pl. 4 f. 2* (*Xylodon versiporus* var.).

Xylodon versiporus var. *microporus* E. Komar. 1959 (BMs 12): 252 *fs. 4, 5* (U.S.S.R., White Russia); fide Domański 1969 (APo 38): 267.

SISTOTREMA Fr.

1821, not ∼ Pers. per Nocca & Balb. 1821 [1960 (Pe 1): 281]. — Monotype: *Sistotrema confluens* Pers.

Nota bene.—Most species of the present emendation of *Sistotrema* are not polyporoid and are not listed in this publication.

alboluteum (Bourd. & G.) Bond. & S. 1941. — *Poria* Bourd. & G. 1925 (France); *Trechispora* D. P. Rog. 1944. — Bourd. & G. 1928: 657 *f. 181* (*Poria*).

eluctor Donk 1967 (Pe 5): 102 (Finland).

M.—*Trechispora onusta* P. Karst. sensu Bres. 1908 (Am 6): 41 (*Poria*). — Bres. 1908 (Am 6): 41; Bourd. & G. 1928: 658 (*Poria*); D. P. Rog. 1944 (M 36): 80 *f. 1*; Wakef. 1952 (TBS 35): 39 (*Trechispora*), Donk 1967 (Pe 5): 101, notes.

SKELETOCUTIS Kotl. & P.

1958 [1960 (Pe 1): 282]. — Holotype: *Polyporus amorphus* Fr. per Fr.

Special literature.—Overholts, *1917*.

amorphus (Fr. per Fr.) Kotl. & P. 1958. — *Polyporus* Fr. 1815 (Sweden) (d.n.) per Fr. 1821; *Bjerkandera* P. Karst. 1879; *Leptoporus* Quél. 1886; *Polystictus* Gillot & Luc. 1890; *Tyromyces* Murrill 1918; *Polystictoides* Lázaro 1916; *Gloeoporus* Killerm. 1928. — Secr. 1833 M. 3: 116, at least as to var. A; Gillet 1874–90 P.: *pl. 555/459*; Bres. 1903 (Am 1): 74; Vuyck 1911–15 (Fb 24): *pl. 1885*; Lloyd 1915 (LMW 4, Ap.): 331; Overh. 1917 (M 9): 263 *pls. 12, 13 (Polyporus)*; Murrill 1918 (M 10): 109 *pl. 6 f. 5 (Tyromyces)*; Bourd. & G. 1928: 549; Konr. & M. 1932 I. 5: *pl. 431 f. 1 (Leptoporus)*; Donk 1933: 166; Pilát 1937 (ACE 3): 153 *f. 39, pls. 80–82 (Gloeoporus)*; Nannf. & Du R. 1952: 243 *pl. 125*; Overh. 1953: 359 *pl. 25 fs. 151, 152, pl. 26 f. 159, pl. 128 fig. (Polyporus)*; H. Jahn 1963 (WPb 4): 63; G. Cunn. 1965: 113 *(Gloeoporus)*.

Boletus irregularis Sow. 1815: *pl. 423* (England) (d.n.); fide Fr. 1830 (Li 5): 701. — *Polyporus* (Sow.) per Pers. 1825, not ∼ Underw. 1897; *Bjerkandera* P. Karst. 1889.

Polyporus aureolus Pers. 1825 (France); fide Pers., l.c. (*Polyporus amorphus* cited as a syn.); fide Donk 1933: 166, 167. — Pers. 1825: 60.

Polyporus laneus Pers. 1825: 112 *pl. 17 f. 3* (Switzerland); fide Lloyd 1910 (LMW 3): 472 & Donk 1967 (Pe 5): 91. — *Poria* Lloyd 1910, ("*laurens*"; Syn.: n.v.p.).

Polyporus kymatodes Rostk. 1830 (StP 4): 51 *pl. 24* (Germany/Poland); fide Donk 1972 (PNA 75): 171. — *Bjerkandera* P. Karst. 1879; *Leptoporus* Quél. 1886, Pilát 1937 misapplied; *Poria* [Höhn. apud] Lloyd 1908 (nom. nud.: n.v.p.); *Polystictus* Big. & Guill. 1913; *Coriolus* Bourd. & G. 1925, misapplied; *Agaricus* E. Krause 1932; *Tyromyces* Pilát 1938 (syn.: n.v.p.), Bond. & S. 1941, not ∼ Donk 1933. — Sensu Fr. 1884 I. 2: 82 *pl. 183 f. 1* = ?; sensu Bres. apud Bourd. & G. → *Tyromyces balsameus*; sensu E. Krause → *Polyporus cymatoides* (0). — V.s.: "*cymatodes* (Sacc. 1888), "*kimatodes*", "*kymathodes*".

Polyporus armeniacus Berk. 1836: 147 (Scotland), not ∼ Berk. 1856; fide Cooke 1885 (G 13): 87 (var.); Reid & Austw. 1963 (GN 18): 308 ("almost certainly").

Polyporus roseoporus Rostk. 1848 (StP Fs. 27–28): 23 *pl. 12* (Germany/ Poland), not *P. roseoporus* (Pat.) . . . 18, not ∼ (Lloyd) J. Rick 1938; fide Fr. 1863 M. 2: 339 (var.).

Polyporus albo-aurantius Veull. in Roum. 1883 (Rm 5): 19 (nom. nud.), 46 [repr. 1891 (ASL 17): 288] (France; fide Bres. 1903 (Am 1): 74 & Lloyd 1915 (LMW 4, Ap.): 375. — *Polyporus* Veull. in Roum. 1883 F.g.: No. 2403 [repr. 1883 (Rm 5): 19] (nom. nud.).

Bjerkandera mollusca P. Karst. 1887 (Rm 9): 9 & 1887 (Mfe 4): 80 (Finland). — *Polyporus* Sacc. 1888. — Sensu Lloyd 1915 (LMW 4, Ap.): 101 (syn.); Bourd. & G. 1925 (BmF 41): 130 (*Leptoporus amorphus* var. C.). — Cf. Donk 1967 (Pe 5): 98.

Polyporus horaci Velen. 1922: 651 [see Pilát 1948: 249 for Latin translation] (Czechoslovakia); fide Pilát 1937 (ACE 3): 153.

Polyporus pertenuis Velen. 1922: 651 [see Pilát 1948: 249 for Latin translation] (Czechoslovakia), not ∼ (Kalchbr. apud Thüm.) Lloyd 1912 (n.v.p.), not ∼ (Murrill) Sacc. & Trot. 1912; fide Pilát 1937 (ACE 3): 153.

Gloeoporus amorphus var. *vasilkovii* Bond. 1940 (U.S.S.R., Mari) —

Skeletocutis amorphus var. Domański & al. 1967. — Bond. 1953: 252 *fs. 57, 58*. — The small, ovoid spores (3–3.5 × 1.5–1.75 μ) as described and depicted by Bond. suggest a sp. distinct from *Skeletocutis amorpha* with cylindrical and slightly curved sp., (3–)4–5 × 1–1.5 μ (Overh. 1953: 359, 3–4 × 0.7–1 μ). — V.s. *"vassilkovii"*.

Tyromyces pini-glabrae Murrill 1940 (BTC 67): 65 (U.S.A., Florida); fide Lowe & Murrill apud Overh. 1953: 428. — *Polyporus* Murrill 1940 (nom. altern.).

M.—*Boletus abietinus* Anon. [and Dicks.] sensu DC. 1815: 40; fide Fr. 1828 E. 1: 89.

M.—*Poria vitrea* Pers. sensu Quél. apud Bourd. & G. 1925 (BmF 41): 130 (*Leptoporus amorphus* forma [unnamed]). — Bourd. & G., l.c., 1925 & 1928: 549 (*Leptoporus amorphus* forma).

Gloeoporus **acidulus** M. Bond 1970 (NSn 6): 144 *f. 2* (U.S.S.R., European Russia).

Polyporus **uralensis** (Pilát) Killerm. 1943. — *Leptoporus* Pilát 1932 (U.S.S.R., Russia); *Gloeoporus* Pilát 1937, Bond. & S. 1941. — Pilát 1932 (BmF 48): 11 *f. 3: 1, pl. 1 fs. 1–3* (*Leptoporus*); 1937 (ACE 3): 156 *f. 40, pl. 83* (*Gloeoporus*); Murašk. 1939: 82 *f. 6* (*Leptoporus*); Kartav. 1967: 97 (*Gloeoporus*).

SPONGIPELLIS Pat.

1887 [1960 (Pe 1): 284]. — Monotype: *Polyporus spumeus* (Sow.) per Fr.
? *Somion* Adans. 1763 (d.n.) [1960 (Pe 1): 283]. — Lectotype: *Agaricum squamosum album, superne villosum* ... Mich. ≡ *Hydnum occarium* Batsch per Fr.
Irpiciporus Murrill 1905 [1960 (Pe 1): 230]. — Holotype: *Irpex mollis* B. & C.

SPECIAL LITERATURE.—David, *1969a*; Kotlaba & Pouzar, *1965*.
Spongipellis pachyodon: Ebert, *1940*.
Spongipellis schulzeri: Bondarcev & Ljubarskij, *1938*; Igmándy, *1957a*, *1957b*; H. Lohwag, *1931b*.

delectans (Peck) Murrill 1907. — *Polyporus* Peck 1884 (U.S.A., Ohio); *Leptoporus* Pat. 1900 (nom. nud.: n.v.p.). — Morg. 1885 (JCi 8): 99 *pl. 1* (*Polyporus*); Murrill 1907 (NAF 9): 38 (*Spongipellis*); Overh. 1914 (AMo 1): 99; 1915 (AMo 2): 701, 708 *pl. 29 f. 14b*; Lloyd 1915 (LMW 4, Ap.): 325 *f. 667*; Lowe 1942: 70; Overh. 1953: 320 *pl. 19 f. 113, pl. 116 f. 632, pl. 129 fig.* (*Polyporus*); Al. David 1969 (BlL 38): 192 *fs 1–6, tpl. 1*, with cult. char.; Thind & Ratt. 1971 (IPh 24): 52 *fs. 8–11, 24* (*Spongipellis*).
Trametes krekei Lloyd 1919 (LMW 5, L. 69): 12 (U.S.A., Iowa); fide Overh. 1953: 320, 321.
Leptoporus bredecelensis Pilát 1937 (lacking Latin descr.: n.v.p.), 1953 (U.S.S.R., Ukraine); fide Al. David 1969 (BlL 38): 192, 196. —*Spongipellis* Bond. 1953 (lacking valid ref.: n.v.p.), Domański & al. 1967. — Pilát

1937–8 (ACE 3): 240 *f. 87, pls. 159–161* (*Leptoporus*); Domański & al. 1967 (FpG 3): 96 *f. 25, pl. 8 f. 2* (*Spongipellis*).

foetidus (Velen.) Kotl. & P. 1965. — *Polyporus* Velen. 1927 (Czechoslovakia), not ∼ Lloyd 1915 (n.v.p.); *Leptoporus* Pilát 1937; *Spongipellis* Bond. 1953 (incomplete ref.: n.v.p.). — Pilát 1938–9 (ACE 3): 244 *f. 90, 97, pls. 152, 158* (*Leptoporus*).

pachyodon (Pers.) Kotl. & P. 1965. — *Hydnum* Pers. 1825: Fr. 1832 (France); *Sistotrema* Fr. 1838; *Irpex* Quél. 1888, Lloyd 1906 (*"pachylon"*); *Lenzites* Pat. 1900; *Trametes* Pilát 1939; *Irpiciporus* Kotl. & P. 1957; *Radulomyces* M. P. Christ. 1960. — Bres. 1897 (AAR III 3): 101; 1908 (Am 6): 42; Lloyd 1922–4 (LMW 7): 1159 *pl. 225 f. 2306* & cf. p. 1287; Bourd. & G. 1928: 574 (*Irpex*); Pilát 1939–40 (ACE 3): 326 *f. 141, pl. 219* (*Trametes*); Al. David 1969 (BlL 38): 196, cult. char. (*Spongipellis*).
   ? *Hydnum occarinum* Batsch 1783: 113 (d.n.) per Fr. 1821 .— [≡ *Agaricum squamosum, album, superne villosum, inferne denticulis longioribus, & planis praeditum* Mich. 1729: 122 *pl. 64 f. 3* (Italy)]. — Sensu Secr. 1833 = ? — "Squamosum" in Micheli's phrase name means 'with imbricated caps'.
   ? *Hydnum paleaceum* Thore 1803: 491 (France). — *Sistotrema* (Thore) per Pers. 1825; *Irpex* Fr. 1828; *Hydnum* Duby 1830; *Xylodon* O.K. 1898. — The substratum (rotten coniferous trunk) does not agree.
   Hydnum acutum Pers. 1825: 179 (Switzerland).
   Sistotrema dermatodon Pers. 1825: 195 (Switzerland); fide Donk 1967 (Pe 5): 89. — *Hydnum* Duby 1830.
   Irpex crassus B. & C. 1849 (HJB 1): 236 (U.S.A., North Carolina); fide Lloyd 1919 (LMW 5, L. 69): 11 & Bres. 1920 (Am 18): 70. — *Xylodon* O.K. 1898. — Morg. 1887 (JCi 10): 14.
   Irpex mollis B. & C. 1849 (HJB 1): 236 (U.S.A., South Carolina); fide Murrill 1905 (BTC 32): 471 & 1907 (NAF 9): 15 (*Irpex crassus* cited as a syn.). — Reported from Europe by Bizzoz. 1885 F.v. 1: 104.
   Irpex heterodon Sacc. 1873 (ASv 2): 106 *pl. 7 fs. 16–19* (Italy); fide Bres. 1897 (AAR III 3): 101.
   Hydnum schestunovii Nikol. 1949 (U.S.S.R., Caucasia); fide Nikol. 1964 (NSn): 171. — *Hericium* Nikol. 1961. — Nikol. 1949 (BMs 6): 87 *fs. 1, 2* (*Hydnum*); 1961: 238 *fs. 181, 182* (*Hericium*).
   M.—*Sistotrema lacteum* Fr. sensu Pat. 1886 T.a. 1: 200 *f. 455*.

schulzeri (Fr.) Bourd. & G. 1925. — *Polyporus* Fr. 1874 (Yugoslavia, Slavonia), not ∼ Kalchbr. 1868; *Ungulina* Lohw. 1931, misapplied; *Leptoporus* Pilát 1937, misapplied; ≡ *Polyporus irpex* S. Schulz. 1866: 42 (nom. nud.: n.v.p.) ex E. Krause 1925; *Leptoporus* Igmándy 1957 (incomplete ref.: n.v.p.). — Kalchbr. 1877: 53 *pl. 34 f. 1*, descr. partly incorrect; S. Schulz. 1880 (ÖbZ 30): 108 (*Polyporus*); Bourd. & G. 1925 (BmF 41): 117 & 1928: 534 *f. 149* (*Spongipellis*); Igmándy 1957 (EFK):

67 *fs. 1–5* (*Leptoporus irpex*); Donk 1972 (PNA 75): 171, notes. — Sensu Lohw. → *Tyromyces chioneus?*

Spongipellis litschaueri Lohw. 1931 (Austria); fide Donk 1972 (PNA 75): 172, 185. — *Leptoporus* Pilát 1938, not ∼ Pilát 1932; *Polyporus* Bond. apud Bond. & Ljub. 1938. — Lohw. 1931 (ArP 75): 297 *pl. 18, pl. 19 fs. 3–5* (*Spongipellis*); Pilát 1937–9 (ACE 3): 241 *f. 88, pls. 162–164, pl. 165 f. a*; Igmándy 1957 (BK 47): 101 *fs. 1–3* (*Leptoporus*); Kotl. & P. 1975 (ČM 11): 216, 223 *2 figs.*; 1965 (ČM 19): 70, 75 *pl. 5* (*Spongipellis*).

spumeus (Sow. per Fr.) Pat. 1900. — *Boletus* Sow. 1799 [fide Overh. 1915 (AMo 2): 701–704, text only] (England) (d.n.); *Polyporus* (Sow.) per Fr. 1821; *Boletus* Spreng. 1827; *Bjerkandera* P. Karst. 1882; *Inodermus* Quél. 1886; *Leptoporus* Quél. 1888, Pilát 1937; *Spongipellis* Pat. 1887 (nom. nud.; n.v.p.); *Polystictus* Big. & Guill. 1913; *Tyromyces* Imaz. 1943. — Hornem. 1823 (Fd 10 / F. 30): 12 *pl. 1794?* (*Boletus*) [cf. Overh. 1915 (AMo 2): 703]; Fr. 1828 E. 1: 84; Bres. 1897 (AAR III 3): 72; Overh. 1914 (AMo 1): 99; 1915 (AMo 2): 701, 707 *pl. 24 fs. 10, 11, 14a*; Lloyd 1915 (LMW 4, Ap.): 304 *fs. 641, 642* (*Polyporus*); Bourd. & G. 1928: 533 (*Spongipellis*); Shope 1931 (AMo 18): 345 *pl. 28 f. 2* (*Polyporus*); Pilát 1937–8 (ACE 3): 237 *f. 86, pls. 155–157* (*Leptoporus*); Lowe 1942: 68 (*Polyporus*); Yen 1950 (AUL 6): 32, 80 *tpl. 4 fs. F–I*, cult. char. (*Spongipellis*); Overh. 1953: 318 *pl. 19 f. 114, pl. 21 f. 126, pl. 97 f. 552, pl. 132 fig.* (*Polyporus*); Al. David 1969 (BlL 38): 197 *fs. 5, 6*, cult. char. of American cultures (*Spongipellis*). — Sensu J. Schroet. → *Tyromyces fissilis.*

Polyporus suberosus Fr. 1821, not ∼ (Bull. per St-Am.) Chev. 1826, not ∼ Wettst. 1885; fide Fr. 1828 E. 1: 84. — [≡ *Boletus suberosus* L. sensu Wahl. 1820: 457 (Sweden)]; ≡ *Boletus pulvinatus* Wahl. 1826, typonym. — Sensu Krombh. → *Buglossoporus pulvinus.*

Spongipellis occidentalis Murrill 1907 (U.S.A., New York); fide Bres. 1926: 79 & Overh. 1953: 318. — *Polyporus* Sacc. & Trott. 1912. — Lloyd 1915 (LMW 4, Ap.): 327 *f. 671* (*Polyporus*).

TRAMETES Fr. (114)

1836 [1960 (Pe 1): 288; 1962 (Pe 2): 209]. — Lectotype: *Polyporus suaveolens* (L.) per Fr. [sensu Fr.].

Artolenzites Falck 1909 [1960 (Pe 1): 187]. — Lectotype: *Lenzites repanda* (Pers.) Fr.

Pseudotrametes Bond. & S. ex Sing. 1944 [1960 (Pe 1): 275]. — Holotype: *Trametes gibbosa* (Pers. per Fr.) Fr. — (114).

Haploporus Bond. 1953, not ∼ Bond. & S. ex Sing. 1944 (Polyporaceae) [1962 (Pe 2): 205]. — Holotype: *Trametes ljubarskyi* Pilát.

SPECIAL LITERATURE.—David, *1967a*; Gilbertson, *1961*; Vandendries, *1934c*.

Irpex foliaceo-dentatus: Domański, *1970e*; Nikolajeva, *1949*.

Trametes gibbosa ("*Pseudotrametes*"): Gay, Hutchinson, & Taggart, *1959*; Hemmi & Ikeya, *1939*; Pirk & Tüxen, *1957*; Schwantes, *1962*.

Trametes hoehnelii: Domański, *1970c*.

Trametes ljubarskyi: David, *1966*.

Tyromyces pseudohoehnelii.—Komarova, *1959*.

Trametes suaveolens: Aye, *1931*; Birkinshaw, Bracken, & Findlay, *1944*; Darley & Christensen, *1943*; Enslin, *1784, 1798*; Hirt, *1932*; Schlesinger, *1838*.

cervina (Schw.) Bres. 1903. — *Boletus* Schw. 1822 (U.S.A., North Carolina); *Polyporus* Steud. 1824: Fr. 1828, not ∼ Pers. 1825, not ∼ (Quél.) Sacc. 1895; *Polystictus* Cooke 1886, not ∼ (Quél.) Big. & Guill. 1913; *Microporus* O.K. 1898; *Trametes* Bres. 1903, not ∼ (Pers.) Lloyd 1910; *Coriolus* Bond. 1953; *Coriolellus* Kotl. & P. 1957. — Sensu Fr. 1828 E. 1: 92 (*Polyporus*); Bres. 1903 (Am 1): 81; 1908 (Am 6): 39; Bourd. & G. 1928: 594; Pilát 1939 (ACE 3): 293 *f. 119, pl. 202* (*Trametes*); Bond. 1953: 493 *f. 127, pl. 137 fs. 5–7, pl. 157 f. 2* (*Coriolus*); Domański 1965 (FpG 2): 192 *pl. 55 f. 3, pl. 56 f. 1* (*Coriolellus*).

Sistotrema symphyton Schw. 1822: 101 (U.S.A., North Carolina); fide Lloyd 1913 (LMW 4, L. 50): 8 = *Polystictus biformis.* — *Polyporus* Fr. 1828.

Irpex epiphyllus Schw. 1832: 164 (U.S.A., Pennsylvania); fide B. & C. 1856 (JAP II 3): 217 = *Polyporus biformis*; Bres. 1908 (Am 6): 39 ("forma *resupinata* e vetustate dentata").

Polyporus biformis Fr. apud Kl. 1833 (Li 8): 486, in part ("Americ. bor.") & Fr. 1838: 475; fide Bres. 1903 (Am 1): 81 & 1916 (Am 14): 223; & cf. Berk. 1841 (AM 7): 452. — *Polystictus* Fr. 1851; *Bjerkandera* P. Karst. 1882; *Coriolus* Pat. 1897, misapplied; *Microporus* O.K. 1898; *Trametes* Pilát 1939, misapplied. — Emend. Fr. 1838: 475 (*Polyporus*); 1851 (NAu III 1): 84/68 (*Polystictus*); Overh. 1914 (AMo 1): 95; D. Baxt. 1949 (PMi 33): 25 *pl. 7*; Overh. 1953: 328 *pl. 9 fs. 50–52, pl. 128 fig.* (*Polyporus*). — Sensu Berk. 1839 → *Hirschioporus pargamenus*.

Polyporus candidulus Lév. 1846 (ASn III 5): 301 (France); fide Donk, and Pegl. in litt.

Polyporus molliusculus Berk. 1847 (LJB 6): 320 (U.S.A., Ohio); from descr.; fide Overh. 1953: 328 = *Polyporus biformis.* — *Polystictus* Fr. 1851; *Microporus* O.K. 1898; *Coriolus* Murrill 1914.

Polyporus carolinensis B. & C. 1849 (HJB 1): 102 (U.S.A., South Carolina); fide Fr. 1851 (NAu III 1): 84/68 = *Polystictus biformis.*

Polyporus scariosus B. & C. apud Berk. 1872 (G 1): 53 ("*scarrosus*") (U.S.A., North Carolina); fide Murrill 1907 (NAF 9): 26 = *Coriolus biformis.* — *Polystictus* Cooke 1886; *Microporus* O.K. 1898.

? *Polyporus balsamiferae* Cooke 1878 (TEd 13): 133. — ≡ *Polyporus biformis* var. *populi-balsamiferae* Kl. 1833 (Li 8): 486 (Canada, Saskatchewan); Berk. 1838 (AM 3): 392. — I have also thought of *Hirschioporus subchartaceus* (0), a species closely related to *H. pargamenus*, which Kl. confused with *Polyporus biformis* [= "*Trametes*" *cervinus*], but the pores seem to be too large: 'majusculus' (Kl.) and 'rather large' (Berk.).

? *Polyporus pachylus* Berk. ex Sacc. 1888 (SF 6): 149 (Canada); fide Lloyd 1915 (LMW 4, Ap.): 383, cf. *Polyporus biformis*.

Trametes populina Bres. apud Sacc. 1896 (Mal 10): 262, not ∼ (Schum. per Fr.) Fr. 1849; fide Bres. 1903 (Am 1): 81. — *Coriolus* Murrill 1948. — [≡ *Polyporus populinus* (Schum.) ex Fr. sensu S. Schulz. apud Kalchbr. 1877 (Yugoslavia, Slavonia)]. — Kalchbr. 1877: 57 *pl. 37 f. 1b* (as *Polyporus* "*populinus Sch.*", treated as a syn. of *P. vulpinus* Fr. [sensu Kalchbr.]); Bres. 1897 (AAR III 3): 90.

M.—*Polyporus vulpinus* Fr. sensu Kalchbr. 1877: 56 *pl. 37*, in part, as to *f. 1b* (lower fig.); fide Kalchbr., l.c. = *Polyporus populinus* (Schum.) ex Fr. sensu S. Schulz.; fide Bres. 1903 (Am 1): 81.

M.—*Boletus populinus* Schum. sensu S. Schulz. in litt. & MS., 1880 (*Polyporus*). — S. Schulz. 1880 (ÖbZ 30): 109 (*Polyporus*). → *Polyporus vulpinus* Fr. sensu Kalchbr. 1877, in part, → *Trametes populina* Bres. apud Sacc.

flavescens Bres. 1903 (Poland). — *Trametes* Jørst. 1937 ("*flavicans*"; error: n.v.p.); *Coriolellus* Bond. & S. 1941; *Daedalea* Aosh. 1967; *Diplomitoporus* Domański 1970. — Bourd. & G. 1928: 597; Pilát 1939 (ACE 3). 296 *f. 121, pl. 203 f. b* (*Trametes*); Kotl. & P. 1959 (ČM 13): 34 *fig*;. M. P. Christ. 1960 (DbA 19): 371 *f. 372*; Domański 1963 (Mob 15): 315 *f. 14*; 1965 (FpG 2): 182 *pl. 52 f. 1* (*Coriolellus*); Al. David 1967 (Nca 94): 560 *tpl. 1 f. 3, tpl. 2 f. b*, cult. char. (*Trametes*); Domański 1970 (APo 39): 191 *fs. 1, 2a, 3a, b, 4–6* with cult. char. (*Diplomitoporus*).

Polyporus winogradowii Bond. 1912 (TlR 37): 9 *pl. 2 f. 4* (U.S.S.R., European Russia); fide Bond. 1953: 510.

Irpex **foliaceo-dentatus** Nikol. 1949 (U.S.S.R., Caucasus). — Nikol. 1949 (BMs 6): 85 *fs. 1–3* (*Irpex*); Domański *1970c*, with cult. char. (*Coriolus*).

Coriolus **genistae** (Bourd. & G.) Bourd. & G. 1928 (**115**). — *Coriolus hoehnelii* subsp. *C. genistae* Bourd. & G. 1925; ≡ *Poria vulgaris* var. *pileata* Bourd. & L. Maire 1920 (France). — Bourd. & G. 1928: 569 *f. 164*.

gibbosa (Pers. per Fr.) Fr. 1838 (**114**). — *Merulius* Pers. 1795 (Germany) (d.n.); *Daedalea* Pers. 1801 (d.n.) per Fr. 1821; *Polyporus* Kumm. 1871, not ∼ Pers. 1825, not ∼ Bl. & Nees 1826; *Lenzites* Hemmi apud Hemmi & Ikeya *1939*; *Pseudotrametes* Bond. & S. 1941 (generic name n.v.p.), Sing. 1944. — Gillet 1874–90 P.: *pl. 656/474*; Boud. 1904–11: 83 *pl. 162*; Bourd. & G. 1928: 589; Pilát 1939 (ACE 3): 289 *f. 116, pl. 198, pl. 199 f. b* (*Trametes*); Švarcm. 1964: 566 *fs. 250, 251* (*Pseudotrametes*); Al. David 1967 (Nca 94): 558 *fs. 1, 2, tpl. 2 f. a*, cult. char. (*Trametes*). — Sensu Tratt. → *Daedalea torulosa* Pers. (**0**); sensu Wahl. → *Climacocystis borealis*; sensu Purt. 1821: 248 *pl. 14* → *Daedalea quercina*, fide Berk. 1836: 131.

Agarico-suber scalptum Paul. 1793 T. 2: 76 (descr.), Ind. (name) (France) (d.n.). — ≡ *Agaricus tectulum* Paul. 1812–35: *pl. 2 fs. 2–4* (d.n.?).

Boletus sinuosus Sow. 1799: *pl. 194* (England) (d.n.); fide Fr. 1821: 338.
— *Daedalea* (Sow.) per S. F. Gray 1821; *Boletus* Lenz 1840.

Trametes kalchbrenneri Fr. apud Kalchbr. 1868 (MtK 5): 264 *pl. 4 f. 2*
[repr. 1879 (H 8): 117] (Hungary, now Czechoslovakia); fide Bres. 1897
(AAR III 3): 91. — *Daedalea* Lübstorf 1896; *Agaricus* E. Krause 1934.

Bulliardia virescens Lázaro 1916 (RMa 14): 843 / 1917: 155 (Spain);
fide Bres. apud Trott. 1925 (SF 23): 449 ("vetusta").

hoehnelii (Bres. apud Höhn.) Pilát 1939. — *Polyporus* Bres. apud Höhn.
1912 (Austria); *Coriolus* Bourd. & G. 1925; *Polystictus* Killerm. 1943.
— Höhn. 1912 (SbW 121): 344; Bres. 1920 (Am 14): 58 (*Polyporus*);
Bourd. & G. 1928: 568 *f. 163* (*Coriolus*); Pilát 1939 (ACE 3): 270 *f. 107*,
pls. 185, 186; H. Jahn 1967 (WPb 6): 159 *fig.* (*Trametes*); Domański
1970c, with cult. char. (*Coriolus*). — V.s. (by Lloyd): *Polyporus* "*Hohenia-lus*", "*Hoehnelialus*", "*Hohnelianus*".

Polyporus rufopodex Haglund ex Romell 1912 (SbT 6): 641 *fig. in part*
(Sweden); fide Bourd. & G. 1928: 568.

Polyporus scaber Velen. 1922: 657 [see Pilát 1948: 252 for Latin transla-
tion] (Czechoslovakia), not ∼ Bres. 1920; fide Pilát 1939 (ACE 3): 270.
— Velen. 1925 (MP 2): 73, 75. — Apparently the spores were wrongly
described as globose, 2–3 μ.

M.—*Polyporus epileucus* Fr. (**O**) sensu Lloyd 1915 (LMW 4, Ap.):
309, in part: excl. of American collection depicted in *f. 649*; fide Lloyd,
l.c. = *Polyporus* "*Hohenialus*" [= *hoehnelii*]. — Lloyd (LMW 7): 1260
pl. 279 f. 1255, very poor.

ljubarskyi Pilát 1937 (U.S.S.R., Russia, Siberia). — *Haploporus* Bond. &
S. 1941 (n.v.p.), Bond. 1953. — Nikol. 1940 (TSR 4): 413 *fs. 27, 28*
(*Trametes*); Bond. 1953: 47, 523 *f. 139, pl. 151 fs. 1–4*; Švarcm. 1964:
570 *fs. 252, 253* (*Haploporus*); Pouz. 1966 (Fgp 1): 365; Al. David 1966
(BmF 82): 504 *fs. 1–8*, with cult. char. (*Trametes*).

nigrescens Bres. 1905 (Am 5): 163 (Italy, Tirol), not ∼ Lázaro 1916.
— *Fomitopsis* Bond. & S. 1941. — Nikol. 1940 (TSR 4): 408 *f. 2?*
(*Trametes*); Bond. 1953: 619? (*Fomitopsis*), with note.

Polyporus **onychoides** Egeland 1913 (Norway) (**116**). — *Tyromyces* Ryv.
1967. — Egeland 1913 (NMN 51): 91 (*Polyporus*); Ryv. 1967 (Bly 25):
207 (*Tyromyces*).

Tyromyces **pseudohoehnelii** Bond. & Kom. apud E. Komar. 1959 (U.S.S.R.,
White Russia). — Bond. & Kom. apud E. Komar. 1959 (DAb 3): 507
f. 1 (**117**).

Polyporus **semisupinus** B. & C. apud Berk. 1872 (U.S.A., "New England")
(**118**). — *Tyromyces* Murrill 1907, Bond. & M. Bond. 1960 (incomplete
ref.: n.v.p.); *Leptoporus* Pilát 1937; *Aporpium* Bond. 1953. — Lowe 1934

(PMi 19): 146 *f. 5B, pl. 15 f. 2* (*Polyporus*); Pilát 1937–8 (ACE 3): 212 *f. 67, pl. 138, pl. 139 f. a* (*Leptoporus*; Lowe 1942: 64; Overh. 1953: 376 *pl. 17 f. 102, pl. 18 fs. 108, 109, pl. 95 f. 541, pl. 106 f. 589, pl. 124 f. 675, pl. 132 fig.* (*Polyporus*); Domański 1965 (FpG 2): 153 *f. 54, pl. 33 f. 3, pl. 42* (*Tyromyces*); Lowe 1966: 126 *f. 114* (*Polyporus*). — Sensu Lloyd → *Incrustoporia semipileata.*

Polyporus pachycheiles Ell. & Ev. 1894 (PAP): 322 (U.S.A., New Jersey); fide Lowe 1934 (PMi 19): 146 & Overh. 1953: 376, 377. — Lowe, op. cit. *f. 5C*, spores.

Polyporus pallescens Romell 1911 (**119**), not ∼ Fr. per Fr. 1821 (**16**); fide Lundell 1953 (LNF 43–44): 4 No. 2104 & cf. Overh. 1953: 377. — *Coriolus* Pilát 1932 ("Karsten, sensu Romell . . . nec Fries!"), misapplied; [≡ *Polyporus pallescens* Fr. per Fr. sensu P. Karst., specimen in herb. (Finland)]. — Romell 1911 (ABS 11³): 19 *pl. 1 f. 5.* — Sensu Pilát (f. *resupinata*) → *Coriolus subradula* (**0**).

M.—*Polyporus pallescens* Fr. sensu P. Karst. → *Polyporus pallescens* Romell.

suaveolens (Fr.) Fr. 1838. — *Polyporus* Fr. 1828 (Sweden), not ∼ (L.) per Fr. 1821; *Daedalea* E. Krause 1928 ("Fr."), 1930 ("Linn.") ("*Daedaleus*"). — Fr. 1874: 584; Gillet 1874–90 P.: *pl. 658/473* (*Trametes*); J. Schroet. 1888: 475 (*Polyporus*); Boud. 1904–11: 83 *pl. 163*; Bourd. & G. 1928: 589; Konr. & M. 1930 I. 5: *pl. 445*; Pilát 1940 (ACE 3): 292 *f. 118, pl. 200*; Overh. 1953: 143 *pl. 85 fs. 485, 486, pl. 99 f. 563, pl. 100 f. 568, pl. 125 fig.*; Jo. Erikss. 1958 (Sbu 16¹): 143 *f. 45*; Westh. 1971 (Bo 10): 226 *fs. 23, 24*, with cult. char. (*Trametes*); Donk 1971 (PNA 74): 19, notes.

Boletus discoideus Dicks. 1793 P.c. 3: 21 (England) (d.n.); fide Fr. 1832^Ind.: 57.

Polyporus itoi Lloyd 1924 (LMW 7): 1274 *pl. 288 f. 2816* (Japan); fide Imaz. 1943 (Apg 13): 255.

Trametes radiata Burt 1931 (AMo 18): 475 (U.S.S.R., Russia, Siberia), not ∼ (Sow. per Fr.) Fr. 1849; fide Nikol. apud Bond. 1953: 521 (*Trametes suaveolens* f. *gibbosiformis* Nikol. apud Bond., l.c., lacking Latin descr.: n.v.p.).

M.—*Boletus suaveolens* L. sensu Enslin 1784 (*Boletus*) & Fr. 1821 (*Polyporus*). — Enslin 1784; Wulf. 1789 (CoJ 2): 147 (*Boletus*); Fr. 1821: 366; Hornem. 1825 (Fd 11 / F. 31): 12 *pl. 1849* (*Polyporus*). → *Polyporus suaveolens* Fr.

? M.—*Boletus salicinus* Bull. sensu Sow. 1790: *pl. 227* & cf. discussion to *pl. 228.*

M.—*Boletus suberosus* L. (**0**) sensu Bolt. 1791: 162 *pl. 162*; Pilát (ACE 3): 292 ("*tuberosus*"; error & syn.: n.v.p.). — Under *Polyporus suberosus.* Secr. 1833 M. 3: 105 (as a sp. of *Boletus*: n.v.p.) refers to Bolt. without actually excluding the type, which may have been the intention.

TRUNCOSPORA Pilát

1953 [1960 (Pe 1): 290]. — *Truncospora* Pilát 1941 & 1942 (n.v.p.). — Lectotype: *Polyporus ochroleucus* Berk.

    Pyrofomes Kotl. & P. 1964 (FR 69): 140. — Holotype: *Polyporus demidoffii* Lév.
M.—*Ungulina* Pat. sensu Kotl. & P. 1957 (ČM 11): 168, excl. of type = *Ungulina* Kotl. & P. (n.v.p.) (**O**).

    SPECIAL LITERATURE.—von Schrenk, *1900* (*Polyporus demidoffii*); Torrend, *1910* (*T. ochroleuca*); Wakefield, *1915* (*Polyporus demidoffii*):

    Polyporus **demidoffii** Lév. 1842 (U.S.S.R., Ukraine). — *Trametes* P. Karst; 1882; *Fomes* Cooke 1885; *Scindalma* O.K. 1898; *Xanthochrous* Pat. 1897 (nom. nud.: n.v.p.), 1900; *Inonotus* Pilát 1937; *Phellinus* Bond. & S. 1941, Pilát 1942; *Fulvifomes* Murrill 1948; *Pyrofomes* Kotl. & P. 1964, — Lév. 1842 D. 2: 92 [*pl. 3*] (*Polyporus*); Shope 1931 (AMo 18): 385 *pl. 35 f. 3* (*Fomes*); Pilát 1937 (BmF 53): 88 (*Inonotus*); 1942 (ACE 3): 514 *f. 243, pl. 341*; Bond. 1953: 398 *f. 96, pl. 183 f. 1* (*Phellinus*); Lowe 1957 F.: 62 *f. 45* (*Fomes*); Kotl. & P. 1964 (FR 69): 140 (*Pyrofomes*).

    Polyporus juniperinus Schrenk 1900 (U.S.A., Tennessee), not ∼ (Murrill) Sacc. & Trott. 1912; fide Lloyd 1912 (LMW 4): 523 = *Fomes demidoffii* (mentioned as a syn.). — *Fomes* Sacc. & Syd. 1902; *Pyropolyporus* Murrill 1903; *Fulvifomes* Murrill 1914. — Schrenk 1900 (BVP 21): 9 *pls. 1–4, pl. 7 fs. 1–7, 11* (*Polyporus*); Lloyd 1912 (LMW 4): 522; Hedgc. & Long 1912 (M 4): 110 *pl. 64 f. 1, pl. 65 fs. 3, 6, 9*; Lloyd 1915 (LMW 4, F.), 232; Wakef. 1915 (BmI): 102 *fig.*; Overh. 1953: 97 *pl. 70 fs. 410–412*: *pl. 126 fig.* (*Fomes*).

    Pyropolyporus earlei Murrill 1903 (BTC 30): 116 (U.S.A., New Mexico). fide Lloyd 1912 (LMW 4): 523 = *Fomes juniperinus.* — *Fomes* Sacc. & D. Sacc. 1902; *Phellinus* A. Ames 1913. — Hedgc. & Long 1912 (M 4): 111 *pl. 64 fs. 4–6, pl. 65 fs. 2, 5, 8*; D. Baxt. 1949 (PMi 33): 27 (*Fomes*).

    ochroleuca (Berk.) S. Ito 1955 (**120**). — *Polyporus* Berk. 1845 (West Australia); *Trametes* Cooke 1891, Bres. apud Torrend 1910, not ∼ Pat. 1914, not ∼ Sacc. 1917; *Fomes* Pat. 1898; *Ungulina* Pat. 1900; *Truncospora* Pilát 1942 (generic name n.v.p.), 1953 (incomplete ref.: n.v.p.); *Fomitopsis* Imaz. 1943, G. Cunn. 1948, Bond. 1953 (incomplete ref.: n.v.p.); *Poria* Kotl. & P. 1959; *Heterobasidion* G. Cunn. 1965. — Lloyd 1911 (LMW 3, L. 33): 2; 1915 (LMW 4, Ap.): 311 *f. 651* (*Polyporus*); 1917 (LMW 5): 714 *f. 1070* (*Fomes*); Clel. 1935: 220 (*Trametes*); Boed. 1940 (BBu III 16): 368 (*Polyporus*); G. Cunn. 1965: 145 *f. 29, pl. 1 f. b* (*Heterobasidion*).

    Polyporus compressus Berk. 1845 (LJB 4): 53 (West Australia); fide Lloyd 1915 (LMW 4, F.): 279 ("effete") & Bres. 1916 (Am 14): 223. — *Fomes* Cooke 1886; *Scindalma* O.K. 1898.

    Polyporus detritus Berk. 1856 (HJB 8): 197 (Brazil); fide Bres. 1916 (Am 14): 224.

    ? *Polyporus havannensis* B. & C. 1868 (JLS 10): 310 (Cuba); Cooke

1878 ("*havanensis*"); fide Cooke 1891 (G 19): 99. — *Trametes* Murrill 1907.

Trametes ungulata Berk. 1872 (JLS 13): 165 (South Australia), not ∼ (Lloyd) Yas. 1918; fide G. Cunn. 1949 (BPZ 81): 13 & 1965: 145, 147. — *Polyporus* Cooke 1885, not ∼ (Schaeff. per St-Am.) Balbis 1828.

Trametes scrobiculata Berk. 1877 (G 6): 70 (Australia, Victoria); fide G. Cunn. 1949 (BPZ 81): 13 & 1965: 145, 146.

Polyporus leveillei Pat. 1891 (Rm 13): 137, not ∼ Cooke 1878; cf. Pat. 1890 (BmF 6): xxi & fide Lloyd 1915 (LMW 4, Ap.): 381 ("old, effete"). — [≡ *Favolus crassus* Lév. "mscr. Herb. Mus. Par." (Abyssinia)].

Trametes ochroleuca var. *lusitanica* Torrend 1910 (Portugal) **(120)**. — Torrend 1910 (BSp 4): 36; 1913 (Bro 11): 68.

Trametes varia Lloyd 1922 (LMW 7): 1114 *pl. 195 f. 2089* (Australia, Tasmania); fide G. Cunn. 1949 (BPZ 81): 13 & 1965: 145.

? *Polyporus junctus* Lloyd 1924 (LMW 7): 1317 *pl. 311 f. 3028* (Japan); cf. Lloyd, l.c., "it is *Polyporus ochroleucus*, excepting as shown in the figure".

TYROMYCES P. Karst. **(121)**

1881 [1960 (Pe 1): 290]. — Lectotype: *Polyporus chioneus* Fr. per Fr.

Postia Fr. 1874 [1960 (Pe 1): 273], not ∼ Boiss. & Blanch. 1875 (Compositae). — Lectotype: *Polyporus lacteus* Fr. — **(121)**.

Leptoporus Quél. 1886 [1960 (Pe 1): 236]. — Lectotype: *Polyporus mollis* (Pers.) per Fr.

Persooniana Britz. 1897 [1960 (Pe 1): 251]. — Monotype: *Persooniana albocana* Britz. **(122)**.

Spongiporus Murrill 1905 [1960 (Pe 1): 284]. — Holotype: *Polyporus leucospongia* Cooke & Harkn.

? *Aurantiporus* Murrill 1905 [1960 (Pe 1): 189] **(121)**. — Holotype: *Polyporus pilotae* Schw.

Hemidiscia Lázaro 1916 [1960 (Pe 1): 233]. — Lectotype: *Polyporus lacteus* Fr.

Ptychogaster Corda 1838 (nom. anam.) [1962 (Ta 11): 96] **(123)**. — Monotype: *Ptychogaster albus* Corda.

SPECIAL LITERATURE.—General: Martirosjan, *1963*.

"House fungi": Bondarcev, *1936a*; Hennings, *1888*.

Tyromyces albellus & *T. lacteus*: Aoshima & Kobayasi, *1966*; Arita & Sakamoto, *1966*; Bondarcev, *1941*.

Tyromyces alborubescens: Jacquiot, *1965* (n.v.).

Tyromyces balsameus: Aoshima, *1953b*; Hubert, *1929*; Kotlaba & Pouzar, *1968a*.

Tyromyces caesius: Buchwald, *1941b*; Harmsen, *1954*; Lloyd, *1909b*

Polyporus croceus: Černý, *1960*; Jahn, *1967b*; Kotlaba & Pouzar *1966b*.

Leptoporus ellipsosporus: Romagnesi, *1944*.

Tyromyces floriformis: Bondarcev, *1949b*.

Tyromyces gloeocystidiatus: Kotlaba & Pouzar, *1956*, *1964b*; Lowe & Lundell, *1956*; Orlicz, *1971*.

Tyromyces guttulatus: Bergstädt, *1970*; Davidson, Christensen & Darley, *1946*; O. Fidalgo, *1959a*.

Tyromyces kmetii: Bondarcev, *1949a*; Kotlaba & Pouzar, *1965*.

Tyromyces lowei: Domański, *1964a*.

Tyromyces ptychogaster: Cornu, *1876*; Ludwig, *1880a, 1880b*; Meyer, *1922*; Stevenson, *1878*; Tulasne, *1864*.

Tyromyces spraguei: Pilát, *1936b*; Weir, *1923b, 1927*.

Tyromyces stipticus (*T. albidus*): Bondarcev, *1949b*; Domański, *1960c*; Kauffman, *1926*.

Tyromyces minusculoides: Lowe, *1957b*.

Tyromyces undosus: Brotzman & Gilbertson, *1967*.

Tyromyces wynnei: Hansen, *1969*; Jacquenoud, *1968*; Lundell & Pilát, *1936*; Miersch, *1965*.

Ptychogaster: Cornu, *1876*; Davidson, Christensen, & Darley, *1946*; R. & O. Falck, *1937*; O. Fidalgo, *1959a*; Hennings, *1888*; Ludwig, *1880a, 1880b*; Tulasne, *1865*; Ulbrich, *1941*.

Persooniana **albocana** Britz. 1897 (BCb 71): 88 [*pl. 694 f. 90*] (Germany) (**122**). — *Hydnum* Sacc. & Syd. 1899.

alborubescens (Bourd. & G.) Bond. 1953. — *Phaeolus albosordescens* subsp. Bourd. & G. 1925 (France); *Leptoporus* Pilát 1938; *Polyporus* Badet 1934 (n.v.p.); *Spongipellis* Maire 1945; *Phaeolus* R. Heim 1957 (n.v.p.). — Bourd. & G. 1928: 556 *f. 159R* (*Phaeolus albosordescens* subsp.); Fallahyan 1964: 14 *tpl. 1*, cult. char.; Al. David 1969 (Nca 96): 217, cult. char. (*Phaeolus*).

apalus (Lév.) Bond. 1953 (**O**) sensu Bourd. & G. (**127**). — Bourd. & G. 1928: 566 (*Coriolus kymatodes* subsp.). — Cf. *Polyporus* [*Tyromyces*] *henningsii*.

balsameus (Peck) Murrill 1914. — *Polyporus* Peck 1878 (U.S.A., New York); *Polystictus* Cooke 1886; *Microporus* O.K. 1898; *Coriolus* Murrill 1907. — Murrill 1907 (NAF 9): 21 (*Coriolus*); McCallum 1928 (BAC II 104): 9 *pl. 7 fs. 2, 3*; Lowe 1942: 65; D. Baxt. 1949 (PMi 33): 17 *pl. 5*; Overh. 1953: 356 *pl. 17 fs. 96, 97, pl. 18 fs. 106, 107, pl. 96 f. 547, pl. 113 f. 621, pl. 128 fig.* (*Polyporus*); Kotl. & P. 1968 (ČM 22): 121, 128 *fs. 1–3* (*Tyromyces*).

Polyporus crispellus Peck 1885 (RNS 38): 91 (U.S.A., New York); fide Lowe 1934: 71 & Overh. 1953: 356, 358. — *Tyromyces* Murrill 1907.

Tyromyces kymatodes Donk 1933, not ∼ (Rostk.) Bond. & S. 1941. — [≡ *Coriolus kymatodes* (Rostk.) Bourd. & G. sensu Bres. apud Bourd. & G. 1925 (Italy)]. — Donk 1933: 154.

Leptoporus alma-atensis Pilát 1937 (U.S.S.R., Kazachstan); fide Pilát 1938 (ACE 3): 214, 215 = *Leptoporus kymatodes* [sensu Pilát]; fide Kotl. & P. 1968 (ČM 22): 128. — Pilát 1937 (BmF 52): 307 *pl. 3 fs. 1, 2*.

M.—*Polyporus kymatodes* Rostk. sensu Bres. apud Bourd. & G. 1925 (*Coriolus*); fide Pilát 1938 (ACE 3): 214 (*Polyporus balsameus* cited as a syn.); fide D. Baxt. 1949 (PMi 33): 17. — Bourd. & G. 1928: 565 (*Coriolus*); Pilát 1937–8 (ACE 3): 214 *f. 69 pls. 131–133, pl. 134 f. 1* [see remarks by Kotl. & P. 1968 (ČM 22): 128]; Wakef. 1952 (TBS 35): 34 (*Leptoporus*); Donk 1972 (PNA 75): 176, notes. → *Tyromyces kymatodes* Donk.

caesius (Schrad. per Fr.) Murrill 1907. — *Boletus* Schrad. 1794 (Germany) (d.n.); *Polyporus* (Schrad.) per Fr. 1821; *Boletus* S. F. Gray 1821; *Postia* P. Karst. 1881; *Bjerkandera* P. Karst. 1881; *Leptoporus* Quél. 1886; *Tyromyces* Murrill 1907; *Cyanosporus* Lloyd 1909 (n.v.p.); *Polystictus* Big. & Guill. 1913. — Bres. 1897 (AAR III 3): 71 & 1903 (Am 1): 73; Overh. 1915 (AMo 2): 705, 708; 1915 (WUS 3¹): 37 *pl. 2 f. 8* (*Polyporus*); Bourd. & G. 1928: 540; Pilát 1937 (ACE 3): 171 *f. 44, pls. 92–95, 100* (*Leptoporus*); Lowe 1942: 74; Overh. 1953: 292 *pl. 17 f. 103, pl. 18 fs. 104, 105, pl. 128 fig.*; Siepm. 1969 (NH 18): 188 *tpl. 2 f. 2, tpl. 4 f. 3, tpl. 5 f. 3 in part,* cult. char. (*Polyporus*).

? *Agaricus caesius* Dubois 1803: 178 (France (d.n.). — *Agaricus* Dubois per Dubois 1833.

Polyporus gossypinus Moug. & Lév. apud Lév. 1843 (ASn III 9): 123 (France); fide Bres. 1916 (Am 14): 224. — *Polystictus* Cooke 1886; *Daedalea* Quél. 1886; *Leptoporus* Pat. 1897; *Microporus* O.K. 1898; *Fomes* Big. & Guill. 1930 ("*grossypinus*").

Bjerkandera ciliatula P. Karst. 1887 (Mfe 14): 80 (Finland); fide Lowe 1956 (M 48): 100. — *Polyporus* Sacc. 1888. — P. Karst. 1887 I. 2: 12 / 1888 (ASf 16): 526 *pl. 10 f. 54* (*Bjerkandera*). — Sensu Bourd. & G. = *Tyromyces lacteus* [sensu Bourd. & G.] (forma).

Polyporus caesiocoloratus Britz. 1893 (BCb 54): 103 [*pl. 635 f. 145*] (Germany); fide Killerm. 1922 (Dba 15): 70 & Donk.

? *Tyromyces caesiosimulans* Atk. 1908 (Am 6): 61 (U.S.A., New York) (nom. conf.); fide Lowe 1942: 74, probably based on a specimen of *Polyporus caesius* contaminated with another fungus.

M.—*Boletus albidus* Schaeff. sensu Sow. 1799: *pl. 226*; fide Pers. 1801: 526.

M.—*Boletus caeruleus* Schum. sensu Lázaro 1916 (RMa 14): 576 / 1917: 78 (*Hemidiscia* "*coerulescens*").

carpatorossicus (Pilát) Bond. 1953 (lacking Latin descr.: n.v.p.). — *Leptoporus* Pilát 1937 (U.S.S.R., Ukraine) (lacking Latin descr.: n.v.p.). — Pilát 1937–8 (ACE 3): 169, 226 *f. 79, pl. 140 f. b* (*Leptoporus*).

chioneus (Fr. per Fr.) P. Karst, 1881. — *Polyporus* Fr. 1815 (Sweden) (d.n.) per Fr. 1821; *Bjerkandera* P. Karst. 1881; *Leptoporus* Quél. 1886; *Polystictus* Gillot & Luc. 1890; *Ungularia* Lázaro 1916. — Fr. 1815 O. 1: 125; 1874: 546 (*Polyporus*); Murrill 1907 (NAF 9): 35 (*Tyromyces*); Overh. 1915 (AMo 2): 697, 706 *f. 3* (*Polyporus*); Donk 1972 (PNA 75): 290, notes. — Sensu Quél. → *Incrustoporia semipileata*; sensu R. Fr. → *Tyromyces lacteus.*

Boletus candidus Pers. 1801 (Germany) (d.n.), not ∼ Roth 1797 (d.n.); fide Fr. 1838: 453, 1874: 546 and Donk 1972 (PNA 75): 290. — *Boletus* Pers. per Steud. 1824; *Polyporus* Pers. 1825, not ∼ (Roth per Pers.) Fr 1838, not ∼ (Speg.) Lloyd 1913 (n.v.p.). — Pers. 1801: 524 (*Boletus*); 1825: 51 *pl. 15 fs. 4, 5* (*Polyporus*).

Polyporus albellus Peck 1878 (U.S.A., New York), not ∼ Mass. 1901; fide Murrill 1907 (NAF 9): 35 & Donk 1972 (PNA 75): 287. — *Leptoporus* Bourd. & L. Maire 1920; *Tyromyces* Bond. & S. 1941. — Lloyd 1915 (LMW 4, Ap.): 294, ? in part (*Polyporus*); Bourd. & L. Maire 1920 (BmF 36): 83; Bourd. & G. 1928: 543 *f. 153* (*Leptoporus*); Overh. 1933: 21; 1953: 299 *pl. 18 f. 110, pl. 119 fs. 647, 648, pl. 128 fig.* (*Polyporus*); Aosh. & Kob. 1966 (RTm 5): 13 *f. 1, pl. 9 f. 1* (*Tyromyces*).

? *Bjerkandera pellita* P. Karst. 1895 (H 34): 7 (Finland). — *Polyporus* Sacc. 1896, Sacc. & Syd. 1899, not ∼ G. Meyer per Fr. 1821. — Lowe 1956 (M 48): 102 could not find the type. The original descr. reads like *Tyromyces chioneus*. It gives the spores as "elongatae, rectae vel curvulae, 2-guttulatae, 2–3 = 0.5–1 mm", which was changed into ". . . 2–3 = 1–2 mm" by P. Karst. 1898 T. 3: 11 / 1903 (BFi 62): 75.

M.? — *Polyporus lacteus* Fr. sensu Fr. in herb. K; fide Romell 1911 (ABS 11[3]): 15 = *P. lacteus* [sensu Romell]. — Romell, l.c.; 1926 (SbT 20): 11; Bres. 1931 (BIm 20): *pl. 985 f. 2* ?; H. Jahn 1963 (WPb 4): 42, in part, *Abb. 22* (*Tyromyces*), fide H. Jahn 1969 (WPb 7): on cover-page opposite p. 112 = *Tyromyces albellus.*

M.—*Polyporus tephroleucus* Fr. sensu Fr. in herb. K. — Romell 1911 (ABS 11[3]): 24 *pl. 1 f. 4.* — Cf. Donk 1972 (PNA 75): 300, note.

M.—*Polyporus trabeus* Rostk. (**O**) sensu Lloyd 1915 (LMW 4, Ap.): 301 *f. 638* at least in part; fide Overh. 1953: 300 = *Polyporus albellus* (as to fig.).

? M.—*Polyporus schulzeri* Fr. sensu Lohw. 1931; cf. Donk 1972 (PNA 75): 172. — Lohw. 1931 (ArP 75): 309 *pl. 19 fs. 2, 7, 8*; Kotl. & P. 1965 (ČM 19): 76.

Polyporus **croceus** (Pers.) per Fr. 1821 (**121**). — *Boletus* Pers. 1796 (Germany) (d.n.); *Polyporus* Fr. 1815 (d.n.); *Boletus* (Pers. per Fr.) Spreng. 1827; *Inonotus* P. Karst. 1882; *Inodermus* Quél. 1886; *Ochroporus* J. Schroet. 1888; *Phaeolus* Pat. 1900; *Polystictus* Big. & Guill. 1913; *Aurantiporus* Murrill 1920, 1939, Kotl. & P. 1957; *Poria* Clel. & Rodw. 1929, misapplied, not (error for) ∼ Lloyd 1922; *Hapalopilus* Bond. & S. 1941 ("Donk"). — Bres. 1903 (Am 1): 74; Lloyd 1915 (LMW 4, Ap.): 332 (*Polyporus*); Bourd. & G. 1928: 557; Pilát 1937 (ACE 3): 146 *f. 37, pl. 76* (*Phaeolus*); Lowe 1942: 53; R. W. Davids. 1942 (TUS 785): 28 *f. 4G, pl. 2 f. C*, cult. char.; Overh. 1953: 384 *pl. 47 f. 285, pl. 52 f. 311, pl. 129 fig.* (*Polyporus*); Fallahyan 1964: 21 *tpl. 3* with cult. char.; Černý 1966

(ČM 20): 90 *fs. 1–5 (Phaeolus)*; Kotl. & P. 1966 (ČM 20): 99, 103 *f. 2, pls. 11, 12*; H. Jahn 1967 (WPb 6): 145 *fig. (Hapalopilus)*; Al. David 1969 (NCa 96); 218, cult. char. *(Phaeolus)*.

Polyporus pilotae Schw. 1832 (U.S.A., North Carolina); fide Lloyd 1910 (LMW 3, L. 29): 7, Bres. 1912 (Am 10): 496, & Overh. 1953: 384. — *Aurantiporus* Murrill 1905. — Murrill 1905 (BTC 52): 487 *(Aurantiporus)*; Overh. 1914 (AMo 1): 115 *(Polyporus)*.

Polyporus pini-canadensis Schw. 1832: 157 (U.S.A., Pennsylvania); fide Murrill 1905 (BTC 32): 487 = *Aurantiporus pilotae*; fide Overh. 1953: 384, 385. — *Fomes* Cooke 1885; = *Scindalma pini* O.K. 1898.

Polyporus hypococcinus Berk. 1847 (LJB 6): 319 (U.S.A., Ohio); fide Murrill 1905 (BTC 32): 487 = *Aurantiporus pilotae*; fide Lloyd 1910 (LMW 3, L. 29): 7 & 1915 (LMW 4, Ap.): 380, & Overh. 1953: 384.

Polyporus castanophilus Atk. 1902 (JM 8): 118 (U.S.A., North Carolina); fide Murrill 1907 (NAF 9): 72 = *Aurantiporus pilotae*; fide Overh. 1953: 384.

destructor (Schrad. per Fr.) Bond. & S. 1941 **(O)** sensu Bourd. & G. *(Leptoporus)* **(128)**. — Bourd. & G. 1928: 546 *f. 155*, excl. of vars. *(Leptoporus)*.

Leptoporus destructor var. *pileatus* Bourd. & G. 1925 (BmF 41): 128 (France).

Leptoporus **ellipsosporus** (Pilát apud Lund. & Pil.) Romagn. 1944. — *Polyporus wynnei* forma Pilát apud Lund. & Pil. 1936 (England). — Pilát apud Lund. & Pil. 1936 (SbT 30): 232 *(Polyporus wynnei* f.); Pilát 1937–8 (ACE 3): 233 *f. 84 (Leptoporus wynnei* f.); Romagn. 1944 (BmF 60): 88 *(Leptoporus)*.

Leptoporus **epileucinus** Pilát 1937 (lacking Latin descr.: n.v.p.) (Czechoslovakia, now U.S.S.R., Ukraine). — Pilát 1937–8 (ACE 3): 229 *f. 81, pl. 141*.

fissilis (B. & C.) Donk 1933. — *Polyporus* B. & C. 1849 (U.S.A., North Carolina); *Spongipellis* Murrill 1907. — Murrill 1907 (NAF 9): 39 *(Spongipellis)*; Lloyd 1915 (LMW 4, Ap.): 319; B. O. Dodge 1916 (M 8): 14 *pl. 176 (Polyporus)*; Donk 1933: 153 *(Tyromyces)*; Pilát 1937–8 (ACE 3): 227 *f. 80, pls. 144–146 (Leptoporus)*; Lowe 1942: 71; R. W. Davids & al. 1942 (TUS 785): 30 *f. 4J, pl. 2 f. F*, cult. char.; Overh. 1953: 321 *pl. 19 f. 117, pl. 21 f. 129, pl. 114 f. 624, pl. 129* fig. *(Polyporus)*. — Cf. Bres. 1926: 76.

Polyporus albosordescens Romell 1912; fide Lloyd 1915 (LMW 4, Ap.): 320, 375. — *Phaeolus* Bourd. & G. 1925; *Leptoporus* Pilát 1932. — Romell 1912 (SbF 6): 637 *f. 1 (Polyporus)*; Bourd. & G. 1928: 555 *f. 159S*; Fallahyan 1964: 18 *tpl. 2*, with cult. char.; Al. David 1969 (Nca 96): 216 *fs. 3, 4*, cult. char. *(Phaeolus)*.

Polyporus fusco-mutans Lloyd 1922 (U.S.A., Ohio); fide Overh. 1953: 321, 322. — Lloyd 1922–3 (LMW 7): 1158 *pl. 223 f. 2291*.

? *Polyporus cavernosus* Velen. 1922: 640 [see Pilát 1948: 244 for Latin translation] (Czechoslovakia); fide Pilát 1938 (ACE 3): 227. — Described as 'effused'.

Polyporus pomaceus Velen. 1922: 645 [see Pilát 1948: 246 for Latin translation] (Czechoslovakia); not ∼ (Pers. per S. F. Gray) Pers. 1825; fide Pilát 1938 (ACE 3): 227.

M.—*Boletus spumeus* Sow. sensu J. Schroet. 1888: 471 (*Polyporus*); fide Pilát 1938 (ACE 3): 227. — Quél. 1888: 384, cf. Bourd. & G. 1928: 534 (*Leptoporus*).

M.—*Polyporus rubiginosus* Fr. sensu Bres. 1879 (AAR III 3): 72; fide Bres. apud Romell 1912 (SbT 6): 636 = *Polyporus albus* Huds. [sensu Bres. = *Polyporus albosordescens*].

M.—*Boletus albus* Huds. (**O**) sensu Bres. 1903 (*Polyporus*); fide Bres. apud Romell 1912 (SbT 6): 636 = *Polyporus albosordescens*. — Bres. 1903 (Am 1): 73; 1931 (BIm 20): *pl. 978* (*Polyporus*).

floriformis (Quél. apud Bres.) Bond. & S. 1941. — *Polyporus* Quél. apud Bres. 1884; *Leptoporus* Quél. 1886; *Coriolus* Quél. 1888; *Polystictus* Big. & Guill. 1913; *Cladomeris* Lázaro 1916. — Bres. 1884 F.t. 1: 61 *pl. 68*; Quél. 1886 (Crf 14²): 450 (*Polyporus*); Bourd. & G. 1928: 546 (*Leptoporus albidus* subsp.); Bres. 1931 (BIm 20): *pl. 975* (*Polyporus*); Konr. & M. 1932 I. 5: *pl. 431 f. 2* (*Leptoporus*); Lowe 1934 (PMi 19): 144 *f. 5A, pl. 15 f. 1*; 1942: 65; Overh. 1953: 252 *pl. 18 f. 112, pl. 129 fig.* (*Polyporus*).

Bjerkandera subsericella P. Karst. 1884 (Mfe 11): 136 (Finland); fide Lowe 1956 (M 48): 103. — *Polyporus* Sacc. 1888.

Polyporellus albulus P. Karst. 1894 (H 33): 15 (Finland); fide Lowe 1956 (M 48): 116 ("a pale centrally stipitate form"). — *Polyporus* Sacc. 1895.

Polyporus tabulosus Velen. 1922: 650 [see Pilát 1948: 248 for Latin translation] (Czechoslovakia); fide Pilát 1938 (ACE 3): 208.

Polyporus subsericeo-mollis Romell 1926 (SbT 20): 17 (Sweden); fide Kotl. & P. 1964 (ČM 18): 217. — *Leptoporus* Pilát 1937; *Tyromyces* Jo. Erikss. 1958, misapplied; Donk 1972 (PNA 75): 176, notes. — Sensu Romell (as to many specimens determined by him) → *Poria šimanii*, cf. Lowe 1966: 74, in obs.; sensu Lowe apud Lowe & Lund. 1956 (PMi 41): 23 = *Polyporus trabeus* [sensu Bres. = *Tyromyces leucomallelus*], acystidiate condition.

Polyporus **fodinarum** Velen. 1922 (Czechoslovakia) (**129**). — *Leptoporus* Pilát 1937. — Velen. 1922: 640, wrong spores [see Pilát 1948: 244 for Latin translation] (*Polyporus*); Pilát 1937–8 (ACE 3): 199 *f. 57, pls. 116–119* (*Leptoporus*).

? *Boletus hybridus* Sow. 1800: *pl. 289* (England) (d.n.). — *Polyporus* (Sow.) per B. & Br. apud Berk. 1860; *Poria* Cooke 1886; ≡ *Polyporus erosus* Pers. 1825 (**130**). — Sow. 1800: *pl. 289*; 1803: *pl. 387 f. 6* (*Boletus*). — Sensu Berk. in herb., Lowe → *Poria vaillantii*.

Byssus globosus Scop. 1772 P.s.: 93 *pl. 6* (Hungary, now Czechoslovakia) (nom. anam.) (d.n.).

M.—*Poria vaporaria* Pers. (**O**) sensu Mez 1908 (*Prolyporus*). — Mez 1908: 84 *fs. 24–28*; Kallenb. 1935 (ZP 14): *pls. 9, 10, 13, 14* & 1936 (ZP 15): 14, 36, 73, at least in part (*Polyporus*).

M.—*Boletus vaillantii* DC. sensu Pilát 1927 (Sčz 2): 472, 506, in part: *pl. 9 fs. 13–16, pl. 10 fs. 18, 19* (*Poria*); fide Pilát 1938 (ACE 3): 199.

M. ?—*Boletus destructor* Schrad. sensu Bond. 1953: 205 *f. 51: 19, pl. 56, pl. 187 f. 2*; 1956: 46 *fs. 13, 14* (*Tyromyces*), at least in part: illustrations.

fragilis (Fr.) Donk 1933. — *Polyporus* Fr. 1828 (Sweden), not ∼ Velen. 1922; *Bjerkandera* P. Karst. 1881; *Leptoporus* Quél. 1886; *Polystictus* Big. & Guill. 1913; *Spongipellis* Murrill 1915; *Fomes* Bres. 1926 (error of printing: n.v.p.). — Fr. 1828 E. 1: 86; 1838: 451; 1884 I. 2: 81 *pl. 182 f. 2*; Shope 1931 (AMo 18): 341 *pl. 24 f. 4*; Overh. 1953: 274 *pl. 23 fs. 142, 143, pl. 116 f. 635, pl. 130 fig.*; Lundell 1959 (LNF 53–54): 8 No. 2618 (*Polyporus*); Domański 1964 (Ffg 10): 86; Kotl. & P. 1964 (ČM 18): 213, 217 *fs. 2–4* (*Tyromyces*). — Sensu Bres. apud Bourd. & G. → *Tyromyces leucomallelus*.

Polyporus weinmannii Fr. 1838; fide Donk 1972 (PNA 75): 301. — *Postia* P. Karst. 1881; *Bjerkandera* P. Karst. 1881, 1882 ("*Veinmanni*"); *Inodermus* Quél. 1886; *Daedalea* Quél. 1888; *Spongipellis* Pat. 1900. — [≡ *Polyporus labyrinthicus* Fr. sensu Weinm. 1836 (U.S.S.R., European Russia)]. — Weinm. 1836: 313 (*Polyporus labyrinthicus*); sensu Bres. in herb. & Höhn. 1913 (SbW 122): 286 (*P. weinmannii*).

? *Polyporus keithii* B. & Br. 1875 (AM IV 15): 30 (Scotland); cf. Reid & Austw. 1963 (GN 18): 308, "the type could be interpreted as *P. mollis* Fr. [= *P. fragilis* sensu Pilát, Overholts]". — Reid & Austw., l.c.

Polyporus vermiculus Veull. apud Roum. 1883 (Rm 5): 46 [repr. 1891 (ASL 17): 288] (France); fide Bres. 1920 (Am 6): 68 = *Polyporus mollis* [sensu Bres.]; fide Donk 1972 (PNA 75): 299.

Spongipellis sensibilis Murrill 1912 (M 4): 93 (U.S.A., Washington); fide Overh. 1953: 274, 275. — *Polyporus* Murrill 1912 (nom. altern.).

Polyporus cavinae Velen. 1922: 648 [see Pilát 1948: 248 for Latin translation] (Czechoslovakia); fide Pilát 1937 (ACE 3): 176 (for *P.* "*Kavinae* Vel. . . . in herb.").

Daedalea sistotremoides Velen. 1926 (MP 3): 102 *f. 3* (Czechoslovakia), not ∼ (A. & S. per Schw.) Fr. 1832, not ∼ Beeli 1930; fide Pilát 1937 (ACE 3): 176.

M.—*Polyporus labyrinthicus* Fr. (**O**) sensu Weinm. 1836: 313. → *Polyporus weinmannii* Fr.

M.—*Boletus mollis* Pers. sensu Bres. 1920 (Am 18): 59, 67 (*Polyporus*); fide Donk 1972 (PNA 75): 294, note.

fumidiceps Atk. 1908 (U.S.A., New York). — *Polyporus* Sacc. & Trott. 1912. — Overh. 1915 (AMo 2): 706, 709 *pl. 23 f. 6*; Lowe 1942: 66; Overh.

1953: 305 *pl. 20 f. 122, pl. 118 f. 640, pl. 130 fig.* (*Polyporus*); E. Komar. 1960 (DAb 4): 132 *f. 2a* (*Tyromyces*). — Perhaps confirmation of Komarova's record for Europe is wanted.

guttulatus (Peck) Murrill 1907 (**133**). — *Polyporus* Peck 1883; ≡ *Polyporus maculatus* Peck 1874 (U.S.A., New York), not ∼ Berk. 1851. — Overh. 1914 (AMo 1): 100; Pilát 1937–8 (ACE 3): 204 *f. 60, pl. 123* (*Leptoporus stipticus* f.); Lowe 1942: 77; Overh. 1953: 286 *pl. 20 fs. 123–125, pl. 118 fs. 643, 644, pl. 130 fig.* (*Polyporus*); Bergstädt 1970 (MMH 14): 96 (*Tyromyces*).

Tyromyces substipitatus Murrill 1912 (M 4): 96 (U.S.A., Washington); fide Overh. 1953: 286, 287. — *Polyporus* Murrill 1912 (nom. altern.).

M.—*Polyporus alutaceus* Fr. (**O**) sensu Lloyd 1915 (LMW 4): 301; fide Lloyd, l.c. (*Polyporus guttulatus* cited as a syn.).

Polyporus **henningsii** Bres. apud Sacc. 1891 (Germany) (**131**), not ∼ Lloyd 1912. — Bres. apud Sacc. 1891 (SF 9): 167; P. Henn. 1899 (VBr 40): 126 *fs. 1c, b on p. 177*, as a syn. of *Polyporus vaillantii*.

Ptychogaster rubescens Boud. 1887 (France) (nom. anam.) (**125**). — *Ceriomyces* Sacc. 1888; *Oligoporus* Bref. 1888. — Boud. 1887 (JBM 1): 10 *pl. 1 f. 2* (*Ptychogaster*); Bref. 1888 U. 8: 136 *pl. 8 fs. 41–50*; P. Henn. 1889 (VBr 30): v (*Oligoporus*); 1896 (Vbr 37): 6; 1899 (VBr 40): 127 *fs. 8a, 9 on p. 177*; R. W. Davids. & al. *1946*; Fidalgo *1958* (*Ptychogaster*). — Sensu O. & R. Falck → *Tyromyces ptychogaster*.

M.—*Boletus vaillantii* DC. sensu P. Henn. 1893 (Gfl 42): 579 (*Polyporus*). — P. Henn. 1896 (VBr 37): 6 (*Polyporus*).

M.—*Poria vaporaria* Pers. (**O**) sensu P. Henn. 1899 (*Polyporus*). — P. Henn. 1899 (VBr 40): 125 *fs. 1–9 on p. 177* (*Polyporus*). — In part → *Polyporus henningsii* Bres.

M.—*Polyporus lacteus* Fr. sensu P. Henn. 1899 (VBr 40): 126 *fs. 1b, 5 on p. 177* [as a form of "*Polyporus vaporarius* (Pers.) Fries" sensu P. Henn.].

kmetii (Bres.) Bond & S. 1941. — *Polyporus* Bres. 1897 (Hungaria, now Czechoslovakia); *Leptoporus* Pilát 1937. — Bres. 1897 (AAR III 3): 70; Höhn. 1912 (SbW 121): 345, in obs. (*Polyporus*); Pilát 1937–8 (ACE 3): 231 *f. 82, pl. 137* (*Leptoporus*); Kotl. & P. 1965 (ČM 19): 74, 77 *pl. 6* (*Tyromyces*).

Polyporus ferro-aurantiacus Romell 1911 (Sweden); fide Höhn. 1912 (SbW 121): 345. — Romell 1911 (ABS 11³): 13 *pl. 1 f. 1*.

lacteus (Fr.) Murrill, 1907, not ∼ Murrill 1948 (n.v.p.). — *Polyporus* Fr. 1821 (Sweden); *Bjerkandera* P. Karst. 1881; *Postia* P. Karst. 1881; *Leptoporus* Quél. 1886; *Polystictus* Big. & Guill. 1913; *Hemidiscia* Lázaro 1916; *Daedalea* E. Krause 1925; *Spongiporus* Aosh. apud Aosh. & Kob. 1966. — Fr. 1821: 359 (*Polyporus*). — Sensu Lowe in litt., said to be apparently a species distinct from *Tyromyces tephroleucus* of this Check list; some of the references cited under that species below (as *Polyporus* *lacteus* sensu Fr. 1838) might belong here, for instance 'Aosh. & Kob. 1966'. It might appear desirable to treat the basionym as a nomen dubium and ambiguum. — Sensu Fr. 1838 → *Tyromyces tephroleucus*; sensu Fr. in herb. K, Romell → *Tyromyces chioneus*; sensu P. Henn. → *Polyporus* [*Tyromyces*] *henningsii*; sensu E. Krause = *Merulius tremellosus* (**O**), fide E. Krause 1928 B.r.: 54.

leucomallelus Murrill 1940 (U.S.A., Florida). — *Polyporus* Murrill 1940 (nom. altern.). — Murrill 1940 (BTC 67): 63.

? *Tyromyces newellianus* Murrill 1940 (BTC 67): 63; fide Lowe in litt. — *Polyporus* Murrill 1940 (nom. altern.).

Tyromyces gloeocystidiatus Kotl. & P. 1964 (Czechoslovakia); fide Lowe in litt. — Kotl. & P. 1964 (ČM 18): 207 *f. 1, tplate f. 1, pls. 15, 16*; H. Jahn 1964 (WPb 4): 45 *fs. 2c, 24, 35*; Domański 1964 (Ffg 10): 81 *f. 1*; 1965 (FpG 2): 135 *pl. 37*; Lowe 1966: 74 *f. 52*; Orl. 1971 (AmW 7): 15, 25 *fs. 1–6*; Mal. & Bert. 1971 (Apb 8): 28 *f. 5*.

M.—*Polyporus trabeus* Rostk. (**O**) sensu Bres. 1908; fide Kotl. & P. 1964 (ČM 18): 207, 216. — Bres. 1908 (Am 6): 37 (*Polyporus*); Bourd. & G. 1928: 541 (*Leptoporus*); Lowe & Lund. 1956 (PMi 41): 21 *fs. 1–3* (*Polyporus*); M. P. Christ. 1960 (DnA 19): 351 *f. 348* (*Chaetoporellus*); Domański 1963 (Mob 15): 313 (*Tyromyces*).

M.—*Polyporus fragilis* Fr. sensu Bres. apud Bourd. & G. 1925 (BmF 41): 123 (*Leptoporus trabeus* subsp.); fide Reid & Austw. 1963 (GN 18): 306 = *Polyporus trabeus* sensu Lowe. — Bourd. & G. 1928: 542 *f. 151* (*Leptoporus trabeus* subsp.); Domański 1961 (Ffg 7): 207 *f. 5* (*Leptoporus*). — The first descr. under this name is perhaps by Höhn. 1913 (SbW 122): 281, after a collection so named by Bres.; no colours of fruitbody mentioned.

M.—*Polyporus undosus* Peck sensu Pilát 1937–8 (ACE 3): 189 *f. 51, pls. 110, 111* (*Leptoporus*), in part; fide Kotl. & P. 1964 (ČM 18): 208. — Domański 1961 (Ffg 7): 208 *f. 4* (*Leptoporus*).

M.—*Poria notata* Overh. sensu Lowe 1946: 69, in part: on coniferous substrata; fide Lowe & Lund. 1956 (PMi 41): 73 = *Polyporus trabeus* [sensu Lowe & Lund.].

M.—*Polyporus albobrunneus* Romell sensu Kotl. & P. 1956 (ČM 10): 59 *2 figs.* (*Tyromyces*) fide Kotl. & P. 1964 (ČM 18): 208.

M.—*Leptoporus lowei* Pilát sensu Lundell 1959 (*Polyporus*); fide Lundell 1959 (LNF 53–54): 9 No. 2619 (*Polyporus trabeus* sensu Bres. cited as a syn.); fide Kotl. & P. 1964 (ČM 18): 208. — Lundell, l.c.

lowei (Pilát) Domański 1964. — *Leptoporus* Pilát 1937 (lacking Latin descr.: n.v.p.); *Polyporus* Lowe 1942 (lacking valid ref.: n.v.p.); *Tyromyces*, Bond. 1953 (lacking valid ref.: n.v.p.); Lundell 1959 (lacking valid ref.: n.v.p.), misapplied; *Leptoporus* Pilát 1953 (U.S.S.R., Ukraine). — Pilát 1937–8 (ACE 3): 168, 205 *pl. 129* (*Leptoporus*); Lowe 1942: 78 (*Polyporus*); Domański 1964 (Ffg 10): 81 *fs. 2, 3* (*Tyromyces*). — Sensu Lundell → *Tyromyces leucomallelus*.

M.—*Polyporus trabeus* Rostk. (**0**) sensu Lowe 1934: 84 [presumably only 'in part']; fide Pilát 1938 (ACE 3): 205.

Polyporus **marianii** Bres. apud Bres. & Cav. 1900 (Italy). — *Leptoporus* Pilát 1937. — Bres. & Cav. 1900 (NGi 7): 313 *pl. 11*; 1931 (BIm 20): *pl. 988.*

minusculoides (Pilát) Bond. 1953 (lacking valid ref.: n.v.p.). — *Leptoporus* Pilát 1937 (lacking Latin descr.: n.v.p.), 1953 (Czechoslovakia, now U.S.S.R., Ukraine); *Polyporus* Lowe 1957 (lacking valid ref.: n.v.p.). — Pilát 1937 (ACE 3): 193 *pl. 109* (*Leptoporus*); Lowe 1957 (PMi 42): 37 *f. 1* (*Polyporus*).

mollis (Pers. per Fr.) P. Karst. 1881. — *Boletus* Pers. 1795 (Germany) (d.n.); *Daedalea* Fr. 1815 (d.n.), not ∼ Sommerf. 1826, not ∼ Velen. 1922; *Polyporus* (Pers.) per Fr. 1821, not ∼ (Sommerf.) Lundell 1932; *Boletus* Wahl. 1826, not ∼ Vitt. 1844 (Boletales); *Bjerkandera* P. Karst. 1881; *Leptoporus* Quél. 1886, Pilát 1937; *Fomes* Big. & Guill. 1913, misapplied; *Polystictus* Big. & Guill. 1913, not ∼ (Pat.) Sacc. & Trott. 1912. — Pers. 1795 (ABU 15): 22 / 1796 O. 1: 22 (*Boletus*); sensu Fr. 1838: 454; 1884 I. 2: 81 *pl. 182 f. 3* ?, fide Overh. 1953: 278; Romell 1912 (SbT 6): 639 *f. 2*; Lloyd 1915 (LMW 4, Ap.): 318 *f. 658*; Overh. 1953: 277 *pl. 23 fs. 137, 138, pl. 130 fig.* (*Polyporus*); Kotl. & P. 1959 (ČM 13): 27 *2 figs.* (*Tyromyces*). — Sensu A. & S. (var. α) → *Climacocystis borealis*; sensu Rostk. → *Polyporus erubescens*, see this sp., below; sensu R. Hartig → *Phaeolus schweinitzii* sensu Bres. → *Tyromyces fragilis.*

? *Polyporus gelatinosa* Weinm. 1826 (SPR 2): 101 (U.S.S.R., European Russia); fide Weinm. 1836: 317 ("fungus junior").

Polyporus erubescens Fr. 1838, at least as to type; fide Donk 1972 (PNA 75): 293. — *Ischnoderma* P. Karst. 1879; *Placodes* Quél. 1886; *Fomes* Big. & Guill. 1913; *Leptoporus* Bourd. & G. 1925; *Tyromyces* Donk 1933, Bond. & S. 1941; [≡ *Polyporus mollis* (Pers.) per Fr. sensu Rostk. 1830 (StP 4): 53 *pl. 25* (Germany/Poland)]. — Sensu Bres. 1920 (Am 18): 59 ("*rubescens*") & 1931 (BIm 20): *pl. 985 f. 1* (*Polyporus*); Bourd. & G. 1928: 542 *f. 152* (*Leptoporus*). — Sensu Killerm. (with "?"), Kotl. & P. → *Fomitopsis pinicola.*

Bjerkandera alborosea P. Karst. 1889 (H 28): 366 (Finland); fide Lowe 1956 (M 48): 100. — *Polyporus* Sacc. 1891.

Polyporus mollicomus Britz. 1897 (BCb 71): 57 [*pl. 656 f. 209*] (Switzerland). — *Polystictus* Big. & Guill. 1913.

Tyromyces smallii Murrill 1907 (NAF 9): 32 (U.S.A., Florida); fide Lloyd 1915 (LMW 4, Ap.): 386 & Overh. 1953: 277. — *Polyporus* Sacc. & Trott. apud Sacc. 1912.

Polyporus pini-ponderosae Long 1917 (PNP 1): 3 (U.S.A., New Mexico); fide Overh. 1953: 277.

pseudoalbidus M. Bond. 1970 (NSn 6): 142 *f. 1* (U.S.S.R., White Russia).

ptychogaster (F. Ludw.) Donk 1933. — *Polyporus* F. Ludw. 1880 (Germany). — *Leptoporus* Pat. 1900, Pilát 1937; *Oligoporus* R. & O. Falck 1937. — F. Ludw. 1880 (ZgN III 5): 430 *pls. 13, 14* (*Polyporus*); Donk 1933: 153 (*Tyromyces*).

Oligoporus ustilaginoides Bref. 1888 (Germany); fide Bref. 1888 U. 8: 134 = *Polyporus ptychogaster* (cited as a syn.) & Donk. — Bref. 1888 U. 8: 126 *pl. 7 fs. 23–25, pl. 8 fs. 26–33*.

Ptychogaster fuliginodes (Pers. per Steud.) Donk 1972 (**124**). — *Trichoderma* Pers. 1801 (Germany) (nom. anam.) (d.n.); *Strongylium* Ditm. 1809 (d.n.), misapplied; *Trichoderma* Pers. per Steud. 1824. — Donk 1972 (PNA 75): 170, notes. — Sensu Ditm. 1809 & 1816 (StP 1): 77 *pl. 38* (*Strongylium*) = *Reticularia lycoperdon* Bull. (Myxomycetes).

Ptychogaster albus Corda 1838 (nom. anam.) (Czechoslovakia) (**124**); fide F. Ludw. 1880 (ZgN III 5): 430 (anamorphosis). — *Ceriomyces* Sacc. 1888. — Corda 1838 I. 2: 24 *pl. 12 f. 90*; Tulasne *1865*; L. Tul. 1872 (ASn V 15): 228, 235 *pl. 12 fs. 1–4*; Lloyd 1909 (LMW 3, P.I.): 31 *f. 265*, wrong conidia (*Ptychogaster*); Ulbr. 1941 (NBe 15): 579 (*Ceriomyces*). — Sensu Richon → *Ceriomyces richonii*, see this sp. below.

Ceriomyces richonii Sacc. 1888. — [≡ *Ptychogaster albus* Corda sensu Richon 1879: 216 *pl. 1 f. 2* (n.v.) [repr. Roum. 1879 (Rm 1): 132] (France)].

? *Myceliophthora fusca* Doyer 1927 (MCS 10): 33 *fs. 1–3* (from culture; unknown origin); fide R. & O. Falck 1937 (HF 12): 48, 58 = *Ptychogaster rubescens* [sensu R. & O. Falck].

Ptychogaster flavescens R. & O. Falck 1937 (lacking Latin descr.: n.v.p.) (nom. anam.) (presumably Germany). — *Ceriomyces* Ulbr. 1941 (incidental mention: n.v.p.). — R. & O. Falck 1937 (HF 12): 1, 11, 59 *f. 6, pl. 2 f. 7*: "bisher nur unterschieden [von *P. rubescens* sensu O. & R. Falck] durch die Färbung der jungen Sporenbeläge." — Cf. (**124**).

M.—*Ptychogaster rubescens* Boud. sensu R. & O. Falck 1937 (HF 12): 1, 58, 59 *fs. 1, 3–5, 8–17, 19–25, pl. 1, 2 fs. 1–6, 8, pl. 3 f. 1, pl. 3b f. 5.* — Cf. (**124**).

Polystictus **revolutus** Bres. 1920 (Italy) (**132**). — *Leptoporus* Bourd. & G. 1925; *Poria* Bres. apud Bourd. & G. 1925 (syn.: n.v.p.); *Agaricus* E. Krause 1933; *Polyporus* E. Krause 1934. — Bres. 1920 (Am 18): 35 (*Polystictus*); Bourd. & G. 1928: 548 *f. 157*; Pilát 1937–8 (ACE 3): 210 *fs. 65, 94, 95, pl. 124 f. a* (*Leptoporus*).

spraguei (B. & C. apud Berk.) Murrill 1907. — *Polyporus* B. & C. apud Berk. 1872 (U.S.A., "New England"). — *Leptoporus* Pilát 1937. — Murrill 1907 (NAF 9): 33 (*Tyromyces*); Overh. 1914 (AMo 1): 101; 1915 (WUS 3¹): 40; Lloyd 1915 (LMW 4, Ap.): 305 (*Polyporus*); Murrill 1919 (M 11): 103 *pl. 6 f. 4* (*Tyromyces*); Lowe 1942: 68; R. W. Davids. & al. 1942 (TUS 785): 38 *pl. 5 f. H, pl. 2 f. O*, cult. char.; Overh. 1953: 311 *pl. 24 f. 145, pl. 42 f. 255, pl. 116 f. 633, pl. 132 fig.* (*Polyporus*).

Polyporus sordidus Cooke 1886 (G 15): 20 (U.S.A.), not ∼ Lév. 1844,

not ∼ (Berk. ex Fr.) Murrill 1908 (n.v.p.); fide Murrill 1907 (NAF 9): 33 &
Lloyd 1915 (LMW 4, Ap.): 386.

Polyporus castaneae Bourd. & G. 1925 (France); fide Weir 1927 (Ph 17):
340 & Lowe 1961 (PMi 46): 207. — *Grifola* Pilát 1934; *Tyromyces* Kotl. &
P. 1957. — Bourd. & G. 1928: 522 *f. 148* (*Polyporus*); Pilát 1936 (BmF 52):
100 *pl. 1*; 1936 (ACE 3): 52 *f. 12, pl. 25, pl. 26 f. a* (*Grifola*).

stipticus (Pers. per Fr.) Kotl. & P. 1959 (**134**). — *Boletus* Pers. 1801
(Germany) (d.n.); *Polyporus* (Pers.) per Fr. 1821; *Boletus* Spreng. 1827;
Bjerkandera P. Karst. 1882; *Leptoporus* Quél. 1886; *Polystictus* Big. &
Guill. 1913. — ("*Stypticus*") Romell 1917: 277; 1926 (SbT 20): 18; Kallenb.
1926 (ZP 5): *pl. 7 fs. 1, 2* & 1927 (ZP 6): 58 (*Polyporus*); Pilát 1938
(ACE 3): 202, in part (*Leptoporus*); Siepm. & Zycha 1968 (NH 15): 565
pl. 79 f. 3, pl. 80 f. 2, pl. 82 f. 2, cult. char. (*Polyporus*). — Sensu Fr.
1884 I. 2: 80 *pl. 81 f. 2* (*Polyporus*) = ? (variously interpreted); sensu
Bres. → *Dichomitus squalens*.

Boletus albidus Schaeff. 1774 (Germany) (d.n.), not ∼ Pers. 1801
(d.n.), not ∼ (Pers. per S. F. Gray) Zant. 1822, not ∼ Roques 1832
(Boletales); fide Romell 1926 (SbT 20): 3, 5 (*Polyporus stipticus* cited as a
syn.). — *Polyporus* (Schaeff.) per Trog apud Fr. 1838, non/an ∼ Chev.
1837 (n.v.p.), not ∼ Saut. 1869; *Boletus* Lenz 1840, not ∼ Roques 1832
(Boletales); *Trametes* Fr. 1849, not ∼ (Fr.) Killerm. 1928; *Bjerkandera*
P. Karst. 1882; *Polystictus* Cooke 1886, not ∼ Mass. 1892; *Leptoporus*
Quél. 1886; *Microporus* O.K. 1898; *Fomes* Big. & Guill. 1913; *Tyromyces*
Donk 1933; *Agaricus* E. Krause 1933. — Secr. 1833 M. 3: 116 ?; Fr. 1838:
475; Bres. 1903 (Am 1): 74; Romell 1912 (SbT 6): 635; 1926 (SbT 20):
2 (*Polyporus*); Bourd. & G. 1928: 545 (*Leptoporus*); Donk 1933: 151
(*Tyromyces*); Lowe 1934 (PMi 19): 141 *pl. 14 fs. 1–3*; 1934: 85 (*Polyporus*);
Konr. & M. 1935 I. 5: *pl. 430 f. 2*; Domański *1960c* (*Leptoporus*). — Sensu
With. → *Abortiporus biennis*; sensu Sow. → *Tyromyces caesius*; sensu
Wahl. → *Bjerkandera fumosa*?

? *Polyporus immitis* Peck 1884 (U.S.A., New York); fide Lowe 1934:
86 & 1942: 79 = *Polyporus albidus* sensu Romell, but cf. Overh. 1953:
426 & (**134**). — *Leptoporus* Pilát 1937; *Tyromyces* Bond. 1953. — C. H.
Kauffm. 1926 (M 18): 28; Lowe 1934: 86; 1942: 78; Overh. 1953: 288
pl. 18 f. 111, pl. 96 fs. 546, 549, pl. 101 f. 569, pl. 112 f. 615, pl. 130 fig.

Bjerkandera acricula P. Karst. 1888 (Rm 10): 73 (Finland); fide Lowe
1956 (M 48): 100. — *Polyporus* Sacc. 1891.

Bjerkandera colliculosa P. Karst. 1890 (H 29): 177 (Finland); fide
Lowe 1956 (M 48): 101.

Polyporus candidissimus Velen. 1922: 645 [see Pilát 1948: 245 for
Latin translation] (Czechoslovakia), not ∼ Schw. 1832; fide Pilát 1938
(ACE 3): 202.

Polyporus perdurus Velen. 1922: 646 [see Pilát 1948: 247 for Latin
translation] (Czechoslovakia); fide Pilát 1938 (ACE 3): 202.

? *Coriolus maublancii* Pilát 1932 (U.S.S.R., Russia, Siberia); fide Pilát 1938 (ACE 3): 211 = *Leptoporus immitis.* — *Polyporus* Killerm. 1943. — Pilát 1932 (BmF 48): 17. — Type from *Populus tremulae.*

M. ?—*Polyporus alutaceus* Fr. (**O**) sensu Bres. in herb. & litt. — Fide Bres. apud Romell 1926 (SbT 20): 3, 6 = *Polyporus albidus*; & cf. Bourd. & G. 1928: 545, in obs.

Polyporus **subtestaceus** Bres. 1905 (Hungary, now Czechoslovakia). — *Phaeolus* Bourd. & G. 1925; *Hapalopilus* Bond. 1953 (incomplete ref.: n.v.p.), 1963. — Bres. 1905 (Am 3): 162 (*Polyporus*); Bourd. & G. 1928: 557 *f. 159T* (*Phaeolus*).

Polyporus **testaceus** Fr. 1838: 453 (Sweden), not ~ Lév. 1846. — *Bjerkandera* P. Karst. 1882; *Leptoporus* Quél. 1886; *Polystictus* Gillot & Luc. 1890. — Sensu Bres. 1920 (Am 18): 58 (*Polyporus*); Bourd. & G. 1928: 552 (*Leptoporus*). — Fr. 1874: 545 refers here *Polyporus rutilans* Pers. sensu Rostk. 1837 (StP 4): 75 *pl. 36* ("var. margine sulcato-crenato"), of very doubtful identity; Bres., l.c., cited the same plate for the *P. testaceus* "meo sensu".

tephroleucus (Fr.) Donk 1933. — *Polyporus* Fr. 1821 (Sweden), not ~ (Berk.) Bres. 1926; fide Bourd. & G. 1928: 540 = *Leptoporus lacteus* [sensu Bourd. & G.] (forma). — *Boletus* Spreng. 1827; *Bjerkandera* P. Karst. 1881; *Leptoporus* Quél. 1886; *Polystictus* Big. & Guill. 1913, not ~ (Berk.) Cooke 1886. — Fr. 1821: 360; 1838: 452; Rostk. 1830 (StP 4): 55 *pl. 26*, fide Fr. 1838: 452 "bona" [?] (*Polyporus*); Bourd. & G. 1928: 540 (*Leptoporus lacteus* forma 2); Konr. & M. 1935 I. 5: *pl. 430 f. 2* (*Leptoporus*). — Sensu Fr. in herb. K, Romell → *Tyromyces chioneus.*
   Bjerkandera melina P. Karst. 1887 (Mfe 14): 80 (Finland); fide Lowe 1956 (M 48): 101 = *Polyporus tephroleucus* Fr. sensu Overh. [in part]; fide Donk; & cf. Romell 1911 (ABS 11[3]): 16, "authentic specimens. . . seem to agree with [*Polyporus chioneus* sensu R. Fr.]". — *Polyporus* Sacc. 1888; *Tyromyces* Bond. & S. 1941. — Bourd. & G. 1928: 340 (*Leptoporus lacteus* forma 3). — Referred by Bres. 1920 (Am 18): 67 to *Polyporus lacteus* (forma).
   Bjerkandera simulans P. Karst. 1888 (Rm 10): 73 & 1888 (Mfe 16): 20 (Finland); cf. Lowe 1956 (M 48): 103, "appears to agree" with *Polyporus tephroleucus* Fr. sensu Overh. [in part]. — ≡ *Polyporus karstenii* Sacc. 1891. — P. Karst. 1889 I. 3: 4 / 1891 (ASf 18): 104 *pl. 64*, very poor.
   Bjerkandera cinerata P. Karst. 1890 (Mfe 16): 103; fide Lowe 1956 (M 48): 100–101 = *Polyporus tephroleucus* Fr. sensu Overh. [in part]. — *Polyporus* Sacc. 1891; *Leptoporus* E. Krause 1935 (syn.: n.v.p.); [≡ *Bjerkandera tephroleuca* (Fr.) P. Karst. sensu P. Karst. 1890 (Finland)]. — Sensu Bourd. & G. → *Bjerkandera fumosa.*
   M.—*Polyporus lacteus* Fr. sensu Fr. 1838: 453; 1874: 546; 1884 I. 2: 81 *pl. 182 f. 2*; sensu Bres. apud Lloyd 1915 (LMW 4, Ap.): **317** *f. 656*

(*Polyporus*); Bourd. & G. 1925 (BmF 41): 121; 1928: 539 (*Leptoporus*); Lundell 1946 (LNF 27–28): 7 No. 1315, in part: as to distributed specimens (*Polyporus*).

M.?—*Polyporus chioneus* Fr. sensu R. Fr. in herb., cf. Romell 1911 (ABS 11³): 16.

M.—*Bjerkandera ciliatula* P. Karst. sensu Bourd. & G. 1925 (BmF 41): 121 (*Leptoporus lacteus* forma 1). — Bourd. & G. 1928: 539 (*Leptoporus lacteus* forma 1).

undosus (Peck) Murrill 1907. — *Polyporus* Peck 1883 (U.S.A., New York); *Leptoporus* Pat. 1900 ("*undatus*"), Pilát 1937. — Lloyd 1915 (LMW 4, Ap.): 318 *f. 657*; Lowe 1934 (PMi 19): 146 *f. 5D, pl. 14 fs. 4, 5*; 1942: 64; Overh. 1953: 331 *pl. 16 fs. 94, 95, pl. 19 f. 116, pl. 132 fig.* (*Polyporus*); Domański 1964 (Ffg 10): 86 (*Tyromyces*); Lowe 1966: 77 *f. 57* (*Polyporus*); Brotzm. & Gilb. 1967 (Myp 33): 33 *fs. 1–5*; H. Jahn 1969 (WPb 7): 98 *fig.*, distribution in Europe (*Tyromyces*). — Sensu Pilát, in part → *Tyromyces gloeocystidiatus*.

Tyromyces pseudotsugae Murrill 1912 (M 4): 95 (U.S.A., Washington); fide Overh. 1953: 331, 332. — *Polyporus* Murrill 1912 (nom. altern.).

wynnei (B. & Br.) Donk 1933. — *Polyporus* B. & Br. 1859 [repr. Lund. & Pil. 1936 (SbT 30): 229] (Great Britain); *Polystictus* Cooke 1886; *Leptoporus* Bourd. 1894; *Microporus* O.K. 1898; *Merisma* Torrend 1924; *Fibuloporia* Bond. & S. 1941 (generic name n.v.p.), Bond. 1953. — B. & Br. 1859 (AM III 3): 358 (*Polyporus*); Bourd. 1932 (BmF 48): 225 (*Leptoporus*); Donk 1933: 156, excl. of forma? (*Tyromyces*); Lund. & Pil. 1936 (SbT 30): 229 *pls. 1, 2* (*Polyporus*); Pilát 1937–8 (ACE 3): 232 *f. 83, pls. 142, 143* (*Leptoporus*); L. Hansen 1969 (BT 64): 242 (*Fibuloporia*); Grög. 1970 (MMH 14): 98 (*Tyromyces*).

NOTES

Abortiporus

(**1**).   The European species *Abortiporus biennis*, which is perhaps not specifically different from its North American counterpart *A. distortus* (**2**), in more than one respect shows an astonishing range in variation. The configuration of the hymenophore is so variable that the species was repeatedly described as new during the period that highly artificial definitions of the genera of Hymenomycetes were still accepted; the so-called novelties were accommodated in *Polyporus, Daedalea, Sistotrema, Irpex,* and *Hydnum.*

Above all, the range of variation of the shape of the fruitbody is perhaps unmatched among the polypores. As remarked by Bourdot & Galzin (*1928*: 576): "*D[aedalea] biennis* est si variable qu'il n'y a pas de forme qui lui soit propre: stipité infundibuliforme, à chapeaux libres ou connés, spathuliforme ou sessile flabellé, dimidié, imbriqué. Nous [avons trouvé des] formes plus ou moins résupinés, à hyménium supère, dans lesquelles l'espèce est plus difficile à reconnaître." They described the most reduced form (*Poria terrestris* Pers.) as "étalé interrompu, mince, formé de pores oblongs, tendres et fugaces . . ."!

Moreover the species produces secondary spores. In north-western Europe these are sometimes almost (but perhaps never completely) absent in the 'normal' hymenium of well-developed stalked fruitbodies. The opposite extreme is that they are formed profusely in fruitbodies of their own. These are quite different from typical basidiferous ones and have been placed in the genera for imperfect states *Ceriomyces* (**0**) or *Ptychogaster.* In this case, too, all gradations between the extremes may be met with. Somewhere in between perfectly 'mixed' fruitbodies result. Since it would seem forced to consider the production of the secondary spores as an 'abnormal' condition in the basidiferous fruitbody I have abstained from drawing a line between 'normal' basidiferous fruitbodies and those in which a more advanced production of secondary spores is to be found.

(**2**).   It is now common practice to include *Abortiporus distortus,* described from North America, in *A. biennis,* whether or not as a distinct variety with its own geographical distribution. It seems to differ in the somewhat smaller pores; perhaps also in the less wide range of variation of the shape of the fruitbody; and in not producing typical (imperfect) *Ptychogaster* states in nature. Overholts (*1953*: 224–226) merely stated "chlamydospores often present in distorted forms". His remark that "a ptychogastric form is known in Europe under the name *Ceriomyces terrestris* Schulzer" strongly suggests that such a form was not known to him from North America. Fidalgo (*1969*: 162) made a similar remark; he regards *A. distortus* as a variety and calls it *Heteroporus biennis* var. *flabelliformis* ("Mont.") Fid. As pointed out by Donk (*1971a*: 2), this is a misapplied name. Fidalgo stated that in this taxon "abnormal forms corresponding to those of the extra-American variety [= *Abortiporus biennis*] were not found, neither have imperfect forms ever been recorded." On the other hand in North America monstrous fruitbodies may reach considerable dimensions in the form of more or less spherical balls (Buller, *1922a*). These are apparently not a chlamydosporous imperfect state; at any rate so far nothing to suggest the contrary has been written. Such fruitbodies are the basis of *Polyporus abortivus* var. *subglobosus* Peck. Although at the moment I have no overriding reason for doubting the specific identity of the two I have not entered either *A. distortus* or the synonyms based on material from the two Americas referred to it under *A. biennis*. These will be found in the "List of omitted names".

In addition to the synonymy related to *Abortiporus distortus* it will be noticed that names based on all other extra-European collections are also omitted from the synonymy of *A. biennis*. This is because our knowledge of *Abortiporus* in the palaeotropics and the regions south of them is still poor. Careful microscopical examination is needed to avoid confusion with *Diacanthodes novo-guineensis* (P. Henn. apud Schum. & Hollr.) Fid. For this reason the names *Polyporus platyporus* Berk. (**O**), *P. anthelminthicus* (**O**), *P. proteiporus* (**O**), and *Ptychogaster rufescens* (**O**) are omitted.

Albatrellus

(3). Niemelä (*1970*) recorded *A. syringae* not only from its type locality in Estonia but also from Finland and Sweden. The several habitats he mentioned suggest that the relation with *Syringa* is merely incidental. After careful comparison with the North American counterpart, *A. peckianus*, he concluded that the two should be kept apart as distinct species, although the differences appear rather trivial and may prove to be of questionable importance. He overlooked a paper by Malençon (*1966*).

Malençon described a fungus from Tirol that he considered probably identical with *Polyporus* [*Albatrellus*] *peckianus*. Comparison of the description and figures with the specimens of the same collection of *A. peckianus* studied by Niemelä (Canada, Quebec, leg. J. H. Ginns 1336) seems to indicate a remarkable degree of likeness. A few items (like spore size and colour of cap) however suggest instead that the Tirolean collection should also be referred to *A. syringae*.

Antrodia

(4). *Antrodia* is the same genus as *Coriolellus* as defined by workers primarily interested in cultural characters (Sarkar, *1959b*). As Sarkar defines it the group would appear to be homogeneous, but recently doubt has been raised as to the constancy of some of the characters she emphasized. One of the main features of the genus is the presence, among other things, of a certain type of thick-walled hyphae that are differentiated to form nodose-septate hyphae with irregularly thickened walls or with scattered thick-walled, refractive areas on walls (code symbol 9 of Nobles, *1945*: 1107). This type of hyphae is most readily observed in cultures and often not easily demonstrated in dried fruitbodies (5). Sarkar described both the irregularly thickened walls and the refractive projections on them for *Coriolellus* [*Antrodia*] *malicola* as well. Domański (*1966c*: 608) was unable to observe them in any of his strains, including two used by Sarkar. This species and *A. ramentacea* have been found to be homothallic, in contradistinction to the other species more closely investigated in cultures; those appeared to be heterothallic and bipolar (Sarkar, *1959b* and Domański, *1966c* for *A. malicola*; Domański, *1969a* for *A. ramentacea*). It should be remarked however that in a later publication, when contrasting *Antrodia* with *Diplomitoporus*, Domański (*1970b*: 206) categorically attributes to *Antrodia* "the bipolar type of interfertility".

Aoshima (*1967*) transferred all of the species of *Coriolellus* studied more extensively except one to *Daedalea* sensu stricto by emphasizing the thickened generative hyphae with clamps and irregular thickening of the walls (5). In my opinion this implies an underrating of certain characters of the fruitbody. I am not prepared to follow him in this respect. The exception alluded to is *Poria* [*Antrodia*] *sinuosa* (**10**), which Sarkar has found to be a species sharing all the characters deemed essential for the genus *Coriolellus* except that its small, allantoid spores are a unique feature within the genus.

In *Antrodia serialis* the hyphal construction of the fruitbody is described by Sarkar (*1959b*: 1259 *f. 29d*) as somewhat more complex than in the other species; she reports that in the trama of the dissepiments of the tubes the skeletals "are frequently branched and contorted". These look like 'binding hyphae' and would make the context trimitic, but in my opinion this is an insufficient reason for the introduction of a distinct genus.

(**5**).   The occurrence of clamped and irregularly thick-walled generative hyphae mentioned under (**4**) is not restricted to *Daedalea quercina* and *Antrodia*. This feature is best observed in cultural mats. It has been found to be widely distributed among polypores and at present there is no strong evidence that it is *per se* indicative of close relationship. In her latest paper Nobles (*1971*: 180, 182, 183) listed the character (her code symbol 9) for both 'oxidase-negative' and 'oxidase-positive' species. In the former category she listed under her Key Code 1.3.8.9 *Fomes* [*Fomitopsis*] *Cajanderi*, *Daedalea quercina*, six species of *Coriolellus* [= *Antrodia*] inclusive of *C.* [*A.*] *malicola* (**4**), *Poria placenta*, *P. xantha*, and *P. alpina* (perhaps not correctly identified), as well as two extra-

European species, *"Polyporus" durus* Jungh. and *"P." lilacino-gilva* Berk.; and under her Key Code 1.3.9 (hyphae differentiated to form 'fiber hyphae' lacking), *Osteina obducta, Poria oleracea, P. placenta, P. sericeomollis, Tyromyces* [*Poria*] *albo-brunneus, T. stipticus, T. tephroleucus* [= ?], and *T. undosus.*

In the other category ('oxidase-positive') she listed *Skeletocutis amorphus, Polyporus* [*Tyromyces*] *albellus* [= *T. chioneus*], *Climacocystis borealis,* three species of *Poria* referable to *Incrustoporia* (*P. stellae, P. subincarnata, P. tschulymica*), as well as *Leptotrimitus* [*Incrustoporia*] *semipileatus* and the extra-European *Poria odora* (Peck) Sacc.

This kind of hyphae is very much in evidence in the fruitbodies of *Osteina obductus* and *"Tyromyces" undosus.* The introduction of the genus *Cartilosoma* (based on *"Trametes" subsinuosa*) was largely but indirectly induced by this kind of hyphae in the fruitbody; its type species appears on this Check list as *Antrodia ramentacea.*

These observations tend to stress the necessity for caution. When hyphae of this kind are used by the taxonomist to aid in the delimitation of generic taxa they should be strongly supported by other characters as well. In this connection see also some remarks on *Incrustoporia* (Donk, *1971b*: 37).

(6). The core of *Antrodia* is a complex whose exact limits and the probable ranks of whose constituent taxa are as yet insufficiently understood. It consists of *Antrodia serpens, A. albida,* and *A. heteromorpha,* all three described by Fries from Sweden, and of *Trametes sepium,* described from North America. In his last comprehensive work Fries (1874) placed both his cap-producing species in *Lenzites* (which implies that well-developed fruitbodies tend to form a lamellate hymenophore); one (*A. albida*) was thought to grow on frondose wood, the other (*A. heteromorpha*) on coniferous wood. The third of Fries's species (*A. serpens*), with resupinate fruitbody and growing on frondose wood, was referred to *Trametes.*

Scandinavian authors like Jørstad [1937 (KnS 1936[10]): 31 fs. 7–9] and Lundell [1941 (LNF 21–22): 6 No. 1011a] thought that *A. heteromorpha* also occurred on frondose wood and suggested that the true *A. albida* might be no more than a form of *A. heteromorpha.* In this connection, however, it may be recalled that Bourdot [1932 (BmF 48): 227] dit not identify true *A. heteromorpha* from Sweden with what he had previously called *Trametes albida,* of which he had studied numerous collections from France. This could be taken as an indication that *Trametes albida* sensu Bourd. is not the same species Jørstad and Lundell had in mind. Other authors think that *A. serpens* and *A. albida* are not specifically different and maintain the two as infraspecific taxa (7). As to *Trametes sepium,* it has been either maintained as a good species or reduced to the synonymy of *A. albida.* Finally, *Trametes salicina* Bres., with resupinate fruitbody on frondose wood and relatively small pores, has also been included in this complex by some authors but Domański (*1966c*: 607) compared it with *Coriolellus* [*Antrodia*] *malicola*.

If all, or some, of the above-mentioned taxa are really to be treated as good species it is fundamental to work out 'new' characters that are more reliable than the 'old' ones. This will be by no means an easy task. To give an example of the difficulties ahead Sarkar (*1959b*: 1258) differentiated *Coriolellus* [*Antrodia*] *sepium* (8) by "Cystidioles present: basidia usually with four, but occasionally with two sterigmata with definite thickening of the walls at the base of some basidia and cystidioles ..." from *Coriolellus* [*Antrodia*] *heteromorphus*, "Cystidioles absent; basidia consistently thin-walled with four sterigmata . ..." Yet Domański (*1965b*: 183 *f. 63*) depicts not only cystidioles for *C. heteromorphus* but also somewhat thick-walled basidia and cystidioles.

While awaiting renewed studies in this group I have retained all the above-mentioned taxa as distinct species. *Trametes sepium* is treated as a doubtful European species for which descriptions based mainly on North American material are cited. It seems to be close to *A. albida* sensu Bourd. & G. from southern Europe.

In connection with *Antrodia albida* all references to descriptions based on Eastern Asian material have been left out [Imaz. 1939 (JJB 15): 302 *fs. 1, 2*; S. Ito 1955: 227 *f. 175*; &c. (*Trametes*)] and also possible synonyms based on material from Japan: compare *Coriolellus kusanoi* (**O**), *Irpiciporus tanakae* (**O**), *Lenzites yoshinagae* (**O**), and *Daedalea kusanoi* Murrill sensu Lloyd (**O**).

(7). Should *Antrodia albida* and *A. serpens* be considered conspecific it would appear that the correct name (or its basionym) is *Trametes* [*Antrodia*] *serpens*. As far as I am aware, of the two specific names involved and simultaneously validly published the one first reduced to a synonym of the other is *T. albida*; Bourdot & Galzin (*1925*: 167) made *T. albida* a subspecies of *T. serpens*.

(8). European mycologists often consider *Trametes sepium* to be merely a synonym of *Antrodia albida* but the correct interpretation of the latter species is still rather doubtful (6, 7). It is not unlikely that *A. albida* itself (exclusive of *Trametes serpens*, *T. salicina*) is not interpreted consistently in Europe and that (i) apart from a taxon related to *A. serpens* (ii) specimens of *A. heteromorpha* have also been referred to it. In that case it might turn out that *Trametes sepium* will have to be identified with taxon (i) but in my opinion this still needs further study. For this reason *T. sepium* is included separately on the Check list, but with a question mark; the accompanying references are all to descriptions and illustrations based on North American material except for the one to Bakshi & al. (*1958*) that deals with specimens from India.

It was Bresadola [1903 (Am 1): 81, as *Trametes*] who recorded *T. sepium* from Europe, although he mentioned some differences with North American material. Later in life [Bresadola, 1932 (BIm 21): text to pl. 1024 f. 2 (*Trametes sepium*) & *op. cit.*, text to pl. 1022 f. 1 (*T. albida*)] he thought that *T. sepium* was probably merely a straw-coloured form of *T.* [*Antrodia*] *albida*.

Should it prove necessary to regard *T. sepium* as a species different from *Antrodia albida* the correct name to be used for it will have to be examined anew. Overholts [1923 (M 15): 214, 221] concluded that *Polyporus favescens* Schw. (1832) and *Polyporus rhododendri* Schw.[7] (1832) are the same species as *T. sepium*; both names are of an earlier date; both were described as 'resupinates'. In later work Overholts (*1953*) seems to have forgotten these names. Lowe (*1966*) does not mention them either.

(**9**).   Fries used the name *Trametes serpens* in three different senses.

(i)   *Trametes serpens* Fr. 1849: 324, for a species collected by Lindblad. This was reduced to the synonymy of *Daedalea mollis* "Wormsk." (Fries, 1863 M. 2: 256). 'Wormsk.' is apparently an error for 'Sommerf.', resulting from a previous reference by Fries of a collection by Wormskjold from Kamčatka to *Daedalea mollis* Sommerf. (as "*β. D. membranacea*" Fries, 1828 E. 1: 72, a subspecific designation that he later dropped, cf. Fries, 1874: 585 under *Trametes mollis*). This is *Datronia mollis*.

(ii)   *Trametes serpens* Fr. 1863 M. 2: 255, 349, 355. In the index of the same work (pp. 349, 355) this was stated to be a *lapsus calami* for *T. serialis*.

(iii)   *Trametes serpens* (Fr. per Fr.) Fr. 1874: 586 is a recombination of *Daedalea* [*Antrodia*] *serpens* Fr. per Fr.

(**10**).   The current interpretation of *Polyporus sinuosus* was founded by Bresadola [1903 (Am 1): 78], who referred to it a specimen that in his opinion agreed completely with Fries's figure (1884 I. 2: 88 *pl. 190 f. 1*) drawn from "exempl. original. sicca". Bourdot & Galzin (*1928*: 672) gave a full description of it in this sense.

Romell (*1926*: 17) was not satisfied that the above conception was correct and thought that "perhaps the true *Pol. sinuosus* is rather the resupinate state of *Trametes subsinuosa* or of *Lenzites heteromorpha*, both of which are rare in Sweden," since it was "raro" according to Fries (1821: 382 & 1874: 576) and *Poria sinuosa* sensu Bres. is very frequent in Sweden. Moreover, Romell thought that Fries's figure contradicted Bresadola's interpretation. On the other hand there is "an authentic specimen in Fries' herbarium [that] seems to confirm the general inter-

---

[7]   Von Schweinitz (1832: 158) compared his species with '*Polyporus contiguus*'. It may be recalled in this connection that *Boletus contiguus* var. *dimidiatus* A. & S. 1805 (d.n.) is rather *Antrodia serialis* than *Phellinus contiguus* (cf. Donk, *1971b*: 27).

pretation", and stability in nomenclature will best be served by accepting Bresadola's interpretation.

The taxonomic position at the generic level is still under discussion (4).

(**11**). When Persoon (1825: 107) accepted *Polyporus sinuosus* Fr. he made a clerical error: instead of copying Fries's phrase and description he associated the name with an adaptation of the phrase and description of *Polyporus callosus* Fr. described by Fries on the same page and preceding *P. sinuosus*. This did not escape Fries (1828 E. 1: 120): "*P. sinuosum* l.c. p. 107. evidentem nostram sistere *P. callosi*, descriptionem, verbotenus transcriptam". This error could be interpreted as the introduction of a typonym of *Polyporus callosus*, viz. *Polyporus sinuosus* Pers., not *P. sinuosus* Fr., but I prefer to look on it for what it actually is, 'an unintentional error'.

Bjerkandera

(**12**). Some years ago Pouzar (*1966c*: 370) denied *Bjerkandera* generic status and reduced it to a subdivision of *Tyromyces*. The monomitic context with clamps apparently turned the scale. It is also evident that Kotlaba & Pouzar (*1966b*: 103) were on the verge of doing the same with *Hapalopilus*, of which they stated that it differed from *Tyromyces* practically only by the presence of the extracellular pigment which produces such a spectacular change of colour in KOH solution. It seems to me that they attach so much weight to the term monomitic that it is detrimental to the correct appraisal of other features.

The present genus *Bjerkandera* owes its scope to Murrill, who (in addition to the anoderm cap and other features) emphasized the more or less smoke-coloured hymenium at maturity. This is not a very impressive character. It was Ames (*1913*: 236, 239) who provided a much better one: "Hymenophore arising from a distinct layer." This layer is thin but quite typical. It always preceeds the formation of the tubes and on section is much darker coloured than the context of the cap. The species of *Bjerkandera* have a strong tendency to develop sterile portions devoid of hymenophore; nevertheless in such places the layer is formed and is easily noticeable because of its colour and its polished to shiny surface. The presence of this layer, in combination with other features, induced Donk (1933: 160) to introduce Murrill's conception of *Bjerkandera* in European mycology. I still believe the genus as thus defined to be a usable generic taxon.

Pilát [1937 (ACE 3): 149] combined *Bjerkandera* with *Gloeoporus*, which I prefer to keep apart, even though Domański (*1966b*) thought that *Gloeoporus* is not merulioid but has sterile edges along the dissepiments (at least as far as the European species are concerned). Its hyphal make-up and consistency show important differences; it also lacks the dark layer found in *Bjerkandera*.

(13). European and North American mycologists now admit only two species. These are easily distinguishable by the colour of the context of the dissepiments, best seen in sections transferred to a drop of KOH solution; it is the same colour as the dark layer **(12)** in *Bjerkandera adusta* and the same colour as the context of the cap (hence, paler than the dark layer) in *B. fumosa*. These two species are far fewer in number than those accepted before Bresadola; Bourdot & Galzin reduced the chaos to some degree of order. I have long been fascinated by the question as to how Fries came to admit about a dozen species instead of the two currently accepted. In the following notes **(14, 15)** I have tried to sketch what happened. Species varying greatly in the shape of the fruitbody were often stumbling-blocks to Fries, especially where this feature is coupled with considerable variation in the shape of the pores, the sets of main characters on which his groups of polypores were based. He put too much faith in his own scheme of classification.

The dark layer between the context of the cap and the hymenophore escaped his notice almost completely. This is not surprising with *Bjerkandera adusta*, where it is the same colour as the hymenophore. In *B. fumosa*, where it is much easier to observe, however, he noticed it only once **(15)**.

(14). Very much the same as with *Bjerkandera fumosa* **(15)** also happened with *B. adusta*. In this case Fries distinguished among at least five species (and subspecies).

Tribus *Merisma* sect. *Lenti.—Polyporus candidus?*
Tribus *Apus* A. *Anodermei* sect. *Lenti.—Polyporus scanicus, P. adustus, P. carpineus, P. crispus.*

Of these, *Polyporus crispus* survived the longest as a taxon distinct from *P. adustus*, especially in the U.S.A., where Overholts kept it alive for a long time, suppressing it only in his latest work (*1953*: 365). This form was maintained mainly because of the crisped margin of the cap and the somewhat larger and unequal pores, which tend to become labyrinth-like. *Polyporus carpineus* is another form that is often easily recognizable.

(15). The species that Fries (1874) accepted as distinct and that are listed on the Check list as belonging to *Bjerkandera fumosa* are the following:

Tribus *Merisma*.
  Sect. *Caseosi.—Polyporus alligatus* (cf. Donk, *1971b*: 27).
  Sect. *Suberosi.—Polyporus imberbis, P. salignus, P. holmiensis.*
Tribus *Apus* A. *Anodermei.*
  Sect. *Carnosi.—Polyporus pallescens?* **(16)**.
  Sect. *Lenti.—Polyporus albus* (*Boletus salicinus* Bull.), *P. fumosus.*

Of some of these taxa it is not altogether certain that they really belong to the *B. fumosa* complex, but I doubt whether many errors have crept in. From their distribution over the genus it may be concluded that Fries over-emphasized the variations in the 'shape' and consistency of the context. It is not surprising that Fries's later determinations were not always consistent and perhaps even erratic. The 'species' that went into *Polyporus* tribus *Merisma* Fr. owe their position to the fact that *B. fumosa* often forms a common 'stroma' from which the caps arise. What is especially remarkable is that Fries noticed the dark line seen in sections between the tube-layer and the context of the cap in only one of his taxa, viz. *P. salignus* [subsp.] **P. holmiensis*! Its description contains: "pileis ... *linea fusca ab hymenophoro limitatis*" (italics as in the original). In several of the above mentioned 'species' Fries did not even indicate the usually characteristic darkening of the pores but merely called them whitish or white.

I do not wish to go so far as to state that *B. fumosa* is a thoroughly homogeneous, though variable, species; it may well be that the current circumscription has been drawn up too widely and that in the future it will be possible to distinguish between a few taxa at a level perhaps even higher than 'subspecies'. Especially on *Salix* some forms occur that are worthy of further study. In this connection attention is drawn to the fungus depicted by Bulliard as *Boletus salicinus* (*pl. 433 f. 1*) and diversely interpreted.

(16). When publishing *Polyporus pallescens*, Fries cited *Boletus pelloporus* Bull. sensu Sow. 1799: *pl. 230* ["250"] as "optime". This plate is difficult to place but I suggest that it might represent sessile fruitbodies of *Abortiporus biennis*. Romell (*1911*: 19–20) concluded that Fries's species "seems to be identical with *Pol. fumosus* ... as far as authentic specimens in the Kew herbarium are concerned." The original description does not seem to contradict this.

Polyporus pallescens as published by Kalchbrenner [1868 (MtK 5): 262] has occasionally been treated as a species distinct from *P. pallescens* Fr. This is not correct; it represents an application of *P. pallescens* Fr. and the accompanying Latin description strongly suggests that the collection was also *Bjerkandera fumosa*.

For *Polyporus pallescens* Romell ("Karst."), see (119).

Boletopsis

(17). I have found it difficult to make up my mind about the number of species to be recognized within this genus. I have seen few specimens in fresh condition and these were typical *Boletopsis leucomelaena*. Most authors now recognize two taxa which are taken either to be good species or else colour forms of a single species. Murrill [1903 (BTC 30): 431] conceived them as species with different geographical distribution, one (*Polyporus griseus*) North American, one (*P. leucomelas*) European; he maintained that "the two are certainly distinct." Lloyd (*1911*: 77–79) claimed that the taxon with pale caps occurred in North America and Europe (called *Boletopsis griseus* below) while that with a much darker cap (*B. leucomelaena*) is a rare plant occurring in Europe only.[8] This would rather point to two different species. It has proved however that Lloyd's assertion does not hold true. The dark-coloured taxon does occur in North America. During a trip in northern Idaho I saw a few fresh (and some very young) fruitbodies; if these had been found in Europe I would unhesitatingly have referred them to *B. leucomelaena*. Dr. K. A. Harrison (personal communication) also insists that he found this dark taxon in the U.S.A. This shows that Overholts's description, excerpts from which will be quoted below, was drawn up without knowledge of this dark form.

The two species are not always easy to separate from one another, at least when their identity is to be judged from published descriptions and illustrations, not only because these are often poor, but also because the pale form becomes darker when old. According to Overholts (*1953*: 228) the colour of the cap of North American material of *B. griseus* is "ivory-white or with a darker center when young and growing, but soon becoming gray and then smoky umber and darkening even more on drying (especially in old sporophores) . . .." For *B. leucomelaena* the colour of the cap is described by Boudier (1911: 71) as "gris fuligineux plus ou moins foncé" and by Bourdot & Galzin (*1928*: 519) as "isabelle foncé gris bistré, brunissant au toucher, . . . à la fin entièrement bistré". These indications suggest some overlapping in colour. Moreover, other descriptions indicate that the colours of both may also vary in other respects.

Boletopsis appears to be rather common in southern Bavaria; this resulted in its occurrence under at least four different names in Britzelmayr's work; three of these names were given to 'new' species. Of two other species that have sometimes been referred to here no spores were mentioned and Britzelmayr's brief descriptions and sketchy drawings are in my opinion insufficient for their acceptance as surely belonging to this group; these were named *Polyporus subsquamosus* "Linn." ("nicht als Speisepilz bekannt") and *P. dapsilis* Britz. (**O**). The descriptions of

---

[8]   *Polyporus leucomelas* is the name that for some time was used in the U.S.A. to indicate *P. griseus*; it is usually accompanied by descriptive matter taken at least partly from descriptions by European authors. For instance the description of the spores ("cylindric-fusoid, pale brown, 10–12 × 4–5 μ") was copied from Massee (1892 B.F. 1: 229) and is patently wrong; compare McIlvaine (1900: 480) and Hard (1908: 391). "*Polyporus violaceus* apud Hard", as mentioned in a note by Saccardo & Trotter [1912 (SF 21): 255], is an error for '*Polyporus leucomelas*' (Hard, l.c.).

the other species taken together pretty well cover the whole colour gamut
of *Boletopsis*. In the following recapitulation of Britzelmayr's species
now referred to *Boletopsis* I have copied not only the colours but also the
taste and smell (**18**) as mentioned in his descriptions:—

Polyporus involutus Britz.: Hut weisslich bis isabellfarben, auch mit gelbbräunlichen Flecken. Stiel weiss, gelblich. Fleisch weiss. Sporenstaub weiss. Sporen
gelblich. Geschmack nicht unangenehm.

Polyporus conspicabilis Britz.: Hut grau, graubläulich. Stiel schmutzig grau.
Fleisch blass grauviolet. Porenschicht weisslich. Sporen gelblich. Geruch angenehm.

Polyporus formatus Britz.: Hut braunrot, braungrau und ins Weissliche verblassend. Stiel oben weisslich, nach unten bräunlich, rotbraun, am Grunde dunkler.
Fleisch weiss, sehr blass violettbräunlich, unten im Stiel grauschwärzlich. Porenschicht weiss, weisslich, später schmützig isabellfarben. Sporen farblos. Geruch
und Geschmack nicht unangenehm.

Polyporus leucomelas Pers.: [Colours of cap and stalk not mentioned in the text,
but dark.] Fleisch weisslich, angebrochen sich rötlich färbend. Porenschicht weiss,
rötlich grauweiss. Sporen gelblich, blass goldgelb. Geschmack nicht gerade unangenehm, aber etwas zusammenziehend, herb.

Of these forms the first would be *Boletopsis griseus* in the young stage
before the colours had turned darker; the last two would be good *B.
leucomelaena*; while the second would appear almost intermediate between
the two usually accepted taxa, but perhaps it could still be referred to
B. leucomelaena in a state before the colours had turned darker. Bresinsky &
Stangl (*1968*: 77), who live in the same region as Britzelmayr, reduce
all the above forms (and also *P. dapsilis*) to a single species, which they
call *Boletopsis subsquamosus*.

Bresadola [1931 (BIm 19): *pls. 947, 949*] recognized two species, a
conclusion based on personal collections in Italy: "*Polyporus subsquamosus*: affinis *Polyporo leucomelano* Pers. sed pileo nunquam atro et
stipite breviore satis distinctus." There are indications that typical
Boletopsis griseus (*Polyporus subsquamosus* in the sense of Bresadola)
is rare in Sweden, if it occurs there at all. Lundell [1946 (LNF 27–28): 4
No. 1309] does not include *Polyporus subsquamosus* in the then current
sense of European authors in the synonymy of *Polyporus leucomelas*.
He knew only the dark form and admitted that he had never seen such
pale and giant specimens as those illustrated by Fries (1863 S.S.: *pl. 53*)
under the name *Polyporus subsquamosus*. This also implies that he never
saw such almost pure white fruitbodies as Overholts mentioned for
Boletopsis griseus and as Britzelmayr depicted as *Polyporus involutus*.
On the other hand in the text accompanying the plate just-mentioned
Fries remarked that *P. subsquamosus* was often also more whitish (translated from the Swedish).

Because of insufficient experience in the field I have no personal opinion
to offer. That mycologists of repute (like Bresadola) maintained two
taxa, however, and that certain other authors who came across typical
B. leucomelaena do not know the whitish form has induced me to list

two species of *Boletopsis*. This is also done to encourage critical examination of this point of view.

(**18**).   *Boletopsis griseus* (usually under the name *Polyporus subsquamosus*) is often supposed to be an edible species. The odour seems to vary; there are indications that it may be pleasant in young and vigorous fruitbodies and less so in old ones. The taste, as given by Malençon [1956 (BmF 71): 274] for *B. griseus* from Morocco, is "douce puis amarescente" and "très amère et immangeable après cuisson", after Maire *in litt*. As to *B. leucomelaena*, this was annotated by Konrad & Maublanc (1935 I. 5: *pl. 422*) as "Comestible. — Pas fameux, goût amer, ne vaut pas la cuisson. Nous l'avons essayé." B. Hennig (1927: no. 289) wrote: "nicht gerade wohlschmeckend". As to North America I have not come across any detailed not-copied statements about the taste and edibility of *B. griseus*.

Malençon raises doubts about the specific identity of the species he described from North Africa as *Boletopsis griseus* and the European species that has been called *Polyporus subsquamosus*: "Sans doute existe-t-il en Europe une forme de [*Boletopsis*] . . . comestible . . .." This assumption is not well tenable: the reputation of *P.* [*Boletopsis*] *subsquamosus* sensu auctt. or *Boletopsis grisea* as an edible species is due to the fact that Fries placed *P. subsquamosus* sensu Wulf. (= *Polyporus carinthiacus* Pers.) by error in the synonymy of *P. subsquamosus*. The species described by von Wulfen is not a species of *Boletopsis* but like *Polyporus subsquamosus* itself (cf. Donk *1969a*: 255) it is *Albatrellus ovinus*, which is rated high as an edible species.

(**19**).   As to the correct name for the taxon with a whitish cap while young I have stated my reasons for rejecting both the names *Polyporus subsquamosus* and *P. carinthiacus* (Donk, *1969a*: 255). The use of the earliest name available as basionym (viz. *P. griseus*) is accepted here on the tentative assumption that the North American fungus which received this name is identical with the European fungus that has often erroneously been called *P. subsquamosus*; the two were identified by Bresadola.

(**20**).   When Fries in 1821 revalidated *Polyporus subsquamosus* (=*Albatrellus ovinus*) he associated it with two varieties which both represent forms of *Boletopsis*. Variety α ("pileo . . . fuligineo-nigricantibus") is simply the true *Boletopsis leucomelas* itself.

The other variety, "β. *P*[*olyporus*] *repandus*", "pileo cinereo", shows that Fries himself had seen specimens of a species of a pale *Boletopsis* ("v.v."). He first thought that this was the same as a species of Micheli (*pl. 70 f. 2*, now identified with *B. leucomelaena*) but later he was not sure about this, as can be concluded from a remark under *Polyporus leucomelas* in 1838 (p. 429). Variety β was raised to the rank of a species in 1879 by Karsten as *Polyporus repandus*. Does this taxon belong to *Polyporus* [*Boletopsis*] *griseus* Peck (1872) or is it a rather pale state of *Boletopsis leucomelaena*? Compare also (17).

Ceriporia

(21). *Ceriporia bresadolae* has also been recorded from North America (Arizona) by Gilbertson & Lowe [1962 (PMi 47): 168 *pl. 1 f. 3, pl. 2 fs. 5C, D*]. Besides resupinate fruitbodies their material also includes effuso-reflexed and sessile ones. Although these authors compared their material with "isotype" material of *C. bresadolae* (which Lowe *1966*: 34 later reduced to *Poria* [*Ceriporia*] *purpurea*) I hesitate to enter a reference to their description on the Check list. The French authors of the species appear to have seen many collections ("toujours récolté dans un rayon très limité, au-dessus de Carbassas, Causse Noir"); nevertheless no mention was made of any other condition of the fruitbody except the resupinate.

The North American record is in any case very interesting since it demonstrates that *Ceriporia* need not always be completely 'resupinate' (effused).

(22). *Polyporus rhodellus* is a highly controversial species. One interpretation was founded by Bresadola [1897 (AAR III 3): 80] and propagated by Bourdot & Galzin (*1928*: 662), who provided a good description. Lundell [1946 (LNF 27–28): 14 No. 1328] did not accept this interpretation as correctly named but introduced the name *Poria excelsa* for it. To him *Polyporus rhodellus* was a dubious species. This thesis he defended as follows:—

"It has always been a mystery what Fries meant by his *Polyporus rhodellus* and there is no Swedish tradition about this species (comp. e.g. Romell, Sv. Bot. Tidskr. 20 p. 17): There is no material left in Hb. Fries, and the figure cited by Fries (Bulliard tab. 442 D) gives no clue, as it is too vague. The illustration in Fries, Icon. sel. tab. 189: 2, shows without any doubt *Poria purpurea* (Fr.) Cke, and it is rather possible that the original description of *Polyporus rhodellus* is based on young specimens of that species. It has thus been necessary to give a new name to the species that Bresadola interpreted as Fries's species."

Romell (*1926*: 10, 17) thought that *Polyporus rhodellus* might have been *Poria eupora* (= *Chaetoporus nitidus*). The basis for this was Fries's reference of an unpublished picture of this species by H. A. von Post to *P. rhodellus*. In view of Fries's description (1821: 380), which contains, *inter alia*, "... digito tritus evanescens ..."this is unacceptable. Romell, however, referred to Fries's own figure (1884 I. 2: 88 pl. 189 f. 2) as probably representing *Poria purpurea*. This shows that Bresadola [1897 (AAR III 3): 81], Romell, and Lundell were in agreement about this figure published after Fries's death.

A new development was started by Lowe [1959 (Ll 21): 106; 1962

(PMi 47): 185; *1966*: 30] when he took *Poria viridans* in a still more inclusive sense than had been done before (for instance by Bourdot & Galzin, *1928*: 661) by including *Poria excelsa* (to mention only one of the European synonyms he added recently). For this emendation he simultaneously adopted the name *Poria rhodella* "in the sense of Bresadola". Therefore it may be concluded that if he had decided to keep *Poria viridans* (in the circumscription of Bourdot & Galzin) and *P. excelsa* (= *P. rhodella* in the sense of Bresadola) apart he would not have dropped the name *Poria viridans*. No arguments were advanced either to defend the inclusion of *Poria excelsa* or to disprove Lundell's conclusion that *P. rhodella* was a nomen dubium.

(**23**). Since no reliable authentic material remains my effort to form my own opinion about the original *Polyporus rhodellus* leaves little alternative but to reproduce Fries's first published and at the same time most extensive description:—

"...*Polyporus rhodellus*, effuso-elongatus membranaceus subfugax roseus, margine tenui subpubescente, poris subrotundis obtusis. / In ligno fagineo Aug. / Obs. Membrana 2–4 unc. longa, laevigata, tota albido-rosea, tacta obscurior subincarnata, arcte adnata tenuis, ut digito trita evanescat, margine tenui irregulari pubescente; sed non byssino. Tubuli haud lin. longi subrecti. Pori mediocres. Diversi sunt et *Polyp. sanguin.* et *P. incarnatus*, mox describendus".—Fries (1818 O. 2: 261).

To my surprise this for its time rather full account calls to mind the most common species of *Ceriporia* in Europe, viz. *C. viridans*. The observation "...ut digito trita evanescat..." rules out such species as *Poria placenta* (Fr.) Cooke, *P. gilvescens* Bres., and *P.* [*Rigidoporus*] *sanguinolenta*, the last two often also becoming reddish when bruised but with a much firmer consistency. The substratum ("in ligno fagineo") and the length of the tubes rule out *Poria bresadolae*, another of the species that reddens when touched. The words "...tota albido-rosea, tacta obscurior subincarnata..." suggest the form of *C. viridans* that received the name *Physisporus inconstans* P. Karst. [1887 (Rm 9): 10; "...albus, fractus dilute violascens..."]. I cannot wholeheartedly share Lundell's conclusion that the name *Polyporus rhodellus* is a nomen dubium. If it is concluded that *Polyporus rhodellus* is the correct basionym for *C. viridans* in the broad sense of Bourdot & Galzin, whether with the inclusion of *C. excelsa* (*Poria rhodella* sensu Bres.) or not, it might still be advisable to rule out this earlier name as a nomen ambiguum. In any case a more careful consideration of the question as well as a renewed and detailed taxonomic investigation of the whole of the complex to which *C. viridans* belongs is needed before the name *Polyporus rhodellus* is again brought into circulation for a species that does not include *C. excelsa*.

The original publication of *C. viridans* mentions the colour thus: "...primum album, exsiccatus pallide viridans ..." and "...perfectly white at first, but in drying assuming a delicate pale green, with a honey-like tinge in parts . . .." The present conception in Europe includes many colour forms.

Coltricia

(**24**). *Coltricia focicola* was described from North America. As re-described by Coker "it is usually if not always confined to burnt-over woods or ash piles. While found as far north as New York, it is much more common in the southern states." Moreover he remarked that in America *C. perennis* does not have the habit of growing on charcoal heaps, while *C. focicola* does. His description and illustrations also show that *C. focicola* has much larger pores than *C. perennis*. Overholts added that it is a short-stalked species the cap of which soon weathers to gray or ash colour and that it is both zonate and radiate-striate. He also stated that it has longer spores that are much narrower in proportion to their length than in either *C. perennis* or *C. cinnamomea*. In contradistinction to Coker, Overholts, like other American authors, found *C. perennis* "often where fires have been kindled"!

Like Overholts, many European authors report *C. perennis* as common on burnt-over places, and I have seen quite a number of collections from such spots; in general habit they suggested *C. focicola* but had the smaller spores of *C. perennis*. In agreement with other European mycologists I refer such forms (not without some hesitation) to *C. perennis*. This leaves the question open as to whether or not *C. focicola* has also been found in Europe. If it has it must be rare. Its inclusion on the present Check list is based on the report of it from "Carpatorossia" (Hungary) by Pilát as *Polystictus perennis* f. *focicola* (B. & C.) Pilát [1942 (ACE 3): 581]. He did not describe the spores.

(**25**). It is rather obvious that there is great similarity between *Coltricia montagnei* and *Cycloporus greenei*, especially when the hymenophore is left out of consideration; it is 'polyporoid' in the former and 'cyclomycetoid' in the latter. Gilbertson (*1954*) has combined the two taxa into a single species by reducing *Cyclomyces greenei* to the rank of a variety. This solution appears acceptable provided it is realized that the variety is geographically limited to the extra-European territory of the species; in Europe the clearly cyclomycetoid state of the hymenophore is unknown, in contrast to the U.S.A. and Japan, where both forms have been reported.

Lloyd also referred two other cyclomycetoid taxa to *Cyclomyces greenei*, viz. *Cyclomyces turbinatus* Berk. (India) (**O**) and *Xanthochrous javanicus* Pat. (Indonesia) (**O**). There is little doubt but that they are closely related to *Coltricia montagnei*, but a more careful study of this question is still needed.

(**26**). I refer *Polyporus casimiri* Velen. and *P. baudyšii* Velen. to *Coltricia cinnamomea*, although with a good measure of hesitation. This reference is based not only on the original descriptions and on drawings of the spores and their measurements but also on the assumption that it is not very likely that two species with smooth spores other than those admitted on the Check list might also occur in Europe.

Pilát [1942 (ACE 3): 580] referred *P. baudyšii*, with doubts and without comment, to *Polystictus* [*Coltricia*] *perennis*. He reduced *Polyporus casimiri* to the rank of a form of *Polystictus perennis* (*op. cit.*, p. 581). His conception of the latter species was so inclusive that it covers the whole of the genus *Coltricia* of this Check list. I am convinced that this is untenable and that several good species are involved. A renewed study of the two species published by Velenovský is suggested.

Coriolus

(**27**). *Coriolus zonatus* has been a source of much confusion; misdeterminations in the herbaria are numerous. Although it does not always seem to be easily separable from *C. versicolor*, little exact evidence for fusing the two species has been offered. On the contrary Jahn (*1963*: 79) has pointed out that the distribution areas do not overlap completely; in Europe *C. zonatus* has a more pronounced northern limit than *C. versicolor*. Moreover *C. zonatus* is more selective in its hosts, *Betula* and *Populus* being its favourites. British authors have understood the species only poorly.

It would seem as though North American authors have neither a consistent nor a correct idea of *Polyporus zonatus*. (i) Thus Overholts (*1953*: 344) stated that *Polyporus zonatus* "in my earlier papers" is a form of *Polyporus* [*Coriolus*] *versicolor*. (ii) Later he (Overholts, *1953*: 348) believed that the conception held by Romell, Karsten, and Bresadola covers a form intermediate between *Polyporus pubescens* and its variety *grayi*. (iii) At the same time he also thought that his interpretation of *P. velutinus* "might better be referred to *Polyporus zonatus* Fries, but not in the sense of Romell and Karsten" (Overholts, *1953*: 355). His final description and his figures of what he called *P. velutinus* do suggest certain forms of *Coriolus zonatus*. According to Lowe & Gilbertson [1962 (M 53): 489] the "*zonatus* Fries of American authors" is *Polyporus* [*Coriolus*] *velutinus*, which they distinguish from *P. pubescens*. They do not include *P. zonatus* in their synopsis of the Polyporaceae of the western U.S. and Canada. I believe that there is little reason to ascribe an erroneous interpretation to Romell and Karsten, as Overholts did.

It also seems doubtful whether Cunningham (1965: 167) correctly interpreted the material from New Zealand and Australia that he described and listed as *Trametes zonata*. I have deleted all references not only to his treatment under that name but also to the synonyms he referred to it. These are briefly mentioned in the "List of omitted names" as *Polyporus bireflexus* B. & Br. ex Cooke (**O**), *P. orbiculatus* Colenso (**O**), and *P. ruforugosus* Lloyd (**O**).

(28). *Boletus [Coriolus] ochraceus* Pers. is recognizable from its first description as belonging to *Coriolus zonatus*; this was somewhat amplified in 1801. Fries (1874: 568) made it a direct synonym of *Polyporus [Coriolus] zonatus*, which explains why various authors have treated *Boletus ochraceus* as a variety of *P. zonatus*. Persoon himself was not sure whether it was really distinct from *Coriolus versicolor* and in later work (1825: 72) he reduced it to the rank of a variety of that species. This uncertainty about the specific distinction between the two species is still shared by some contemporary mycologists. The inclusion by Maire [*apud* Maire & Pol. 1940 (AAt 1): 78] of *Boletus ochraceus* in *Coriolus hirsutus* as a variety is surprising: it was perhaps induced by Bresadola's conception. Donk reported that specimens which the latter author had referred to the Persoonian species agreed with what he called *Coriolus versicolor* f. *subhirsutus* Donk (1933: 181). The form of *Polyporus versicolor*, which Overholts (*1953*: 344) referred to as *Polystictus ochraceus*, as well as *Polystictus ochraceus* sensu Lloyd [1920 (LMW 6): 979], perhaps coincide with it.

Daedalea (4)

(29). The precise identity of the remarkable kind of mycelial growth called *Xylostroma giganteum* is still something of a puzzle. I cannot improve on the following quotation from Sowerby (1802: text to pl. 358):—

"It should seem that this fungus may have given rise to the use of *Agaricus Chirurgorum*, Pharmac. Edin. and *Agaricus querneus*, Pharmac. Gener. As Ray says it was used by the country people of Ireland to cure wounds; and thus I think the true styptic Agaric should be the *Agaricus querneus*, Linn. tab. 181 [= Sow. *pl. 181* as *Agaricus quercinus* L. ≡ *Daedalea quercina*] and not *Boletus [Phellinus] igniarius*, tab. 132, which I believe never has been found growing on the oak, although *Boletus [Fomes] fomentarius*, [Sow.] tab. 133, which nearly resembles it, sometimes does ... *Agaricus quercinus* nearest resembles the Oak Leather in delicate fibrous texture, and may be readily cut into slices and freed from impurities. The other two are of a less delicate colour, and require more preparation." — And compare Ray 1686 H. 1: 110 (*Fungus coriaceus Quercinus haematodes* Breyne) and 1724: 25 ("... Oak-Leather Hibernis").

This and some other suggestions made in early literature led me to assimilate *Xylostroma giganteum* with a sterile mycelial state known for *Daedalea quercina*. This state answers well to Tode's original description, which does not mention the host species. However he cited as synonyms *Fungus amplissimus* Scop. (**0**) and "Oak-Leather Raj ...." The second citation specifically introduced oak as substratum.

Another suggestion of relationship is to *Fomes fomentarius*. I have found large mycelial sheets formed by this species in the stem of an old beech in Scotland. Such pieces may well have been the basis for "*Agaricus coriaceus faginus haematoidis* Gagnebin" Haller (1768 H. 3: 143 no. 2290).

Dana (*1771*) described still another xylostromatoid mycelium from *Larix* as *Agaricus, seu Boletus pelliceus* $\equiv$ *Agaricum pelliculum* Latourr. (**O**). I am unaware of any identification of this mycelial growth.

Another mycelial production in the form of very extensive leathery sheets was found by Dr. H. H. Burdsall in northern Michigan. I saw them in fresh condition and could identify them as being a product of *Stromatoscypha fimbriatum* (**O**), a species usually referred to the 'Cyphellaceae'.

Fuckel [1873 (Jna 27–28): 86] thought that *Xylostroma corium* "Rbh." [$\equiv$ *X. giganteum*] was a state of *Polyporus xylostromatis* Fuck., a species that has been identified with *Perenniporia medulla-panis*.

Daedaleopsis

(**30**). There is a pronounced tendency to refer all names based on European specimens of this genus to a single species. This variable species is represented three times in Fries's "Systema", viz. as *Daedalea confragosa*, *D. rubescens*, and *D. suaveolens*. Of these simultaneously published names *D. rubescens* and *D. suaveolens* were reduced to the synonymy of *D. confragosa* before similar but reverse reductions were made; this makes *Daedalea confragosa* the basionym of the correct name.

Some authors still prefer to distinguish between two European species, *Daedaleopsis confragosa* and *D. tricolor*, and I have followed this trend. One reason is that typical *D. tricolor* seems to have a distribution area of its own and a different preference for hosts. Another reason is to keep the attention of mycologists who regularly come across the two taxa focused on the problem.

(**31**). The enormous variation of the hymenophore of *Daedaleopsis confragosa*—from typically lamellar to strictly poroid—has been the main reason for the many names proposed for this species and the various genera to which the forms were assigned. This variation is not always displayed by fruitbodies from different mycelia; it may even occur in the hymenophore of a single fruitbody or in fruitbodies growing closely together. Other variations also exist. It seems legitimate to ask whether or not all the forms found in Europe (apart from *D. tricolor*) really belong to a single species. I have decided on the single species but am not prepared to go further and reduce unconditionally a long series of species described from outside Europe to this species, as is done at present. What is needed are extensive studies on a world wide scale. Modern Japanese authors distinguish between several species in this complex in their country. (It would seem as though they are misapplying the name *D. confragosa*.) In the "List of omitted names" no less than nearly a score of names based on extra-European material are mentioned as having been placed in the synonymy of *D. confragosa*.

(32). Some authors who admit *Lenzites* [*Daedaleopsis*] *tricolor* as a distinct species (following Bresadola) also refer to it forms with trametoid hymenophore. Therefore it is not surprising that several synonyms have been attributed to this species. Those based on extra-European material are mentioned only on the "List of omitted names".

Fomitopsis

(33).   As it appears on the Check list *Fomitopsis* is much reduced but it is still heterogeneous. There are only a few true species and these are held together by the whitish to so-called bright but not brown coloured, dimitic (by skeletals) to incipiently trimitic context of the cap; the colourless, acyanophilous, smooth spores; the tendency of the hymenophore to become layered; the presence of a 'crust' covering the cap; and the lack of an imperfect state referable to *Oedocephalum*; they are brown-rot fungi ('oxidase-negative') in contradistinction to the typical species of *Trametes*, which are white-rot fungi.

Fomitopsis cytisina is a strongly deviating element that does not fit well in any of the existing European genera. It seems to deserve a distinct genus to be shared with some tropical species.

Other elements that have been included in *Fomitopsis* are species with a more or less pinkish to pinkish-vinaceous context; the European representatives are *Fomitopsis cajanderi* and *F. rosea*. They have been thought to be closely related to each other or even barely distinguishable on the specific level. Recent studies however indicate that possibly they should not even be included in the same genus. While (according to data furnished by Nobles, *1971*: 179) *F. rosea* would appear to be closely related to the type species of *Fomitopsis*, recent studies by Nobles (*1971*: 180) and Van der Westhuizen [1971 (Bo 10): 208 *f. 18*] show that *F. cajanderi* shares important characters with *Daedalea* sensu stricto, such as, for instance, Nobles's code symbol 9 (4). Removal of this second species from *Fomitopsis* seems unavoidable but whether it should be included in *Daedalea* or in a closely related group (well represented in the tropics), characterized *inter alia* by the pinkish colour of the context and pores, must still be worked out. In my card index this group is filed as 'Rhodoporus'; this is not a 'name' but merely a convenient indication of a group whose status is still undecided.

(34).   Donk (*1973*: 223) concludes that Fries introduced a *Polyporus marginatus* (1838) of his own, distinct from *P. marginatus* (Pers.) per Fr. (1821); after some initial hesitation he has also concluded that the two are not specifically different and that both are to be included in *Fomitopsis pinicola*. Given the often very incomplete references to basionyms by older authors it is not always easy (and sometimes impossible) to decide unequivocally on the basionyms, whether 'marginatus Fr. 1838' was

intended or 'marginatus (Pers.) Fr. 1821'. On the Check list I have tried
to register the complete set for 'marginatus Fr. 1838' at the generic
level but have not tried to work out the 'parallel' set for 'marginatus
(Pers.) per Fries 1821'. The recombinations of the latter set apparently
all date from later than the corresponding recombinations of the former set.

Funalia

(**35**). *Funalia* is accepted in this publication because it is convenient
and is appealing as acceptable for the European species. However the
main distribution is presumably in the tropics, where the type species
was collected (Guinea) as well. If one also takes into consideration the
mixed lot of extra-European species that is now occasionally assembled
under the name *Coriolopsis* it would appear that the limits between the
two are far from settled.

Recently Domański (*1968*) rejected the name *Funalia* because "he
had no herbarium material or culture at his disposal, and not even a
satisfactory description of *P[olyporus] funalis*", the tropical type species
of the generic name. For this reason he preferred to revive *Trametella*,
based on *Trametes hispida* [= *Funalia gallica*]. Not until it is shown
that *Polyporus funalis* is a nomen dubium or that this species is not
congeneric with *Funalia gallica* in the sense accepted here will I abandon
the earlier name *Funalia*, which is now currently applied to two European
species.

(**36**). I accept two species for Europe, both taken from *Trametes*:
Funalia gallica (often called *Trametes hispida*) (37) and *F. trogii*. The
former in particular is extremely plastic; there is considerable variation,
especially as regards the size and shape of the pores, and to a lesser degree
the colour of the flesh and the hirsuteness of the cap, so that the question
arises whether or not it is homogeneous in its present European conception.
For some time the two species just mentioned were considered to be
varieties of a single taxon.

In view of certain discrepancies between European and North American
descriptions and material and considering the uncertainty of the exact
delimitations of the two species, on the Check list proper I have refrained
from listing synonyms attributed to them based on material collected
outside Europe and adjacent North Africa. Those names will be found
included in the "List of omitted names"; they are *Trametes fergussonii*
(Berk.) ex Corda, *Irpex grossus* Kalchbr., ? *Trametes lindheimeri* B. & C.
apud Berk., *T. peckii* Kalchbr. apud Peck, ? *Polyporus proteus* Berk.,
Polyporus pulcher Speg. ((≡ *Polystictus celottianus* Sacc. & Manc. apud
Sacc.), *Polyporus sciurinus* Kalchbr. apud Thüm., *Trametes stuppea* Berk.,
Daedalea trametes Speg. (≡ *Trametes daedalea* Speg.), and *Trametes
tucumanensis* Speg.

(37). The use of the name *Polyporus gallicus*, introduced for *Boletus favus* L. sensu Bull. from France, needs some explanation. Bulliard's plate shows a fruitbody with big and rather regular hexagonal pores not exactly matched by any specimen I have come across. Lloyd [1912 (LMW 4): 520 *f. 517*] published a photograph of a fruitbody with big but more irregular pores from France itself; this comes close to Bulliard's picture. Considering that the pores in the present European conception are extremely variable in shape and size; that no other European species could be compared with any confidence with Bulliard's plate; and that some of Bulliard's objects have been found to be unsatisfactorily rendered on his plates, I am strongly inclined to follow Bresadola [1897 (AAR III 3): 90]: ". . . valde probabiliter . . . *Boletus favus* Bull. tab. 421 (certe nimis artificiose depicta), species a nemine intellecta et Mycologicis gallicis quoque ignota, huc pertinet . . ..''; [1908 (Am 6): 39] "*Boletus favus* Bull. tab. 421 absque dubio ad hanc speciem [*Trametes favus* sensu Bres.] pertinet; icon sat bona, tantum pori solito latiores delineantur.''

Ganoderma

(38). Most species of this genus are easily recognizable as such through their highly characteristic type of spores (Coleman, *1927*; Furtado, *1962*; Heim, *1963*). *Amauroderma* Murrill is very closely related but it is not found in Europe; it lacks the truncate-depressed apex of the dried up *Ganoderma* spore. The number of European species is low in comparison with those of North America and the tropics but the confusion has been impressive. The following notes indicate a considerable number of gaps in our knowledge together with controversial problems; these need further detailed attention. Compare also Donk (*1969b*).

It has appeared more and more unsatisfactory to separate *Elfvingia* (type, *Ganoderma applanatum*) from *Ganoderma* sensu stricto (type, *G. lucidum*), so that the two genera have now been re-united.

(39). The present restricted interpretation of *Ganoderma applanatum* in Europe still covers a variable species. For instance the initial colour of the surface of the cap (before it becomes heavily dusted with spores, as it often does) and the thickness and hardness of the crust show considerable variation. At one extreme stands a form with a whitish-grey and comparatively thick and hard crust; at the other extreme a form with a brown crust, thin enough to be indented with the thumb-nail. The former has been called *Ganoderma leucophaeum* and also *G. megaloma*. Both these names were based on North American types. If this form deserves a specific name at all it should perhaps be derived from one published earlier, viz. *Polyporus stevenii* Lév. (Crimea), as basionym, but Kotlaba & Pouzar (*1971*: 99) consider this as possibly the oldest name from Europe for *Ganoderma adspersum*, a suggestion I cannot accept.

Patouillard (*1889a*: 73) reported *G. leucophaeum* for Europe (France and Switzerland); he described the spores as "lisses, . . . 8–9 × 5 *μ*" and considered it "bien distinct de *G. applanatum* [sensu Patouillard] par sa couleur blanche et ses spores lisses". He was followed by Bresadola [1897 (AAR III 3): 73–74] who listed it from 'Hungary' (Czechoslovakia) and who emphasized the following differences: "A *Ganoderma*[*te*] *applanato* Pers. [sensu originario!] distinguitur crusta albida, initio lactea et contextu brunneo." Under *G. applanatum*, which Bresadola then called *G. rubiginosum*, he remarked, "A *Ganodermate leucophaeo* Mont. distinguitur cute rubiginosa initio [pulvere] copioso concolore obtecta et substantia cinnamomea." Although the spores of the second species were given as 8–10 × 5.5–6 *μ* (slightly too big) these characterizations seem to cover the two forms under discussion. The context of the 'leucophaeum' form ('brunneus') may indeed be of a darker colour than cinnamon. Some other early European records are by Bresadola & Saccardo [1897 (Mal 11): 249] for Italy, by Rolland [1898 (BmF 14): 86] for Corsica and by Torrend [1913 (Bro 11): 63] for Portugal.

Lloyd (*1915a*: 263, 264) also admitted two species; to him *Fomes leucophaeus* has a hard, pale or white crust and it was considered to be the common form in the United States; his *Fomes applanatus* has a brown, rather soft crust when fresh and he considered it to be a species frequent in Europe but rare in the United States. (As suggested by Atkinson, *1908b*: 183, Lloyd at first "evidently confused *Fomes reniformis* Morgan with *Fomes applanatus*" but in later work he kept the two apart.) The impression that the 'leucophaeum' form is much more common in North America than in Europe is confirmed by Overholts's description (*1953*: 98): "Sporophore . . . grayish or grayish black, sometimes brown, covered with a crust which is usually hard and horny but which may at times be thin enough to be indented with the thumb-nail . . .."

Contemporary mycologists, in both Europe and North America, have found it impossible to separate these two forms. They follow Atkinson (*1908b*), who subjected them to a detailed investigation and concluded that they were conspecific. Atkinson's discussion will need slight adjustment because he was not aware that Bresadola and Patouillard had called two different species *G. applanatum*. One of these, *G. adspersum*, has indeed the bigger and more distinctly 'echinulate' spores also mentioned by these authors but Atkinson—correctly—denied them for the true *G. applanatum*.

(**40**). *Ganoderma applanatum* has been made a receptacle for quite a number of closely related or more or less similar species from all parts of the world. The resulting mess has not yet been sorted out, although a brave attempt was made by Humphrey & Leus (*1931*; *1932*). The synonyms entered on the Check list proper and on the "List of omitted names" are only those that are based on collections from the northern temperate region.

(**41**). A flat radiating fungous growth has been found repeatedly on the pore-surface of *Ganoderma applanatum* and some related species occurring in the tropics. Lloyd [1915 (LMW 4): 538 *f. 742*] first called it *Sebacina dendroidea* (B. & C. apud B. & Br.) Lloyd; consequently when it turned out not to belong to *Sebacina* (Tremellaceae) he adopted the name *Institale bombacina* (Fr.) Fr. for it [Lloyd 1920 (LMW 6): 917]; this is undoubtedly a misnomer of a magnitude characteristic of his methods. He reported it from Venezuela (type of *Hymenochaete dendroidea* B. & C.), Ceylon, Madagascar, and the U.S.A. What might be considered a similar growth was described from Yugoslavia as *Septocylindrium lindtneri* W. Kirschstein (*1936*).[9] The true nature of this fungous growth is still open to discussion.

Petch [1912 (APe 5): 280, 281] reported that the collection described from Ceylon as *Hymenochaete dendroidea* [Berkeley & Broome, 1873 (JLS 14): 69, in the original publication stated to occur "on dead wood"] represented mycelial growth on which the spores of the substratum (*Ganoderma* sp.) had been deposited and which was parasitized by *Hypomyces chrysostomus* B. & Br. He also mentioned that *Reticularia apiospora* B. & Br. [1873 (JLS 14): 82, type from Ceylon] consisted of the thallus of '*Hymenochaete dendroidea*' containing the spores of a species of *Ganoderma* (which he called *Fomes australis*) and the conidia of *Hypomyces chrysostomus*.

According to Lloyd (op. cit., 1920) Bourdot showed that it was not a species of *Sebacina* and "that it has spores directly on the hyphae". Bourdot also suggested that it was an outgrowth of the *Ganoderma* fruit-body, an "expansion mycéliale conidifère." Lloyd doubted this. A similar theory was elaborated by Pilát [1942 (ACE 3): 491] for *Septocylindrium lindtneri*. He took this growth to be a chlamydospore-producing outgrowth of the substratum itself, and called it *G. applanatum* "f[orma] *Septocylindrium Lindtneri*".

Corner (*1968*: 102) stated that *Thelephora dendroidea* is a mass of germinating spores of *G. applanatum*; he examined the type at Kew.

When Kirschstein described *Septocylindrium lindtneri* he considered it to be an imperfect fungus parasitic on the fruitbody of the ganoderma but apparently living mainly on the fallen spores of the host. He described the conidia for his species as colourless and formed in long chains, of variable length (8–16 × 3 μ), straight and cylindrical, and 1–4 celled.

Apparently Kirschstein came the nearest to the correct evaluation of the problem, at least as far as European material is concerned. A study of excellent material collected in Germany shows that the growth over the pore surface consists of hyaline hyphae without clamp connections at the septa, in my opinion almost certain proof that this mycelium does

---

[9] Kirschstein calls the 'host' *Ganoderma lucidum* but Pilát (see below) referred it to *G. applanatum*.

not belong to the host, with its clamped generative hyphae. I also observed the spores described by Kirschstein, although not *in situ*; they occurred in addition to a considerable number of spores of *Ganoderma*.

A renewed study of this question is much needed.

(**42**). *Ganoderma adspersum* and *G. pfeifferi* might perhaps appear closely related. They can be distinguished easily by their crust: thick, hard, and opaque in the former, thinner and covered by a thin varnish-like film (wrinkling upon drying) in the latter, and also quite distinct in structure. However fruitbodies that have the crust removed are practically indistinguishable. Both are variable species. It is remarkable to find that authors who know these two species well usually come across both of them in their native regions.

(**43**). As will be shown below it is of importance to decide whether or not the name '*Fomes advena* Quél.' was validly published. Its publication was highly irregular. In "Les Champignons du Jura et des Vosges", as this work appeared in the periodical, Quélet [1872 (MMb II 5): 278, 330] published a personal interpretation of "*P[olyporus] Resinosus.* (Schrad.) F[ries]", which is an application of *Polyporus resinosus* (Schrad.) per Fr. [= *Fomitopsis pinicola*]. On page 330 (in the index) the species is listed as "Polyporus XIX. 2 [= 1]. Resinosus", 'XIX, 1', being a reference to the accompanying figure 1 on plate 19; the explanation of the figure on the plate runs "1. *Fomes Advena*". In the reprint of the work it will be found that in the index (Quélet, 1872: 318) the entry reads "[*Polyporus*] XIX. 1 [!]. *Resinosus* v. *Advena*".[10] I do not know whether or not the instalment of the periodical and the reprint were issued simultaneously; the odds are that the reprint was published a short time later. Both the plate and the reprint make it clear that in 1872 Quélet managed to publish three names for the same species, viz. *Polyporus resinosus*, *Fomes advena*, and *Polyporus resinosus* var. *advena*. I am as yet not quite certain whether it will prove necessary to treat the first two as alternative names.[11] Were the question to be answered in the affirmative this would make *Fomes advena* a validly published name.

According to a later remark by Quélet [1892 (Crf 20²): 469] "le nom de *advena* [était] donné selon l'avis de Fries".

---

[10] The meaning of 'v.' is not quite evident. For a while I thought that it stood for 'vel', which would result in '*Polyporus resinosus* or *P. advena*'. However, from a remark by Niel [1893 (BRo III 28): 49] who had sent a specimen to Quélet and had confirmed the determination "*Polyporus resinosus* (Schrad.) Fr. var. *advena* (Quélet, Champignons du Jura et des Vosges, p. 262, pl. XIX, fig. 2)", the correct meaning appears to be 'var.'.

[11] I am also not quite certain that the plates were issued at the same time as the text in the periodical. In the unbound copies that I was able to consult the plates were in a separate wrapper; perhaps such sets of plates were prepared for the smaller-sized reprint.

In case the name '*Fomes advena* Quél. 1872' is to be accepted as validly published the next puzzle to be solved is which species Quélet described in 1872 as *Polyporus resinosus*. Is it *Ganoderma resinaceum* or *G. pfeifferi*? Both names were published at a later date. Quélet gave the colour of the context as "*pâle puis chatain*" (italics as in the original); 'pâle' points amongst other species to *G. resinaceum*, 'chatain', to *G. pfeifferi*. Later on Quélet (1888: 400, as *Placodes resinosus*) dropped the qualification 'pâle' and called the flesh 'chatain'. He added, 'Spore ovoïde (0^{mm}01–12), grenelée, brune", which supports the placing of his interpretation in *Ganoderma*. Still later Quélet himself [1892 (Crf 20^2): 469] identified his interpretation of *Polyporus resinosus* with *G. pfeifferi* but simultaneously he insisted on using "*Placodes resinosus*, Schrad., Spec. (Bol. no. 33). Très bien décrit par Schrader comme un champignon particulier au hêtre", a conclusion I cannot follow.

Through the kindness of Dr. J. A. Nannfeldt I was able to study the material Quélet sent to Fries. A specimen (a piece of a fruitbody broken into several fragments) in Fries's herbarium is annotated in Fries's handwriting "*Polyporus flaviporus* / Gallia. Quélet" (**O**) and also has a label "*Polyporus flaviporus* / Gallia / L. Quélet. / e coll. E. Fries" (presumably written when Fries's herbarium was incorporated in UPS). Lundell determined the specimen as *Ganoderma cupreo-laccatum* (1944) (**O**) and later added a note that the correct name seemed to be *Polyporus (Ganoderma) pfeifferi* Bres. (1954). Both Dr. Nannfeldt and I think that this is correct. The second item in Fries's herbarium is a watercolour drawing annotated as follows, "*Polyporus (Fomes) advena* Quél. / Gallia austr.-orient. / Quélet". It shows a fruitbody with yellowish pore-surface and whith uniformly dark-coloured context and tube-layers; I have no hesitation in also referring this figure to *G. pfeifferi*. It resembles Quélet's published picture but it is not a copy. All this reveals that '*Fomes advena* Quél.' belongs to *G. pfeifferi*.

If the name *Fomes advena* were to be accepted as validly published by means of a protologue consisting of Quélet's description of *Polyporus resinosus* and the figure labelled '*Fomes advena*' then the lectotype must be the fruitbody depicted. If valid publication is considered to be effected by the figure only (this in case the plate was not simultaneously published with the text) then the fruitbody depicted on the published plate must be accepted as the holotype.

(**44**). It will be observed that few references to descriptions and illustrations of *Ganoderma lucidum* based on North American material have been mentioned on the Check list. This is due to the far too inclusive conception of this species imposed over a long period by Overholts (*1953*: 208). Under the denomination *Fomes lucidus* he admitted a number of taxa that are either not matched by European material or that are referred to other species by contemporary European mycologists. One of these elements is *G. sessile* Murrill, which corresponds to *G. resinaceum* in Europe. Another is *G. zonatum* Murrill, treated by Overholts as a variety; it occurs on palms and is not found in Europe. It appears to be a good species [cf. Steyaert, 1967 (BBr 37): 473 *fs. 4A–C, 14, 15*].

It looks as though 'typical' *G. lucidum* is rare in North America, at least as far as forms on hardwoods are concerned. Murrill [1908 (NAF 9): 118] did not admit *G. lucidum* at all. Other authors like Atkinson (*1908a*: 335; using the specific name *G. pseudoboletus*) and Haddow (*1931*: 33, 39–42) held to the commonly accepted European conception of *G. lucidum* except that they widened it by including *G. tsugae* Murrill. Atkinson was not convinced that the typical European form occurred in North America ("also in N.A.?"). Haddow cited very few collections from hardwoods from that continent and stated that *G. lucidum* occurs "in America most commonly on Hemlock", thus referring to the *G. tsugae* component.

Even in the restricted sense adopted here for *G. lucidum*, this species shows considerable variation in Europe, where it occurs on a wide range of hosts, mostly hardwoods. One form occurs on *Abies* (*G. pseudoboletus* var. *montanum* Atk.); it should apparently not be confused with *G. valesiacum* (**47**). This form very likely corresponds to Kreisel's "Boreale Rasse" (*1963b*: 52). For *G. tsugae*, see also (**45**).

When Curtis published the (devalidated) name *Boletus lucidus* he cited *Boletus rugosus* Jacq. as synonym. Therefore some authors consider it to be a mere substitute name for the latter (compare Dandy & Stearn apud A. Stevenson 1961: 401). This would make Austria the type locality. It would be more satisfactory to take the specimen depicted by Curtis as lectotype (England, near Peckham).

Some authors have made *G. lucidum* a very inclusive species, with many synonyms based on specimens from all over the world. Such a wide conception is certainly unacceptable. Working out the taxonomy of this group will prove to be a major undertaking. On the Check list proper and on the "List of omitted names" only synonyms based on European specimens are recorded; an exception is made for *G. tsugae*.

(**45**). It was mentioned in the previous note that those American authors who kept to a stricter conception of *G. lucidum* than that of Overholts felt compelled to merge it with *G. tsugae*. Atkinson kept *G. tsugae* distinct as a variety but Haddlow did not wish to make even this distinction. *Ganoderma tsugae* was originally described from *Tsuga*, the usual host in North America, where it is common. Overholts (*1953*: 211) also recorded it from *Picea* and *Pinus*. Moreover he seemed disposed to completely equate Haddow's conception of *G. lucidum* with his own of *G. tsugae* and also to accept records from hardwoods.

It will be necessary for European mycologists to decide what to do about *G. tsugae*, whether to (i) include it in *G. lucidum*, as Atkinson and Haddow did, (ii) keep it distinct from the latter species or (iii) follow Pilát and include it in *G. valesiacum* (47). On the Check list I have listed *G. tsugae* as possibly belonging to *G. lucidum*; the references cited have been limited to descriptions based on North American collections.

(**46**). Although there is a tendency to consider *Ganoderma resinaceum* as not sharply distinguishable from *G. lucidum*, therefore as conspecific with it, I believe that it is preferable to keep the two apart. This opinion is based solely on collections from north-western Europe and does not take into account the various forms that appear to occur in southern Europe. For the latter see Bresadola (*G. resinaceum* var. *martelli* Bres. 1892 F.t. 2: 31 *pl. 139*) and Bourdot & Galzin (*1928*: 610).

This conception apparently agrees with that of Haddow (*1931*); he studied material from both Europe and North America and called the species *G. sessile* Murrill. Evidently he was not aware of the existence of the earlier published name *G. resinaceum*.

Steyaert (*1967*: 197) circumscribed *G. lucidum* in a broad sense and evidently included *G. resinaceum* but without mentioning the latter name. He is now convinced, however, that the two are different species.

A number of synonyms based on types from the tropics are left off the Check list proper and the "List of omitted names".

(**47**). *Ganoderma valesiacum* is said to be a species that occurs in Europe only on *Larix*. German authors now identify it with *G. tsugae*, apparently following Pilát [1942 (ACE 3): 483]. In my copy of the reprint of the paper in which *Ganoderma tsugae* was published Bresadola wrote "= *Gan. valesiacum* Boud.?", after the former name. I have no personal opinion about this matter because I have not sufficient knowledge of *G. valesiacum*. The original description in combination with the fact that later Bresadola accepted it as a distinct species suggest that the problem needs renewed and detailed study.

(**48**). In my opinion the correct identity of *Polyporus roburneus* Fr. (**0**) is still open to discussion. Fries (1828 E. 1: 106) at first referred his material to *Polyporus [Heterobasidion] annosus* but later (Fries, 1838: 464) he described it as a distinct species, *Polyporus roburneus*. No details about the type collection were mentioned except that it grew on oak in Sweden. For this reason it is difficult to decide whether or not it is still in existence as an 'authentic' specimen. The dark context ("obscure badia l. cochalatae coloris in fungo vegeto"), the presence of a crust on the cap ("superficie glaberrima, quasi laccata, tamen opaca"), the colour of the spore print, emphasized by being printed in italics (*"sporidiis purpureofuscis"*), as well as the remark "Omnium forsan una cum *P. australi* . . ." (in the description of 1828), all taken together, point to a species of *Ganoderma*. The one species of this genus that comes to mind is *G. adspersum* (compare "superficie . . . tuberculoso-scrobiculata, absque zonis distinctis" in the description of *P. annosus*), but several other features refute this suggestion ("Duritie et pondere fere lapideus"; "interna substantia tota durissima"; "poris . . . umbrino-purpureis"). I have now given up trying to think up another and more likely suggestion. According to Bourdot & Galzin (*1928*: 619) Lloyd originally thought of *"Ganoderma laccatum"* [= *G. pfeifferii*], but subsequently he rejected this view.

Fries's figures (1884 I. 2: *pl. 184 f. 1*) of *Polyporus roburneus* published posthumously agree well with his original description. Bresadola [1897 (AAR III 3): 74] suggested that they represented *Fomes nigricans* sensu Bres. = black-crusted *Fomes fomentarius*: "Icon Friesii l.c. optima, sed pori fere semper versus lucem micore argenteo-carneo praediti, ideoque, meo sensu, nec *Polyp. roburneus* Fr. specifice differt."

The next interpretation was published by Lloyd [1910 (LMW 3, L. 28): 4] in connection with a specimen sent to him by A. Weidmann, Bohemia: *"Fomes roburneus*, teste Bresadola. It was on birch and has same context color and texture as *igniarius*, but has a smooth *black* crust. It has abundant cystidia." This suggests, first, that the specimen was named by Bresadola and, secondly, that Bresadola had apparently changed his mind, perhaps because of a specimen he had studied that had earlier been named by Fries. A few years later Lloyd (*1915a*: 246) claimed that he had unearthed the "type" at Kew. It is more likely that 'type' should be revised to 'an authentic specimen'; it is not unlikely that Bresadola had studied it previously. The specimen was said to be a slightly laccate form of *Fomes nigricans* (which is a vague determination); "it is exactly the same, excepting there is a slight resinous exudation on the crust and the setae are quite abundant. The pore mouths are strongly silverny, glancing." Mr. L. Ryvarden informed me that the Kew specimen is a slice and that it is sterile, with short stout setae 12–23 × 6–8 μ; he believes it to be a young fruitbody of a species of *Phellinus* (cf. *igniarius*). Lloyd also remarked that Fries's "Icones 184 has no resemblance whatever to his specimens, and I believe it represents *Fomes roseus*." Weidmann's specimen mentioned above was said to be "typical". The original *Polyporus roburneus* was recorded from oak ("ad ligna durissima *Quercus*"). Compare also Lloyd 1922 (LMW 7): 1155 & 1923 (LMW 7): *pl. 220 f. 2270*.

There are still other interpretations. *"U[ngulina] roburneus* Fr." of Patouillard (1900: 102) has "Trame pâle, très dense." According to Bourdot & Galzin (*1928*: 618) *"Pol. roburneus* Fr. était le nom que donnait Quélet dans ses déterminations au *Ph. robustus."*

It will be noticed that most of the interpretations mentioned above passed over the recorded colour of the spores in silence. If this were legitimate (of which I am not at all convinced) it would still be difficult to come to a decision. In view of the many interpretations (belonging to such genera as *Fomes, Fomitopsis, Phellinus,* and *Ganoderma*) and the uncertainty as to whether the Kew specimen is really the type it seems appropriate for the moment to reject the name *Polyporus roburneus,* as it is a nomen ambiguum et dubium.

Gloeophyllum

(**49**). The circumscription adopted here includes three genera now often accepted in Europe, viz. *Gloeophyllum* sensu stricto, *Phaeocoriolellus*, and *Osmoporus*. Studies of cultural characters by David (*1968*) support this emendation.

Gloeophyllum (sensu stricto) is, from a world-wide point of view, the largest element, with two European species, *G. sepiarium* and *G. abietinum*. The hyphal structure of the fruitbody is best known for the former species [cf. Van der Westhuizen 1971 (Bo 10): 175 *fs. 5, 6*]. It is trimitic; the generative hyphae are differentiated into thin-walled and sclerified hyphae.

Phaeocoriolellus, with one species, *G. trabeum*, has been found to be dimitic. The fruitbody lacks binding hyphae; the binding system, admittedly not well developed, consists of sclerified generative hyphae [Van der Westhuizen 1971 (Bo 10): 180 fs. 7 8]. This genus was introduced because Teston (*1953b*: 91; as *Trametes*) had reported only generative hyphae and skeletals, in contradistinction to *G. sepiarium* and *G. abietinum* for both of which she also reported binding hyphae. Van der Westhuizen [1971 (Bo 10): 185] believed that "this difference is of fundamental importance in the anatomy of the carpophores [of *Lenzites trabea* and *L. sepiaria*], which therefore cannot be regarded as congeneric."

Osmoporus includes *G. odoratum*, from which *G. protractum* may or may not be specifically distinct. Teston (*1953b*: 91; as *Trametes*) found only generative and skeletal hyphae in *G. odoratum*; M. E. P. K. Fidalgo (*1962*; as *Osmoporus*) reported for the same species a dimitic hyphal construction without binding hyphae but with sclerified generative hyphae in older parts of the context of the cap. This led Van der Westhuizen [1971 (Bo 10): 185] to conclude that *G. trabeum* and *Osmoporus* "are thus congeneric and if the genus *Osmoporus* Sing. is accepted as valid, *Lenzites trabea* should be transferred to it."

The above brief analysis shows that there is considerable conformity between the three composing groups, both in cultural characters and in the construction of the fruitbody, the sole difference advanced being the presence of binding hyphae in *Gloeophyllum* sensu stricto in contradistinction to *Phaeocoriolellus* and *Osmoporus*. The same facts from which this single strongly stressed difference is extracted also admit of a contrary interpretation: the system of binding hyphae in *Gloeophyllum* sensu stricto is rather poorly developed while at least in *Phaeocoriolellus* there is sufficiently leeway to speak of a small quantity of incipient binding hyphae. I believe it premature to separate the three groups and await supporting characters before admitting them as distinct genera.

Gloeoporus

(**50**). This small genus has been combined with *Bjerkandera* by Pilát [1937 (ACE 3): 149] and the type species of *Skeletocutis* by Killermann

(1928: 202). The hyphal construction of these three genera and several other details, however, are so different that there is little reason for keeping them together.

L. Hansen (*1956*: 253–255) concluded that the tube-layer of *Gloeoporus dichrous* was 'merulioid' with the hymenium continuous over the edge of the tubes as in *Poria* [*Merulius*] *taxicola* (**O**). Van der Westhuizen (*1971*: 170) also concluded that the hymenium is continuous over the edges of the pores. This would remove *Gloeoporus* from the Polyporaceae. Domański (*1966b*), however, came to the opposite conclusion, viz. that the edges of the tubes were sterile, as in *Poria pannocincta*, a species that Eriksson transferred to *Gloeoporus*.

Although the association of *G. dichrous* and *G. pannocinctus* in a single genus is here tentatively accepted it must be pointed out that according to Nobles the former is oxidase-positive (*1965*: 1122; *1971*: 181) and the latter apparently a brown-rot fungus (typically oxidase-negative; *1971*: 179). More recently Van der Westhuizen (*1971*: 172–173) reviewed the pertinent evidence published in connection with *G. dichrous*; this species appears to behave inconsistently in the production of oxidase reactions but he concluded that it causes a white rot. Much of this kind of research was based on extra-European collections which may not have been conspecific with the typical European taxon (**51**).

I cannot agree with Lowe (*1966*: 72) that the type species of *Podoporia* P. Karst. (*P. confluens* P. Karst.) (**O**) "may be the same plant" as *Poria* [*Gloeoporus*] *pannocincta*. I have seen the type but am not prepared to make any other suggestion about its correct identity.

(**51**). The precise circumscription of *Gloeoporus dichrous* is far from settled. After a few random tests (including tests on North American material referred to *G. dichrous*) I refrain not only from listing any description based on specimens collected outside Europe but also from adding specific synonyms based on extra-European collections. The genus needs very thorough revision. All the species that were or have been referred to *Gloeoporus dichrous* are mentioned on the "List of omitted names".

Grifola

(**52**). Pilát's emendation of this genus hsa been shrinking rapidly owing to the exclusion of such elements as *Bondarzewia*, *Meripilus*, *Laetiporus*, *Leptoporus castaneae* (= "*Tyromyces*" *spraguei*), and *Osteina*. The principal character indicated for the separation of *Flabellopilus* Kotl. & P. = *Meripilus* P. Karst. was the absence of clamps. This feature needs further study; according to Nobles (*1965*: 1101; code symbols 5 + 6) the lack of clamps is not absolute, at least as far as cultures are concerned. On the other hand there are a number of other differences between the type species of *Grifola* (*G. frondosa*) and *Meripilus* (*M.*

giganteus). Taking these together the two taxa still appear worthy of recognition as distinct genera. One of the features I have in mind is the strongly pronounced fissility of the flesh in *M. giganteus*; another is the pale ochraceous spore print.

(**53**). Recently Pouzar (*1966c*: 360) transferred *Grifola umbellata* to the genus *Polyporus* because of the hyphal system; he considered it to be dimitic by binding hyphae, which reminded him of *Polyporus squamosus*. I am not always much impressed by the taxonomic importance of binding hyphae as a generic character. Moreover Nobles (*1971*: 188) concluded that on the evidence of cultural studies *Grifola umbellata* should be excluded from *Polyporus* s. str. My retention of this species in *Grifola* means that at the moment I am not willing to suggest any alternative solution.

Incrustoporia

(**53bis**). **Incrustoporia percandida** (Mal. & Bert.) Donk, *comb. nov.*; basionym, *Poria percandida* [J. L.] G. Malençon & R. Bertault *in* Acta phytotax barcinon. 8: 30, 35 *f. 6.* 1971. — Al. David (*1971*: 95) used this new combination but without validly publishing it. Not only is the replaced synonym not clearly indicated but no full and direct reference with page is given to the original publication (Code, Art. 33). The publication of the basionym as "*Poria* (*Incrustoporia*) *percandida*" does not constitute valid publication of the combination *Incrustoporia percandida* either.

Recently I made several collections of this species in Morocco.

Inonotus

(**54**).   The circumscription of *Inonotus* used here does not differ much from the genus as emended by Donk (1933: 240) and monographed by Pegler (*1964*). This should not be taken as evidence that I consider the genus to be homogeneous. That the group will eventually be broken up into a number of smaller genera I do not doubt. Actually for some time I contemplated doing this but it became more and more obvious that a fair knowledge of the European species is definitely not a sufficient basis for effecting this task. As in so many other groups, the European species form only a minority.

When in the future an attempt is made to break up the present genus I have little doubt that two available generic names, now listed as synonyms of *Inonotus*, will have to be restored; they are *Inodermus* Quél. (lectotype, *Inonotus radiatus*) and *Xanthoporia* Murrill (holotype, *I. andersonii*). The former name is discussed below from a nomenclative point of view (**55**).

On an earlier occasion Donk (1933: 241) divided the Dutch species over three sections. These are:

(i) *Inonotus* sect. *Inonotus* (formerly called *Inonotus* sect. *Euinonotus* Donk). Species: *Inonotus cuticularis, I. hispidus.*

(ii) *Inonotus* sect. *Inoderma* (P. Karst.) Donk (*"Inodermus"*). Species: *Inonotus radiatus, I. nodulosus* (**56**), *I. polymorphus* Rostk. sensu Bourd. & L. Maire (**57**), and presumably also *I. weirii.*

(iii) *Inonotus* sect. *Dryadeus* Donk. Sole species: *Inonotus dryadeus.*

I would now add one more section taken from section *Inonotus* and give it a name: —

(iv) *Inonotus* sect. **Phymatopilus** Donk, *sect. nov.* Basidioma e corpore fibroso-granulato myceliato formatum; superficies anodermatica. Sporae distincte brunneo-coloratae. Setae nullae. Typus: *Inonotus dryophilus* (Berk.) Murrill. — Etymology: φῦμα, a growth or tumor; πῖλος, cap. The European species are mentioned below (**59**).

(v) With one exception the rest of the European species could be temporarily regarded as forming a group characterized by yellow or sulphur-yellow spore-print. Available names are: *Xanthochrous* sect. *Resupinati* Pat. 1900: 1 (lectotype, *Polyporus andersonii*), *Poria* sect. *Xanthochrous* Bourd. & G. 1925 (BmF 41): 247 (monotype, *Poria obliqua*), and *Xanthoporia* Murrill. Species: *Inonotus andersonii, I. glomeratus, I. niduspici, I. obliquus.*

The exception is the resupinate species

(vi) *Inonotus subiculosus* (**61**).

(**55**).   If it were to be decided that *Inonotus* sect. *Inoderma* (P. Karst.) Donk must be treated as a distinct genus then the establishment of the correct name for the genus is something of a minor nomenclative puzzle. There are three published names to be considered: *Inoderma* P. Karst. 1879, *Inodermus* Quél. 1886, and *Mensularia* Lázaro 1916, all of which can be taken as based on *Polyporus radiatus* (Sow.) per Fr. If *Inoderma* P. Karst. had not been preoccupied by the earlier homonym *Inoderma* (Ach.) S. F. Gray 1821 (a genus of lichens), *Inodermus* Quél. would have to be treated as merely a superfluous orthographical variant of *Inoderma* P. Karst., differing slightly in termination. Since *Inoderma* P. Karst. is illegitimate however *Inodermus* appears to be available precisely because of the difference in termination; *Inoderma* (Ach.) S. F. Gray and *Inodermus* Quél. are not likely to be confused since they are applicable to unrelated taxa, viz. a genus of lichens and one of Hymenomycetes (cf. Art. 75 of the "Code", the *Peponia-Peponium* Example). The other generic names ending in —a, *Inoderma* Kütz. 1883 (Algae) and *Inoderma* Berk. 1881 (Ascomycetes) cannot affect the above conclusion, not only because they are generic names given to systematically widely different taxa but also because they drop out as later homonyms, viz. illegitimate names that are permanently unavailable.

In this connection I would mention such generally accepted substitute names as *Melanoleuca* Pat. (for *Melaleuca* Pat.) and *Elmerina* Bres.

(for *Elmeria* Bres.). If the names between brackets had not been pre-occupied by earlier homonyms the substitute names would have to be regarded as mere orthographic variants and as such would not be available.

It is not impossible that an attempt will be made to brand *Inodermus* Quél. as 'superfluous', hence illegitimate (impriorable), because one of its original species [*Inonotus hispidus* (Bull. per Fr.) P. Karst.] is the lectotype of the generic name *Inonotus* P. Karst. published at an earlier date. This selection of the type species of *Inonotus* however was made long after Quélet had introduced *Inodermus*. One interpretation of Art. 63 of the "Code" accepts this as an argument against censuring *Inodermus* Quél. as illegitimate (impriorable). Personally I ignore much of the ruling on 'superfluous' names in its present formulation as a traumatic spot in the "Code" that ought to be cured as soon as possible.

(**56**). It is with considerable hesitation that I maintain the separation of *I. nodulosus* as a satellite species from *I. radiatus*; I have done so in order to attract attention to this taxonomic problem.

The species received two names in the same year, viz. *Polyporus nodulosus* Fr. and *P. polymorphus* Rostk. (I have not tried to establish the relative dates.) A study of the protologue of the latter name reveals that the fungus grew on *Fagus* and that it formed true caps (not merely nodules); these facts, supported by the rest of the protologue, leave no doubt in my mind about its correct identity. Pilát [1942 (ACE 3): 559] arrived at the same conclusion these many years ago. It is remarkable that *P. polymorphus* could have been so long misapplied to the strictly resupin-ate species that I now call *Polyporus* [*Inonotus*] *polymorphus* sensu Bourd. & L. Maire for want of a correct name (**57**).

Moravec [1954 (ČM 8): 93; 1960 (ČM 14): 29] records *Dialonectria cosmariospora* (Ces. & De Not.) Moravec, a parasitic pyrenomycete, as growing on *Inonotus nodulosus* only. The conception of this host species favoured by some Czech authors, like Pilát (l.c.), also includes *I. polymor-phus* sensu Bourd. & L. Maire but I have not seen any collection of *Dialonectria cosmariospora* growing on it. Perhaps a more precise know-ledge of the preference of this species of *Dialonectria* may help to decide on the status to be ascribed to the taxa of *Inonotus radiatus* complex. The substrata given on labels of works of exsiccati and in other publica-tions are often wrongly named. The determination of the substrata is not always easy because the fruitbodies are often represented in the collections as fragments (usually merely resupinate portions) and old and more or less blackened by rotting. Examples: Rabenhorst (*in* F.e. No. 459, as *Cosmospora coccinea*) gave "In lignis putridis". Roumeguère (*in* F.g. No. 2755, as *Nectria cosmariospora*) wrote "Sur le *Polyporus ferruginosus* pourissant." This induced the Tulasnes (1865 S.C. 3: 93) to correct this to "crescit in hymenio . . . *Polypori* . . ."; they themselves found the species in beech-wood, which would seem to indicate *Inonotus nodulosus*. However

Saccardo (S.F. 2: 508) merely copied Roumeguère. When Munk [1957 (DbA 17[1]): 50] stated "apparently not rare on old fruit bodies of *Polyporus radiatus*" he may have had in mind *Inonotus nodulosus*, which in Denmark is common on *Fagus*.

(**57**). The fungus that Bourdot & L. Maire first described under the name *Xanthochrous polymorphus* and which Bourdot & Galzin treated as a subspecies of *X*. [*Inonotus*] *radiatus* is now often accepted in Europe under the name *Inonotus polymorphus* as a good species (cf. Jahn, *1965c*). The use of the epithet 'polymorphus' however is due to a misinterpretation (**56**) and must be abandoned for the fungus with the conspicuous tramal setae in the dissepiments.

It might appear that a correct basionym exists, viz. *Polyporus dentifer* Velen., but the original description is not completely convincing and the type has as yet not been located for confirmation (Drs. F. Kotlaba & Z. Pouzar *in litt.*, Dec. 14, 1970).

Polyporus salebrosus Lasch (**0**) has been mentioned in this connection.

(**58**). *Inonotus radiatus* is usually broadly conceived but there is a growing tendency to admit more than one species in Europe. The most extreme authors admit not only *I. radiatus* but also *I. nodulosus* and *I. polymorphus* sensu Bourd. & L. Maire (**57**). As pointed out elsewhere the name *Polyporus polymorphus* has been misapplied; in my opinion it is a synonym of *Polyporus nodulosus* (**56**).

Even when these three taxa are admitted on the specific level a residue remains that still needs further study. This applies perhaps (again as far as Europe is concerned) especially to part of what is now called *I. radiatus* var. *resupinatus* (Bourd. & G.) Pegl. (after the exclusion of resupinate *I. nodulosus* and *I. polymorphus* sensu Bourd. & G.).

The division of *I. radiatus* into three species on this Check list does not mean that I am satisfied that this is correct: it is merely a working hypothesis. Outside Europe there are still some taxa that are now included in *I. radiatus* but perhaps deserve autonomous status; for instance there are *Polyporus cucullatus* and *P. illinoisensis* [= *Inonotus radiatus* var. *cephalanthi* (Overh.) Pegl.], both described from North America.

(**59**). The *Inonotus rheades* complex is far from well understood in Europe. Lloyd lumped the whole of it in one species under the incorrect name *Polyporus corruscans*, while Bourdot & Galzin admitted four subspecies. I have followed Pegler's modification (*1964*) of the treatment by the French authors; he admitted three species, each corresponding to one of Bourdot & Galzin's subspecies, except that he combined their *Xanthochrous rheades* (sensu stricto) and *X. vulpinus* (subspecies) into a single species. This combination is accepted by Kotlaba & Pouzar (*1969*: 169). A nomenclative change of epithet from Bourdot & Galzin's scheme was also necessary because the correct name for *X. corruscans* is *Inonotus dryophilus*.

Another matter in need of careful study by European mycologists is the host range of these species. The rule of thumb followed in northern Europe is that *I. dryophilus* grows on species of *Quercus* and *I. vulpinus* (= *I. rheades*, in part) on species of *Populus*. Bourdot & Galzin (*1928*: 635–637) came to a broader range: *Xanthochrous rheades* (sensu stricto) they reported from *Populus*, *Fagus*, and *Betula*, and *X. vulpinus* from species of *Populus* and from *Fagus*. If simple contraction of these two taxa into one species (*I. rheades* sensu Pegl.) is in order then the species now usually referred to as *I. vulpinus* may on the one hand also be expected to occur in northern Europe on hosts other than species of *Populus* alone. On the other hand Kotlaba & Pouzar (*1969*: 169) stated that "In Czechoslovakia, *Inonotus rheades* is known to occur almost only on dead (only once on living) standing trunks of moderately aged aspen (*Populus tremula*); in one case it was collected on *Populus canescens* (*P. tremula* × *P. alba*) and once in a wound on a branch of living *Salix* cf. *aurita* (quite an exceptional host plant)." They regard the reported occurrence on *Pinus* as surely wrong, while the record of *Rosa* as a host plant was due to an error of determination of the fungus, which turned out to be *I. radiatus*.

The host range of *I. dryophilus* outside Europe is much wider than oak alone.

(**60**). Bresadola supposed that his *Polyporus friesii* corresponded to the *P. fulvus* of Fries. It should be observed however that his description mentions "Hab. ad truncos praecipue *Quercus* et *Populi* in Silva nigra (Dr. Pfeiffer)." There can be little doubt that the type must be selected from Pfeiffer's material; it is also very likely that it represented two species, viz. *Inonotus dryophilus* (the material from *Quercus*) and *I. vulpinus* = *I. rheades* (the material from *Populus*); compare (**59**). Both species have been recorded from the general region of the Black Forest (= Schwarzwald = Silva nigra). Bresadola recorded the spores as "7–9 $\vee$ 4–5 μ"; the length suggests *I. dryophilus*, the breadth, *I. vulpinus*. An inspection of Bresadola's specimens is called for to determine their identity one way or the other. I have followed present usage and list the name as a synonym of *I. rheades*, an important reason being that Bresadola himself listed *Polyporus fulvus* Fr. [= *Inonotus vulpinus*] as a synonym.

(**61**). *Inonotus subiculosus* has been reported from Finland. I have seen several collections from North America and have made an ample collection in Michigan myself; I have also studied a portion of the material on which the European record is based. All the American collections studied had shallow, or rather shallow, tubes with poorly developed hymenium and they had produced spores that were perhaps not fully mature (viz. thin-walled, colourless). The European collection had much longer tubes and also a fully developed (as well as exhausted) hymenium, with apparently more mature spores. The exhausted hymenium is of the honey-comb type; the just-ripened basidia are broadly club-shaped; and the spores are hardly adaxially flattened, with a hyaline apiculus protruding abruptly from the somewhat thickened wall. These spores may have been faintly tinted but of this I am not quite sure (no spore-print seen). As to the correct genus, this may well be *Inonotus*, as long as this is accepted in the wide sense retained for it on this Check list.

Irpex

(62).   The *Irpex lacteus* of this Check list is broadly conceived, irrespective of possible diversity in certain characters of the generative hyphae. From The Netherlands a clamp-bearing collection was described by Maas Geesteranus [1963 (PNA 66): 452 *fs. 11–13*] [12]; and Boidin & Lanquetin [1965 (RM 30): 10] reported for this species the tetrapolar type of interfertility and [*fide* David, 1959 (BlL 38): 200] "un comportement normal", which means that it has uninuclear spores that germinate by producing a mycelium of uninuclear cells, while the "diplonte" is regularly binucleate. Later investigations reported hyphae without clamps for *I. lacteus* (Domański & Orlicz, 1969: 158) while David (l.c.) concluded that *I. tulipiferae* was presumably homothallic, without clamps in polyspermous cultures, and "holocénocytique" (mycelium consisting completely of coenocytic cells). This disparity between the results of Boidin and those of David was apparently the reason why instead of *I. lacteus* the name *I. tulipiferae* was used for European collections.

Irpex sensu stricto of this Check list is retained as conceived by Domański & Orlicz for a broadly conceived species, presumably characterized by coenocytic generative hyphae without clamps.

For almost none of the type collections, most of which have still to be selected, is information available on the presence or absence of clamps. As to the names *Irpex lacteus, I. sinuosus*, and *I. canescens*, where Fries left no types, substitutes for them must be selected; this will require separate study. Preferably such a substitute and illustrative specimen (or 'neotype') of *I. lacteus* should come from southern Sweden (presumably Femsjö); should answer to the first description and picture published by Fries (1818 O. 2: 266 *pl. 6 f. 1*); and should agree with the only substratum mentioned in its first description, "In ramis dejectis *Fagi*." The original material of *I. sinuosus* came from two localities,

---

[12]   Dr. R. A. Maas Geesteranus has come to the conclusion that the two collections on which the description was based represent *Steccherinum ochraceum* (Pers. per Fr.) S. F. Gray. The fruitbodies are small, with barely reflexed margins, whitish, and with irpicoid rather than toothed hymenophore.

"Pluries a Ruthenia misit Cel. Weinmann; equidem unico tantum loco, sed per quinque annos reducem habui supra ramos dejectos coacervatus *Quercus* . . .. (v.v.)." This second locality is again presumably Femsjö.

Irpex tulipiferae is the name now used for *I. lacteus* sensu lato as it occurs in North America. Even though it is there an extremely common species, so far no specimens with clamps have been reported. Although Bresadola [1897 (AAR III³): 101] at first identified *I. tulipiferae* with *I. lacteus*, later he (Bresadola, 1926: 79) apparently treated it as distinct.

Fries (1838: 522) received the type of *I. pallescens* from von Schweinitz, who collected both this and the type of *I. tulipiferae* on *Liriodendron*. Hence it may well belong to the latter, which is clampless.

Irpicodon

(**63**). *Irpex pendulum* is a remarkable species of, as yet, doubtful alliance; it seems worthy of being segregated as a distinct genus. This was already Fries's opinion at an early date: "Optima generis species" (1828 E. 1: 143) and ". . . ut facile peculiaris generis typus" (1874: 620). After Reid & Austwick [1963 (GN 18): 314] and Boidin & Lanquetin [1965 (RM 30): 10] had discovered that the spores were amyloid a generic name was provided, viz. *Irpicodon* Pouz.

Ischnoderma

(**64**). In a recent paper Pouzar (*1971*) defended the Friesian thesis that the *Ischnoderma benzoinum* of the present Check list is composed of two species. He emphasized that it was the colour of the trama of the dissepiments of the tubes in the hard and final stage of development that constituted the main difference: dark brown in *I. benzoinum* sensu stricto and white, whitish, or at most somewhat dirty yellowish in what he calls *I. resinosum*. Moreover from the specimens at his disposal it appeared that (at least in Europe) *I. benzoinum* is confined to coniferous trees, whereas *I. resinosum* occurs only on deciduous trees, mostly on *Fagus*. It would seem to me that many authors knew only the former taxon (Bourdot & Galzin, *1928*: 605, as *Ungulina fuliginosa*; Donk, 1933: 175, as *I. resinosum*).

This revival of the two taxa earlier differentiated by Fries poses a thesis that is warmly recommended for further study. While Pouzar distinguishes two phases in the development of the fruitbody of both taxa (a soft, sappy one and a hard one) it must be pointed out that Fries (1874: 554) emphasized two phases for his *Polyporus resinosus* only, "A. *sitaneus*, pileo carnoso . . . B. *hornotinus* pileo suberoso . . .."

Those following Pouzar should reconsider the name to be used for what he prefers to call *I. resinosus*. This was discussed by Donk (*1971a*: 8); he rejected the name for the whole complex because it was simply a misapplication by Fries that did not exclude the type (Art. 55); the type is represented by the name *Boletus resinosus* Schrad., which both Pouzar (*1971*: 18) and Donk (*1971a*: 8) refer to *Fomitopsis pinicola* as a synonym. Had Fries excluded the type he would have introduced a 'new' name based on a different type (Art. 48). The same argument applies to *Boletus fuscus* Pers. (= *I. resinosum* sensu Pouz.); according to Pouzar, Wahlenberg, in revalidating *Boletus fuscus* Pers., misapplied it to the other taxon (= *I. benzoinum* sensu Pouz.).

The correct names and their synonyms of the two taxa defended by Pouzar would appear to be as follows (with the 'original' substrata added): —

(i) *Ischnoderma benzoinum* (Wahl.: Fr.) P. Karst.—" . . . ad caudices *abietinos* putridos . . .."

? *Boletus fuliginosus* Scop.—". . . in Truncis prope radices."

Polyporus morosus Kalchbr.—"In pinetis . . . in trunco *Abietis piceae* reciso . . .."

Polyporus pini-silvestris Allesch.—The name leaves little doubt about the substratum.

(ii) *Ischnoderma fuscum* (Pers. per Wahl.) John Doe (nomen eventuale). —"Passim ad truncos *fagi sylvat.*" (Pers.); ". . . ad radices denudatus *Abietinum* . . ." (Wahl.).

Boletus velutinus Vahl.—". . . in truncis *fagi* . . .."

Polyporus guttatus Weinm.—"Ad truncos *Alni* etc."

M.—*Boletus resinosus* Schrad. per Fr. sensu Fr. 1821 emend. Fr. 1828.— "Major in arboribus frondosis, minor in pinetis. . . . Ad truncos" (1821), "In trunci *Fagi, Aceris, Alni* etc." (Fries, 1874: 554); ≡ *Ischnoderma resinosum* Donk, not ∼ (Fr.) P. Karst.

Laetiporus

(**65**). *Laetiporus sulphureus* is a common species in Europe and temperate North America, where it shows not only a wide range of variation in shape and colour but also in host. Some authors who received and named polypores from all parts of the world extended their conception of this species by also including taxa described from tropical Asia and South America; of these the best known is *Laetiporus miniatus* (Jungh.) Over. (**0**) from the eastern Asian tropics; it also appears to have a wide colour range. However, I am not convinced that it should be merged in *L. sulphureus*. (I myself have collected it in West Java, Indonesia.) For this and other reasons names based on taxa from the southern hemisphere and occasionally referred to *L. sulphureus* have not been entered on the Check list; they are mentioned briefly on the "List of omitted names".

Lenzites

(**66**). In many respects *L. warnieri* strongly resembles *Daedalea* *quercina* and several mycologists (among others Bresadola) have even reduced the species to the rank of a mere lenzitoid form of *D. quercina*. On the other hand under the name *L. reichardtii*, Igmándy (*1962*) and David (*1967b*) have found that it deviates considerably from *D. quercina*, for instance in cultural characters, and these characters agree much better with *Lenzites betulina*. In contradistinction to *D. quercina* the species is, *inter alia*, 'oxidase positive' and lacks clamped hyphae with irregularly thickened walls (Nobles's code symbol 9). On the other hand the fruitbody shows considerable differences vis-à-vis *L. betulina*, especially in the surface of the cap. However a great deal of information on tropical lenzitoid species is needed before changes in the taxonomic position can be proposed.

Meripilus

(**67**). Much of the confusion of *Meripilus giganteus* with species of *Grifola* has arisen because the fruitbodies are generally too big to be depicted natural sized, and too intricately built, inducing reduced and schematical rendering on many of the classical plates. A meticulous reading of the accompanying text is necessary to find out the real dimensions; often these will point in the direction of *M. gigantea* in cases where a fruitbody considerably exceeds about one or two feet in diameter. A description of the flesh might also reveal misidentifications, as in the case of *P. frondosus* sensu Rostk.: "Das Fleisch des Hutes ist fest, lederartig, elastisch, weiss, und lässt sich in unendliche sehr feine Fasern zertheilen"; this is certainly *M. gigantea*.

Compare also (**52**).

Oligoporus

(**68**). The genus *Oligoporus* is in my opinion still on trial. It was introduced by Brefeld for three *Ptychogaster* states, two of which he knew to be connected with polypores; of the third (*Ptychogaster rubescens*) he knew no perfect state. Brefeld was prejudiced in favour of the co-occurrence of the imperfect states. Had he been followed his genus would have grown into a bigger one at the expense of *Tyromyces* (**123**) and *Poria* but his view failed to draw support.

The type species of Brefeld's generic name is *Oligoporus farinosus*, which Donk identified with *Polyporus rennyi*. When Pouzar (*1967*: 206) transferred *Polyporus rennyi* to his new genus *Strangulidium* he emphasized the 'suburniform' (rather, utriform) basidia and the cyanophilous sporewall, the latter a character strongly emphasized—perhaps with some bias—in the work of Kotlaba and Pouzar. This implied that the other species of polypores known to produce *Ptychogaster* states in nature did not possess these two characters, but this was not stated positively.

The fruitbody of *Oligoporus rennyi*, when not associated with its imperfect state, is effused, but on vertical substrata fruitbodies associated with the *Ptychogaster* state this state may produce raised edges with tubes formed on the underside. If the genus is not accepted the perfect state could be placed in an inclusive genus *Poria* or in *Strangulidium*, which is based on *Poria sericeo-mollis*. Pending further research I prefer to maintain *Oligoporus* for a single species.

Onnia

(**69**).　The generic status of the *Polyporus tomentosus* complex and its precise circumscription is still open to discussion. I have restricted *Onnia* to this complex only. It has been included in *Coltricia* but the differences are that its fruitbody (which may vary from stalked to sessile or even pulvinate) is less definitely stalked; the duplex context of the cap is more or less pronounced (the lower layer becomes firm to hard and even almost woody and the soft upper layer varies in its degree of development from a mere tomentum to a thick layer); big haplosetae (hymenial) are present; the spores are smaller and paler (colourless); and through mycorrhizal association or parasitically the species are related to conifers.

Other genera that should be taken into consideration before autonomous generic status is assigned to *Onnia* are *Phaeolus* and *Inonotus*; here these are not discussed further. Authors who favour an inclusive genus *Inonotus* will perhaps find it not too easy to define clear cut characters for these three genera and they may be inclined to combine them into one.

In some recent European publications the genus *Onnia* is called *Mucronoporus* on the assumption that *Polyporus circinatus* is the type species of this generic name. It was recently pointed out by Donk (*1971a*: 12) however that the correct lectotype of the name *Mucronoporus* is *Polyporus gilvus*; this makes of *Mucronoporus* a synonym of *Phellinus*, *Onnia* thus becoming the correct generic name.

(**70**).　The next problem connected with *Onnia* (**69**) is the division of its contents, the *Polyporus tomentosus* complex, into taxa, together with the rank to be assigned to them. Although a few mycologists and phyto-pathologists (Gosselin, *1944*) have claimed that they are able to distinguish between two groups without the use of a microscope others insist that the only tangible difference is in the setae (Whitney, *1962*). These bodies may all be straight or else all to many of them distinctly curved to hooked. This difference was first appreciated by Ellis & Everhart (*1889a*) and it was particularly emphasized by Jørstad & Juul (*1939*) and Haddow (*1941*). It has become current practice to admit two taxa on this basis but there is still no concensus about the rank to be assigned to them. Most modern authors now admit two varieties but some prefer species, in which case they distinguish between *Polyporus tomentosus* (straight setae) and *P. circinatus* [which, assuming it to be broadly conceived, should be called *P. triqueter*] (hooked setae). Recently Eriksson & Strid (1939: 136) "have become convinced that *Inonotus triqueter* is a species of its own." Another difficulty has been to find out which name should go with which type of setae; this problem was reviewed by Haddow (*1941*), who concluded that the forms with curved setal tips belonged to *Polyporus circinatus*.

Much of the confusion in this connection was due to Lloyd (*1908*: 2–4); originally he ascribed straight setae implicitly to both *P. tomentosus* and *P. circinatus*,[13] supposing that the main difference (of the stalked forms), as indicated by Fries, lay in the degree of development of the tomentum on the cap, poor in *P. tomentosus* and strong in *P. circinatus*. This gave rise in North America to the conception of a species with straight setae (*P. tomentosus*) to which for some time the name *P. circinatus* was applied and for which the name *P. tomentosus* var. *americanus* Jørst. & Juul [1939 (MnS 6): 434] was introduced.

Of the two species recognized here *Onnia triqueter* (*O. circinata*) shows the greater variability in the shape of the fruitbodies; the shape ranges from centrally stalked (developing from buried roots or growing on duff from conifers) over sessile to even pulvinate (on stumps and trunks). According to Jørstad it would seem that in Norway *O. circinata* is always found with sessile fruitbodies or with a not clearly marked lateral stalk. This variation in shape has given rise to several specific names that now live on in the names of varieties and forms. The taxonomic value of these subdivisions, based primarily on the shape of the fruitbody, is as yet not well understood.

This separation of *Onnia* into only two species however may appear to be too simple a solution.

Oxyporus

(71). The genera *Oxyporus* and *Rigidoporus* as conceived by Donk (*1967*: 73–74) form two parallel series which agree in monomitic context and the lack of clamps in the fruitbody. (Clamps have been recorded in cultures for a few species.) This conformity induced Pouzar (*1966c*: 366–369) to combine the two under the name *Rigidoporus*. The differences between the two genera must still be formulated more precisely than has been done so far. Cultural characters should be helpful. In this respect the species referred to *Rigidoporus* appear to have a number of characters in common among which there is one expressed by Nobles's (*1971*: 184) code symbol 10 ("Hyphae differentiated to form cuticular cells, closely packed together to form a pseudoparenchyma.") This last character is negative in *Oxyporus populinus*, the type of the generic name *Oxyporus*.

Oxyporus will perhaps appear to be composed of two or three groups

---

[13] Moreover, he described the spores of European *P. circinatus* as $12 \times 7\ \mu$, which is much too big and certainly wrong.

the taxonomic rank of which is as yet uncertain. The typical subseries differs from the rest in a number of details of the cystidia. As far as is known the '*Poria*' group consists of species with non-amyloid hyphae and one with weakly amyloid hyphae. The latter has been placed in a group of its own, viz. *Rigidoporus* subgen. *Neooxyporus* Pouz. (*1966c*: 368); according to Nobles this species (*Oxyporus latemarginata*) agrees with *Rigidoporus* in forming 'cuticular cells' in cultures (code symbol 10).

The poria-like *Oxyporus* species are all supposed to be cystidiate but the number of cystidia may vary considerably to the point where they have even been reported to be absent. Of two putative synonyms of *O. latemarginata* (viz. *Poria consobrinoides* and *P. reticulatomarginata*) no cystidia have been reported. This raises the question of the importance of the character of 'cystidia present' in connection with the poria-like series in *Oxyporus*. In *Rigidoporus* cystidia may be present or absent, depending on the species, and they may vary in number in certain species.

On comparing *Chaetoporus* [*Oxyporus*] *philadelphi* (cystidia present) with *Poria millavensis* (cystidia lacking) a possibly significant similarity is apparent and I have been considering the transfer of *P. millavensis* to the neighbourhood of the other poria species in the genus *Oxyporus*. In any case, with our present incomplete knowledge the introduction of the genus *Riopa* for a species (*R. davidii*) lacking cystidia may have been premature, since in other respects *Riopa* seems to closely resemble the poria species of *Oxyporus*. Pending more information on these fungi it seems advisable to assign *Riopa* either to the rather indefinitely conceived genus *Oxyporus* of this Check list or (as is done in these pages) to *Poria*, the container for unclassified resupinate polypores.

(72).   The correct name for the highly characteristic type species of the generic name *Oxyporus* is still in doubt. There are two rival epithets that have been used almost indiscriminately; these were validly published in the following combinations, viz. *Polyporus populinus* (Schum.) per Fr. 1821 and *Polyporus connatus* Weinm. 1826. Provided both names were based on the same species there should be no hesitation in accepting *P. populinus* as the correct basionym.

The trouble is that there is no absolute certainty about the identity of both taxa with the species under discussion. The name *Polyporus populinus* was a recombination of *Boletus populinus* Schum., although Fries (1821: 367) made some reservations. His citation of the devalidated basionym was: "*Bol.* Schum. Sael. p. 384, varietas videtur" and he listed the name in the index (1821: 519) as "[*Polyporus*] *populinus* (Schum.?)". It is clear that he hesitated, but in any case he did not explicitly exclude Schumacher's taxon, the original illustration of which was later published by Hornemann in the "Flora danica" (*pl. 1791*). After the publication of the plate (of which it is likely that Fries had seen the original before 1821) Fries (1838: 472) had no longer any scruples; this time he cited,

"Schum.! n. 1951." Still later (Fries, 1863 M. 2: 254) he rejected this point of view and remarked "*P. populinus* Fr. (noster primitus lignosus, hinc diversus a *B. populino* Schum.)." The context of the fruitbody changed in his descriptions from "carnoso-suberosus" (1821) to "suberoso-lignosus" (Fries, 1874: 564).

The original description of *Polyporus connatus* Weinm. runs: "Char. spec. Effuso-reflexus, seriato-confluens; pileo lateritio, velutino, subzonato; poris minutis, subrotundis albis. / Descr. Substantia suberoso-coriacea, tenax, elastica, alba, pallida: odore acidulo. Pileo 3–8 lin. lati, 1–2 pedis longus, 2–3 unc. et ultra latus, imbricato-connatus. / Patria: Ad truncos semiemortuos *Aceris platanoidis* sero autumno. Ante *Polyp. hirsutum* Fr. [1821:] pag. 376. ponendus." Fries (1828 E. 1: 92) saw a dried specimen from the type locality and considered it to be intermediate between *Polyporus neesii* Fr. (to which in 1828 he appended Weinmann's species as a variety) and *P. populinus*, "ab utroque forsan diversus ob odorem peculiarem acidulum." Later in connection with *P. connatus* Fries (1874: 563) remarked "[*P.*] *connatus* Weinm. Ross. p. 332 [= 1836: 332], in *Acere* quoque lectus, cum nostro convenit excepto pileo lateritio". A very good illustration of the *Polyporus connatus* of modern mycology was published posthumously under the name "*Polyporus connatus* Weinm.?" (Fries, 1884 I. 2: *pl. 185 f. 2*). The question mark shows that he continued to doubt the identity of Weinmann's species with his own interpretation.

One of the striking features of to-day's interpretation of *Polyporus populinus* (*P. connatus*) is that on section the tubes are very often many-layered. It is remarkable to find that this feature appeared only at a late stage in Fries's work and then only in his accounts of *P. connatus*. In 1838 it is stated of this species, "Perennans, reviviscens, sed non vere stratosus"; not until 1874 do we read, "intus porisque *stratosis*" (italics as in the original). Until the end (1874: 564) *Polyporus populinus* remained in a group, "Perennantes . . ., *sed non stratosi* . . .." After Secretan (1833 M. 3: 113) had given a good description of the now current interpretation of *P. populinus*, Fries (1838: 472) remarked "Uterque intus candidus nec stratosus, hinc Secr. n. 60 carne alutacea (nisi e succo arboris sit imbuta) recedit." From these notes it follows that eventually to Fries *P. populinus* was a species with one-layered tubes and growing only on *Populus*; and *P. connatus* a different species with many-layered tubes and growing on various kinds of trees, although *Acer* was originally mentioned as its only host.

To-day's *P. populinus* may be found with well-developed fruitbodies but still having one-layered tubes after a rapid initial growth; in such specimens the context of the cap is rather soft and more or less water-soaked ('carnoso-suberosus'). I have collected it on *Populus* in this condition in the dunes along the Dutch coast. As a result I am not yet prepared to deny hat (i) the fungus of Schumacher, (ii) that of Fries of 1821 named

P. populinus, and (iii) *P. connatus* sensu Fr. might all represent the single species to which these names are now applied. On the other hand I am not prepared to say the same of the original *P. connatus.* Keeping in mind that Fries saw material forwarded by Weinmann however it is difficult (without the study of an authentic specimen) completely to exclude the possibility of identity with the *P. populinus* of modern authors. On the Check list, in accordance with a long tradition, the name *Polyporus populinus* is retained as basionym for the correct name.

Should this conclusion not be accepted it will not be easy to decide on the correct name. *Polyporus neesii* Fr. 1821 is a nomen dubium (**O**). A sure name is *Trametes secretanii* Otth 1866, the next in succession to be taken into consideration.

(**73**). Seemingly when Fries (1838) published *Polyporus ravidus* he based this name on "*Bol. heteroclit.* Sowerb. t. 367." His original description (protologue) contains no other references, and no specimens or localities are cited. However, on comparing the description of *P. ravidus* with the (devalidated) protologue of *Boletus heteroclitus* Sow. (**O**) there can be no doubt that Fries's description was drawn up from a different source. In later work Fries (1874: 566) repeated his original description practically unaltered, but with the addition "*Sistotrema cinereo-lutescens* Pers. Myc. Eur. 2. p. 205. ... in Gallia". This seems to be a possible clue to the identity of the hypothetic type collection: a specimen collected in France near Angers by Guépin on which Persoon had previously based *Sistotrema *lutescens* Pers. 1825: 205.[14] Its description and that of Fries's *P. ravidus* show undeniable similarity. When Bresadola published his interpretation of *P. ravidus* he must have seen material either of the *Polyporus ravidus* as named by Fries or of Persoon's *Sistotrema lutescens.* There is no type of *Polyporus ravidus* in Fries's herbarium. Nor have I been able to locate the type of *Sistotrema lutescens* in Persoon's herbarium.

Should it be decided to accept the above, provisional, thesis, viz. that, perhaps by some error, in 1838 Fries did not mention the type of *P. ravidus* and that in fact he published a new name (typonym) for *Sistotrema lutescens,* it will be necessary to restore the earlier-published epithet of *Sistotrema lutescens* Pers. in a new combination. What must be done first however is to establish the basis of Bresadola's interpretation of *Polyporus ravidus* as published in 1903.

---

[14]   The asterisk indicates (Person, 1801: x) "obscure species, of which either the [taxonomical] position is doubtful, or which are still in need of a more exact investigation" (translated from the Latin). Such species were thus clearly set apart from varieties and subspecies. Fries took the asterisk to indicate a variety of the species preceding it in Persoon's work; therefore he should have mentioned it as '*Sistotrema cinereum* var. *lutescens*' but by some error he wrote "*Sistotrema cinereo-lutescens*".

(74). The status of *Poria pearsonii* as a distinct species, different from *P. corticola*, needs further consideration. Lowe (*1966*: 19) considered them to be synonyms. If they are different the morphological differences are slight. A rather long set of specimens collected by the author of the species and preserved in BPI were all found on wood of *Abies* and I hesitate to identify them unconditionally with *P. corticola*. If the two are kept apart comparison of *P. pearsonii* with *P. vicina* Bres. (U.S.A., Washington) is recommended; Gilbertson [1956 (Ll 19): 66, 84] regarded the latter taxon as a synonym of *P. corticola*. Its three syntypes were all found on stems of *Abies grandis*.

According to Nobles (*1958*: 901), cultures of *P. pearsonii* have fiber hyphae and clamped septa and appear to be 'oxidase-negative'. Donk (*1967*: 74) suggests that this determination is incorrect.

Parmastomyces

(75). The protologue of *Tyromyces mollissimus* Maire [1935 (BAN 36): 37 *f. 7bis*; Algeria] strongly suggests that this may be an earlier name for *Parmastomyces kravtzevianus* or that Maire's species is one closely related to it. The type of *Tyromyces mollissimus* could not be located.

Perenniporia

(76). I have followed Lowe (*1966*: 110–111) in tentatively treating *Poria tenuis* and *P. pulchella* as together forming a single species. This meant investigating which name was first reduced to the synonymy of the other. For the present I leave the question unanswered; Lowe treats *Poria pulchella* as a variety of *P. tenuis*. The reason why this is not done on the Check list proper is to be found in the answer to quite a different question: Are the two taxa really conspecific? It may be pointed out that Lowe records *Poria tenuis* var. *pulchella* from "throughout the U.S. and in Europe," while *Poria tenuis* [var. *tenuis*] is said to occur "throughout Canada and throughout the U.S.", thus with no mention of Europe. Lowe is convinced that "they are not the same thing" (oral communication, Dec. 1969) and this has induced me to list *Poria pulchella* as European and *P. tenuis* as a questionable synonym.

The specific delimitations within the genus *Perenniporia* are far from settled; Lowe (*1966*: 111) concluded that "[*Poria tenuis*] and its variety intergrade with *P. medulla-panis* and *P. subacida* and the validity of the taxonomic concepts of these species and their synonyms is by no means clear." This and the current preparation of a European monograph on the group will explain much of the unsatisfactory treatment on this Check list, on which I preferred not to enter several personal and other unpublished conclusions.

Phellinus

(77).   This is a big and heterogeneous group. The number of European species is high if compared with that of other genera but it must be called small in comparison with the extra-European genera, many of which occur almost only in the tropics. No attempt at a new classification should be made without a fair knowledge of the whole group. Moreover such an attempt should be preceded by improved terminology for the hyphal elements and the setae in all their variations.

(78).   It would appear that in the names "*P.*" *salicinus*, "*P.*" *conchatus*, and "*P.*" *versatilis*, as published by Bourdot [1898 (RBo 11): 231], the 'P.' does not stand for '*P[lacodes]*' but rather for '*P[hellinus]*', the generic name that was not mentioned in full presumably by error. This is accepted here as having been the case.

(79).   *Phellinus gilvus* is now considered to be widely distributed over the tropics and to occur there abundantly, as a species that varies greatly, mainly in the shape of the fruitbody. It becomes rarer towards the temperate regions and is recorded from Europe by an older generation of mycologists from only a few localities: France (Le Mans, Gironde, Pyrenees), Spain (Catalonia), and Italy (Piemonte, Lombardia, Sardinia). Some or all of these collections may have been incorrectly named; confusion with *Phellinus torulosus* in particular would not be unexpected. European material has served as types of two names that have subsequently been reduced to the synonymy of *Phellinus gilvus*: viz. *Placodes fucatus* Quél. and *Polyporus marcuccianus* Lloyd ex Trott.

Fries (1874: 548) was the first to record *P. gilvus* from Europe [15] because he thought that *Boletus impuber* Sow., described from England, might belong to it but in my opinion this identification is not acceptable.

In accepting the absence of *Phellinus gilvus* in Europe (I have not yet studied any of the European collections attributed to it) I also assume as a thesis that (even in a broad circumscription) it is specifically distinct from *P. torulosus*, a point that has been questioned by a few leading American mycologists. Thus Overholts (*1953*: 72), who recorded *P. torulosus* from North America, stated that he "can only express the hypothesis . . . that *F[omes] torulosus* may be a thick form of *P[olyporus] licnoides*", a taxon often reduced to subspecific rank under *Phellinus gilvus*. Lowe (*1957a*: 51, 90) went a step further and stated that the North American material reported by Lloyd and Overholts as *Fomes torulosus* "appear to be perennial specimens of *Polyporus licnoides* Mont." He even went so far as to make the more general statement that tropical

---

[15]   It is true that Wallroth (1833: 589) included *Polyporus gilvus* in his flora but this was because he thought a European species of *Hymenochaete* to be a form of it, "m[onstrum] *thelephoreum* Wallr."

P. licnoides is the annual form of *Fomes torulosus* from southern Europe and subtropical America. It seems worth while to re-examine the question whether these American authors interpreted *Phellinus torulosus* consistently and correctly or not.

(80). Those who accept *Phellinus gilvus* as a European species (79) are confronted with still another problem: What is the exact scope of this versatile species? Some authors have made it very inclusive. Others divide it into several species; in addition to *P. gilvus* they also admit as autonomous *Polyporus hookeri* Lloyd (India), *Phellinus licnoides* (Mont.) Pat. & Har. (French Guiana), and so on. Pilát and, more recently, O. & M. E. P. K. Fidalgo [1968 (MNY 17²): 9], accepted the inclusive conception and brought together as synonyms not only all the names that had been reduced at one time or another to *P. gilvus* but also several that had merely been suspected of being related to it often on the questionable authority of Lloyd. In this way some species crept into the synonymy that do not even belong to *Phellinus*. I am not prepared to do the same and have left off the "List of omitted names" all synonyms, reliable or suggested, that are not based on European material. The number of these completely omitted names (not counting the recombinations and other isonyms) amounts to about two dozen.

(81). The *Phellinus igniarius* of many modern authors has grown out into a big conglomerate in which they have accumulated a considerable number of not only pileate but also resupinate forms; for a note on the latter, see (82). A few European authors have been engaged in a more critical study of the complex and they are now agreed not only that certain resupinate taxa should be excluded (*Phellinus laevigatus, P. rhamni*) but also that *P. tremulae* deserves treatment as a distinct species. This still leaves an imposing residue. Some authors divide this into a few more species, *P. igniarius* and *P. nigricans* and a third, often called *P. trivialis* (cf. Donk, *1971e*), whose taxonomic status is as yet not firmly settled. I accept *P. nigricans* on this Check list. The delimitation by Bourdot & Galzin (*1928*: 618, as a subspecies) and by Jahn (*1963*: 98, as *Phellinus trivialis*) is too wide. The correct typification and nomenclature as well as the specific status of a narrowly conceived *P. trivialis* is still to be worked out. The latest segregate is *P. lundellii*.

Populus tremula harbours at least two species, sometimes on the same tree, or more often on different trees, viz. *Phellinus tremulae* and a form of *P. igniarius* with black, soon strongly rimose, crusted surface; the latter seems to agree with *P. igniarius* f. *tremulae* E. Komar. (not validly published).

(82). A fact similar to that mentioned for *Phellinus robustus* (86) exists in connection with *Phellinus laevigatus*; this is often considered to be only a resupinate form of *P. igniarius*. Fries described it with a thin fruitbody ("vix lin. crasso"), "ad corticem arborum v.c. *Betulae*." Fruitbodies may however become thick at the middle, up to about 10–20 mm. Such thick portions recall certain 'resupinate', almost cushion-shaped conditions of certain pileate species of *Phellinus*. Bresadola [1897 (AAR III 3): 79] regarded *P. laevigatus* as barely autonomous; he preferred to consider it a form of *Fomes fulvus* [sensu Bres.] = *Phellinus pomaceus*. This view was firmly rejected by Bourdot & Galzin (*1928*: 624), but it must be recalled that their conception of *Phellinus laevigatus* was a different species that (at least in part) is now called *P. rhamni*. Nevertheless most European authors do not now doubt the autonomy of *P. laevigatus*; they consider it to be a species that in Europe is confined almost exclusively to *Betula*.

With this pronounced preference for *Betula* in Europe it seems almost a matter of course to hesitate before referring to *P. laevigatus* a number of other resupinate taxa described from other parts of the world and found on other substrata. Overholts (1942: 59; *1953*: 63) listed only one synonymous taxon, *Fomitiporella betulina* Murrill. Lowe (*1966*: 156) goes much further; not only does he not uphold any real distinction between *Poria* [*Phellinus*] *laevigata* and 'resupinate' *Fomes igniarius*, but to the *F. igniarius* complex he also refers *Poria pereffusa* (Murrill) Sacc. & Trott. (**O**) and *Poria prunicola* (Murrill) Sacc. & Trott. (**O**), both of which Overholts (1942: 61, 62) treated as distinct. These as well as a few other names that have been reduced to *Phellinus laevigatus* will here be found only on the "List of omitted names".

(83). *Phellinus pini* was described twice, in two successive years, both times as *Boletus pini*, first by Thore (1803), then by Broterus (1804). In 1821 Fries revalidated the name of the second author; hence *Polyporus pini* (Brot.) per Fr. American authors, in the period before the introduction of the later starting-point date for 'Fungi caeteri', as a matter of course preferred the name by Thore published earlier. This will explain such incorrect authors' citations as 'Thore ex Fr.' and 'Thore ex Lloyd'. Because both Fries and Lloyd had in mind *Boletus pini* Brot. as basionym, it would seem correct to transcribe such citations into 'Fr.' and 'Lloyd' and, in the case of new combinations, to consider these shortened authors' citations as indirect references to 'Brot. ex Fr.' and 'Brot.'. These corrections proved necessary to disentangle the synonymy (recombinations) of the two devalidated homonyms.

As to the references to descriptions and illustrations, I have not tried to sort them out over these two competing names but cite them all under their 'correct' name, *Phellinus pini* (Brot. per Fr.) A. Ames.

(84). *Phellinus pini*, as it occurs in Europe, is readily separable into two (and perhaps three) taxa that have been assigned different rank by different authors. Among European mycologists there is now a strong tendency to regard *P. pini* and *P. abietis* as distinct species. This thesis was recently defended at some length by Jahn (*1967a*: 85). Eriksson & Strid (*1969*: 136) came to the same conclusion: "After having seen [*Phellinus abietis*] numerous times in nature we have come to the opinion that it is a good species, distinct from *P. pini*." The correct name for *P. abietis*, however is, in my opinion, *Phellinus chrysoloma* (cf. Donk, *1971a*: 39).

Other authors, like Pilát and his followers, and especially North American mycologists, like Overholts and Lowe, all of whom have studied material from more extensive areas or from different parts of the world (other than Europe), tend to conceive *P. pini* as an increasingly inclusive species. This may be correct but I am far from certain that it really is so; a more likely conclusion is that *P. pini* sensu lato is a species aggregate which, besides the main species, includes a number of satellite species, each with its own area of distribution and its own preference for hosts. Starting from this point of view I have omitted from the Check list as synonyms all specific names based on material collected outside Europe. These are *Phellinus piceinus* (Peck) Pat. (**O**), *Daedalea vorax* Harkn. (**O**), *Cryptoderma yamanoi* Imaz. (**O**).

Pilát [1942 (ACE 3): 521] recorded one collection of what he called *Phellinus pini* var. *abietis* f. *micropora* Pilát from Tirol, a form with small pores ("pores très petits, 0.2–0.3 mm de diam., arrondis à arrondis anguleux, rarement un peu allongés"). Small-pored forms might also be expected from northern Fennoscandia. They should be compared with *P. piceinus* (**O**).

(**85**).   The precise host range of *Phellinus pomaceus* still requires the attention of collectors. The usual hosts are various species of the genus of Rosaceae trib. Pruneae, *Prunus* L. in a wide sense, including *Amygdalus* L., *Cerasus* Mill., *Persica* Mill., *Padus* Mill., and *Laurocerasus* Duham.; the fungus has been reported from all of these subdivisions.

It has also been recorded from *Pyrus* L. (including *Malus* Mill.), *Sorbus* L. (Kotlaba, *1966*), and *Crataegus* L., genera of Rosaceae trib. Pomeae. Some of this second set of records may have been based on *Phellinus igniarius*, which includes a 'form', especially well-pronounced on apple (*Pyrus malus* L.), that superficially resembles certain fruitbodies of *P. pomaceus*. In this connection the following remark by Murrill [1903 (BTC 30): 113] is worth mentioning: "In an orchard near Mauritzberg, Sweden, where *P*[*yropolyporus*] *igniarius* was abundant on apple trees, *P*[*yropolyporus*] *fulvus* [= *Phellinus pomaceus*] was confined to the stumps and dead or dying trunks and branches of plum trees."

Other records of this second set of hosts (i.e. on *Crataegus*) could have been based on *P. robustus*. The colour of the context of *P. pomaceus* lies between the somewhat darker colour of *P. igniarius* and the much brighter colour of *P. robustus*.

Finally, quite a number of records of *P. pomaceus* mention the occurrence of this species on hosts that do not belong to the Rosaceae, such as *Corylus avellana* L. (Bourdot & Galzin, *1928*: 619, "coudrier"), *Syringa* L. (Kreisel, 1961: 136), *Vitis vinifera* L. [Malençon & Delécluse, 1937 (BMa 17): 136], *Cytisus laburnum* L., *Juglans regia* L., *Rhamnus frangula* L. (*Frangula alnus* Mill.), and *Morus* sp. [Pilát, 1942 (ACE 3): 513–514]. I would suggest re-examination of the collections on which these records are based before definitely including them in *P. pomaceus*. Thus the material from *Vitis* should be compared with *P. punctatus* (**86**).

A long time was needed before European mycologists started to accept *Phellinus pomaceus* as a good species and even now misdeterminations are not rare. Misgivings about its autonomous status were voiced as late as 1942 by Pilát [(ACE 3): 513]. The confusion in France can easily be proved by an analysis of Quélet's conceptions (1888: 399). What he called *Placodes igniarius* is very likely *Phellinus robustus* ("chair ... brun rouillé"). His *Placodes igniarius* var. *fulvus* (Scop. per Fr.) Quél. is mainly *Phellinus igniarius* ("chair ... brun foncé), but it also included *Phellinus pomaceus* ("prunier"), which he called *Placodes igniarius* var. *pomaceus* (Pers. per S. F. Gray) Quél. ("chair brun fauve") as well. In confusing typical *Phellinus igniarius* with *P. robustus*, Quélet was following an old French tradition (cf. Donk *1971e*: 411).

I assume that the epithet 'pomaceus' in this case really means 'of orchards' rather than 'of apple trees'. Persoon's first description (1799 0.2: 5) does not explicitly mention apple as a host. He wrote, "Hab. ad *Pyros*, praesentim vero ad caudices *Pruni domesticae* aut *Cerasi* . . .." At that period Persoon (1807: 40) called the apple *Pyrus malus*, but by '*Pyrus*' he may also have meant *Pyrus communis*, the pear or a number of other species which he admitted to the genus. The name *Prunus* he used for a rather inclusive genus which for instance included *Prunus domestica* and *P. spinosa*. Konrad & Maublanc (1927 I. 5: text to pl. 459) considered the type form of *P. pomaceus* as growing on 'pommier'; the form from 'prunier' (*Prunus domesticus*) and 'cerisier' (*P. avium* presumably) they identified with Persoon's form 'prunastri'. This typification of *Phellinus pomaceus* was apparently caused by their interpretation of the meaning of the specific epithet. In later work Persoon (1801: 538) gave as the typical habitat merely, "in pomariis [orchards!] ad caudices arborum." I prefer to stress "praesentim vero ad caudices *Pruni domesticae* aut *Cerasi*" and select as 'typical' a specimen from *Prunus domesticus* answering to Persoon's first description. Later Persoon (1801: 538) segregated a form '*prunastri*', which he characterized as "pileo subresupinato crasso laeui truncato". There is a specimen in Persoon's herbarium (L 910.263–397) labeled "*Polyporus pomaceus. Boletus.* / Gallia Ad truncos praesentim *Prunorum* et *Cerasorum*." It is typical of the species.

This raises the question whether *Phellinus pomaceus* consists of genetically different taxa or not. Is there a form on Pomeae (*Pyrus, Sorbus, Crataegus*) distinguishable from one on Pruneae (*Prunus* sensu lato)?

Is the form on *Crataegus* worth distinguishing? (What Baxter [1925 (AJB 12): 563] described under the name *Fomes pomaceus* f. *crataegi* had a completely resupinate fruitbody; its author recorded it not only from *Crataegus* but also from *Prunus americana* and *P. persica*). Typical material on *Prunus domesticus* often develops sessile fruitbodies, but occasionally the position on the substratum also gives rise to 'subresupinate' fruitbodies on the same tree. Is this *Phellinus pomaceus* var. *prunastri* (Pers. ex S. F. Gray) Pat.? Or is there a more consistently and typically resupinate form that deserves this name, like a form from *Prunus spinosa* mentioned by Pilát (l.c.), of which Jahn (*1963*: 103) stated that "die Frk. meist kleiner und vorwiegend resupinat bleiben: f. *prunastri* (Pers.) ...."? The specimens pictured on Konrad & Maublanc's plate (l.c.) I would refer to typical, pileate *P. pomaceus* rather than to something deserving of the special epithet 'prunastri', which taxon should be 'subresupinate'. *Phellinus prunicola* (**O**), described from North America, is a distinct species allied to *P. laevigatus*. It is now usually referred to *Phellinus laevigatus* as a form, but this is questionable.

(**86**). Many species of *Phellinus* have strikingly variable fruitbodies; these may range from effused or resupinate over effuso-reflexed to sessile. This has led to much-needed lumping, but on the other hand to insufficiently-founded extrapolation. Thus the conceptions of both *Phellinus robustus* and *P. igniarius* have been extended by some authors to also include a number of taxa with consistently effused (resupinate) fruitbodies. As far as Europe is concerned it would appear that lumping has gone too far.

The thesis I subscribe to is also defended by Jahn: although *Phellinus robustus* may occasionally be effused it is certainly not conspecific with the consistently effused *Phellinus punctatus*. American authors have referred many names first to *Poria punctata* but then to *Fomes robustus*, according to a contrary thesis. In a few cases the types were redescribed extensively enough for tentative inclusion under *P. punctatus*; these are *Fomitiporia laminata* Murrill and *F. obliquiformis* Murrill. In many other cases however the original (and a few supplementary) descriptions are either inadequate to justify a personal opinion or else they seem even positively to exclude inclusion under *P. punctatus*.

(**87**). Bourdot & Galzin (*1928*: 623) reported the occurrence of *Phellinus friesianus* [= *P. punctatus*] on *Vitis*: "Sur vigne: le champignon attaque les ceps dans le Midi et les tue assez vite; il est mal développé, en nodules souvent à peine porés, assez rare." Assuming that Gard (*1922*, as *Fomes igniarius* var. *viticidus* Gard) and Rezende-Pinto (*1940*, as *Poria viticola* Lázaro) were describing the same fungus it is very likely that this is also the fungus mentioned by Bourdot & Galzin: cystidia absent or very rare and then tawny, pointed, swollen at the base, 15–25 μ long (Gard) and spores globose, 7 μ in diameter (Rezende-Pinto).

In a second paper Rezende-Pinto (*1942*) concluded that *Poria viticola* (Schw.) Cooke, *Poria viticola* Lázaro, and *Poria friesiana* Bres. all belonged to the same species, which he called *Poria viticola* (Schw.) Sacc. Here this conclusion is partially accepted (identification with *P. friesiana*), but the identification with von Schweinitz's species is incorrect. The latter is now considered to be quite distinct from *Phellinus punctatus*. The identity of *Poria viticola* Lázaro with the poria on *Vitis* under discussion seems quite likely and for the time being I accept it.

(88). The history of true *Polyporus rimosus* Berk. (**O**) is intricate and will not be worked out here. Over a considerable part of the tropics as well as in North America and eastern Europe the name came to be mis-applied to a different species, the correct basionym of which has been accepted as being *Pyropolyporus robiniae* Murrill. The confusion of the two species, apparently together with one or two others, has been intense. In the first place therefore I have selected descriptions that were drawn up by authors who lived in the regions where *Phellinus rimosus* does not occur, viz. the northern hemisphere north of the tropics.

In view of Patouillard's description and Lloyd's identification (apparently after study of the type) with "*Fomes rimosus*" there seems to be little possible doubt that *Xanthochrous tuniseus* is an earlier-published name for the species often misnamed *Fomes rimosus* (**O**) and now also known as *Phellinus robiniae*.

On this basis and referring the species to the genus *Phellinus* it would seem obligatory to introduce a new combination of Patouillard's specific epithet 'tuniseus' with this generic name. I refrain from doing so however because in my opinion in the near future *Phellinus* will fall apart and then the present species will almost certainly be removed from it and placed in a revived genus *Fulvifomes* Murrill, of which *Phellinus robiniae* is the type species.

(89). *Phellinus robustus* is generally conceived as a species mainly infecting *Quercus*, but its occurrence in Europe on a rather limited number of other angiosperms is now generally accepted. On some of these hosts the fruitbody may look rather distinct from that on oak, but this seems to be due chiefly to its smaller size and the much thinner branches from which it develops, resulting in a different attachment. The most striking example of this is the form on *Hippophaë*.

Phellinus hartigii is undoubtedly closely related to *P. robustus*. It can most easily be distinguished by its hosts, which are species of conifers (*Abies, Picea, Pinus*). Many mycologists doubt the autonomy of *P. hartigii* and reduce it to a form or variety of *P. robustus*. However, authors who have paid special attention to this question have upheld the two as distinct species; compare K. Lohwag (*1937*), who made a study of the fruitbodies, and Jacquiot (*1960*), who based his conclusion on cultural evidence. Since the question can hardly be called settled the two are maintained as distinct species.

Both of the above species may occur with effused fruitbodies more or less closely resembling certain taxa with consistently effused fruitbodies. It is either incorrect or premature to refer this last mentioned set to *Phellinus robustus* (**86**).

Polyporus

(**90**). I have accepted *Polyporus arcularius* (described from Italy) in the sense of Bourdot & Galzin's *Leucoporus arcularius* var. *strigosus* and assumed that it is specifically different from the *P. brumalis* of this Check list (= *Leucoporus arcularius* var. *scabrellus* Bourd. & G., as "*scabellus*") notwithstanding the fact that the French authors found the two taxa to be "reliées par trop d'intermédiaires". *Polyporus brumalis* is far more uniform in northern Europe than one would suspect from the description by Bourdot & Galzin based on French material. It is suggested that Italian mycologists pay renewed attention to this problem.

Although I have accepted that Kreisel interpreted *P. brumalis* correctly (Donk, *1969a*: 242), the words "pileo . . . margine ciliato" in Persoon's original description have left an uneasy feeling in my mind; they could be taken as pointing to *P. arcularius* rather than *P. brumalis*.

Polyporus arcularius has been variously, but mostly very inclusively, conceived. In my (tentative) opinion it has also been misapplied to the complex that I now call *P. alveolarius* (Bosc) per Fr. All this has led to the reduction of a number of names to the synonymy of *P. arcularius*; some of these names were based on collections of *P. alveolarius*. It is difficult to reassess the correct status of these presumed 'synonyms' without renewed study; for this reason those based on material collected outside Europe and North America outside the tropics have not been entered on the "List of omitted names" (unless they have been entered on it for other reasons). The following enumeration mentions such names as have been reduced by various authors to the synonymy of *P. arcularius*; many of these were compiled by Pilát [1936 (ACE 3): 75–76].

Polyporus aemulans B. & C. (Cuba).

For *Polyporus agariceus*, see (**O**).

Polyporus armitii F. Muell. & Kalchbr. apud Cooke (Australia),

Polyporus binnendykei Kurz ex Cooke (Java),

Polyporus ciliaris Mont. (Madagascar),

Polyporus collybioides Kalchbr. ex Cooke (Australia),

Polyporus cremoricolor Berk. (India),

Polyporus maculatus Berk. = *P. squamoso-maculatus* Sacc. (India), not ∼ Peck, not ∼ (Cooke) ex Lloyd (n.v.p.?),

Polyporus obscurus Kalchbr. apud Kalchbr. & Thüm. (Mongolia),

Polyporus penningtonii Speg. (Argentina),

Polyporus similis Berk. (Brazil).

For *Polyporus umbilicatus* Jungh., see (**O**).

(**91**).   From the description there can be little doubt but that *Polyporus rhombiporus* Pers. is the same as *Leucoporus arcularius* var. *strigosus* Bourd. & G., which I take as representing typical *P. arcularius*. The original description is rather full and differs in only one respect from that of Bourdot & Galzin, viz. in the fact that these authors do not mention small scales on the cap, while Persoon wrote "squamulis fuscis adpressis." However small scales are mentioned by Konrad & Maublanc [1935 I. 5: *pl. 428 f. 2*; "chapeau ... plus ou moins squamuleux ou granuleux ..."]. Precisely because of these scales *P. floccipes* comes to mind and this is strengthened by "margine strigosis ... villo rudi hirtus"; this is also typical however of the fungus described by Bourdot & Galzin (*1928*: 532; "a bords fortement hispides et ciliés de poils rigides, fauvatres, assez longs"); they were certainly not describing what I call *P. floccipes* because their fungus has smaller, medium-sized spores. The same can be said of Konrad & Maublanc's conception. Compare also Donk (1933: 135), who wrote about a specimen in Persoon's herbarium that could be the type "Der gewimperte Rand sowie die Sporen weisen auf *P. arcularius.*"

Persoon hesitated to identify *P. rhombiporus* with *P. arcularius* (with which he compared it) because in Micheli's figure the pores were drawn very schematically and crudely; of the fruitbody in the figure he wrote" ... poros sistit fere quadrangulares et multo majores, et hinc minore in copia (fere 30)".

The name *P. rhombiporus* was preceded by an asterisk, for the meaning of which see foot-note 14.

(**92**).   It is with considerable hesitation that I maintain *Polyporus intermedius* Rostk. as a synonym of *P. arcularius*. Fries (1874: 526) suggested this identity and later authors accepted it as settled. If they are right then the specimens figured must have been old, weathered fruitbodies. Another suggestion is *P. floccipes*. For a very brief moment I thought of *Gyrodon lividus* (Bull. per Fr.) Sacc. (Boletales) because of the habitat, "auf alten Erlenwurzeln". Both of these last two suggestions I find untenable.

(**93**).   The name *Polyporus brumalis* has often been used for an extremely broadly conceived species, but it has also been misapplied to what is called *P. ciliatus* on this Check list; compare Kreisel (*1963a*) and Donk (*1969a*). The synonyms referable to this complex are numerous. Many of them could be placed but others are so insufficiently known that they could not be disposed of with any degree of certainty; this latter category will be found mentioned only in the 'List of omitted names'.

(94). *Polyporus ciliatus* is now generally assumed to be a species with considerable variation in its fruitbodies; for instance their size ranges from minute (cap a few mm in diam.) to large (cap up to about 10 cm in diam.). Some of the synonyms now accepted could however represent elements that might perhaps better be treated as distinct species, for instance *P. vernalis*. I have reluctantly given up any hope of separating *P. ciliatus* sensu stricto from *P. lepideus*; the two are connected by such a gradual series of intermediates that I feel compelled to follow Kreisel in considering them merely as forms. On the other hand I have kept *P. corylinus* and *P. incendiarius* (96) separate until more is known about them.

(95). *Polyporus latiporus* Britz. is difficult to place. Killermann [1922 (Bba 15): 66] made it a synonym of *Polyporus elegans* [sensu auctt. = *P. varius* s. str.]; this cannot be correct because of the "sich bald abnützenden Faserschuppen" on the cap and its big spores ($14 \times 4 \mu$). Bresinsky & Stangl (*1968*: 78) referred it to *Polyporus forquignonii* [= *P. floccipes*]. This is a much more likely suggestion, although neither for the margin of the cap nor for the base of the stalk are whitish-hyaline bristles mentioned while the base of the stalk is black. I would tentatively assume that Britzelmayr described and depicted an old fruitbody with worn off surfaces. In an occasional fruitbody of *P. floccosus* the very base of the stalk may become black; this also seems to have been the case in this instance.

(96). Neither the correct status and synonymy nor the precise relationship of *Polyporus incendiarius* is as yet well established. Pilát [*1936a*: 15; 1936 (ACE 3): 64] reduced it to a form of *Polyporus brumalis* (which species he conceived in a very broad sense) and this was accepted by Bondarcev (*1953*: 463). The few and incomplete descriptions published by European authors would suggest that so far this reduction is a mere guess and that it needs further study. Other authors, like Muraškinsky (1940: 2), reduced *P. incendiarius* to infraspecific rank under *P. elegans* [= *P. varius*]. The identification with *P. brumalis* var. *albus* Allesch. is still doubtful.

The original material came from the Ukraine in southern Europe while Pilát and Muraškinsky based their descriptions and opinions on collections from various localities in Siberia—which raises the question of identity with the type.

Pilát [1937 (ACE 3): 104] also suggested that *P. incendiarius* might turn out to be the same as the North American *Polyporus albiceps* Peck (**O**), a species which Overholts (*1953*: 250–251) compared with *P. admirabilis* Peck (**O**), suggesting that it might represent only a depauperate condition of the latter. Both have white or cream-coloured caps when fresh and also in other respects agree with *P. incendiarius*.

Polyporus corylinus, described from Italy, also has a white-coloured cap.

As far as the descriptions go, the black layer that develops on the stalk of *Polyporus* subgen. *Melanopus* (Pat.) Maubl. is not much in evidence, if present at all, in either *P. incendiarius* or *P. albiceps*. The size of the spores mentioned by Pilát for *P. incendiarius* exclude it from *Polyporus* subgen. *Polyporus*. The proper place for these species might be in *Polyporus* subgen. *Leucoporus* (Quél.) Maubl.

(**97**). *Polyporus michelii* is maintained on this Check list as a species distinct from *P. squamosus*; it seems to differ in its snow-white somewhat silky looking cap, which becomes squamulose without the scales becoming conspicuous by turning dark; its smaller size; its preference for willows (*Salix* spp.); and its southern distribution. I have seen it in the field only once, and then quite old, so that I am not in a position to form a well-founded personal opinion about its true status. The species was upheld by Maire [1933 (TrB 15): 36], so that it should not be rashly reduced to a mere form of *P. squamosus* by those who do not really know it.

(**98**). *Polyporus mori* is another name for the species appearing in much of the European and North American literature as *Favolus europaeus* or as *Favolus/Polyporellus alveolaris* (often erroneously spelt 'alveolarius'). The latter name has no relation to the *Polyporus alveolarius* of this Check list.

After collecting a long series of abundant material in Canada and the U.S.A. of what Overholts (*1953*: 156) called *Favolus alveolarius* I became hesitant about accepting unconditionally the identity of this material with the *Polyporus mori* of Europe. In North America a small-pored form (*Favolus striatus*; *Hexagona micropora*) is common and appears to be scarcely important taxonomically; I was unable to match it with any of the (rather few) European collections that I saw. Among the European specimens there occur fruitbodies of a bigger size and with much bigger pores than I found in any of the American collections. This will explain why I have left out all names based on American (and other non-European) collections from the synonymy of *Polyporus mori*; they have been relegated to the 'List of omitted names'.

If the North American fungus is considered to be a distinct species its name under *Polyporus* should be derived from *Favolus canadensis* as basionym. A careful study of the correct circumscription of *Polyporus mori* is recommended.

(**99**). *Polyporus squamosus* is a very variable species that, in Europe, is the cause of a wood decay of stems of various living deciduous trees. Under suitable conditions however infested dead and even worked wood is still capable of producing crops of fruitbodies over a period of several years. It is on this second type of substratum and more particularly when exposure to day-light is strongly reduced (as is the case in sheds, cellars, caves, and the like) that more or less abnormal fruitbodies develop; these have received a long series of names: *Polyporus dissectus* Letell. (habitat not mentioned), *P. squamosus* f. *erectus* Bres., *P. squamosus* var. *glaber* Graff in part, *P. helopus* Har. & Pat., *P. infundibuliformis* Rostk. = *P. rostkowii* Fr., *P. pallidus* S. Schulz., *Boletus polymorphus* Bull. in part, *B. rangiferinus* Bolt., and others. In these cases the stalk is usually relatively long and tends to be erect; it may even become branched or very long without developing caps. If one wishes to use special names for these abnormal fruitbodies, forms that produce caps (often tending to become infundibuliform) may be called 'forma *erecta* Bres.'.

(**100**). The black rind on the stalk in *Polyporus squamosus* may be limited to the very base of the stem in short-stemmed fruitbodies developing on living trees or else it may cover the entire, sometimes long stalk in the more or less 'aberrant' forms (**99**). I am inclined to regard it as a feature indicating the sclerotium-like nature of the black-rinded portions. Sometimes a distinct sclerotium-like body with black rind is formed first before the fruitbody proper develops from it. Usually the first stage is strongly suppressed, the two stages following each other without interruption, leaving the black rind at the base of the stalk as the only indication of the sclerotioid initial. Sometimes several fruitbodies (with black stipes) may develop from a common sclerotioid stroma or base (*P. helopus*). The specimen Bolton described and figures under the name *Boletus rangiferinus* had "a base or pedestal consisting of brown fungous tubercles, the size of hazel nuts, adhering together at their sides, and forming an irregular knotty surface on the upper side; on the under side, where it is adhered to the wood, it is smooth and a little hollow". Each of the abnormal fruitbodies seems to represent a 'tubercle' of this kind that did not stop at this stage but developed further into a fruitbody. This tendency to form black-rinded sclerotia is emphasized here in connection with a discussion on the status of *P. tuberaster* (**103**). It is interesting to find that Campbell & Munson (*1936*) suggested that the black 'lines' (plates) developing in wood form the limiting layer or rind of sclerotium-like bodies buried in attacked wood.

When reading descriptions of *P. squamosus* by North American authors I was struck by the habitat mentioned, for instance, by Shope [1931 (AMo 18): 356–357 *pl. 29 fs. 1, 2*] for Colorado: "... generally found growing from the base of cottonwood (*Populus* spp.) stumps." His figures show one somewhat excentrically stalked specimen and another, *in situ*, also practically centrally stalked and arising from the ground, though close to the base of a stem rather than from the stem itself. There are other records of this kind, some of which are mentioned by Graff (*1936*) for Montana, Wisconsin, and Minnesota on a limited number of tree species. In Europe the normal habitat is stems of living or dead trees at some distance from the ground up to very high in the tree; this seems also to be the case in the north-eastern states of the U.S.A. I wonder whether the 'terrestrial' habitat might point to a special form in the mid-western states or whether confusion with *P. tuckahoe* (= *P. tuberaster*) is perhaps involved; Güssow reported this from "popular woods" in Saskatchewan and Manitoba, the neighbouring parts of Canada.

(**101**). The name *Boletus platyporus* Pers. [1794 (NMB 1): 107] is treated here as a typonym of *B. juglandis* Schaeff. This latter name itself was not cited but the plates were: "Schaeff. fung. tab. 101, fig. 2". To my mind there is no possible doubt but that 'fig. 2' is an error for '102'. The citation of the cited figure makes no sense unless it is corrected. Compare *Polyporus platyporus* Secr. (**0**).

(**103**). *Polyporus tuberaster* is in Europe in the first place a Mediterranean species, known especially from Italy and already esteemed by the Romans for its edible fruitbodies. It has also been recorded from a few northern outposts (for instance The Netherlands). It differs from *P. squamosus* (**100**) in its subterranean, big sclerotia from which the (often rather small) fruitbodies arise directly. Its North American counterpart is *P. tuckahoe*, which appeared to be interfertile with *P. tuberaster* from Europe (Vanterpool & Macrae, *1951*).

In (**100**) attention is drawn to the sclerotioid nature of the base of the stalk in *P. squamosus* and to the formation of some kind of true sclerotia by this species when developing from and on dead wood under abnormal conditions. The note also mentions the formation of sclerotium-like bodies buried in the wood. Is it possible that some of these sclerotia may finally become strongly developed, persisting after the substratum from which they originated has disappeared? The sclerotia of *P. tuberaster* are subterranean and differ in their substance by consisting of a mycelial mass interspersed with earth particles. This mingled substance may be caused by its subterranean development; the 'externally' developed sclerotia and sclerotum-like bodies (stalk or stalk-base) of *P. squamosus* obviously lack such interspersions. Thus the question is raised whether after all the two fungi may not belong to the same species. If not, with our present knowledge it would seem difficult to answer the question as to what the specific differences between the fruitbodies (minus the sclerotia) of the two might be.

In connection with finds in Germany several authors [Bergstädt & al., 1969 (MMH 13): 83–88 *fs. 5–7*; Anonymus, *1970*] have commented on the resemblance between the fruitbody of *P. tuberaster* and *P. forquignonii* [= *P. floccipes*]. A closer investigation of the true relationship between these two species seems worth while. It would be interesting for instance to know the results of interfertility experiments between *P. tuberaster*, *P. squamosus*, and *P. floccipes*.

(102)   Modern French authors do not seem to have formed an opinion about *Polyporus tubarius*. The species was based by Quélet on a collection made by Boudier ("sur les racines de Bruyère"); later Quélet published a description and a plate of it of his own, based on a different collection sent to him by Barla ("comme récolté sur du bois mort des Cistes"). The figures published by the two French authors show a strong resemblance. As to the spores, these are stated by Quélet to be punctate, which is almost certainly an error (as in so many other cases); the spores were presumably smooth, with contents including guttules. In other respects Quélet described the spores as 'pruniform', 8–9 μ, while Boudier called them avoid, 10–12 × 5–6 μ [most likely somewhat too big, as is usual with his later spore-measurements: cf. van Brummelen 1969 (Pe 5): 233]. The original description and redescription by Quélet calls the tubes, "très tenues" and the pores, "polygones"; Boudier wrote "Pores . . . petits mais assez longs, plus ou moins denticulés."

Bresadola identified *P. tubarius* with the species he first called *P. tubarius* (1897) and later (among other names) *P. agariceus* and for which I now accept the name *P. alveolarius*. I am not at all certain that Bresadola's identification is correct since (judging from Quélet's and Boudier's descriptions and figures) *P. tubarius* seems to differ in the general shape of the fruitbody (for instance longer and relatively more slender stalk), much smaller pores and longer tubes, and perhaps also more ovoid spores.

Poria

(104).   This genus, as used on the Check list, is decidedly artificial: it comprizes all strictly resupinate species of uncertain affinity, that is, uncertain as far as my personal knowledge and judgment go. For instance *Poria hippophaës* Bres. has been insufficiently described but since it is not unlikely that the type has been preserved it was not transferred to the 'List of omitted names'. Amyloidity and cyanophily of the spore-wall in themselves do not appeal to me as characters that are *ipso facto* of generic importance in resupinate species. This is for instance why *Anomoporia* Pouz. is not yet admitted as a distinct genus; the few species put together under this name may prove to be not mutually closely related. As another example a monomitic context does not make a resupinate whit(ish) polypore automatically a species of the pileate genus *Tyromyces*. It looks as though most porias have 'reduced' rather than 'primitive' fruitbodies.

The name *Poria* itself is retained here arbitrarily for a residual genus of ever-dwindling but still considerable size, irrespective of the correct taxonomic position of the type species. *Boletus medulla-panis* sensu Pers. is now often accepted as lectotype, usually in a conception that was fixed by Bresadola. This species will be found listed under *Perenniporia*.

For a more elaborate discussion on these and other points, see Donk (*1967*: 50–67).

(**105**). *Poria consobrina*. In the protologue Bourdot & Galzin (*1925*: 230) remarked: Même pourriture que *P. mucida* [sensu Bres.] avec lequel on le confond à simple vue; mais sa trame tendre ceracée le sépare bien de ce dernier qui est coriace." This, and other details in the original description, indicate that *P. consobrina* is presumably monomitic in contrast to *Schizopora paradoxa*, which is clearly dimitic, although this has generally been overlooked. Therefore, I distrust both von Höhnel's identification [*in litt. apud* Jaap 1922 (VBr 64): 44, "Bres. in herb."] of it with *Irpex deformis* [= *Schizopora paradoxa*] and that of Lowe [1962 (PMi 47): 186] with *Poria versipora*. On a previous occasion Lowe [1959 (Ll 21): 108] had thought of *Poria myceliosa*. Later he (*1966*: 61, in obs.) wrote that it may be the same species as *Poria mollusca* [sensu Lowe = *Poria mucida* sensu orig.] but at the same time (op. cit., p. 65) he favoured its inclusion in *Poria hypolateritia* Berk. (see below), which he had previously included in *Poria versipora* as a variety.

Lowe [1959 (Ll 21): 101, 108] made of *Poria eyrei* another synonym of *P. hypolateritia*, with the annotation "cystidia reported by Bresadola not found on type". However, more recently he published a figure of the hymenium from the type, "showing variable cystidia" (Lowe, *1966*: *f. 41*).

I am reluctant to accept the identification of these two species with *Poria hypolateritia* (**0**), a name denoting that the fruitbody is brick-coloured on the underside. The type came from India and Bresadola [1912 (Am 10): 504], Petch [1916 (APe 6): 136; 1923: 154, 199, 204 *f. 57, pl. 2 fs. 10, 11*; 1928 (TBS 13): 242],[16] Boedijn [*apud* Schoorel & Boed., 1939 (ATh 13): 12 *fs. 1–3*], and I myself apply the name to a quite different and apparently tropical species that is considered to be the cause of a red-root disease of tea. Interpreted in this sense *P. hypolateritia* appears to be near, or to belong to, *P. vincta* (Berk.) Cooke, an extra-European species. As long as I am not completely convinced that the name *P. hypolateritia* in the sense of Petch is a misapplication I am not prepared to follow Lowe and report the species for Europe ("England, France"); the record for 'England' is apparently based on *P. eyrei*; and the one for 'France' on *P. consobrina*.

All three species are in need of renewed examination.

(**106**). Karsten described *Physisporus flavicans* twice, first in Latin (1896), then in Swedish (1898). This second description is almost a literal translation of the first; a brief remark was added. In the first publication

---

[16] Pegler & Gibson [1972 (CDp): no. 322] referred *Poria hypolateritia* sensu Petch to *P. hypobrunnea* Petch. What they took as the type of *P. hypolateritia* was identified with *Schizopora paradoxa*.

the material mentioned was indicated as "m. Aug.", in the second this became "Oct. 1866 och Aug. 1895".

The two collections thus indicated appear to represent widely different species: the "Aug. 1895" collection belongs to *Poria [Chaetoporus] luteo-alba* (*fide* Lowe, *1956*: 111–112), the "Oct. 1866" collection to a species that until now has been known as *Poria flavicans*, a tradition founded by Romell and later subscribed to by Egeland, Romell, and Baxter. Because in the first description (1896) only a single collection was mentioned explicitly Lowe felt bound to take this as the type. This resulted in *Physisporus flavicans* becoming a synonym of *Poria luteo-alba*; to him *Poria flavicans* sensu Romell became nameless.

The 'first' description, taken as a whole, however, points to this latter species rather than to *Poria luteo-alba*. The statement "Cystidia nulla" is especially significant since *Poria luteo-alba* has easily observable cystidia. For myself I consider the 1866 specimen as the type rather than the one actually selected by Lowe. It would appear that when first publishing the name, Karsten merely committed an error but in his next description he corrected it; the existence of the 'real' type was clearly implied by the first description.

(**107**). *Poria expansa* seems to be much more common in southern Europe (or at least in France and Spain) than in Germany and other more northern countries. It has been known in France under various names, often misapplications. Bourdot & Galzin (*1928*: 686) considered it "le type des champignons destructeurs des charpentes".

Its systematic position has not yet been settled. As far as I know it is a true 'poria', although it has been thought to form caps too; this seems to be due to confusion with the form of *Heterobasidion annosum* that was described as *Boletus cryptarum* Bull., which often produces superimposed and seriate caps. I cannot agree with its transfer to *Fomes* sensu stricto by Domański & Orlicz (*1967a, 1967b*). The best solution I can think of is to place it in a genus of its own. Before considering a new generic name it must first be decided whether *Spongioides* Lázaro (**O**) should be taken up or not. Like that of Quélet and some other French descriptions that of Lázaro mentions caps (in addition to 'resupinate' fruitbodies), apparently as a result of confusion with '*Boletus cryptarum*'.[17]

---

[17] The following is a translation from the Spanish (with apologies):—
Gen. *Spongioides*. Nov. gen. / Sporophore formed of thick folds irregularly superimposed, strongly projecting, undulate and confluent in some spots which appear attached to the trunks. The upper as well as the lower surface are convex, generally the former more so than the latter; but the irregularity of its forms and the extensive development of its horizontal dimensions, often bigger than the vertical or height, remove its forms from the biconical, ungulate and flat-topped [types of fruitbody]. A single tube-layer, with the tubes vertical and rather long (on the average one or two centimeters). Pores small and equal.

Another problem to be resolved is the correct basionym to be used on transference to another genus. Although *Boletus expansus* Desm. (1823) is the earliest name available, so far the possibility that the correct basionym might be *Polyporus unitus* Pers. (1825) cannot be definitely excluded. Fries treated *Polyporus unitus* as an autonomous species in a volume of the starting-point book (1828 E. 1: 116) in contradistinction to *P. expansus*, which he mentioned under *Polyporus* [*Fomes*] *fomentarius* (Fries, op. cit., p. 109), suspecting that it might be a resupinate condition. It is entirely possible however that Fries misapplied the name *P. unitus* when drawing up his description from Norwegian material collected by Sommerfelt.

Until now I have not been able to locate the specimen (type) of *P. unitus* studied by Bresadola (from which he apparently took a fragment now in S) and which is the basis of his conclusion that it represented *Poria megalopora* (= *P. expansa*) but I have not given up hope of eventually unearthing it. Compare also Donk (*1967*: 116).

(**108**). *Polyporus spongiosus* Pers. (**O**) ≡ *Boletus resupinatus* Bolt. was identified with *Poria expansa* and the name reintroduced as *Poria spongiosa* by Quélet [1892 (Crf 20²): 468 *pl. 2 f. 19*], whose description and figure are sufficiently detailed for recognition of the species. The identity of Bolton's taxon however is doubtful. *Polyporus spongiosus* was a mere substitute name for *Boletus resupinatus* Bolt. When renaming Bolton's fungus Persoon had not seen any specimens attributable to it and his account of it was certainly based exclusively on that of Bolton.[18]

Bolton collected his fungus near Halifax in northern England, apparently in the open air "on dry decayed hasle boughs" (branches of *Corylus avellanea*) which vary in diameter (when measured from the plate) from less than 1 cm up to 4 cm thick. Locality and substratum are together a strong indication that it was not *Poria expansa*; this is supported by several other features, such as the colour ("the colour, like all the other

---

Spongioides cryptarum. (Bolt. [!]) Láz. / . . . / Description.—Sporophore distinct or resupinate [?], 10–20 centimeters in diameter, superimposed or assembled, of variable thickness, although never very large, wavy or folded, with the upper surface silky-tomentose, uneven, strongly sinuous, pale brown (pardo) coloured or lurid, becoming paler as time goes on. Flesh spongy or desiccated, rather pale brown. Tubes relatively long, thin, of the same colour as the flesh. Pores small, dirty yellowish. / Habitat.—Frequent on coniferous logs decaying in the dark, especially on those of pines used in mine galeries. / Distribution.—This species can be found in various localities of the northern, western, central and eastern regions of the [Iberian] Peninsula.

Several passages in the original text are not quite clear.

[18]   There can be no doubt about this conclusion, this in contradistinction to Massee's opinion (1892 B.F. 1: 226). Massee wrote: "Saccardo, in Sacc., Syll. Fung. vi. n. 6525, has given the present species [*Boletus resupinatus* Bolt.] as a synonym under *Fomes spongiosus*, Pers.; there is no evidence, however, that the two are identical . . .."

parts, is a rusty brown"). The rather soft context ("coriaceo-spongiosus") and the irregular shape of the fruitbodies with relatively long (one-layered) pores remind me of *Hapalopilus rutilans* but 'rusty brown' hardly applies to this species. Fries (1838: 455) reduced *Polyporus spongiosus* to a subspecies ("*") of *P. nidulans*, which is a synonym of *Hapalopilus rutilans*. In any case I am not prepared to accept the identity of Bolton's fungus with *Poria expansa* and since I do not recognize it with certainty I must treat *Polyporus spongiosus* Pers.: Fr. 1828 ($\equiv$ *Boletus resupinatus* Bolt.) as a nomen dubium.

According to Massee (1892 B.F. 1: 226) "there are specimens in the Kew Herbarium agreeing exactly with Bolton's species, and determined as such by the Rev. M. J. Berkeley." These specimens appear to have furnished the source for some of the features that Massee added to Bolton's description, as for instance, "spores colourless, elliptical, 4 × 1.5 μ" and "the absence of large, coloured cystidia projecting from the hymenial surface."

(**109**). A poria that produces a *Ptychogaster* state was called "*Oligoporus*" *friesii* R. & O. Falck. The authors believed they had found the imperfect state that Fries (1849: 564) had brought into relation with *Climacocystis borealis*; Fries wrote, ". . . *Ptychogaster*; est nempe monstrosa progenies *Polypori borealis*, his diebus a me plene observata." Because it produced a 'twin-fruitbody' consisting of both the perfect and imperfect states the species was placed in *Oligoporus* but it was concluded that the perfect state could not be *Polyporus borealis*.

This association by Fries of a *Ptychogaster* state with *Climacodon borealis* must still be confirmed. I have carefully searched through modern literature for additional records but have found no reliable data. I am reluctant to assume that Fries misnamed the perfect state.

Oligoporus friesii appears to be different from *Ptychogaster fuliginoides* = *Tyromyces ptychogaster* because of the structure of its imperfect state and because the chlamydospores are of a somewhat different shape. The fruitbody of this state is radially fibrous (the strands are said to be 'koremienartig') but the interior is not lacunose, as in the *P. fuliginoides* complex. On page 45 (R. & O. Falck, *1937*) the chlamydospores are given as measuring 5–6 × 4.5–5 μ but on page 59 as 6–8 × 5–6.5 μ. The species differs from *Oligoporus rennyi* for instance in the colour of the chlamydospores: "Macroskopisch erscheinen [die Chlamydosporen] im Zusammenschluss junger Anlagen erst hell, später dunkel rahmfarben, zuletzt tabakbraun gefärbt."

New collections and information on several characters are needed before the status and systematic position of *O. friesii* can be worked out.

I have interpreted this species as forming resupinate fruitbodies. This will explain its inclusion in the artificial genus *Poria*. The original description contains *inter alia*: "Die von uns gefundenen Früchte . . . besitzen eine deutliche dem Substrat ansitzende Fruchtplatte. Konsolen . . . oder Hutbildungen fanden wir nicht.''

(110). Few species of *Poria* have been so diversely interpreted as *Poria incarnata* Pers. This may be the main reason why by now the name has been almost completely dropped. It is not improbable that in the near future, once the material in Persoon's herbarium has been exhaustively analyzed, the identity of his species will be securely re-established. Even so, however, if its identity with *Poria placenta* is confirmed, it would seem not unreasonable to treat the name as a nomen ambiguum.

On comparing Fries's earliest and personal description (1818 O. 2: 262) and the abbreviated description in his "Systema" (1821: 379) with Persoon's first description a number of discrepancies appear, but there is no indication that Fries wished to exclude the original fungus; in any case he acknowledged Persoon's authorship without any restriction. As is so often true in similar instances it is impossible, I believe, to form a definite opinion about the species that Fries used in 1821 for emendating the Persoonian description. At one time Fries (1832[Ind.]: 149) referred *Poria incarnata* Pers. to *Polyporus isabellinus* [(Schw.) Fr.], at the same time maintaining a *Polyporus* "*incarnatus* Fr. [Syst. myc.] 1. 379. El. 1. 119". This may be taken as the birth of a new but homonymous name, *Polyporus incarnatus* Fr., non (Pers.) ex Fr. Nevertheless in subsequent work Fries restored Persoon as the author (1849: 322; and cf. 1884 I. 2: 87) or excluded only Persoon's conception of 1825, apparently based on parasitized *Antrodia serialis* (cf. Donk, *1971b*: 26).

It would seem as though Fries had no fixed opinion about his own conception of 1818 & 1821. In his first description the fruitbody is effused and immarginate but later he wrote "ambitu passim solvitur, sed pileum non formare potest" (Fries, 1838: 484) and "Margo superior passim reflexus" (Fries, 1874: 573). The published figure, referred to *Polyporus incarnatus* in posthumous work (Fries, 1884 I. 2: 87 *pl. 189 f. 1*), bears the name "*Polyporus helvolus* n. sp." (written in ink) on the original drawing; the name "*Polyporus incarnatus*" (in pencil) was added later but neither name was written by Fries. Romell (*1911*: 13) thought that the figure probably represented *Polyporus rixosus* [= *Chaetoporus collabens*], "though the margin is shown too radiate."

In 1821 Fries excluded the *Boletus incarnatus* as described by von Albertini & von Schweinitz (1805: 258; "Vix A.S. p. 258 hujus loci") but in 1828 (E. 1: 199) he changed his mind and stated that "*Boletus incarnatus*, A.S. p. 258" belonged there after all: "revera hujus loci, forma optima explicata." Accompanying the figure discussed above is the remark "Nominis auctor A.S. (vel potius forte Pers.) nec Fr." This may explain why some authors ascribe the name *Polyporus* (or *Poria*) *incarnatus* to 'A. & S.', but this is no more than an indirect reference to the original name published by Persoon, unless the latter is simultaneously and expressly excluded. Persoon renamed the 'A. & S.' conception *Polyporus niskiensis*. There is no acceptable suggestion available as to the identity of this species.

In connection with *Polyporus incarnatus* Persoon committed an error of his own. In the "Mycologica europaea" (1825: 98), under this name he all at once described a different fungus, as mentioned above. However, his description in the "Synopsis" (1801: 546) clearly reflects the original fungus and not what he was later to describe under the name *Polyporus incarnatus*.

From Karsten's herbarium material and descriptions it is evident that during a considerable period he identified *Poria incarnata* with *Poria* [*Merulius*] *taxicola* (Pers.) Bres. (**O**); compare Donk [1962 (Pe 2): 229], who concluded that it was this interpretation that had served as the basis of the generic names *Caloporus* P. Karst. 1881 = *Caloporia* P. Karst. 1893. In Donk's opinion *Merulius taxicola* is a species of *Merulius* Fr. rather than a resupinate member of the Polyporaceae. It is not unlikely that in later years Karsten changed this interpretation [1896 (H 35): 44 & 1898: 12 / 1903 (BFi 62): 76 (*Caloporia*)] but I find it difficult to make a pertinent suggestion that can be relied on.

Bresadola's interpretation as published by Bourdot & Galzin (*1925*: 253; *1928*: 689) agrees fairly closely with Persoon's original description. There were some followers, but still more recently even these have preferred the use of the name *Poria placenta* Fr. By calling this conception *Ceriporia incarnata* "(Bres. non Fr. nec al.)" Parmasto (1963: 112 *f. 2*) published a new (but not validly published) name, *Ceriporia incarnata* Parm.

(**111**). The publication of *Polyporus sericeo-mollis* Romell and its subsequent treatment by the author was a matter in which no fewer than at least four species were involved. A brief recapitulation may be useful.

(i) The name was based on a specimen from Nattavara (Sweden). It was pointed out by Lowe (*1966*: 84–85) that "the type has very abundant cystidia. By some mischance Romell did not mention them in his original description, nor did he correct this omission in later writings. His identifications and sketches with his specimens amply show that he knew cystidia were present in this specimen." Lowe did not record chlamydospores for this restricted conception.

(ii) The second element Romell mentioned in the observations to *Polyporus sericeo-mollis*; it was associated with a chlamydosporous state of "sulphureous or pallid colour." For some time this was taken to be the true *P. sericeo-mollis*, but Donk (*1971c*: 210) identified it with *Oligoporus farinosus* Bref. and *Ptychogaster citrinus* Boud., both of which in turn he referred to *Oligipororus rennyi*.

(iii)   A third element, found at Femsjö (Sweden), was taken to be a variety "or perhaps distinct species." Like the preceding it was also associated with a chlamydosporous state, but of a different colour, "of about isabelline color, still more like small specimens of *Ptychogaster albus* [= *P. fuliginoides*]." For the present this element may be tentatively referred to *Tyromyces ptychogaster*.

(iv)   Finally, in later work Romell included specimens of a fourth species that he later called *Polyporus subsericeo-mollis* Romell. This was referred to *Tyromyces floriformis* (Quél. apud Bres.) Bond. & S. by Kotlaba & Pouzar (*1964*: 217). For a note on the somewhat irregular publication of Romell's name for this element, see Donk (*1972a*: 176).

Pycnoporus

(**112**).   By almost common consent this generic name is now restricted to a small group of species with fruitbodies of an eye-catching red colour (**114**). Two members (*P. cinnabarinus* and *P. sanguineus*) have been confused and often considered to belong to a single variable species. It is now generally agreed that the two taxa are distinct; this conclusion is supported by the fact that in Europe *P. sanguineus* (**0**) has not yet been found. In a monograph on the genus by Nobles & Frew (*1962*) a third species is recognized, viz. *P. coccineus* (**0**), from countries bordering on the Pacific or Indian Oceans. Certain synonyms now placed under *P. cinnabarinus* and *P. sanguineus* may belong to this third species.

Schizopora

(**113**).   Bourdot & Galzin (*1925*: 237; *1928*: 681) described an "état conidifère" of *Poria mucida* [sensu Bres.] = *S. paradoxa*. It was associated with what was "probablement *Irpex obliquus*," which they reduced to a form of their *P. mucida*. On the basis of this apparently chlamydo- (or, rather, aleurio-)sporous state that was found rarely "en bordure de la partie porée peu développée" R. & O. Falck (*1937*: 58) transferred "*Poria mucida* Bourdot et Galzin," with a question mark, to their new genus *Multiporus* (name not validly published), remarking that it was perhaps identical with *Multiporus chlamydoformans* R. & O. Falck. This seems improbable in view of the size of the conidia described by Bourdot & Galzin (8–12 × 5–7 μ), while those of *Multiporus chlamydoformans* were stated to be 10–15 × 9–12 μ.

I have not come across any reliable confirmation of the existence in nature of a chlamydosporous state of *Schizopora paradoxa*, which is an extremely common species throughout Europe. The question fully deserves special attention.

Trametes

(**114**). The unfinished classification of the polypores makes the acceptance of a number of artificial genera unavoidable; *Trametes* of this publication is one of them, *Fomitopsis* (**33**), *Tyromyces* (**121**) and *Poria* (**104**) are other examples. Even the definition of the generic character is artificial and arbitrary, especially where cultural characters are emphasized. As a working thesis I have assumed that the only representative in Europe taxonomically typical of *Trametes* is *T. suaveolens*. All other species are included here not only because I do not know where else to place them without more precise additional knowledge about various characters but also because the status of a number of proposed generic taxa (a few based on extra-European species) has not as yet been worked out. In any case all the species concentrated here under *Trametes* have a pale coloured, di- or trimitic, not-fleshy context, and lack sceletocystidia or conspicuous hymenial cystidia (but compare *Tyromyces pseudo-hoehnelii*); all of them possess typical skeletals; these features are shared by *Fomitopsis* (**33**). In both genera the generative hyphae possess clamps, in '*Trametes*' apparently with a sole exception, viz. *Polyporus onychoides* (**116**). The genus *Haploporus* is maintained as distinct.

Pycnoporus (**112**) is excluded mainly because of a contemporary fashion. The colour of the fruitbody is showy and has therefore been taken as important, but it should be remembered that white strains occur not only in cultures but also in nature, at least in the tropical and subtropical, extra-European *Pycnoporus sanguineus* (**O**). Of this species Murrill [1948 (Ll 10): 268] reported that "specimens in bright sunlight may bleach to pure white." If the red colour were less emphasized separation from the more typical species of '*Trametes*' would be more difficult.

Pseudotrametes, which is not excluded, is accepted by several European authors; so far it consists of a single variable species, *Trametes gibbosa*. It is conceivable that this species will be found to be related to certain tropical forms, such as *Daedalea* [*Trametes*] *palisotii* (**O**). In case these species are to be regarded as congeneric and as representing a distinct genus then an earlier generic name is available: *Artolenzites* Falck.

(**115**). The name *Coriolus genistae* appeared in literature for the first time as a synonym when *Poria vulgaris* (Fr.) Cooke "Var. ? B. *pileata* (*Polyporus genistae*, B. et G., in herb.)" was described in a paper by Bourdot & L. Maire [1920 (BmF 36): 84]. Later (Bourdot & Galzin, *1925*: 145) the name was published as that of a subspecies: *Coriolus hoehnelii* [subsp.] "*C. genistae*". Eventually Bourdot & Galzin (*1928*: 569) published *Coriolus genistae* as a species distinct from both *Poria vulgaris* (Fr.) Cooke [sensu Bres. = *P. romellii*] and *Coriolus* ["*Trametes*"] *hoehnelii*. They reported it from "genêts", the substratum from which the specific epithet was derived; collections from the various other substrata mentioned were considered to be the same species ("identique"). This shows that the lectotype must be a collection from *Genista*. From notes kindly supplied by Dr. J. L. Lowe a collection from Bourdot's herbarium (PC) marked "sur genêt, base de pins. Env. de Epinal, Vosges, legit Galzin Dec. 1904" (Bourdot no. 3975) accords well with the prerequisites; it is herewith selected as lectotype. Some descriptive notes on this specimen follow:—

Sporophore effuso-reflexed, up to 1 cm forward, 0.2 cm thick. Cap convex, the upper surface cream, appearing appressed-fibrillose, rather azonate and rather smooth, the margin resinous-translucent. Pore surface dark cream, the pores rounded, averaging 4 per mm, the dissepiments thin with entire edges. Context pale cream, homogeneous; texture leathery, up to 0.1–0.2 mm thick. Tubes dark to pale cream, coriaceous in texture, up to 1 mm long.

Hyphae of context of 2 types; the generative thin-walled, rarely simple-septate, usually nodose-septate (cross-walls with clamps), 2–3 μ in diam.; the skeletal hyphae nonseptate, thick-walled to solid, 2–4 μ in diam. Cystidia none; hyphal pegs present. Spores oblong, dorsally flattened to slightly depressed, 3–4 $\times$ 1.5–2 μ.

One other collection from Bourdot's herbarium studied by Dr. Lowe agrees with the lectotype (Bourdot 6032; "sur genêt carbonisé. Vergnas, Aveyron, leg. Galzin 3914–17, 20 I 1909"); it also has a small amount of clamped hyphae as well as thick-walled to solid hyphae and spores of the same type. A third specimen ["sur genêt et bruyère. Chateau Charles (Allier), leg. H.B.] has thin-walled hyphae, 2.5–3.5 μ in diam., but these lack clamps, while the spores (3–4 $\times$ 1.5–2 μ) are more ellipsoid-ovoid. This last specimen may appear to differ specifically and it should be compared with *Polyporus onychoides* (**116**).

Pilát [1939 (ACE 3): 271] reduced *Coriolus genistae* to the synonymy of *Leptoporus* ["*Trametes*"] *semisupinus*, but this should be reconsidered. As Dr. Lowe remarked in his notes, the spores of *Coriolus genistae* are perhaps too slender for that species. It may also be mentioned that Bourdot & Galzin (*1928*: 569) compared their material with "*Polyporus pallescens* Karst. Rom., Hym. of Lappl., p. 19" (**119**), a taxon which is now supposed to belong to "*Trametes*" *semisupinus*. They found it distinguishable by the hyphal structure, which they described for their *C. genistae* (apparently in error) thus: "les hyphes sont toutes similaires, flexueuses."

In view of the above I hesitate to reduce *Coriolus genistae* to a mere synonym of "*Trametes*" *semisupinus*, although it seems quite near it and may be a 'variety'. I follow Bourdot & Galzin in not reducing it to *Poria vulgaris* (Fr.) Cooke sensu Bres. (= *P. romellii*]: "nous n'avons pas encore vu de forme de passage entre les deux plantes."

(**116**). The status of *Polyporus onychoides* is still undecided. The description suggests *Polyporus* ["*Trametes*"] *semisupinus* but the lack of clamps may appear to be an important difference. This conclusion is based on the following notes taken from the type collection. Dr. J. L. Lowe's notes run:—

Context hyphae closely and parallel-arranged, rarely branched, thick-walled, rarely to frequently septate, lacking clamps, 3–4 μ in diam.; also a variable amount of very thin-walled, septate hyphae 2–3(–3.5) μ in diam. with the cross-walls always thin and inconspicuous. — In a letter (January 6, 1970) he added, *inter alia*: When I studied the type, I evidently thought *P. onychoides* could be a synonym of *P. semisupinus*; but the presence of simple-septate hyphae in the context eliminates that possibility.

The lack of clamps is also underlined in a note by Mr. L. Ryvarden (*in litt.*, December 3, 1970):—

I have examined the type microscopically very thoroughly and can state the following: The hyphae are dimitic. The skeletal hyphae are thick-walled to almost solid with a diameter of 3–4.5 μ and without septa as far as I could see, and I concentrated on that point: no septa at all. The generative hyphae are thin-walled and with a few simple cross-walls. I was unable to detect a single clamp during an examination of about one hour. . . . The type is small and in bad condition; it looks very much like a small specimen of *T. semisupinus*. As far as I can judge, the only really good separating character is the absence of clamps on the generative hyphae.

According to a suggestion by Dr. Lowe *Polyporus onychoides* may be the same species as the clampless element among Bourdot's material of *Coriolus genistae* (**115**).

(**117**). The specific epithet of *Tyromyces* [*"Trametes"*] *pseudohoehnelii* suggests a degree of similarity with *Trametes hoehnelii*; this is due, for instance, to the surface of the cap rather than the size of the fruitbody, which is much smaller than in the other species. In this and various other respects the description recalls *Tyromyces* [*"Trametes"*] *semisupinus*, but the presence of conspicuous cystidia separates the two and directs attention to *Tyromyces semisupiniformis* Murrill [1912 (BTC 39): 148 (Mexico)] as redescribed by Lowe [1961 (PMi 46): 200 *f. 1*; as *Polyporus*]. Comparison of the two taxa is suggested.

The hyphal structure of the *Tyromyces pseudohoehnelii* was not described but it may be dimitic by skeletals ("hyphae 2,5–4 μ in diam., cum fibulis raris"). According to Lowe's description the same seems to be true for *T. semisupiniformis*. He mentioned the presence of clamps in "a very small amount of thin-walled . . . hyphae" in the latter species; this does not suggest relationship with *Rigidoporus*, as he had inferred from the cystidia.

The original description does not clearly indicate the holotype, and the name *Tyromyces pseudohoehnelii* would seem not to be validly published, unless it is based on a single collection.

(**118**). Under the name *Polyporus semisupinus* I have brought together a few taxa that may not at all belong to a single species, but if the conception is heterogeneous a thorough analytical study will be needed to unravel the complex. For instance Lundell identified *Polyporus pallescens* Romell (**119**) [not ∼ Fr. per Fr.; cf. (**16**)] with *P. semisupinus* and Pilát does the same with *Coriolus genistae*, but Bourdot & Galzin (*1928*: 569) evidently hesitated to consider *P. pallescens* Romell and *Coriolus genistae* as conspecific (**115**). On the whole the small spores are not very carefully described so that it is not always clear whether or not there exists any real difference between the spores of the taxa here reduced to *P. semisupinus*.

Lowe (*1966*: 127) wrote that his conceptions of *Polyporus semisupinus* and the *Poria byssina* of Eriksson [= *Poria romellii*] "do not seem to differ". I think that there are just enough differences (Donk, *1967*: 84–85), including the shape of the spores, to keep the two apart until further contra-indication is presented.

(**119**).　Romell (*1911*: 19) introduced a *Polyporus pallescens* "Karst." thus: "Karsten, . . . has found this plant at Mustiala in Finland . . .. His other collections so named belong to *Pol. velutinus* [= *Coriolus pubescens*], as also Fuckel's plant in Fungi rhen. 1379. − It is not *Pol. pallescens* of Fries . . .." What Romell did (according to the "Code") was to publish a new, but homonymous, species, based on the cited specimen of Karsten: "at Mustiala in Finland growing on the hymenium of *Polyp. fomentarius*". This taxon should be called *Polyporus pallescens* Romell (rather than 'Karst.'), not *P. pallescens* Fr. per Fr. 1821. Romell gave no reference to a publication by Karsten and it is misleading to cite Karsten's treatment of *Bjerkandera pallescens* (Fr.) P. Karst. 1882 (BFi 37): 36, as Lundell [1953 (LNF 43–44): 4 No. 2104] did when he cited Romell's species as a synonym of *Polyporus* ["*Trametes*"] *semisupinus*. In the publication referred to by Lundell Karsten did not even list the specimen just mentioned.

Romell's description of his conception of *Polyporus pallescens* "Karst." mentions the fruitbody as ". . . nunc totus resupinatus, usque ad 8 cm. vel ultra effusus . . ., nunc breviter (1–10 mm.) reflexus . . .." Lundell (l.c.) refers Romell's species to *Polyporus semisupinus*, which also suggests that the type is pileate, a conclusion that agrees not only with Karsten's determination of it as *P. pallescens* Fr. (another pileate species) but also with Karsten's inclusion of the latter species in *Bjerkandera* (which he instituted for pileate species). This conclusion raises the question as to whether the resupinate material included by Romell in his conception might, at least in part, not be *Poria vulgaris* (Fr.) Cooke sensu Bres. = *P. romellii*. Romell no longer mentioned his species in later publications; he did not take it into consideration in 1926 when discussing *Poria byssina* Romell ("Pers.") and *P. vulgaris*. According to Lowe (*1956*: 119) "most of Romell's type material is now at the Riksmuseet in Stockholm; these are the resupinate condition of *P*[*olyporus*] *semisupinus*."

Truncospora

(**120**).　The occurrence of *Truncospora ochroleuca* in Europe is still somewhat doubtful. The species is apparently not uncommon in Australia and New Zealand and the eastern tropics. It has also been reported from some of the warmer countries of America. Lloyd (*1915b*: 312) records it as "a frequent plant" from Africa as well; if it also occurs in north-western Africa the records from Portugal may not be altogether un-expected. *Trametes ochroleuca* var. *lusitanica* Torrend (*1910*: 36) was described from this last-mentioned country, but no microscopical details were given. In a later report Torrend [1913 (Bro 11): 68] stated that he had not seen spores but he recorded "conidiis sporiformibus numero-sissimis 14–18 × 8–9 μ"; in size these 'conidia' may well have been the basidiospores of *Truncospora ochroleuca*.

Other material from Portugal was originally reported as *"Fomes" scutellatus* (**O**); it was later called *Trametes [Truncospora] ohiensis* (**O**) and finally listed by Pinto-Lopes (*1953*: 206–208) under *Ungulina [Truncospora] ochroleuca*. It should be kept in mind, however, that Pinto-Lopes reduced the North American species *Trametes ohiensis* as such to the synonymy of *Ungulina ochroleuca* (apparently an incorrect decision) and that he did not remedy the lack of microscopical details already conspicuous in Torrend's records of *Fomes scutellatus* and *Trametes ohiensis*. Other records of *"Ungulina" ochroleuca* were published by Farinha (*1957*: 20) and Almeida & al. (*1964*: 161), again with no description, except that Farinha mentioned that the species is distinguished by its thick-walled and truncate spores. This applies more or less to all species of *Truncospora* and does not really single out *T. ochroleuca*.

All and all, not only is the occurrence of *Truncospora ochroleuca* in Europe still in want of confirmation, but a renewed study of the distribution of the species on a world-wide scale is also needed. Moreover, a detailed study of its difference from *T. ohiensis* is desirable, particularly in connection with tropical American collections. Compare, for instance, *Fomes ohiensis* var. *macrosporus* A. Teix. [1948 (Bra 8): 78 *pl. 5*], from Brazil, and *Polyporus turbinatus* Pat. [1891 (Rm 13): 137], from Venezuela; in both taxa the recorded dimensions of the spores are comparable to those of *T. ochroleuca*. It is conceivable that other taxa now referred to *T. ochroleuca* and based on American material ought not to be included in this species (*Polyporus havannensis* B. & C.).

Tyromyces

(**121**).   This genus is now accepted by most modern European mycologists, though not always with precisely the same contents. A rather broad version of it is also called *Leptoporus*, a name particularly entrenched in France because it was used by Bourdot & Galzin. An important core is formed by species with pileate, white (immutable or discolouring) fruitbodies with soft and often somewhat watery, cheesy or chalky context. The interpretation of these species and their synonymies forms a thorny bunch of riddles.

It looks as though the supporters of the genus consider it sufficiently homogeneous. With very few exceptions the better known species have a monomitic context, and this feature has been made an important item in the generic character. Some authors include a number of strictly resupinate species with a context of this kind. So much weight has been attached to the term monomitic that there is a tendency to overemphasize it. An example is the incorporation of *Bjerkandera* (**12**) by Pouzar (*1966c*: 370). *Hapalopilus* barely escaped also being included in it.

During the last years my faith in the homogeneity of the genus has been shattered. This is the reason it is treated on this Check list as an essentially artificial genus and why I restricted it to pileate species by referring the strictly 'resupinate' species to *Poria*, another artificial genus. Some species have a dimitic context with skeletals or have an even more complicated hyphal make-up, for instance *Tyromyces spraguei* (supposed to be the same as *Polyporus castaneae* Bourd. & G.); other similar species are now excluded, like *Incrustoporia semipileata* and the *"Trametes" semisupina* group. The context of the cap of one species is monomitic, but the context of the dissepiments is dimitic with skeletals: *Tyromyces chioneus* (*T. albellus*). As far as is known clamps are typical of all the species except one, *T. mollis*.

Most species produce brown rot (and are 'oxidase-negative'); others produce white rot ('oxidase-positive'). The resurrection of the genus *Spongiporus* Murrill by Aoshima (*apud* Aoshima and Kobayashi, *1966*; Japanese text without summary) seems to be due to the stressing of the oxidase-negative character for it. The type species of the name *Spongiporus* is *Tyromyces leucospongia* (Cooke & Harkn. apud Cooke) Bond. & S., a species first described from North America and not occurring in Europe; according to Overholts (*1953*: 290) and Bakshi & al. [1958 (CJB 36): 363] it is known to produce a "brown carbonizing cubical" rot. Aoshima transferred to this originally monotypic genus *Spongiporus* what he understood to be *T. lacteus* and also (Aoshima, *1967*: 1) *Antrodia sinuosa* (**10**).

As was earlier pointed out (Donk, *1960*: 273) the correct name for the genus in its now current circumscription is *Postia* Fr., but restoration of this name would not only require a considerable number of recombinations but it would also deprive phanerogamists of the name *Postia* Boiss. & Blanch. (Compositae), a later homonym. Donk suggested that there is much in favour of rejecting the fungus-name *Postia* Fr. in order to save *Tyromyces* P. Karst., at least until the taxonomy of that genus has been improved. Pouzar (*1966c*: 370) was of the same opinion. A formal proposal for conservation meets with various difficulties, one of which is that the exact identity of the type species of the name *Postia* Fr. (*Polyporus lacteus*) is far from settled (cf. Donk, *1972b*: 294). In fact it is tempting to conclude that the identity of *P. lacteus* is still in doubt so that both this specific name and the name *Postia* would have to be treated as nomina dubia. In case Aoshima hit the nail on the head in regard to his use of the name *Polyporus lacteus* the correct name for what he called *Spongiporus* (1905) would become *Postia* (1874).[19]

Tyromyces chioneus (*T. albellus*) might appear worthy of generic segregation not only on account of its hyphal make-up (briefly sketched above) but also because of several cultural characters that it shares with *Incrustoporia* (5). The trouble is that such a genus would quite likely claim the generic name *Tyromyces* and this in turn would necessitate deciding on the correct name for the rest, a path full of pitfalls. *Tyromyces mollis* (lacking clamps) might conceivably also be placed in a distinct genus, which would have to be called *Leptoporus* Quél. (1886). It has already been set apart in *Tyromyces* subgen. *Leptoporus* (Quél.) Pouzar (*1966c*: 370).

Tyromyces of this Check list also contains a distinct group with obovoid to ellipsoid spores and a pronouncedly fibrous context with the fibers inclining toward the surface of the cap; examples are *T. fissilis, T. alborubescens*; to these I would add *Polyporus croceus*. Bourdot & Galzin entered all three in their conception of *Phaeolus* (and not in *Leptoporus*). I have been tempted to transfer the group to *Spongipellis*, a genus with much the same macroscopic and also nuclear and cultural characters, apparently differing mainly in its thin-walled spores, at least if the emendation of *Spongipellis* by Kotlaba & Pouzar (*1965*: 76) is followed. If considered homogeneous and worthy of being set apart as a genus of its own the name *Aurantiporus* Murrill is available; this name would be very inapt for the group as a whole.

These few and incomplete notes will suffice as a forewarning that at the generic level no stability can be expected for some time to come. Renewed attempts at solving the riddles of the identity of *Polyporus lacteus* (in connection with the generic name *Postia*) and of the type of *Persooniana* Britz. (**122**) are necessary.

The problems at the specific level are also numerous. The following, together with previously published notes (Donk, *1972b*) on separate specific taxa, are ample demonstration of the existence of numerous uncertainties within the limits of the genus *Tyromyces*.

(**122**). The genus *Persooniana* Britz. was published with one new species, *P. albocana*. It was differentiated from *Irpex* Fr. (sensu lato) because "sowohl der Furchtkörper als auch das Sporenlager nicht leder-

---

[19] Since *Polyporus lacteus* is a lectotype an improved typification might change the application or assumed impriorability (as a nomen dubium) of *Postia*. This would require evidence that the selection of *P. lacteus* was not in agreement with the "Code". The two other species of the three from which *P. lacteus* was chosen (cf. Donk, *1960*: 274) are *Polyporus trabeus* Rostk. sensu Fr. (**O**) and *P. weinmannii*, both much debated names of uncertain application, although the latter seems identifiable with *Tyromyces fragilis* (cf. Donk, *1972b*: 301).

artig, sondern weich fleischig ist." I studied the protologue in connection with *Irpicodon pendulus* and *Irpex lacteus* carefully but was forced to conclude that *Persooniana albocana* differs from both these species.

This I believe leaves only one other solution, viz. that *P. albocana* is a species, or form of a species, of *Tyromyces* in which the hymenophore is more or less typically irpicoid, as in the original *T. lacteus*, the original description of which contains: "... pileo carnoso, poris inaequalibus lacerisve. [Pileus] ... mollis .... Pori plani, flexuosi, sed valde subtiles & profundi, in dentes omnino Sistotremoideos lacerati. ..." I have also seen *T. caesius* with perfectly irpicoid hymenophore. The spores of *Persooniana albocana* support this general suggestion. They are drawn as allantoid, a shape that is hardly expressed by the published measurements, $4 \times 2\ \mu$. It would seem premature to suggest the specific identity of Britzelmayr's species with a particular species of *Tyromyces*; this should wait until the species of that genus with allantoid, small spores has been worked out more thoroughly.

(123). The form-genus *Ptychogaster* is essentially 'artificial'. It is not correlated with a particular natural group of perfect fungi; the corresponding basidiferous states are found among several families of the Aphyllophorales, but this may be due to some extent to the somewhat indefinite generic character attributed to it. As far as the European polypores are concerned *Ptychogaster* states have been recorded for certain species of *Fistulina*, *Ganoderma*, *Abortiporus*, *Oxyporus*, *Laetiporus*, *Oligoporus*, *Poria* and *Tyromyces*. The rival generic name *Ceriomyces* must be dropped because it was presumably not based on a fungus state but rather on old galls produced by wasps (Donk, *1972a*: 165).

The conidia are of the chlamydospore-type (thick-walled) and are of endogenous origin. This last condition often escapes observation since the chlamydospores may quickly disintegrate.

Bourdot & Galzin (*1928*) mentioned the following species of *Tyromyces* as capable of producing *Ptychogaster* fruitbodies.

Tyromyces stipticus (as *Leptoporus albidus*), "quelquefois précédé ou accompagné d'un *Ptychogaster* (*P. rubescens* Boud.?) plutôt roussâtre que rougeâtre." Compare *P. rubescens* sensu F. & O. Falck (**124**).

Tyromyces floriformis (as *Leptoporus albidus* subsp. *floriformis*), "forme pétaloide, ... accompagnée d'un *Ptychogaster* ...," followed by a description.

Tyromyces apalus sensu Bourd. & G. (as *Coriolus kymatodes* subsp. *apalus*). "Ce champignon se développe soit totalement à l'état de *Ptychogaster*, soit partie *Ptychogaster* et partie *Polyporus*, soit uniquement *Polyporus*." This is succeeded by the description and the remark "Ce *Ptychogaster* parait répondre à *P. rubescens* Boud., mais il est plus franchement rougeâtre et moins gros que celui qui accompagne *L. albidus* et *L. floriformis*."

To these must be added:

Tyromyces ptychogaster, with the chlamydosporous state *Ptychogaster fuliginoides*, formerly called *Ptychogaster albus*. The species was apparently unknown to Bourdot & Galzin, unless they included it in their *Leptoporus albidus* (= *Tyromyces stipticus*), listed above.

Chlamydospore formation in nature among the Polyporaceae (exclusive of Fistulinaceae, Hymenochaetaceae, and Ganodermataceae), but not amounting to the production of narrowly attached *Ptychogaster* fruitbodies, was recorded by Bourdot & Galzin for the following taxa, in which the imperfect states are effused.

Leptoporus [*Tyromyces*] *revolutus*: "Il est ... fréquemment accompagné de conidies en tas plus ou moins épais ...." Apparently this chlamydosporous state had not yielded the typical (narrowly attached) *Ptychogaster* fruitbodies known to the French authors. They compared it with the next species but decided that the two were different.

Leptoporus destructor subsp. *sericeo-mollis* sensu Bourd. & G. = *Oligoporus rennyi*, the conidial state of which is called *Ptychogaster citrinus* (cf. Donk, *1971c*: 210).

Oxyporus obducens (as *Coriolus connatus* subsp. *obducens*).

Poria metamorphosa Fuck. (treated as a form of *Poria aneirina* by Bourdot & Galzin). I refer *Sporotrichum aureum* here.

Schizopora paradoxa (as *Poria mucida* "Etat conidifère"), see (113).

R. & O. Falck (*1937*) paid special attention to *Ptychogaster* states. The new taxa they described are *Ptychogaster flavescens*, mentioned elsewhere (124); and *Oligoporus friesii* (109) and *Multiporus chlamydoformans*, listed on this Check list as a species of *Poria*. Their interpretation of *Ptychogaster rubescens* will be found scrutinized in (124). Compare also (125).

It is mentioned above (1) that *Abortiporus biennis* apparently invariably produces chlamydospores in the hymenium or even in separate fruitbodies that have been described as a *Ptychogaster*.

(124). The name *Ptychogaster albus* Corda is now generally used in Europe for an imperfect state in which the 'chlamydospores' (often present as 'aleuriospores') finally end by forming a dark brown powdery mass more or less loosely held together by worn out hyphal strands. The darkening of these spores is rapidly progressive. When using the name *P. albus* for these states it is implied that Corda described a young stage in which the spores had not really started to ripen ("die Sporen ... bilden ein ... blasses Pulver"); the outside of the fruitbody is the last portion to turn brown. The epithet 'albus' indicates that no change of colour was visible on the outside. The question is whether all the states of the same structure as *P. albus* (lacunose by radiating hyphal strands and plates forming elongate cavities which become filled with spores and with a dark brown final colour of the spores) do not belong to the same species. The earliest available name for the complex is *Trichoderma fuliginoides* Pers. ≡ *Ptychogaster fuliginoides* (Pers. per Steud.) Donk (*1972a*: 170). Under this last name I have assembled the following taxa. This conspecificity, tentatively assumed, is evidently in need of reconsideration through studies in nature and in culture.

Trichoderma fuliginoides Pers. 1801.—'pulvere rutilo'. The spores in mass are now very dark brown in the type specimen. 'Fuliginoides' refers to a certain similarity with the genus *Fuligo* Haller (Myxomycetes) rather than to the colour of the spores.

Ptychogaster albus Corda 1838.—"Sporen ... bilden ein ... blasses Pulver", & "sporis ochraceo-argillaceis".

Polyporus ptychogaster F. Ludw. 1880.—Spores 'braun'.

Oligoporus ustilaginoides Bref. 1888 (may be considered a substitute name for *Ptychogaster albus*).—The final stage disintegrates "in eine braun verstäubende Sporenmasse".

Ceriomyces richonii Sacc. 1888.—"... pulpe ... blanche d'abord, puis couleur canelle ... spores ... [4–6 μ], jaune nankin ..." [repr. 1879 (Rm 1): 132].

Polyporus sericeomollis Romell (111), "a variety (or perhaps distinct species)" Romell 1911 in obs.—Associated with "a fibrous-pulveraceous *Ptychogaster* of about isabelline color ... like small specimens of *Ptychogaster albus*." — Romell *1911*: 22) also included in the same species a chlamydosporous state "of sulphureous or pallid color", for which see under *Oligoporus rennyi* (**68**).

Ptychogaster rubescens Boud. sensu F. & O. Falck 1937.—"... der Beginn der Sporenbildung ... durch eine rosafarbene Verfärbung zu erkennen" and "... Beläge färben sich erst rötlich, dann rötlich-gelb bis rötlich braun, schliesslich braun bis dunkelbraun". Compare (**125**). In

Ptychogaster flavescens R. & O. Falck 1937 is "diese erste Färbung hellgelblich" and "... Beläge färben sich erst gelblich, dann lehmgelb, schliesslich braun". — "In grösseren Mengen zusammenliegend sehen die älteren Sporen beider Ptychogasterarten sehr ähnlich aus. Sie erscheinen als braune, rötlich-braune oder schokoladefarbene Lager."

All the taxa listed above were collected in nature. *Ptychogaster rubescens* sensu R. & O. Falck and *P. flavescens* alone were not only found in nature [cf. R. & O. Falck 1937 (HF 12): *f. 6* and table on p. 11] but also in buildings; they were stated to be causal organisms of rot "des technisch verwerteten Coniferenholz" (p. 1).

(**125**). The correct assignment of *Ptychogaster rubescens* Bourd. to a perfect state is in my opinion still problematic. Boudier himself found this chlamydosporic state in greenhouses, "parmi le *Polyporus vaporarius* fréquent dans les serres ... [et qui se tache] de rouge par le toucher ou par le froissement." It is difficult to suggest which polypore Boudier had in mind. The true *Poria vaporaria* (**O**) is insufficiently known; what Fries called *Polyporus vaporarius* is now often considered to be a form of *Poria* [*Antrodia*] *sinuosa*, a species that does not turn reddish. I am acquainted with only one white poria that may be common in greenhouses in northern France and that turns red, viz. *Rigidoporus sanguinolentus*. A connection between this and *Ptychogaster rubescens* is not very likely.

This conclusion shifts attention to the case described by Hennings, who found a *Ptychogaster* state in greenhouses in Berlin associated with a (presumably) polymorphic fungus that is tentatively called *Polyporus [Tyromyces] henningsii* Bres. (**131**) on this Check list. *Tyromyces apalus* sensu Bourd. & G. (**127**) is another species producing a *Ptychogaster* state that 'seems to answer' to *P. rubescens*.

Another suggestion has been voiced by Davidson *& al.* (*1946*), who directed attention to the great similarity between cultures of two isolates identified as *Ptychogaster rubescens* and *Polyporus [Tyromyces] guttulatus*; this suggestion has been supported by O. Fidalgo (*1959a*). So far such a connection does not seem to have been established. The correct determination of the cultures named *Ptychogaster rubescens* is open to doubt; moreover as far as I am aware the two states have not yet been found together in nature and *Polyporus guttulatus* has not yet been reported from greenhouses in Europe. It would seem as though we are confronted with different *Ptychogaster* states with similar cultural characteristics.

What R. & O. Falck (*1937*) called *Ptychogaster rubescens* perhaps agrees better with *Ptychogaster fuliginoides* (**124**), which state I take to be conspecific with *P. albus* and *Tyromyces ptychogaster*.

(**126**). Among the many difficulties and problems still to be straightened out in *Tyromyces* are those connected with a taxonomically mixed group of species that may be indicated as the 'house fungi': species that occur in dwellings, mines, hothouses, and the like. They are of considerable economic importance and this presages inextricable confusion in literature. In fact it could not be worse. In order to compress the subject into the trammels of this Check list I have cut many knots without much personal knowledge; for this I tender my sincerest apologies.

I have admitted four species: *Tyromyces apalus* sensu Bourd. & G. (**127**), *T. destructor* sensu Bourd. & G. 1928 emend. (**128**), *Polyporus fodinarum* (**129**), and *Polyporus henningsii* (**131**). In this connection species with only strictly resupinate fruitbodies are not taken into consideration; they are referred to *Poria*.

(**127**). Polyporus *apalus* Lév. (**0**) has been variously interpreted. According to Bourdot & Galzin (*1928*: 566) Bresadola was of the opinion that Boudier considered the true *P. apalus* to be the same as *Polyporus kymatodes* sensu Bres. [= *Tyromyces balsameus*]. Lloyd (*1915b*: 375) could find no type; he added "almost passed out of European tradition excepting locally in France, where the specimens are *Polyporus [Bjerkandera] fumosus*." Finally, Bresadola [1931 (BIm 20): text to pl. 985 f. 2] listed *P. apalus* as a synonym of *Polyporus [Tyromyces] lacteus*, which seems to be contradicted by the original description.

Bourdot & Galzin applied the name to an infrageneric taxon which they appended as a subspecies to *Coriolus kymatodes* sensu Bres. [= *Tyro-*

myces balsameus]. The fungus they described also developed a *Ptychogaster* state (**123**). Their description of the basidiferous state agrees in many respects with Léveillé's original description but there are also serious discrepancies. The type of *Polyporus apalus* was found "ad truncos quercuum caesos", a quite different habitat from that of Bourdot & Galzin's fungus: "sur vieilles planches de sapin, dans une serre un peu chauffée et humide." I am far from convinced that their fungus was correctly named.

Still another matter is the presumed relationship of *Tyromyces apalus* sensu Bourd. & G. with *T. balsameus*. Again judging from Bourdot & Galzin's description, I am not prepared to accept close specific relationship between the two. For *T. balsameus* North American authors do not mention a *Ptychogaster* state. It looks as though *T. apalus* sensu Bourd. & G. is in need of a new name.

The habitat of the single collection thus far reported suggests that this may be a 'house fungus' (**126**). Its Ptychogaster state recalls *Ptychogaster rubescens* (**125**). *Polyporus henningsii* (**131**) was also collected in a hothouse and also in connection with a *Ptychogaster* state that has been identified with *P. rubescens*.

(**128**). What is *Polyporus destructor*? So many interpretations have been published that a reproduction and inspection of Schrader's original description may not be altogether superfluous as the best approach we have to the lost type:

Boletus c. *Acaules* α. *Carnosi*. "*B. destructor*, albidus, pileo inaequali rugoso, poris subrotundis obtusis. / Hab. in domibus recenter extructis et humori obnoxiis, quas secus ac *Merulius vastator* [= *Serpula lacrimans*] destruit. / Fungus variae figurae, magis vel minus explanatus, superficie rugosa, glabra. / Pori breves, minuti, subrotundi, obtusi. / Substantia fibrosa, per siccitatem, facile in pulverem farinaceam conterenda. / Color totius fungi albidus. / Odor gravis, sed haud ingratus."

This description contains so much information that it is surprising that no interpretation can be found in literature that really agrees with it entirely. When Fries (1821: 359) revalidated the name as *Polyporus destructor* he merely rewrote what Schrader had already written and there can be no doubt that he had not yet shown that he had a personal interpretation; the type of the revalidated name continues to be represented by Schrader's original account. It is true that Fries added "v.v.", but if the fungus he had in mind was different it made no real change in the phrase and description. It is not unlikely that what he associated with Schrader's fungus was something different; he did not find it "ad 'ligna putrida domestica'" (this is the way he cited the habitat of the original fungus) but "in silvis ad truncos udos pineos . . . per Annum."

When many years later Fries (1838: 454) redescribed *P. destructor* he emphasized the following characters: "pileo aquose-carnoso fragile effuso-reflexo *fuscescenti-albido, intus zonatus, poris longis* subrotundis dentatis lacerisque albidis." (Italics as in the original.) No habitat. Still later Fries (1874: 547) also printed "aquose carnoso" in italics and gave the habitat as "Ad ligna, praecipue domestica et fabrefacta, quae more *Merulii lacrymantis* emollit et destruit." He referred *Polyporus alutaceus* Fr. (**O**) sensu Rostkovius [1830 (StP 4): 57 *pl. 27*] here. Mez (1908: 110 *f. 42*) claimed to have found such a fungus.

I feel compelled to interpret Schrader's fungus as 'resupinate'; although it is stated to be variously shaped ("fungus variae figurae") the only further elucidation is in the "magis vel minus explanatus" immediately following.[20] The table below contrasts Schrader's and Fries's fungi:

Schrader	Fries 1838, 1874
Fruitbody variable but flattened out. "Color totius fungi albidus."	Fruitbody effuso-reflexed. Cap 'fuscescente-albidus'; pore surface 'albidus'.
Flesh fibrous, when dry easily rubbed to a mealy powder.	Flesh watery-fleshy, zonate.
Tubes short, with obtuse edges.	Tubes long, with toothed and torn dissepiments.

Even if it is assumed that Schrader's fungus need not be consistently resupinate there can be little doubt but that more than likely the two are different species, as Bresadola [1908 (Am 6): 37–38] had concluded.

Another interpretation is that of Bourdot & Galzin (*1928*: 546 *f. 155*). They described the fungus from a single collection with sessile fruitbodies and an abnormal hymenophore: "pores cériomycétoïde et altéré par un parasite". This condition recalls *Poroptyche candida* G. Beck, another house fungus, but with resupinate fruitbodies and perhaps similar basidiospores and also with a spongy (myriadoporous) hymenophore, but its rope-like mycelium apparently allying it differently.

In connection with a French specimen that he referred to *P. destructor* and which is referable to *Poria resupinata* (not to *Leptoporus destructor* sensu Bourd. & G.!) Bresadola (l.c.) suggested: "Species haec, quoad formam valde variabilis, saepe resupinata, mollis, friabilis, praecipue si in locis suffucatis obvia, meo sensu, tantum forma *Polypori albidi* Schaeff. qui tantum sub diu et optime evolutus tenax et lignosis est." Romell (*1926*: 4) came to the same conclusion after he was told by R. Falck that *P. albidus* [= *Tyromyces stipticus*] also occurs in houses. This last claim still needs verification. It should be remembered however that Romell stipulated that his interpretation of *P. stipticus* was not that of Bresadola, who for some time used this name to indicate *Dichomitus squaleus*. I do

---

[20] Schrader placed his species in a group that he indicated as "Acaules". It contained both pileate and resupinate species: "Pileo dimidiato . . ., altero latere horizontaliter inserte . . .."

not recall having come across an explicit record of a *Tyromyces* house
fungus in which an intensely bitter taste of the fruitbody is mentioned,
a character typical of *Tyromyces stipticus* (*T. albidus*) (**134**).

Other authors (Bondarcev, *1953*: 205; Domański, *1965b*: 139) equated
P. destructor (inclusive of Bourdot & Galzin's interpretation of 1928)
with *Leptoporus fodinarum*; in my opinion this is not readily acceptable (**128**).

The above review has been written to stimulate renewed study of the
'white' house fungi. In order to focus attention on the problem connected
with the identification of *P. destructor* this species as conceived by
Bourdot & Galzin is entered on the Check list as distinct. It would seem
reasonable to keep in mind the possibility that Schrader's fungus might
also occur in the forest, as would be the case if it really could be proved
identical with *Tyromyces stipticus*.

(**129**). *Polyporus fodinarum* Velen. is entered on this Check list as
a third white 'house' fungus of the genus *Tyromyces* (**126**). This is done
on the authority of Pilát, who claimed to have seen a specimen named
by Velenovský and who gave it its present circumscription. At first he
called it *Poria vaillantii* and listed *Polyporus henningsii* = *Polyporus
vaporarius* sensu P. Henn. (**131**) as a synonym. When accepting the
name *Polyporus fodinarum* as that of a good species he dropped this
synonym; instead he listed *Polyporus destructor* "auctorum p. max. p.
non Schrader nec Fries".

The species as conceived by Pilát is remarkable in its ability to develop
extensive superficial mycelial growth with drooping portions not only in
damp mines and caves but also in houses. This mycelial state commonly
assumes striking and bizarre shapes; often it resembles the mycelium of
Serpula lacrimans (Wulf. per Fr.) J. Schroet.; the two may have been
confused when not found with well-developed fruitbodies. I would suggest
that *Polyporus vaporarius* sensu Mez belongs here. To my knowledge no
free-living chlamydosporous state has thus far been reported, as it has
been for *Polyporus henningsii*, which also seems to produce a different
kind of superficial mycelial state.

Polyporus fodinarum is placed here in *Tyromyces* rather than in *Poria*
on account of Pilát's description, from which I conclude that the fruitbody
may occasionally be pileate ("receptacle résupiné ou formant de petits
chapeaux . . . aussi latéraux"). The mycelia Mez and Pilát depicted look
very similar, but this does not hold true for the fruitbodies they associated
with them. The question arises whether *Tyromyces fodinarum* is a homogeneous conception or not. The correctness of the name itself may also be
doubted (**130**). Compare also *Boletus fodinalis* Humb. (**O**).

(**130**). Attention is drawn here to the existence of a possible earlier
name for *Leptoporus fodinarum* of Pilát (**129**), viz. *Polyporus erosus*

Pers. ≡ *Boletus hybridus* Sow. The latter forms spectacular superficial mycelia ("spreading in large patches, forming more or less broad ramifications, often inosculating, of a cottony substance like [*Boletus (Serpula) lacrimans*]") and it may produce pileate fruitbodies with rather long pores (as figured by Sowerby, *pl. 289*). On a later plate (Sowerby, *pl. 387 f. 6*) a long, mycelial 'stalactite' is depicted; it bears an orbicular disc of pores at the lower (free) end.

The original account (*pl. 289*) would seem to furnish sufficient evidence for not equating *Boletus hybridus* with *Poria vaillantii*, in contradistinction to the following citation: "An unlocalized specimen named [*Poria hybrida*] by Berkeley in Kew herbarium has been selected as a neotype by Prof. J. L. Lowe, according to his annotation of the packet. This collection . . . is *Poria vaillantii* (DC. ex Fr.) Cooke."—Reid & Austwick [1963 (GN 18): 309].

(**131**). Another species of house fungi that produces both resupinate and pileate fruitbodies is *Polyporus henningsii*; it has obovoid spores of 3–4 × 2 μ, hence apparently different from those of *Tyromyces destructor* sensu Bourd. & G. It was described by Bresadola as a species with a flabelliform and stalked fruitbody, but according to Hennings (who had sent the material that became the type) such a form made up part of a long series of fruitbodies varying from completely resupinate to laterally or even more or less centrally stalked forms (the latter 'cyathoid'). Here it is tentatively assumed that all these various forms really represent a single species. I am not prepared to identify this species with *P. destructor* sensu Bourd. & G. for several reasons. Certain microscopical characters (spores) that Bresadola indicated for *Polyporus henningsii* support my view and further a superficial system of rhizomorphic mycelial strands from which the fruitbodies originate has not been recorded for *P. destructor*.

Hennings himself [1899 (VBr 40): 125–126] identified the strictly resupinate states with *Poria vaillantii*, which he in his turn identified with *Poria vaporaria*. The latter species is insufficiently known; it has often (incorrectly) been considered to be the same as *Poria vaillantii* (cf. Donk, *1967*: 117). Since Hennings's drawing showed that the mycelial strands produce both resupinate and sessile fruitbodies I do not accept his identification with *Poria vaillantii* either; no modern author who described the current interpretation of *Poria vaillantii* has ever made any mention of pileate to stalked fruitbodies. (Hennings did not study the fungus microscopically.) Yet I must confess that his drawing of the resupinate fruitbodies is very suggestive of *Poria vaillantii*, though again assuming the conspecificity of the whole range of fruitbodies the microscopical details published by Bresadola for *P. henningsii* would also seem to exclude identification of the two taxa.

For the *Ptychogaster* state, see (**123**).

Modern authors, following Pilát, identify Hennings's fungus and

Polyporus destructor sensu Bourd. & G. with *Polyporus fodinarum*; this still needs confirmation (**128**).

(**132**).   *Tyromyces cerifluus* (**O**) was known from a single collection made in South Carolina, U.S.A., when Lowe [1961 (PMi 46): 205] identified it with the European *Polystictus* [*Tyromyces*] *revolutus*. On comparing descriptions a number of discrepancies can be noted; these may appear to be unimportant when additional collections cause the American species to become better known; nevertheless for the time being the existing descriptions induce me to retain *Polystictus revolutus* as a distinct species. The strongly disjunct areas of distribution and the lack of knowledge of the existence of an accompanying imperfect state for the American fungus also seem to support cautious treatment.

It has also been suggested to me that *Polystictus revolutus* be the same species as *Tyromyces ptychogaster*. It is true that according to Bourdot & Galzin the species is often accompanied "de conidies en tas plus ou moins épais" but this quotation does not in the least suggest the *Ptychogaster* fruitbodies of often quite considerable size and globular shape which are typical of *T. ptychogaster*. Moreover, *P. revolutus* is recorded from a diversity of hosts, all of which are frondose trees, while *T. ptychogaster* I have invariably found on coniferous wood. Finally, the spores of *P. revolutus* are "subcylindriques, un peu déprimées latéralement" and as drawn by Bourdot & Galzin, different from the more ovoid and wider spores of *T. ptychogaster*.

(**133**).   Both the status of *Tyromyces guttulatus* as a good species and its occurrence in Europe need some comment. It was described from North America. According to Overholts (*1953*: 288), "*P*[*olyporus*] *immitis* intergrades considerably with *P. guttulatus*, which is best distinguished by the thin, applanate, and typically spotted pileus." It is now often assumed that *P. immitis* Peck is the same as *Tyromyces stipticus* (which species has also been called *T. albidus*), an identification that is recommended for renewed critical study (**134**).

Lloyd (*1915b*: 301) was perhaps the first author who reported *Polyporus guttulatus* from Europe, although he adopted an old Friesian name for it, viz. *Polyporus alutaceus*. He did this because "the European plant, as named by Bresadola, appears to me to accord with Fries' description and the figure that he cites." Fries's description mentioned by Lloyd is not the original one; the figure that Fries cited is that of *Polyporus epixanthus* Rostk. (**O**). Neither the original description of *P. alutaceus* (**O**) nor Rostkovius's figure even remotely suggests *Tyromyces guttulatus* as I know it from North America. According to Bresadola (*in litt. apud* Bourd. & G., *1928*: 545) his conception of *Polyporus alutaceus* "est très voisin de *L*[*eptoporus*] *albidus* [= *Tyromyces stipticus*] et en est peut-être une simple variété bien développée et jaunissante." This differential description is hardly, if at all, suggestive of *T. guttulatus*.

Romell (*1926*: 9) was not sure whether *Polyporus guttulatus* occurs in Sweden; Bourdot & Galzin (*1928*) do not report it; but Pilát [1938 (ACE 3): 204] mentioned a Czechoslovakian collection that he pronounced to be absolutely identical with North American collections [as *Leptoporus stipticus* f. *guttulatus* (Peck) Pilát]. I have seen a few European collections that I would refer here.

(**134**). *Polyporus immitis* is a disputed species. At first it was confused with, and referred to [Murrill 1907 (NAF 9): 39], *Spongipellis galactina* (**O**). Kauffman (*1926*) pointed out that this interpretation was erroneous; his redescription suggests a species close to, or perhaps conspecific with, *Tyromyces stipticus* (*T. albidus*). Although I repeatedly collected material of this complex both in Europe and in Michigan, U.S.A., at present I do not wish to commit myself as to the correct status of *Tyromyces immitis*.

Lowe (1934: 86; 1942: 79) was the first to accept the identity of the two taxa. On the other hand Pilát [1937–8 (ACE 3): 211 *f. 66, pl. 130*] keeps the two apart; in his key he contrasted them as follows:—

Leptoporus stipticus: Chair de saveur amère, adstringente. Sur bois de conifères.
Leptoporus immitis: Chair sans saveur perceptible. Sur bois d'arbres feuillus.

This characterization is somewhat surprising because the epithet 'immitis' denotes an unpleasant taste or smell which the author of the name described as "a subacid odor." The suspicion arises that Pilát did not really taste the collections (from Czechoslovakia and the U.S.A.) which he referred to *Leptoporus* [*Tyromyces*] *immitis*, not only because he did not mention any taste in his description but also because it may safely be assumed that the North American collections cited were bitter. Overholts (*1953*: 288) attributed to *Polyporus immitis* "a very bitter taste in both fresh and dried condition." It would thus seem that the two taxa do not differ in taste. Neither Peck nor Kauffman (l.c.) who re-introduced the species, seem to have tasted it.

The host range may offer a more positive difference. The *Tyromyces immitis* of North American authors prefers deciduous trees. The type was found on a species of *Fraxinus* ("decaying ash trunks"); other species of *Fraxinus* and apple, *Acer*, and *Betula* have been recorded as additional substrata, although Overholts also reports it from various coniferous trees. This behaviour is at variance with the European *T. stipticus*, which is known to nearly all authors from this continent from coniferous trees only. An exception is Romell (*1926*: 9), who stated that he had also found it on *Betula* and *Sorbus*. On the other hand Lundell [1946 (LNF 27–28): 8 No. 1316] reports that he found *T. stipticus* to be rather common in Femsjö (where Fries lived in his youth) "always on spruce".

Compare also Lowe (1934: 141 *pl. 14 fs. 1–3*) and Overholts (*1953*: 426).

LIST OF OMITTED NAMES

The names on the following list have been omitted from the preceding Section, viz. the Check list proper. The list enumerates in alphabetical order names given to, or brought into connection with, European polypores but for various reasons thought to be unfit for incorporation in the preceding enumeration.

Many of these names (i) are not validly published; they are nomina nuda in the sense of the "Code". Identifiable devalidated names however have been included in the Check list proper. (ii) Other names included here cover taxa that are so insufficiently known that they amount to nomina dubia. (iii) Still other names are (or were) either so diversely interpreted that they should be dropped (cf. Art. 69 of the "Code"; nomina ambigua) or else are illegitimate (not available) in the light of Art. 70 (nomina confusa).

A considerable number of names are preceded by an asterisk (*); these are based on type material collected outside Europe. The reason for the inclusion of these names is (iv) that they are or for some time were identified on erroneous or, at least in my opinion, (still) dubious grounds with species thought to occur in Europe. In the case of some of these European species the number of these putative or doubtful synonyms is so considerable that it was convenient to leave them out altogether. In the "Notes" brief mention of this is in each case made separately. (v) A small category comprises aliens that were reported incidentally from Europe, usually from hot- or green-houses. Of a few of them the 'type locality' is Europe.

Genera like *Irpex*, *Merulius*, and *Sistotrema* were for a long time the recipients of polypores as well as non-polypores; as a rule specific names bestowed on this second category, of non-polypores, have been left out.

Mention of a family or order following a specific or generic name indicates that the species is not considered to belong to the 'polypores'. It will not be found treated elsewhere in this book.

The addition of '(O)' to a name indicates that the name is treated more fully elsewhere in the present list.

For a discussion on names published by Secretan, see Donk in Taxon **11**: 170–173. 1962.

abietina, Daedalea, Secr. 1833 M. 2: 496 (Switzerland), not ∼ (Bull.) Fr. 1818 (d.n.); [≡ *Daedalea abietina* (Bull.) Fr. sensu Secr., excl. of type].—Nomen dubium. Fr. 1874: 495 (Secr. "n. 17") referred this to *Lenzites* [*Antrodia*] *heteromorpha*, but Secretan's descr. does not agree.

abietinum, Sistotrema, Secr. 1833 M. 2: 501 (Switzerland) (nom. nud.: n.v.p.).—*Sistotrema violaceum* Pers. "(excl. syn.)" is cited as a syn. of "Var. A", the description of which suggests *Hirschioporus fusco-violaceus.*

abietinus, Boletus, Cumino 1805: 224 (Italy) (d.n.), not ∼ Anon. 1790 (d.n.) per Purt. 1821, not ∼ Pers. apud J. F. Gmel. 1792 (d.n.) & (Pers. apud J. F. Gmel. per Fr.) S. F. Gray 1821; ≡ *Polyporus cumini* Pers. 1825.—Nomen dubium. Referred by Fr. 1832[Ind.]: 55 to *Polyporus stereoides* (**O**).

abnormis.—"*D[aedalea] cinerea* Fr. (vel *abnormis* Kill.)" Killerm. 1940 (DrG 21): 75 *tpl. 31 f. 20* (lacking Latin descr.: n.v.p.) (Germany).– Nomen dubium. *Daedalea cinerea* Fr. = *Cerrena unicolor.*

abortivus, Polyporus,* Peck 1881 (BG 6): 274 (U.S.A., Illinois); *Daedalea* Pat. 1900.—This is conspecific with *Abortiporus distortus* (O**) fide Murrill 1904 (BTC 31): 422, but like the latter taxon it has been reduced further to the synonymy of *Polyporus [Abortiporus] biennis,* for instance by Bres. 1926: 79 and Overh. 1953: 224. – Cf. (2). – Descr.: Peck 1885 (RNS 38): 90.

acanthoides, Cladomeris, Big. & Guill. 1909: 410 (syn.: n.v.p.); [≡ *Clado-meris acanthoides* (Bull. per Mérat) Quél. sensu Fr. (*Polyporus*), excl. of type, (Sweden)].– This name was listed by error as a syn. of *Clado-meris [Bondarzewia] montana;* simultaneously *Boletus acanthoides* Bull. [= *Meripilus giganteus*], which represents the 'type', was treated as a distinct sp. Fries's conception of *B. acanthoides* belongs to *Abortiporus biennis.*

acerinus, Polyporus, Opiz 1852: 137 (Czechoslovakia) (nom. nud.: n.v.p.).— Nomen dubium.

acidula, Bjerkandera, (Fr.) P. Karst. 1882 (Mfe 9): 62; *Polyporus stipticus* var. Fr. 1828 E. 1: 86 (Sweden); *Boletus* Mussat 1901 (syn.: n.v.p.); *Polyporus* Romell 1912 (incidental mention: n.v.p.).—Romell 1926 (SbT 20): 19 referred this to *Polyporus [Tyromyces] stipticus,* but this seems doubtful. Fries separated this taxon because of its deviating taste, "sapor acidulus . . ., sapore haud stiptico", although he thought it not specifically distinct ("certe tamen varietas"). – *Boletus acidulus* (Fr.) Mussat 1901 (SF 15): 62 is an error for *Polyporus stipticus β. acidulus* Fr. 1828 E. 2: 86 ≡ *Polyporus stipticus* [var.] "B." *acidulus* Fr. 1838: 453 ("b."), 1874: 546, Sacc. 1888 (SF 6): 113, the 'B.' being taken for the abbreviation of '*Boletus*'.

**admirabilis, Polyporus,* Peck 1899 (BTC 26): 69 (U.S.A., Maine); *Poly-stictus* Lloyd 1914 (nom. nud.: n.v.p.).—This is another species (cf. also *P. albiceps*) that should be compared with *Polyporus incendiarius.* – Syn.: *Polyporus underwoodii* Murrill apud Peck 1906 (BNS 105): 27; Murrill 1907 (NAF 9): 61. – Descr.: B. O. Dodge 1916 (M 8): 7 *pl. 173, pl. 174 f. 2;* Overh. 1953: 249 *pl. 37 f. 233, pl. 128 fig.*

**adpressa, Poria,* Murrill 1920 (M 12): 85 (Cuba).—Referred by Lowe

1947 (Ll 10): 45 to *Poria versipora* [= *Schizopora paradoxa*] and 1966: 65 to *Poria hypolateritia* (**0**).

affinis, Polyporus, Boud. in herb.—Fide Lloyd 1905 (LMW 4, Ap.): 375 = *Polyporus* [*Ganoderma*] *resinaceus.*

**agariceus, Polyporus,* (Kön.) ex Berk. 1843 (AM 10): 371 (Ceylon); *Boletus* Kön. in herb.; *Favolus* Lév. 1844; *Leucoporus* Pat. 1900; *Polyporellus* Pilát 1936.—Correct application not yet quite certain, cf. Petch 1916 (APe 6): 89, Donk 1969 (Pe 5): 238, 1971 (PNA 74): 17. − Sensu Bres. (European specimens) → *Polyporus alveolarius* (Bosc) per Fr.

agaricicola, Polyporus, F. Ludw. 1882 (H 21): 145 (Germany); 1883 (ZPf 1): 209 ("*agaricinicola*"); *Poria* Sacc. 1888.—Abnormal growth on the surface of the cap of *Amanita pantherina* (DC. per Fr.) Krombh., fide F. Ludw. 1908 (FwN): 116 and Malkovský *1931. Daedalea parasitica* (**0**) is a similar abnormalty.

agaricoides, Boletus, Thunb. → *Daedalea thunbergii* Fr. (**0**).

alba, Daedalea, Pers. 1801 (d.n.), not ∼ Pers. 1828; [≡ *Agaricus daedaleis Sinibus minoribus* Batt. 1755: 72 *pl. 38 fs. E, F* (Italy)].—Nomen dubium. Pers. 1828: 5 (with doubt) and Fr. 1832^Ind.: 81 referred it to *Daedalea saligna* [= *Bjerkandera fumosa*].

alba, Daedalea, Pers. 1828, not ∼ Pers. 1801 (d.n.); [≡ *Boletus albus* Huds. sensu Bolt. 1788: 78 *pl. 78* (England)]; ≡ *Polyporus albus* Wettst. 1888, typonym, misapplied?, not ∼ (Huds.) per Fr. 1838.—Nomen dubium. Bolt. (1791: Ind. gener.) soon referred his interpretation to *Boletus salicinus* Bull. [≡ *Bjerkandera fumosa*]; Fr. 1874: 544 placed it under *Polyporus salignus* [≡ *Bjerkandera fumosa*], leading Wettst. to replace the name *P. salignus* by *P. albus.* Both Pers. and Wettst. apparently excluded the type of *Boletus albus* Huds., although not explicitly so. I tend to agree with Berk. 1836: 133, who thought that Bolton's fungus could be *Daedalea* [*Abortiporus*] *biennis.*

alba, Mensularia, Lázaro 1916 (RMa 14): 738 / 1917: 124 (Spain).—Nomen dubium. *Tyromyces* sp.?

**albertinii,* [*Polyporus*], Lloyd 1912 (LMW 3, S.P.): 160 *f. 460* (Australia, Queensland); *Phaeolus* D. Reid 1963.—The type was referred to *Polyporus* [*Phaeolus*] *schweinitzii* by Cooke in herb.; Bres. 1926: 26 thought it to be close to this sp.; G. Cunn. 1965: 201 referred it to *Inonotus rheades* [misnamed]; D. Reid 1963 (KB 17): 277 *f. 1: 5* (descr.) treated it as a distinct sp. of *Phaeolus.* It differs from the type sp. of *Phaeolus* by the presence of "setal hyphae" (macrosetae) in the context of the dissepiments of the tubes and by the strongly coloured spores.

"*albescens* Quél." as cited by Killerm. 1922 (Dba 15): 87 as a syn. of *Polystictus* [*Coriolus*] *zonatus* "f. *lutescens* Bres." It is not clear what specific or other name Killerm. had in mind. I could not locate the name in Quélet's publications. The epithet haunts the literature in various combinations: *Polystictus zonatus* var. *albescens* [J. Rick 1906 (Am 4): 310 & Theiss. 1911 (DAW 83): 239; "Quél."; no descr., "det.

Bres."], *Polystictus versicolor* var. *albescens* [Torrend 1909 (Bro 8): 132; "Quél."; no descr.]; "*Polyporus albescens* Quél. (sec. Killerm.)" cited as a syn. of *Polyporus zonatus* Quél. (Bond. 1953: 486). Possibly it originated as a printing or reading error for 'lutescens'.

albescens, Polystictus, Lázaro 1916 (RMa 14): 750 / 1917: 136 (Spain).— Nomen dubium. *Coriolus pubescens?*

*albiceps, *Polyporus,* Peck 1900 (BTC 27): 19 (U.S.A., New Hampshire); *Polyporellus* Pilát 1932.—This has been brought into relation with *Polyporus incendiarius* (**96**). – Descr.: Overh. 1953: 250 *pl. 37 f. 224, pl. 115 fs. 627, 628.*

*albida, *Daedalea,* Schw. 1822: 93 (U.S.A., North Carolina), not ∼ Fr. 1821; ≡ *Daedalea discolor* Fr. 1828; *Striglia* O.K. 1891.—Fide Murrill 1905 (BTC 32): 86, 87 & Overh. 1953: 120 = *Agaricus |Daedalea [Daedaleopsis] confragosa.* – Descr.: Fr. 1828 E. 1: 68 (*Daedalea discolor*). – *Daedalea discolor* Fr. sensu Kl. → *Lenzites klotzschii* Berk. (**0**) & *L. discolor* Fr. (**0**). – Cf. (**31**). – *Daedalea discolor* was reported for Switzerland by Trog 1857 (MiB): 39.

albidofuscus, Polyporus, Secr. 1833 M. 3: 67 (Switzerland) (as a sp. of *Boletus*: n.v.p.).—Almost certainly *Polyporus badius*, rather than *P. varius.* Referred by Fr. 1874: 536, with doubt, to *P. petaloides* [= *P. varius*].

albidus, Polyporus, Chev. 1837: no. 40 var. α (as a var. of *P.* [*Albatrellus*] *cristatus*: n.v.p.) (France), not ∼ (Schaeff.) per Trog apud Fr. 1838, not ∼ Saut. 1869.

albidus, Polyporus, Saut. 1869 (H 8): 41 (Austria), not ∼ Chev. 1837 (n.v.p.), not ∼ (Schaeff.) per Trog apud Fr. 1838.—Nomen dubium. Keissl. 1917 (AW 31): 110 referred this to *Polystictus albidus* [= *Tyromyces stipticus*], a suggestion in no way supported by the descr.

albocarneus, Polyporus, Velen. 1922: 642 [see Pilát 1948: 245 for Latin translation] (Czechoslovakia).—Fide Pilát 1943 (ACE 3): 394 = *Poria [Merulius] taxicola* (**0**).

albomarginatus, Polyporus, Opiz 1852: 137 (Czechoslovakia) (nom. nud.: n.v.p.).—Nomen dubium.

alboniger, Polyporus, Lloyd in herb. (Australia, Tasmania) ≡ *Polyporus atromaculatus* Lloyd ex G. Cunn. [= *Incrustoporia semipileata*], fide G. Cunn. 1950 (PNW 75): 217.

albopallescens, Poria, Bourd. & G. 1925 (BmF 41): 216 (France); *Trechispora* D. P. Rog. 1944 (M 36): 79.—Fide Lundell 1947 (LNF 29–30): 11 No. 1415a = *Sistotrema muscicola* (Pers.) Lundell ('Corticiaceae'), a species usually not *Poria*-like. – *Poria albolutescens* var. *albopallescens* Bourd. & G. 1914 (BmF 30): 253 (France) (nomen subnudum), must perhaps be taken as basionym.

*albovestidus, *Polystictus,* Lloyd 1923 (LMW 7): 1192 *pl. 236 f. 2401* (South Australia); *Coriolus* G. Cunn. 1950.—Referred to *Trichaptum [Hirschioporus] pargamenum* by G. Cunn. 1965: 100.

album, Agarico-igniarium, Paul. 1793 T. 2: 85 (descr.), Ind. (name) (d.n.).—Name based on various sterile mycelia; lectotype, "Celle qu'on voit ici sous la figure 1" = presumably sterile mycelium of *Serpula lacrimans* (**O**), "a pris naissance sur un mât de navire". Some of the other pieces described may well have been sterile mycelia of polypores, such as *Xylostroma giganteum.* – The figures designed to accompany the descr. were published at a later date as *Pyreium giganteum* Paul. 1812–35: *pl. 6 fs. 1–3.* (generic name n.v.p.), of which Murrill 1903 (JM 9): 89 assumed that it was a recombination of *Xylostroma giganteum* Tode.

albus, Agaricus.—Occasionally [Harz 1868 (BSM 41): 3; P. Magn. 1905: 183] cited as a botanical syn. of *Polyporus* [*Agaricum*] *officinalis.* In my opinion it is merely the name of a drug and it was so used by T. Martius 1845 (BdN 23): 102–103.

albus, Boletus, Huds. 1762: 496 (England) (d.n.), not ∼ Schaeff. 1774 (d.n.), not ∼ Pers. 1818 (d.n.), &c.; *Polyporus* (Huds.) per Fr. 1838, misapplied, not ∼ Wettst. 1888; *Leptoporus* Quél. 1886, misapplied; *Bjerkandera* P. Karst. 1889; *Polystictus* Big. & Guill. 1913, misapplied.—According to Donk 1971 (PNA 74): 21 this is not only a nomen dubium but also a nomen ambiguum. – Sensu Lightf., Bolt. = spp. incertae sedis; sensu Fr. → *Bjerkandera fumosa*; sensu Bres. → *Tyromyces fissilis.*

albus, Boletus, Schrank 1789: 615 (Germany) (d.n.), non/an ∼ Huds. 1762 (d.n.), not ∼ Schaeff. 1774 (d.n.), not ∼ Pers. 1818 (d.n.), &c.— Nomen dubium.

albus, Polyporus, Wettst. → *Daedalea alba* Pers. 1828 (**O**).

alneum, Sistotrema, Secr. 1833 M. 2: 504 (Switzerland).—Nomen dubium. Secretan cited *Hydnum argutum* Fr. as a syn. Referred by Fr. 1838: 523 to *Irpex obliquus* [= *Schizopora paradoxa*, forma].

alni, Polyporus, Sorok. 1892: 142 *pl. 22 | f. 230* (U.S.S.R., Caucasia), not ∼ Velen. 1922; *Fomes* Sacc. 1895; *Scindalma* O.K. 1898.—Nomen dubium. Lloyd 1915 (LMW 4, F.): 278 thought of *Fomes* [*Fomitopsis*] *roseus* or *F.* [*Fomitopsis*] *pinicola*. Compare also Bond. 1953: 618.

alni, Polyporus, Velen. 1922: 650 [see Pilát 1948: 249 for Latin translation] (Czechoslovakia), not ∼ Sorok. 1892.—Referred by Pilát 1938 (ACE 3): 185 to *Leptoporus* [*Tyromyces*] *lacteus* in a very inclusive sense. Velen. described the spores as ovoid-globose, 4–5 μ in diam.

alpinus, Polyporus, Saut. 1876 (H 15): 33 (Germany).—Nomen dubium. Insufficiently described; referred by Pilát 1936 (ACE 3): 87, 93 to *Polyporellus* [*Polyporus*] *squamosus* f. *rostkowii* (Fr.) Pilát, but in my opinion this is an unacceptable guess. Saut. 1877 (H 16): 72 published a second descr. based on a different collection, which may well have been another species.

alutaceus, Polyporus, Fr. 1821: 360 (Sweden); *Boletus* Spreng. 1827, not ∼ Fr. 1821 (syn., error for *B. alutarius* Fr.: n.v.p.); *Bjerkandera*

P. Karst. 1882; *Leptoporus* Quél. 1886; = *Polyporus delens* E. Krause 1930.—Nomen dubium v. ambiguum. Cf. Romell 1926 (SbT 20): 4. — Sensu Rostk. 1830 (StP 4): 57 *pl. 27* [= *Polyporus griseus* (Wint.) Pilát (**O**)] = ?; referred by Fr. 1838: 454 to *Polyporus* [*Tyromyces*] *destructor* var. b, but his interpretation of *P. destructor* (**O**) is an enigma; cf. also Romell 1926 (SbT 20): 9–10, Bourd. & G. 1928: 546, 547. — Sensu Britz. 1887 (BnS 29): 277 [*pl. 601 f. 31*], as *Polyporus destructor* f. *alutaceus* (Fr.) Britz. = ? (no descr.). — Sensu Bres. → *Tyromyces stipticus*, & cf. Romell 1926 (SbT 20): 3, 6, Lundell 1946 (LNF 27–28): 8 No. 1316, in obs. Fries's descr. is too ambiguous firmly to support this interpretation. Moreover, the specific autonomy of *Polyporus alutaceus* sensu Bres. has been doubted, even by Bresadola himself: cf. Bourd. & G. 1928: 545 in obs., ". . . *P. alutaceus* Fr., qui, selon M. Bresadola (in litt.), est très voisin de *L*[*eptoporus*] *albidus* [= *Tyromyces stipticus*] et en est peut-être une simple variété bien développée et jaunissante." Lundell 1946 (LNF 27–28): 8 was inclined to accept Bresadola's interpretation: "There are two and only two species described by Fries in Syst. myc. that may refer to [*Polyporus* [*Tyromyces*] *stipticus*], viz. *P. stipticus* and *P. alutaceus*. Fries attributed, to be sure, the latter name to old specimens of *P. stipticus*, in which the surface of the pileus had turned yellow." — Sensu Lloyd 1915 (LMW 4, Ap.): 301–302 → *Tyromyces guttulatus*, at least in part. Lloyd not only thought that Bresadola had applied the name to European material of *Polyporus guttulatus*, but he also accepted it as the correct name of this North American species; at the same time he reported it from Europe. The autonomous status of *Tyromyces guttulatus* accepted in this Check list still needs careful investigation (**133**). — Romell 1926 (SbT 20): 4 suggested *Polyporus velutinus* [= *Coriolus pubescens*], but this is not supported by the protologue. — The valid publication of the name *Polyporus delens* depends on the ref. (no descr.). The name was presumably intended as a substitute for *P. alutaceus* sensu Rostk., but since Fries's original conception was not explicitly excluded, the ref. "*Pol. alutaceus* St. 10, 27" must be regarded as an indirect ref. to *P. alutaceus* Fr. itself. Later E. Krause 1935 (ANM II 9): 13 referred his material to *Polyporus imberbis* [= *Bjerkandera fumosa*], but it is doubtful whether this correction is correct.

amanitoides, Daedalea, P. Beauv. → *Lenzites palisotii* (Fr.) Fr. (**O**).

**amara, Poria,* Burt in herb.—Fide Bres. 1926: 80 = *Poria* [*Perenniporia*] *medulla-panis* (forma). In my opinion 'amara' is an error for 'omaema' and apparently its use a misapplication of the name *Poria omaema* [= *Perenniporia subacida*].

amarissimus, Polyporus, Post in herb.—Fide Romell 1926 (SbT 20): 3, 6 = *Polyporus* [*Tyromyces*] *stipticus*.

**amesii, Polyporus,* Lloyd 1915 (LMW 4, Ap.): 309 & 1915 (LMW 4, L. 60): 9 (U.S.A., New York) (binomial introduced for a subdivision of a sp.: n.v.p.).—Considered by its author to be a 'marked collection' and 'a marked deviation' of *Polyporus* [*Bjerkandera*] *fumosa*; & cf. Overh. 1953: 367.

Amphitretia Hill 1751 (pre-Linnaean name) [1960 (Pe 1): 185].—Apparently non-hymenomycetous.

amplissimus, Fungus, Scop. 1772 P.s.: 116 *pl. 44* (Hungary, now Czechoslovakia) (nom. anam.?) (d.n.).—Nomen dubium. Tode 1790: 36 referred this to *Xylostroma giganteum* (29), but the mode of origin appears to be quite different.

**amurensis, Fomes*, Bond. in herb. (U.S.S.R., Russia, Siberia).—Fide Murašk. 1940: 10, a name given to a collection of *Ungulina* [*Fomitopsis*] *cytisina*.

**angolensis, Polyporus*, Lloyd 1920 (LMW 6): 997 *pl. 163 f. 1801* (Angola).— Lloyd compared this with *Polyporus* [*Laetiporus*] *sulphureus*, and Pilát 1936 (ACE 3): 41 listed it with a question mark as a syn. of *Grifola* [*Laetiporus*] *sulphurea*. The sp. was described insufficiently.

angustata, Lenzites, S. Schulz. in MS. & litt.—Fide Fr. 1874: 494 = *Lenzites* [*Daedaleopsis*] *tricolor*, a determination refuted by S. Schulz. who published his sp. as *Lenzites bresadolae* (**O**).

aniseus, sulfureo-albescens, & sulfurescens, Poria, Romell in herb.— Herbarium names mentioned by Jo. Erikss. 1949 (SbT 43): 20–22 as syns. of *Poria xantha*.

Anisomyces Pilát 1936, 1940 (lacking Latin descr.: n.v.p.), not ∼ Theiss. & Syd. 1914 (Ascomycetes) [1960 (Pe 1): 186]. − Monotype: *Trametes* [*Gloeophyllum*] *odorata*.

annosus, Boletus, P. Magn. 1905: 180 ("Swartz") (syn.: n.v.p.).—An error for *Polyporus* [*Fomitopsis*] *pinicola* "B. *hornotinus et annosus*" Fr. 1874: 561, in part.

**annularis, Polyporus*, Fr. sensu Overh. 1953: 103 *pl. 63 f. 374* (*Fomes*).— Overh. used the wrong name for the sp. he described; he used it in error for *Fomes annularis* Lloyd 1912 (LMW 4, L. 40): 6 (Union of S. Africa). The same error was made when *P. annularis* Fr. was transferred to *Ganoderma* by R. L. Gilb. apud Lowe & Gilb. 1962 (M 53): 505 ("*annularis*"). The species should be called *Ganoderma annulare* (Lloyd) Boed. 1940 (BBu III 16): 391; Humphr. & Leus 1931 (PJS 45): 509, 528, 533, 565 (incidental mention: n.v.p.). − Kotl. & P. 1971 (ČM 25): 94, 99 listed "*Fomes annularis* Fr. sensu Overholts" as a possible syn. of *Ganoderma adspersum*, going by Overholts's descr.

**antarctica, Fistulina*, Speg. 1887 (BCó 11): 20 (Argentina).—Referred to *Fistulina hepatica* by Bres. 1916 (Am 14): 222 but Sing. 1969 M.a.: 380 considers it a good sp.

**anthelminticus, Polyporus*, Berk. 1866 (GCh): 753 (Burma). − This was compared by Berk. with *Polyporus rufescens*, which resulted in the name being listed as a doubtful syn. of *Heteroporus* [*Abortiporus*] *biennis* by Pilát 1937 (ACE 3): 115. − Cf. (2).

apalus, Polyporus Lév. 1848 (ASn III 9): 124 (France), not ∼ Berk. 1843; *Polystictus* Cooke 1886, Sacc. 1888 (*"hapalus"*); *Leptoporus* Pat. 1900 (*"hapalus"*); *Coriolus kymatodes* subsp. *C. apalus* Bourd. & G. 1928, misapplied; *Oligoporus* R. & O. Falck 1937 ("syn. *Coriolus apalus* Bourdot et Galzin"); *Coriolus* R. & O. Falck 1937 (syn.: n.v.p.); ≡ *Tyromyces leveilleanus* Bond. apud Bond. & S. 1941 (Am 39): 52 (nom. nud.: n.v.p.), *Polyporus* Bond. 1953 (syn.: n.v.p.), fide Bond. 1953: 714. – Sensu Bourd. & G. → *Tyromyces apalus* sensu Bourd. & G. (**127**); sensu Kallenb. → *Oligoporus rennyi*.

apiospora, Reticularia*, B. & Br. 1873 (JLS 14): 82 (Ceylon).—Cf. (41**).

Aporpium Bond. & S. ex Sing. 1944 [1958 (Ta 7): 166]; holotype, *Poria canescens* P. Karst. (**0**) (Tremellaceae).

appendiculatus, Polyporus*, B. & Br. 1873 (JLS 14): 48 (Ceylon).— Referred to *Laetiporus miniatus* (0**) by Over. 1925 (IFm 12): 1, 4. Lloyd 1912 (LMW 3, S.P.): 154 considered *Polyporus* [*Laetiporus*] *miniatus* not specifically distinct from *Polyporus* [*Laetiporus*] *sulphureus*. – Descr.: Petch 1916 (APe 6): 94, 125.

applicatus, Agarico-suber, (Batsch) Paul. 1793 → *Agaricus resupinus* Paul. (**0**).

arboreus, Boletus, Sow. 1802: *pl. 346* (d.n.) per Purt. 1821 (England); ≡ *Xylomyzon pelliceum* Pers. 1825.—Not a polypore; apparently *Serpula* sp.

**arctica, Trametes*, Berk. in herb. (Canada); Lloyd 1915 (LMW 4, F.): 225, 283 (as a form of *Trametes carnea* [sensu Lloyd]: n.v.p.); *Trametes carnea* var. Pilát 1940 (ACE 3): 321 (lacking Latin descr.: n.v.p.).—Fide Weir 1923 (Rh 25): 216 = *Trametes subrosea*; fide Lowe 1957 F.: 70 = *Fomes cajanderi*. Erroneously referred to *Fomes* [*Fomitopsis*] *roseus* by Overh. 1953: 56.

**arcticus, Polyporus*, Fr. 1838: 479 ("Kamtschatka"); *Polystictus* Cooke 1886; *Microporus* O.K. 1898.—Nomen dubium. Murrill 1907 (NAF 9): 28 designated the specimen from Kamčatka as type ("not found") and concluded that the sp. was "evidently near *C*[*oriolus*] *nigromarginatus* [= *C. hirsutus*] or *C.* [*Hirschioporus*] *abietinus*"; judging from the descr. both suggestions seem doubtful. The other specimen mentioned by Fr. is presumably *Hirschioporus pargamenus*; cf. Berk. 1841 (AM 7): 452.

**arctostaphyli, Fomes*, Long 1917 (U.S.A., Arizona).—Overh. 1953: 60, 62 referred this to *Fomes* [*Phellinus*] *igniarius* in the broad sense he accepted. However, specimens from Arizona are held to represent a distinct sp. by R. L. Gilb. (in herb.) and judging from some of his collections I would agree. – Descr.: Long 1917 (PNP 1): 2; Zeller 1934 (M 26): 299 *fs. 4, 5*; D. Baxt. 1952 (PMi 37): 96 *pl. 1 f. 2* (*Fomes*).

areolatus, Boletus, Batsch → *Boletus coriaceus* Batsch 1786 (**0**).

argenteum ("*Sistotrema argenteum*").—Fide Fr. 1828 E. 1: 94 = *Polyporus* [*Coriolus*] *versicolor*. – Error for *Polyporus argyraceus* Pers. [= *Cerrena unicolor*]? Fr., l.c., wrote, "Etiam sub *Sistotrem. argentei* nomine accepi."

argillaceus, Polyporus, Cooke 1878 (G 7): 1 (U.S.A., California), not ∼ (Murrill) Overh. 1926; *Poria* Cooke 1886.—Fide Teix. & Rog. 1955 (M 47): 413 = *Aporpium caryae* (**O**).

armoraceus, Polyporus, Hook. f. in herb. (India).—Referred to *Polyporus crispus* [= *Bjerkandera adusta*] by Berk. 1851 (HJB 3): 82. – This is apparently the same as "*P*[*olyporus*] *armoracius* Berk.! Ined." Fr. 1851 (NAu III 1): 53/37 (nom. nud.: n.v.p.) ex Sacc. 1888 (SF 6): 102. Fr. considered it to be close to *Polyporus secernibilis* (**O**).

asbestinus, Boletus, Scop. 1770: 148 (Hungary, now Czechoslovakia) (d.n.).—Fr. 1828 E. 1: 128 referred this to *Fistulina hepatica*, but the original descr. does not agree at all with this identification.

atripes, Polyporus, S. Schulz. 1866: 42 (Yugoslavia, Slavonia) (nom. nud.: n.v.p.).—Nomen dubium.

atrofuscus, Polyporus, Velen. "in herb.", not ∼ (Schaeff.) Secr. 1833.—Fide Pilát 1942 (ACE 3): 501 = *Phellinus torulosus*.

atrorufus, Boletus, Pers. 1800: 107 & Fr. 1821: 369 (syn.).—An error for *Boletus atrofuscus* Schaeff. [= *Coriolus versicolor*].

aurantiacum, Sistotrema, Secr. 1833 M. 2: 503 (Switzerland).—Nomen dubium.

aurantiacus, Polyporus, Rostk. 1838 (StP 4): 119 *pl. 58* (Germany/Poland), not ∼ Lasch 1853, not ∼ Peck 1874; *Physisporus* P. Karst. 1887; *Poria* Sacc. 1891; *Hapalopilus* Bond. & S. 1941 ["(Karst.)"].—Nomen dubium, fide Donk 1967 (Pe 5): 80. – Sensu Bres. → *Poria salmonicolor*.

aurantiacus, Tyromyces, (E. Komar.) E. Komar. 1964; *Tyromyces albellus* forma E. Komar. 1959 (U.S.S.R., White Russia).—Apparently lacking indication of type, hence both names n.v.p.? – Descr.: E. Komar. 1959 (BMs 12): 254 *f. 6* (*Tyromyces albellus* f.); 1964: 89 *f. 27* (*Tyromyces*).

aurantius, Mucor, Bull. 1790: *pl. 504 f. 5* & 1791 H.: 103 (nom. anam.) (d.n.); *Aegerita* DC. 1805: 72 (d.n.), Fr. 1832: 423 ("*aurantiaca*"; syn.: n.v.p.).—This was reduced to the synonymy of *Sporotrichum aureum* Link [≡ *S. aurantiacum* Fr. = imperfect state of *Poria metamorphosa*] by Link 1824: 15, in which he was followed by Fr. 1832: 423. It is certainly a different sp. from the modern interpretation of *S. aurantiacum*. No recent interpretation known to me.

aurea, Lenzites, Velen. 1930 (MP 7): 18, 19 (Czechoslovakia).—Velen. compared this with *Lenzites flaccida* [= *L. betulina* forma] but his descr., although incomplete, strongly suggests to me *Phyllotopsis nidulans* (Pers. ex Fr.) Sing. Not mentioned by Pilát 1936–42 (ACE 3).

aurea, Poria, Peck 1890 (U.S.A., New York); *Leptoporus* Pat. 1900; *Chaetoporellus* Bond. 1953.—Recorded from Europe by Pilát 1933 (H 73): 31 *f. 1, pl. 1*, 1941 (ACE 3): 401 *f. 171, pl. 254*, but according to Lowe 1966: 68 these records were based "on specimens similar to the type except in the intensely orange-red color of the context, which is very different from the much paler context in the American plant." More recently Dr. J. L. Lowe (oral communication, Dec. 1969) told me that as far as he knew *Poria aurea* had not been found in Europe. Descriptions drawn up from Pilát's collections [Pilát, l.l. c.c.; Domański 1965 (FpG 2): 120 *f. 42*] do not show the remarkable two- to many-rooted bases of the cystidia revealed by the figures and mentioned in the descriptions based on American material: Overh. 1919 (BNS 205–206): 74 *pl. 3, pl. 4 fs. 1, 2*; Lowe 1966: 67 *f. 44* (*Poria*).

aureum, Acrosporium, Pers. 1822: 25 (nom. anam.); *Oidium* Link 1824 (excl. of *Trichoderma aureum* Pers., cf. Link 1824: 59), Steud. 1824: Fr. 1832: 429 (excl. of *Trichoderma aureum* Pers., cf. p. 418); *Torula* Corda 1829; *Oospora* Wallr.; [≡ *Trichoderma aureum* Pers. sensu Link 1809 (MBe 3): 18 *pl. 1 f. 29* (*Oidium*), excl. of type, (Germany)].— S. Hugh. 1958 (CJB 36): 811 cited this sp. in the synonymy of *Trichoderma aureum* Pers. [= imperfect state of *Poria metamorphosa*] in my opinion incorrectly so because, by the exclusion of the type of *Oidium aureum* (Pers.) Link sensu Link 1809, the 'new' names *Acrosporium aureum* Pers., *Oidium aureum* Link 1824: Fr. 1832, &c. came into being, all based on the same new type, viz. Link's material of 1809. Cf. also Donk 1962 (Ta 11): 91–92.

aureus, Agaricus, Graff 1936 (M 28): 159 (syn.: n.v.p.); [≡ *Agaricus aureus flabelli effigie* Batt. 1755: 68 *pl. 37 fs. A, B* (Italy)].—A syn. of *Polyporus squamosus.*

aureus, Polyporus, Berk. 1843 (AM 10): 373 ("Without habitat"), not ∼ Secr. 1833 (n.v.p.), not ∼ Lloyd 1917; *Trametes* Sacc. 1888.—Nomen dubium. P. Cout. 1919: 57 suggested that a collection from Welwitsch ("Fung. Lusit. n. 129 − Febr. 1844") found in Portugal (of which he mentioned some features taken from Welwitsch's notes) is perhaps a portion of the type. This is not likely since Welwitsch's specimen was collected in 1844 and the species was described in 1843.

aureus, Polyporus, Secr. 1833 M. 3: 153 (Switzerland) (as a sp. of *Boletus*: n.v.p.), not ∼ Berk. 1843, not ∼ Lloyd 1917.—Nomen dubium. Secr. cited *Boletus lacrymans* [≡ *Serpula lacrimans*] (0) sensu Bolt. 1789: 167 *pl. 167 f. 1*; figure and descr. by Bolt. are poor, but it is not impossible that Bolton's determination was correct. However, the descr. by Secr. does not agree at all with *Serpula lacrimans*; he apparently described some abnormal condition of a fungus found "dans un vivier des bains soufres sur les planches . . . qui en formaient l'encaissement."

auricoma, Polyporus, Lév. ("in Herb.") ex Cooke 1886 (G 15): 2b (Marquesas Is.); *Poria* Cooke 1886 (nom. nud., not definitely accepted: n.v.p.), Lowe 1963 (incomplete ref.: n.v.p.).—Lowe 1966: 86 stated that "*P. auricoma* (Lév.) Cooke, *P. chlorina* Massee, and *P. mellea* (Berk. & Br.) Cooke . . . are very similar [to *Poria* [*Ceriporia*] *reticulata*] and very probably synonyms." − Descr.: Lowe 1963 (M 55): 458 (*Poria*).

auricomum, Ozonium, Link 1809 (MBe 3): 21 (Germany) (nom. anam.) (d.n.) per Link 1824: 138, type of *Ozonium* Link 1809 (d.n.) per Fr.

1821, a genus of sterile mycelia (Deuteromycetes). − Pilát 1927 (Sčz 2): 481, 483 applied the name *O. auricomum* to mycelial growth of *Trametes* [*Gloeophyllum*] *odorata*, which is evidently a misapplication. For a more current interpretation, see Donk 1962 (Ta 11): 94. The correct 'specific' name must still be worked out.

auricula-major, Conchites, Paul. → *Fungoides hyosotis* Paul. (**O**).

auriculatus, Boletus, Vill. 1779: 56 (*auticulatus*) (France) (d.n.); = *Boletus normalis* Vill. 1798 (d.n.).—Nomen dubium.

**australis, Polyporus*, Fr. 1828 E. 1: 108 ("in insulis Oceani pacifici"); *Fomes* Cooke 1885; *Placodes* Quél. 1886, *Ganoderma* Pat. 1888. − Nomen dubium. − Sensu Pat. 1889, at least in part, → *Ganoderma applanatum*; sensu Fr. in herb. K. → *G. adspersum*. − Cf. Donk 1969 (PNA 72): 276.

**avellanei-albus, Tyromyces*, Murrill 1939 (BTC 65): 657 (U.S.A., Florida).— Lowe apud Overh. 1953: 426 remarked, "appears to agree with *Polyporus* [*Tyromyces*] *fumidiceps*."

**badius, Polyporus*, Berk. 1841 (AM 7): 462 (Arctic N. America is an error), not ∼ (Pers. per S. F. Gray) Schw. 1832, not ∼ (Berk.) Lév. 1846, not ∼ Jungh. ex Bres. 1912; *Fomes* Cooke 1885 (nom. nud.: n.v.p.), 1886, typonym; *Scindalma* O.K. 1898; *Phellinus* G. Cunn. 1965.—Murrill 1903 (BTC 30): 111 referred this sp. to *Pyropolyporus* [*Phellinus*] *igniarius*, which is an error, fide Murrill 1908 (NAF 9): 111, Lloyd 1915 (LMW 4, F.): 249, and Lowe 1957 F.: 29.

balloui, Polyporus, (Lloyd) Lloyd 1915; *Polyporus rufescens* forma Lloyd 1914 (U.S.A., New York); *Spongiosus* Torrend 1924.—Lloyd 1915 (LMW 4, L. 58): 7 listed *Polyporus* [*Abortiporus*] *tropicalis* [= *P. fractipes* (**O**)] as a syn.; Overh. 1953: 224 referred it to *Polyporus biennis* [sensu Overh. = *Abortiporus distortus* (**O**)]; and Fid. 1969 (Ri 4): 150, 153, 162, 167 to *Heteroporus biennis* var. *flabelliformis* ("Mont") Fid. [sensu Fid. = *Abortiporus distortus*; see Donk 1971 (PNA 74): 2]. − V.s.: "*balloni*". − Cf. (2). − Descr.: Lloyd 1923 (LMW 7): 1191 *pl. 236 f. 2395* (*Polyporus*).

balticum, Sistotrema, A. Dietr. apud P. Karst. 1882 (BFi 37): 83 ("Heimar i Estland").—Nomen dubium. A sp. with stalked, brown-black fruitbody.

**bambusarum, Poria*, J. Rick 1937 (Bro 6): 146 (Brazil).—Referred by Lowe 1966: 163 to *Poria* [*Phellinus*] *punctata* but in view of the presence of setae, previously recorded by Lowe 1963 (M 55): 472, this determination is doubtful.

**bankeri, Inonotus*, Lloyd 1910 (LMW 3): 475 ("McGinty"; not accepted: n.v.p.); *Polyporus* Lloyd 1915 ("McGinty"; syn.: n.v.p.); [= *Hydnum strigosum* Sw. sensu Schw. 1832: 162, excl. of type, (lacking descr.) (U.S.A.)].—Fide Lloyd, l.c., 1915 = *Polyporus* [*Inonotus*] *hispidus*, "with large split pores".

barbigonus, Polyporus, Opiz in herb.—Fide Pilát 1942 (ACE 3): 592 = *Trametes* [*Coriolus*] *hirsuta*. Error for 'barbizonus'?

**barteri*, [*Polyporus*], Berk. in herb. (type locality not mentioned).—Fide

Cooke 1886 (G 15): 53 = "variety of *biformis*, Fr. [= *"Trametes"* *cervina*]."

bartholomaei, *Polyporus*, Peck 1896 (BTC 23): 418 (U.S.A., Kansas); *Tyromyces* Murrill 1907.—Referred by Lloyd to *Polystictus velutinus* [sensu Lloyd = *Coriolus* sp.], but cf. Overh. 1953: 254, who thought of *Polyporus* [*Abortiporus*] *fractipes* (**0**).

bathyporus, *Polyporus*, Rostk. 1838 (StP 4): 121 *pl. 59* (Germany/Poland); *Poria* Cooke 1886. – Nomen dubium. Pilát 1942 (ACE 3): 592 referred it, with doubt, to *Inonotus radiatus* f. *nodulosus* Donk [= *Inonotus nodulosus*] but the protologue does not support this suggestion.

batschii, *Polyporus*, Pers. → *Boletus coriaceus* Batsch 1786 (**0**).

bicolor, *Lenzites*, Fr. 1851 (NAu III 1): 43/27 (Mexico), not ∼ Falck 1909.—Referred to *Agaricus* [*Daedaleopsis*] *confragosus* by Murrill 1905 (BTC 32): 86, 87. – Cf. (**31**).

bicolor, *Polyporus*, Wallr. 1833: 583 (Germany), not ∼ Pers. apud Gaud. 1827 (n.v.p.), not ∼ Jungh. 1838.—Nomen dubium. This was described as resupinate and, therefore, should not be compared with sessile *Polyporus varius* as was done by Sacc. 1888 (SF 6): 84.

bicolor, *Poria*, Ell. & Langl. in herb. (U.S.A., ? Louisiana), not ∼ Bres. ex Theiss. 1911.—Fide Bres. 1920 (Am 18): 68 = *Poria* [*Rigidoporus*] *nigrescens*.

bifasciata, *Lenzites*, Cooke & Mass. apud Cooke 1892 (G 21): 37 (Australia, Victoria); *Cellularia* O.K. 1898.—Referred by G. Cunn. 1950 (PNW 75): 243 first to *Daedalea* [*Gloeophyllum*] *subferruginea* (Berk.) G. Cunn., an extra-European sp., later (G. Cunn. 1965: 252, 253) to *Gloeophyllum abietinum*. The former identification was retracted by G. Cunn. himself; the latter is doubtful as his descr. of *G. abietinum* does not suggest this sp.

bireflexus, *Polyporus*, B. & Br. ("MSS.") ex Cooke 1882 (G 10): 101 (Australia, Queensland); *Polystictus* Cooke 1886; *Microporus* O.K. 1898.—G. Cunn. 1965: 167, 168 referred this to *Trametes* [*Coriolus*] *zonata*. His interpretation of the latter sp. needs confirmation.

boganidensis, *Daedalea*, J. Boršč. 1856: 142 (U.S.S.R., Siberia).—Nomen dubium. This was referred to *Trametes* [*Antrodia*] *serpens* by Killerm. 1943 (Am 41): 244, perhaps for no other reason than that its author compared it with that species. The descr. hardly suggests the identification.

boltonii, *Hydnum*, Spreng. → *Boletus obliquus* Bolt. (**0**).

bombacina, *Institale*, (Fr. apud Weinm.) Fr. sensu Lloyd.—See under *Septocylindrium lindtneri* (**0**) and (**41**).

bonariensis, *Daedalea*, Speg. 1902 (ABA III 1 / 8): 52 (Argentina).—Bres. 1916 (Am 14): 230 referred this to *Polyporus* [*Abortiporus*] *biennis* [sensu lato] and Speg. 1926 (BCó 28): 370 followed him ("forma truncicola merismoideo-frondosa, hymenio daedaloideo"). Fid. 1969 (Ri 4): 150, 162 made it a syn. of *Heteroporus biennis* var. *flabelliformis* ("Mont.")

Fid. [sensu Fid. = *Abortiporus distortus* (**O**); see Donk 1971 (PNA 74): 2],
apparently without having seen the type. – Cf. (**2**).

borealis, *Fomes*, Lloyd 1915 (LMW 4, F.): 247 (Canada, Ontario).—
Insufficiently described; no re-description available. Lowe 1957 F.: 56
referred it to a very inclusively conceived *Fomes* [*Phellinus*] *igniarius*.
The poor descr. contains, "Pileus . . . with a thin, pale, smooth, hard
crust, variegated with darker spots". This, in combination with the
host (*Betula*), would perhaps suggest *Phellinus nigricans* s. str.

botryites = *botryoides*

botryoides, *Boletus*, Humb. 1793: 103 *pl. 3 f. 9* ("*botryoeides*" & "*botryites*")
(Germany) (d.n.) per Steud. 1824; *Polyporus* Pers. 1825, not ∼ Lév.
1846.—Nomen dubium; approximate systematic position in doubt,
apparently not a polypore or else a monstrosity, found in mines.
Recorded from Austria by Wettst. 1885 (VW 35): 362 (no descr.).

botulatus, *Polyporus*, Secr. 1833 M. 3: 80 (Switzerland) (nom. nud. & as
a sp. of *Boletus*: n.v.p.). – As to var. A. possibly *Phellinus robustus*.

boucheanus, *Favolus*, Kl. 1833 (Li 8): 316 *pl. 5* (Germany); *Polyporus*
Fr. 1838; *Polyporellus* P. Karst. 1882; *Cerioporus* Quél. 1886.—Nomen
dubium & ambiguum. Interpreted in different ways, none of which is
satisfactorily supported by the original figures: cf. Donk 1969 (Pe 5):
241. – Sensu Lloyd → *Polyporus floccipes*; sensu Bres. 1915 → *Polyporus
alveolarius* (Bosc) per Fr.; sensu Berk. *in* Lea 1849: 58 (lacking descr.),
fide Morg. 1886 (JCi 9): 5, = *Favolus* [*Polyporus*] *canadensis* (**O**).

brachyporus, *Boletus*, Pers. "ined."—Referred by Fr. 1828 E. 1: 63 ("sec.
specimen a Cel. Chaillet missum, *Boletus brachyporus* Pers.! ined.
dictum . . .") to *Xylomyzon crustosum* Pers. 1825, which he listed as
syn. of *Merulius serpens* Tode per Fr. ('Corticiaceae').

brachytubus, *Polyporus*, S. Schulz. "Manuscript, p. 760, 1869" (unpublished:
n.v.p.) (Hungary). – Fide Igmándy 1970 (Aph 5): 287, 288 = *Phellinus
robustus*.

bresadolae, *Lenzites*, S. Schulz. 1885 (H 24): 142 ("Hungaria et Slavonia");
lectotype, depicted by Kalchbr. 1875: *pl. 30 f. 4a* as *Lenzites* [*Daedaleopsis*] *tricolor*. Previously published as *Lenzites angustata* S. Schulz.
(nom. nud.: n.v.p.) (**O**), apud Kalchbr. 1875: 49 (syn.: n.v.p.).—Nomen
dubium.

bresadolae, *Polyporus*, S. Schulz. 1885 (H 24): 145 (Yugoslavia, Slavonia),
not ∼ (Bourd. & G.) Lowe apud Gilb. & Lowe 1962 (n.v.p.); *Polystictus* Sacc. 1888; *Microporus* O.K. 1898.—Nomen dubium.

brevipes, *Polyporus*, Opiz 1855 (Lo 5): 87 (Czechoslovakia).—Nomen
dubium.

brevipora, *Poria*, Speg. 1899 (ABA 4): 172 (Argentina).—This was
referred to *Porotheleum* [*Stromatoscypha*] *fimbriatum* (**O**) by Bres. 1916
(Am 14): 228, but according to Lowe 1966: 47 it "appears to be better
referred to the genus *Merulius* ['Corticiaceae']."

brevizonatus, *Polyporus*, Harz 1868 (BSM 41): 18 (as a "Form" of *Polyporus*

[*Agaricum*] *officinalis*: n.v.p.) ("von unbekannter Herkunft").—Nomen dubium. Insufficiently described.

**brisbanensis*, [*Polyporus*], B. & Br. in herb. (Australia, Queensland).— Listed by Cooke 1886 (G 15): 53 as a syn. of *Trametes* [*Truncospora*] *ochroleuca*; & cf. Lloyd 1915 (LMW 4, Ap.): 376, G. Cunn. 1965: 266.

britzelmayri, Lenzites, Killerm. 1925 (Dba 16): 31 (Germany).—Nomen dubium. Insufficiently described. Based on a personal collection (lecto-type) and *Agaricus flaccidus* Bull. sensu Britz. 1893 (BCb 54): 100 [*pl. 557 f. 6*]. The latter is certainly not a sp. of *Lenzites*, but a true agaric.

**brownii, Elfvingia*, Murrill 1915 W.P.: 29 (U.S.A., California); *Fomes* Lloyd 1924 (syn.: n.v.p.), Sacc. & Trott. apud Trott. 1925; *Ganoderma* R. L. Gilb. apud Lowe & Gilb. 1962. – Kotl. & P. 1971 (ČM 25): 94, 99 listed this sp. as a possible syn. of *Ganoderma adspersum*, going by Overholts's descr. – Descr. Humphr. & Leus 1931 (PJS 115): 531, 565 *pl. 6, pl. 23 f. 1, pl. 25 f. 3, pl. 29 f. 2, pl. 30 f. 2, pl. 31 f. 3, pl. 33 fs. 9–12* (*Ganoderma applanatum* var.); Overh. 1953: 104 *pl. 78 f. 444, pl. 125 fig.* (*Fomes*).

brunneo-adherens, Poria*, Clel. & Rodw. 1929 (PTa 1928): 42 (South Australia).—Referred by G. Cunn. 1965: 200 to *Fuscoporia* [*Phellinus*] *laevigata*, but cf. (82**).

brunneus, Polyporus, Opiz 1852: 136 (Czechoslovakia) (nom. nud.: n.v.p.), not ∼ Pers. 1825, not ∼ Schw. (n.v.p.).—Nomen dubium.

**brunneus, Polyporus*, Schw. in herb. (U.S.A.), not ∼ Pers. 1825, not ∼ Opiz 1852 (n.v.p.).—B. & C. 1856 (JAP II 3): 214 reported on it as follows, "This appears to be the same with *P. crocatus* Fr., which is very near to, if not identical with *P. cupreus*, Berk." Referred by Lloyd 1913 (LMW 4, L. 50): 8 to *Polyporus* [*Inonotus*] *radiatus*; & cf. Lloyd 1915 (LMW 4, Ap.): 376.

brusinae, Polyporus, S. Schulz. 1886 (H 25): 9 (Yugoslavia, Slavonia); *Polystictus* Sacc. 1888; *Microporus* O.K. 1898.—Nomen dubium. *Coriolus* sp.?

bulbipes, Polyporus*, Fr. 1846 Pr.: 135 (Australia), not ∼ G. Beck 1889; *Trametes* Fr. 1848 (n.v.p.?); *Polystictus* Fr. 1851; *Microporus* O.K. 1898; *Xanthochrous* Pat. 1900.—Referred by Bres. 1916 (Am 14): 233 to *Polyporus connatus* Schw. [≡ *Coltricia focicola*], and by Pilát 1942 (ACE 3): 580, 582 to *Polystictus perennis* f. *cinnamomeus* (Jacq. per S. F. Gray) Pilát [≡ *Coltricia cinnamomea*]. – *Polyporus* [*Coltricia*] *oblectans* (O**) is an earlier name for this extra-European sp., which is taken here as a sp. of its own.

bullosus, Polyporus, Fr. apud Weinm. 1826 (nom. nud.: n.v.p.), 1836: 336 (U.S.S.R., European Russia); *Physisporus* P. Karst. 1882; *Poria* Cooke 1886, Quél. 1886.—Nomen dubium. Sensu Bres. in herb. and Bourd. & G. 1928: 675 = *Poria lenis*. Fide Dománski 1964 (APo 33): 171 specimens named by Pilát as *Poria calcea* var. *bullosa* (Fr. apud Weinm.) Pilát belong to *Poria xantha* f. *pachymeris* Jo. Erikss. *q.v.*

byssinoides, ciliato-byssina, & *subpallescens, Polyporus,* Romell in herb.—
Herbarium names mentioned by Jo. Erikss. 1949 (SbT 43): 7 as syns.
of *Poria byssina* Romell [≡ *P. romellii* Donk].

byssinus, Boletus, Schrad. 1794: 172 *pl. 3 f. 1* (Germany) (d.n.); *Poria*
Fr. 1832 (syn.: n.v.p.), Secr. 1833 (as a sp. of *Boletus*: n.v.p.), (Schrad.)
per Quél. 1888 misapplied; *Physisporus* Cost. & Duf. 1891, misapplied;
Polyporus E. Krause 1930, misapplied; *Tyromyces* Bond. 1953 (incomplete ref.: n.v.p.; "Pers."), misapplied.—Fide Fr. 1821: 506, Pers.
1825: 108 & Donk 1959 (Pe 1): 81, 82 = *Polyporus fimbriatus* ≡ *Stromatoscypha fimbriatum* (**O**). – Sensu Quél. → *Cristella molleusca*; sensu
Romell ("Pers.") → *Poria romellii*; sensu E. Krause 1930 B.r.: 97 = ?

byssinus, Fungus, Scop. 1772 P.s.: 113 *pl. 40* (Hungary, now Czechoslovakia) (d.n.); *Polyporus* (Scop.) per Pers. 1825, not ∼ (Schrad. per
Quél.) E. Krause 1930.—Nomen dubium.

byssoides, Polyporus, S. Schulz. 1858 (VW 7): 140 (without descr.) ≡ *Merulius* [*Phlebia*] *rufus* Pers. per Fr. ('Corticiaceae').

caespitosus, Irpex, S. Schulz. 1866: 40 (Yugoslavia, Slavonia) (nom. nud.:
n.v.p.).—Nomen dubium.

**caespitulans, Polyporus,* Schw. "olim".—Cited by Schw. 1832: 156 as a
syn. of *Polyporus* [*Coriolus*] *pubescens.* This may not be correct.

calceus, Polyporus, (Fr. ex Pers.) Schw. 1832, not ∼ B. & Br. 1873;
[≡ *Polyporus vulgaris* "*β. P. calceus*" Fr. 1821: 381 (Sweden)]; *Poly-
porus vulgaris* var. Fr. ex Pers. 1825; *Poria* Cooke 1886, Bres. 1908,
not ∼ (B. & Br.) Cooke 1886; *Physisporus* P. Karst. 1893; *Amyloporia*
Bond. & S. 1941 (generic name n.v.p.), Sing. 1944.—Nomen dubium,
fide Donk 1967 (Pe 5): 85. Variously applied. – Sensu Bres. 1897 →
Poria lenis; sensu M. P. Christ. → *Incrustoporia subincarnata*; sensu
Bond. & S. → *Poria xantha.*

Caloporia P. Karst. → *Caloporus* P. Karst. (**O**).

Caloporus P. Karst. 1881 [1960 (Pe 1): 192], not ∼ Quél. 1886; ≡ *Caloporia*
P. Karst. 1893 [1960 (Pe 1): 192].–Fide Donk 1962 (Pe 2): 227 the
monotype, "*C*[*aloporus*] *incarnatus* (Alb. et Schw.)", as interpreted
by Karsten, = *Merulius taxicola* (**O**). – (**110**).

cameleon, Polyporus, Chev. 1837: no. 40 var. *β* (as var. of *P. cristatus*:
n.v.p.) (France) = *Albatrellus cristatus.*

**canadensis, Favolus,* Kl. 1832 (Canada): Fr. 1832.—This has been
identified with *Hexagona*/*Favolus alveolaris* and *Favolus europaeus*
[= *Polyporus mori*] by North American mycologists like Murrill 1904
(BTC 31): 327; Lloyd 1909 (LMW 3, P.I.): 19; Overh. 1953: 157.
Cf. however (**98**). – Descr.: Kl. 1832 (Li 7): 197; Berk. 1839 (AM 3):
380; Morg. 1886 (JCi 9): 5; Hard 1908: 430 *f. 359*; Overh. 1914 (AMo 1):
148 (*Favolus canadensis*); A. Ames 1913 (Am 11): 241 *pl. 13 f. 67*;
Overh. 1953: 156 *pl. 80 f. 453, pl. 102 fs. 575, 576, pl. 125 fig.* (*Favolus
alveolaris*).

**canaliculatus, Polyporus,* Pat. 1898 (BmF 14): 153, not ∼ Overh. 1941;

Leptoporus Pat. 1900; *Baeostratoporus* Bond. & S. 1941; *Flaviporus* Bond. 1953 (incomplete ref.: n.v.p.).—Fide Bres. 1912 (Am 10): 494 = "*Polyporus*" *pusiolus* Ces. Incorrectly cited by C. & D. Over. 1922 (BBu III 4): 63 & Pilát 1938 (ACE 3): 220 as a syn. of *Fomes/ Leptoporus rufoflavus* [≡ *Flaviporus brownei*], presumably because Lloyd 1915 (LMW 4, F.): 278 wrote, "Compared by author to *Fomes rufo-flavus*."

candicans, Daedalea, P. Karst. 1911 (Ttk 12^{1,2}): 110 (U.S.S.R., 'Transbaikal').—Fide Lowe 1956 (M 48): 105 = *Daedalea* [*Daedaleopsis*] *confragosa* (forma). − Cf. (**31**).

candida, Poroptyche, G. Beck 1888 (VW 38): 657 *figs.* (Austria).—Nomen dubium & anam.: hymeniophore abnormal, sponge-like. A resupinate species as far as described, found in a court-yard on soil. The scanty descr. to some degree recalls *Poria vaillantii* ("subtus mycelii ramis funiformibus solo indefinite sed arcte affixus") but the spores as drawn are too narrow. Type of the name *Poroptyche* G. Beck, which is listed elsewhere in this publication as a syn. of *Poria*.

candidus, Gloeoporus, Speg. 1884 (ASa 17): 70 (Paraguay); *Leptoporus* Pat. 1900, not ∼ (Roth per Pers.) Quél. 1886; *Polyporus* Lloyd 1913 (nom. nud. & as a form of [*P.*] "*conchoides*"; n.v.p.), not ∼ (Pers. per Steud.) Pers. 1825, not ∼ (Roth per Pers.) Fr. 1838.—Referred by Bres. 1912 (H 53): 74 to *Gloeoporus dichrous*; by Lloyd 1915 (LMW 4, Ap.): 376 to *Polyporus* [*Gloeoporus*] *conchoides*; by Bres. apud Speg. 1919 (BCó 23): 411 to *Gloeoporus thelephoroides* (var.); but later Speg. 1926 (BCó 28): 365 claimed that it was a distinct sp. − Cf. (**51**).

canescens, Fomes, Bres. apud Torrend 1910 F.e.: No. 16 (Portugal) (nom. nud.: n.v.p.).—This name appears on a list of names printed on a single unpaged sheet, apparently distributed with Torrend's "Fungi selecti exsiccati. . . . Series I–IV. N. ^{os} 1–100" (1910), but also inserted as a folder in the periodical "Brotéria" (1912). The name appears as "16. *F*[*omes*] *canescens* Bres.", without any descr. I have seen only one packet of this number (BPI): the name is not to be found on its label, which bears the determination "*Poria Friesiana* Bres.", written in ink. This replacement of the original name is in agreement with later-published notes, which referred *Fomes canescens* to *Poria friesiana* [= *Phellinus punctatus*] and *Fomes* [*Phellinus*] *igniarius*, by Bres. apud Torrend and by Torrend 1913 (Bro 11): 64, 65.

canescens, Poria, P. Karst. 1887 (Rm 9): 10 (Finland); *Aporpium* Sing. 1944.—Fide Teix. & Rog. 1955 (M 47): 410 = *Aporpium caryae* (**O**).

capricida, Boletus, Rox. Clem. 1864: 64 (Spain) (nom. nud.: n.v.p.).—Since the author of this name did not distinguish between the genera *Boletus* and *Polyporus* this may be a name given to a polypore. In any case a nomen dubium.

carbonarius, Polyporus, Fr. 1821; *Polystictus* Cooke 1886; *Leucoporus* Quél. 1886; *Microporus* O.K. 1898; [≡ *Polyporus lignosus*, & *caespi-*

tosus, infundibulum imitans, superne nigricans, inferne cum pediculo albus, areolis carbonariis innascens Mich. 1729: 131 *pl. 70 f. 6* (Italy)].— Nomen dubium. The habitat and the figure suggest a species of *Coltricia*, particularly *C. perennis* or *C. focicola*, but the colours (see Micheli's phrase above) do not agree. I would suggest that Micheli entered his species in *Polyporus* instead of *Erinaceus* (= *Hydnum* sensu lato) by error and that he had before him a species of *Phellodon* P. Karst., e.g. *P. melaleucus* (Sw. per Fr.) P. Karst.

carbonarius, Tyromyces, Murrill 1912 (M 4): 94 (U.S.A., Washington); *Polyporus* Murrill (nom. altern.), not $\sim$ Fr. 1821.—Referred by Bres. 1926: 79 to *Polyporus albidus* [= *Tyromyces stipticus*], but Overh. 1953: 296 treated it as a distinct sp.

carnea, Trametes, Wettst. 1886 [1887?] (SbW 94): 64 (Austria), not $\sim$ (Bl. & Nees) Lloyd 1915.—Nomen dubium. Perhaps an alien ("in caldario").

carneo-albus, Irpex, Fr. 1838: 521 (Sweden); *Xylodon* O.K. 1898.—Nomen dubium. Sensu Velen. 1922: 741 = *Leptoporus* [*Tyromyces*] *mollis*, fide Pilát 1942 (ACE 3): 594.

carneus, Polyporus, Bl. & Nees 1826: 14 *pl. 3 fig.*: Fr. 1828 E. 1: 99 (Indonesia, Java); *Fomes* Cooke 1885; *Scindalma* O.K. 1898; *Ungulina* Pat. 1900; *Trametes* Lloyd 1915, Pilát 1932 & 1939, misapplied; *Fomitopsis* Imaz. 1943.—A species of the eastern tropics. – Sensu Ravenel, Berk. 1872 → *Fomitopsis cajanderi*. Also recorded for England by B. & Br. 1873 (AM IV 11): 342, almost certainly in error. – For some time this species was also confused with *Fomitopsis roseus*.

carpineus, Polyporus, S. Schulz. 1866: 41 (Yugoslavia, Slavonia) (nom. nud.: n.v.p.), not $\sim$ (Sow. per S. F. Gray) Trog 1844.—Nomen dubium.

carpini, Irpex, S. Schulz. 1866: 40 (Yugoslavia, Slavonia) (nom. nud.: n.v.p.).—Nomen dubium.

carteri, Poria, (Berk.) ex Cooke sensu G. Cunn. 1948 (BPZ 73) 10 *tpl. 1 f. 7*.—Referred to *Fuscoporia* [*Phellinus*] *punctata* by G. Cunn. 1965: 215, but this determination is doubtful. For a brief descr. of *Poria carteri* sensu stricto, see Lowe 1963 (M 55): 472.

caryae, Polyporus, Schw. 1832 (U.S.A., Pennsylvania); *Poria* Cooke 1886 = *Aporpium caryae* (Schw.) Teix. & Rog. (Tremellaceae). – Descr.: Teix. & Rog. 1955 (M 47): 410 *fs. 1–9* (*Aporpium*).

castaneus, Boletus, Web. 1787: 13 (Germany) (d.n.), not $\sim$ Bull. 1786 (d.n.) per Fr. 1821, not $\sim$ Fr. 1835 ("Müll.").—A sp. of Boletales, perhaps *Gyroporus castaneus* (Bull. per Fr.) Quél.: the descr. deviates on one point, "Stipes … reticulatim". In any case not *Polyporus* [*Ganoderma*] *lucidus*, with which Fr. 1821: 533 identified it.

castaneus, Polyporus, Fr. 1821: 369; = *Boletus populneus* Pollini 1816 H.: 34 (Italy) (d.n.); *Boletus* Pollini per Pollini 1824, not $\sim$ Web. 1787 (d.n.), not $\sim$ Bull. per Fr. 1821, not $\sim$ Fr. 1835 ("Müll."); *Scindalma* O.K. 1898; *Fomes* Sacc. 1916; = *Boletus veronensis* Spreng. 1827.— Nomen dubium. Lloyd's suggestion 1915 (LMW 4, F.): 278 that "the description seems to be *Fomes annosus*" is certainly incorrect. – Descr.: Pollini 1824: 611 *pl. 2 f. 5* (*Boletus populneus*). – Sensu Rostk. = ?; sensu Britz. (citing "Rostk.") → *Inonotus rheades* (*I. vulpinus*).

castanicola, Polystictoides, Lázaro 1916 (RMa 14): 834 / 1917: 146 (Spain); *Polystictus* Sacc. & Trott. apud Trott. 1925.—Nomen dubium.

cavernatus, Phellinus, Haracsi, Erdészeti növénykórtan 275. 1969 (nom. nud.). – Fide Igmándy 1970 (Aph 5): 298 = *Phellinus pilatii*.

ceciliae, Polyporus, Roum. 1888 F.e.: No. 4420 (France); *Poria* Sacc. 1888.—Nomen dubium.

cellaris, Boletus, Chaill. "in litt."; *Polyporus dryadeus* var. *cellaris* (Chaill.) ex Fr. 1828 E. 1: 108 (Switzerland).—Nomen dubium. *Phellinus contiguus* or *Poria expansa?*

cellulosus, Boletus, O. F. Müll. 1776 (Fd 4 / F. 12): 6 *pl. 716 f. 1* (presumably Denmark) (d.n.), not ∽ Lightf. 1778 (d.n.); *Boletus* O. F. Müll. per Wahl. 1826.—Nomen dubium. Wahl. 1826: 961 used the name to replace *Polyporus vulgaris* Fr. [= *Poria lenis*] after Fries 1821: 381 had declared the two to be synonymous.

celottianus, Polystictus, Sacc. & Manc. apud Sacc. → *Polyporus pulcher* Speg. (**O**).

cephalotes, Poria, Pers. 1799 O. 2: 15 (Germany) (d.n.); *Polyporus* (Pers.) per Pers. 1825.—Nomen dubium. Found in a mine; apparently an abnormal growth-form.

cerasi, Daedalea, S. Schulz. in MS. & in litt.—Fide Fr. 1874: 494 = *Lenzites* [*Daedaleopsis*] *tricolor*, of which it is "forma hymenio poroso insignis", fide Kalchbr. 1875: 49.

cerasi, Odontia, Pers. 1799 O. 2: 16 (Germany) (d.n.); *Sistotrema* Pers. 1801 (d.n.); *Hydnum* Poir. 1808 (d.n.); *Polyporus* (Pers.) per Fr. 1821, misapplied, not ∽ Richon 1889 (**O**); *Hydnum* Mérat 1821; *Sistotrema* Schw. 1822 (n.v.p.), Chev. 1826; *Irpex* Fr. 1828, misapplied; *Coriolus* Pat. 1900 ("Fr."); = *Hydnum* [*Hyphoderma*] *radula* Fr. ('Corticiaceae'). – Sensu Fr. 1821, 1828 → *Schizopora paradoxus*, fide Fr. 1838: 523 (sub *Irpex p.*) & cf. Donk 1967 (Pe 5): 86; sensu Rostk. 1838 (StP 4): 125 *f. 61* (*Polyporus*), referred by Bres. 1897 (AAR III 3): 77 to "*Polystictus*" *fibula* [sensu Bres. = *Coriolus hirsutus* forma], but in my opinion Rostkovius's plate and descr. are insufficient for determining the correctness of this suggestion.

cerasi, Polyporus, Richon 1889: 98 (France), not ∽ (Pers.) per Fr. 1821.— Nomen dubium. In certain respect the descr. suggests *Phellinus pomaceus*, but "tubes assez grands" appears to disagree.

cerasicola, Irpex, S. Schulz. 1866: 41 (Yugoslavia, Slavonia) (nom. nud.: n.v.p.).—Nomen dubium.

ceratoniae, Pseudofomes, Lázaro 1916 (RMa 14): 586 / 1917: 87 *pl. 9 f. 19* (Spain); *Fomes* Sacc. & Trott. apud Trott. 1925.—Nomen dubium. Cf. Bres. apud Trott. 1925 (SF 23): 388, "Videtur affinis *Fomes* [*Phellinus*] *Ribis* (sic!)."

Ceratophora Bond. & S. 1941 (Am 39): 54 (lacking Latin descr.: n.v.p.), not ∼ Humb. 1793 (d.n.) per Corda 1842 [1960 (Pe 1): 195]; [≡ *Cerato- phora* Humb. per Corda sensu Bond. & S., l.c., excl. of type]; holotype, *Trametes* [*Gloeophyllum*] *odorata* (Wulf. per Fr.) Fr.—*Ceratophora* Humb. per Corda (nom. anam.) was emended also to include the perfect state; this automatically shifted the type from the imperfect to the perfect state and in this way introduced a 'new' generic but homonymous name.

cerea, *Poria*, Scop. 1772 P.s.: 104 *pl. 22 f. 2* (Hungary, now Czechoslovakia) (d.n.), not ∼ (Berk.) Cooke 1886; ≡ *Polyporus cerinus* Pers. 1825.— Nomen dubium. Found in a mine: apparently an abnormal growth- form. – Sensu Hoffm. → *Polyporus cereus* Pers. [= *Rigidoporus undatus*].

cerifluus, Polyporus, B. & C. apud Berk. 1872 (U.S.A., South Carolina); *Tyromyces* Murrill 1907.—See (**132**). – Descr.: Lowe 1961 (PMi 46): 205 (*Polyporus*).

cerinus, *Polyporus*, Pers. → *Poria cerea* Scop. (**0**).

Ceriomyces Batt. 1755: 62 *pl. 24 f. A* (not a generic name), not ∼ Corda 1837 (gall), not ∼ Murrill 1909 (Boletales) [1960 (Pe 1): 195].—A monoverbal specific name for *Polyporus tuberaster*.

Ceriomyces Corda 1837 [1960 (Pe 1): 196], not ∼ Murrill 1909 (Boleta- ceae); monotype, *Ceriomyces fischeri* Corda (**0**).—Presumably based on a gall, cf. Donk 1972 (PNA 75): 165.

chailletianus, *Polyporus*, Pers. in herb. (Switzerland) (n.v.p.) ≡ *Polyporus aureolus* Pers.—Fide Donk 1933: 167 = *Gloeoporus* [*Skeletocutis*] *amorphus*.

chlorina, *Poria*, Mass. 1906 (BmI): 93 (Christmas I.).—See under *Poly- porus auricoma* (**0**). At first Lowe 1962 (PMi 47): 184 referred this to *Poria mellea* (**0**). G. Cunn. 1950 (PNW 75): 246, 1965: 63 listed it as a syn. of *Poria versipora* [= *Schizopora paradoxa*].

ciliatus, *Boletus*, Hornem. 1806 (Fd 8 / F. 22): 7 *pl. 1297* (Denmark) (d.n.), not *Polyporus ciliatus* Fr. 1818 (d.n.) per Fr. 1821.—This is referable to *Polyporus* subgen. *Leucoporus* (Quél.) Maubl. but it is difficult to be more precise. Fr. 1821: 348 referred it to *Polyporus brumalis* var. b. *vernalis* (not *P. vernalis*) as "forma singularis".

cinctus, *Polyporus*, Velen. "in herb."—Fide Pilát 1937 (ACE 3): 162 = *Gloeoporus* [*Bjerkandera*] *fumosus*.

cinereus, *Boletus*, Schum. 1803: 390 (Denmark) (d.n.), not ∼ Pers. 1801 (d.n.) per Pers. 1825.—Nomen dubium. Lind 1913: 392 listed this as a syn. of *Polyporus* [*Antrodia*] *serialis*, but the protologue hardly supports this identification.

cinereus, Polyporus, Schw. 1832: 159 (U.S.A., Pennsylvania), not ∼ Lév. 1846.—Nomen dubium. Cf. Teix. & Rog. 1955 (M 47): 414.

cinnabarinum, Sistotrema, Schw. 1822: 102 (U.S.A., Georgia); *Hydnum* Spreng. 1827: Fr. 1828; *Acia* P. Karst. 1879.—According to B. & C.

1856 (JAP II 3): 217, "This is . . . only a peculiar condition of *Pol.* [*Pycnoporus*] *cinnabarinus*"; but cf. Lloyd 1911 (LMW 3, L. 35): 3 and Rog. & Mart. 1958 (M 50): 307–308. A nomen dubium. – *Hydnum rubrum* Pers. 1825 was made a syn. of *H. cinnabarinum* and in this way the latter sp. came to be recorded for Europe by Fr. 1874: 615.

*cinnamomeus, *Irpex*, Fr. 1838: 524 (North America); *Hydnochaete* Pat. 1900, not ∼ J. Rick 1959 (n.v.p.) = *Hydnochaete olivacea* (Schw.: Fr.) Banker, an extra-European sp.—See under *Ceratium ferrugineum* Wallr. (**O**).

circulatus, Polyporus, Velen. 1922: 920 [see Pilát 1948: 255 for Latin translation]; ≡ *Polyporus zonatus* Velen. 1922: 671 (Czechoslovakia) (simultaneously published but rejected name: n.v.p.), not ∼ (Nees) per Fr. 1821.—Nomen dubium. Referred by Pilát 1937 (ACE 3): 110, with doubt, to *Polyporellus varius* var. *podlachicus* [≡ *Polyporus podlachicus*], but the original descr. does not support this disposition.

circumscriptus, Polyporus, Lév. ("in Litt., De Guernisac") apud Crouan 1867: 65 (France).—Nomen subnudum & dubium.

citrinus, Oligoporus, R. & O. Falck 1937 (HF 12): 58, 59 (lacking Latin descr.: n.v.p.).—This is a new name for *Ptychogaster citrinus* Boud. (nom. anam.) plus the basidiferous (perfect) state (from France) described by Boud. but not included by him under *Ptychogaster citrinus* [cf. Donk 1971 (Pe 6): 212]. The inclusion of this state by O. & R. Falck, implied *inter alia* by the use of the combination *Oligoporus citrinus*, automatically created a new specific name (Code, Art. 59) for a taxon based on the perfect state described by Boud. (type), and later by Bref. under the name *Oligoporus farinosus*, which was cited as a syn.

citrinus, Polyporus, Chev. 1837: no. 40 var. δ (as a var. of *P.* [*Albatrellus*] *cristatus*: n.v.p.) (Germany), not ∼ (Plan. pcr S. F. Gray) Pers. 1825.

clavata, Helvella, Schaeff. → *Boletus elvela* Batsch. (**O**).

clavatus.—"*Polyporus clavatus* of Fries" Hussey *c.* 1847 I. 1: *pl. 46* in obs. (syn.: n.v.p.); [≡ *Polyporus sulphureus*, "Var. . . . b) clavatus, undique porosus: Sowerb. [*t. 135*]" Fr. 1838: 450, unnamed var. (England)]. – Fr. did not publish the specific name, nor did he validly publish the varietal epithet 'clavatus'. Hussey used the supposed specific name as basionym for a new name, *Polyporus sulphureus* var. *clavatus* Hussey, l.c., which is based on the monstrosity depicted by Sow., l.c.

*clelandii, *Lenzites*, Lloyd 1919 (LMW 6): 887 *pl. 130 f. 1535* (Australia, New South Wales).—Referred to *Gloeophyllum trabeum* by G. Cunn. 1965: 253. It should be remembered that G. Cunn. did not distinguish between *G. trabeum* and *G. striatum*.

cobelliana, Lenzites, (Sacc.) Sacc. 1882 (syn.: n.v.p.); *Lenzites cinnamomea* subsp. *L. cobelliana* Sacc. 1882 (Mi 2): 529 (Italy).—Fide Sacc. 1882 (Mi 2): 682 = *Lenzites crocata* [= *Gloeophyllum sepiarum*].

*coccineus, *Polyporus*, Fr. 1851 (NAu III 1): 67/51 (Marquesas Is.);
Fomes Cooke 1885; *Scindalma* O.K. 1898; *Trametes* Pat. 1906; *Poly-stictus* Lloyd 1916 (as a form of *P. sanguineus*: n.v.p.), Doidge 1950;
Pycnoporus Bond. & S. 1941.—This has been referred to *Trametes*
[*Pycnoporus*] *cinnabarina* sensu lato by, for instance, G. Cunn. 1965:
169, but cf. (112).

cocos, *Poria*, F. A. Wolf, see *Sclerotium cocos* (O).

cocos, Sclerotium, Schw. 1822: 56 (U.S.A., North Carolina) (nom. anam.);
Pachyma Fr. 1822.—A basidiferous state (but perhaps not the full-grown 'perfect' state) is now known as *Poria cocos* F. A. Wolf 1922
(JMS 38): 134 *pls. 34–37*; *Polyporus* Over. 1927 ("[Fries] Weber").
Here records of *Sclerotium cocos* from Europe are supposed to be based
on the sclerotia of *Polyporus tuberaster* (or closely related spp.); such
records are by Otth 1866 (MiB 1865): 170, Prilleux *1889, 1890*, and Hein-richer & Elsler *1910*, as *Pachyma cocos*. – Now often incorrectly cited as
Poria cocos (Schw.) Wolf; but the name for the perfect state dates from
1922, as cited above. According to Ginns 1969 (M 60): 1219 *Merulius
albus* Burt 1917 (AMo 4): 334 *f. 18* is a synonym. It antedates the
name *Poria cocos*.

cognatus, Polyporus, (Bres.) Speg. 1925; *Polyporus rheades* var. Bres.
1920 (Am 18): 34 (Argentina).—Speg. 1925 (BCó 28): 371 considered
this a distinct sp., but more recently Pegl. 1964 (TBS 47): 188 treated
it as a var. of *Inonotus rheades*.

columbiensis, Polyporus, Berk. 1842 (LJB 1): 454 (U.S.A., Oregon).—This
was listed by Pilát 1936 (ACE 3): 76 as a syn. of *Polyporellus* [*Polyporus*]
arcularius [sensu lato]. According to Lloyd 1912 (LMW 3, S.P.): 178,
"The type is a little discolored frustule that tells nothing and which
should never have been named." Not mentioned by Overholts *1953*.
The type locality is Columbia River in Oregon (cf. Lloyd, l.c.) not in
South Carolina, as was stated by Murrill 1907 (NAF 9): 58. – Descr.:
Murrill 1904 (BTC 31): 35; 1907 (NAF 9): 58.

compactum, *Hydnum*, Pers. 1800: 57 (Germany) (d.n.) per Fr. 1821,
misapplied; *Hydnellum* P. Karst. 1879, Nikol. 1954; *Calodon* P. Karst.
1882; *Phaeodon* J. Schroet. 1888; ≡ *Hydnellum compactum* (Pers.
per Fr.) P. Karst. (Thelephoraceae). – Sensu Inz. → *Abortiporus
biennis*.

complanatus, *Boletus*, Latourr. 1785: 39 (d.n.).—This is a binomial name
provided for *Polyporus sessilis, convexo planus, inferne albidus, superne
fulvis, discoloribus zonis* Haller 1768 H. 3: 143 (no. 2289), which judging
from both the descr. and the syns. cited is a very compound species.

concentricus, *Polyporus*, Pers. → *Boletus latus* Sow. (O).

conchoides, Gloeoporus, Mont. 1842 C.: 385 *pl. 15 f. 1* (Cuba); *Leptoporus*
Pat. 1900; *Polyporus* Lloyd 1915.—Referred by Bres. 1912 (H 53):
74 to *Gloeoporus dichrous*, cf. Overh. 1953: 363 *pl. 26 fs. 157, 158*
(*Polyporus*) for descr. (correct interpretation?). "In American tradition",

fide Lloyd 1915 (LMW 4, Ap.): 330 = *Polyporus [Gloeoporus] dichrous.* – Cf. (**51**).

confluens, Podoporia, P. Karst. 1892 (H 31): 297 (Finland, now U.S.S.R., European Russia); *Poria* Sacc. 1895.—Nomen dubium. Lowe 1956 (M 48): 116 thought of *Polyporus [Gloeoporus] pannocinctus*; in my opinion this is incorrect.

**confragosa, Trametes,* Lloyd 1912 (LMW 4, L. 42): 15 ("a . . . form of *Daedalea [Daedaleopsis] confragosa"*; n.v.p.) ex Hara 1927: 392 (Japan).—Fide Imaz. = *Daedaleopsis nipponica* Imaz. 1943 (BTS 6): 78 for descrs. of which see S. Ito 1955: 251 *f. 196* and Imaz. & Hongo 1957 C. I. [1]: 113 *pl. 53 f. 294.*

conglobatus, Polypilus, P. Karst. 1904 (ÖfF 46[11]): 2 (Finland); *Polyporus* Sacc. & D. Sacc. 1905 (SF 17): 108, not ∼ Berk. 1845.—Nomen dubium. Lloyd 1912 (LMW 3, S.P.): 161 remarked, "description reads much like *distortus* [= *Abortiporus distortus* (**O**)]." I would have suspected *Abortiporus biennis,* but Lowe 1956 (M 48): 116, who saw the type, concluded, "It is unknown to me."

constrictus.—*Boletus albidus* var. "*β. Boletus constrictus.* Bolt. funguss." Pers. 1801: 516 (syn.: n.v.p.), not *Boletus constrictus* Pers. 1801 (d.n.).— An error for *Boletus 'substrictus'* Bolt. [= *Polyporus ciliatus*].

**contrarius, Fomes,* (B. & C.) ex Cooke 1886 (G 15): 21; *Polyporus* B. & C. "in Herb." (n.v.p.); *Fomes* Cooke 1885 (G 14): 19 (nom. nud.: n.v.p.); *Scindalma* (B. & C. ex Cooke) O.K. 1898; *Rigidoporus* Murrill 1907; ≡ *Polyporus contrarius* B. & C. apud B. & Br. 1883 (TLS II 2): 60 *pl. 11 fs. 2–4* (Cuba), earlier typonym.—Lloyd 1915 (LMW 4, F.): 279 wrote, "appears to me to be probably subresupinate *Fomes annosus*", which explains the inclusion of the name *Fomes contrarius* Cooke by Pilát as a syn. (with query) of *Fomes [Heteroporus] annosus.* (The type is not from Brazil as stated by Lloyd.) – This is an extra-European sp. Descr.: Lowe 1957 F.: 86 *f. 68 (Fomes).*

contrarius, Polyporus, B. & C. apud B. & Br. → *Fomes contrarius* (B. & C.) ex Cooke (**O**).

cookei, Lenzites,* Berk. 1876 (G 4): 161 (U.S.A., New York); *Cellularia* O.K. 1898.—Fide Bres. 1897 (AAR III 3): 92 = *Trametes rubescens* [= *Daedaleopsis confragosa*]. – Cf. (31**).

corallinus, Boletus, Humb. 1793: 98 *pl. 3 f. 11* (Germany) (d.n.); *Boletus* Humb. per Steud. 1824; *Polyporus* Pers. 1825.—Nomen dubium.

cordonii → *gordonii* (**O**).

cordylina, Poria,* G. Cunn. 1947 (BPZ 72): 23, 39 *f. 17* (New Zealand).—Fide Teix. & Rog. 1955 (M 47): 411, 414 = *Aporpium caryae* (O**).

coriacea grisea, Daedalea, Secr. 1833 M. 2: 488 (Switzerland) (double epithet: n.v.p.) = *Lenzites betulina.*

coriacea subtusrufa, Daedalea, Secr. 1833 M. 2: 486 (Switzerland) (nom. nud. & double epithet: n.v.p.).—Fide Fr. 1874: 493 ("Secret. n. 6")

= *Lenzites betulinus*. – Secr. cited *Agaricus coriaceus* Bull. [= *Lenzites betulinus*] as a syn. of his var. A.

coriaceus, Boletus, Batsch 1783 (d.n.), not ∼ Scop. 1772 (d.n.) per Bergam. 1823, not ∼ Huds. 1778 (d.n.), not ∼ Batsch 1786 (d.n.); = *Helvella infundibuliformis* Schaeff. 1774: 110 [*pl. 277*] (Germany); *Helotium* (Schaeff.) per S. F. Gray 1821; *Leotia* Fr. 1822; = *Boletus milleporeus* Batsch 1789 (d.n.).—Nomen dubium. Fries 1822: 26 accepted the species sensu Sow. 1798: *pl. 153* (*Helvella infundibuliformis*). The discomycetous nature of this fungus is also very doubtful, both in its original sense and in that of Sowerby.

**coriaceus, Boletus*, Batsch 1786: 177 *pl. 24 f. 127* (type locality unknown, perhaps tropical Africa), not ∼ Scop. 1772 (d.n.) per Bergam. 1823, not ∼ Huds. 1778 (d.n.), not ∼ Batsch 1783: 105 (d.n.); *Polyporus* D. Dietr. 1847; not ∼ (Batsch per Bergam.) Endl. 1830; = *Boletus areolatus* Batsch 1789 (d.n.); = *Polyporus batschii* Pers. 1825.—Referred by Fr. 1832[Ind.]: 57 to *Polyporus* [*Coriolus*] *zonatus*. This cannot be correct. If one assumes that the fungus was collected in Germany then the only possibility would be *Daedaleopsis confragosa*. The protologue, however, calls the fruitbody "membranaceus" and the figure strongly suggests "*Hexagonia*" cf. *discopoda* Pat. & Har. described from tropical Africa. The fact that Batsch stated "locum vero regionis . . . ignoro" in my opinion makes this last suggestion almost certain.

**corium, Bornetina*, Mang. & Viala 1903 (CrP 136): 398, 1699 (Israel) (nom. anam.).–Reported from, but apparently not yet found in, Europe: cf. Donk 1971 (PNA 74): 27, 28.

corium.—"*Polyporus corium* Pers., Mycol. Eur. vol. 2, p. 60, 1825" G. Cunn. 1950 (BPZ 83): 2 (syn.: n.v.p.), not ∼ Kunze (n.v.p.), an error; = *Thelephora incarnata* var. *corium* (Pers.) per Pers. 1822: 131 = *Merulius corium* (Pers. per Pers.) Fr. ('*Corticiaceae*').

**corium, Polyporus*, Kunze apud Cooke 1882 (G 10): 103 (nom. nud.: n.v.p.) & 1886 (G 14): 114 (syn.: n.v.p.) (Australia, New South Wales), not ∼ (Pers.) G. Cunn. (n.v.p.).—Referred by Cooke, l.c., 1886 to *Poria* [*Phellinus*] *ferruginosus*. Fide G. Cunn. 1965: 268 = "*Trichaptum*" *flavum* (Jungh.) G. Cunn., an extra-European sp.

corni, Polyporus, Velen. 1925 (MP 2): 98 (Czechoslovakia).—Referred by Pilát 1942 (ACE 3): 511 to *Phellinus igniarius* subsp. *trivialis* [= *P. igniarius?*] (forma). The original descr. is very incomplete.

cornucopiae, Poria, Hoffm. 1797–1811 V.s.: 14, in obs. (Germany) as a var. of *Poria scutata*). – This seems to be a growth form of *Poria scutata* = *Heterobasidion annosum*. Found in mines.

**corrugata, Daedalea*, Kl. 1833 (Li 8): 481 (Canada); *Lenzites* Berk. 1872; *Striglia* O.K. 1891; *Cellularia* O.K. 1898.—Fide Murrill 1905 (BTC 32): 86 = *Agaricus* [*Daedaleopsis*] *confragosa*; according to Lloyd 1913 (LMW 4, L. 44): 10 a thin form of the southern U.S.A. of *Daedalea* [*Daedaleopsis*] *confragosa*, a somewhat surprizing statement for a sp.

described from boreal North America. – Descr.: Berk. 1839 (AM 3): 382 (*Daedalea*). – Cf. (**31**).

corticola salicis, *Poria*, Secr. 1833 M. 3: 174 (Switzerland) (as a sp. of *Boletus* & double epithet: n.v.p.).—Nomen dubium. Secr. identified his fungus with *Polyporus* [*Oxyporus*] *corticola* var. b *fagineus* Fr. 1821: 385; Fr. 1874: 580 accepted it as *Polyporus* [*Oxyporus*] *corticola* var. d *salicinus* Fr.

corylicola, *Polystictus*, Lázaro 1916 (RMa 14): 751 / 1917: 137 (Spain).— Nomen dubium. *Coriolus* sp.?

costaricensis, *Fuscoporella*, Murrill 1907 (Costa Rica); *Poria* Sacc. & Trott. 1912.—Referred by Lowe 1966: 163 to *Poria* [*Phellinus*] *punctata*, but cf. (**86**).

cotyledoneus, *Polyporus*, Speg. in herb. & nom. prov. (Argentina).—Listed by Speg. 1925 (BCó 28): 375 as a syn. of *Polyporus* [*Truncospora*] *ochroleuca*, fide Bres. in litt.

crassidens, [*Sistotrema?*], Delastre, "Lettre . . . à J. B. Mougeot".—Fide Quél. apud Moug. & Ferry 1887: 499 = *Sistotrema* [*Spongipellis*] *pachyodon*.

crassus, *Polyporus*, Fr. 1838: 451 (Germany); *Xylopilus* P. Karst. 1882; *Fomes* Cooke 1885, not ∼ (P. Karst.) E. Komar. 1964 (n.v.p.); *Scindalma* O.K. 1898.—Nomen dubium. Cf. Fr. 1874: 543: "Structura et color exacte *P*[*olypori*] *fomentarii*, ut hujus lusum maxime abnormem suspicor, licet saepius conformis sit lectus."

crataegi, *Lenzites*, Berk. 1847 (LJB 6): 323 (U.S.A., Ohio); *Cellularia* O.K. 1898.—Fide Peck 1878 (RNS 30): 73–74 = *Daedalea* [*Daedaleopsis*] *confragosa* (var.). – Cf. (**31**).

cremaceus, *Ceriomyces*, P. Henn. 1899 (VBr 40): 132 (Germany) (nom. anam.).—Needs redescription. Perhaps an alien, found in the palmhouse of the former botanical garden, Berlin. According to Ulbr. 1941 (NBe 15): 579, 583 this was referred by Killerm. in herb. B to *Ceriomyces* [*Ptychogaster*] *albus* [= *Ptychogaster fuliginoides*], but the descr. disagrees. – A doubtful spedies of *Ptychogaster*, cf. ". . . mit einer dünnen, faserigen, seidigglänzenden Oberhaut bedeckt . . .."

cremoriflorus, *Polyporus*, Crag. 1884 (nom. prov.: n.v.p.); [= *Polyporus fissus* Berk. sensu Crag. 1884 (BWb 1): 22 (U.S.A., Kansas)].—Cf. *Polyporus varius* (*P. elegans* sensu Fr.).

cretaceus, *Polyporus*, S. Schulz. 1878 (Fl 61): 11 (Europe), not ∼ Lloyd 1915.—Nomen subnudum & dubium. Previously published?

crispum, *Hydnum*, Schaeff. 1774: 97 [*pl. 147*] (Germany) (d.n.), not ∼ Scop. 1772 (d.n.); *Hydnum* Schaeff. per Fr. 1821; *Irpex* Fr. 1838.— Nomen dubium, fide Maas G. 1960 (Pe 1): 353. – Sensu Inz. → *Abortiporus biennis*.

crispus, *Cantharellus*, Pers. 1794 (d.n.); *Merulius* Pers. 1800 (d.n.), not ∼ (Bull.) J. F. Gmel. 1792 (d.n.); *Cantharellus* Pers. per Fr. 1821; *Merulius* Schleich. 1821; *Trogia* Fr. 1861; *Plicaturopsis* D. Reid 1964; ≡ *Plicatura crispa* (Pers. per Fr.) Rea 1922. – Descr.: D. Reid 1964 (Pe 3): 130 *fs. 49, 50 (Plicaturopsis).* – *Favolus europaeus* [≡ *Polyporus mori*] sensu Velen. 1922: 696 *f. 119* = *Plicatura faginea* (Schrad. per J. Schroet.) P. Karst. [= *P. crispa*], fide Pilát 1936 (ACE 3): 85.

*cristata, *Trametes,* Cooke 1882 (G 10): 132 (Australia, Queensland); *Polystictus* Cooke 1886; *Microporus* O.K. 1898.—This was referred by Bres. 1897 (AAR III 3): 90, 1908 (Am 6): 39 to *Trametes lutescens* forma *umbrina* and *T. favus* [both names, sensu Bres., = *Funalia gallica*]. More recently the species has been treated as distinct e.g. by Lloyd and G. Cunn. – Cf. (36).

cristatus, Boletus, Baumg. 1790: 634 (d.n.) & *B. cristatus* J. F. Gmel. 1792: 1435 (d.n.), not ∼ Gouan 1762 (d.n.), not ∼ Schaeff. 1774 (d.n.) & (Schaeff. per Fr.) Pollini 1824; [≡ *VIII. Boletus . . .* Gled. 1753: 75 (Germany)].—Nomen dubium. Gleditsch's Latin phrase and German description taken together do not evoke any suggestion in my mind; neither *Albatrellus cristatus* nor *Grifola frondosa* agree, cf. "Ein . . . Sammet-hafter Erd-Bülz, mit Purpur-farbenen kraus-faltig zertheilten Lappen . . .."

cristatus, Boletus, J. F. Gmel. → *Boletus cristatus* Baumg. (**0**).

cristatus, Boletus, Gouan 1762: 540 (France) (d.n.), not Baumg. 1790 and J. F. Gmel. 1792 (d.n.), not ∼ Schaeff. 1774 (d.n.) and (Schaeff. per Fr.) Pollini 1824.—Nomen dubium. Fr. 1821: 355 referred this to *Polyporus [Grifola] frondosus.* As published Gouan's species consisted of several vars.; the type var. was identified with "*Agaricus seu fungus ramosus cristatus medius.* Barr. Icon. 1271" which was also cited when *Polyporus barrelieri* Viv. [= *Grifola frondosa*] was introduced. Gouan recorded it for Montpellier, France, "Magnitudo capitis humani. Albus." The white colour does not agree with the European spp. of *Grifola.* It might suggest a sp. of *Sparassis,* although the inclusion in *Boletus* ("pileus . . . subtus porosus") hardly supports this suggestion. The citation of Barrelier's figure under *Merisma cristatum* Pers. [= *Sebacina incrustans* (Pers. per Fr.) Tul., Tremellaceae] by Lapl. 1894: 353 is an error.

cristatus, Boletus, "Plan."—An error by Fr. 1832[Ind.]: 57 for *B. 'crustatus'* [= *Ganoderma lucidum*].

*cristula, *Polyporus,* Kl. ("in Herb.") ex Berk. 1839 (AM 3): 387 (India?).— Sacc. 1888 (SF 6): 245 appended this as a subsp. to *Polyporus [Pycnoporus] cinnabarinus,* with the remark, "Affinis, nisi idem cum *Pol. cinnabarino.*" This was caused by a remark by Fr. 1851 (NAu III 1): 75/59.

*croceopallens, *Gloeoporus,* Bres. 1912 (Am 10): 506 (Indonesia, Java); *Polyporus* Lloyd 1915 (n.v.p.).—Considered by Lloyd 1915 (LMW 4, Ap.): 331 to be a mere form of *Polyporus [Gloeoporus] dichrous.* – Cf. (**51**).

*croceus (Fr.) ex P. Karst. 1859: 39, apud Nyl. & Sael. 1859: 102; [≡ *Polyporus nitidus* "b. croceus. Schwein.! in litt." Fr. 1828 E. 1: 117 (apparently not an infragenetic name: n.v.p.?) (U.S.A.)]; *Polyporus* *nitidus* var. *croceus* Mont. 1854 ("Schwz."; lacking further references, no descr.: n.v.p.), typonym.—Nomen dubium. Fries's "b. croceus" has been interpreted not as a one-word phrase (as I do), but as an infraspecific epithet. Schw. 1832: 158 listed this taxon as "*P[olyporus]* *nitidus*, Fr. n. 8, Pers. 122, olim *croceus*, L. v. S.*" This use of 'croceus' should most probably not be confused with *Boletus croceus* Pers. as previously used by Schw. 1822: 96 for a sp. that he placed in *Boletus* c) *Apodes*; hence it had a pileate fruitbody in contradistinction to the taxon described by Fr., which is 'resupinate'. Overh. 1923 (M 15): 213 evidently confused these two taxa when he stated of "*Boletus* [!] *croceus* Schw.*" 1822: 96 that no type was preserved in PH and the Michener collection. — Fries's taxon was reported from Finland under the name *Polyporus croceus* "Schwein." by P. Karst. (ll. cc.) with ref. to Fries's publication; it may be assumed that he misapplied the name to *Chaetoporus collabens*.

**crocicolor, Polyporus,* J. B. Ell. "mss. at Kew".—Fide Lloyd 1915 (LMW 4, Ap.): 377 = *Polyporus aurantiacus* Peck [= *Pycnoporellus fulgens*].

**cruentatus, Polyporus,* Mont. 1854 (ASn IV 1): 129; 1856: 161 (French Guinea); *Poria* Cooke 1886.—Bres. 1916 (Am 14): 228 thought that this was a resupinate form of *Polyporus* [*Gloeoporus*] *dichrous*, but Wakef. 1934 (BmI): 249 denied this and accepted it as a good sp.; Murrill 1921 (M 13): 96 and Lowe 1963 (M 55): 458 & 1966: 37 referred it to *Poria spissa* (Schw. apud Fr.) Cooke, an extra-European sp.

**crustacea, Laschia,* Jungh. 1838: 75 ["fig. 40" (n.v.)] (Indonesia, Java); *Polyporus* Lév. 1844; *Hymenogramme* Sacc. & Cub. apud Sacc. 1887; *Aschersonia* O.K. 1898 ("*crustata*"); *Poria* Cooke 1886, Bres. 1910.—This was recorded from France by Richon 1889: 99 as "*Polyp. crustaceus* Lév.", without descr.; I take this to be an erroneous determination; compare Ryv. 1972 (Pe 7): 18.

cryptarum, Agaricus, P. Beauv. 1806 (AMP 8): 346 *pl. 57 fs. 2, 3* (France).— Nomen dubium. A poria (or poria states) found in cellars, caves, and mines, insufficiently described for certain recognition; referred by DC. 1815: 38 to *Boletus* [*Poria*] *vaillantii*. May be a nomen confusum: it is not unlikely that states of more than one sp. were thought to be conspecific and developing from each other, but cf. also *Polyporus* [*Tyromyces*] *fodinarum*.

cryptarum, Spongioides, (Bull. per Fr.) Lázaro sensu Lázaro → *Spongioides* (**O**).

**cubensis, Lenzites,* B. & C. 1868 (JLS 10): 303 (Cuba), an extra-European sp.—Pilát 1940 (ACE 3): 327, 328 made this a variety of *Trametes* [*Lenzites*] *betulina*, but this identification is doubtful. According to Bond. & Ljub. 1964 (NSn): 182 collections referred to *Trametes betulina* var. *cubensis* (B. & C.) Pilát are "*Lenzites*" *acuta* Berk., another extra-European sp.

cumini, Polyporus, Pers. → *Boletus abietinus* Cumino (**O**).

cupreolaccatus, Polyporus, Kalchbr. (in litt. ad A. von Kerner, 1882)
apud Wettst. 1885 (ÖbZ 35): 81 (syn.: n.v.p.); Ferd. & W. 1943 (lacking
Latin descr.: n.v.p.); *Fomes* Ferd. & Jørg. 1939: 332 (lacking descr.:
n.v.p.); *Ganoderma* Igmándy 1968 (lacking Latin descr.: n.v.p.);
≡ *Polyporus laccatus* Kalchbr. apud Wettst. [= *Ganoderma pfeifferi*]. –
Descr. Ferd. & W. 1943: 95 *fig.*

**curreyanus, Polyporus,* Berk. ex Cooke 1886 (G 15): 20 (New Zealand);
Poria G. Cunn. 1947.—Bres. 1916 (Am 14): 224 referred this to *Polyporus*
[Gloeoporus] dichrous, and Lloyd 1915 (LMW 4, Ap.): 377 to *Polyporus*
[Bjerkandera] adustus. It is treated by G. Cunn. 1965: 50 (*Poria*; with
descr.) as distinct from both spp. and as strictly resupinate. – Cf. (**51**).

**curtisii, Polyporus,* Berk. ("MSS.") apud B. & C. 1849 (HJB 1): 101
(U.S.A., South Carolina); *Fomes* Cooke 1885, not ∼ Lloyd 1922 (n.v.p.);
Scindalma O.K. 1898; *Ganoderma* Murrill 1908.—At first taken to be
conspecific with *Ganoderma pseudoboletus* [= *G. lucidum*] by Murrill
1902 (BTC 29): 602, 604, but later correctly accepted by him, Murrill
1908 (NAF 9): 120, as a distinct sp. of *Ganoderma.* Extra-European.

cuticularis, Boletus, Thore 1803: 487 (France) (d.n.) ex Duby 1830, not ∼
Bull. 1789 & ∼ (Bull. per Fr.) Mérat 1821.—Nomen dubium.

Cyanosporus Lloyd 1909 (not accepted by publishing author: n.v.p.)
[1960 (Pe 1): 202]; monotype, *Polyporus* [*Tyromyces*] *caesius.*

cymatoides, Polyporus, E. Krause 1928 B.r.: 53; [≡ *Polyporus kymatodes*
Rostk. sensu E. Krause 1925 (ANM II 1): 129, excl. of type, (lacking
descr.) (Germany)].—Nomen dubium. Apparently not validly published:
"Entspricht Fr. t. 183, 1 gut, Habitus wie *serialis* St[urm = Rostk.]
17, 49, Farben fast wie *hirsutus* St. 16, 44. Aber nicht *P. kymatodes*
St. 10, 24."

daedalea, Trametes, Speg. → *Daedalea trametes* Speg. (**O**).

dapsilis, Polyporus, Britz. 1887 (BnS 29): 274 [*pl. 591 f. 3*] & 1910 (BbC
26): 208 (Germany).—According to Killerm. 1922 (Dba 15): 62, "dürfte
[*Polyporus*] *melanopus* od. [*P.*] *varius* sein"; both guesses are undoubtedly far off the mark. Referred by Bres. apud Killerm. 1925
(Dba 16): 118 to *Polyporus subsquamosus* [sensu Bres. = *Boletopsis*
grisea] (**17**). I have thought it might be either *Albatrellus ovinus* or
A. similis, but could not convince myself one way or another. No
spores mentioned. – Rather, a nomen dubium.

debilis, Polyporus, Wallr. 1833: 600 (Germany).—Nomen dubium. Referred
by Fr. 1838: 430 to *Polyporus brumalis,* but the original descr. is quite
insufficient for correct determination.

**decipiens, Polyporus,* Schw. 1832: 157 (U.S.A., Pennsylvania); *Polystictus*
Fr. 1851; *Hansenia* P. Karst. 1879; *Microporus* O.K. 1898; *Coriolus*
Pat. 1900.—Nomen dubium. Schw. stated "Ambit inter *P. versicolorem*
et *abietinum.*" Murrill 1907 (NAF 9): 27–28 and Lloyd 1913 (LMW 4,
L. 50): 8 could not find the type or any authentic specimen. The species

has been referred to *Coriolus versicolor, Polystictus ochraceus* [= *Coriolus zonatus*], *Coriolus pavonius* (Hook.) Murrill (the last sp. extra-European). Not mentioned by Overh. *1953.* Murrill thought that the description called for a plant near *Coriolus prolificans* [= *Hirschioporus pargamenus*].

deformis, Boletus, Schaeff. 1774: 90 [*pl. 264*] (Germany) (d.n.) per Steud. 1824; *Polyporus* Fr. 1832, 1838; *Fomes* Cooke 1885; *Placodes* Quél. 1886; *Scindalma* O.K. 1898.—Nomen dubium. Pers. 1800: 106 thought that this might be a var. of *Boletus* [*Albatrellus*] *ovinus*. The sp. was accepted by Fr. 1838: 441, after he had supposed for a while (Fr. 1832[Ind.]: 147) that it was the correct name for *Polyporus* [*Albatrellus*] *pes-caprae*. The original plate suggests something abnormal; the context was indicated as "lignosus". A polypore overgrown by a mould?

delens, Polyporus, E. Krause → *Polyporus alutaceus* Fr. (**0**).

*delicatissima, Daedalea, Speg. 1890 (ABA 6): 175 (Argentina).—Bres. 1916 (Am 14): 230 found the type to be a mixture ("[*Pol.*] *aculeifer* Berk. et *Pol. biennis* Bull. commixti") while Speg. 1926 (BCó 28): 370 apparently considered the last element to be typical (lectotype), accepting it as a syn. of *Polyporus* [*Abortiporus*] *biennis* [sensu lato] as "forma truncicola dimidiato-sessilis hymenio daedaloideo". Fid. 1969 (Ri 4): 150, 162, 165 made it a syn. of *Heteroporus biennis* var. *flabelliformis* ("Mont.") Fid. [sensu Fid. = *Abortiporus distortus* (**0**); see Donk 1971 (PNA 74): 2]. – Cf. (**2**).

*delicatus, Polyporus, B. & C. apud Berk. 1872 (G 1): 37 (U.S.A., Alabama); *Favolus* A. Ames 1913.—Referred by Lloyd 1912 (LMW 3, S.P.): 146 to *Polyporus* [*Abortiporus*] *fractipes* (**0**) and Overh. 1953: 254 was inclined to accept this disposition, but Fid. 1969 (Ri 4): 121, 150, 154, 166 identified it with *Heteroporus biennis* var. *flabelliformis* [sensu Fid. = *Abortiporus distortus* (**0**); see Donk 1972 (PNA 74): 2].

*demissus, Polyporus, Berk. 1845 (LJB 4): 52 (West Australia); *Merulius* Pat. 1908.—This has been referred to *Polyporus* [*Bjerkandera*] *adustus* by Cooke and G. Cunn. 1949 (BPZ 81): 2, to *Polyporus* [*Bjerkandera*] *fumosus* by Lloyd 1915 (LMW 4, A.): 377, and to *Gloeoporus thelephoroides* (**0**) by G. Cunn. 1965: 111, 112. – Cf. (**51**).

*dendroidea, Hymenochaete, B. & C. apud B. & Br. 1873 (JLS 14): 69 (Ceylon). – Cf. (**41**).

dense-lamellata, Lenzites, Opiz, "n.n." (n.v.p.)—Fide Pilát 1940 (ACE 3): 339 = *Gloeophyllum trabeum*.

dentatus, Polyporus, Velen. 1922: 650 [see Pilát 1948: 248 for Latin translation] (Czechoslovakia).—Referred by Pilát 1938 (ACE 3): 185 without comment to *Leptoporus* [*Tyromyces*] *lacteus* in a very inclusive sense.

destructor, Boletus, Schrad. 1794: 166 (Germany) (d.n.); *Polyporus* (Schrad.) Fr. 1815 (d.n.); *Polyporus* (Schrad.) per Fr. 1821; *Boletus* Spreng. 1827; *Bjerkandera* P. Karst. 1882; *Leptoporus* Quél. 1886; *Polystictus* Gillot & Luc. 1890; *Cladomeris* Lázaro 1916; *Fibroporia* Parm. 1968; = *Poria destruens* S. F. Gray 1821.—Nomen dubium & ambiguum (128). — Sensu Fr. 1838, Mez 1908: 110 *f. 42* (*Polyporus*) = ?; sensu Bres., in part → *Poria resupinata*; sensu Bond. → *Tyromyces fodinarum*?

destruens, Boletus, S. F. Gray → *Boletus destructor* Schrad. (**O**).

destruens, Leptoporus, Pilát 1937–8 (ACE 3): 197, 198 *f. 57* (lacking Latin descr.: n.v.p.).—This name is the precursor of *Leptoporus* [*Tyromyces*] *fodinarum,* which replaced it farther on in the book.

detmoldiense, Ganoderma, Kreis. & Jahn in herb. (Germany).—Fide H. Jahn 1963 (WPb 4): 86 = *Ganoderma europaeum* [= *G. adspersum*].

Dictyoporus F. Clem. → *Retiporus* Endl. (**O**).

difformis, Boletus, Willd. ("Plant. Cryptog. ined.") ex Humb. 1793: 102 (Germany) (d.n.).—Nomen dubium. Found in mines: perhaps an abnormal growth.

difformis, Poria, Scop. 1772 P.s.: 105 *pl. 26* (Hungary, now Czechoslovakia) (d.n.); ≡ *Polyporus voluta* Pers. 1825.—Nomen dubium. This was referred to *Polyporus* [*Coriolus*] *hirsutus* by Fr. 1832[Ind.]: 149.

discolor, Daedalea, Fr. → *Daedalea albida* Schw. (**O**).

discolor, Lenzites, Fr. → *Lenzites klotzschii* Berk. (**O**).

**discolor, Polyporus,* Kl. 1833 (Li 8): 483 (Mauritius).—Regarded as a syn. of *Polyporus* [*Laetiporus*] *sulphureus* by Lloyd 1912 (LMW 3, S.P.): 156. This may be incorrect.

**dispar, Polyporus* Kalchbr. in Thüm. 1878 (Fl 61): 441 (Australia, Victoria); *Polystictus* Sacc. 1888; *Microporus* O.K. 1898.—Referred at first to *Polystictus* [*Hirschioporus*] *pargamenus* by Bres. 1903 (Am 1): 76, but later to "*Polyporus*" [*Hirschioporus*] *elongatus* Berk., an extra-European species.

dissitus, Polyporus,* B. & Br. 1873 (JLS 14): 48 (Ceylon).—Referred by Lloyd 1915 (LMW 4, Ap.): 378 and Bres. 1916 (Am 14): 224 to *Polyporus* [*Bjerkandera*] *adustus.* However Petch 1916 (APe 6): 94, 100, 119 identified it with "*Polyporus*" *secernibilis* Berk. (O**) and his brief descriptive note would not at once exclude a sp. of *Gloeoporus.* He referred the *P. secernibilis* "of Lloyd" to "*Polyporus*" *introfuscus* Petch (Ceylon).

**distortus, Boletus,* Schw. 1822: 97 (U.S.A., North Carolina); *Polyporus* Steud. 1824; Fr. 1828; *Daedalea* Pat. 1900; *Abortiporus* Murrill 1904; *Heteroporus* Bond. & S. 1941.—This has been referred to *Polyporus*/ *Daedalea* [*Abortiporus*] *biennis,* for instance, by Schw. 1832: 160, Bres. 1926: 79, and Overh. 1953: 224; but cf. (2). — Descr.: Murrill 1907 (NAF 9): 64 (*Abortiporus*); Lloyd 1912 (LMW 3, S.P.): 158 *f. 458*; Overh. 1914 (AMo 1): 105 & 1915 (WUS 3¹): 84 *pl. 1 f. 3*; Lloyd 1916 (LMW 4): 549 *f. 753* (*Polyporus*); &c. Other descrs. under misapplied names: Buller *1922a* (*Polyporus rufescens*); Fid. 1969 (Ri 4): 149 *fs. 18–37, 51* (as *Heteroporus biennis* var. *flabelliformis*). Cf. Donk 1971 (PNA 74): 2 on the epithet 'flabelliformis'. — Sensu Pilát, in part (as to European specimens), → *Abortiporus biennis,* typical variety.

doidgei, *Polystictus*, Lloyd 1924 (LMW 7): 1294 (nom. nud.), 1329 and 1925 (LMW 7): *pl. 314 f. 3039* (South Africa).—Lloyd, l.c., remarked, "We would put it in . . . '*Microporus*' "; G. Cunn. 1965: 166 referred it to *Trametes* [*Coriolus*] *versicolor*.

dolosa, *Thelephora*, Lév. 1844 (ASn III 2): 209 (Indonesia, Java).— Referred by Bres. 1912 (H 51): 318 to *Gloeoporus dichrous* and later, Bres. 1912 (H 53): 74, to *Gloeoporus conchoides* (**O**). – Cf. (**51**).

dorsale, *Ganoderma* (Lloyd) Torrend 1920; *Polyporus* Lloyd 1917 (LMW 5): 658 *f. 939* (Brazil).—Steyaert 1967 (BBB 100): 198 tentatively reported this sp. from a garden in Antwerp (Belgium); he considered it an alien. – Syn.: *Polyporus perturbatus* Lloyd (**O**).

dryinus virescens, *Polyporus*, Secr. 1833 M. 3: 137 (Switzerland) (as a sp. of *Boletus* & double epithet & nom. nud.: n.v.p.).—Nomen dubium.

dryophila, *Fomitoporia*, Murrill 1907 (NAF 9): 8 (U.S.A., Mississippi); *Poria* Sacc. & Trott. 1912; *Fuscoporia* G. Cunn. 1948.—Referred by Lowe 1949 (Ll 11): 167 & Overh. 1953: 87 to resupinate *Fomes* [*Phellinus*] *robustus* but cf. (**86**). A re-description of the type is needed. Cf. D. Baxt. 1937 (PMi 22): 282 *pl. 33*; 1952 (PMi 37): 100 *pl. 8* (*Fomitoporia*).

dubius, *Boletus*, Plan. 1788 I.F.: 27 (Germany) (d.n.), not ∼ Retz. 1795 (d.n.), not ∼ Fr. (n.v.p.).—Nomen dubium. Fr. 1821: 370 referred this to *Polyporus* [*Hirschioporus*] *abietinus*. In his copy of Planer's booklet Persoon wrote, "Est *Merulius serpens* Tode" [sensu Pers.]; I am reminded of *Merulius tremellosus* Schrad. per Fr. (both, 'Corticiaceae').

dubius, *Boletus*, Retz. 1795: 318 (d.n.), not ∼ Plan. 1788 (d.n.), not ∼ Fr. (n.v.p.); ≡ "*Merulius*" *crispatus* O. F. Müll. 1777 (Fd 4 / F. 12): 6 *pl. 716 f. 2* (d.n.) per Fr. 1821 ('Corticiaceae').

dubius, *Polyporus*, Fr. "in litt." (type locality?); Lloyd 1914 (LMW 4, L. 52): 10 (matter of record: n.v.p.), not ∼ Plan. 1788 (d.n.), not ∼ Retz. 1795 (d.n.).—"It appears to be the plant called [*Tyromyces*] '*lacteus*' now". It is not clear what sp. Lloyd thought was called '*lacteus*' at that time; cf. Donk 1972 (PNA 75): 289.

dussii, *Myriadoporus*, Pat. 1889 (BmF 5): 85 (Martinique).—Referred by Bres. 1890 (Rm 12): 104 to *Fomes fomentarius* as a myriadoporous form. Perhaps *Fomes sclerodermeus* (Lév.) Cooke, an extra-European sp.

earleae, *Fomitiporia*, Murrill 1907 (NAF 9): 9 (U.S.A., Mississippi); *Poria* Sacc. & Trott. 1912.—Referred by Lowe 1949 (Ll 11): 167 & 1957 F.: 56 to *Poria* [*Phellinus*] *punctata* which he considered to be resupinate *Fomes* [*Phellinus*] *robustus* but cf. (**86**). A re-description of the type is needed.

earlei, *Lenzites*, Murrill 1908 (NAF 9): 128 (Jamaica).—Lloyd 1917 (LMW 5): 709 referred this to *Lenzites tenuis* Lév. (*q.v.*), which on the base of Japanese material so named by Lloyd was referred to

Lenzites [*Trametes*] *gibbosa*, together with *L. earlei*. The original fungus is likely to be a syn. of *"Trametes" palisotii* (**O**).

eburneus, Polyporus, Wallr. 1833: 599 (Germany).—At first Fr. 1838: 448 considered this to be *Polyporus lobatus* [sensu Fr. = ?] "status senilis leuciticus", but later Fr. 1874: 541 thought that it could perhaps be *Polyporus osseus* [= *Osteina obductus*]. My suggestion is white-weathered specimens of *Polyporus varius*.

echinata.—"[*Lenzites*] *echinata*" in 1894 (BmF 9): xxxiii is an error for '*Lycoperdon echinata*'; cf. 1894 (BmF 10): 100.

**effusa, Daedalea,* Speg. 1909 (ABA 19 / III 12): 274 (Argentina).— Referred by Bres. 1916 (Am 14): 229 to *Poria* [*Phellinus*] *contigua*.

effusa, Trametes, S. Schulz. 1866: 41 (Yugoslavia, Slavonia) (nom. nud.: n.v.p.).—Fide S. Schulz. 1880 (ÖbZ 30): 108 ≡ *Polyporus schulzeri* Kalchbr. [= *Funalia gallica*]. Cf. Donk 1972 (PNA 75): 166.

effusus, Boletus, Latourr. 1785: 39 (d.n.); [≡ *Polyporus crustaceus effusus, farinosus albus* Haller 1768 H. 3: 139 no. 2272 (Switzerland) ≡ *Polyporus effusus, crustaceus albus* Haller 1769: 201].—Nomen dubium. Haller's sp. was referred to *Polyporus* [*Perenniporia*] *medulla-panis* by Fr. 1821: 380, apparently because Haller cited *Agaricum album, terrestris, medullam panis referens* Mich. 1729: 121 *pl. 63 f. 2* as being an illustration of the sp. he had in mind. Jacq. also cited Micheli's sp. when he published *Boletus medulla-panis,* cf. Donk 1960 (Pe 1): 266, who doubted whether it was the same as Jacquin's fungus. It is also doubtful, to say the least, whether Haller correctly identified his insufficiently described sp. with Micheli's.

ellipticus, Polyporus, Secr. 1833 M. 3: 154 (Switzerland) (as a sp. of *Boletus* & nom. nud.: n.v.p.).—Nomen dubium. Secr. cited *Odontia cerasi* Pers. (**O**) as a syn. of his var. A.

elvela, Boletus, Batsch 1783: 105 (d.n.); ≡ *Helvella clavata* Schaeff. 1774: 100 [*pl. 149*] (Germany) (d.n.); *Spathularia* (Schaeff.) per Sacc.; = *S. flavida* Pers. per Fr. (Discomycetes), fide Fr. 1821: 491.

**endoxantha, Fistulina,* Speg. 1921 (BCó 25): 21 (Chile).—Fide J. E. Wright 1961 (BaB 9): 225, 227 = *Fistulina hepatica* (var.) but Sing. 1969 M.a.: 381 considers it a good sp.

epidryphloea, Daedalea, E. Krause 1928 B.r.: 55 ("*Daedaleus*") (Germany). —Nomen dubium. Presumably a sp. of *Merulius* Fr. sensu lato ('Corti-ciaceae'), hence excluded.

epigaeus, Polyporus, S. Schulz, 1866: 41 (Yugoslavia, Slavonia) (nom. nud.).—Nomen dubium.

epigeus.—"*epigeus* [, *Polyporus*], Europe, Link. Specimens in his herbarium Berlin are *Polyporus* [*Skeletocutis*] *amorphus* . . . ."—Lloyd 1915 (LMW 4, Ap.): 378. – This quotation implies the specific name '*Polyporus epigeus*', which in its turn must be connected with *Polyporus amorphus* b. *epigeus* Fr. 1818 O. 2: 259 ≡ *Polyporus* [*Skeletocutis*] *amorphus*. Not *Polyporus epigaeus* S. Schulz. 1866 (n.v.p.).

epileucus, Polyporus, Fr. 1838: 452 (Sweden); *Bjerkandera* P. Karst. 1882; *Leptoporus* Quél. 1886; *Polystictus* Gillot & Luc. 1890; *Coriolus* Bond. 1953.—Nomen dubium. Variously interpreted: cf. Romell 1926 (SbT 20): 10 and Bourd. & G. 1928: 543, in obs. − Sensu Morg. ("of Morgan's flora" [?]) = *Polyporus [Tyromyces] spraguei,* fide Lloyd 1915 (LMW 4, Ap.): 305; sensu Peck 1885 (RNS 38): 91 (*Polyporus epileucus* var. *candidus* Peck) = *Polyporus immitus* [= *Tyromyces stipticus*? (134)] fide Overh. 1953: 289; sensu Lloyd, in part, → "*Trametes*" *hoehnelii.*

epiphegum, Polyporus, Pers. 1825: 88, epithet preceded by an asterisk; [≡ *Boletus cryptarum* Bull. sensu Schum. 1803: 389 (Denmark)].— Nomen dubium. Fr. 1828 E. 1: 109 thought this was a cryptophilous form of *Polyporus nigricans* (p. 128); if one relies on the descr. this is a highly improbable guess.

epiphylla, Poria, Pers. 1799 O. 2: 15 (Germany) (d.n.); *Polyporus* Fr. 1815 (d.n.); *Polyporus molluscus* var. (Pers.) per Pers. 1825; *Polyporus* Wettst. 1888.—Nomen dubium. Referred by Fr. 1821: 384 to *Polyporus molluscus* (forma) [sensu Fr. = *Poria* sp. indet.] and taken up by Wettst. as the correct name for this species.

epitheleus, Polyporus, Fr. 1852 (ÖVS 9): 129 (Sweden); *Poria* Cooke 1886.—Nomen dubium. "Fungus maxime abnormis".

epixanthus, Polyporus, Rostk. 1830 (StP 4): 63 *pl. 30* (Germany/Poland). —Nomen dubium. Fr. 1832[Ind.]: 146, 1838: 453 referred this to his *Polyporus alutaceus* (**O**). Lloyd 1915 (LMW 4, Ap.): 302 thought the plate a "fairly good" representation of his interpretation of *P. alutaceus,* which he identified with *Polyporus [Tyromyces] guttulatus.* − Kallenb. 1927 (ZP 6): 59 suggested *Polyporus [Laetiporus] sulphureus.*

eurystegon, Polyporus, E. Krause 1929 B.r.: 78 (Germany).—Nomen dubium. E. Krause 1930 B.r.: 97 soon referred this to *Polyporus byssinus* (Schrad. per Quél.) E. Krause [= *Stromatoscypha fimbriatum* (**O**)], but his original descr. does not support him.

**eutelea, Trametes,* Pat. 1912 (BmF 28): 144 (Mauretania); *Poria* Trott. 1925.—Lowe 1963 (M 55): 465 referred this to *Trametes [Antrodia] serialis,* but the original descr. does not seem to support him.

**everhartii, Mucronoporus,* Ell. & Gall. 1889 (JM 5): 141 *pl. 12* (U.S.A., New Jersey); *Xanthochrous* Pat. 1898; *Scindalma* O.K. 1898; *Pyropolyporus* Murrill 1903; *Fomes* Schrenk & Spauld. 1909, Lloyd 1909 (n.v.p.), Overh. 1914; *Phellinus* A. Ames 1913, Pilát 1942; *Fulvifomes* Murrill 1914.—Certain collections from Czechoslovakia were named *Phellinus everhartii* by Černý; these were later referred to *Phellinus pilatii,* cf. Kotl. 1968 (ČM 22): 174. Thus far not reported from Europe with certainty. − Descr.: Hirt *1930*; Overh. 1953: 82 *pl. 69 fs. 402, 403, pl. 73 f. 425 pl. 107 f. 593, pl. 127 fig.*; Lowe 1957 F.: 34 *f. 19* (*Fomes*). − Special literature: Hirt, *1930*.

fagineus, Boletus, Schrad. apud J. F. Gmel. 1792: 1435, Schrad. 1794:

161 (Germany) (d.n.), not $\sim$ With. 1776 (d.n.).—Nomen dubium. Fr. 1838: 456 referred this to *Polyporus albus* (Huds.) per Fr. [sensu Fr. = *Bjerkandera fumosa*].

favolus albus, *Polyporus*, Secr. 1833 M. 3: 51 (Switzerland) (as a sp. of *Boletus* & double epithet & nom. nud.: n.v.p.).—Fide Fr. 1874: 533 = *Polyporus michelii*.

favolus juglandis, *Polyporus*, Secr. 1833 M. 3: 49, 615 (Switzerland) (as a sp. of *Boletus*; double epithet: n.v.p.) ≡ *Boletus juglandis* Schaeff. [= *Polyporus squamosus*].

**favus*, *Boletus*, L. 1759: 1349 & 1763: 1645 (China) (d.n.) per Spreng. 1827, in part; *Hexagonia* Hariot 1891 (nom. prov.: n.v.p.), not $\sim$ Quél. 1888; = *Polyporus* [*Scenidium*] *sinensis* Fr. (extra-European). —Variously misapplied by European authors: sensu O. F. Müll. (with "?") → *Polyporus squamosus*; sensu "Nonnull." (which became "Quorund." in Fr. 1832$^{Ind.}$: 58), fide Fr. 1821: 367, = *Polyporus* [*Coriolus*] *hirsutus*. This seems to refer to Liljebl. 1798: 453, who admitted a *Boletus favus* from Sweden, whose descr. suggests *Coriolus hirsutus*. He added as a syn. "*Bolet. hirsutulus?*", which perhaps stands for *B. hirsutulus* J. F. Gmel. ⩾ *B.* [*Coriolus*] *hirsutus*. – Sensu Bull. → *Funalia gallica*.

fendleri*, *Polyporus*, B. & C. 1868 (JLS 10): 317 (Venezuela); *Polystictus* Cooke 1886; *Microporus* O.K. 1898; *Poria* Lowe 1947.—Fide Lowe 1966: 132 = *Aporpium caryae* (O**). – V.s.: "*fendzleri*".

**fergussonii*, *Polystictus* (Berk.) ex Cooke 1886 (G 15): 23 (South Africa); *Trametes* Berk. in herb. (n.v.p.); *Microporus* O.K. 1898.—Referred by Lloyd 1910 (LMW 3, M.): 68 and Bres. 1916 (Am 14): 229 to *Trametes hispida* [= *Funalia gallica*]. – Cf. (36).

**ferox*, *Poria*, Long & Baxt. apud D. Baxt. 1940 (PMi 25): 149 *pl. 4* (U.S.A., Arkansas, fide Dr. J. L. Lowe MS. notes).—For some time Lowe 1959 (Ll 21): 105 thought that this might be a form of *Trametes* [*Antrodia*] *serialis*, but later Lowe 1966: 94 *f. 75* treated it as a distinct sp.

farinaceus, *Boletus*, Pers. in herb. (Europe); ≡ *Polyporus versiporus* var. *farinosus* Pers. 1825: 106.—Fide Donk 1933: 226 = *Polyporus versiporus* [= *Schizopora paradoxa*].

farinaceus, *Polyporus*, Pers. 1825: 120, epithet preceded by an asterisk (Europe).—Nomen dubium.

farlowii*, *Polyporus*, Lloyd 1915 (LMW 4, Ap.): 363 *f. 697* (U.S.A., Arizona).—A sp. of *Inonotus*. Long 1945 (Ll 8): 235 cited *Polyporus* [*Inonotus*] *munzii* (O**) and *Inonotus schinii* (**O**) as syns. Overh. 1953: 421 apparently did not know of Long's paper (*1945*); he wrote "certainly it might be surmised that the species is synonymous with *Polyporus munzii*." – Descr.: Long, l.c. ("setae . . . straight or catclaw shaped"); Overh., l.c.

fasciatus, *Polyporus*, Paul. 1812–35 *pl. 164 fs. 5, 6* ("*fascietus*") (type locality not indicated; presumably France).—Nomen dubium. Lév.

1855: 90 referred this to *Polyporus* [*Coltricia*] *perennis*; no other suggestion comes readily to mind. The figures may be fakes that perhaps found inspiration in *Fungus lignosus, fasciatus* Vaill. 1727: 61 *pl. 12 f. 7*, which however is a sp. of *Lactarius*.

**favillaceus, Polyporus*, B. & C. apud Frost 1869 (PB 12): 78 (nom. nud.); apud Berk. 1872 (G 1): 53 (U.S.A., "New England").—Nomen dubium. Bres. 1896 (H 35): 283 thought of *Polyporus crispus* [= *Bjerkandera adusta*] resupinate form; Lowe 1947 (Ll 10): 49 at first concluded that it "seems scarcely determinable", but later, 1959 (Ll 21): 111, he referred it to *Polyporus* [*Hirschioporus*] *abietinus*.

**ferreus, Polyporus*, Berk. 1847 (LJB 4): 502 (Ceylon), not ~ Pers. 1825; *Fomes* Cooke 1885; *Scindalma* O.K. 1898; *Ganoderma* Over. & Steinm. 1923, misapplied; *Trametes* G. Cunn. 1950.—This was referred to *Fomes fomentarius* by Bres. 1890 (Rm 12): 104 but afterwards he rejected this identification. "*Fomes ferreus*" is a distinct species of the eastern tropics, the correct genus and name of which are not yet quite certain. It is not a sp. of *Ganoderma*.

ferruginea, Daedalea, Schum. 1803: 373 (Denmark) (d.n.); *Daedalea* Schum. per Fr. 1821; *Striglia* O.K. 1891.—Nomen dubium.

ferrugineum, Ceratium, Wallr. 1833: 305 (Germany); *Xylodon* O.K. 1898.—Nomen dubium. Fr. 1838: 524 thought that this differed little from *Irpex* [*Hydnochaete*] *cinnamomeus* Fr. [= *Hydnochaete olivacea* (Schw.) Banker] "forma suffocato-byssina". The sp. with which Fr. compared it is extra-European so that the suggested identity appears unacceptable. Wallroth's fungus is apparently non-polyporaceous and might be some underdeveloped state.

fibula, Boletus, Sow. 1803: *pl. 387 f. 8* (England) (d.n.); *Polyporus* Secr. 1833 (as a sp. of *Boletus*: n.v.p.); *Polyporus* (Sow.) per Fr. 1838, misapplied; *Polystictus* Fr. 1851, misapplied; *Bjerkandera* P. Karst. 1882, misapplied; *Coriolus* Quél. 1886, misapplied; *Microporus* O.K. 1898; = *Polyporus peltatus* Pers. 1825, not ~ Pers. 1825, not ~ Fr. 1851.— Nomen dubium. – Sensu Fr., Bres. *Coriolus hirsutus*.

filamentosus, Boletus, Humb. 1793: 95 *pl. 3 f 14* (Germany) (d.n.); *Boletus* (Humb.) per Steud. 1824; *Polyporus* Pers. 1825.—Nomen dubium. Apparently an abnormal growth, found in mines.

fimbriata, Daedalea, Schleich. 1821: 57 (Switzerland) (nom. nud.: n.v.p.).— Nomen dubium. Any connection with *Hydnum* [*Steccherinum*] *fimbriatum* (Pers.) DC. sensu DC.?

fimbriata, Poria, Pers. 1794 (NMB 1): 109 / 1797 T. 29 (Germany) (d.n.); *Boletus* Pers. 1801 (d.n.), not ~ Bull. 1785 (d.n.) per St-Am. 1821; *Porotheleum* Fr. 1818 (d.n.); *Polyporus* (Pers.) per Fr. 1821, not ~ Fr. 1830: Fr. 1832, not ~ (Bull. per St-Am.) Gillet 1878; *Porotheleum* Fr. 1832; *Poria* Lloyd 1910 (syn.: n.v.p.), 1917; = *Polyporus fimbriatus supinus* Secr. 1833 M. 3: 164 (double epithet & as a sp. of *Boletus*:

n.v.p.); ≡ *Stromatoscypha fimbriatum* (Pers. per Fr.) Donk ('Cyphellaceae' or Schizophyllaceae). – Sensu Quél. → *Cristella mollusca*.

fimbriatum, Sistotrema, Pers. sensu DC 1815: 37 (*Hydnum*).—The descr.
was drawn up from material sent in by de Chaillet and from a sample
of "*Boletus abietinus*. Schleich. cent. exs. n. 93, non Pers.", specimens
from material that Pers. 1825: 77 later referred to *Polyporus dolosus*
[= *Hirschioporus abietinus*]. In his copy of de Candolle's book Persoon
crossed out rather vigorously not only the ref. to his own *Sistotrema*
[*Steccherinum*] *fimbriatum* (a non-polyporaceous, resupinate 'hydnum')
but also the one to Schleicher's material.

fimbriatus supinus, Polyporus, Secr. → *Poria fimbriata* Pers. (**O**).

**fimbriporus, Polyporus*, Schw. 1832: 155 (U.S.A., Pennsylvania).—Lloyd
    1913 (LMW 4, L. 50): 6 thought that this might perhaps be *Polyporus*
    [*Tyromyces*] *fragilis*, which is not suggested by the description. Not
    mentioned by Overh. *1953*.

firmus, Boletus, Spreng. 1807: 23 (Germany) (d.n.).—Nomen dubium.
    Fr. 1821: 380 referred this to *Polyporus* [*Perenniporia*] *medulla-panis*,
    but the too brief phrase hardly suggests this (". . . subtus zonatus
    fuscus . . ."). *Heterobasidion annosum*?

fischeri, Ceriomyces, Corda 1837 (StP 3): 133 *pl. 61* (Czechoslovakia).—
    Donk 1972 (PNA 75): 165 suggested that this may be a gall caused
    by a wasp.

fissus, Polyporus, Berk. sensu Crag. – *Polyporus cremoriflorus* Crag. (**O**).

Flabellaria Chev. 1826 (nom. prov.: n.v.p.), not ∼ Pers. 1818 (n.v.p.)
    (Schizophyllaceae) [1960 (Pe 1): 214]; ≡ *Polyporus* trib. *Merisma*
    Fr. [= *Grifola*].

flabellatus, Polyporus, Schulz. & Bres. apud S. Schulz. 1885 (H 24): 145
    (Yugoslavia, Slavonia).—Nomen dubium. At first Quél. 1890 (Crf 18²):
    512 referred this, with doubt, to *Coriolus helveolus* [sensu Quél. =
    Buglossoporus pulvinus], but afterwards he dissociated it from this
    species (which he then called *Caloporus fuscopellis*). Lloyd 1912 (LMW 3,
    S.P.): 156 thought of abnormal *Polyporus* [*Laetiporus*] *sulphureus*. The
    published descr. does not support these identifications.

flabelliformis, Boletus, (Batsch) Sow. 1799: text to pl. 231 (syn.: n.v.p.);
    Boletus suberosus var. Batsch 1789: 117 *pl. 41 f. 226* (Germany) (d.n.).—
    Sow. listed the specific combination as a syn. of *Boletus carpineus* Sow.
    [= *Bjerkandera adusta* forma].

flabellum, Boletus, Hoffm. 1797–1811 V.s.: 33 *pl. 18 f. 1* (Germany).—
    Nomen dubium. Fruitbody laterally attached, white, thin, flaccid,
    subcoriaceous when dry, irregular in shape; on rotten wood in mines.
    Apparently an abnormal growth.

flammea, Auricula, Graff 1936 (M 28): 155, 159 (syn.: n.v.p.) ≡ (an error
    for) *Auricula flammea Malchi* Sterb. 1675 & 1712: 105 *pl. 13* [= *Poly-
    porus squamosus*].

flavescens, Polyporus, Secr. 1833 M. 3: 155 (Switzerland) (as a sp. of

Boletus: n.v.p.), not ∼ Mont. 1856.—Nomen dubium. The collection described by Secr. (leg. Chaill.) was said by its collector to have been used for the descr. of *Hydnum obliquum* DC. 1815: 36 (**O**), although the two descriptions do not readily suggest this.

flavicans, Trametes, Jørst. ("Bres.") ≡ (an error for) *Trametes flavescens* Bres.

flavidus, Polyporus, Peck → *P. peckianus* Cooke (**O**).

flaviporus, Fomes, Quél. in herb. (France).—Fide Lloyd 1915 (LMW 4, F.): 267, 280 = *Fomes laccatus* (Kalchbr. apud Wettst.) Sacc. [= *Ganoderma pfeifferi*].

flavovirens, Polyporus, Romell in herb. (Sweden), D. Baxt. 1938 (PMi 23): 290 *pl. 3 f. 2* (incidental mention, lacking Latin descr.: n.v.p.).—Nomen dubium. Resupinate.

floriformis, Boletus, Pers. 1801: 522 (as var. of *B. cristatus*) ≡ (an error for) *Boletus flabelliformis* Schaeff. [= *Albatrellus cristatus*].

fluctuosus, Polyporus, Weinm. 1836: 337 (U.S.S.R., European Russia).— Nomen dubium. Fr. 1838: 484 thought of *Polyporus [Perenniporia] medulla-panis*, but it is difficult to know precisely what sp. he had in mind under the latter name.

fodinalis, Boletus, Humb. 1793: 99 *pl. 2 f. 8* (Germany) (d.n.) per Steud. 1824; ≡ *Polyporus pera* Pers. 1825.—Nomen dubium. Found in mines. von Humboldt's figure shows the fruitbody as though it were erect and had a superior tube-layer (that is, with the tubes opening upward), with the text in agreement ("superne poroso"); he also stated that the fruitbody was infundibuliform. The picture makes the impression that the tube-layer was convex; it is perhaps also likely that in reality the fruitbody was growing downward, in which case it rather suggests a fruitbody depicted by Sow. *pl. 387 f. 6* as *Boletus hybridus* Sow. (= *Polyporus [Tyromyces] fodinarum?*). "Substantia carnoso mollis"!

foetens, Polyporus, Velen. 1922: 650 [see Pilát 1948: 248 for Latin translation] (Czechoslovakia), not ∼ J. Rick 1935.—Referred by Pilát 1938 (ACE 3): 185 without comment to *Leptoporus [Tyromyces] lacteus* in a widely inclusive sense. Velen. described the spores as ellipsoid, 4–5 μ!

foetidus, Polyporus, Lloyd in herb. (U.S.A., Ohio), not ∼ Velen. 1937.— Fide Lloyd 1915 (LMW 4, Ap.): 305 = *Polyporus [Tyromyces] spraguei*.

foliacea, Poria, Scop. 1772 P.s.: 104 *pl. 22 f. 1* (Hungary, now Czechoslovakia) (d.n.); *Polyporus* (Scop.) per Pers. 1825, not ∼ Jungh. (n.v.p.).—Nomen dubium. Found in a mine; apparently an abnormal growth.

fomentarius, Irpex, Mont. 1856: 174 (France); *Xylodon* O.K. 1898.—Nomen dubium.

fractipes, Polyporus, B. & C. apud Berk. 1872 (U.S.A., South Carolina); *Grifola* Murrill 1904, misapplied; *Petaloides* Torrend 1924; *Polyporus* Bond. & S. 1941; *Abortiporus* Bond. apud E. Komar. 1956 (incomplete

ref.: n.v.p.); *Spongipellis* E. Komar. 1964 (incomplete ref.: n.v.p.);
Heteroporus Fid. 1969.—Reported from Europe (White Russia) by
E. Komar. 1956 (VBb 2): 125 *fs. 1–3*, 1959 (BMs 12): 251 *fs. 1, 2*
(*Abortiporus*), and 1964: 120 *f. 43, pl. 12 f. 52* (*Spongipellis*), but according to Fid. 1969 (Ri 4): 171 the descr. and figures do not support the
correctness of the record.—Descr.: Lloyd 1912 (LMW 3, S.P.): 131,
191; Overh. 1953: 253 *pl. 35 fs. 207, 208, pl. 36 f. 220, pl. 129 fig.*
(*Polyporus*), Fid. 1969 (Ri 4): 169 *fs. 38–48, 52* (*Heteroporus*). – Syns.
(basionyms only): *Polyporus humilis* Peck 1874 (U.S.A., New York);
Polyporus bartholomaei Peck (U.S.A., Kansas)? (**O**); *Fomes cremeotomentosus* P. Henn. 1904 (Brazil); *Abortiporus tropicalis* Murrill 1910
(Jamaica); and cf. *Polyporus delicatus* (**O**). – Sensu Murrill 1904
(*Grifola* → *Polyporus* [*Albatrellus*] *peckianus* (**O**), fide Murrill 1920
(**M** 12): 11.

fragilis, Polyporus, Velen. 1922: 651 [see Pilát 1948: 249 for Latin translation] (Czechoslovakia), not ∼ Fr. 1828.—Pilát 1937 (ACE 3): 176
referred this to *Leptoporus* [*Tyromyces*] *fragilis* without any comment.
His interpretation of *L. fragilis* is not homogeneous. Both *Tyromyces
fragilis* and *T. gloeocystidiatus* apparently do not agree because for
Velenovský's (translated) descr. contains "totus vivus et exsiccatus
pure albus" and "sporis subglobosus, 4 μ".

fraxinicola, Polyporus, S. Schulz. 1882 (RjA 64): 179 (Yugoslavia, Slavonia)
(nom. nud.: n.v.p.), not ∼ Velen. (n.v.p.).—Nomen dubium.

fraxinicola, Polyporus, Velen. "in herb.", not ∼ S. Schulz. 1882 (n.v.p.).
—Fide Pilát 1937 (ACE 3): 162 = *Gloeoporus* [*Bjerkandera*] *fumosus*.

friesii, Poria, Romell 1926 (SbT 20): 24 (nom. prov.: n.v.p.).—New name
suggested for *Poria vaporaria* Pers. sensu Fr. [= *Antrodia sinuosa*].

frondosus, Polyporus, Secr. 1833 M. 3: 57 (Switzerland) (as a sp. of *Boletus*:
n.v.p.), not ∼ (Dicks.) per Fr. 1821.—Nomen dubium: see Donk 1971
(Pe 6): 205.

fucatus, Placodes, Quél. 1887 (Crf 15²): 487 *pl. 9 f. 7*; 1888: 399 (France);
Fomes Sacc. 1891; *Scindalma* O.K. 1898; *Polyporus* Pat. apud Rolland
1890; *Xanthochrous* Pat. 1900; *Chaetoporus* Romell 1901; *Boudiera*
Lázaro 1916; *Lazaroa* Gonz. 1917. – Fide Bres. 1920 (Am 18): 67
= *Polyporus* [*Phellinus*] *gilvus* (**O**) but cf. also (**79**). The protologue calls
to mind such forms as *Phellinus torulosus* f. *subsalicinus* Bourd. & G.
1928: 620.

fugax, Poria, Pers. 1805 I.p.: 36 *pl. 16 f. 2* (Germany) (d.n.); *Polyporus*
(Pers.) per Steud. 1824; *Poria* Lloyd 1910 (incidental mention: n.v.p.).
—Nomen dubium.

fulgens, Poria, (Fr.) Cooke 1886; *Polyporus nitidus* *P. *fulgens* Fr. 1874:
574; [= *Polyporus micans* (Ehrenb.) per Fr. sensu Rostk. 1838 (StP 4):
129 *pl. 63* (Germany/Poland)].—Nomen dubium. This has been associated with *Polyporus* [*Chaetoporus*] *nitidus* (from the start), *Poria
eupora* [= *Chaetoporus nitidus*], and *Polyporus medulla-panis* var.

pulchella (Schw.) Bres. apud J. Rick 1898 (ÖbZ 48): 137 [≡ *Perenniporia pulchella*].

fuligineo-albus, *Polyporus*, Trog 1857 (MiB): 38 (Switzerland).—Nomen dubium.

fuligineus, *Boletus*, Pers. → *Boletus polyporus* Bull. (**O**).

fulvus, *Boletus*, Scop. 1772: 469 (Yugoslavia, Carniola) (d.n.), not ∼ Schaeff. 1774 (d.n.), not ∼ Willd. 1787 (d.n.); *Polyporus* (Scop.) per Fr. 1838, not ∼ Fr. 1863; *Boletus* Lenz 1840, not ∼ Willd. per Pollini 1824; *Trametes* Fr. 1848, 1849, and Otth 1866 (nom. nud.: n.v.p.), Fuck. 1870, not ∼ (Fr.) Kalchbr. 1868; *Fomes* Gillet 1877; *Phellinus* Quélet 1886; *Ochroporus* J. Schroet. 1888; *Inonotus* P. Karst. 1889; *Placodes* Bourd. 1894; *Xanthochrous* Pat. 1897; *Scindalma* O.K. 1898; *Pyropolyporus* Murrill 1903; *Mensularia* Lázaro 1916.—Nomen dubium, fide Donk 1971 (PNA 74): 419. – Sensu Pers., Bres. → *Phellinus pomaceus*; sensu Fr. 1838, 1874 → *Inonotus rheades*, & cf. *Polyporus fulvus* Fr. 1863 [= *Inonotus rheades*; sensu Quél., J. Schroet. → *Phellinus hartigii*, sensu Velen. → *Fomes fomentarius*.

fulvus, *Boletus*, Willd. 1787: 391 (Germany) (d.n.) per Pollini 1824, not ∼ Scop. 1772 (d.n.), not ∼ Schaeff. 1774 (d.n.).—Nomen dubium. Willd. cited Mich. 1729: 121 *pl. 62* as a syn., which represents the type of *Mison* Adans. ≡ *Scindalma* [Hill] O.K., typonym, = *Phellinus*. Referred by Fr. 1821: 375 to *Polyporus* [*Phellinus*] *igniarius* sensu latissimo. Willdenow's descr. is too indefinite to place his fungus with any degree of certainty.

Fungoides Tourn. 1700 (pre-Linnaean name) [1960 (Pe 1): 219], see *Fungoides hyosotis* Paul.

fusca, *Scalaria*, Lázaro 1916 (RMa 14): 741 / 1917: 127 (Spain); *Fomes* Sacc. & Trott. apud Trott. 1925.—Nomen dubium. See also *Scalaria*. (**O**).

fuscatus, *Polyporus*, Lázaro 1916 (RMa 15): 96 / 1917: 188 (Spain), not ∼ (Fr. per Fr.) Cooke 1878, not ∼ Lloyd 1920.—Nomen dubium. Bres. apud Trott. 1925 (SF 23): 369 thought of *Polyporus arcularius* ("juvenilis"). Another possibility is *P. alveolarius*. However, the stalk was said to be lateral and the cap semicircular in outline, which would be abnormal for both these species.

fuscidulus, *Boletus*, Schrad. apud J. F. Gmel. 1792 & Schrad. 1794: 153 (Germany) (d.n.); *Polyporus* (Schrad.) per Fr. 1838; *Polyporellus* P. Karst. 1879; *Leucoporus* Quél. 1886.—Nomen dubium. Fr. 1821: 349 thought that this might be *Polyporus leptocephalus* [= *P. varius*]. Later Fr. 1838: 431 accepted it as a distinct species, but it is difficult to decide whether or not he interpreted it correctly. Sacc. 1916: 960 changed the name of Fries's conception to *Polyporus substrictus* because Fr. had cited this last name as a syn. Pilát 1936 (ACE 3): 64, 67 listed *P. fuscidulus* with doubt as a syn. of *Polyporellus* [*Polyporus*] *brumalis* var. *vernalis* (Fr.) Pilát. – Cf. Lloyd 1912 (LMW 3, S.P.): 171.

*fuscogilvus, Polyporus, Schw. 1832: 157 (U.S.A., Pennsylvania) (syn.: n.v.p.).—Fide Berk. 1872 (G 1): 52 = Polyporus [Inonotus] radiatus.

fuscopallidus, Polyporus, Saut. 1878 (Msa 18): 180 (Austria).—Nomen dubium.

*fuscovelutinus, Xanthochrous, Pat. 1908 (BmF 24): 6 (U.S.A., Louisiana). —Referred by Lloyd 1915 (LMW 4, Ap.): 379 to Polyporus [Inonotus] cuticularis; and by Murrill 1915 S.P.: 41 to Inonotus ludovicianus (Pat.) Murrill (O), an extra-European sp. The original descr. suggests the latter disposition as the most probable.

fuscus, Polystictoides, Lázaro 1916 (RMa 14): 755 / 1917: 141, 293 (Spain); Polystictus Sacc. & Trott. apud Trott. 1925 (syn.: n.v.p.); Polyporus Sacc. & Trott. apud Trott. 1925, not ∼ Lév. 1846, not ∼ (Pers. per Wahl.) Lloyd 1915.—Nomen dubium.

*galactinus, Polyporus, Berk. 1847 (LJB 6): 321 (U.S.A., Ohio); Spongipellis Pat. 1900; Leptoporus Pat. 1900, Pilát 1937.—Reported by Pilát 1937 (BmF 53): 85, 1938 (ACE 3): 225 pl. 140 f. a (Leptoporus) for Europe from Yugoslavia on the basis of a single collection with fruitbodies "un peu déformés-ptychogastérique" and spores formed "sur les hyphes des dissépiments, comme conidies, en grand nombre"; no basidia. Apparently an abnormal collection and a doubtful record. – No syns. of this sp. are mentioned on this list. – Descr.: Overh. 1953: 317 pl. 20 fs. 119–121, pl. 130 fig. (Polyporus).

*gausapata, Trametes, (B. & Rav.) ex Cooke 1891 (G 19): 102; [Polyporus] B. & Rav. in herb. (U.S.A., South Carolina), Cooke 1886 (nom. nud.: n.v.p.)—In selecting the lectotype, material from the first-cited locality ("United States") has been chosen: Lowe 1957 F.: 88 indicated as lectotype "Ravenel No. 1327 at K" and identified it as "Trametes" carbonaria (B. & C. apud Berk.) Overh., stating simultaneously that another American specimen ("Ravenel No. 2922 at K") is "Irpex" farinaceus (Link) ex Fr.; both extra-European spp. I do not know on which specimen (or specimens) previous assessments of the name were based. Murrill 1905 (BTC 32): 368 referred Trametes gausapata to Porodaedalea [Phellinus] pini, for which see (84), but according to Wakef. apud Haddow 1938 (TBS 22): 185 the original collection is not this sp.

gelsorum, Polyporus, Fr. 1821: 377 (incidental mention: n.v.p.) ex ?Steud. 1824 ("†"), ex Fr. 1832^Ind.: 146 & 1838: 469; Boletus Pollini (as var. of B. igniarius: n.v.p.); Fomes Cooke 1885, Bizzoz. 1885; Placodes Quél. 1886; Scindalma O.K. 1898; [≡ Agaricus stratosus Gelsorum Batt. 1755: 70 pl. 37 f. D (Italy)]; Agaricus stratosus Fr. 1821: 377 (syn.: n.v.p.).—Nomen dubium & ambiguum. Cf. Bres. 1891 (BSb 9): 30, ". . . est species valde dubia, quia interum non inventa. Valde probabiliter ad Polyporum [Heterobasidion] annosum vel ad Polyporum [Rigidoporum] ulmarium, non vero ad Polyp. [Phellinum] igniarium adscribenda est." I would add as a probability *Ganoderma* cf. *applanatum*. Pers. 1825: 81 made it a var. of *Polyporus* [*Fomes*] *fomentarius*. – Sensu Martelli → *Inonotus hispidus*. – Batt., l.c., wrote, "Hic apud Mich. ord. 2. n. 4. pag. 118. prostat"; this ref. became in Fr. 1838: 469–470, "Mich. gen. p. 118 n. 4." It is likely that he had in mind Mich. 1729: 118 no. 3 (*Agaricum Gelsis, seu Moris adnascens ex obscuro ferrugineum, inferne album, tenuissime, & densissime perforatum, foraminulis rotundis*). Later Fr. 1874: 562 compounded the confusion by changing 'n. 4' into "no. 7", which refers to a quite different fungus, "*Agaricum Gelsis, seu Moris adnascens, squamosum & lignosum, sordide luteum ubique, superne subhirsutum, inferne tenuissime, & breviter perforatum, foraminulis rotundis* Mich. 1729: 118. All this will explain to some extent the confusion around *Polyporus gelsorum*.

*Gemmularia Rafin. 1819 (JPC 89): 106 & 1820 (JBD 1): 243 (nom. anam.) per Steud. 1824 [1962 (Ta 11): 85]; = *Tucahus* Rafin. 1830, 1833 [1962 (Ta 11): 101]; = *Rugosaria* Rafin. 1833 (nom. prov.: n.v.p.) [1962 (Ta 11): 98]; lectotype, *Gemmularia rugosa* Rafin. (**O**).

geophilus, *Polyporus*, S. Schulz. 1866: 42 (Yugoslavia, Slavonia) (nom. nud.: n.v.p.).—Nomen dubium.

*geotropus, *Polyporus*, Cooke 1884 (British Guiana); *Fomes* Cooke 1885; *Scindalma* O.K. 1898; *Rigidoporus* Imaz. apud S. Ito 1955.—Bres. 1920 (Am 18): 68 and other authors have identified this with *Fomes* [*Rigidoporus*] *ulmarius*, q.v., but Lowe [1955 (M 47): 217; 1957 F.: 80, 81 (*Fomes*), 1963 (PMi 48): 173 (*Polyporus*)] considered the type distinct. – Cf. *Polyporus glabrescens* Berk. (**O**).

gibbus, *Polyporus*, S. Schulz. "Mpt., p. 772. 1869" (n.v.p.).—Igmándy 1968 (Aph 3): 353 referred this, with doubt, to *Ischnoderma resinosum* [sensu Fr. = *I. benzoinum* sensu lato].

*gilvoides, *Trametes*, Lloyd 1912 (LMW 4): 520 *f. 516* (U.S.A., Florida); *Phellinus* Imaz. 1943; *Inonotus* Teng 1964.—This was referred to *Fomes tenuis* [= *Phellinus viticola*] by Overh. 1953: 67, 69, but Lowe 1954 (M 46): 492 stated that it "appears to be a distinct species". – Descr.: Lloyd, l.c.; Bres. 1920 (Am 18): 61, 62; Lloyd 1924 (LMW 7): 1216 *pl. 280 f. 2766* (*Trametes*).

*gilvus, *Phellinus*, (Schw.) Pat. 1900 (**79, 80**); *Boletus* Schw. 1822 (U.S.A., North Carolina); *Polyporus* Steud. 1824: Fr. 1828; *Mucronoporus* Ell. & Ev. 1889; *Chaetoporus* Romell 1901; *Hapalopilus* Murrill 1904; *Fomes* Speg. 1898, not ∼ Lloyd 1912.—The notes cited above give the reasons why I hesitate to enter this species on the list of European polypores. – Descr. (all drawn up from North American material): Berk. 1839 (AM 3): 389; Overh. 1914 (AMo 1): 117; Lloyd 1915 (LMW 3, Ap.): 346; Hirt *1926, 1928*; Bourd. & G. 1928: 621; Corner 1932 (TBS 17): 78; Overh. 1953: 401 *pl. 48 fs. 292, 293, pl. 51 f. 309, pl. 130 fig.* (*Polyporus*). – Cf. also *Placodes fucatus* (**O**) and *Polyporus marcuccianus* (**O**). – Special literature: Hirt, *1926, 1928*.

glabrata, Lenzites, Opiz, "n.n."—Fide Pilát 1940 (ACE 3): 339 = *Gloeophyllum trabeum.*

**glabrescens, Daedalea,* Berk. 1877 (JLS 16): 39 (Australia, New South Wales); *Striglia* O.K. 1891; *Lenzites* G. Cunn. 1950.—Referred by Lloyd 1914 (LMW 4, L. 53): 10, in obs., as a form to *Daedalea quercina,* certainly in error. Bres. 1916 (Am 14): 230 considered it a form of "*Elmerina*" *vespacea* (Pers.) Bres. while G. Cunn. 1965: 178 referred it to "*Daedalea*" *beckleri* (Berk.) G. Cunn., both extra-European spp.

**glabrescens, Polyporus,* Berk. 1839 (AM 3): 391 (Mauritius); *Trametes* Fr. 1851; *Fomes* Cooke 1885; *Scindalma* O.K. 1898.—Cf. Bres. 1912 (Am 10): 496, "Forte tantum varietas *Fomitis ulmarii* Fr."; and Bres. 1916 (Am 14): 240, ". . . *Trametes Marchionicae* Mont., cum quo valde probabiliter identica". Lloyd 1915 (LMW 4, Ap.): 280 thought of *Fomes* [*Rigidoporus*] *geotropus* (Cooke) Cooke.

glaucus, "*Leptoporus*".—An error in 1958 (BsM 1957): 25 for '*Leptotus*' *glaucus* (Batsch per Fr.) Maire (Agaricales).

gordonii, Polyporus, "Berkeley, nomen nudum", Lloyd 1915 (LMW 4, Ap.): 379 (Scotland?) (incidental mention: n.v.p.); Pilát 1936 (ACE 3): 64 ("*Cordonii*") (syn.).—According to Lloyd, l.c., "type a remnant. Probably *Polyporus brumalis.*"

**gossypium, Poria,* Speg. 1898 (ABA 6): 169 (Argentina); *Fibroporia* Parm. 1968.—Referred by Bres. 1916 (Am 14): 229 to *Poria vaillantii,* but Lowe 1966: 118 accepted it as a good sp. that he also recorded for Europe, presumably because he listed *Leptoporus destructor* var. *resupinatus* Bourd. & G. = *L.* [*Poria*] *resupinatus* as a syn.

**gratus, Polyporus,* Berk. 1854 (HJB 6): 163 (India); *Polystictus* Cooke 1886; *Microporus* O.K. 1898.—According to Lloyd 1915 (LMW 4, Ap.): 379, "seems same as *Polyporus* [*Tyromyces*] *floriformis*".

gratzianus, Polyporus,* Murrill 1945 (QFA 8): 197 (n.v.) (U.S.A., Florida). —Lowe apud Overh. 1953: 426 thinks that it "does not appear to differ from *P*[*olyporus*] [*Abortiporus*] *biennis.*" Cf. (2). Fid. 1969 (Ri 4): 150, 153, 162 referred it to *Heteroporus biennis* var. *flabelliformis* ("Mont.") Fid. [sensu Fid. = *Abortiporus distortus* (O**); see Donk 1971 (PNA 74): 2], without having seen the type.

grisea, Bulliardia, Lázaro 1916 (RMa 14): 841 / 1917: 153 (Spain); *Daedalea* Sacc. & Trott. apud Trott. 1925.—Nomen dubium. Cf. Bres. apud Trott. 1925 (SF 23): 449, ". . . species haec vix forma *Daed.* [*Cerrenae*] *unicoloris* habenda."

griseopora, Trametes, Lázaro 1917 (RMa 15): 371 / 1917: 285 (Spain.) —Nomen dubium.

griseus, Polyporus, (Wint.) Pilát 1938 (syn.: n.v.p.), not ~ Peck 1873, not ~ Bres. 1912; *Polyporus destructor* var. Wint. 1882 (RKF 1): 433; [= *Polyporus alutaceus* Fr. sensu Rostk. 1830 (StP 4): 57 *pl. 27* (Germany/Poland)].—Pilát 1938 (ACE 3): 183 (& cf. p. 200) referred this to

Leptoporus [Incrustoporia] semipileatus, which cannot be correct. – See further under *Polyporus alutaceus*.

griseus violaceus, *Polyporus*, Secr. 1833 M. 3: 157 (Switzerland) (as a sp. of *Boletus* & double epithet: n.v.p.).—Fide Fr. 1838: 479 ("n. 107") = *Polyporus [Hirschioporus] abietinus*.

**grossus*, *Irpex*, Kalchbr. 1881 (G 10): 57 (South Africa); *Xylodon* O.K. 1898.—Fide Bres. 1920 (Am 18): 70 = *Trametes [Funalia] gallica* [sensu lato]. The original descr. states, "substantia ligneo-patens" ['-palente', fide Sacc. 1888 (SF 6): 487]. – Cf. (36).

guttatus, *Polyporus*, Preuss 1846 (BCR 1): 202 (Germany) (nom. nud.: n.v.p.), not ∼ Weinm. 1826.—Nomen dubium.

Gymnoderma Humb. 1793: 109 (nom. anam.?) (d.n.) per Steud. 1824 [1957 (Ta 6): 72], monotype, *Gymnoderma sinuatum* Humb. (**O**).

Gyrophana Pat. 1897 [1958 (Fu 28): 9] ≡ *Gyrophora* Pat. 1887, basionym, not ∼ Ach. 1803 (Gyrophoraceae, Lichenes) [1958 (Fu 28): 10]; lectotype, *Merulius lacrimans* (Coniophoraceae) (**O**). Correct name, *Serpula* (**O**).

Gyrophora Pat. → *Gyrophana* Pat.

haematodes, *Polyporus*, Fr. "in litt." (type locality not mentioned), not ∼ Rostk. 1838.—Cited by Fr. 1851 (NAu III 1): 55/39 as a syn. of *Polyporus hypococcinus* [= *Polyporus [Tyromyces] croceus*].

haematodes, *Polyporus*, Rostk. 1838 (StP 4): 127 *pl. 62* ("*haematodus*") (Germany/Poland), not ∼ Fr. (n.v.p.); *Poria* Egeland 1912.—Nomen dubium. Referred by Fr. 1874: 573 to *Polyporus rufus* [→ *Boletus rufus* Schrad. (**O**)], by Bres. 1903 (Am 1): 76 to *Poria [Merulius] taxicola* (**O**).

**haematoxyli*, *Pyropolyporus*, Murrill 1903 (BTC 30): 117 (Jamaica); *Fomes* Sacc. & D. Sacc. 1905.—Referred with some reservations to *Fomes [Phellinus] robustus* by Lowe 1957 F.: 55. – Descr.: Murrill 1908 (NAF 9): 111.

hapalus = apalus.

**hartmannii*, *Polyporus*, Cooke 1884 (G 12): 14 (Australia, Queensland); *Caloporus* R. Heim *1962*, presumably misapplied.—Reported by R. Heim 1962 (RM 27): 93 from near Paris, France, as an alien, but I doubt the correctness of his determination. He described the spores as "phaséoliformes en profil frontal, . . . de 5–6 × 3.9–4.2 μ . . ."; this does not agree with the descriptions based on Australian material as published by Clel. 1935: 211 *f. 15* ("spores narrow, oblique, mummy shaped, . . . 7 to 9 × 2.5 to 3.5 μ") and G. Cunn. 1965: 79 *f. 10* ("spores narrowly cylindrical or fusiform, . . . 7–9 × 3–3.5 μ"). – Apparently also different from *Albatrellus syringae*.

heloporus = holoporus

Henningsomyces O.K. 1898 [1951 (Re 1): 212].—A genus of 'Cyphellaceae', formerly occasionally included in the Polyporaceae.

hepaticus, *Ceriomyces*, Seyn. 1890 (BbF 37): 112 (nom. nud.; n.v.p.), non/an ∼ Sacc. 1888; = chlamydosporous state of *Fistulina hepatica*.

heteroclitus, *Boletus*, Bolt. 1791: 164 *pl. 164* (England) (d.n.); *Polyporus* (Bolt.) per Fr. 1821; *Boletus* S. F. Gray 1821, Purt. 1821.—Nomen dubium. None of the suggestions about the identity made so far are acceptable to me. Cf. Fr. 1828 E. 1: 78, "Ambiguus ille *P. heteroclitus* Bolt. num hinc [*P. alligatus*], an *P. squamoso* affinior?" Lloyd 1915 (LMW 4, Ap.): 380 did not question that it was *Polyporus* [*Albatrellus*] *cristatus*; he also stated that Cooke referred *Polyporus rufescens* [= *Abortiporus biennis*] here. The original descr. states that the upper surface is thickly covered with a rough hairy pile or shag, the young fruitbody a golden colour, changing to orange brown with age, and the colour of the pores varying with age from a pale bright yellow to an orange brown; internal substance white, leathery. – Sensu Sow. 1802: *pl. 367* is certainly different; Fr. 1838: 475 referred this fungus to *Polyporus* [*Oxyporus*] *ravidus*, in my opinion an unlikely disposition.

heteroporus, *Polyporus*, Pers. "in Litt." (n.v.p.), not ∼ Mont. 1841, not ∼ Fr. apud Quél. 1872.—Referred by Cooke 1886 (G 15): 25 to *Poria pinguedinea* (**O**).

heufleri, *Polyporus*, S. Schulz. 1866 (nom. nud.), 1879 (VW 28): 431 ("*Huefleri*") (Yugoslavia, Slavonia) (nomen subnudum)—Nomen dubium. Insufficiently described.

Hexagona Bond. 1953: 47, 524 (lacking Latin descr.: n.v.p.), not ∼ Pollini 1816 (d.n.) per Fr. 1835; [≡ *Hexagonia* Pollini per Fr. 1835 F.s.: 339, 1838: 496 ("*Hexagona*") sensu Bond., l.c., excl. of type, viz. *Polyporus mori*, cf. Bond. 1953: 446]; holotype, *Hexagonia* [*Apoxona*] *nitida*.

hippocastani, *Polyporus*, Saut. 1869 (H 8): 41 (Austria).—Nomen dubium.

hirsutulus, *Boletus*, see under *Boletus favus* (**O**).

hirsutus, *Boletus*, Batsch 1783 (d.n.), not ∼ Scop. 1772 (d.n.), not ∼ Latourr. 1785 (d.n.), not ∼ Wulf. 1788 (d.n.); [≡ *Polyporus Alpinus, cinereus* . . . Mich. 1729: 130 *pl. 71 f. 2* (Italy)]; ≡ *Polyporus tessulatus* Fr. 1818 (d.n.) per Fr. 1821; *Boletus* Pollini 1824.—Insufficiently described and crudely depicted. Nevertheless I would suggest that this may well be *Strobilomyces floccopus* (Vahl per Fr.) P. Karst. (Boletales). Quél. 1888: 405 referred *P. tessulatus* to *Caloporus subsquamosus* [sensu Quél. = *Boletopsis grisea*].

hirsutus, *Boletus*, Latourr. 1785: 39 (d.n.), not ∼ Scop. 1772 (d.n.), not ∼ Batsch 1873 (d.n.), not ∼ Wulf. 1788 (d.n.); [≡ *Polyporus sessilis, convexo planus, anulis versicoloribus, poris albis tenuissimis* Haller 1768 H. 3: 140 no. 2283 (Switzerland)].—Nomen dubium. Latourr. gave no descr., but only a ref. to Haller's non-binomial sp. Fr. 1821: 368 referred this to *Polyporus* [*Coriolus*] *zonatus* (sub forma b). Haller's descr. and the syns. cited show that he had no clear conception of his sp.; he made it a strongly heterogeneous one, lacking *Coriolus hirsutus* as a constituent element.

**hirsutus*, *Trametes*, Lloyd 1924 (LMW 7): 1319 (Australia, Victoria) (nom. nud.: n.v.p.) [cf. Stev. & Cash 1936: 144], not ∼ (Wulf. per Fr.)

Pilát 1939.—Fide Lloyd, l.c., "a trametoid form of *Polyporus* [*Coriolus*] *hirsutus*".

hirsutus ramealis, *Polyporus*, Secr. 1833 M. 3: 131 (Switzerland) (as a sp. of *Boletus* & double epithet: n.v.p.).—Apparently a sp. of *Coriolus*, perhaps a form of *C hirsutus*.

hirtus juglandis, *Polyporus*, Secr. 1833 M. 3: 86 (Switzerland) (as a sp. of *Boletus* & double epithet: n.v.p.).—Fr. 1874: 552 referred this to *Polyporus* [*Spongipellis*] *spumeus*. I would suggest *Tyromyces fissilis*.

hispida, *Lenzites*, Lázaro 1916 (RMa 14): 847 / 1917: 159 (Spain).—Nomen dubium. *Lenzites betulina?*

hispidus, *Boletus*, Rox. Clem. 1864: 64 (Spain) (syn.: n.v.p.), non/an ∼ Bull. 1784 (d.n.) & ∼ (Bull. per Fr.) St-Am. 1821.—This was reduced (apparently by the editor of the paper, Colm.) to *Polyporus* [*Coriolus*] *hirsutus*. There is no descr., only the indication, "En los troncos de morera." This was perhaps not a new name but merely a record of *Boletus* [*Inonotus*] *hispidus* Bull.

holoporus, *Polyporus*, Pers. 1825: 107 [1822: *pl. 6 fs. 3, 4*] (Germany); Fr. 1828 E. 1: 117 ("*heloporus*"; syn.); *Irpex* Sacc. & Trav. 1910. —Nomen dubium. Referred by Fr. 1828 E. 1: 117 to *Polyporus* [*Poria*] *xanthus*. Considered to be a form of *Poria* [*Antrodia*] *sinuosa* by Bourd. & G. 1928: 672, who relied on the descr. only. The epithet should apparently be corrected into 'heloporus', as was done by Fr., l.c., cf. Persoon's remark, ". . . tubulos decumbentes glabros fere solos constituens." No type could be traced in Persoon's herbarium.

hornotinus, *Boletus*, P. Magn. 1905: 180 ("Swartz") (syn.: n.v.p.).—An error for *Polyporus marginatus* [status] "B. hornotinus et annosus" Fr. 1874: 561, in part [= *Fomitopsis pinicola*].

hornotinus, *Boletus*, (Fr.) Mussat 1901 (SF 15): 65 (syn.: n.v.p.).—An error for *Polyporus resinosus* [status] "B." *hornotinus*" Fr. 1874: 554, Sacc. 1888 (SF 6): 137, the 'B.' being taken for the abbreviation of '*Boletus*'.

hrabatii, *Polyporus*, Opiz 1855 (Lo 5): 87 (Czechoslovakia).—Nomen dubium.

huefleri = *heufleri*

**hyalinus*, *Polyporus*, Berk. 1859 Ta.: 255 (Australia, Tasmania); *Poria* Cooke 1886.—Nomen dubium, fide Lowe 1966: 136, but cf. G. Cunn. 1965: 46 who gives a full descr. Here mentioned because J. Rick 1960 (Ih 7): 271 referred *Poria byssopora* J. Rick [= *P. xylostromatoides*] to this sp. as a var.

hybridus, *Boletus*, "Willd. plant cryptog. ined."—A herbarium name (n.v.p.) mentioned by Humb. 1793: 99 in an observation on *Boletus fodinalis* (**O**).

hydnoides, *Boletus*, Gaterau 1789: 191 (France) (d.n.).—Nomen dubium.

hygrophanus, *Polyporus*, Romell 1926 (nom. prov.: n.v.p.) (Germany). —This is *Tyromyces gloeocystidiata* fide Kotl. & P. 1964 (ČM 18): 207,

217 [*T. leucomallelus*]. Lundell 1946 (LNF 27–28): 8 concluded that the illustration Fr. 1884 I. 2: *pl. 181 f. 2* matches a sp. related to *Polyporus* [*Tyromyces*] *fragilis* and "without doubt identical with *P. trabeus* sensu Bres. [= *Tyromyces leucomallelus*] and with *P. hygrophanus* Romell ad interim [1926 (SbT 20): 18]." Romell, l.c., wrote in connection with certain specimens, "It may be a distinct species (provisionally called *Pol. hygrophanus* in my herbarium)."

hymenopus, Daedalea, F. Brig. *1847* (Italy).—Nomen dubium. Sacc. 1916: 955 referred this to *Polyporus rufescens* [= *Abortiporus biennis*], but several details of the descr. contradict this disposition.

hyosotis, Fungoides, Paul. 1812–35: *pl. 185 fs. 1, 2* (generic name n.v.p.) = *Conchites auricula-major* Paul. 1793 (d.n.).—This sp. has been identified with *Polyporus varius* and *P. melanopus*, but Donk 1960 (Pe 1): 219 suggests that it may be one of the large Pezizaceae (Discomycetes).

**hyperborea, Poria*, Berk. in herb. (Canada) & cf. Cooke 1886 (G 14): 114 (nom. nud.), 1886 (G 15): 27, Sacc. 1886 (SF 6): 325 (incidental mention: n.v.p.).—Fide Bres. 1926: 80 = *Polyporus* [*Hirschioporus*] *pargamenus* (resupinate).

**hypocitrinus, Polyporus*, Berk. 1876 (JLS 15): 50 (Brazil).—Considered to be a form of *Polyporus* [*Tyromyces*] *lacteus* by Bres. 1916 (Am 14): 225, but cf. Speg. 1925 (BCó 28): 374.

hypocrateriformis, Boletus, Schrank 1789: 621 (Germany) (d.n.).—Nomen dubium. This was referred to *Polyporus brumalis* by Fr. 1821: 348, but the original descr. is too incomplete for certain determination. A sp. with white fruitbody.

hypolateritia, Poria* (Berk.) ex Cooke 1886 (G 15): 24; *Polyporus* Berk. in herb. (India); *Schizopora* (Berk. ex Cooke) Parm. 1968.—At first this was referred by Lowe 1962 (PMi 47): 186 to *Poria versipora* [= *Schizopora paradoxa*]; later he treated it as a distinct, but closely related, sp. (Lowe 1966: 65) but I doubt his interpretation (105**). Its being recorded for Europe is because Lowe considered *Poria eyrei* and *P. consobrina* to be syns.

**hypopolius, Polyporus*, Kalchbr. apud Cooke 1882 (G 10): 20 (Australia, Queensland).—Fide G. Cunn. 1950 (PNW 75): 228 & 1965: 147 = *Fomitopsis/Heterobasidion hemitephra* (extra-European). Also cited as a syn. of *Heterobasidion annosum* because Lloyd 1915 (LMW 4, Fo.): 281 thought that the "description indicates *Fomes annosus*."

impolitus, Polyporus, Fr. 1851 (NAu III 1): 58/42 (Costa Rica).—A nomen dubium. According to Lloyd 1915 (LMW 4, Ap.): 381 "no type exists." – Reported from Portugal by Thüm. 1878 (JSL 6): 243, a record almost certainly based on a misidentification. In later Portuguese literature sometimes listed as *Boletus impolitus* Fr. (Boletaceae), a quite different sp.

impuber, Boletus, Sow. 1799: *pl. 195* ("Bull.") (d.n.); *Polyporus* (Sow.)

per Pers. 1825.—The specific epithet was ascribed to Bull.; perhaps Sow. had *Boletus 'imberbis'* Bull. in mind, although this suggestion is not very likely. If treated as a new species, then the name could better be qualified as nomen dubium. – During the course of his life Fr. made various suggestions; in 1874: 548 he referred Sowerby's fungus to *Polyporus [Phellinus] gilvus*, an unacceptable identification (**79**).

incarnatus, Polyporus, Fr. 1832^Ind.: 149 (Sweden), not ∼ (Pers.) per Fr. 1821.—Nomen dubium. Cf. also (**110**).

**induratus, Polyporus*, Peck 1879 (RNS 31): 37 (U.S.A., New York) [repr. 1881 (H 20): 141], not ∼ Berk. 1875, not ∼ Lloyd 1918; *Myriadoporus* Peck 1884 (nom. prov.: n.v.p.); *Poria* Cooke 1886.—Pat. 1889 (BmF 5): 84 referred this to *Polyporus [Oxyporus] obducens* (myriadoporous), Murrill 1920 (M 12): 80 made it a doubtful syn. of *Poria [Perenniporia] subacida*. Not mentioned by Lowe 1966. – Descr.: Overh. 1919 (BNS 205–206): 82 *pl. 6 fs. 6–7, pl. 7 fs. 1–3 (Poria)*.

informis, Boletus, Cumino 1805: 223 (Italy) (d.n.) per Pollini 1824; *Polyporus* Pers. 1825.—Nomen dubium. Referred by Fr. 1838: 455 to *Polyporus [Tyromyces] croceus*, and 1874: 540 to *Polyporus lobatus* [sensu Fr. = *Albatrellus cristatus*?]. The descr. is insufficient to determine the identity, I believe.

infularis = insularis

infundibuliformis, Helvella, Schaeff. → *Boletus coriaceus* Batsch 1783 (**0**).

infundibuliformis, Polyporus, "Roth".—Cited by Fr. 1832^Ind.: 146; "Roth" is an error for 'Rostk.' – A syn. of *Polyporus squamosus*.

infundibulum, Agaricus, Paul. 1812–35: *pl. 4 f. 2* (d.n.?); [≡ Agaric-iris en entonnoir Paul. 1793 T. 2: 81 (France)].—Nomen dubium. Lév. 1855: 4 called this *Polyporus [Coriolus] versicolor* var. *infundibuliformis* Lév. However, the protologue is so poor that nothing definite can be suggested. It is even doubtful whether it is a polypore at all.

inodora, Trametes, Fr. 1849 (nom. nud.: n.v.p.), 1863 M. 2: 273 (Sweden); *Daedalea* E. Krause 1925.—Nomen dubium. Pilát 1934 (BmF 49): 264 reduced this to the rank of a form of *Trametes suaveolens* [sensu Fr.], differing only in the lack of the odour of anise. Fries also emphasized other differences: "... sed minor, firmior, semper alba (pori non fuscescens), poris minoribus, rotundis et odore nullo distincta." – Fr. 1884 I. 2: *pl. 191 f. 1* reminds me of *Trametes gibbosa* (small fruitbody) by the shape and size of the pores.

insularis, Polyporus, Pers. → *Boletus terrestris* Sow. (**0**).

intermedius, Polyporus, Sing. 1922 (PK 5): 265 (Germany), not ∼ Rostk. 1837.—Nomen dubium. Sing. 1962: 159 referred this to *Polyporus arcularius*; in the original descr. he thought it to be intermediate between *P. arcularius* and *P. ciliatus*. The descr. is insufficient to decide on its identity, but cf. *P. brumalis*; Singer's conceptions of *P. arcularius* (of 1922 and of 1962) and *P. ciliatus* (of 1922) are not clear to me.

internus, *Polyporus*, Schw. 1832: 159 (U.S.A., Pennsylvania); *Poria* Cooke 1886.—Referred by Bres. 1926: 30 to *Poria monticola* [= *Poria placenta*]; but it is not clear what conception Bres. had of the latter sp. Overh. 1923 (M 15): 215 could not find a type in herb. Schw. and in the Michener collection, but a portion of it may be at K.

interruptus, *Boletus*, Pers. "Mscr." ≡ *Polyporus frustulatus* Pers. = *Rigidoporus undatus*.—Cited by Fr. 1828 E. 1: 93 as a syn. of *Polyporus* [*Antrodia*] *serialis* (forma).

interruptus, *Polyporus*, B. & Br. 1873 (JLS 14): 55 (Ceylon); *Poria* Cooke 1886.—This name is mentioned because Petch 1919 (APe 7): 5 here referred *Poria aquosa* Petch, which Lowe listed as a syn. of *Poria xylostromatoides*.

inversus, *Boletus*, "[Gunn.] Fl. N. 540", 1779: 249 (d.n.), not ∼ Vill. 1789 (d.n.) per Steud. 1824 or per Pollini 1824.—Nomen dubium. Retz. cited "Fl. N. 540" = J. E. Gunnerus, Fl. norv. 2. "1772" [1776]; I have been unable to locate this name in the cited work.

inversus, *Boletus*, Vill. 1789: 1042 (France) (d.n.), not ∼ Retz. 1779 (d.n.); *Boletus* Vill. per Steud. 1824 or per Pollini 1824.—Nomen dubium. Referred by Fr. 1821: 381 to *Polyporus vulgaris* [sensu typo = *Poria lenis*], which means scarcely anything.

inversus, *Polystictus*, Lázaro 1916 (RMa 14): 746 / 1917: 132 (Spain). —Nomen dubium. *Coriolus hirsutus*?

investiens, *Polyporus*, Desm. in herb. Pers. (n.v.p.).—Fide Pers. in herb. = *Boletus abietinus* [sensu] DC. [= *Skeletocutis amorphus*], as mentioned by Lloyd 1915 (LMW 4, Ap.): 381.

irpicoides, *Daedalea*, P. Henn. 1898 (BJ 25): 501 (New Guinea) was cited in error by Bond. 1953: 478 as a syn. of *Cerrena unicolor* var. *irpicoides* (Bres. ex Bourd. & G.) Bond.

isabellina, *Lenzites*, Lloyd 1922–3 (LMW 7): 1156 *pl. 220 f. 2274* (Philippine Is.) (n.v.p.).—Fide Lloyd, l.c., a form of *Lenzites betulina*.

isabellinum, *Chromosporium*, Ell. & Sacc. apud J. B. Ell. 1885 N.A.F.: No. 1391 ("ined."; n.v.p.) (U.S.A., New Jersey).—Fide Ell. & Ev. 1890 (JM 6): 79 = *Mucronoporus* [*Inonotus*] *andersonii*, spore print; "on old *Polyporus* (*obliquus*)? and on rotten wood adjacent" is mentioned on the original label. Cf. Donk 1971 (PNA 74): 8 for similar cases.

isidioides, *Polyporus*, Opiz 1852: 136 (Czechoslovakia) (nom. nud.: n.v.p.), not ∼ Berk. 1843.—Nomen dubium.

isopora, *Daedalea*, (Pers.) E. Krause 1928 ("*Daedaleus*"); *Xylomyzon* Pers. 1825: 33 *pl. 16 fs. 1, 2* (France); *Merulius* Duby 1830; = *Merulius rufus* Pers. per Fr. ('Corticiaceae') fide Fr. 1828 E. 1: 63 ("*isopyrum*") & Bres. 1903 (Am 1): 83.

jamaicensis, *Fomitiporia*, Murrill 1907 (NAF 9): 11 (Jamaica); *Poria* Sacc. & Trott. 1912.—Referred by Lowe 1949 (Ll 11): 167 & 1957 F.: 56 to *Poria* [*Phellinus*] *punctatus*, which he regarded as resupinate

Fomes [*Phellinus*] *robustus* but cf. (86). A re-description of the type
is needed.

**jamaicensis, Inonotus,* Murrill 1904 (BTC 31): 597 & 1908 (NAF 9):
88 (Jamaica); *Polyporus* Sacc. & Trott. 1912.—Lloyd 1915 (LMW 4,
Ap.): 381 was inclined to refer this to *Polyporus* [*Inonotus*] *cuticularis*,
but it is a distinct sp. – Descr. : D. Reid 1955: 12 *f. 1*; Pegl. 1964
(TBS 47): 190 *f. 2: 10* (*Inonotus*).

**japonica, Lenzites,* B. & C. 1858 (PAA 4): 121 (Japan); *Cellularia* O.K.
1898.—Bres. 1916 (Am 14): 221 referred this to *Lenzites betulina* "f.
biennalis"; Imaz. 1943 (BTS 6): 73 and S. Ito 1955: 232 consider it
to be a syn. of *Trametes acuta* (Berk.) Imaz. [= *Lenzites acuta* Berk.],
an extra-European sp.

**javanicus, Xanthochrous,* Pat. 1897 (ABu, S. 1): 113 *pl. 24 fs. 8–10*
(Indonesia, Java); *Cyclomyces* Sacc. & Syd. 1899.—This was referred
to *Cyclomyces greenei* [≡ *Polyporus* [*Coltricia*] *montagnei* var. *greenei*
(Berk.) R. L. Gilb.] by Lloyd 1917 (LMW 5): 633.

**jezoensis, Daedalea,* Yamano 1930 (Goryorin No. 25): 70 *fs. 1, 3, 7, 8*
(n.v.) (Japan); *Fomes* Tochinai & Kamei 1933 (n.v.); *Cryptoderma*
Imaz. 1943 ("*yezoense*").—See under *Cryptoderma yamanoi.*

*jodiolens, gilvescens, gilvotacta, farinella, microspora, minutosinuosa,
radiosa, senescens, suecia, Poria,* Romell in herb., and *jodoformicus,
lunaris, tactus, Polyporus,* Romell in herb.—Herbarium names mentioned
by Jo. Erikss. 1949 (SbT 43): 7–15 as syns. of *Poria lenis.*

**junghuhnii, Lenzites,* Lév. 1844 (ASn III 2): 180 (Indonesia, Java);
Cellularia O.K. 1898.—This has been referred to *Lenzites guineensis,*
which in its turn has been referred to *Lenzites betulina*; cf. Fr. 1851
(NAu III 1): 44–45/28–29, "Non igitur dubitamus esse *L. Junghuhnii*
Lev. . ., quam Cel. Junghuhn ipse pro *L. betulina guineensi* habuit,
quae lamellis acie subvelutinis modo recedit." Bres. treated it as a
distinct sp. from *L. betulina.*

**juniperina, Fuscoporia,* Murrill 1907 (NAF 9): 4 (U.S.A., Louisiana);
Poria Sacc. & Trott. 1912.—Referred by Lowe 1949 (Ll 11): 168 to
Poria [*Phellinus*] *punctata* and 1957 F.: 56 to resupinate *Fomes*
[*Phellinus*] *robustus,* but cf. (86). Redescription of the type is needed.
Cf. D. Baxt. 1941 (PMi 26): 111 *f. 4* (*Fuscoporia*); 1952 (PMi 37):
103, 104 (*Fomes robustus* forma).

**juniperinus, Agaricus,* Murrill 1905 (BTC 32): 85 (U.S.A., Kansas);
Daedalea Murrill 1908; ≡ *Daedalea kansensis* J. B. Ell. in herb. [cf.
Lloyd 1906 (LMW 2, L. 10): 3, 1911 (LMW 3, L. 38): 11].—This sp.
was referred by Bres. 1920 (Am 18): 64 as a form to *Daedalea quercina,*
but I believe that it can be better kept apart. It may be referred to
Antrodia. – Lloyd 1911 (LMW 3, L. 38): 10, 11 recorded it from Russia,
with the remark, "This is the first record of the plant in Europe".
The record needs verification; modern Russian authors do not report
it from Europe.

kansensis, Daedalea, J. B. Ell. → *Agaricus juniperinus* Murrill (**O**).

kansensis, Trametes*, Crag. 1884 (BWb 1): 24 [repr. 1884 (JM 1): 28] (U.S.A., Kansas).—Nomen dubium. Murrill 1908 (NAF 9): 125 referred this, with doubt, to *Daedalea quercina*. Cf. also Overh. 1953: 124. The type grew "on dry oak planks", which would exclude *Daedalea juniperina*. This should not be confused with *Daedalea kansensis* J. B. Ell. in herb. ≡ *D*. [*Antrodia*] *juniperina* (O**).

karelicus, Polyporus, Jač. 1913 O.G. 1: 635 (n.v.).—Nomen dubium, fide Bond. 1953: 619, who thought of a sp. of *Coriolus*, perhaps even *C. versicolor*.

kauffmannii, Favolus*, Lloyd 1916 (LMW 5): 614 *f. 869* (U.S.A., Michigan) ("I think it is a variation of *Favolus europaeus*": n.v.p.).—Fide Overh. 1953: 157 = *Favolus alveolarius* (O**). I am not convinced that this is correct, judging from the figure only.

**kerensis, Polyporus*, Passerini 1875 (NGi 7): 181 [repr. Martelli 1886: 132] (Abyssinia).—Lloyd 1915 (LMW 4, Ap.): 381 thought that the descr. "reads like *Polyporus* [*Truncospora*] *ochroleucus*", an unlikely guess.

kirchneri, Favolus, Wallr. ("in litt.") apud L. Kirchn. 1856 (Lo 6): 246 (Czechoslovakia).—Nomen dubium. *Polyporus mori?*

klotzschii, Lenzites*, Berk. 1841 (AM 7): 452; *Cellularia* O.K. 1898; [≡ *Daedalea discolor* Fr. sensu Kl. 1833 (Li 8): 481 (Canada) (lacking descr.)]; ≡ *Lenzites discolor* Fr. 1851, typonym.—Fide Murrill 1905 (BTC 32): 86, 87 = *Agaricus* [*Daedaleopsis*] *confragosa*. − Cf. (31**).

kusanoi, Coriolellus*, Murrill 1909 (M 1): 165 (Japan); *Trametes* Sacc. & Trott. 1912.—Imaz. 1939 (JJB 15): 302 referred this to *Trametes* [*Antrodia*] *albida*, but Aosh. 1967 (TmJ 8): 3 considers it a syn. of *Daedalea* [*Antrodia*] *tanakae* (O**). − Descr.: Nohara 1910 (BMT 24): (8) *fs. 4, 6d* (*Coriolellus*); & see next entry.

kusanoi, Daedalea*, Murrill sensu Lloyd 1924–5 (LMW 7): 1331 *pl. 317 f. 3054*.—Referred to *Trametes* [*Antrodia*] *albida* by Imaz. 1939 (JJB 15): 302, but his conception perhaps coincided with *Irpicoporus* [*Antrodia*] *tanakae* (O**).

Labirinthia S. Schulz.—Published as a syn. in the combination *Labirinthia truncata* (**O**).

laburni, Polyporus, Opiz 1852: 136 (Czechoslovakia) (nom. nud.: n.v.p.). —Nomen dubium.

labyrinthica, Lenzites, Quél. & Schulz. apud S. Schulz. 1885 (H 24): 141 (Yugoslavia, Slavonia); *Cellularia* O.K. 1898.—Nomen dubium.

labyrinthicus, Polyporus, Fr. → *Sistotrema spongiosum* Schw. (**O**).

lacerum, Porotheleum, Fr. 1818 O. 2: 273 (Sweden) (d.n.) per Fr. 1828. —Fide Lloyd 1917 (LMW 5): 740 = *Porotheleum fimbriatum* [≡ *Stromatoscypha fimbriatum* (**O**)].

lacrimans, Boletus, Wulf. 1781 (MaJ 2): 111 *pl. 8 f. 2* ("*lacrymans*") (Austria) (d.n.); *Merulius* Schum. 1803 (d.n.); *Xylophagus* Siemss. 1809 (n.v.) (d.n.); *Merulius* (Wulf.) per Fr. 1821 ("*lacrimans*"); *Boletus*

Purt. 1821; *Gyrophora* Pat. 1887 (nom. nud.: n.v.p.); *Gyrophana* Pat. 1900; *Sesia* P. Karst. 1899; *Xylomyzon* E. Krause 1932; = *Merulius lacrymabundus* Link 1833; = *Serpula lacrymans* (Wulf. per Fr.) J. Schroet. 1888 (Coniophoraceae).

lacrimans, Ptychogaster, P. Henn. in herb.—P. Henn. 1889 (VBr 30): v rejected this name in favour of *Ptychogaster rubescens* (**125**).

lacrymabundus, Merulius, Link → *Boletus lacrimans* Wulf. (**0**).

lacrymans, Polyporus, Saut. 1876 (H 15): 150 (Austria); *Poria* Cooke 1886.—Nomen dubium.

**lamelliformis, Irpex,* Lloyd 1917 (LMW 5): 715 *f. 1073*; [= *Irpiciporus noharae* Murrill sensu Lloyd 1916 (LMW 5): 601 *f. 851* (*Irpex*) (Japan)]. —Considered to be a syn. of *Hirschioporus fuscoviolaceus* by Imaz. 1944 (JJB 20): 286, from which sp. it seems to differ in some respects.

lanatus pruni, Polyporus, Secr. 1833 M. 3: 152 (Switzerland) (as a sp. of *Boletus* & double epithet: n.v.p.).—Nomen dubium.

langloisii, Fomitiporia,* Murrill 1907 (NAF 9): 9 (U.S.A., Louisiana); *Poria* Sacc. & Trott. 1912.—Referred by Bres. 1926: 30 and Lowe 1949 (Ll 11): 167 to *Poria friesiana* [= *Phellinus punctatus*] (86**). A redescription of the type is needed.

lapidum, Boletus, Raddi 1807: 361 (Italy) (nom. nud.).—To judge from the epithet alone, this may be *Polyporus tuberaster*. Since Raddi did not add "nob." to the name it may perhaps have been previously published by another author.

laricis, Boletus, Schleich. 1821: 56 (Switzerland) (nom. nud.), not ∼ Rubel 1778 (d.n.) per Roques 1821.—Nomen dubium. This was described by Secr. 1833 M. 3: 160 as *Polyporus* [*Poria*] *vulgaris* var. B. (unnamed var.).

laricis, [Trametes?],* Jač. in litt. (U.S.S.R., Russia, Siberia) = *Xanthochrous pini* var. *laricis* (Jač.) ex Pilát 1934 (BmF 49): 272 *pl. 23 fs. 1, 2*. —Fide Pilát 1942 (ACE 3): 521 = *Phellinus pini* var. *abietis* (P. Karst.) Pilát [= *Phellinus chrysoloma*] (forma). – Cf. (84**).

lateralis, Polyporus, Pers. apud Gaud. 1827: 175 (New Guinea, Rawak I.); *Melanopus* Pat. 1927 ("*lateris*").—Lloyd 1912 (LMW 3, S.P.): 185 ("*lateratus*") wrote, "In my opinion, it is a lateral stemmed form of [*Polyporus*] *elegans* [= *P. varius*] of Europe." To be verified.

latus, Boletus, Sow. 1803: *pl. 387 f. 9* (England) (d.n.); *Polyporus* (Sow.) per Steud. 1824, not ∼ Berk. 1839; = *Polyporus concentricus* Pers. 1825, not ∼ Cooke 1880.—Nomen dubium. Cf. Fr. 1828 E. 1: 122.

**lavendulus, Polystictus,* Lloyd 1922 (LMW 7): 1121 *pl. 201 f. 2142* (U.S.A.: Minnesota) (as a form of *P. pargamenus*: n.v.p.).—Fide Overh. 1953, 336, 338 = *Polyporus* [*Hirschioporus*] *pargamenus*, "spores reported by Lloyd ... are undoubtedly foreign".

laxus, Polyporus, Otth apud Trog 1857 (MiB): 39 (Switzerland).—Nomen dubium.

lazaroi, Polyporus, Trott. → *Polyporus parvulus* Lázaro (**0**).

*leoninus, Polyporus, Kl. 1833 (Li 8): 486 (East Indies); *Polystictus*
Fr. 1851; *Microporus* O.K. 1898; *Coriolus* G. Cunn. 1950; *Trametes*
Imaz. 1952; *Spongipellis* Teng 1964; ≡ *Funalia leonina* (Kl.) Pat. 1900.
—When Karsten 1859: 30 published the name *Polyporus fibrillosus*
[= *Pycnoporellus fulgens*] he cited *Polyporus leoninus* (extra-European),
with doubt, as a syn. The latter sp. is widely different.

Leptopora Rafin. 1808 (d.n.) [1960 (Pe 1): 236], not *Leptoporus* Quél.
1886; ≡ "Leptostroma. Rafin." Reichenb. 1828 (error; incidental
mention: n.v.p.).—Nomen dubium.

Leptostroma "Rafin." → *Leptopora* Rafin. (**0**).

*leucogriseus, Boletopsis, Malenç. 1956 (nom. prov.: n.v.p.); [≡ *Boletopsis
grisea* (Peck) Bond. & S. sensu Malenç. 1956 (BmF 71): 272 (Morocco)]
(**18**).

leucoloma, Polyporus, Pers. 1825: 65; Fr. 1828 E. 1: 94 ("*leucoma*"; syn.);
[≡ *Agaricus cinereus margine albo* Batt. 1755: 69 *pl. 35 f. B* (Italy)].
—Nomen dubium. Fr. 1821: 368 referred this to *Polyporus* [*Coriolus*]
zonatus (forma).

leucomelaena, Polystictoides, Lázaro 1916 (RMa 14): 832 / 1917: 145
("*leucomelas*") (Spain); *Polystictus* Sacc. & Trott. apud Trott. 1925.
—Nomen dubium.

leveilleanus, Tyromyces, Bond. apud Bond. & S. → *Polyporus apalus* (**0**).

ligoniformis, Polyporus, Bon. 1876 (H 15): 76 (Germany).—Nomen dubium.
I would suggest, without conviction, *Buglossoporus pulvinus*.

lilacino-roseus, Polyporus, S. Schulz. "Mpt., p. 763. 1869" (n.v.p.).—Fide
Igmándy 1968 (Aph 3): 225 = *Fomes* [*Fomitopsis*] *roseus*.

*lilacinus, Boletus, Schw. 1822: 100 (U.S.A., North Carolina); *Polyporus*
Steud. 1824: Fr. 1832.—Nomen dubium. Referred by Fr. 1828 E. 1:
118 & 1838: 484 to *Polyporus* [*Ceriporia*] *purpureus*. Overh. 1923 (M 15):
216 reported that no type existed.

lilacinus, Polyporus, (Batsch per Opiz) D. Dietr. 1847 ("*lilacineus*");
Helvella lilacina Batsch 1786: 187 *pl. 25 f. 131* (Germany) (d.n.); =
Stereum purpureum (Pers. per Fr.) Fr. ≡ *Chondrostereum purpureum*
(Pers. per Fr.) Pouz. ('Stereaceae').

*lindheimeri, Trametes, B. & C. apud Berk. 1872 (G 1): 66 (U.S.A., Texas);
Polystictus Cooke 1886; *Microporus* O.K. 1898; *Polyporus* Murrill 1905
(incidental mention: n.v.p.).—Fide Murrill 1908 (NAF 9): 79 & Lloyd
1910 (LMW 3, M.): 65, 68 = *Funalia/Polystictus stuppea* which in its
turn has been referred to *Funalia hispida* = *F. gallica*, but cf. Bres.
1926: 25, "a *Trametes stuppea* . . . prorsus est diversus." − For figure
of type, see Lloyd, op. cit., *f. 355* (as *Polystictus stuppeus*). − Cf. (**36**).

lindtneri, Septocylindrium, W. Kirschst. 1936 (ZP 15): 118 *pl. 15 f. 2*
(Yugoslavia) (nom. anam.).—This has been considered a chlamydosporous state of *Ganoderma applanatum* by Pilát 1942 (ACE 3): 491,
but this conclusion is still open to considerable doubt. Similar growths
were described as *Hymenochaete dendroidea* B. & C. apud B. & Br.,

Reticularia apiospora B. & Br., and *Institale bombacina* (Fr. apud Weinm.) Fr. sensu Lloyd. − Cf. (**41**).

linearisporus, Polyporus, Velen. 1922: 654 [see Pilát 1948: 250 for Latin translation] (Czechoslovakia).—Referred by Pilát 1938 (ACE 3): 185, without comment, to *Leptoporus* [*Tyromyces*] *lacteus* in a very inclusive sense.

lloydii, Fomitiporia,* Murrill 1907 (NAF 9): 10 (U.S.A., Ohio); *Poria* Sacc. & Trott. 1912.—Referred by Bres. 1926: 80 and Lowe 1949 (Ll 11): 167 to *Poria friesiana* [= *Phellinus punctatus*], cf. (86**). A redescription of the type is needed.

lobatum, Xylometron, Paul. → *Agarico-suber lobulatum* Paul. (**0**).

lobulatum, Agarico-suber, Paul. 1793 T. 2: 78 (descr.), Ind. (name) [*pl. 3 f. 1*] (France) (d.n.); ≡ *Xylometron lobatum* Paul. 1812–35 (generic name n.v.p.).—Nomen dubium.

loennbohmii, Poria,* P. Karst. 1904 (ÖfF 46 [11]): 8 (U.S.S.R., Baikal region).—Referred to *Daedalea* [*Daedaleopsis*] *confragosa* by Lowe 1956 (M 48): 121. − Cf. (31**).

**lucidus, Ptychogaster,* Lloyd 1920 (LMW 5): 699 *f. 1046* (Brazil) (nom. anam.).—Lloyd considered this to be an imperfect state of *Polyporus* [*Ganoderma*] *lucidus,* but this is certainly wrong. The presence of "large setal hyphae" observed by R. W. Davids. & al. 1942 (M 34): 145 refer it to the Hymenochaetaceae.

**ludovicianus, Xanthochrous,* Pat. 1908 (BmF 24): 6 (U.S.A., Louisiana); *Polyporus* Sacc. & Trott. 1912; *Inonotus* Murrill 1915, Bond. & S. 1941, not ∼ Murrill 1939.—Lloyd 1915 (LMW 4, Ap.): 381 referred this to *Polyporus* [*Inonotus*] *cuticularis,* but it appears to be a good sp. − Descr.: Overh. 1953: 414 *pl. 44 f. 267, pl. 45 fs. 268, 269, pl. 130 fig.* (*Polyporus*); Pegl. 1964 (TBS 47): 190 *f. 2: 12* (*Inonotus*).

lunata, Poria, Romell 1926 (SbT 20): 12, 20 (syn.: n.v.p.).—Fide Romell, l.c., "*Pol. lenis* Karst, as understood in Hym. Lapp." [≡ *Poria lenis*].

luteiporus = *luteoporosus*

luteocinereus, Polyporus, Britz. 1897 (BCb 71): 57 [*pl. 659 f. 230*] (Germany). —Nomen dubium. Britz., l.c., thought it was related to *Polyporus crispus* [= *Bjerkandera adusta*] but this cannot be correct. It may be *Coriolus hirsutus* or *Funalia trogii* but in certain details the protologue contradicts both suggestions.

luteolamellosa, Lenzites, Opiz in herb.—Fide Pilát 1942: 607 = *Gloeophyllum sepiarium.*

luteoluteus, Polyporus, Lloyd 1911 ("McGinty"; not accepted: n.v.p.) = *Scutiger laeticolor* Murrill [= *Albatrellus confluens*].

luteopora, Poria, Bond. 1940 (U.S.S.R., European Russia); *Asterostromella* Bond. & S. 1941; *Vararia* Sing. 1943, Bond. 1953.—Descr. and illustration by Bond. 1953: 592 *f. 160* (*Vararia*) are barely suggestive of a sp. of *Vararia* P. Karst. ('Corticiaceae'). According to Parm. 1971: 87,

145, "Probably synonymous with *V. granulosa*; the type is not preserved." "*Vararia granulosa* (Fr.) Laurila" is a misapplied name.

luteoporosus, Polyporus, Opiz 1855 (Lo 5): 87 (Czechoslovakia).—Nomen dubium. Not a nom. nud. as stated by Pilát 1942 (ACE 3): 607 ("*luteiporus*"). *Coriolus versicolor?* – Listed by Sacc. 1888 (SF 6): 149 as *P.* "*luteiporus*".

luteosubulatum, Sistotrema, Secr. 1833 M. 2: 502 (Switzerland).—Nomen dubium. Found in mines; apparently an abnormal condition.

*luteolus, *lutescens, Lenzites,* H. & P. Syd. 1900 (MHB 4): 1 (Argentina).—Singer 1930 (BbC 46): 78 stated, "Sicherlich nur Abart von [*Lenzites*] *betulina* (wie auch Bresadola meint)" but this is not readily suggested by the descr. Bres. may have had in mind *Lenzites betulina* var. *lutescens* Sacc. 1873, which is a different taxon.

lutescens, Poria, Lázaro 1917 (RMa 15): 370 / 1917: 284 (Spain).—Nomen dubium.

lutescens, Trametes, Lázaro 1916 (RMa 14): 520 & 1917 (RMa 15): *pl. 10 f. 23* / 1917: 71 *pl. 10 f. 23* (Spain), not ∼ (Pers. per Nocca & Balb.) Bres. 1896.—Nomen dubium. Cf. Bres. apud Trott. 1925 (SF 23): 435, "E diagnosi videtur *Tram. flavescens* Bres." Lázaro himself differentiated it from *Trametes* [*Funalia*] *trogii*, which sp. is another possibility to be kept in mind.

lyallii, Lenzites,* Berk.—I have been unable to locate the place of publication. Murrill 1905 (BTC 32): 87 mentioned it and stated its origin ("from Vancouver"); he thought that it "seems to be a form" of *Agaricus* [*Daedaleopsis*] *confragosa,* but the type was in poor condition. – Cf. (31**).

lychneus, Fomes, Lázaro 1916 (RMa 14): 666 / 1917: 105 (Spain).—Nomen dubium.

lycoperdiformis, Polyporus, Pers. → *Poria pulvinata* Scop. (**O**).

**macounii, Polystictus, Lloyd* 1914 (LMW 4, L. 53): 6, 9 (Canada, British Columbia) (as a "var. of *versicolor*": n.v.p.).—Fide Overh. 1953: 342, 344 = *Polyporus* [*Coriolus*] *versicolor.*

macowanii, Polyporus,* Kalchbr. 1881 (G 10): 54 (South Africa).— Referred by Bres. 1890 (BmF 6): xl ("*Mac Ovani*") to *Polyporus* [*Gloeoporus*] *dichrous* and by Lloyd 1915 (LMW 4, Ap.): 382 to *Polyporus* [*Bjerkandera*] *adustus.* According to Wakef. apud Doidge 1950 (Bo 5): 527, the poor type suggested *Polyporus dichrous* rather than *P. adustus.* – *Polyporus macounii* Peck drifted into the company of *Polyporus dichrous* as a var. through a clerical error by Sacc. 1888 (SF 6): 126, perhaps because it was confused with *P. macowanii*; it is a syn. of *Phellinus ferruginosus,* – Cf. (51**). – V.s.: "*marcowanii*".

macrodon, Hydnum, Pers. sensu Fr. → *Hydnum nodulosum* Fr. (**O**).

macroporus, Polyporus, S. Schulz. 1866: 42 (Yugoslavia, Slavonia) (nom. nud.: n.v.p.), not ∼ Lév. 1848.—Nomen dubium.

macrorrhizus, Boletus, Ach. "in litt. (Sweden?).—Fide Fr. 1821: 384 = *Polyporus* [*Poria*] *vaillantii.*

macrospora, Trametes, S. Schulz. 1866: 41 (Yugoslavia, Slavonia) (nom. nud.: n.v.p.).—Nomen dubium.

*_madagascarensis, Polyporus_, Lloyd 1913 (LMW 4, L. 48): 6 (Madagascar) (as a var. of *Polyporus* [*Gloeoporus*] *dichrous*: n.v.p.). — Cf. (**51**).

malum, Agarico-carnis, Paul. 1793 T. 2: 99 (descr.), Ind. (name) (France); ≡ *Dendrosarcos rutilenis* Paul. 1812–35: *pl. 13/10* (generic name n.v.p.). —Nomen dubium. Referred by Lév. 1855: 7 to *Polyporus* [*Inonotus*] *hispidus*, but there is very little in the protologue to substantiate this guess.

manubriatus, Polyporus, Secr. 1833 M. 3: 89 (Switzerland) (as a sp. of *Boletus*: n.v.p.), not ∼ Lév. apud Zoll. 1854.—Nomen dubium.

marcuccianus, Polyporus, Lloyd 1915 (LMW 4, Ap.): 348 ("Southern Italy or one of the Mediterranean islands") (as a form of *P. gilvus*: n.v.p.) ex Trott. 1925; [≡ *Boletus fulvus* Scop. sensu Marcucci ?1867 U.: No. 70 (*Polyporus*) (Italy, Sardinia)]. — *Phellinus torulosus*? — Cf. (**79**).

*_marmoratus, Polyporus_, B. & C. 1858 (PAA 4): 122 (Nicaragua); *Fomes* Cooke 1885; *Scindalma* O.K. 1898; *Elfvingiella* Murrill 1920 (nom. prov.?), 1948.—Fide Lowe 1957 F.: 43 = *Fomes sclerodermeus* (Lév.) Cooke, a species of the neotropics, often erroneously identified with "*Boletus*" *fasciatus* Sw. Listed by various authors as a syn. of *Fomes fomentarius*. — Descr.: Overh. 1953: 94 *pl. 68 fs. 397, 398* (*Fomes marmoratus*); Lowe 1957 F.: 43 *f. 28* (*Fomes sclerodermeus*); A. Teix. 1962 (Ri 1): 73 *tpl. 3* (*Fomes fasciatus*).

martellii, [*Polyporus*], (Bres.) Lloyd 1915 (LMW 4, Ap.): 382 (syn.: n.v.p.); *Ganoderma resinaceum* var. Bres. 1892 F.t. 2: 31 *pl. 137* (Italy). —As yet not treated as a sp. distinct from *G. resinaceum*.

*_maxonii, Fomitiporia_, Murrill 1907 (NAF 9): 11 (Costa Rica); *Poria* Sacc. & Trott. 1912.—Referred by Lowe 1949 (Ll 11): 167, 1957 F.: 56 to *Poria* [*Phellinus*] *punctata* which in turn was referred to *Fomes* [*Phellinus*] *robustus* (resupinate), but cf. (**86**). A redescription of the type is needed.

medullaris, Daedalea, Purt. 1821: 250 (England) (d.n.).—Purt. himself identified his fungus with *Daedalea asserculorum* (Schrad.) Pers. [= *Gloeophyllum abietinum*] as distributed by Moug. & Nestl. 1815 S.C.: No. 491; in addition he mentioned *Boletus angustatus* Sow. [*Daedaleopsis confragosa*] as a possible synonym. Purton's descr. hardly suggests *Gloeophyllum abietinum* to me. Perhaps to be regarded as a substitute name for '*Daedalea asserculorum*', but Purton's specific epithet was suggested by his own specimen: "I have therefore adopted a new name which is more applicable to the character of my specimen."

melanostroma, Poria, Secr. 1833 M. 3: 176 (Switzerland) (as a sp. of *Boletus*: n.v.p.).—Nomen dubium and perhaps also a nomen confusum. Secr. himself thought of the possibility that this was a species of poria growing over another fungus, "la *Tremella glandulosa*?"

melleus, Polyporus, B & Br. 1873 (JLS 14): 53 (Ceylon); *Poria* Cooke

1886, not ~ Bond. & Ljub. 1959 (n.v.p.).—See under *Polyporus auricomus* (**0**). – Descr.: Petch 1916 (APe 6): 134; Boed. 1940 (BBu III 16): 383, colour when fresh; Lowe 1962 (PMi 47): 184 (*Poria*).

mellinus, Polyporus, Pers. 1825: 96 (Europe).—Nomen dubium. Referred by Fr. 1828 E. 1: 120 to *Polyporus [Antrodia] sinuosus*. I could not find the type in Persoon's herb.

membrana, Boletus, Batsch → *Helvella pileus* Schaeff. (**0**).

membranacea, Poria, Scop. 1772 P.s.: 106 *pl. 28 f. 2* (Hungary, now Czechoslovakia) (d.n.); ≡ *Polyporus scopolii* Pers. 1825.—Nomen dubium. Found in a mine: apparently an abnormal growth.

Merulioporia Bond. & S. 1941 (n.v.p.), 1943 [1958 (Fu 28): 12], not *Meruliporia* Murrill 1942 (Coniophoraceae); ≡ *Meruliopsis* Bond. apud Parm. 1959, Bond. 1961; holotype; *Poria taxicola* (**0**).

Meruliopsis Bond. apud Parm.; Bond. → *Merulioporia* Bond. & S. (**0**).

Merulius Fr. 1821 [1958 (Fu 28): 10], not ~ [Haller] St-Am. 1821 (Cantharellaceae); lectotype, *Merulius tremellosus* Schrad. per Fr. ('Corticiaceae').—Now excluded from the Polyporaceae.

metallicus, Polyporus, Lloyd 1921 (LMW 6): 1099 (". . . of course only a form . . . of *Polyporus lucidus*: n.v.p.) (U.S.A., New York).—This has been referred to *Ganoderma/Polyporus lucidum* by Lowe 1934: 125 and Overh. 1953: 208, which leaves it undecided whether it is *G. resinaceum* or *G. lucidus*.

**micheneri, Polyporus*, Berk. (in herb.) (U.S.A.).—Fide Cooke 1885 (G 13): 115 = *Polyporus [Ischnoderma] benzoinus*.

micropora, Hexagonia*, Murrill 1904 (U.S.A., Maine); *Favolus* Sacc. & D. Sacc. 1905 ("*microsporus*").—Referred to *Hexagonia striatula* [= *Favolus [Polyporus] canadensis* (0**)] by Murrill 1907 (NAF 9): 48 and to *Favolus alveolaris* [= *Polyporus mori*] by Overh. 1953: 157, but cf. (**98**). – Descr.: Murrill 1904 (BTC 31): 328 (*Hexagonia*); Lloyd 1909 (LMW 3, P.I.): 19 *f. 257* (*Favolus*).

**mikadoi, Polyporus*, Lloyd 1912 (LMW 4, L. 43): 3 (Japan); *Xanthochrous* Murašk. 1940; *Inonotus* Bond. 1953 (incomplete ref.: n.v.p.).—This was referred to *Polyporus [Inonotus] cuticularis* by Lloyd 1915 (LMW 4, Ap.): 360 and others, but the sp. is now treated as distinct. – Descr.: Lloyd 1915 (LMW 4, Ap.): 360 *f. 695* (*Polyporus*); Kawam. 1954 I. 1: 125 *f. 115* (*Polyporus cuticularis* var.); Pegl. 1964 (TBS 47): 190 *f. 2: 13* (*Inonotus*).

milleporeus, Boletus, Batsch → *Boletus coriaceus* Batsch 1783 (**0**).

Milleporus Pfeiff. 1874 (incidental mention: n.v.p.) [1960 (Pe 1): 244]; [≡ *Boletus* "subordo" *Milleporei* Batsch 1783: 101 (inadmissible term denoting rank)]; lectotype, *Polyporus lacteus* Batsch [in part, = *Polyporus tuberaster*]. – A taxon introduced for the stalked polypores.

**miniatus, Polyporus*, Jungh. 1838: 68 (Indonesia, Java), not ~ Liboš. ex Steud. 1824: Fr. 1832; *Polyporellus* P. Karst. 1879; *Laetiporus* Over. 1925, Bond. 1953 (incomplete ref.: n.v.p.); *Tyromyces* Teng 1964.

—Sometimes regarded as belonging to *Laetiporus sulphureus*. If not conspecific, it is certainly closely related. — Descr.: Bres. 1912 (H 53): 52 (*Polyporus*); Over. 1925 (IFm 7): 1 *fig.*, *pl. 12* (*Laetiporus*). — The name *Polyporus miniatus* Jungh. is illegitimate as a later homonym. Cf. also *Polyporus discolor* Kl. — (**65**).

**minima, Trametes*, Berk. in MS. (presumably U.S.A.).—Fide Lloyd 1919 (LMW 5): 850 *f. 1421 = Trametes [Antrodia] sepium* (forma); he did not validly publish the binomial because he applied it to a "form of *Trametes sepium*".

minusculus, Polyporus, Boud. 1902 (France); *Leptoporus* Big. & Guill. 1913 (syn.: n.v.p.), Bourd. & G. 1925; *Polystictus* Big. & Guill. 1913. —Apparently an alien found only once in a greenhouse. According to notes on the type made by Dr. J. L. Lowe the hyphae are of one type, very thin-walled, abundantly septate without clamps, and occasionally branched. These and other characters indicate that the species may not belong to *Leptoporus [Tyromyces]*. — Descr.: Boud. 1902 (BmF 18): 141 *pl. 6 f. 3.* (*Polyporus*).

**minutipora* Rodw. & Clel. 1929 (PTa): 17 (Australia, New South Wales). —Referred by G. Cunn. 1965: 57, 58 to *Poria lenis*. — Descr.: Rodw. & Clel., l.c., Clel. 1935: 238.

**miyabei, Irpex*, Lloyd 1923 (LMW 7): 1126 (nom. nud.: n.v.p.), 1175 *pl. 228 f. 2336* (Japan).—Nomen dubium. Referred, with doubt, to *Irpex lacteus* by S. Ito 1955: 254, but Lloyd's miserable protologue seems to contradict this.

molluscus, Boletus, Fr. 1821: 384 (not definitely accepted: n.v.p.); [≡ *Boletus molluscus* Pers. sensu Nees 1816: 223 *pl. 11 f. 223* (Germany)]. —Nomen dubium. Nees described this fungus from collections he had studied himself and for which "Herr Sturm" (as "St.", cf. p. 329) had drawn the figures. He referred his material to *Boletus molluscus* Pers. but at the same time indicated his doubt ("?") as to this identification. By adopting *Polyporus molluscus* (Pers.) per Fr. and at the same time separately listing "*B. molluscus* Nees Syst. *f. 223.* poris rubris" however, he introduced a new taxon, without validly publishing it (mentioned incidentally as a matter of record).

molluscus, Merulius, Fr. sensu Berk. 1860: 255.—Fide Romell 1911 (ABS 11³): 30 = *Polyporus haematodes* Rostk. [sensu Romell = *Merulius taxicola* (**O**)].

molluscus sebaceus, Polyporus, Secr. 1833 M. 3: 168 (Switzerland) (as a sp. of *Boletus* & double epithet: n.v.p.).—Nomen dubium.

Monka Adans. 1763 (d.n.) [1960 (Pe 1): 245] = *Verpa* Sw. per Fr. (Discomycetes).

mori, Boletus, Pollini in herb., not ∼ (Pollini per Fr.) Pollini 1824.—Fide Sacc. 1873 (ASv 2): 99 = *Polyporus [Inonotus] hispidus*. This may have been published as *Boletus flavus* Pollini (n.v.p.) ≡ *Polyporus pollinii* Heufl. = *Inonotus hispidus*.

*_morincola, Hydnum_, Schw. ("olim"), in herb. (U.S.A., Pennsylvania). —Listed as a syn. of _Irpex sinuosus_ [cf. _I. lacteus_] by Schw. 1832: 164.

morinus, Boletus, Gaterau 1789: 190 (France) (d.n.).—Nomen dubium. Descr., "en sabot de cheval, d'un jaune rougeâtre à sa surface et parsemée de poils, les bords d'un jaune clair; les pores jaunes. / Sur les mûriers."

mucidus, Boletus, Scop. 1770: 149 (Hungary, now Czechoslovakia) (d.n.), not ∼ (Pers.) Pers. 1801 (d.n.), not ∼ (Pers. per Fr.) Pollini 1824. —Nomen dubium. Referred by Fr. 1828 E. 1: 78 to _Polyporus alligatus_ [= _Bjerkandera fumosa_] and listed as a syn. of _Heteroporus [Abortiporus] biennis_ by Fid. 1969 (Ri 4): 116 ("_muscidus_").

Mucilago Hoffm. 1796 [1960 (Pe 1): 245], presumably ≡ _Mucilago_ Mich. 1729 (pre-Linnaean name) ≡ _Mucilago_ [Mich.] Adans., a name presumably based on Myxomycetes.

Muciporus Juel 1897 (nom. conf.) [1957 (Ta 6): 84]; ≡ _Myxoporus_ F. Clem. 1902 (nom. nud.: n.v.p.); lectotype, _Muciporus corticola_ (Fr.) Juel sensu Juel.—Based on a resupinate polypore, _Oxyporus corticola_, overgrown by a species of _Tulasnella_ J. Schroet. (Tulasnellaceae), hence a nomen confusum.

Mucronoporus Domański apud Domański & al. 1967 (FpG 2): 316 (lacking Latin descr.: n.v.p.); [≡ _Mucronoporus_ Ell. & Ev. sensu Domański, l.c., in part: excl. of type]. – Lectotype: _Polyporus circinatus_ (Fr.) Fr. – Cf. (**69**) and Donk 1971 (PNA 74): 12.

Multiporus R. & O. Falck 1937 (lacking Latin descr.: n.v.p.) [1960 (Pe 1): 246]; monotype, _Multiporus [Poria] chlamydoformans_ R. & O. Falck. – A second species was referred to this genus with a question mark.

*_multisetosus, Polyporus_, Lloyd 1920 (LMW 6): 976 (Australia).—Erroneously referred to _Phellinus robustus_ by G. Cunn. 1965: 227, 228.

*_munzii, Polyporus_, Lloyd 1922 (LMW 7): 1163 (U.S.A., California); _Inonotus_ R. L. Gilb. 1969.—This was taken by Pegl. 1964 (TBS 47): 185 as a thick form with very rare setae of _Inonotus cuticularis._ However, Overh. and more recently R. L. Gilb. 1969 (SwN 14): 125 treated it as a distinct sp. Long 1945 (Ll 8): 235, 237 listed it as a syn. of _Polyporus [Inonotus] farlowii_ (extra-European). – Descr.: Overh. 1953: 425 _pl. 124 fs. 672, 673, pl. 132 fig._ (_Polyporus_).

murinus, Boletus, (Pers.) C. & T. Nees 1820: clxxii (d.n.); _Boletus subtomentosus_ var. "_B. murinus_" Pers. 1799 O. 2: 10 (d.n.).—_Boletus substrictus_ Bolt. [= _Polyporus ciliatus_] was cited as a syn. by error. There is no reason to assume that Persoon's var. is not a taxon of the Boletaceae. Fries 1838: 431 cited "_B. murinus_ Nees" as a syn. of _Polyporus fuscidulus_ (**O**).

muscicola, Polyporus, Wettst. 1885 (VW 35): 563 [repr. 1886 (BCb 27): 86] (Austria); _Polystictus_ Sacc. 1891; _Microporus_ O.K. 1898.—Nomen dubium. The descr. suggests _Tyromyces wynnei_, though there are several discrepancies.

muscidus = *mucidus*

Myxoporus F. Clem. → *Muciporus* Juel (**O**).

**nambui, Polyporus*, Yas. "mss. name" (Japan).—Lloyd 1919 (LMW 5):
843 ("*Nambai*") mentioned this as a syn. of *Polyporus greenei* Lloyd
[≡ *P. greenei* Yas. = *Polyporus* [*Coltricia*] *montagnei* var. *greenei*
(Berk.) R. L. Gilb.], which was based on the same collection.

nanus discoides, Polyporus, Secr. 1833 M. 3: 157 (Switzerland) (as a sp.
of *Boletus* & double epithet: n.v.p.).—Fide Fr. 1838 479 ("n. 106")
= *Polyporus* [*Hirschioporus*] *abietinus*.

neesii, Polyporus, Fr. 1821: 370 (Sweden); *Trametes* Fr. 1848 (n.v.p.),
1849; *Fomes* Cooke 1885; *Placodes* Quél. 1888; *Scindalma* O.K. 1898;
Polystictus Rick. 1918.—Nomen dubium. Bres. 1897 (AAR III 3):
76–77 remarked in connection with *Fomes* [*Oxyporus*] *populinus*,
"Forte huc etiam *Fomes neesii* ducendus." The substratum (cf. Fr. 1874:
564), "ad ramos dejectos *Fagi*", hardly supports this guess.

nepevnyi, Trametes, Opiz 1852: 147 (Czechoslovakia) (nom. nud.: n.v.p.).
—Fide Pilát 1942 (ACE 3): 609 = *Trametes gibbosa*.

nieskyensis = *niskiensis*

nigrescens, Fomes, Lloyd 1915 (LMW 4, F.): 237, 288 ("Klotzsch") (as
a form of *Fomes fomentarius*: n.v.p.).—Introduced for *Polyporus
nigricans* Fr. 1838: 466 (in part, viz. as to "e Scotia Klotzsch", which
specimen represents black-crusted *Fomes fomentarius*, fide Lloyd, l.c.

nigrescens, Trametes, Lázaro 1916 (RMa 14): 523 / 1917: 74 *pl. 1* (Spain),
not ∼ Bres. 1905.—Nomen dubium. The protologue calls to mind
Fomitopsis cystisina, but it also shows discrepancies, for instance as to
the pores.

**nigroaspera, Trametes*, Lloyd 1924 (LMW 7): 1272 *pl. 285 f. 2793* (Japan)
("A thick form of *Daedalea* [*Daedaleopsis*] *confragosa*": n.v.p.). –
Cf. (**31**).

nigroporus, Fomes, Lázaro 1916 (RMa 14): 662 / 1917: 101 (Spain).
—Nomen dubium. *Phellinus* sp.?

**nigropurpurascens, Polyporus*, Schw. 1832: 156 (U.S.A., North Carolina);
Leptoporus Pat. 1928.—Referred by Fr. 1851 (NAu III 1): 55/39 &
Bres. 1912 (H. 53): 74 to *Polyporus* [*Gloeoporus*] *dichrous*. – Cf.
(**51**).

nigrozonatus, Polyporus, Saut. 1876 (H 15): 33 (Austria); *Polystictus*
Cooke 1886; *Microporus* O.K. 1898.—Nomen dubium. Keissl. 1917.
(AW 31): 111 thought that this might be *Daedalea* [*Cerrena*] *unicolor*
but the descr. does not support this.

nisetiensis = *niskiensis*

niskiennis = *niskiensis*

niskiensis, Polyporus, Pers. 1825: 258; [≡ *Boletus incarnatus* (Pers.) Pers.
sensu A. & S. 1805: 258 (Germany)].—Nomen dubium. The original
descr. is too brief; it may perhaps faintly suggest *Poria placenta*, but
to assume this as a reasonable identification would be going too far.

— V.s.: the epithet has been often misspelt, *"niskiennis"*, *"Nisetiensis"*. *"Nieskyensis"*. — (**110**).

nitens, Ganoderma, Big. & Guill. 1913: 360 ("Batsch") (syn.: n.v.p.) ≡ *Boletus nitens* var. *crocatus* Batsch. ≡ *Ganoderma nitens* Lázaro, typonym [= *G. lucidum*]. — Cf. also Donk 1971 (PNA 74): 10.

niveus, Boletus, Jullien apud Vill. 1789: 1040 (France) (d.n.).—Nomen dubium. Apparently not a polypore.

**nivosus, Polyporus*, Berk. 1856 (HJB 8): 196 (Brazil).—For this sp. see Lloyd 1915 (LMW 4, Ap.): 310 *f. 650* but also Bres. 1926: 78.—This sp. is mentioned here because Morg. 1885 (JCi 8): 101 misinterpreted it; according to Lloyd 1915 (LMW 4, Ap.): 316 he applied the name to *Polyporus* ["*Trametes*"] *semisupinus*, which Lloyd confused with *Polyporus* [*Incrustoporia*] *semipileatus* Peck.

nodulosum, Hydnum, Fr. 1874: 616; *Acia* P. Karst. 1879, Pilát 1925 misapplied; *Dryodon* Quél. 1886, Cejp 1928 misapplied; *Odontia* Cost. & Duf. 1895; *Hericium* Nikol. 1956 (incomplete ref.: n.v.p.), misapplied; [≡ *Hydnum macrodon* Pers. per Fr. sensu Fr. 1863 M. 2: 279 (Sweden)]. —Nomen dubium. Fide Lundell 1914 (LNF 21–22): 13 No. 1019, in obs., "a juvenile polypore, either *Trametes* [*Antrodia*] *serialis* (Fr.) Fr. or *Polyporus radiatus* Sow. ex Fr. var. *nodulosus* (Fr.) [*Inonotus nodulosus*]." — Sensu Pilát = *Dentipellis fragilis* (Pers. per Fr.) Donk (Hericiaceae), fide Donk 1962 (Pe 2): 233, 234; sensu Bourd. & G. = *Mycoacia stenodon* (Pers.) Donk ('Corticiaceae').

noharae, Irpiciporus, Murrill sensu Lloyd → *Irpex lamelliformis* Lloyd (**O**).

normalis, Boletus, Vill. → *Boletus auriculatus* Vill. (**O**).

notarisii, Polyporus, Berk. in herb.—Mentioned by Lloyd 1915 (LMW 4, Ap.): 383, "nomen nudum. Not 'described' and no specimen found by me." — Is this *Sistotrema notarisii* [= *Abortiporus biennis*]?

Nothotrechispora Sing. 1944 (n.v.p.) [1960 (Pe 1): 247], presumably = *Phlebiella* P. Karst. (n.v.p.) ('Corticiaceae').

**novae-angliae, Polyporus*, B. & C. apud Berk. 1872 (G 1): 51 ("New England"); *Fomes* Lloyd 1906, &c. (n.v.p.).—Nomen dubium. Referred with doubt to *Pyropolyporus* [*Phellinus*] *igniarius* by Murrill 1908 (NAF 9): 103, certainly by error, fide Lloyd 1915 (LMW 4, F.): 283 and Lowe 1954 (M 46): 493, 1957 F.: 89.

**oblectans, Polyporus*, Berk. 1845 (LJB 4): 51 (West Australia); *Polystictus* Cooke 1886; *Xanthochrous* Pat. 1897; *Microporus* O.K. 1898; *Coltricia* G. Cunn. 1948.—Referred by Bres. 1890 (BmF 6): xlii, Murrill 1904 (BTC 31): 343, and Lloyd to *Polystictus/Coltricia cinnamomea* and in consequence of this to *Polystictus* [*Coltricia*] *perennis* f. *cinnamomeus* (Jacq. per S. F. Gray) Pilát by Pilát 1942 (ACE 3): 579, 582. — Descr.: Clel. 1935: 217 (*Polystictus*); G. Cunn. 1965: 191 *pl. 4 f. a* (*Coltricia*).

obliquum, Hydnum, DC. 1815: 36 (Switzerland) (d.n.), not ∼ Schrad. 1794.—Nomen dubium. DC. cited only one syn. and then with doubt

("*Sistotrema obliquum*. Alb. & Schwein. n. 780?" [≡ *Hydnum obliquum* Schrad.]), hence, *H. obliquum* DC. and *H. obliquum* Schrad. may be considered to be based on different types (different taxa). – Mentioned in connection with *Polyporus flavescens* Secr. (**O**).

obliquus, Boletus, Bolt. 1788: 74 *pl. 74* (England) (d.n.); *Sistotrema* C. & T. Nees 1820 (d.n.); *Boletus* Bolt. per Purt. 1821, not ∼ (Pers. per Fr.) Wahl. 1826; ≡ *Hydnum boltonii* Spreng. 1827.—Nomen dubium. At first referred to *Merulius [Serpula] lacrimans* (Wulf.) ex Fr. (forma) (Coniophoraceae) by Fr. 1821: 328 ("ipso teste") and afterwards by Fr. 1838: 515 to *Hydnum squalinum* Fr. ("e descr."), itself a nomen dubium.

obliteratus, Boletus,* Schw. "in Hook. Herb."—Fide B. & C. 1856 (JAP II 3): 214 = *Porotheleum pezizoides* (Schw.) Schw. [= *Stromatoscypha fimbriatum* (O**)].

obtusus, Polyporus,* Berk. 1839 (AM 3): 390 (North America); *Trametes* Berk. 1841, Lohw. 1931; *Daedalea* Neuman 1914; *Leptoporus* Pilát 1937. – Pilát 1939 (ACE 3): 242–243 and Kotl. & P. 1957 (ČM 11): 233 suggested that *Spongipellis litschaueri* [= *S. schulzeri*] might belong to this sp., but later Kotl. & P. 1965 (ČM 19): 75 decided in favour of keeping the two taxa specifically distinct. – The correct name is still debatable, cf. also under *Sistotrema spongiosum* (O**) and *Boletus unicolor* Schw. (**O**). – Descr.: Lloyd 1915 (LMW 4, Ap.): 323 *f. 666*; Lohw. 1931 (APr 75): 305 *fs. 1, 2, pl. 19 fs. 1, 6*; Overh. 1953: 322 *pl. 21 fs. 127, 128, pl. 131 fig.* (*Polyporus*).

obversus, Polyporus, S. Schulz. 1880 (Fl 63): 80 (Yugoslavia, Slavonia. —Nomen dubium. – Schulzer von Müggenburg promised a more detailed Latin description, which, as far as I am aware, was not published. – Special literature: Schulzer von Müggenburg, *1880c.*

**ochracea, Daedalea,* Lloyd 1914 (LMW 4, L. 49): 12 (as a form of *Daedalea [Cerrena] unicolor*: n.v.p.) (U.S.A., Massachusetts).

**ochracea, Lenzites,* Lloyd 1922 (LMW 7): 1106 *pl. 187 f. 2023* (Brazil) ("A form of *Lenzites flaccida* for me": n.v.p.), non/an ∼ Lloyd 1922, simultaneously published. – *Lenzites flaccida* = *L. betulina* forma.

**ochraceo-stuppeus, Polystictus,* Lloyd 1916 (LMW 5, L. 63): 11 (Australia, New South Wales).—Fide G. Cunn. 1965: 114, 115 = *Gloeoporus [Bjerkandera] adusta.*

ochraceus tremulae, Polyporus, Secr. 1833 M. 3: 132 (Switzerland) (as a sp. of *Boletus* & double epithet & nom. nud.: n.v.p.).—Nomen dubium. Referred by Fr. 1874: 568 ("n. 81") to *Polyporus [Coriolus] zonatus.*

**ochrohirsutus, Polystictus,* Lloyd 1923 (LMW 7): 1233 *pl. 262 f. 2591* (as a form of *Polystictus hirsutus*: n.v.p.) (Japan).

**odontoporus* Kalchbr. in herb. K. (unknown locality).—Lowe 1966: 19 thinks that this "is probably a synonym" or *Poria [Oxyporus] latemarginata*; G. Cunn. determined a specimen named by Cooke from Australia (New South Wales) first [1950 PNW 75): 234] as *Poria* *versipora* [= *Schizopora paradoxa*], then (1965: 276) as *Steccherinum ochraceum* (Pers. per Fr.) S. F. Gray ("Hydnaceae").

odorus zonatus, *Polyporus*, Secr. 1833 M. 3: 106 (Switzerland) (as a sp. of *Boletus* & double epithet: n.v.p.).—This was identified by Secr. with Bull. *pl. 310* = *Boletus suaveolens* L. sensu Bull. [= *Daedaleopsis confragosa*].

ohiensis*, *Favolus*, B. & Mont. apud Mont. 1856: 171 (U.S.A., Ohio). —This has been referred to *Hexagonia alveolaris* by Murrill 1904 (BTC 31): 327 and to *Favolus europaeus* by Lloyd 1909 (LMW 3, P.I.): 19, both names being now placed in the synonymy of *Polyporus mori* but cf. **(98).

ohiensis*, *Trametes*, Berk. 1872 (G 1): 66 (U.S.A., Ohio), (Berk.) Bres. "in litt." apud Torrend 1910 (as a "forme" of *Fomes scutellatus*: n.v.p.), Bres. apud Torrend 1913; *Ungulina* Pat. 1900, Murašk. 1940, Kotl. & P. 1957; *Ganoderma* Romell 1901, Coker, 1927; *Fomes* Murrill 1903; *Fomitopsis* Bond. & S. 1941; *Truncospora* Pilát 1942 (generic name n.v.p.), 1953 (incomplete ref.: n.v.p.); *Poria* Kotl. & P. 1959, not *P. ohioensis* (Murrill) Sacc. & Trott. 1912 (*"ohiensis"*).—Reported from Portugal by Torrend 1913 (Bro 11): 67, 68, who had previously referred the material to *Fomes* [*Fomitopsis* ?] *scutellatus* **(0). The identification with *Truncospora ohiensis* is considered here to be improbable and the latter sp. is therefore not reported for Europe; but cf. *Truncospora ochroleuca*. – Descr.: Coker 1927 (JMS 43): 133, 134 *pl. 14, pl. 22 fs. 10, 11* (*Ganoderma*); Overh. 1953: 44 *pl. 69 f. 404, pl. 93 f. 523, pl. 126 fig.*; Lowe 1957 F.: 82 *f. 63* (*Fomes*).

oleae, *Polyporus*, Panizzi 1886 (NGi 18): 65 (Italy).—Nomen dubium. Pilát 1942 (ACE 3): 610 suggested *Grifola frondosa* or *G.* [*Meripilus*] *gigantea*. Perhaps *Laetiporus sulphureus*?

olivaceozonatus, *Polyporus*, Secr. 1833 M. 3: 136 (Switzerland) (as a sp. of *Boletus*: n.v.p.).—Nomen dubium.

orbicularis, *Daedalea*, Bagl. 1865 (Cci 2): 265 (Italy); *Striglia* O.K. 1891. —Nomen dubium.

orbicularis, *Polyporus*, Saut. 1876 (H 15): 150 (Austria), not ∼ Berk. 1839, not ∼ Velen. 1922.—Nomen dubium. Judging from the poor descr. I would suggest *Polyporus mori* but the substratum does not agree. Keissl. 1919 (AW 31): 111 thought of *Polyporus arcularius* but this cannot be correct. Lloyd 1912 (LMW 3, S.P.): 176 *f. 474* (fig. magnified?) published an interpretation that must also be dismissed as incorrect. (Pilát 1936 (ACE 3): 75, 78 saw in Lloyd's interpretation a form of *P. arcularius*.)

**orbiculatus*, *Polyporus*, Colenso "a herbarium name" (New Zealand). —Fide G. Cunn. 1965: 167, 168 = *Trametes* [*Coriolus*] *zonatus*. His interpretation of this sp. needs confirmation.

orthoporus, *Polyporus*, Pers. 1825: 91, in obs. (Switzerland) (not definitely accepted as a distinct sp.: n.v.p.).—Fide Donk 1967 (Pe 5): 102 = *Polyporus* [*Rigidoporus*] *undatus.*

ostrea, *Boletus*, Hoffm. 1797–1811 V.s.: 27 *pl. 15 f. 2* (Germany) (d.n.).
—Nomen dubium. Fruitbody of imbricate, sessile caps, white, not
typically zonate; pores becoming somewhat grey; context soft, somewhat fleshy, flexible. On rotten wood, and if found in a mine (which
seems likely) perhaps based on an abnormal growth.

ostrea, *Poria*, Scop. 1772 P.s.: 105 *pl. 24* (Hungary, now Czechoslovakia).
(d.n.); *Polyporus* (Scop.) per Pers. 1825.—Nomen dubium. Found in
a mine: apparently an abnormal growth.

oxyporus, *Polyporus*, Saut. 1876 (H 15): 150 (Austria); *Fomes* W. Cooke
1949 (syn.: n.v.p.).—Nomen dubium. Bres. 1920 (Am 18): 68 referred
this to *Fomes* [*Oxyporus*] *populinus*. The descr. does not support this
identification.

**Oxyuris* Lloyd 1915 (n.v.p.), not ∼ Linstow 1907 (n.v.) [1960 (Pe 1):
249]; type, *Fomes* [*Phellinus*] *pachyphloeus* Pat. (extra-European).

**Pachyma* Fr. 1822: 242 (nom. anam.) [1962 (Ta 11): 94]; lectotype,
Sclerotium cocos Schw. (**O**).

pachyus, *Polyporus*, Rostk. 1848 (StP Fs. 27–28): 9 *pl. 5* (Germany/
Poland).—Nomen dubium. Fries 1863 M. 2: 339 remarked on the
discrepancy in colour as indicated in the text and on the plate; he
thought of *Polyporus* [*Phellinus*] *contiguus*, a sp. about which he had
no clear idea. Bourd. & G. 1928: 739 suggested, with doubt, *Ungulina*
[*Heterobasidion*] *annosa*.

**palisotii*, *Daedalea*, Fr. 1821: 335; *Lenzites* Fr. 1838; *Trametes* Imaz.
1952; ≡ *Daedalea amanitoides* P. Beauv. 1805 (Nigeria, Oware) (d.n.);
Lenzites (P. Beauv.) per Hariot 1891; *Cellularia* O.K. 1898 ("amani-
todes"); *Daedalea* Murrill 1907.—The name "5. *Lenzites Palisoti* Fr."
appears on the same list, mentioned under *Fomes canescens* (**O**) in such
a manner that one would conclude that the corresponding material
(Torrend, Fungi sel. exs. No. 5) had been collected in Portugal: it was
actually reported from that country by S. Câm. 1956: 60 on the base
of this distribution number. A package of it (BPI) gives "Timor", an
island south of the Moluccas, as the locality where it was collected.

**pallidofulva*, *Daedalea*, Berk. 1847 (LJB 6): 322 (U.S.A., Ohio); *Striglia*
O.K. 1891; *Sesia* Murrill 1904, misapplied; *Gloeophyllum* Murrill 1905,
misapplied.—Fide Murrill 1908 (NAF 9): 126 = ?*Daedalea aesculi* (Fr.)
Murrill [sensu Murrill = *Trametes ambigua* Berk. = ? "*Trametes*"
palisotii (Fr.) Imaz. (**O**)]; fide O. & K. Fid. 1967 (M 58): 868 = *Daedalea
elegans* Spreng. ex Fr. [= "*Trametes*" *palisotii* (Fr.) Imaz.]. — Sensu
Bres., Murrill 1904, 1905 → *Gloeophyllum trabeum*.

pallidomicans, *Polyporus*, Britz. 1897 (BCb 71): 58 [*pl. 658 f. 232*], 1909
(BbC 26): 211 (Germany).—Rather a nomen dubium. Killerm. 1922
(Dba 15): 86 referred this to *Polystictus* [*Coriolus*] *zonatus* as a form.
I think this is the best suggestion available, although the substratum
("an Weiden") would be exceptional for it. The determination by
Bresinsky & Stangl 1969 (ZP 34): 78 (with doubt) as *Phellinus conchatus*

is not acceptable to me; Britz. gave the context as white; further the
spores are quite different.

palmatus, Polyporus, Saut. 1876 (H 15): 151 (Austria), not ∼ (Hook.)
Berk. 1839; = *Polystictus subpalmatus* Sacc. 1888.—Nomen dubium.

**pampeana, Daedalea*, Speg. 1898 (ABA 6): 175 (Argentina).—Bres. 1916
(Am 14): 230 referred this to *Polyporus [Abortiporus] biennis*, and
Speg. 1926 (BAA 28): 370 followed ("forma terrestris stipitata hymenio
sublenzitoideo"). Fid. 1969 (Ri 4): 150, 162, 165 made it a syn. of
Heteroporus biennis var. *flabelliformis* ("Mont.") Fid. [sensu Fid. =
Abortiporus distortus (**O**); see Donk 1971 (PNA 74): 2]. Cf. (2).

papilio melanopus, Polyporus, Secr. 1833 M. 3: 62 (Switzerland) (as a sp.
of *Boletus* & double epithet & nom. nud.: n.v.p.).—Fide Fr. 1838:
439 ("n. 13") = *Polyporus melanopus*.

papyraceus, Boletus, Schrank 1789: 618 (Germany) (d.n.), not ∼ Schw.
1822; *Polyporus* (Schrank) per Wettst. 1888, not ∼ Fr. 1828, not ∼
(Schw.) Schw. 1832.—Nomen dubium. This was referred to *Polyporus
vulgaris* Fr. [sensu stricto = *Poria lenis*] by Fr. 1821: 381 (as *Boletus
"papyrae"*), therefore Wettst. revived it as the correct name for *P.
vulgaris*.

papyrae, Boletus, Fr. → *Boletus papyraceus* Schrank (**O**).

papyrina, Daedalea, (Bull. per St-Am.) E. Krause 1928 ("*Daedaleus*");
Auricularia Bull. 1788: *pl. 402* (France) (d.n.); *Thelephora* DC. 1805
(d.n.) per St-Am. 1821; *Auricularia* Mérat 1821; *Merulius* Quél. 1888;
= *Merulius corium* (**O**).

**paradoxa, Bresadolia*, Speg. 1883 (ASa 16): 277 (Paraguay).—This was
referred to *Polyporus squamosus* ("vetustus, poris laceratis") by Bres.
1916 (Am 14): 222. Sing. and J. E. Wright apud Sing. 1962: 161 compared it with *Polyporus discoideus* B. & C., but it is not clear what these
authors had in mind by this name (apparently covering a mixture,
judging from the syns. given by Sing. 1962: 159). Cf. also Lloyd
1923 (LMW 7): 1191 *pl. 235 fs. 2389, 2390*.

paradoxus, Hydnogaster, P. Henn. in herb. B.—Fide Ulbr. 1941 (NBe 15):
(incidental mention) the fruitbody developed after "*Oligiporus*" [in
this case = *Ptychogaster rubescens*] in a hothouse at Berlin. He thought
that it "vielleicht auch nur eine abweichende Fruchtkörperbildung von
Poria vaporaria darstellt." For '*Poria vaporaria*' sensu P. Henn. one
should in this case perhaps read resupinate *Polyporus [Tyromyces]
henningsii*.

Paramyces Oehm 1937 (BbC A 57): 253–258, 261 (nom. anam.; nom.
prov.: n.v.p.).—Introduced in connection with two imperfect states,
viz. of *Polyporus squamosus* and of *Polystictus [Coriolus] hirsutus*, but
no species were assigned to the genus *Paramyces*, which was introduced
as such in anticipation of the description of imperfect states that do
not form basidiferous fruitbodies anymore.

parasitica, Daedalea, Velen. 1922: 690 [see Pilát 1948: 262 for Latin translation] (Czechoslovakia).—Fide Malkovský *1931* this represents abnormal growth on the surface of the cap of *Amanita spissa* (Fr.) Opiz (Agaricales). Cf. also *Polyporus agaricicola* F. Ludw. (**O**).

parasiticum, Hydnum, L. 1763: 1648 (Europe) (d.n.) per Schleich. 1821; *Agaricus* E. Krause 1932 B.r.: 141.—Nomen dubium: insufficiently described. Fide Schrad. 1794: 180 the name *Hydnum parasiticum* had been used by Willd. and/or Timm for *Hydnum decipiens* [= *Hirschioporus fuscoviolaceus*]. As a result of this identification *H. parasiticum* "Timm" was also used for *Hirschioporus abietinus.*

parvula, Ungularia, Lázaro 1916 (RMa 14): 671 / 1917: 110 (Spain); *Polyporus* Sacc. & Trott. apud Trott. 1925, not ∼ Schw. 1832, &c. —Nomen dubium.

parvulus, Polyporus, Lázaro 1916 (RMa 14): 671 / 1917: 110 (Spain), not ∼ Schw. 1832, &c.; ≡ *Polyporus lazaroi* Trott. 1925.—Nomen dubium. Presumably a sp. of *Polyporus* subg. *Melanopus* (Pat.) Maubl. Cf. Bres. apud Trott. 1925 (SF 23): 369, "probabiliter est *Pol. melanopus.*" Other spp. of the subgenus also come to mind.

patella, Boletus, Humb. 1793: 97 (Germany) (d.n.) per Steud. 1824; *Polyporus* Pers. 1825.—Nomen dubium.

**pauperculus, Polyporus,* Speg. 1889 (BCó 9): 435 (Brazil).—Referred by Bres. 1916 (Am 14): 226 to *Polyporus brumalis* (sensu Bres. = *P. ciliatus*] "juvenilis".

pavichii, Irpex, Kalchbr. 1868 (nom. nud.: n.v.p.) ex Fr. 1874: 621 (Yugoslavia, Croatia); *Xylodon* O.K. 1898.—Nomen dubium. I compared Kalchbrenner's descr. and figures with the protologue of *Irpex foliaceo-dentata* but was forced to conclude that the two taxa could hardly be the same. Apparently does not agree either with *Irpex lacteus.* Cf. *Hirschioporus pargamenus?* – Descr.: Kalchbr. 1877: 60 *pl. 37 f. 2.*

pavonium, Agaricum, Latourr. 1785: 38 (France) (nom. nud.: n.v.p.). —The foot-note "An *Boleti versicoloris* var.?" would suggest a form of *Coriolus versicolor,* although the sp. was described in *Agaricum* rather than *Boletus.*

**peckianus, Polyporus,* Cooke 1878; *Polystictus* Sacc. 1888; *Microporus* O.K. 1898; *Grifola* Murrill 1914; *Scutiger* Parm. 1962; *Albatrellus* Niemelä 1970; ≡ *Polyporus flavidus* Peck 1873 (BBf 1): 61 & 1874 (RNS 26): 68 (U.S.A., New York), not ∼ Berk. 1854; *Polystictus* Cooke 1886. – Sensu Malenç. → *Albatrellus syringae.* – Cf. (3). – Descr.: Lowe 1942: 34; Overh. 1953: 218 *pl. 28 f. 171, pl. 32 f. 195, pl. 36 f. 221, pl. 131 fig.* (*Polyporus*); Niemelä 1970 (Abf 7): 54 *fs. 3, 5* (*Albatrellus*).

**peckii, Trametes,* Kalchbr. apud Peck 1881 (BG 6): 274 (U.S.A., Dakota). —Reduced to the synonymy of *Trametes hispida* [= *Funalia gallica*] by modern North American authors, for instance, Overh. 1953: 147. – The flesh was described as "substance wood-colour", according to Peck's translation [repr. Murrill 1905 (BTC 32): 356], and cf. Bres. 1920

(Am 18): 69 who referred it to *Trametes gallica* var. *trogii* (Berk.) Bres. [= *Funalia trogii*]. − Cf. (**36**).

*pectinatus, *Polyporus*, Kl. 1833 (Li 8): 485 (India); *Fomes* Gillet 1884; *Phellinus* Quél. 1886, misapplied; *Scindalma* O.K. 1898; *Xanthochrous* Pat. 1900; *Mucronoporus* Romell 1901; *Pyropolyporus* Murrill 1907; *Boudiera* Lázaro 1916, misapplied; *Lazaroa* Gonz. 1917; *Porodaedalea* Aosh. 1966 (incomplete ref.: n.v.p.).—An extra-European sp. Descr.: Lowe 1957 F.: 17 *f. 1* (*Fomes*). − No taxonomic syns. of this taxon are included in this list. − Sensu Fr., in part, Quél. → *Phellinus ribis*.

pelliceum, *Xylomyzon*, Pers. → *Boletus arboreus* Sow. (**0**).

pelliculum, *Agaricum*, Latourr. 1785: 39 (nom. anam.) (d.n.); [≡ '*Agaricus, seu Boletus pelliceus*' Dana *1771* (Italy?), a pharmaceutic rather than a botanical name]. − Cf. (**29**).

pellitus, *Polyporus*, S. Schulz. 1866: 41 (Yugoslavia, Slavonia) (nom. nud.: n.v.p.), not ∼ G. Meyer per Fr. 1821, not ∼ (P. Karst.) Sacc. 1896. —Nomen dubium.

Pelloporus Bond. & S. 1941 (Am 39): 54 ("Quél. sens. str.", but excl. of type) (lacking Latin descr.: n.v.p.); monotype, *Polyporus corrugis* Fr. [≡ *Ischnoderma trogii*].

pelloporus, *Cladomeris*, Big. & Guill. 1909: 410 ("Sow."; syn.: n.v.p.); [≡ *Boletus pelloporus* Bull. sensu Sow. 1799: *pl. 230*, excl. of type, (England)].—Cited by Big. & Guill. as a syn. of *Cladomeris* [*Bjerkandera*] *imberbis* sensu Quél. but here Sowerby's plate is referred to *Abortiporus biennis*.

pellucidus, *Boletus*, With. 1796: 309 (England) (d.n.); *Polyporus* (With.) per Steud. 1824.—Nomen dubium. With. compared his fungus with *Boletus* [*Albatrellus*] *ovinus* Schaeff. *pl. 122* (excl. of *pl. 121*), a form apparently induced by a spell of dry weather, but there is little in his descr. to make this comparison understandable.

peltata, *Poria*, Scop. 1772 P.s.: 104 *pl. 23* (Hungary, now Czechoslovakia) (d.n.); *Polyporus* (Scop.) per Pers. 1825, not ∼ Pers. 1825, not ∼ Fr. 1851.—Nomen dubium. Found in a mine: apparently an abnormal growth. Cf. Fr. 1828 E. 1: 87.

peltatus, *Polyporus*, Pers. → *Boletus fibula* Sow. (**0**).

pelviformis.—Pers. 1825: 86 cited a *Poria pelviformis* Scop. of which he stated that Hoffm. had reduced it to a var. of *Poria scutata* Scop. What actually happened is that Hoffm. 1797–1811 V.s.: 14 mentioned (but did not name) a variety of *Poria scutata* Scop. [= *Heterobasidion annosum*] which he described as being 'pelviformis' and depicted on *pl. 10 upper fig.* as *Poria scutata*. As far as I know there is no '*Poria pelviformis* Scop.'

*pendula, *Daedalea*, Berk. 1855 N.Z.: 180 *pl. 150 f. 4* (New Zealand); *Striglia* O.K. 1891.—At first this was identified with *Lenzites* [*Cerrena*] *unicolor* by G. Cunn. 1949 (BPZ 81): 21, but later G. Cunn. 1965: 97, 282 referred it to *"Trichaptum" venustum* (Berk.) G. Cunn., an extra-European species.

penetralis, Polyporus, W. G. Sm. 1875 (JBL 13): 98 *pl. 162 fs. 4–8* (England); *Polyporellus* Pilát 1936.—An alien found in a nursery.

peponius, Favolus*, (B. & C.) Lloyd 1917 (as a form of *Favolus europaeus*: n.v.p.); *Polyporus boucheanus* var. B. & C. 1853 (AM II 12): 432 (U.S.A., South Carolina).—Cf. *Favolus* [*Polyporus*] *canadensis* (O**).

pera, Polyporus, Pers. → *Boletus fodinalis* Humb. (**O**).

perdurans, Polyporus*, Kalchbr. apud Kalchbr. & Cooke 1880 (G 9): 1 (Australia, Tasmania).—Referred by Bres. 1890 (BmF 6): xliii to *Polyporus bulbipes* Fr. [≡ *Coltricia oblectans* (O**)]. Listed as a syn. of *Polystictus perennis* f. *cinnamomeus* (Jacq. per S. F. Gray) Pilát [*Coltricia cinnamomea*] by Pilát 1942 (ACE 3): 580, 582.

pereffusa, Fomitiporia*, Murrill 1907 (U.S.A., Pennsylvania); *Poria* Sacc. & Trott. 1912.—Referred by Lowe 1949 (Ll 11): 166 & 1966: 156 to *Poria* [*Phellinus*] *laevigatus*, which was simultaneously considered to be resupinate *Fomes* [*Phellinus*] *igniarius*, but cf. (82**); fide Niemelä 1972 (Abf 9): 50 microscopically "hardly distinguishable" from *Phellinus laevigatus*. − Descr.: Overh. 1942: 61; Lowe 1946: 81; D. Baxt. 1948 (PMi 32): 207 *pl. 9 f. 1* (*Poria*); Niemelä, l.c. (*Fomitiporia*).

**pergamena, Trametes*, Lloyd 1917 (LMW 5, L. 65): 9 (U.S.A., Montana) ("a thick ... form of *Polystictus* [*Hirschioporus*] *pergamenus*": n.v.p.).

perplexus, Polyporus*, Peck 1896 (RNS 49): 33 (U.S.A., New York); *Inonotus* Murrill 1904, misapplied; *Polystictus* A. Ames 1913.—Lloyd 1918 (LMW 5): 755 suggested *Polyporus* [*Inonotus*] *rheades* [sensu lato] (59**) but so far the precise identity does not seem to have been settled. The substratum of the type is trunks of *Fagus*. − Sensu Murrill → *Inonotus cuticularis*; sensu Lloyd 1912 → *Onnia triqueter*.

**pertenuis*, [*Polyporus*], (Kalchbr. apud Thüm.) Lloyd 1912 (incidental mention: n.v.p.); *Polyporus varius* var. Kalchbr. apud Thüm. 1877 (BSM 52): 144–145 (U.S.S.R., Russia, Siberia).—Fide Pilát 1936 (BbC 56): 64 & 1937 (ACE 3): 103 = *Polyporellus picipes* [= *Polyporus badius*] (forma).

perturbatus, Polyporus*, Lloyd 1918 (LMW 5, L. 68): 11 (Brazil); *Ganoderma* Torrend 1920.—Mentioned by Steyaert 1967 (BBB 100): 198 as a syn. of *Ganoderma dorsale* (O**). − The type material was distributed by J. Rick as *Polyporus formosissimus* Speg. which is the name of a different species.

pertusus, Polyporus, Pers. 1825: 103 (France); *Poria* Bres. 1903.—Nomen dubium. Referred by Fr. 1828 E. 1: 124 ("quoad descript.!") to *Polyporus* [*Oxyporus*] *corticola* var. *quercinum* Fr. The collection that Bres. 1903 (Am 1): 80 named *Poria pertusa* was insufficiently described.

petropolitana, Daedalea, (Fr. apud Weinm.) E. Krause 1928 (*"Daedaleus"*); *Merulius* Fr. apud Weinm. 1836: 348 (U.S.S.R., European Russia); *Xylomyzon* E. Krause 1934; = *Plicatura nivea* (Sommerf. apud Fr.) P. Karst. (Schizophyllaceae?).

pezizoides, *Boletus*, Schw. 1822: 100 (U.S.A., North Carolina); *Polyporus* Steud. 1824; *Porotheleum* Schw. 1832.—Fide Donk 1959 (Pe 1): 81, 83 = *Stromatoscypha fimbriatum* (**O**).

Phaeolus Sart. & M. 1921: 22 (lacking ref. and descr.: n.v.p.), not ∼ Pat. 1900; [≡ *Phaeolus* Pat. sensu Sart. & M., l.c., excl. of type]; monotype: *Polyporus* [*Hapalopilus*] *rutilans*.

phalliodorus, *Polyporus*, S. Schulz., "Mpt., p. 780. 1869" (n.v.p.).—Fide Igmándy 1968 (Aph 3): 352 and Tort. & Jel. 1969 (Abc 28): 382 = *Polyporus cadaverinus* [= *Buglossoporus pulvinus*?].

Physisporinus P. Karst. 1889 [1960 (Pe 1): 256]; monotype, "*Poria vitrea* Pers." sensu P. Karst.—Nomen dubium.

piceae, *Trametes*, A. Möll. 1914 (ZFJ 46): 204 (Germany) (not definitely accepted: n.v.p.), not *T. picei* Yamano 1930; = *Phellinus chrysoloma*.

picei*, *Trametes*, Yamano 1930 (Goryōrin No. 25): 69 *fs. 2, 5* (n.v.) (Japan), not *T. piceae* A. Möll. 1914 (n.v.p.).—See under *Cryptoderma yamanoi* (O**).

piceinus*, *Polyporus*, Peck 1889 (RNS 42): 121 (U.S.A., New York); *Polystictus* Sacc. 1891; *Microporus* O.K. 1898; *Phellinus* Pat. 1900; *Trametes* Peck 1901.—Bres. 1920 (Am 18): 68 referred this to *Trametes abietis* [= *Phellinus chrysoloma*], but cf. (84**). –Descr.: Peck 1901 (RNS 54): 169; Lloyd 1915 (LMW 4, F.): 277 *f. 276* (*Trametes*); & cf. Haddow 1928 (TBS 22): 186.

pictus, *Boletus*, Ehrenb. 1818: 19, 31 (Germany) (d.n.), not ∼ K. F. Schultz 1806 (d.n.), not ∼ Peck 1872.—Nomen dubium. Insufficiently described. Referred by Fr. 1821: 348 to *Polyporus brumalis* (forma).

pilatii, *Poria*, Bourd. 1932 (BmF 48): 230 *pl. 25* ("Tchéchoslovaquie", now U.S.S.R., Ukraine); *Aporpium* Bond. 1953.—Fide Teix. & Rog. 1955 (M 47): 411, 414 = *Aporpium caryae* (Schw.) Teix. & Rog. (Tremellaceae).

pileatus, *Boletus*, (Schaeff.) Lapl. 1894: 475 (incidental mention: n.v.p.); ≡ *Boletus ramosissimus* var. *pileatus* Schaeff. 1774: Ind. secundus [*pl. 111*] (Germany) = *Grifola umbellata*.

pileus, *Helvella*, Schaeff. 1774: 111 [*pl. 281*] (Germany) (d.n.); ≡ (by lectotypification) *Boletus membrana* Batsch 1783 (d.n.).—Fr. 1821: 348 referred Schaeffer's plate 281 to *Polyporus brumalis*, certainly by error. The German descr. gives the context as 'wachsartig', the Latin, as 'cartilagineus', while no tubes were mentioned. In my opinion a species of *Peziza* (modern emendation), cf. *Peziza cerea* Sow. ex Mérat.

pinguedinea, *Poria*, (Gaill.) ex Cooke 1886 (G 15): 3; *Polyporus* "Gaill. in Herb. Desm.; in Herb. Berk. ex Desmazières, No. 2888" [no locality mentioned by Cooke, l.c.; Sacc. 1888 (SF 6): 312 suggested ". . . in Gallia (?)"].—Nomen dubium and subnudum: "We fail to find any description of this. It has large, very irregular, often oblique pores, and is wholly whitish" (Cooke, l.c.). Cooke referred here "*Polyporus heteroporus*, Pers. in Litt."

pininus, *Polyporus*, E. Krause 1930 B.r.: 98 (Germany); *Polystictus*

E. Krause 1934.—Nomen dubium. This was equated with only a part of *Polyporus* [*Coriolus*] *hirsutus* (Wulf.) per Fr. and in any case with the exclusion of its type, mentioned as "*Bol. hirsutus* [Wulf. in] Jacq. coll. II S. 149". The original descr. may be rendered thus: ". . . die oberseits ganz schwarze Form ([*Polyporus hirsutus*] var. F. [Fr. 1874: 568]). Aber in der Jugend is dieselbe hellgrau, dem *P. abietinus* sehr ähnlich. . . . An Nadelholzrinde. Dez.–Mai." Later the following information was given: "Alle von November bis April gesammelten sind ganz schwarz, ein in September gefundener mit noch weichem Rande ist etwas bunt, doch schon durch blauschwarze kahle Zonen auffällig. Poren werden bald orange. An Nadelholz" (E. Krause 1934 M.B.: 18). Next year followed: "Im Dezember erschien die Nadelholzform des Versicolorkreises in Menge. Die frischen Hüten sind braun oder gelbbraun, ziemlich gleichfarbig mit dunkleren kahlen Zonen und blassem Rande, eigentümlich matt (glanzlos). Die Poren sind weiss, trocknend bald orange" [E. Krause 1935 (ANM II 9): 14]. However, E. Krause 1936 (ANM II 10): 53 soon indicated that this last descr. must be excluded from his conception of *Polyporus pininus*; the specimens were not found on coniferous wood but on stumps of birches; he referred them to *P. multicolor* [= *Coriolus zonatus*] but this name was perhaps misapplied.

piniperda, *Boletus*, Link 1804: 243 (Portugal) (nomen subnudum; d.n.); *Polyporus* (Link) per Colm. 1867; *Fomes* Sacc. 1893.—Nomen dubium. Not really described. Referred by some authors to *Trametes* [*Phellinus*] *pini* (for instance by P. Cout. 1919: 57), but others suspected that it might be *Fomes ungulatus* [= *Fomitopsis pinicola*] (cf. Torrend 1913 (Bro 11): 65). Cf. also Pinto-L. 1953 (RCL 3): 205–206. – The basionym has also been ascribed to "Hoffm. ex Link". Portuguese authors cite from the French edition of Link's book.

**pinsitus*, *Polyporus*, Fr. 1828 (Brazil); *Polystictus* Fr. 1851; *Microporus* O.K. 1898; *Coriolus* Pat. 1900; *Trametes* O. & K. Fid. 1957.—Listed by Govi 1969 (MBo 20): 42 *fs. 65–67, pl. 10 f. b* (*Coriolus*) from Italy. I have seen a representative fruitbody of this record and refer it to *Coriolus versicolor*. The correct basionym for the American sp., of which *Coriolus pinsitus* is a syn., is *Polyporus villosus* (Sw.) per Fr., an earlier published name for this highly variable and much-named sp. – Descr.: Overh. 1953: 326 *pl. 4 fs. 19, 20, pl. 5 f. 27, pl. 132 fig.* (*Polyporus pinsitus*); O. & K. Fid. 1967 (M 59): 833 *fs. 1, 2* (*Polyporus villosus*).

pithysus, *Polyporus*, (Fr.) Secr. 1833 M. 3: 171 (as a sp. of *Boletus*: n.v.p.); *Polyporus nitidus* var. *pithisus* Fr. 1818 O. 2: 263 (Sweden).—Nomen dubium. – Sensu Secr. = ?

pithyus, *Boletus*, Chaill. "in litt."—Fide Fr. 1828 E. 1: 85 = *Polyporus* [*Climacocystis*] *borealis* (var.).

**placentaeformis*, *Polyporus*, Berk. "in Herb."; *Polystictus* (Berk.) ex

Cooke 1886 (G 15): 24 ('British North America'); *Microporus* O.K. 1898.—Nomen dubium. Cf. Murrill 1907 (NAF 9): 28: "The small type specimens ... resemble forms of *C*[*oriolus*] *nigromarginatus* [= *C. hirsutus*] or *C. pubescens*, ... but the pores are much too large for the former and the surface too hirsute for the latter."

planus, Polyporus, Wallr. 1833: 602 (Germany), not ∼ Peck 1879.—Nomen dubium. Referred by Pilát 1936 (ACE 3): 64 to *Polyporellus* [*Polyporus*] *brumalis* [sensu lato], but the species is said to be terrestrial and the description is too vague.

**platyporus, Polyporus,* Berk. 1851 (HJB 3): 81 (India), not ∼ (Pers.) Fr. 1815 (d.n.), not ∼ Secr. 1833 (n.v.p.).—This has been cited as a syn. of *Heteroporus* [*Abortiporus*] *biennis* by Pilát 1937 (ACE 3): 115 because Lloyd 1912 (LMW 3, S.P.): 162 thought it might be "*Polyporus rufescens* form *heteroporus*". − Cf. (2).

platyporus, Polyporus, Secr. 1833 M. 3: 50 (Switzerland) (as a sp. of *Boletus*: n.v.p.), not *Boletus platyporus* Pers. 1794 (d.n.), not ∼ Berk. 1851; = *Polyporus squamosus.* This is "*Polyporus*" [≡ *Boletus*] *platyporus* "Pers. Syn. f. p. 521 (pro parte)" exclusive of the type, which is cited under *P. favolus juglandis* Secr. 1833 M. 3: 49 as "Schaeff. t. 101–102. *B. Juglandis*" (**101**).

pleuropus, Daedalea, Sacc. 1891 (SF 9): 200 (Germany); Lind. & Syd. 1917 T. 5: 190 (incidental mention: n.v.p.).—Nomen dubium. In my opinion this specific name is the result of a misrepresentation of the original text [1831 (IO): 1011]: "... Steinheim ... zeigte ... einen in Hamburgs Nähe gefundenen Pilz, *Daedalea, Pleuropus,* [follows a brief Latin descr.]." What was actually published was an unnamed sp. of '*Daedalea* sect. *Pleuropus*'.

**plumbea, Daedalea,* Lév. 1846 (ASn III 5): 302 (U.S.A., New York). —Nomen dubium. Not mentioned by Overh. 1953. Murrill 1905 (BTC 32): 95, 1907 (NAF 9): 125 referred this with doubt to *Agaricus/Daedalea quercina.*

Podoporia P. Karst. 1892 [1960 (Pe 1): 259]; monotype, *Podoporia confluens* P. Karst. (**0**).—Nomen dubium. The name has been misapplied: *Podoporia* sensu Donk → *Rigidoporus,* 'resupinate' species.

polygonius = *polyzonus*

polymorpha, Daedalea, Opiz 1852: 119 (Czechoslovakia) (nom. nud.: n.v.p.).—Fide Pilát 1942 (ACE 3): 612 = *Gloeophyllum sepiarium.*

polyporus, Boletus, Bull. 1789: *pl. 469* & 1791 H.: 331 (France) (d.n.), not ∼ Retz. 1769 (d.n.); *Boletus* Bull. per St-Am. 1821, Purt. 1821; ≡ *Boletus fuligineus* Pers. 1801 (d.n.), not ∼ Fr. 1835; *Polyporus* (Pers.) per Fr. 1821; *Albatrellus* S. F. Gray 1821; *Boletus* Spreng. 1827; *Leucoporus* Quél. 1890 (incidental mention: n.v.p.); *Agaricus* E. Krause 1933.—Nomen dubium. For a long time treated as a distinct species. Pilát 1936 (ACE 3): 64, 69 considered it a form of *Polyporellus* [*Polyporus*] *brumalis* [sensu lato]. This is doubtful, although *Polyporus lepideus* [= *P. ciliatus* forma] comes to mind. Bulliard's protologue gives the species as terrestrial. I have compared it with *Polyporus politus* Fr. (cf. Fr. 1884 I. 2: 79 *pl. 179 f. 2*), a form of *Albatrellus confluens*, but the colour disagrees; and with *Polyporus melanopus*, but in this case too there is much that disagrees. Cf. also Quél. 1872 (MMb II 5): 268 (*P. fuligineus*) and a remark by Bourd. & G. 1928: 530–531 under *Leucoporus* [*Polyporus*] *brumalis* f. *crassior* Bourd. & G. — Sensu Lenz 1840: 93, (*Boletus*) = ?

polyporus, Boletus, Retz. 1769 (SVH 30): 253 (Sweden) (d.n.), not ∼ Bull. 1789 (d.n.) per St-Am. 1821, Part. 1821, not ∼ Fr. 1835; *Polyporellus* Murrill 1903 (generic name not accepted: n.v.p.); *Polyporus* Murrill 1904, presumably misapplied.—Nomen dubium. Referred by Fr. 1821: 348 to *Polyporus brumalis* (forma), but there is little in the original descr. to substantiate this claim. Original descr.: "stipitatus, perennis, pileo subhemisphaerico, subtus planiusculo fulvo"; Mich. 1729: *pl. 70 f. 9* listed as a syn. — Sensu Murrill = *Polyporus brumalis* and other species confused with it.

polyporus perennis, Fungus, Paul. 1793 T. 2: 361 (descr.), Ind. (name) [*pl. 164 fs. 1, 2*] (France) (double epithet); ≡ *Polyporus umbilicatus* Paul. 1812–35 (d.n.?), not ∼ Jungh. 1838, &c.—If typified by the depicted specimen (as is done here), then to me a nomen dubium, perhaps *Polyporus lepideus* = *P. ciliatus* forma. Lév. 1855: 90 referred this to *Favolus* [*Polyporus*] *arcularius* but Paul. mentioned the pores as "très-fins" and depicted them as radially elongate. Among the synonyms cited by Paulet are Mich. 1729: *pl. 70 f. 7, Boletus* [*Coltricia*] *perennis*, and *B. lacteus* Batsch [= *Polyporus brumalis*].

polyposus, Polyporus, Pers. 1825: 102; [≡ *Agaricum album, terrestre medullam panis referens* Mich. sensu Batt. 1755: 66 *pl. 32 f. A* (Italy)]. —Nomen dubium. It is with considerable hesitation that I suggest that this might represent an abnormal condition of *Heterobasidion annosum*, developing fruitbodies on the soil surface, but connected to buried tree-roots by 'roots'.

Polysticta 'Fr.' [1960 (Pe 1): 264].—Sometimes listed as a generic name, but perhaps never validly published as such; ≡ *Polyporus* subgen. *Polysticta* Fr., based on two effused polypores, *Polyporus* [*Oxyporus*] *corticola* Fr. and *P.* [*Ceriporia*] *reticulata* (Hoffm.) per Fr. Not *Polystictus* Fr. 1851.

Polystictus Bond. 1953: 43, 419 (lacking Latin descr. or validating ref.: n.v.p.), not ∼ Fr. 1851 ("*Polyporaceae*") [1960 (Pe 1): 265, in obs.]; [≡ *Polystictus* Fr. sensu Sing. 1944 (M 36): 68, in obs. (provisional emendation) & Bond., l.c., excl. of type); holotype, *Polyporus tomentosus* Fr.

**polyzonus, Polyporus,* Pers. apud Gaud. 1827: 171 (New Guinea); *Polystictus* Cooke 1886, Sacc. 1888 ("*polygonius*", error); *Microporus* O.K. 1898 ("*polygonius*"); *Trametes* Petch 1916; *Daedalea* C. & D. Over.

1922, non/an ∼ Pers. 1828; *Coriolus* Imaz. 1943.—Pers. wrote "n'est vraisemblement qu'une simple variété de *polyporus lutescens* (Mycol. Europ.)". Referred by Pat. 1914 (LPB 6): 2248 (as a var.) and Lloyd 1929 (LMW 6): 1089 (as a form) to *Polystictus [Coriolus] hirsutus*, but this needs further study. − Descr.: Bres. 1913 (H 53): 65; Kawam. 1954 I. 1: 148 *f. 131: 2 (Polystictus)*.

pomariorum, Agaricus, Dubois 1803: 179 (France) (d.n.) per Dubois 1833.—Nomen dubium.

populina, Ungularia, Lázaro 1916 (RMa 14): 670, 1917 (RMa 15): 378 / 1917: 109, 292 (Spain); *Polyporus* Sacc. & Trott. apud Trott. 1925, not ∼ (Schum.) per Fr. 1821.—Nomen dubium. *Fomes fomentarius?*

populneus, Boletus, Pollini → *Polyporus castaneus* Fr. (**O**).

Poria [Hill] Adans. 1763 (d.n.) [1960 (Pe 1): 265], not ∼ Pers. per S. F. Gray 1821, not *Porium* Hill 1751 (pre-Linnaean name).—A name introduced for polypores in general.

Poria Bond. 1953: 156 ["(Fr.) Karst."] (lacking Latin descr. or valid ref.: n.v.p.), excl. of types of *Polyporus* "series" [subgen.] *Poria* Fr. 1851 (NAu III 1): 70/54 [≡ *Poria* Pers. per S. F. Gray 1821] and of *Poria* P. Karst. 1881 (Rm 3 / No. 9): 19 [= *Poria* Pers. 1794, in part.] − Cf. Donk 1960 (Pe 1): 268–269.

porinoides, Trametes, Lázaro 1917 (RMa 15): 372 / 1917: 286 (Spain). —Nomen dubium.

**poripes, Polyporus,* Fr. 1821 (NAu III 1): 48/32 (U.S.A., North Carolina); *Grifola* Murrill 1904, misapplied?—A nomen dubium, fide Lloyd 1911 (LMW 3, O.): 90. − Sensu Ravenel, Murrill → *Albatrellus cristatus*.

Porium Hill 1751 (pre-Linnaean name) [1960 (Pe 1): 270], not *Poria* [Hill] Adans. 1763 (d.n.); not ∼ Pers. per S. F. Gray 1821, not ∼ P. Karst. 1881.—In the main this genus is the same as *Polyporus* Mich. 1729 (pre-Linnaean name), introduced for stalked polypores.

Porodon Fr. 1821 (nom. nud.: n.v.p.), 1851 (nom. nud. & prov.: n.v.p.) [1962 (Pe 2): 209]; holotype: *Polyporus acanthoides* (Bull. per Mérat) Fr. sensu Fr. [= *Abortiporus biennis*].

Porothelium Fr. 1818 (d.n.); *Porotheleum* (Fr. per Fr.) Fr. 1825: Fr. 1828 & 1832, not *Porothelium* Eschw. (Lichenes) [1951 (Re 1): 217]; lectotype, *Boletus fimbriatus* (Pers.) Pers. (**O**).—Often referred to the Polyporaceae and even in recent times included in *Poria*. European spp. referred to *Porotheleum* are not listed except when they actually belong to the European 'polypores'.

**proliferus, Polystictus,* Lloyd 1908 (LMW 3, P.I.): 8 *f. 202* (U.S.A., Ohio) (as a "form" of *P. perennis*: n.v.p.).—Fide Overh. 1953: 387 = *Polyporus [Coltricia] perennis*.

proteiporus, Polyporus,* Cooke 1883 (G 12): 15 (Australia, Queensland). —This has been referred to *Polyporus rufescens* [= *Abortiporus biennis*] by Lloyd 1912 (LMW 3, S.P.): 162 and to *Polyporus/Heteroporus [Abortiporus] biennis* by G. Cunn. 1965: 82, 83. − Cf. (2**).

proteus, Boletus, Bolt. 1788: 66 *pl. 66* (England) (d.n.).—Nomen dubium. Referred by Fr. 1821: 381 to *Polyporus vulgaris* Fr. [sensu stricto = *Poria lenis*] and by Pers. 1825: 100 to *Polyporus [Perenniporia] medulla-panis.*

**proteus, Polyporus*, Berk. 1843 (LJB 2): 514 ("414") (South Africa), not ∼ Kalchbr. apud Cooke 1882; *Trametes* Fr. 1848, G. Cunn. 1949 misapplied; *Polystictus* Cooke 1886; *Microporus* O.K. 1898; *Osmoporus* G. Cunn. 1965, misapplied.—Referred by Bres. 1926: 79 and in earlier publications to *Trametes hispida* [= *Funalia gallica*] sensu lato. – Misapplied by Lloyd; also by G. Cunn. 1965: 249, cf. D. Reid 1967 (TBS 50): 165. – Cf. (36).

provectus, Polyporus, Chev. 1837: No. 40 var. ε *plate f 2* (as a var. of P. [*Albatrellus*] *cristatus*: n.v.p.) (France).

**proxima, Lenzites*, Berk. 1876 (G 4): 162 (U.S.A., New York); *Cellularia* O.K. 1898.—Referred to *Daedalea* [*Daedaleopsis*] *confragosa* (var.) by Peck 1878 (RNS 30): 74. – Cf. (31).

prunastri, Polyporus, (Pers. ex S. F. Gray) Bret. & Niel 1894 (BRo III 29): 135 ("Alb. et Schw."; syn.: n.v.p.); *Fomes* Lloyd 1908 (syn.: n.v.p.); [= *Boletus fomentarius* var. "*B. prunastri*" Pers. 1801: 538 (Germany)]; *Boletus fomentarius* var. *prunastri* (Pers.) ex A. & S. 1805 (d.n.); *Boletus pomaceus* var. Pers. ex S. F. Gray 1821.—Fide Pers. 1825: 85 = *Polyporus [Phellinus] pomaceus* (var.).

pruni, Polyporus, Bertol. 1878 (NGi 10): 383 (Italy) (nom. prov. & subnudum), not ∼ Wittr. (n.v.p.).—Fide Sacc. 1916: 999 = *Fomes fulvus* [sensu Sacc. = *Phellinus pomaceus*].

pruni, Polyporus, Wittr. in herb., not ∼ Bertol. 1878 (n.v.p.).—Fide Nannf. 1967 (WPb 6): 131 = *Phellinus pomaceus.*

**prunicola, Fomitiporia*, Murrill 1907 (U.S.A., Maine); *Poria* Sacc. & Trott. 1912; *Fuscoporia* Aosh. 1950.—Referred to *Poria [Phellinus] laevigata*, which was simultaneously considered to be resupinate *Fomes [Phellinus] igniarius*, by Lowe 1949 (Ll 11): 166, 1966: 156, but cf. (82). Niemelä 1972 (Abf 9): 57 considers it a distinct sp. – Descr.: Overh. 1931 (M 23): 118 *pl. 12 fs. 2, 6, pl. 13 f. 8*; 1942: 62; Lowe 1946: 81 (*Poria*); Aosh. 1950 (RFJ 46): 160 *tpl. 1 f. 3* (*Fuscoporia*); D. Baxt. 1952 (PMi 37): 98 *pl. 7 f. 1* (*Poria laevigata* f.); Niemelä l.c. (*Fomitiporia*).

pseudoacaciae, Polyporus, Kalchbr. in manuscr.—Fide Igmándy 1965 (EFE): 207 = *Gloeoporus [Bjerkandera] fumosus.*

pseudoboletus, Hydnum, DC. 1815: 37 (Switzerland) per Fr. 1821; *Xylodon* O.K. 1898.—Nomen dubium. Referred by Pers. 1825: 77 to *Polyporus dolosus* [= *Hirschioporus abietinus*] and by Fr. 1828 E. 1: 147 to *Irpex deformis* [= *Schizopora paradoxa*]. If the substratum ("les bois de chêne dénudés d'écorce") was correctly identified the first suggestion is not tenable; the second is the more likely but far from certain.

**pseudolacteus, Tyromyces*, Murrill 1940 (BTC 67): 65 (U.S.A., Florida); *Polyporus* Murrill 1940 (nom. altern.).—According to Lowe apud

Overh. 1953: 428, "appears to agree with *Polyporus* [*Tyromyces*] *fumidiceps.*"

pseudo-osseus, Polyporus, J. Schroet. 1893 (JsC 70, Naturw. Abth.): 80 (former Prussian Silesia).—Nomen dubium.

Pseudopelloporus Lázaro 1916 (nom. nud. & error: n.v.p.) [1960 (Pe 1): 224, in obs.] ≡ *Heteroporus* Lázaro = *Abortiporus.*

pulchellus, Polyporus, Sacc. 1873 (ASv 2): 98/50 *pl. 7 fs. 11–15* (Italy), not ∼ Schw. 1832; *Polystictus* Sacc. 1888; *Microporus* O.K. 1898. —Nomen dubium; even Sacc. 1916: 1021 confessed, "rimane quindi un po'dubbia." – Referred by Lloyd 1915 (LMW 4, Ap.) 384 to *Polyporus* [*Tyromyces*] *henningsii* ("picture is surely same thing"); this is contradicted by the original descr., cf. "pileo . . . fusco-zonato; . . . sporis minimis rotundis . . ., circ. 3 µ diam."

**pulcher, Polyporus,* Speg. 1880 (ASa 10): 129 ("*pucher*") (Argentina), not ∼ Fr. 1830; *Trametes* Speg. 1899; ≡ *Polystictus celottianus* Sacc. & Manc. apud Sacc. 1888; *Microporus* O.K. 1898; *Polyporus* Speg. 1899 (syn.: n.v.p.).—Fide Bres. 1916 (Am 14): 223 = *Trametes hispida* [in part, = *Funalia gallica*]. – Cf. (36).

pulchrum, Xylomyzon, Pers. 1825: 32 *pl. 14 f. 1* (Switzerland); *Merulius* Duby 1830.—Fide Romell 1911 (ABS 11³): 30 and Lundell 1941 (LNF 21–22): 3 No. 1004 = *Polyporus haematodes* Rostk. [sensu Romell = *Merulius taxicola* (**O**), fide Donk 1967 (Pe. 5): 104].

pulvinata, Poria, Scop. 1772 P.s.: 106 *pl. 28 f. 1* (Hungary, now Czechoslovakia) (d.n.); *Boletus* Humb. 1793 (d.n.); Fr. 1821 (incidental mention: n.v.p.); *Polyporus* (Scop.) per Steud. 1824; ≡ *Polyporus lycoperdiformis* Pers. 1825.—Nomen dubium. Found in a mine and was very likely a monstrous growth form of some polypore.

pulvinato-griseus, Polyporus, Secr. 1833 M. 3: 114 (Switzerland) (as a sp. of *Boletus*, nom. nud.: n.v.p.).—Nomen dubium.

punctatus, Boletus, With. 1776; [≡ *Fungus arboreus Lobis rubellis, diversimodè figuratis & punctatis* Ray 1696: 20, 1724: 23 (England)]. —Nomen dubium. This may not even be a polypore.

puniceus, Polyporus,* Kalchbr. 1882 (Rm 4): 96 *pl. 29 f. 4* ("*punicus*") ("Ins. Ocean. pacifici"); *Polystictus* Sacc. 1888; *Microporus* O.K. 1898.—Fide Bres. 1890 (BmF 6): xliv = *Polystictus* [*Pycnoporus*] *sanguineus* (q.v.). Referred by G. Cunn. 1965: 169 to *Trametes* [*Pycnoporus*] *cinnabarina* sensu lato, but cf. (112**). – To be compared with *Trametes* [*Pycnoporus*] *coccinea* Fr. 1851 (extra-European) (**O**).

purpurascens, Boletus, DC. 1815: 41 (France) (d.n.) per Steud. 1824, not ∼ Pers. 1796 (d.n.), not ∼ Hook. 1822, not ∼ Rostk. 1844; *Polyporus* Pers. 1825, not ∼ (Hook.) Fr. 1838, not ∼ Stahlschmidt 1877.—Fide Fr. 1828 E. 1: 58 = *Merulius corium* (**O**). The monotype was found by de Candolle himself. Fries's identification is apparently based on a specimen sent from France by Guépin; it is supported by the original descr.

purpurascens, Trametes, B. & Br. 1879 (AM V 3): 210 (England).—Cf.

Bres. 1916 (Am 14): 240, "Vix dubie abortus speciei jam notae, ideoque species delenda."

purpurea, *Trametes*, Cooke 1882 (G 10): 121 (Japan).—Referred to *Lenzites* [*Daedaleopsis*] *tricolor* ("f. trametoidea") by Bres. 1916 (Am 14): 229. — Cf. (31).

purpureo-rufus, *Polyporus*, Post in herb. (Sweden).—Fide Romell 1926 (SbT 20): 15 = *Polyporus* [*Ceriporia*] *purpureus*.

pusillus, *Trametes*, Lloyd 1918 (LMW 5): 774 *f. 1165* (U.S.A., Minnesota), not ∼ Ell. & Gall., Lloyd 1915 (n.v.p.).—Overh. 1953: 354, 355 referred this to *Polyporus* [*Coriolus*] *velutinus*, which, in his interpretation, might be *Coriolus zonatus*.

queletii, *Daedalea*, S. Schulz. 1885 (H 24): 145 (Yugoslavia, Slavonia); *Striglia* O.K. 1891.—Nomen dubium.

queletii, *Lenzites*, S. Schulz. 1885 (H 24): 142 (Yugoslavia, Slavonia). —Nomen dubium. If this is *Lenzites warnieri* then the descr. contains several errors.

quercina, *Ungularia*, Lázaro 1916 (RMa 14): 674 / 1917: 113 (Spain); *Polyporus* Sacc. & Trott. apud Trott. 1925, not ∼ (Schrad.) per Fr. 1838.—Nomen dubium. Cf. *Fomes fomentarius* or *Phellinus* sp.

querneus, *Polyporus*, Berk. 1861 (GCh): 121.—Berk. mentioned this name rather casually in a comparison with *Fistulina hepatica*. This may be taken as a lapsus calami for *Polyporus* 'quercinus' [= *Buglossoporus pulvinus*].

radians, *Polyporus*, Lloyd 1923 (LMW 7): 1186 (nom. nud.: n.v.p.). —This is apparently an error for *Polyporus* [*Inonotus*] *radiatus*. According to G. Cunn. 1965: 278 the specimens involved (New Zealand, leg. G. Cunn.) are *Inonotus nothofagi* G. Cunn. (extra-European).

radiata, *Lenzites*, (Peck) Lloyd 1924 (as a form of *Lenzites betulina*: n.v.p.); *Lenzites betulina* var. *radiata* Peck 1902 (BNS 54): 965 (U.S.A., New York).—Fide Lowe 1942: 97 = *Lenzites betulina*.

radula, *Poria*, Romell 1926 (SbT 20): 16 ["Bres. (non Pers.)"] (nom. prov.: n.v.p.), not ∼ (Pers. per Fr.) Cooke 1886; = *Poria eupora* var. *subfimbriata* Romell, l.c. (Sweden); = either *Chaetoporus separabilimus* or *C. nitidus*. − Romell did not include the 'type' of *Poria radula* sensu Bres., cf. Donk 1967 (Pe 5): 105–106.

ramealis, *Polyporus*, S. Schulz. 1866: 41 (Yugoslavia, Slavonia) (nom. nud.: n.v.p.).—Nomen dubium.

ramosus, [*Polyporus*], Schw., "Published ?".—Fide Lloyd 1912 (LMW 3, S.P.): 157 = *Polyporus* [*Grifola*] *frondosus*.

rangiferinus, *Polyporus*, Pers. 1825: 114; [= *Boletus medulla-panis* Jacq. sensu Sow. 1801: *pl. 326* (England)].—Nomen dubium, cf. Donk 1967 (Pe 5): 93. Referred by Fr. 1828 E. 1: 122 to *Polyporus* [*Poria*] *vaillantii* ("optime"). *Serpula* sp.?

ravenelii, *Merulius*, Berk. 1871 (G 1): 69 (U.S.A., South Carolina). —Fide Lowe 1966: 35 = *Poria* [*Merulius*] *taxicola* (**O**).

*repsoldii, Polyporus, A. Möll. 1897 (BCb 72): 232 (Brazil), Spongiosus Torrend 1924 ("Repsodi").—Lloyd 1912 (LMW 3, S.P.): 159 compared this with Polyporus [Phaeolus] schweinitzii and J. Rick 1960 (Ih 7): 219 ("P. Henn.") reduced it to a var. of that sp.

resinosus, Boletus, Rubel 1779 (MaJ 1): 180 (presumably Austria) (d.n.). —Nomen dubium. An inclusively conceived and insufficiently described sp. The citation of Schaeff. pl. 106 [= Gloeophyllum odoratum] and pl. 136 [= Hapalopilus rutilans] indicates only some of the elements included.

resupinatus, Boletus, Bolt. 1791: 165 pl. 165 (England) (d.n.), not ~ Sw. 1788 (d.n.), not ~ Sow. 1815 (d.n.); Fomes (Bolt.) per Mass. 1892; ≡ Boletus spongiosus Pers. 1801 (d.n.), not ~ Lightf. 1778 (d.n.); Polyporus (Pers.) per Fr. 1821; Boletus Hook. 1821; Poria S. F. Gray 1821; Fomes Cooke 1886; Physisporus Cost. & Duf. 1895; Scindalma O.K. 1898.—Nomen dubium. This has been reduced to Polyporus nidulans [= Hapalopilus rutilans] by Fr. 1838: 455 (as a subsp.) & cf. (108). – Misspellings: Boletus "reticulatus", Hartig 1833: 43 (syn.); Poria "spongia Quél.", D. Baxt. 1950 (PMi 34): 43. – Sensu Quél. 1892 (Poria spongiosa) → Poria expansa; sensu Romell (Polyporus spongiosus) ~ Phellinus nigrolimitatus.

resupinatus, Boletus, Sow. 1815: pl. 424 (England) (d.n.), not ~ Sw. 1788 (d.n.), not ~ Bolt. 1791 (d.n.); ≡ Daedalea vermicularis Pers. 1825; Striglia O.K. 1891.—Nomen dubium. Referred by Fr. 1830 (Li 5): 701 to Daedalea latissima, which is referred to Cerrena unicolor (forma) by Bourd. & G. 1928: 564. As to Sowerby's fungus, from the plate I do not venture to suggest its identity.

resupinatus pseudo-platani, Polyporus, Secr. 1833 M. 3: 147 (Switzerland) (as a sp. of Boletus & double epithet: n.v.p.).—Nomen dubium.

resupinus, Agaricus, Paul. 1812–35: pl. 2 f. 5 ≡ (by lecto-typification) Agaricus applicatus Batsch 1786: 171 pl. 24 f. 125 (d.n.); Agarico-suber Paul. 1793; ≡ Resupinatus applicatus (Batsch per Fr.) S. F. Gray (Agaricales). – Paulet confused this with Gloeophyllum sepiarium. Cf. Donk 1971 (PNA 74): 6.

*reticulata, Auricularia, Fr. 1838: 555 (Brazil); Gloeoporus Fr. 1848, Graff 1922.—Fide Bres. 1896 (H 35): 284 = Gloeoporus conchoides, but cf. (51).

reticulata sebacea, Poria, Secr. → Polyporus sebaceus Fr. (O).

Retiporus Endl. 1836 (syn.: n.v.p.) [1960 (Pe 1): 277]; [≡ Boletus "subordo" Retiporei Batsch 1783: 107 (inadmissible terms denoting rank)]; lectotype, Boletus officinalis Batsch [≡ Agaricum officinale]; ≡ Dictyoporus Clem. 1902 (nom. nud.: n.v.p.).

*retiporus, Polyporus, Cooke 1883 (G 12): 15 (Australia, Queensland). —Lloyd 1912 (LMW 3, S.P.): 154 thought that this "will prove to be only a form" of Polyporus [Laetiporus] sulphureus. Fide Bres. 1916 (Am 14): 227 = Polyporus [Bondarzewia] berkeleyi Fr., an extra-European species.

retiporus, Polyporus, Opiz in herb., not ∼ Cooke 1883.—Fide Pilát 1937
(ACE 3): 88 = *Polyporellus [Polyporus] squamosus* ("Fungus juvenilis").
**rhaponticus, Fomes,* Lloyd 1913 (LMW 4, L. 44): 11, 1915 (LMW 4, F.):
258 (Japan).—Lowe 1957 F.: 55 mentioned this as a probable syn.
of *Fomes [Phellinus] robustus.* S. Ito 1955: 360 took it to be a nomen
dubium. The original descr. disagrees with Lowe's suggestion.
**richardsonii, Irpex,* [Kl. ?] in "Hook. Herb." ("North America", Canada).
—Fide Berk. 1839 (AM 3): 395 = *Irpex [Hirschioporus] fuscoviolaceus.*
**richardsonii, Polyporus,* B. & C. 1856 (JAP II 3): 224 (boreal North
America).—Needs re-examination of type. Murrill 1907 (NAF 9): 28
thought, "Apparently near *C[oriolus] pubescens*", a suggestion hardly
supported by the original description, which, however, is very poor.
**rigidus, Polyporus,* Lév. 1844 (ASn III 2): 189 (Indonesia, Java), not ∼
Berk. in Ravenel 1852 (nom. nud.), not ∼ Lloyd 1924 (n.v.p.); *Polystictus* Cooke 1886, not ∼ (B. & Mont.) Cooke 1886, not ∼ Lloyd 1916;
Microporus O.K. 1898; *Phaeolus* Pat. 1915, misapplied; *Funalia* House
1918; *Trametes* C. & D. Over. 1922.—Sensu Overh. 1953: 308 a North
American sp. of *Rigidoporus* for which he listed *Polyporus [Rigidoporus]
undatus* Pers. 1825 and *P. broomei* Rab. [= *Rigidoporus undatus*] as
syns., both based on specimens of a European sp. I cannot agree with
these identifications and consider Overholts's sp. to be unreported for
Europe. — The application of the name *Polyporus rigidus* by Overh.
seems to have been induced by determinations by Lloyd 1915 (LMW 4,
Ap.): 337 of certain collections found in the U.S.A. The true *P. rigidus*
is apparently a different sp. and has been listed as a syn. of another
extra-European sp., viz. *Rigidoporus zonalis* (O) by Pegl. & Wat. 1968
(CDp): no. 200. Bres. 1916 (Am 14): 227 identified *P. rigidus* with
P. rugulosus Lév., which was listed as another syn. of *Rigidoporus
zonalis* by Pegl. & Wat., l.c. Another interpretation of *P. rigidus* is
by Pat. 1915 (PJS 10): 92 (as *Phaeolus*) = "*Polyporus*" *durus* Jungh.,
also extra-European.
rimosus, Polyporus, Berk. → *Fomes scaber* (Berk.) Lloyd (O).
**robiniophilus, Trametes,* Murrill 1907 (NAF 9): 42 (U.S.A., Virginia);
Polyporus Lloyd 1912, Overh. 1914; *Leptoporus* Pilát 1937.—Reported
from Great Britain by Rea 1932 (TBS 17): 47 (no material in K and
BM); and from Romania by Eliade 1965: 215 without descr. or comment.
— Descr.: Lloyd 1912 (LMW 4, L. 42): 12 (*Polyporus*); 1915 (LMW 4,
Ap.): 314 *f. 653 Trametes*); Overh. 1953: 314 *pl. 12 fs. 68, 69, pl. 95
f. 544, pl. 131 fig.* (*Polyporus*). — Special literature: Kauffman &
Kerber, *1922.*
**robinsoniae, Pyropolyporus,* Murrill 1908 (NAF 9): 108 (Jamaica);
Fomes Sacc. & Trott. 1912.—This was referred to *Fomes [Phellinus]
robustus* by D. Baxt. 1952 (PMi 37): 104 ("ex Bres."), but Lowe 1954
(M 46): 494 identified it with *Fomes zealandicus* (Cooke) Cooke [=
Phellinus laurencii (Berk.) Aosh.] (extra-European).

roburneus, Polyporus, Fr. 1838; Fomes Fr. 1849 (nom. nud.: n.v.p.), Gillet 1877; *Placodes* Quél. 1886; *Scindalma* O.K. 1898; *Ungulina* Pat. 1900; [≡ *Polyporus annosus* Fr. sensu Fr. 1828 E. 1: 106 (Sweden)]. —Nomen dubium & ambiguum, cf. (**48**).

rosae, Polyporus, Velen. 1925 (MP 2): 98 (as a "subspecies" of *P. evonymi*: n.v.p.) (Czechoslovakia).—Fide Pilát 1942 (ACE 3): 528 = *Phellinus ribis* (forma). — *Polyporus ribis* var. *rosae* Velen. 1922: 676, typonym?

rosmarinus, Polyporus, Schwalb 1891: 166 (Austria?).—Nomen dubium.

rotundata, Poria, Roussel 1806: 72 (France) (d.n.).—Nomen dubium.

rotundatus, Polyporus, Lindblad apud Fr. 1874: 554 (Sweden) (nom. prov.: n.v.p.).—Nomen dubium. According to Romell 1909 (M 1): 266 there is no type. He suspected that *P. rotundatus* was *Polyporus [Fomitopsis] pinicola.* He also remarked that, "according to a note, Dr. Lindblad found his fungus 'in cortice vetustiori *Betulae*', while Fries wrote 'ad truncos *Pini.*' "

ruber, Agaricus, Lam. → *Boletus sanguineus* L. (**O**).

ruber, Boletus, Scop. 1772 P.s.: 107 *pl. 30 f. 1* (Hungary, now Czechoslovakia) (d.n.), not ∼ (Lam). Fr. 1821 (n.v.p.); *Polyporus* (Scop.) per Pers. 1825.—Nomen dubium. Found in a mine; may be an abnormal growth.

rubescens, Boletus, Gaterau 1789: 190 (France) (d.n.), not ∼ Trog. 1839. —Nomen dubium.

rubescens, Poria,* Petch 1922 (APe 7): 286 (Ceylon).—Referred by Lowe 1966: 31, 32 to *Poria [Ceriporia] rhodella,* which, sensu Lowe, is a broadly conceived sp. (22**).

rubiginosus, Boletus, Retz. 1769 (SVH 30): 253 (Sweden) (d.n.) per Steud. 1824: 81, not ∼ Schrad. 1794 (d.n.), not ∼ Fr. 1874 (Boletaceae). —Nomen dubium. Sensu Fr. 1818 O. 2: 246 = *Boletus rubiginosus* Fr. 1874: 521 (Boletaceae), not ∼ Retz. Later Fr. 1832[Ind.]: 62 thought of "*Polyp. badius?*" This is difficult to accept; Retzius's sp. was apparently a fleshy bolete, otherwise he would not have cited as a syn. "*Suillus esculentus* [!] ... Mich. 127."

rubiginosus, Polyporus, Wallr. 1833: 587 (Germany), not ∼ Fr. 1838; not ∼ Berk. 1839.—Nomen dubium. This taxon was introduced for a combination of various setae-bearing spp., representing a sp. described in the specific descr. and various others, previously published, appended as syns. and vars. From the descr. I am unable to suggest the identity of the main sp.

rubricus, Polyporus,* Berk. 1851 (HJB 3): 81 (India); *Polyporellus* P. Karst. 1879.—Fide Bres. 1912 (H 53): 52 = *Polyporus [Laetiporus] miniatus* (O**). Referred by Lloyd 1912 (LMW 3, S.P.): 157 to *Polyporus [Laetiporus] sulphureus.*

rubripes, Polyporus, Rostk. 1848 (StP Fs. 17–28): 31 *pl. 16* (Germany/ Poland); *Leucoporus* Quél. 1886.—Nomen dubium. Killerm. 1927 (ZP 6): 135 referred this to *Polyporus arcularius*; Bourd. & G. 1925

(BmF 41): 113 reduced it to a form of *P. brumalis* [sensu Bres. = *P. ciliatus*, forma], but they apparently described something different. Another suggestion, made here, is *P. floccipes*: cf. "am Rande mit weissgelbigen Stacheln besetzt. Die Poren sind . . . gross . . .."

rubromaculatus, Polyporus, Britz. 1891 H.S. 8: 13 *pl. 624 f. 118* (Germany). —This has been referred to *Polyporus [Tyromyces] fragilis* by Killerm. 1922 (Dba 15): 70, 73 and Ade 1923 (ZP 2): 42, but the protologue does not agree well: fruitbody, "weisslich, rötlich, rotfleckig. Fl[eisch] . . . wässerig, fast krokhart . . . . Sp[oren] bald gerade, bald wenig gebogen . . .." As drawn, the spores appear rather very narrowly spindle-shaped. I do not recognize *Tyromyces mollis* in the figures.

rufa, Daedalea, Weinm. 1836: 314 (syn.: n.v.p.).—A herbarium name ("an. 1830. in Sched. adnotavi") for the species actually published by Weinm. 1836: 313 as *Polyporus labyrinthicus* Fr. [sensu Weinm.] → *Polyporus weinmannii* Fr. [= *Tyromyces fragilis*].

rufescens, Bulliardia, Lázaro 1916 (RMa 14): 844 / 1917: 156 (Spain); *Daedalea* Sacc. & Trott. apud Trott. 1925, not ∼ (Pers. per Fr.) Secr. 1833.—Nomen dubium.

**rufescens, Ptychogaster,* Lloyd 1923 (LMW 7): 1195 *pl. 241 f. 2426* (South Australia).—Lloyd remarked: "It is one of those abnormal derivates from a *Polyporus*, probably *Polyporus rufescens* [= *Abortiporus biennis*]." — Cf. (2).

ruficolor, Polyporus, E. Krause 1928 B.r.: 53 (Germany).—Nomen dubium. The original descr. is insufficient for recognition, although *Inodermus radiatus* comes to mind. The author cited Schaeff. *pl. 136* [= *Hapalopilus rutilans*] and Bolt. *pl. 81* [= *Coriolus versicolor*]; later E. Krause 1929 B.r.: 79 identified the sp. with "*Bol. radiatus* Sow. 196 . . . (Nicht *Pol. radiatus* Fr.!)."

**rufitinctus, Poria,* (B. & C.) ex Cooke 1886 (G 15): 25 (Cuba); [*Polyporus*] B. & C. in herb.; *Mucronoporus* Ell. & Macbr. 1896; *Phellinus* Pat. 1900; *Fuscoporia* Murrill 1907; *Fomes* J. Rick 1907; *Trametes* Bres. 1920, misapplied.—Nomen dubium, fide Lowe 1966: 165. Previously Lowe 1954 (PMi 39): 34 referred this doubtfully to *Poria castletonensis* (Murrill) Sacc. & Trott. = *Poria [Phellinus] nicaraguensis* (Murrill) Sacc. & Trott., an extra-European sp. – Sensu Bres. = *Phellinus torulosus*, fide Overh., but Overholts's conception of this sp. is doubtful and apparently erroneous.

**rufo-rugosus, Polyporus,* Lloyd 1924 (LMW 7): 1331, 1925 (LMW 7): *pl. 318 f. 3058* (Australia, Tasmania).—G. Cunn. 1965: 167, 168 referred this to *Trametes [Coriolus] zonatus*. His interpretation of the latter sp. needs confirmation.

rufo-velutinus, Agaricus, DC. 1805: 134 (France) (d.n.) per Mérat 1821; *Daedalea* Duby 1830.—Nomen dubium. Fr. 1821: 340 wrote, "Cf. *Daed. saep.*" [*Gloeophyllum sepiarium*].

rufus, Boletus, Schrad. apud J. F. Gmel. 1792: 1435 & Schrad. 1794:

172 (Germany) (d.n.), not ~ Schaeff. 1774 (d.n.) per Krombh. 1821; *Polyporus* (Schrad.) per Fr. 1821; *Physisporus* Gillet 1877; *Poria* Cooke 1886, Quél. 1886; *Merulius* Sacc. & Lind. 1897 ["(Pers.) Karst."]. —Nomen dubium fide Lundell 1941 (LNF 21–22): 3 No. 1004, in obs., and Donk 1967 (Pe 5): 108. — Sensu Fr. 1874, Romell = *Merulius taxicola* (**O**).

rugosa, Gemmularia*, Rafin. 1819 (JPC 89): 106, 1820 (JBD 1): 243 (nom. anam.) (d.n.) (U.S.A.) per Steud. 1824; *Tucahus* Rafin. 1830, G. F. Web. 1929 (*"Tuckhaus"*; syn.: n.v.p.).—This is now held to be the same as *Pachyma cocos* (O**), see G. F. Web. 1929 (M 21): 122 and Donk 1962 (Ta 11): 101.

**rugosa, Trametes*, B. & Br. 1873 (JLS 14): 55 (Ceylon).—Nomen dubium. Lloyd apud Petch 1916 (APe 16): 109 referred the portion of the type collection in K to *Trametes [Truncospora] ochroleuca*; Petch, l.c., thought that the portion at Peradeniya "appears to be nearer *Hexagona durissima* [B. & Br.; extra-European], but they are scarcely identifiable." Bres. 1916 (Am 14): 240 wrote about *T. rugosa*: *"Trameti Moritzianae* Lév. valde affinis et tantum colore saturatiori distincta. Sterilis inventa, at mihi species valde dubia."

Rugosaria* Rafin. → *Gemmularia* Rafin. (O**).

rugulosus, Polyporus, Lasch in Rab. 1859 F.e.: No. 16, not ~ Lév. 1844. —Wint. 1884 (RKF 1): 416 referred it to *Polyporus [Coriolus] velutinus*. I saw three copies of the type distribution with strongly insect-damaged material which I would refer to *Coriolus zonatus*. The distributed material may be mixed, cf. "Ad trunc. *Carpini* et *Betulae* . . .."

rupiforme, Agarico-suber, Paul. 1793 T. 2: 76 [*pl. 2 f. 1*] (descr.), Ind. (France) (d.n.).—Nomen dubium. Paul. 1812–35: *pl. 2 f. 1* later called this a variety of *Agaricus [Daedalea] quercinus*. The figure suggests *Lenzites betulina* but there is scarcely any indication of zonation of the surface of the cap and Paul. explicitly stated, "la surface supérieure n'est point zonée ou à bandes".

rutilenis, Dendrosarcos, Paul. → *Agarico-carnis malum* Paul. (**O**).

sabinae, Polyporus, (Rox. Clem.) Colm. 1867: 23 (nom. nud.: n.v.p.); *Boletus* Rox. Clem. 1864: 64 (Spain) (nom. nud.: n.v.p.).—Nomen dubium.

saccharinus, Polyporus, B. & C. in herb. (type locality?).—Listed by Cooke 1885 (G 13): 85 as a syn. of *Polyporus [Tyromyces] lacteus*; this means next to nothing.

**saharanpurensis, Polyporus*, P. Henn. 1901 (H 40): 325 (India).—Lloyd 1912 (LMW 3, S.P.): 162 remarked, *"Sahranpurennis* [!] . . . from the description seems to be *[Phaeolus] Schweinitzii"*.

**saitoi, Polyporus*, Lloyd 1924 (LMW 7): 1269 *pl. 282 f. 2770* (Japan), not ~ Lloyd 1925 (n.v.p.).—Lloyd considered this "close to *Polyporus brumalis"*; S. Ito 1955: 321 listed it with doubt as a syn. of *P. picipes* [= *P. badius*].

salebrosus, Polyporus, Lasch in Rab. 1852 Kl.: No. 1606 (Germany), not ~ Lloyd 1912.—Nomen dubium. Pilát 1942 (ACE 3): 557, 558 listed this name ("teste Rechinger") as a syn. of *Inonotus radiatus* var. *nodulosus* [= *I. nodolosus*] and Bond. 1953: 327 transferred it as a syn. (still with the remark "teste Rechinger") to *Inonotus polymorphus* [sensu Bourd. & L. Maire] (**57**). I have been unable to make a determination on the basis of the poor material of two copies of the type-distribution (L); no tramal setae could be located so that I cannot accept the name *P. salebrosus* as basionym for the correct name of *Inonotus polymorphus* sensu Bourd. & L. Maire.

salicinus fumidus, Polyporus, Secr. 1833 M. 3: 107 (as a sp. of *Boletus* & double epithet & nom. nud.: n.v.p.) ≡ *Boletus salicinus* Bull. [= *Bjerkandera fumosa*].

salleana, Poria, (Berk.) ex Cooke 1886 (G 15): 25 (Mexico); *Polyporus* Berk. in herb.—Lowe 1966: 87 writes, "similar to or the same as *Polyporus [Gloeoporus] dichrous*". – Descr.: Lowe 1963 (M 55): 468. – Cf. (**51**).

saloisensis, Poria, (P. Karst.) Bourd. & G. 1925 (syn.: n.v.p.); *Physisporus aurantiacus* var. P. Karst. 1887 (Rm 9): 10 ("*taloisensis*"), 1887 (Mfe 14): 81 (Finland).—Fide Bres. 1903 (Am 1): 77 = *Poria nitida* [sensu Bres. 1903], fide Lowe 1961 (PMi 46): 206 = *Poria salmonicolor*.

salpincta, Polyporus, Cooke 1880 (New Zealand); *Polystictus* Cooke 1886; *Microporus* O.K. 1898; *Coltricia* G. Cunn. 1948.—Lloyd 1912 (LMW 3, S.P.): 160 remarked about this, "probably an abnormal [*Coltricia*] *oblectans*"; this explains why Pilát 1942 (ACE 3): 580, 582 listed it as a syn. of *Polystictus [Coltricia] perennis* f. *cinnamomeus* (Jacq. per S.F. Gray) Pilát. Treated as a distinct sp. by G. Cunn. – Descr.: G. Cunn. 1965: 194 *pl. 4 f. d* (*Coltricia*).

sanguineus, Boletus, Fr. 1821: 372 (incidental mention: n.v.p.), not ~ L. 1759 (d.n.), not ~ With. 1792 (d.n.) per Purt. 1821 (Boletales); not ~ Pers. 1801 ex Secr. 1833 (Boletales); [≡ *Boletus sanguineus* L. sensu Vill. 1789: 1041, excl. of type, (France)]; = *Fistulina hepatica*.

sanguineus, Boletus, L. 1759: 1350 (Surinam) (d.n.) (**112**), not ~ With. 1792 (d.n.) per Purt. 1821 (Boletales), not ~ Pers. 1801 ex Secr. 1833 (Boletales); *Agarico-suber* Paul. 1793 (d.n.); *Xylometron* Paul. 1812–35 (generic name n.v.p.); *Polyporus* G. Meyer 1818 (d.n.); *Polyporus* (L.) per Fr. 1821; *Boletus* Hook. 1822, not ~ Fr. 1821 (n.v.p.), not ~ With. per Purt. 1821, not ~ Pers. per Secr. 1833 (Boletales), not ~ With. 1792 (d.n.) per Fr. 1835 (Boletales); *Polystictus* Fr. 1851; *Trametes* Mart. 1853, Imaz. 1943, not ~ (Kl.) Pat. 1897, not ~ Lloyd 1924; *Microporus* Pat. 1897 (generic name n.v.p.), O.K. 1898; *Pycnoporus* Murrill 1904; *Petaloides* Torrend 1924; *Coriolus* G. Cunn. 1948 (nom. prov.: n.v.p.), 1949, 1950; ≡ *Agaricus ruber* Lam. 1783 (d.n.), not ~ Scop. 1772 (d.n.), not ~ Schaeff. 1774 (d.n.) per Krombh. 1821, not ~ Pers. 1801; *Boletus* Fr. 1821 (syn.: n.v.p.).

—This has been reduced to *Trametes* [*Pycnoporus*] *cinnabarina* by various authors, usually as a form or variety. It is now more often considered distinct from *Pycnoporus cinnabarina*; cf. Nobles & Frew 1962 (CJB 40): 1010 *fs. 78–98, pl. 2 fs. 25–27* (*Pycnoporus*). It is as yet not certain that it occurs in Europe. − Synonyms of this sp. are not mentioned in this list. − Sensu Plan., Vill. = *Fistulina hepatica*; sensu Liljebl. → *Fomitopsis pinicola*.

"*Sarrafini*", *Polyporus*.—An error in Lind. & Syd. 1917 T. 5: 190 for *P.* 'sarazinii' S. Schulz.

saxicola = *taxicola*

**scaber, Fomes* (Berk.) Lloyd 1915; *Polyporus igniarius* var. Berk. 1839 (AM 3): 324 (Van Diemen's Land = Australia, Tasmania); ≡ *Polyporus rimosus* Berk. 1845; *Fomes* Cooke 1885; *Scindalma* O.K. 1898; *Xanthochrous* Pat. 1900; *Phellinus* Pilát 1940, misapplied, Bond. & S. 1941, G. Cunn. 1965; *Pyropolyporus* Teng 1964.—The sp. from Europe and North America that was identified with *Polyporus rimosus* is a different fungus, → *Phellinus robinae* (88). In the sense of its type it is extra-European; cf. Lowe 1957 F.: 22. No taxonomic syns. included in this list.

Scalaria Lázaro 1916 [1960 (Pe 1): 278]; monotype, *Scalaria fusca* Lázaro, a nomen dubium. It is just possible that part of the sp. (on almonds, not represented by the theoretical type) is *Phellinus pomaceus*. Cf. Trott. 1925 (SF 23): 397, "Hoc nov. gen. lazaroanum *Scalaria* . . . abnormitas tantum sistit."

scalaris, Boletus, Schrank 1789: 617 (Germany) (d.n.).—Nomen dubium.

schini, Inonotus*, J. G. Brown 1922 (Sci II 55): 547 (U.S.A., Arizona); *Polyporus* Pilát 1942 ("Overholts"; syn.: n.v.p.).—Pegl. 1964 (TBS 47): 185 referred this to *Inonotus cuticularis* as a thick form with rare setae. Previously Long 1945 (Ll 8): 235 had listed it as a syn. of *Polyporus* [*Inonotus*] *farlowii* (extra-European) and Overh. 1953: 425, as a syn. of *Polyporus* [*Inonotus*] *munzii* (O**), a later name than *P. farlowii*. − Descr.: J. G. Brown *1930*. − V.s.: "*schinus*".

schinus = *schini*.

schönbrunnensis, Polyporus, Kalchbr. 1868 in herb. (Austria). − Fide Fr. apud Kalchbr. 1868 (VW 18): 431 = *Polyporus cyphelloides* Fr. apud Kalchbr. [= *Flaviporus brownei*] "nimis affinis".

schulzeri, Daedalea, Poetsch 1879 (ÖbZ 29): 289 (Austria); *Striglia* O.K. 1891.—Nomen dubium. Collections referred here to were included by Bres. apud Sacc. 1896 (Mal 10): 262 and Bres. 1897 (AAR III 3): 90 in *Trametes populinus* Bres. apud Sacc. [= "*Trametes*" *cervina*]. − Sensu Velen. 1922: 689 = *Gloeoporus* [*Bjerkandera*] *fumosus*, fide Pilát 1937 (ACE 3): 162.

schulzmorlinii, Sistotrema, Spreng. 1806: 377 (Germany) (d.n.) per Steud. 1824.—Nomen dubium.

"*schwartzii* J[oachim]", *Polyporus*, Badet 1934: 145 = (error for) *Polyporus* [*Phaeolus*] *schweinitzii*.

sciurinus, *Polyporus*, Kalchbr. apud Thüm. 1882 (BSM 56^2): 117 (U.S.S.R., Russia, Siberia); *Bjerkandera* P. Karst. 1882; *Polystictus* Cooke 1886; *Microporus* O.K. 1898.—Fide Bres. 1890 (BmF 6): xlvi = *Trametes hispida* [sensu Bres.]. – "... substantia pilei pallide ligneis ... fere alba vetat ...." *Funalia trogii*?

scopolii, *Polyporus*, Pers. 1825 → *Poria membranacea* Scop. (**O**).

scoriatus, *Polyporus*, Pers. in herb.—Fide Lloyd 1910 (LMW 3): 469 = *Fomes evonymi* [= *Phellinus ribis*].

scrobiculatus, *Polyporus*, (P. Karst.) Lloyd 1915 (LMW 4, Ap.): 385 (incidental mention: n.v.p.); *Inonotus radiatus* var. P. Karst. 1882 (BFi 37): 238, 1882 (Mfe 9): 50 (Finland); *Polystictus* Pilát 1942 (syn.: n.v.p.)—Fide Lowe 1956 (M 48): 107 = *Polyporus* [*Inonotus*] *radiatus*.

scutellatus, *Polyporus*, Schw. 1832: 157 (U.S.A., Pennsylvania), not ∼ I. Boršč. 1856; *Fomes* Cooke 1885; *Trametes* Morg. 1886, misapplied, G. Cunn. 1965; *Scindalma* O.K. 1898; *Ungulina* Pilát 1934; *Fomitopsis* Bond. & S. 1941, G. Cunn. 1948.—Reported for Portugal by Bres. 1902 (AAR III 8): 129 and Torrend 1902 (Bro 1): 133; later Torrend 1913 (Bro 11): 67, 68 referred his material to *Trametes* [*Truncospora*] *ohiensis* (**O**). – Apparently not yet reported for Europe where it may be expected perhaps in the north-east of the continent. – Descr.: Overh. 1953: 51 *pl. 61 fs. 364, 365, pl. 126 fig.*; Lowe 1957 F.: 78 *f. 59* (*Fomes*).

sebaceus, *Boletus*, Leyss. 1783: 300 (Germany) (d.n.); *Polyporus* (Leyss.) per Lloyd 1922, presumably misapplied, not ∼ Fr. 1838.—Nomen dubium. The specific phrase runs: "sebaceus acaulis vndulatus late expansus ex albido spadiceus." Cited syn.: "*Agaricum sebaceum vndulatum ex albo spadiceus.* Hall. helv. 2250." Pers. 1801: 543 regarded *B. sebaceus* Leyss. as a var. of *Boletus* [*Tyromyces*] *destructor* (**128**). – Sensu Lloyd 1922 (LMW 7): 1163 (*Polyporus*) = ?

sebaceus, *Polyporus*, Fr. 1838, not ∼ (Leyss.) per Lloyd 1922; = *Poria reticulata sebacea* Secr. 1833 M. 3: 175 (Switzerland) (as a sp. of *Boletus* & double epithet: n.v.p.).—Nomen dubium. Fr. 1874: 580 thought that this was the "primordia" of myxomycetes; this seems incorrect.

sebaceus abietis, *Polyporus*, Secr. 1833 M. 3: 170 (Switzerland) (as a sp. of *Boletus* & double epithet: n.v.p.)—Nomen dubium.

secernibilis, *Polyporus*, Berk. 1847 (LJB 6): 500 (Ceylon).—Sensu Lloyd 1913 (LMW 4, L. 45): 4 & 1915 (LMW 4, Ap.): 329 = "an Eastern form" of *Polyporus* [*Bjerkandera*] *adustus*; but cf. Petch 1916 (APe 6): 94, 119. – Referred to *Gloeoporus crispus* by G. Cunn. 1965: 113, 114, whose interpretation of *Boletus crispus* Pers. [= *Bjerkandera adusta*] is not altogether clear to me.

secretanii, *Scindalma*, O.K. → *Boletus variegatus* Sow. (**O**).

semicircularis, *Boletus*, With. 1776 (d.n.); [= *Fungus arboreus porosus minor absque pediculo, semicircularis* Ray 1696: 336 & 1724: 24 (England)].—Nomen dubium.

semipetiolatus, Boletus, Gaterau 1789: 191 (France) (d.n.).—Nomen dubium. Descr.: "Chapeau en demi-cercle, bai fauve ou lisse, pores petits, d'un jaune blanchâtre; pédicule court et mammelonné. / Sur les troncs de Chêne."

semisanguineus, Polystictus,* Lloyd 1912 (LMW 4, L. 39): 8 (Singapore).—Reduced to *Trametes [Pycnoporus] cinnabarina* sensu lato by G. Cunn. 1965: 169, but cf. (112**). Nobles & Frew 1962 (CJB 40): 1006 referred it, with doubt, to *Pycnoporus coccineus* (extra-European) (**O**).

senescens, Agaricus, Willd. in herb.—Fide Fr. 1830 (Li 5): 513 and 1838: 407 = *Daedalea/Lenzites [Gloeophyllum] abietina.* Fries thought the name had been published by Willd. ("Ber. p. 376"), but Willd. did not publish it. On the page cited *A. antiquus* is described, its descr. points to *Gloeophyllum sepiarium;* the syn. cited by Willd. represents *Daedalea quercina.*

separabilis, Polyporus, Velen. "in herb." (Czechoslovakia).—Listed by Pilát 1937 (ACE 3): 176 as a syn. of *Leptoporus [Tyromyces] fragilis,* but according to Kotl. & P. 1964 (ČM 18): 267, 215 = *Tyromyces gloeocystidiatus [= T. leucomallelus].*

septicus, Polyporus, Anon. 1876 (BbF 23): 336 (nom. nud.) ≡ (presumably an error for) *Polyporus [Tyromyces] stipticus.*

serpens, Daedalea, (Tode per Fr.) E. Krause 1928 (*"Daedaleus"*); *Merulius* Tode 1783 (AhG 1): 355 (Germany) per Fr. 1821; *Xylomyzon* Pers. 1825; *Serpula* P. Karst. 1889; *Byssomerulius* Parm. 1967.—Variously interpreted; the currently accepted interpretation, as *Merulius serpens,* is that of Romell 1911 ('Corticiaceae').

Serpula (Pers.) per S. F. Gray 1821, P. Karst. 1884; *Merulius* sect. Pers. 1801: 496 (d.n.); ≡ *Xylophagus* Link 1809 per Murrill 1903 [1958 (Fu 28): 14]; ≡ *Xylomyzon* Pers. 1818 (nom. nud.) ex Pers. 1825 [1958 (Fu 28): 14]; lectotype, *Merulius destruens* Pers. = *Serpula lacrimans* (**O**).—For a long time this genus (Coniophoraceae) was included in the Polyporaceae.

serpula, Bjerkandera, P. Karst. 1887 (Mfe 14): 79 (Finland); *Polyporus* Sacc. 1888.—Fide Lowe 1956 (M 48): 103, cf. *Coriolus* sp., indet.

sesia, Daedalea, (Scop.) per Pers. 1828; *Merulius* Scop. 1772 P.s.: 101 *pl. 18 f. 1* (Hungary, now Czechoslovakia) (d.n.).—Nomen dubium. Found in a mine: apparently an abnormal growth. The specific epithet is explained by Scop. thus: "*Agaricus de St. Clou.* Vaillant. Paris. p. 3. *Tab. 1 Fig. 1. 2. 3.* huic proximus, quem Cl. Adansonius *Sesiam* vocavit." This *Agaricus de St. Clou* has been generally identified with *Gloeophyllum sepiaria:* it is certainly different from *Merulius sesia,* the diagnosis of which contains as sole colour indication "albus".

sessilis, Polyporus, Graff 1936 (M 28): 155, 159 (syn.: n.v.p.); [= *Polyporus sessilis, convexo planus, poris amplissimis albidis* Haller 1768 H. 3: 140 no. 2278 (Switzerland)].—Haller cited Schaeff. *pls. 101, 102, Boletus juglandis* Schaeff. [= *Polyporus squamosus*] for his sp. and his descr. suggest the same sp. Graff cited '*Polyporus sessilis*' as a syn. of *P. squamosus.*

setulosus, Fomes, Lloyd 1915 (LMW 4, F.) 243; *Phellinus* Imaz. 1943; = *Fomes robustus* var. *setelatus* Lloyd 1912 (LMW 4, L. 42): 11 (Ceylon). —Some authors erroneously referred this to *Phellinus robustus.* Often misapplied. – Descr.: Lowe 1957 F.: 60 *f. 44* (*Fomes*).

setulosus, Polyporus, Killerm. 1940 (DrG 21): 71 *tpl. 31 f. 34* (Germany) (lacking Latin descr.: n.v.p.).—Insufficiently described. Apparently belonging to the *Polyporus brumalis* group.

shiraianus, Polyporus, P. Henn. 1900 (BJ 28): 269 (Japan).—Bres. identified this with *Favolus europaeus* [= *Polyporus mori*], but it may appear to belong to *Favolus* [*Polyporus*] *canadensis* Kl. (**0**) if the two should appear to be different species. – Cf. (**98**).

shiraianus, Polyporus, P. Henn. in herb. (Japan), not ∼ P. Henn. 1900 (BJ 28): 269 (**0**).—Fide Murrill 1907 (NAF 9): 70 = *Pycnoporellus fibrillosus* [= *P. fulgens*]; but cf. Lloyd 1922 (LMW 7): 1108, in obs. Murrill thought he had studied the type of the published homonym, rather than that of the herbarium name. The former is *Favolus europaeus* [= *Polyporus mori*], fide Bres. 1916 (Am 14): 227, or a closely related sp.

sibiricus, Lenzites, P. Karst. 1904 (ÖfF 46^{11}): 3 (U.S.S.R., Baikal region). —According to Lowe 1956 (M 48): 109 a poorly developed, reddish form of *Daedalea* [*Daedaleopsis*] *confragosa.* – Cf. (**31**).

sibiricus, Polyporus, [Murašk.] in herb. (U.S.S.R., Russia, Siberia). —Fide Pilát 1934 (BmF 49): 265 = *Trametes* [*Dichomitus*] *squalens.*

simulans, Polyporus, B. & C. ex Sacc. 1888 (SF 6): 117 ["Cuba", but cf. Lloyd 1915 (LMW 4, Ap.): 386, probably from Ceylon)], not ∼ Błoński 1889, not ∼ Wakef. ex Lloyd 1924; *Polyporus* B. & C. in herb., Cooke 1885 (nom. nud.: n.v.p.); *Trametes* Killerm. 1928 ("Bres."); *Polystictus* Pilát 1937 (syn.: n.v.p.; "*similans*"); = *Bjerkandera subsimulans* Murrill 1907.—Recorded by Maire, Dumée, & Lutz 1903 (BbF 48): ccxxviii from Corsica as "*Phaeolus simulans* (Berk. et Curt.) Pat.", certainly in error. It may be that these authors had *Phaeolus nidulans* ("Pers." [= Fr.]) Pat. [= *Hapalopilus rutilans*] in mind rather than *Bjerkandera simulans* P. Karst. [= *Tyromyces lacteus?*]. The identity of *Polyporus simulans* B. & C. is still a puzzle. Lloyd, l.c., referred it to *Polyporus* [*Bjerkandera*] *fumosus* and Pilát 1937 (ACE 3): 158 listed it as a syn. of this species. Bres. 1926: 27 rejected this identification and stated that it was a species of "*Trametes*" (with descr.).

sinensis, Lenzites, Cooke 1889 (G 17): 75 (China); *Cellularia* O.K. 1898. —Referred to *Lenzites* [*Daedaleopsis*] *tricolor* by Bres. 1916 (Am 14): 221 ("*chinensis*").

sinuatum, Gymnoderma, Humb. 1793: 109 (nom. anam.?) (d.n.) per Steud. 1824.—The genus *Gymnoderma* (**0**) was introduced to accommodate this fungus; it has been listed as belonging to the 'Thelephoraceae', but Donk 1957 (Ta 6): 72 suggested that this might be an abnormal and/ or sterile growth form of some polypore; it was found growing in a mine.

sinuosa, Trametes, "Bres.", an error of Lind 1913: 393 (as syn.) for *T. subsinuosa* Bres. [= *Antrodia ramentacea*].

skeptonymus, Agaricus, E. Krause 1932 B.r.: 142; [≡ *Polyporus vegetus* Fr. sensu E. Krause 1931 B.r.: 122 (Germany)].—Apparently a sp. of *Ganoderma,* but the descr. is too incomplete to guess which one. – The following combines all descriptive matter furnished by the author: Mit dickem Wesen; mit harziger Kruste; heurige Zuwachszone weiss, bald dunkelnd; Röhren kurz, erst weiss, aber bald dunkelbraun; Sporen rostfarben, 9–10 × 8–9 μ. Buche. Cf. Donk 1969 (PNA 72): 281.

sorbicola, Polyporus, Fr. 1874: 570 (U.S.S.R., European Russia); *Poria* P. Karst. 1882.—Fide Bres. 1897 (AAR III 3): 80 = *Poria* [*Merulius*] *taxicola* (**O**). The descr. agrees with this identification; apparently the substratum was incorrectly identified.

sordentulus = *sordulentus*

**sordulentus, Polyporus,* Mont. 1852: 357 (Chile).—Referred to *Polyporus* [*Laetiporus*] *sulphureus* by Lloyd 1912 (LMW 3, S.P.): 154 (forma), but later he seems to have accepted the name [Lloyd 1918 (LMW 5, L. 67): 4, as "*sordentulus*"] when naming specimens (no author, no notes or descr.). Sing. 1962: 300 & 1969 M.a.: 382 transferred the species to *Grifola,* which indicates that it does not belong to *Laetiporus.*

**sororius, Polyporus,* (P. Karst.) Sacc. & Trott. 1912; *Polyporellus caudicinus* subsp. *P. sororius* P. Karst. 1906 (Ttk 8[1]): 61 [repr. 1912 (SF 21): 267] (U.S.S.R., Russia, 'Transbaikal').—Nomen dubium. Undoubtly belonging to the group of *Polyporus squamosus* (which species P. Karst. at that time called *Polyporellus caudicinus*) but it is difficult to make a more precise suggestion from the descr.

sowerbei. – *Daedalea biennis* "*β. D. sowerbei*" Fr. 1821: 332 (published as an infraspecific taxon: n.v.p.); [≡ *Boletus biennis* Bull. sensu Sow. 1799: *pl. 191* (England)]. – This is *Abortiporus biennis.* According to Pilát 1937 (ACE 3): 116 a specimen cited by him as "*Daedalea Sowerbii* ex herb Berkeley in h. Corda, h. N.M.P.*" is also *Heteroporus* [*Abortiporus*] *biennis.*

spadiceus, Polyporus, Fr. "in litt." apud Kalchbr. 1868 (Mtk 5): 263 *pl. 4 f. 1* [repr. 1869 (H 8): 117] (Hungary, now Czechoslovakia), not ∼ Jungh. 1838, not ∼ Berk. 1839, not ∼ Bres. 1915.—Nomen dubium. A resupinate brown-coloured species of *Poria* sensu lato; it was dropped from literature apparently because Fr. did not mention it in his own publications.

spathulatum, Hydnum, Schrad. 1794: 178 *pl. 4 f. 3* (Germany) (d.n.); *Sistotrema* Pers. 1801 (d.n.); *Hydnum* Schrad. per Fr. 1821, not ∼ Schw. 1822; *Sistotrema* Pers. 1825; *Irpex* Fr. 1828; *Xylodon* P. Karst. 1882; *Radulum* Bres. 1903; *Odontia* Litsch. 1939, not ∼ (Schw.: Fr.)

Rea 1922; *Hyphodontia* Parm. 1968.—In my opinion a nomen dubium. The name was revived by Bres. 1903 (Am 1): 89 (as *Radulum*) but was too briefly annotated to determine what he was naming. The now accepted conception was published by L. W. Mill. 1934 (M 26): 25 *pl. 2 f. 3* ("*Radulum?*") who also stated that it was considered to be a form of *Odontia arguta* (Fr.) P. Karst. by Miss E. M. Wakefield. Pilát 1941 (ACE 3): 273, 276 referred the sp. to *Trametes* [*Hirschioporus*] *abietinus* var. "*Xylodon candidum*" [= *Sistotrema candidum* = *Hirschioporus fuscoviolaceus*].

spathulatum.—"*Sistotrema spathulatum* Schw."—According to Overh. 1953: 331 the "types" of this name are *Polyporus tulipiferae* [= *Irpex lacteus*]. It is not clear which name Overh. had in mind for there is no *Sistotrema spathulatum* "Schw." There exist (i) a *Hydnum spathulatum* Schw. 1822: 104: Fr. 1828 E. 1: 139, which Fr. 1838: 517 listed as "*Systotr.* Schwein! Car. n. 993" as (syn.), evidently by error; and (ii) a *Hydnum spatulatum* [!] Schrad. 1794 (d.n.) per Fr. 1821, which was reported from South Carolina as *Sistotrema spathulatum* (Schrad.) Pers. 1801: 553 by Schw. 1822: 102 without descr.

spathulatus, *Polyporus*, Fr. ("in litt.") apud Weinm. 1826 (SPR 2): 101 (nom. nud.) ≡ *Polyporus* [*Climacocystis*] *borealis* var. *spathulatus* (Fr.) ex Fr. 1828 E. 1: 85 (U.S.S.R., European Russia).

spiraeformis, *Polyporus*, Secr. 1833 M. 3: 103 (Switzerland) (as a sp. of *Boletus* & nom. nud.: n.v.p.).—Nomen dubium.

splendens, *Polyporus*, E. Krause 1931 B.r.: 121 (syn.: n.v.p.).—Presumably an error for *P.* [*Spongipellis*] '*spumeus*'.

Spongioides Lázaro 1916 [1960 (Pe 1): 283]; monotype, *Boletus cryptarum* Bull. [sensu Lázaro].—It looks as though at least a good portion of Lázaro's descr. of the type species is derived from *Poria expansa* growing in the dark, especially in mines and the generic name should be kept in mind in case *P. expansa* is placed in a genus of its own **(107)**.

spongiosum, *Sistotrema*, Schw. 1822: 101 (U.S.A., North Carolina); *Hydnum* Spreng. 1827; ≡ *Polyporus labyrinthicus* Fr. 1828: E. 1: 83; *Spongipellis* Pat. 1900.—A somewhat doubtful species not mentioned by Overh. 1953. It was referred together with *Polyporus obtusus* (**O**) to *Spongipellis unicolor* (Schw.) Murrill 1907 (NAF 9): 37; if this is in order then the correct name for the species is *Spongipellis labyrinthicus* (Fr.) Pat. — Mentioned in this list because *Polyporus labyrinthicus* was recorded for Europe by Weinm. 1836: 313; his fungus was renamed *Polyporus weinmannii* Fr. [= *Tyromyces fragilis*].

spongiosus, *Boletus*, Pers. → *B. resupinatus* Bolt. (**O**).

spuma, *Poria*, Hoffm. 1797–1811 V.s.: 21 *pl. 13 f. 2* (Germany) (d.n.). —Nomen dubium. Found in a mine: apparently an abnormal growth.

squalens, *Poria*, Lloyd 1908 (LMW 2, L. 24): 2 ("Russia"), not ∼ (P. Karst.) Lowe 1956.—Nomen dubium (subnudum).

squamatus, Polyporus, Lloyd 1911 (LMW 3, O.): 84 *f. 505* (Hungary), not ∼ Schwalb 1891; *Polyporellus* Pilát 1936, in part.—The type no longer exists. However, I have little doubt that the species was based on *Strobilomyces floccopus* (Vahl per Fr.) P. Karst. (Boletales). – Pilát 1934 (ČčH 14): 103 and 1935 (ČčH 15): 54 ("Kalchbr.") referred this taxon as a form to *Polyporellus* [*Polyporus*] *squamosus*. Soon afterwards Pilát 1936 (ACE 3): 94 accepted Lloyd's sp. and referred *Polyporus boucheanus* sensu Velen. 1922: 665 to it. Fide Pouz. 1966 (Fgp 1): 360 Velenovský's material represents poorly dried fruitbodies of a sp. of *Albatrellus,* viz. *A. ovinus* or *A. similis.*

squamosum, Agaricum, Graff 1936 (M 28): 154, 159 (syn.: n.v.p.); [≡ *Agaricum squamosum Ilicibus . . . innascens . . .* Mich. 1729: 118 (Italy)]. —Listed by Graff as a syn. of *Polyporus squamosus.* This is not acceptable; it has little in common with this sp. – Micheli used 'squamosus' for 'imbricate' in connection with caps.

**squamulosus, Polyporus,* Bres. 1890 (BmF 6): xxxix *pl. 5 f. 1* (San Thomé), not ∼ P. Henn. 1895.—Lloyd 1915 (LMW 4, Ap.): 386 referred this to *Polyporus* [*Tyromyces*] *tephroleucus;* he thought that Bres. had described the spores wrong.

squarrosus, Fomes,* Lloyd 1914 (LMW 4, L. 53): 10 and 1915 (LMW 4, F.): 247 *f. 590* (Australia, Victoria).—This has been taken as a form of *Fomes* [*Phellinus*] *robustus* by J. Rick 1960 (Ih 7): 208; G. Cunn. 1965: 227 included it as a mere syn. in that sp. This disposition is very doubtful. Lloyd 1916 (LMW 5, L. 62): 5 at first referred *F. squarrosus* to *Fomes robinsoniae* Murrill, which according to Lowe 1957 F.: 59 = *Fomes* [*Phellinus*] *zealandicus* (Cooke) Cooke [= *Phellinus laurencii* (Berk.) Aosh.] (extra-European); later Lloyd 1918 (LMW 5, L. 67): 14 made it a syn. of *Fomes* [*Phellinus*] *setulosus* Lloyd (extra-European) (O**).

stalachte & stalachtites = stalactites

stalactites, Poria, Hoffm. 1797–1811 V.s.: 11 *pl. 7* (Germany) (d.n.); Fr. 1828 E. 1: 105 ("*stalachte*") & 1832[Ind.]: 150 ("*stalachtites*") (syn.); *Polyporus* (Hoffm.) per Pers. 1825.—Nomen dubium. Found in mines. Fr. 1828 E. 1: 105 referred this to *Polyporus* [*Fomitopsis*] *roseus* as a monstrosity, but I cannot follow him in this.

starbaeckii, Poria, P. Karst. in herb.—Listed by Egeland 1912 (NMN 49): 370 as a syn. of *Poria* [*Ceriporia?*] *rhodella* Fr. (**22**). Referred by Bourd. & G. 1928: 669 to *Polyporus* [*Poria*] *resinaceus.* Cf. also Lowe 1956 (M 48): 121.

stereoides, Polyporus, Fr. 1818 O. 2: 258 (Sweden) (d.n.) per Fr. 1821; *Bjerkandera* P. Karst. 1882; *Polystictus* Cooke 1886, not ∼ (Berk.) ex Sacc. 1888, not ∼ Lloyd 1918 (n.v.p.); *Coriolus* Quél. 1886; *Trametes* Bres. 1897; *Microporus* O.K. 1898; *Antrodia* Bond. & S. 1941, Bond. 1953. *Datronia* Ryvarden 1967 (incomplete ref.: n.v.p.; "*steroides*"), 1968.—Nomen dubium, fide Donk 1966 (Pe 4): 338, 1971 (PNA 74): 5.

– Sensu Fr. 1884 → *Datronia mollis*; sensu Romell, Ryv. → *Datronia epilobii*.

stictis.—"*Poria stictis* Fries": Opiz 1856 (Lo 6): 211.—Apparently an error.

stillativus, *Polyporus*, Britz. 1893 (BCb 54): 103 [*pl. 628 f. 126*] (Germany). —Nomen dubium. Lloyd 1915 (LMW 4, Ap.): 386 thought of *Polyporus fulvus* [= *Ischnoderma benzoinum*] and Killerm. 1922 (Dba 15): 73 of *Polyporus squamosus*. To me both identifications are unacceptable. It is possible that the spores (spindle-shaped, 10–12 × 2–3 μ) belong to a contamination growing on a rotting fruitbody.

stipitatus.—"[*Boletus*] *stipitatus* Sow.", Fr. 1832[Ind.]: 63 (error, syn.: n.v.p.) ≡ *Boletus spumeus* Sow. *pl. 211*, the plate depicting a stipitate fruitbody, cf. Fr. 1821: 358.

stratosus, *Agaricus*, Fr. 1821 → *Polyporus gelsorum* Fr. (**O**).

striatulus, *Favolus*, Ell. & Ev. 1897 (AN 31): 339 (U.S.A., Delaware); *Hexagonia* Murrill 1907.—This has been referred to *Favolus* [*Polyporus*] *canadensis* (**O**) by Overh. 1914 (AMo 1): 148, 149, a taxon that in its turn has been included in *Polyporus mori*, but cf. (**98**).

striatus, *Agaricus*, Sw. 1788: 148 (Jamaica) (d.n.); *Daedalea striata* (Sw.) per Fr. 1821; *Lenzites* Fr. 1838; *Cellularia* O.K. 1898; *Sesia* Murrill 1904; *Gloeophyllum* Murrill 1905.—This sp. of *Gloeophyllum* was reduced to the synonymy of *G. trabeum* by G. Cunn. 1965: 253, an unacceptable identification. As a result the tropical *Lenzites striata* has been reported from Europe, for instance by Eusebio 1971 (PJS 98): 179. – Descr.: Overh. 1953: 113 *pl. 93 fs. 526, 527, pl. 94 fs. 539, 540, pl. 127 fig.* (*Lenzites*).

striatus, *Boletus*, Humb. 1793: 102 *pl. 4 f. 19* (Germany) (d.n.) per Steud. 1824, not ∼ Hook. 1822; *Polyporus* Pers. 1825.—Nomen dubium.

strigosus, *Boletus*, K. F. Schultz 1806: 488 (Germany, now Poland) (d.n.); *Polyporus* (K. F. Schultz) per Pers. 1825.—Nomen dubium. Perhaps misled by the substratum, Fr. 1821: 375 at first referred this to *Polyporus* [*Phellinus*] *igniarius*, in which he also included *Phellinus pomaceus*. However, Pers. 1825: 57 pointed out that this identification was untenable, and Fr. 1874: 559 changed his mind and wrote "forte *P. vulpinus*". This too is not readily acceptable because the fruitbody was stated to be thin. I would have suggested *Coriolus hirsutus* had Schultz not described the underside as 'cervinus, nec albus': he compared his sp. with *C. versicolor*.

strobiliformis, *Polyporus*, (Dicks.) per Steud. 1824, Loud. 1829, or (Dicks. per Pollini) Steud.; *Boletus* Dicks. 1785 (d.n.) (England), not ∼ Vill. 1789 (d.n. & syn.); = *Strobilomyces floccopus* (Vahl per Fr.) P. Karst. (Boletales).

Strongylium Ditm. apud Link 1809 (NJB 3[1, 2]): 16; Ditm. 1809 (NJB 3[3, 4]): 55; monotype, *Trichoderma fuliginoides* Pers. [sensu Ditm., Link].—The type appears to belong to *Reticularia lycoperdon* Bull., (Myxomycetes); cf. Donk 1972 (PNA 75): 170.

stuppea, Trametes, Berk. 1841 (AM 7): 453 (Canada, British Columbia);
Polystictus Cooke 1886; *Microporus* O.K. 1898; *Funalia* Murrill 1905.
—Murrill 1905 (BTC 32): 356 listed *Trametes peckii* (**O**) as a syn. − For
a picture of the type, see Lloyd 1910 (LMW 3, M.): *f. 355 (Polystictus)*.
− Cf. (36).

subbadia, Poria, Murrill 1921 (M 13): 93 (U.S.A., Alabama).—Fide
Lowe 1947 (Ll 10): 57 = *Poria [Ceriporia] viridans* (sterile).

subchartaceus, Coriolus, Murrill 1907 (NAF 9): 24 (U.S.A., Colorado);
Polystictus Sacc. & Trott. 1912; *Polyporus* Overh. 1915; *Hirschioporus*
Bond. & S. 1941.—Regarded as a syn. of *Polystictus [Hirschioporus]*
pargamenus by some authors, for instance, Lloyd 1922 (LMW 7): 1121.
In my opinion a distinct species, not reported for Europe.

subchioneus, Tyromyces, Murrill 1908 (BTC 35): 406 (Philippine Is.);
Polyporus Sacc. & Trott. 1912; *Fomes* Graff 1914.—Lloyd 1915 (LMW 4,
Ap.): 386 wrote, "I should refer it to *Polyporus [Tyromyces] caesius.*"

subcordatum, Gloeophyllum, Murrill "msn." (U.S.A., New York).—Fide
K. Fid. 1962 (Ri 1): 129 (syn.: n.v.p.) = *Osmoporus odoratus* subsp.
americanus (Overh.) K. Fid. [= *Gloeophyllum protractum*?].

subectypus, Coriolus, Murrill 1907 (NAF 9): 22 (U.S.A., Florida); *Polystic-*
tus Sacc. & Trott. 1912; *Polyporus* Overh. 1953.—Referred by Bres.
1926: 80 to "*Fomes*" [*Coriolus*] *velutinus*. ('Fomes' is due to a typo-
graphical error.) Overh. 1953: 371 (which see for descr.) treated the
sp. as distinct, calling it *Polyporus ectypus* (Murr.) "Bres."; 'Bres.' is
an error because at the place cited Bresadola referred the sp. to *Polyporus*
velutinus.

suberosa, *Trametes*, Quél. 1873 (MMb II 5): 356 ["(Sow.?) Q. n.s."]
(France).—Nomen dubium. Insufficiently described; probably either
Fomitopsis cytisina or *Oxyporus ulmarius*. 'Sow.' of Quélet's authors'
citation refers to *Boletus suberosus* L. sensu Sow. 1800: *pl. 288* =
Polyporus gibbosus Pers. [= *Fomitopsis cytisina*].

suberoso-tomentosum, *Agarico-igniarium*, Paul. 1793 T. 2: 93 (descr.),
Ind. (name) (d.n.) & Paul. 1812–35: *pl. 9 fs. 1–3* as *Pyreium fomen-*
tarium.—Nomen dubium. Referred to *Polyporus [Fomes] fomentarius*
by Lév. 1855: 6; *Boletus fomentarius* L. was one of the spp. that Paul.
cited as a syn.

suberosus, *Boletus*, L. 1753: 1176 (Sweden) (d.n.); *Agaricus* Lam. 1783
(d.n.); *Boletus* L. per Purt. 1821, not ∼ Bull. per St-Am. 1821; *Agaricus*
Boisd. 1828 (nom. nud. & error: n.v.p.), Dubois 1833; *Polyporus*
Wettst. 1885, not ∼ Fr. 1821, not ∼ (Bull. per St-Am.) Chev. 1826;
Piptoporus Murrill 1903; *Fomes* Big. & Guill. 1913 ("Sow."; syn.:
n.v.p.).—Nomen dubium. Fide Fr. 1821: 338 ("ob poros") & 1832[Ind.]:
63 (L. "suec. ed. 1") = *Daedalea [Trametes] gibbosa*; & fide Fr. 1821:
358 (sensu Wulf.) & 1832[Ind.]: 63 (L. "suec. ed. 2. Wulf."))→ *Polyporus*
[*Piptoporus*] *betulinus*. − Sensu Bolt. → *Trametes suaveolens*; sensu
Sow. → *Polyporus gibbosus* [= *Fomitopsis cytisina*]; sensu Wahl. →

Polyporus suberosus Fr. 1821, not ~ L., → *Boletus pulvinatus* Wahl.
= *Polyporus* [*Spongipellis*] *spumeus*, fide Fr. 1828 E. 1: 84.

*subferruginea, Lenzites, Berk. 1854 (HJB 6): 134 (India); *Cellularia*
O.K. 1898; *Gloeophyllum* Bond. & S. 1941, Bond. 1953 (incomplete
ref.: n.v.p.); *Daedalea* G. Cunn. 1950.—When Lloyd 1922 (LMW 7):
1165 referred *Lenzites cinnamomea* [= *L. betulina*] to *Lenzites* [*Gloeophyllum*] *subferruginea*, indirectly he recorded the latter species from
Sweden.

subfusco-flavidus, Polyporus, Rostk. 1848 (StP Fs. 27–28): 21 *pl. 11*
(Germany/Poland); *Physisporus* P. Karst. 1882 ("Chev."); *Poria*
Cooke 1886, Quél. 1886.—Nomen dubium. Variously interpreted.
Sensu auctt. nonn. in herb. → *Poria cinerascens* = *P. lindbladii*; and
cf. Bourd. & G. 1928: 691 for other interpretations, e.g. of Bres. 1897
(AAR III 3): 82 (*Poria*).

subganodermica, Ungularia, Lázaro 1916 (RMa 14): 674 / 1917: 113
(Spain); *Polyporus* Sacc. & Trott. apud Trott. 1925.—Nomen dubium.
Cf. Bres. apud Trott. 1925 (SF 23): 369, "potius *Fomes roburneus*
[(**48**)] habenda". *Ganoderma* sp.?

subintegra, Lenzites, S. Schulz. 1866: 43 (Yugoslavia, Slavonia) (nom.
nud.: n.v.p.).—Nomen dubium. The author cited as a syn. "*Agaricus*
[*Daedaleopsis*] *tricolor* Bull. Tab. 541?"

*sublaevigata, Poria, Clel. & Rodw. 1929 (PTa 1928): 39 (Australia,
New South Wales).—Referred by G. Cunn. 1965: 215 to *Fuscoporia*
[*Phellinus*] *punctata*. The original descr. (apparently abundant setae
present!) does not support this identification.

*subliberatus, Polyporus, B. & C. 1868 (JLS 10): 318 (Cuba); *Poria* Cooke
1886.—Referred by Murrill 1921 (M 13): 87 to *Poria* [*Rigidoporus*]
undata, but Lowe 1959 (Ll 21): 111 stated that the type is sterile and
afterwards, 1963 (M 55): 479, that it "appears to belong to the *Polyporus*
[*Rigidoporus*] *lignosus* Klotzsch complex."

sublobatus, Polyporus, Opiz 1852: 137 (nom. nud.: n.v.p.) (Czechoslovakia).
—Fide Pilát 1937 (ACE 3): 99 = *Polyporellus picipes* [= *Polyporus*
badius]. Opiz cited "*Boletus badius* P. ?" as a syn.

submembranaceus, Polyporus, Saut. 1876 (H 15): 153 (Austria).—Nomen
dubium.

submersus, Irpex, Killerm. 1940 (DrG 21): 67, 68 (Germany).—Nomen
dubium. "An Röhricht in 1 m Tiefe."

submurinus, Polyporellus, P. Karst. 1881 (nom. nud.: n.v.p.); *Polyporellus*
lepideus var. *submurinus* P. Karst. 1882 (BFi 37): 28 (Finland).—Cf.
Polyporus lepideus Fr. = *P. ciliatus* (forma).

*subodoratus, Lenzites, Murrill "msn" (U.S.A., New York).—Fide K. Fid.
1962 (Ri 1): 129 (syn.: n.v.p.) = *Osmoporus odoratus* subsp. *americanus*
(Overh.) K. Fid. [= *Gloeophyllum protractum*?].

subpalmatus, Polystictus, Sacc. → *Polyporus palmatus* Saut. (**O**).

subpapillosa, Boletus, Reb. 1804: 376 (Germany) (d.n.) per Steud. 1824.

—Nomen dubium. The description somehow suggests *Stromatoscypha fimbriatum* (**O**), but the subiculum is perhaps inconsistent with this guess: "Substantia ... coriaceo suberosa 2 lin. fere crassa."

*subradula, *Coriolus*, Pilát 1936 (BmF 51): 366 *pl. 7 f. 3*; *Polystictus* Killerm. 1943; = (by lectotypification) *Coriolus pallescens* f. *resupinatus* Pilát 1932 (U.S.S.R., Russia, Siberia; Zilling 233).—The type collection was determined by Litsch. as *Coriolus pallescens* [= *"Trametes" semisupinus*] but later Pilát decided that it was a different sp.

subroseus, Polyporus, Berk.—A name listed without ref. or descr. from Portugal by Henriq. 1881: 58. This may be a herbarium name: in the introduction to the paper it is stated, "Fungos fere omnes Cl. Berkeley ... praecipue ex imaginibus a me depinctis determinavit." I could trace no further information supplied by Portuguese mycologists.

*subrubescens, *Scutiger*, Murrill 1940 (BTC 67): 277 (U.S.A., Florida); *Polyporus* Murrill 1948 (nom. altern.).—Lowe apud Overh. 1953: 428 wrote that "the type material of this species does not appear to differ from *Polyporus* [*Albatrellus*] *confluens*." The original description does not readily suggest this identity.

subsimulans, Bjerkandera, Murrill → *Polyporus simulans* B. & C. ex Sacc. (**O**).

subspadiceus, Polyporus, Fr. 1818 O. 2: 263 (Sweden) (d.n.) per Fr. 1821; *Physisporus* Gillet 1877; *Poria* P. Karst. 1882.—Nomen dubium. The descr. suggests resupinate *Bjerkandera* cf. *adusta*.

subsuberosus, Polyporus, S. Schulz. 1866: 42 (Yugoslavia, Slavonia) (nom. nud.: n.v.p.).—Nomen dubium.

*subsulphureus, *Polyporus*, P. Henn. 1895 (BJ 22): 89 (incidental mention: n.v.p.), a name mentioned in connection with a collection of *P. sulphureus* f. *substipitatus* P. Henn. (Cameroons). Perhaps a tropical sp. of *Laetiporus*.

subterraneus, Boletus, Scop. 1772 P.s.: 107 *pl. 30 f. 1* (Hungary, now Czechoslovakia) (d.n.); = *Polyporus ruber* Pers. 1825.—Nomen dubium. Found in a mine: apparently an abnormal condition.

subtomentosus, Boletus, L. 1753: 1178 (Sweden) (d.n.) per Fr. 1821; &c. (Boletales).—Sensu Bolt. → *Coltricia perennis*.

subtus-alba, Daedalea, Secr. 1833 M. 2: 487 ("*subtus alba*") (nom. nud.: n.v.p.) (Switzerland) = *Lenzites betulina*.

*subvaporaria, *Poria*, Cooke "in herb. Kew" (Australia, Queensland). —Fide G. Cunn. 1965: 63, 64 = *Poria versipora* [= *Schizopora paradoxa*].

*sulphurea, *Poria*, Killerm. "n. sp. ... in scheda" (U.S.S.R., Russia, Siberia), not ∼ Petch 1922.—Fide Pilát 1932 (BmF 48): 44 = *Poria calcea* var. *xantha* (Fr. per Fr.) Bourd. & G. [= *Poria xantha*]. Afterwards referred to as *Poria selecta* "var. *sulphurea* Romell?" by Killerm. 1943 (Am 41): 251.

sulphureo-flavida, "determined by Prof. Maire" apud Lloyd 1910 (LMW 3,

L. 28): 1 (France) (nom. nud.: n.v.p.).—Nomen dubium. Error for *Poria 'subfusco-flavida'* (**O**)?

sulphureum, Sporotrichum, Grev. 1822 (MWS 4): 69 *pl. 5 f. 3* (Scotland). —Jaap 1922 (VBr 64): 45, under *Poria* [*Inonotus*] *obliqua*, stated, "In Gesellschaft wächst *Sporotrichum sulphureum* Grev., wohl als Konidienform hierzu gehörend." This shows that Jaap found the abundant spore deposit of the polypore; his suggestion that it would be *Sporotrichum sulphureum* is certainly untenable. The 'type' habitat of that species is, "in stercore et in cellariis, toto anno."

sulphureus, Polyporus, Chev. 1837: no. 40 var. *γ plate f. 2* (as a var. of *P.* [*Albatrellus*] *cristatus*: n.v.p.) (France), not ∼ (Bull.) per Fr. 1821.

**tabacinum, Irpex*, B. & C. apud Fr. 1851 (NAu III 1): 106/90, apud Berk. 1873 (G. 1): 102 (U.S.A., South Carolina); *Xylodon* O.K. 1898; *Trametes* Pat. 1900; *Cerrenella* Murrill 1905.—An extra-European sp. (Hymenochaetaceae). Erroneously referred as a var. to *Trametes* [*Hirschioporus*] *abietina* by Pilát 1941 (ACE 3): 273, 275.

tabacinus, Polyporus, Fr. "in litt." apud Weinm. 1826 (SPR 2): 103 (U.S.S.R., European Russia) (nom. nud.), not ∼ Mont. 1835.—Nomen dubium.

taloisensis = saloisensis

**tanakae, Irpiciporus*, Murrill 1909 (M 1): 166 (Japan); *Irpex* Sacc. & Trott. 1912; *Daedalea* Aosh. 1967.—Referred to *Trametes* [*Antrodia*] *albida* by Imaz. 1939 (JJB 15): 302, but Aosh. 1967 (TmJ 8): 3 maintains it as a distinct sp. of *Daedalea* emend. Aosh. [in part = *Antrodia*].

taurinus, Polyporus, Pers. 1825: 37 (Italy).—The descr. suggests a sp. of Boletales. Referred by Fr. 1874: 525 to *Polyporus viscosus* Pers. (**O**).

taxicola, Merulius, Pers. 1825: 32 *pl. 14 fs. 4, 5* (Switzerland); *Poria* Bres. 1897; *Merulioporia* Bond. & S. 1941 (generic name n.v.p.), 1943; *Meruliopsis* Bond. apud Parm. 1959, Bond. 1961; *Ceriporia* E. Komar. 1964 (incomplete ref.: n.v.p.); = *Merulius taxicola* (Pers.) Duby 1830 ('Corticiaceae'). – V.s.: *Poria* "*saxicola*", Torrend 1913 (Bro 11): 67.

tenerrimus, Polyporus, Saut. 1878 (Msa 18): 180 (Austria), not ∼ B. & Rav. apud Berk. 1872.—Nomen dubium.

tenuiparietale, Aporpium, Bond. apud Bond. & S. 1941 (nom. nud.: n.v.p.); *Polyporus* Bond. & S. 1943 (SB¹): 30 (lacking Latin descr.: n.v.p.). —Fide Bond. 1953: 162 = *Aporpium* ["*Trametes*"] *semisupinum*. The descr. of 1943 runs (translated), 'not always resupinate'.

tenuis, Lenzites*, Lév. 1846 (ASn III 5): 122 (Guadeloupe), not ∼ (Berk.) Pat. 1914; *Cellularia* O.K. 1898.—Specimens so named by Lloyd 1917 (LMW 5): 709 from Japan (with *Lenzites earlei* Murrill as syn.) led to this sp. being referred by Hemmi apud Hemmi & Ikeya 1939 (ApJ 9): 14 to *Lenzites* [*Trametes*] *gibbosa*. I doubt the correctness of Lloyd's determination. The original fungus was referred to *Daedalea elegans* Spreng. ex Fr. [= "*Trametes*" *palisotii* (O**)] by O. & K. Fid. 1966 (M 58): 868.

tephrodes, Poria, Fr. in herb. (Norway).—Fide Egeland 1912 (NMN 49): 368, 1914 (NMN 52): 153 = *Poria cinerascens.*

**terebrans, Polyporus,* B. & C. 1868 (JLS 10): 306 (Cuba); *Bjerkandera* Murrill 1907.—Referred by Lloyd 1915 (LMW 4, Ap.): 387, with some reservation, to *Polyporus [Bjerkandera] fumosus.*

terrestris, Boletus, Sow. 1803: *pl. 387 f. 5* (England) (d.n.), not ~ (Pers.) DC. 1815 (d.n.); = *Polyporus insularis* Pers. 1825, Fr. 1832 (*"infularis"*; syn.: n.v.p.), not ~ (Har. & Pat.) Lloyd 1912.—Nomen dubium. Referred, with doubt, to *Polyporus [Poria] bombycinus* by Fr. 1832[Ind.]: 63, 146.

terrestris luteus, Polyporus, Secr. 1833 M. 3: 173 (Switzerland) (as a sp. of *Boletus* & double epithet: n.v.p.).—Nomen dubium. Secr. thought of *Poria terrestris* Pers., but concluded that "ma plante n'a guère de rapport avec celles des auteurs cités. . .."

tessulatus, Polyporus, Fr. → *Boletus hirsutus* Batsch (**0**).

**testaceo-fuscus, Fomes,* Bres. 1912 (Am 10): 498 (Indonesia, Java).—Mentioned by Lowe 1957 F.: 55 as a probable syn. of *Fomes [Phellinus] robustus,* but the original descr. does not support this suggestion.

**texanus, Inonotus,* Murrill 1904 (BTC 31): 597 (U.S.A., Texas); *Polyporus* Sacc. & Trott. 1912.—Referred by Lloyd 1915 (LMW 4, Ap.): 387 to *Polyporus [Inonotus] corruscans* [sensu Lloyd], but other authors maintain the species as distinct from the *I. rheades* complex (no granular basal core in the fruitbody). – Descr.: Schrenk 1914 (AMo 1): 247 *pls. 6, 7;* Overh. 1953: 416 *pl. 132 fig. (Polyporus);* Pegl. 1964 (TBS 47): 188 (*Inonotus*).

**texanus, Pyropolyporus,* Murrill 1908 (NAF 9): 104 (U.S.A., Texas); *Fomes* Hedgc. & Long 1912, Sacc. & Trott. 1912; *Phellinus* A. Ames 1913.—Referred by Lowe 1954 (M 46): 496, 1957 F.: 55 to *Fomes [Phellinus] robustus* in a broad sense. Growing on *Juniperus,* hence perhaps rather *Phellinus hartigii.* – Descr.: Hedgc. & Long 1912 (M 4): 112 *pl. 64 fs. 2, 3, pl. 65 fs. 1, 4, 7;* Lloyd 1915 (LMW 4, F): 242; Overh. 1953: 89 *pl. 62 fs. 369, 370, pl. 127 fig. (Fomes).*

thelephoroides, Boletus,* Hook. 1822: 10 (Peru); *Polyporus* Fr. 1838; *Polystictus* Cooke 1886; *Microporus* O.K. 1898; *Gloeoporus* Bres. apud Speg. 1919, G. Cunn. 1965 misapplied?—At first referred by Bres. 1912 (H 53): 74 to *Gloeoporus dichrous,* but afterwards referred by Bres. 1916 (Am 14): 227 and Lloyd 1915 (LMW 4, Ap.): 331 to *Polyporus/ Gloeoporus conchoides.* – Cf. (51**).

thunbergii, Daedalea,* Fr. 1821: 335; *Lenzites* Fr. 1838; *Lentinus* Sacc. 1891; = *Boletus agaricoides* Thunb. 1784: 347 & apud Murr. 1784: 977 (Japan) (d.n.).—This is now referred to *Daedalea ["Trametes"] palisoti* Fr. (0**). Referred to *Lenzites betulina* by S. Ito 1955: 246, which seems incorrect in view of the descr. of the surface of the cap as 'laevis'. – *Lenzites thunbergii* was recorded for Switzerland by Otth 1871 (MiB 1870): 93 as growing next to *Trametes trabea* Otth [= *Gloeo-*

phyllum trabeum (Pers. per Fr.) Murrill] and "die sich von der *Trametes*, innen und aussen, absolut durch nichts unterschied, als durch das volkommen und rein lamellöse Hymenium."

tiliophila, Tyromyces, Murrill 1907 (NAF 9): 33 (Canada, Ontario); *Polyporus* Sacc. & Trott. 1912.—This was referred by Overh. 1953: 286, 287 to *Polyporus* [*Tyromyces*] *guttulatus,* but Lowe in litt. believes this to be extremely doubtful. It is not bitter in taste, has spots only at the extreme base of the fruitbody, and grows on hardwoods.

**tinctorius, Polyporus,* Quél. apud Malinvaud 1881 (BbF 28): 216 (Algeria); *Placodes* Quél. 1888; *Xanthochrous* Pat. 1897; *Polystictus* Big. & Guill. 1913; *Inonotus* Maire apud Maire & Wern. 1938.—Pegl. 1964 (TBS 47): 184 (apparently type not examined) referred this to *Inonotus hispidus.* − Descr.: Quél. 1888: 399 (*Placodes*); Pat. 1897 (BmF 13): 20, setae said to be lacking (*Xanthochrous*); Lloyd 1915 (LMW 4, Ap.): 365 (*Polyporus*). − Cf. Hazslinszky *1870* in connection with *Polyporus* [*Inonotus*] *hispidus.*

**tisdalei, Scutiger,* Murrill 1943 (Ll 6): 227 (U.S.A., Florida); *Polyporus* Murrill 1943 (nom. altern.).—Lowe apud Overh. 1953: 429 thought this to be "a noncystidiate condition of *Polyporus* [*Abortiporus*] *biennis.* Dr. Murrill does not agree, maintaining that it is a distinct species." Fid. 1969 (Ri 4): 154 did not agree either, but offered no alternative solution.

**tokyoensis, Polyporus,* Lloyd 1915 (LMW 4, Ap.): 302 *f. 639* (Japan); Trott. 1925 ("*tokyvensis*").—Referred to *Tyromyces caesius* (with a question mark) by S. Ito 1955: 278; and to *Tyromyces mollis* by G. Cunn. 1965: 129.

tomentosum, Hydnum, L. 1753: 1178 (Sweden) (d.n.), not ∼ Schrad. 1784 (d.n.); *Hydnum* L. per Fr. 1821; &c.; sensu Pers. 1800: 53 & Fr. 1821 ≡ *Phellodon tomentosum* (L. per Fr.) Banker 1906 (Bankeraceae). − Sensu Oed. → *Coriolus versicolor.*

Tortula Hedw. per Hedw. (Musci frondosi).—Through a complicated error this generic name was brought into connection with a polypore by Pfeiffer 1874 [1960 (Pe 1): 287].

**tortuosa, Daedalea,* Crag. 1884 (BWb 1): 26 [repr. 1884 (JM 1): 28] (U.S.A., Kansas).—Nomen dubium. Referred, with doubt, to *Cerrena unicolor* by Murrill 1908 (NAF 9): 124, but the original descr. hardly supports this guess.

torulosa, Daedalea, Pers. 1828, epithet preceded by an asterisk: [≡ *Daedalea gibbosa* (Pers.) Pers. sensu Tratt. 1806: 150 *pl. 15 f. 30* (Austria)].—Nomen dubium. The surface of the cap is said to be without zones ("Binden"), which may have been the reason why Persoon did not recognize *Trametes gibbosa* in it. Nevertheless the pores suggest this species.

**torulosoides, Fomes,* Lloyd "in litt." (Brazil).—Listed by Torrend 1940: 328 as a form of *Fomes* [*Phellinus*] *torulosus.*

trabens = trabeus (*Polyporus*)

trabeus, Ceriomyces, J. Schroet.—Fide Mez 1908: 249 = *Polyporus* [*Antrodia*] *serialis* Fr. – I have been unable to trace the place of publication of this name.

trabeus, Polyporus, Rostk. 1830 (StP 4): 59 *pl. 28:* Fr. 1832 (Germany/ Poland); *Postia* P. Karst. 1881; *Bjerkandera* P. Karst. 1881; *Leptoporus* Quél. 1886; *Polystictus* Big. & Guill. 1913; *Tyromyces* Parm. 1959, misapplied; *Chaetoporellus* M. P. Christ. 1960, misapplied.—Nomen dubium & ambiguum. Variously interpreted. Romell 1926 (SbT 20): 7 suggested that the original plate might represent *Trametes* [*Dichomitus*] *squalens,* but he did so without conviction. – Sensu Bres. → *Tyromyces leucomallelus;* sensu Lloyd, at least in part → *Tyromyces chioneus;* sensu Fr. = ?, perhaps a mixture, perhaps including *Polyporus albidus* [= *Tyromyces stipticus*], cf. Romell, op. cit., pp. 3, 5; sensu Lowe 1934, in part, → *Tyromyces lowei.* – V.s.: "*trabens*".

trametea, Lenzites, Barbier in herb., ("Quélet"); ≡ *Lenzites tricolor* var. *trametea* Quél. 1888: 368 (France).—The specific combination was cited by Lloyd 1912 (4, L. 42): 2 as a syn. (n.v.p.) of *Daedalea* [*Daedaleopsis*] *confragosa* (forma).

trametes, Daedalea,* Speg. 1880 (ASa 9): 166 (Argentina); *Striglia* O.K. 1891; ≡ *Trametes daedalea* Speg. 1899.—Referred to *Trametes* [*Funalia*] *trogii* by Bres. 1916 (Am 14): 229. – Cf. (36**).

tremellosa, Daedalea, (Schrad. per Fr.) E. Krause 1928 ("*Daedaleus*"); *Merulius* Schrad. 1794: 139 (Germany) (d.n.) per Fr. 1821; *Xylomyzon* Pers. 1822; *Thelephora* L. March. 1828.—*Merulius tremellosus* Schrad. per Fr. is the type of *Merulius* Fr., a genus now excluded from the polypores ('Corticiaceae').

tremulae, Polyporus, Wittr. in herb.—Fide Nannf. 1967 (WPb 6): 131 = *Phellinus tremulae.*

tricolor, Trametes,* Lloyd 1920 (LMW 6): 998 *pl. 163 f 1807* (Japan) ("For me a trametoid form of *Lenzites tricolor*": n.v.p.).—Fide Imaz. = *Daedaleopsis conchiformis* Imaz. 1943 (BTS 6): 77. – Cf. (31**).

trilobatus, Boletus, With. 1776; [≡*Fungus albus minimus trilobatis, sine pediculo, foliis quercinis adnascens* Ray 1696: 18; 1724: 22 (England)].—Nomen dubium. It is not even certain that this fungus had pores, as was accepted when it was placed in *Boletus* [sensu lato].

**trizonatus, Polyporus,* Cooke 1883 (G 12): 17 (Australia, Victoria): *Polystictus* Cooke 1886; *Microporus* O.K. 1898.—Bres. 1926: 80 referred this to *Polyporus* [*Coriolus*] *velutinus;* G. Cunn. 1949 (BPZ 75): 3, 1950 (PNW 25): 245, 248, and 1965: 155, to *Coriolus versicolor, C. zonatus,* and *Microporus flabelliformis* (Kl.) O.K. respectively.

truncata, Labirinthia, S. Schulz. "Mpt. p. 712, 1869" (n.v.p.).—Fide Igmándy 1968 (Aph 3): 354 = *Phaeolus schweinitzii.*

truncigenus, Boletus, (Fr.) Mussat 1901 (SF 15): 68 (syn.: n.v.p.).—An error for *Polyporus intybaceus* [var.] "B." *truncigenus* Fr. 1838: 446, 1874: 539, Sacc. 1888 (SF 6): 96, the 'B.' having been taken for the abbreviation of '*Boletus*'.

*tsuginae, Fomitiporia, Murrill 1907 (NAF 9): 9 (U.S.A., New Hampshire); Poria Sacc. & Trott. 1912.—This has been reduced to Polyporus/Fomes [Phellinus] robustus as a resupinate condition by Bres. 1926: 81, Overh. 1953: 87, and Lowe 1949 (Ll 11): 168, 1957 F.: 56; or to Fomes [Phellinus] hartigii by Englerth 1942 (BFY 50): 17. This needs confirmation. No setae have been reported. – Descr.: Overh. 1931 (M 23): 121 pl. 12 f. 4, pl. 14 fs. 14, 16; Lowe 1946: 77 (Poria).

tuber, Physisporus, P. Karst. 1885 (H 24): 72 (Finland).—Nomen dubium. Cf. Lowe 1956 (M 48): 115.

Tuberaster W. Cooke 1953 ("Boccone") (incidental mention: n.v.p.) [1960 (Pe 1): 290] is not a generic name, as treated by Cooke, but the first word of a non-Linnaean name, Tuberaster Fungos ferens Boccone, which is a syn. of Polyporus tuberaster.

tuberculosa, Poria, Pers. 1795 (ABU 15): 14 / 1796 O. 1: 14 (Germany) (d.n.); Boletus Pers. 1801 (d.n.), not ∼ Baumg. 1790 (d.n.), not ∼ J. F. Gmel. 1792 (d.n.); Polyporus (Pers.) per Pers. 1825, not ∼ Fr. 1821; Boletus L. Morel 1865, misapplied (sensu D.C.); Poria Cooke 1886. –Nomen dubium. No type in herb. Pers. Found in a mine; apparently an abnormal growth. – Sensu DC. → Pachykytospora tuberculosa.

Tucahus Rafin. → Gemmularia Rafin. (O).

*tucumanensis, Polyporus, Speg. 1898 (ABA 6): 162 (Argentina).—Referred to Polyporus similis Berk. (extra-European) (90) by Bres. 1916 (Am 14): 227 and to Leucoporus [Polyporus] brumalis var. vernalis "Quél." by Speg. 1925 (BCó 28): 367.

*tucumanensis, Trametes, Speg. 1898 (ABA 6): 174 (Argentina).—Fide Bres. 1916 (Am 14): 229 = Trametes [Funalia] trogii. – Cf. (36).

*tunetanus, Melanopus, Pat. 1902 (BmF 18): 50 (Tunisia); Polyporus Sacc. & D. Sacc. 1905 (SF 17): 106.—Reported from Spain by Codina Viñas 1925, but his descr. and the original show considerable differences, and I assume that the determination (by Hariot) is incorrect.

tunicatus, Boletus, Schum. 1803: 391 (Denmark) (d.n.).—Nomen dubium. Referred by Fr. 1821: 381 to Polyporus vulgaris [sensu typi = Poria lenis]; and by Secr. 1833 M. 3: 164 to Polyporus fimbriatus supinus [≡ Stromatoscypha fimbriatum (O)].

*turbinatus, Cyclomyces, Berk. 1854 (HJB 6): 166 (India); Xanthochrous Pat. 1900; Cycloporus Bond. & S. 1941.—Referred to Cyclomyces greenei [≡ Polyporus [Coltricia] montagnei var. greenei (Berk.) R. L. Gilb.] by Lloyd 1917 (LMW 4): 633.

"typha (tubulosus)", Polyporus, Secr. 1833 M. 3: 64 (Switzerland) (as a sp. of Boletus: n.v.p.).—Nomen dubium. Apparently belonging to Polyporus subgen. Melanopus (Pat.) Maubl.

typha porosus, Polyporus, Secr. 1833 M. 3: 65 (Switzerland) (as a sp. of Boletus & double epithet & nom. nud.: n.v.p.).—Nomen dubium. Rather (var. A) a sp. of Polyporus subgen. Melanopus (Pat.) Maubl.; Secr. himself thought of Boletus elegans Bull. [= Polyporus varius].

ulicis, Xanthochrous, "Quél." apud Pat. 1900: 101 (nom. nud.: n.v.p.); ≡ *Polyporus ulicis* Boud. [= *Phellinus ribis*]?

ulmarius, Boletus, Gaterau 1789: 190 (France) (d.n.).—Nomen dubium. If the stalks had not been described as "blanchâtres et écailleux" the brief descr. almost suggests *Polyporus squamosus.*

ulmarius, Polyporus, S. Schulz. 1866: 41 (Yugoslavia, Slavonia) (nom. nud.: n.v.p.), not ∼ (Sow.) per Fr. 1821.—Nomen dubium.

umbilicatus, Boletus, Schrank 1789: 621 (Germany) (d.n.), not ∼ Scop. 1772 (d.n.).—Nomen dubium. Referred by Fr. 1821: 348 to *Polyporus brumalis* [sensu lato], but the original descr. is too brief to make this identification convincing.

umbilicatus, Boletus, Scop. 1772: 466 (Yugoslavia, Carniola) (d.n.), Fr. 1832^[Ind.]: 64 ("*umbilicus*"; syn.), not ∼ Schrank 1789 (d.n.): *Boletus* (Scop.) per Spreng. 1827; *Polyporellus* P. Karst. 1889; *Polyporus* Romell 1898, not ∼ Paul. 1812–35 (d.n. ?), not ∼ Jungh. 1838, not ∼ Berk. 1851.—Nomen dubium. Referred by Fr. 1821: 348 to *Polyporus melanopus* var. *cyathoides* [= *P. melanopus*], but this is not supported by Scopoli's descr., fide Donk 1968 (Pe 5): 258.

**umbilicatus, Polyporus,* Jungh. 1838: 72 (Indonesia), not ∼ Paul. 1812–35 (d.n.?), not ∼ Berk. 1851.—To be kept in mind in connection with *Polyporus alveolarius* (Bosc) per Fr., cf. Donk 1969 (Pe 5): 238. Reduced to a very inclusive *Polyporus arcularius* by Lloyd 1912 (LMW 3, S.P.): 179.

umbilicatus, Polyporus, Paul. → *Fungus polyporus perennis* Paul. (**0**).

umbilicus = *Boletus umbilicatus* Scop.

**umbrinus, Daedalea,* Lloyd 1914 (LMW 4, L. 53): 10 (Brazil) (as a var. of *Daedalea quercina*: n.v.p.).—The correct status of this taxon needs further study.

underwoodii, Polyporus, Murrill apud Peck, see *Polyporus admirabilis* Peck (**0**).

ungulaeformis, Lenzites,* B. & C. 1849 (HJB 1): 101 (U.S.A., North Carolina); *Cellularia* O.K. 1898.—Nomen dubium. At first referred by Murrill 1905 (BTC 32): 86, 87, to *Agaricus* [*Daedaleopsis*] *confragosa* (31**); later Murrill 1908 (NAF 9): 127 listed it as a syn. of *Lenzites betulina.* – For a picture of the type, see Lloyd 1918 (LMW 5): *f. 1265* on p. 811.

ungulatus, Fomes, Lázaro 1916 (RMA 14): 663 & 1917 (RMa 15): *pl. 10 f. 24* / 1917: 102 *pl. 10 f 24* (Spain), not ∼ (Schaeff. per St-Am.) Sacc. 1888.—Nomen dubium. Cf. Bres. apud Trott. 1925 (SF 23): 396, "Versimiliter est *Fomes* [*Phellinus*] *igniarius* minor." The descr. is incomplete but as the fungus occurred on *Salix* it suggests this sp.

Ungulina Kotl. & P. 1957 (lacking Latin descr.: n.v.p.), not ∼ Pat. 1900 [repr. 1960 (Pe 1): 292, in obs.]; [≡*Ungulina* Pat. sensu Kotl. & P. 1957 (ČM 11): 168, excl. of type]; holotype, *Polyporus* [*Truncospora*] *ochroleucus* Berk.

*_unicolor, Boletus_, Schw. 1822: 97 (U.S.A., North Carolina); _Polyporus_
Fr. 1838; _Trametes_ Murrill 1906; _Spongipellis_ Murrill 1907.—This was
identified by Murrill 1907 (NAF 9): 37 with _Polyporus_ [_Spongipellis_]
obtusus (**O**), a North American species. The reason for including the
present entry is that Murrill, l.c., referred _Polyporus_ [_Spongipellis_]
schulzeri Fr. here; see Donk 1972 (PNA 75): 175. See also Lloyd 1915
(LMW 4, Ap.): 324.

*_unicolor, Irpex_, Lloyd 1920 (LMW 6): 921 _pl. 145 fs. 1649, 1950_ (Japan)
(as a form of _Daedalea_ [_Cerrena_] _unicolor_: n.v.p.).—Fide S. Ito 1955:
265 = _Cerrena unicolor_.

*_usambarensis, Poria_, P. Henn. 1905 (BJ 38): 108 (Tanganyika).—Bres.
1916 (Am 14): 229 referred this to _Poria_ [_Phellinus_] _ferruginosa_, but
Lowe 1963 (M 33): 472 thought it was a distinct sp.; later Lowe 1966:
150 remarked that it may be the same as _Poria_ [_Phellinus_] _ferrea_.

ustalis, Polyporus, Velen. 1922: 671 [see Pilát 1948: 255 for Latin trans-
lation] (Czechoslovakia).—Needs typification. Cf. _Polyporus lepideus_ =
P. ciliatus (forma).

vaporaria, Poria, Pers. 1794 (ABU 11): 30 (Germany) (d.n.); _Boletus_
Pers. 1801 (d.n.); _Polyporus_ Fr. 1818 (d.n.), misapplied; _Polyporus_
(Pers.) per Fr. 1821, misapplied; _Boletus_ Spreng. 1827; _Physisporus_
Gillet 1878, misapplied; _Poria_ Cooke 1886, Quél. 1886 misapplied;
Trametes Pat., fide R. Heim 1934. _Coriolus_ Bond. & S. 1941; _Tyromyces_
M. P. Christ. 1960, misapplied.—Nomen dubium. Cf. Bres. 1897 (AAR
III 3): 88, "vix dubia cum _Poria vaillantii_ Dec. identica est", but the
original descr. and type prevent acceptance of this identification, fide
Donk 1967 (Pe 5): 117. – Sensu Fr. → _Antrodia sinuosa_; sensu R. Hartig
1878 → _Antrodia serialis_; sensu P. Henn. → _Polyporus_ [_Tyromyces_]
henningsii; sensu Mez → _Polyporus_ [_Tyromyces_] _fodinarum_; sensu Berk.
→ _Schizopora paradoxa_; "sensu Liese", &c. → _Poria placenta_. See also
Poria vaporaria Bres. [= _Antrodia sinuosa_].

variegata, Daedalea, P. Karst. 1911 (Ttk 12[1, 2]): 111 (U.S.S.R., 'Transbaikal'), not ~ Fr. per Fr. 1821.—Considered to be a form of _Daedalea_
[_Daedaleopsis_] _confragosa_ by Lowe 1956 (M 48): 105. – Cf. (**31**).

variegatus, Boletus, Schaeff. 1774: 90 [_pl. 263_] (Germany) (d.n.), not ~
Sow. 1802 (d.n.), not ~ Sw. 1810 (d.n.) per Fr. 1821; _Scindalma_ (Schaeff.)
per O.K. 1898.—Nomen dubium. Referred by Fr. 1821: 353 to _Polyporus_
[_Ganoderma_] _lucidus_. – Sensu Fr. (syn.), O.K. → _Ganoderma lucidum_.

variegatus, Boletus, Sow. 1802: _pl. 368_ (England) (d.n.), not ~ Schaeff.
1774 (d.n.), not ~ Sw. 1810 (d.n.) per Fr. 1821; _Polyporus_ (Sow.) per
Loud. 1829, not ~ Endl. 1830; _Fomes_ Cooke 1885; _Placodes_ Quél. 1886;
Agaricus E. Krause 1934; = _Scindalma secretanii_ O.K. 1898.—Nomen
dubium. Sow. cited as syn. _Boletus versicolor_ Schacff. [= _Hapalopilus_
rutilans] which does not agree. Fr. 1821: 352 referred it to _Polyporus_
varius but evidently incorrectly so. Redescribed by Secr. 1833 M. 3: 99
("_Polyporus_", but as a sp. of _Boletus_: n.v.p.), which induced Fr. 1838:

470 to accept the sp. I am at a loss as to the identity of Sowerby's fungus; Secretan's description suggests *Inodermus nodulosus*.

**variiformis, Polyporus*, Peck 1889 (RNS 42): 122 (U.S.A., New York); *Polystictus* Sacc. 1891; *Microporus* O.K. 1898; *Coriolus* Pat. 1900; *Trametes* Peck 1899; *Coriolellus* Sarkar 1959; *Antrodia* Donk 1966. —Referred by Murrill 1907 (NAF 9): 29 to *Coriolellus [Antrodia] serialis*. Currently considered to be a distinct species; not yet reported from Europe. – Descr.: Overh. 1953: 140 *pl. 88 fs. 504, 505, pl. 125 fig.* (*Trametes*); Sarkar 1959 (CJB 37): 1262 *fs. 58–71, pl. 1 f. 8* (*Coriolellus*).

vavraeanus, Polyporus, Opiz 1852: 136 (Czechoslovakia) (nom. nud.: n.v.p.).—Nomen dubium.

vegetus, Polyporus, Fr. 1838 (Sweden); *Fomes* Fr. 1849 (nom. altern.); *Placodes* Quél. 1886; *Phaeoporus* J. Schroet. 1888; *Ganoderma* Pat. apud Rolland 1890, Bres. 1897 & 1912; *Scindalma* O.K. 1898; *Friesia* Lázaro 1916.—Variously interpreted. Now usually referred to *Ganoderma applanatum*, but the original descr. does not really support this. Fr. himself (1838: 464) compared it with his conception of *Polyporus australis*, which is problematic. Cf. Donk 1969 (PNA 72): 280, note. – Descr.: Fr. 1838: 464; 1884 I. 2: 82 *pl. 183 f. 3*. – Sensu Rick. → *Ganoderma pfeifferi* (a possibility); sensu Velen. → *Phellinus torulosus*.

**vellereus, Polyporus*, Berk. 1842 (LJB 1): 455 (New Zealand): *Polystictus* Fr. 1851; *Hansenia* P. Karst. 1879; *Microporus* O.K. 1898; *Coriolus* Pat. 1923, or perhaps earlier.—Referred to *Polystictus fibula* [sensu Fr. = *Coriolus hirsutus*] by Bres. 1912 (Am 10): 504 and to *Trametes [Coriolus] hirsuta* by G. Cunn. 1965: 172. – Descr.: Lloyd 1921 (LMW 6): 1033 *pl. 172 f. 1877* (*Polystictus*).

**veluta, Hydnum*, Schw. in herb. (U.S.A.).—Fide Schw. 1832: 164 = *Irpex [Hirschioporus] fuscoviolaceus*.

velutina, Bulliardia, Lázaro 1916 (RMa 14): 841 / 1917: 153 (Spain): *Daedalea* Sacc. & Trott. apud Trott. 1925, not ∼ Ces. 1879.—Nomen dubium. Bres. apud Trott. 1925 (SF 23): 450 suggested, "est verisimiliter *Polyporus [Coriolus] velutinus* hymenio lacerato." In view of the substratum ("en la base de los troncos de pino") and some notes by Bourd. & G. 1928: 561 in connection with *Inodermus maritimus* Quél. (". . . une forme de *Tr[ametes] Trogii*, blanchâtre finement véloutée, sans soies rigides, récoltée sur pin . . .") I would make another suggestion, viz. *Funalia trogii*, but without conviction.

velutina, Daedalea, Ces. 1879: 7 ("*vetulina*") (Ceylon), not ∼ (Lázaro) Sacc. & Trott. apud Trott. 1925; *Striglia* O.K. 1891; *Lenzites* Bres. 1916 (syn.: n.v.p.).–Petch 1916 (APe 6): 111 suggested that this was *Lenzites betulina*; and Lloyd 1923 (LMW 7): 1205 referred it to that sp. (form '*L. berkeleyi*'), but Bres. 1916 (Am 14): 221, 230 considered it to be a syn. of *Lenzites junghuhnii* Lév. (**O**).

velutinum.—"[*Hydnum*] *velutinum*. Fl.d. †", Steud. 1824: 205, not ∼ Fr. 1821.—This name is a puzzle. I have thought of accepting it as a

substitute name for *Hydnum tomentosum* L. sensu Oed. 1770 (Fd 3 /
Fasc. 9): 7 *pl. 534 f. 3*, of which Fr. 1821: 406 suggested "potissimum
Polyp. [*Bjerkandera*] *adustus.*" I would refer the figure to *Coriolus
versicolor*.

venetus, Polyporus, Sacc. 1873 (ASv 2): 100/52 *pl. 7 fs. 4–6* (Italy); *Polystictus* Cooke 1886.—Nomen dubium. Perhaps abnormal fruitbodies:
found in a wine-cellar. Pilát 1937 (ACE 3): 124, 125 referred it to *Piptoporus quercinus* [= *Buglossoporus pulvinus*] as a monstrous form, but
this is contradicted by the descr.

venosus, Boletus, Humb. 1792 (ABU 3): 57 (Germany) (d.n.) per Steud.
1824; *Polyporus* Pers. 1825.—Nomen dubium.

vermicularis, Daedalea, Pers. → *Boletus resupinatus* Sow. (**O**).

vernus, Polyporus, S. Schulz. 1870 (VW 20): 177 (Hungary, now Czecho-
slovakia) (nom. nud.: n.v.p.).—Apparently a sp. of *Polyporus* subgen.
Leucoporus (Quél.) Maubl.: the author appended a var., "*β. fascicularis
Schrad.*" = *Boletus fasciculatus* Schrad. [= *Polyporus brumalis*].

veronensis, Boletus, Spreng. → *Polyporus castaneus* Fr. (**O**).

versicolor fusco-purpureus, Polyporus, Secr. 1833 M. 3: 149 (Switzerland)
(as a sp. of *Boletus* & double epithet & nom. nud.: n.v.p.) = *Coriolus
versicolor*.

versicolor impolitus, Polyporus, Secr. 1833 M. 3: 144 (Switzerland) (as a
sp. of *Boletus* & double epithet & nom. nud.: n.v.p.).—*Coriolus versicolor?*

versicolor iris, Polyporus, Secr. 1833 M. 3: 138 (Switzerland) (as a sp. of
Boletus & double epithet & num. nud.: n.v.p.).—Var. A = *Polyporus
[Coriolus] versicolor*.

versicolor olivascens, Polyporus, Secr. 1833 M. 3: 141 (Switzerland) (as a
sp. of *Boletus* & double epithet: n.v.p.).—Nomen dubium. Perhaps a
form of *Coriolus versicolor*.

versicolor pseudo-platani, Polyporus, Secr. 1833 M. 3: 148 (Switzerland)
(as a sp. of *Boletus* & double epithet & nom. nud.: n.v.p.): = *Coriolus
versicolor*.

versicolor psittacinus, Polyporus, Secr. 1833 M. 3: 143 (Switzerland) (as a
sp. of *Boletus* & double epithet: n.v.p.).—*Coriolus versicolor?*

versicolor quercinus, Polyporus, Secr. 1833 M. 3: 145 (as a sp. of *Boletus* &
double epithet & nom. nud.: n.v.p.).—*Coriolus versicolor?*

versicolor ramorum, Polyporus, Secr. 1833 M. 3: 147 (Switzerland) (as
a sp. of *Boletus* & double epithet: n.v.p.); = *Coriolus versicolor*.

versicolor salicinus, Polyporus, Secr, 1833 M. 3: 142 (Switzerland) (as a sp.
of *Boletus* & double epithet & nom. nud.: n.v.p.).—Nomen dubium.

versipellis, Polyporus, Pers. 1825: 96 (Italy).—Nomen dubium. I could not
locate the type specimen; descr. too incomplete for certain identification.

versiporus, Polyporus, Lloyd = *versisporus*.

**versisporus, Polyporus,* Lloyd 1915 (LMW 4, Ap.): 312 (Japan), 1915
(LMW 4, L. 60): 8 ("*versiporus*"); Yas. 1817 (BMT 31): 52 (nom. nud.).

—This was listed as a syn. of *Heterobasidion* [*Truncospora*] *ochroleucum* by G. Cunn. 1965: 145, but Japanese authors now refer it to *Laetiporus versisporus* (Lloyd) Imaz. (extra-European), basionym, *Calvatia versispora* Lloyd, cf. S. Ito 1955: 275.

veselskyi, Polyporus, Opiz in herb.—Fide Pilát 1942 (ACE 3): 623 = *Trametes suaveolens.*

"*vetusta, Daedalea,* Ell. & Harkn. "manuscript name".—Fide Murrill 1905 (BTC 32): 368 = *Porodaedalea pini* (Thore per Pers.) Murrill [= *Phellinus pini* (Brot. per Fr.) A. Ames]. — Cf. (**84**).

**victoriensis, Polyporus,* Lloyd 1921 (LMW 6): 1095 (Australia, Victoria); *Inonotus* Pegl. 1964.—Referred to *Inonotus rheades* by G. Cunn. 1965: 210, but this is not acceptable. — Descr.: Pegl. 1964 (TBS 47): 180 *fs. 1: 18, 2: 31* (*Inonotus*).

villosus, Boletus, With. 1776: 768 (England) (d.n.), not ∼ Huds. 1778 (d.n.), not ∼ Sw. 1788 (d.n.); [≡ *Agaricus villosus & porosus, substantiae coriaceae* Dill. 1719: 193 sensu Ray 1724: 24; ≡ *Fungus arboreus variegato illo Cerasorum, &c. C.B. similis, sed hirsutior, foraminulis etiam majoribus* Ray 1696: 336 (England)].—Nomen dubium.

vinaceo-rosea, Poria,* Rodw. & Clel. 1929 (PTa 1928): 78 (Australia, Queensland).—Referred by Lowe 1962 (PMi 47): 185 to *Poria* [*Ceriporia*]. *rhodella,* which, sensu Lowe, is a broadly conceived sp. (22**).

vinosus, Boletus, Thore 1803: 488 (France) (d.n.).—Nomen dubium.

vinosus juglandis, Polyporus, Secr. 1833 M. 3: 90 (Switzerland) (as a sp. of *Boletus* & double epithet: n.v.p.).—Nomen dubium.

violaceus, Polyporus, Fr. 1818 O. 2: 263 (d.n.) per Fr. 1821: 379, misapplied?, not ∼ (C. E. Mart.) C. E. Mart. 1904; *Boletus* Lenz 1840; *Physisporus* Gillet 1877; *Poria* Cooke 1886, Quél. 1886; *Caloporia* P. Karst. 1893, misapplied; *Merulius* Pat. 1900; *Merulioporia* Bond. 1953; *Meruliopsis* Bond. apud Parm. 1959, Bond. 1961; ≡ *Boletus nitidus* var. *violascens* A. & S. 1805 (Germany) (d.n.); *Polyporus* Secr. 1833 (as a sp. of *Boletus*: n.v.p.); *Polyporus* (A. & S.) per J. Schroet. 1888.—Nomen dubium (sensu originario A. & S.). — Sensu Fr. 1818 = ?; sensu Quél., P. Karst., &c. = *Poria* [*Merulius*] *taxicola* (**0**), fide Bourd. & L. Maire 1920 (BmF 36): 84, & Donk 1962 (Pe 2): 230, &c.; sensu Bres. 1903 (Am 1): 76, a distinct sp. of *Poria* or *Ceriporia?*

violascens, Hypodrys, (Pers.) Streinz 1861 (syn.: n.v.p.); ≡ *Hypodrys hepaticus* var. *violascens* Pers. 1825; [≡ *Fistulina buglossoides* Bull. sensu St-Am. 1821: 546 (France)]; = *Fistulina hepatica.*

viridis hirtus, Polyporus, Secr. 1833 M. 3: 131 (Switzerland) (as a sp. of *Boletus* & double epithet: n.v.p.).—Nomen dubium.

viridiuscula, Poria,* D. Baxt. 1947 (PMi 31): 128 *pls. 4, 5* (U.S.A., Georgia). —At first Lowe 1959 (Ll 21): 107 referred this, with doubt, to *Poria* [*Gloeodoporus*] *pannocincta,* apparently without having seen the type (which for some time was not available), later Lowe 1966: 135 made it a syn. of *Polyporus* [*Gloeoporus*] *dichrous.* — Cf. (51**).

viridi-zonatus, Polyporus, Secr. 1833 M. 3: 135 (Switzerland) (as a sp. of *Boletus*: n.v.p.).—Nomen dubium. *Coriolus* sp.?

viscosus, Polyporus, Pers. 1825: 41 (France): Fr. 1828; *Caloporus* Quél. 1886; *Heteroporus* Lázaro 1916.—Fide Bres. 1920 (Am 18): 68 = *Boletus bovinus* L. per Fr. (Boletaceae).

voluta, Polyporus, Pers. → *Poria difformis* Scop. (**0**).

Vonkhout Endl. 1836 (syn.: n.v.p.), Pfeiff. 1874 (syn.: n.v.p.; *"Vonckhout"*). —The 'generic' name *Vonkhout* has been cited as a syn. of *Daedalea,* but van Sterbeeck's typical species is presumably *Fomes fomentarius,* cf. Donk 1960 (Pe 1): 293.

vorax, Daedalea, Harkn. 1879 (PrP 17): 49 (U.S.A., California) (nomen subnudum).—This was referred to *Porodaedalea pini* (Thore per Pers.) Murrill [= *Phellinus pini* (Brot. per Fr.) A. Ames] by Murrill 1905 (BTC 32): 368 and to *Trametes* [*Phellinus*] *pini* (Brot. per Fr.) Fr. by Haddow 1938 (BTS 22): 185. – Cf. also Haddow, op. cit., p. 190. – Cf. (**84**).

vukasovićiana, Lenzites, S. Schulz. 1887 (H 26): 192 (Yugoslavia, Slavonia). —Nomen dubium. *Lenzites betulina* (forma)?

whetstonei, Favolus, Lloyd ex Trott. 1925; Lloyd 1916 (LMW 5): 615 *f. 870* (U.S.A., Minnesota) (referred to *Favolus europaeus* as "a sport": n.v.p.).—Fide Overh. 1953: 157 = *Favolus* [*Polyporus*] *europaeus* [= *Polyporus mori*], but cf. (**98**). Published in the best of the Lloydian tradition with an extremely poor protologue and a single miserable specimen as the type that should never have been named. A re-examination of the type is suggested.

wirtgenii, Polyporus, Fr. 1838: 483; *Poria* Cooke 1886, Quél. 1886; [= *Polyporus bombycinus* Fr. sensu Wirtg. 1835 (Fl 18): 324 (Germany)].—Nomen dubium.

wranyi, Polyporus, Opiz. "in herb." (presumably Czechoslovakia).—Fide Pilát 1941 (ACE 3): 451 = *Poria vaillantii.*

xoilopus, Polyporus, Rostk. 1828 (StP 4): 23 *pl. 10*: Fr. 1832 (Germany/ Poland); *Pelloporus* Quél. 1886; *Caloporus* Pilát 1931.—Nomen dubium. This is most likely a sp. of *Boletus* sensu lato. At first Killerm. 1922 (Dba 15): 62 thought of young fruitbodies of *Boletus* [*Gyroporus*] *castaneus* Bull. per Fr., but later he believed he had found the real fungus: Killerm. 1929 (H 67): 125 *f. A*, 1928 (Dba 17): 73 *pl. 11 f. 3*, Bres. 1931 (BIm 20): *pl. 961 f. 2* (*Polyporus*). These descriptions also suggest some sp. of Boletales. – V.s.: *"xylopus".*

Xylodon (Pers.) ex S. F. Gray 1821 [1956 (Ta 5): 113, 1963 (Ta 12): 156]; *Sistotrema* sect. "***... (*Xylodon*)" Pers. 1801: 552; lectotype, *Sistotrema quercinum* (Pers.) Pers.—To me a nomen dubium because of the uncertain identity of the type species, which may not be a 'polypore'. – *Xylodon* "Ehrenb." ex P. Karst. 1881 (Afe 2¹): 31 is a monadelphous homonym of which *Irpex* [*Schizopora*] *paradoxus* has been regarded as type species: this makes it a typonym of *Schizopora.* O.K. 1898: 540 implicitly took

Xylodon candidum Ehrenb. [= *Hirschioporus fuscoviolaceum*] as type species and regarded *Xylodon* "Ehrenb." as the correct name for *Irpex* Fr. If validly published as distinct names from *Xylodon* (Pers.) ex S. F. Gray, than both *Xylodon* 'P. Karst.' and *Xylodon* 'O.K.' are impriorable as later homonyms.

Xylomyzon Pers. → *Serpula* (Pers.) per S. F. Gray (**0**).

Xylophagus Link per Murrill → *Serpula* (Pers.) S. F. Gray (**0**).

Xylopilus P. Karst. 1882 [1960 (Pe 1): 295]; monotype, *Polyporus crassus* Fr. (**0**).—Nomen dubium.

xylopus = *xoilopus*

xylostromeus, Polyporus, Pers. 1825: 112 (Europe).—Nomen dubium. Referred by Fr. 1838: 485 ("Pers. . . . n. . . . 159") to *Polyporus vitreus* ("mycelium sterile") and 1874: 577 to the same sp. ("fungus siccus"). Fries's interpretation of *Polyporus* [*Rigidoporus*] *vitreus* in his later publications is discussed by Donk 1967 (Pe 5): 122.

yamanoi, Cryptoderma*, Imaz. 1951 (FPJ 4): 176 (Japan).—Apparently a superfluous name introduced for the combination of *Daedalea jezoensis* Yamano and *Trametes picei* Yamano. It has been referred to *Fomes* [*Phellinus*] *pini* by Lowe 1954 (M 46): 49, 1957 F.: 47. − Cf. (84**). − Descr.: Imaz., l.c.; S. Ito 1955: 367 *f. 262*; Imaz. & Hongo 1957 C.J. [1]: 144 *pl. 67 fs. 397, 398.* − If considered a distinct sp., the corrcet basionym appears to be *Daedalea jezoensis*; the simultaneously published name *Trametes picei* was reduced to the synonymy of the former by Tochinai & Kamei 1933 (ApJ 2): 573 (n.v.).

**yasudai, Fomes*, Lloyd 1915 (LMW 4, F.): 272 *f. 606* (Japan).—According to Lowe 1957 F.: 55 a "probable synonym" of *Fomes* [*Phellinus*] *robustus*. Listed, with doubt, as a syn. of *Phellinus hartigii* by S. Ito 1955: 373.

yezoense = *jezoensis*

**yoshinagae, Lenzites*, Lloyd 1922 (LMW 7): 1108 *pl. 189 f. 2042* (Japan). —Imaz. 1939 (JJB 15): 302 referred this to *Trametes* [*Antrodia*] *albida*, but Aosh. 1967 (TmJ 8): 3 considered it to be synonymous with *Daedalea* [*Antrodia*] *tanakae*.

zollingeri Lév. → *Polyporus zollingerianus* Lév.

**zollingerianus, Polyporus*, Lév. 1846 (ASn III 5): 131 (Indonesia, Java); *Trametes* Sacc. 1888; = *Polyporus zollingeri* Lév. 1846 M.: 120 (nom. nud.: n.v.p.), not ∼ Sacc. 1891.—At first referred to *Trametes lutescens* f. umbrina [= *Funalia gallica*] by Bres. 1897 (AAR III 3): 90 and to *Trametes favus* [sensu Bres. in part = *Funalia gallica*] but later identified with *Polyporus* [*Hirschioporus*] *versatilis* (Berk.) Romell by Bres. 1916 (Am 14): 229, and by Lloyd 1917 (LMW 5): 784 as *Polystictus* [*Hirschioporus*] *versatilis* (Berk.) Fr. (extra-European). Referred to '*Trametes cristata* Cooke' by Pat. 1928 (ACe 1): 11.

**zonalis, Polyporus*, (Kön.) ex Berk. 1843 (AM 10): 375 *pl. 10 f. 5* (Ceylon); *Boletus* Kön. in herb.; *Hansenia* P. Karst. 1879; *Leptoporus* Pat. 1900; *Fomitopsis* Imaz. 1943; *Fomes* Teng 1964; = *Rigidoporus zonalis* (Kön.

ex Berk.) Imaz. 1952.—Recorded for Europe, Portugal, by Alm. & Câm. 1909 (BSb 24): 152 without any description or comment: "In ligno emortuo *Populi* sp., pr. Coimbra, Bemcanta, leg. Octavio Vecchi, junio, 1903." Since it is doubtful that the determination is correct, the species and its numerous synonyms are neigher entered on the Check list proper nor taken further into account on this list. — Recently Al. David 1972 (BmF 87): 417 remarked that *Rigidoporus moeszii* is "identique ou peu différent de *Polyporus* [*Rigidoporus*] *zonalis* Berk." No supplementary information was presented.

*zonata, Daedalea, Schw. 1822: 94 (U.S.A., North Carolina): *Striglia* O.K. 1891.—Referred to *Agaricus* [*Daedaleopsis*] *confragosus* by Murrill 1905 (BTC 32): 86; according to Lloyd 1913 (LMW 4, L. 50): 11, this is the thin, zonate, lenzitoid, southern form of *Daedalea* [*Daedaleopsis*] *confragosa* in the U.S.A. — This species was recorded for Europe by Trog 1850 (MiB): 53 (no descr.) and Saut. 1878 (Msa 18): 149 (no descr.). — Cf. (31).

zonatus, Polyporus, Velen. → *Polyporus circulatus* Velen. (**O**).

zonatus coryli, Polyporus, Secr. 1833 M. 3: 134 (Switzerland) as a sp. of *Boletus* & double epithet & nom. nud.: n.v.p.).—Nomen dubium.

ABBREVIATIONS OF AUTHORS' NAMES

In deciding on abbreviations all groups of plants have been taken into account. This will explain my preference for instance for 'P. Karst.' (= P. A. Karsten, 1834–1917), not 'Karst.' (= G. K. W. H. Karsten, 1817–1908), who started publishing at an earlier date; and for 'Murrill', to avoid confusion with 'Murr.' (= J. A. Murray, 1740–1791).

A few wittingly irregular abbreviations are included as exceptions. These are 'Bond'. (not O. N. Bondarenko), 'J. B. Ell.' (instead of J. B. Ellis), 'Lloyd' (instead of 'C. Lloyd' = C. G. Lloyd), and 'Lowe' (instead of 'J. L. Lowe').

Ach.—E. Acharius
Adans.—M. Adanson
Ade—A. Ade
Afz.—A. Afzelius
All.—C. Allioni
Allen—W. B. Allen
Allesch.—A. Allescher
All. & Schn.—Allesch. & J. N. Schnabl
Alm. & Câm.—J. Verissimo de Almeida & M. Sousa da Câmara
A. Ames
Angerer—J. Angerer
Anon.—Anonymus
Aosh.—K. Aoshima
Aosh. & Furuk.—Aosh. & H. Furukawa
Aosh. & Kob.—Aosh. & T. Kobayasi
Arx—J. A. von Arx
A. & S.—J. B. von Albertini & L. D. von Schweinitz
Atk.—G. F. Atkinson
B. & . . . → Berk.
Badet—M. Badet
Bagl.—F. Baglietto
Bagl. & De-Not.—Bagl. & G. De-Notaris
Bagl. & Razz.—Bagl. & A. Razzore
Bakshi—B. K. Bakshi
Balabán & Kotl.—K. Balabán & F. Kotlaba
Balbis—G. B. Balbis
Bamb.—C. Van Bambeke
Banker—H. J. Banker
Barbier—M. Barbier
Barla—J. B. Barla
Barrel.—J. Barrelier
Bataille—F. Bataille
Batsch—A. J. G. K. Batsch

Batt.—G. A. Battarra
C. Bauh.—C. Bauhin
Bauh. & Cherl.—J. Bauhin & J. H. Cherler
Baumg.—J. C. G. Baumgarten
D. Baxt.—D. V. Baxter
Baxt. & Manis—D. Baxt. & W. E. Manis
P. Beauv.—A. M. F. J. Palisot de Beauvois
G. Beck—G. Beck von Mannagetta und Lerchenau
Beeli—M. Beeli
Bergam.—G. Bergamaschi
Bergeret—J. P. Bergeret
Bergstädt—V. Bergstädt
Berk.—M. J. Berkeley
B. & Br.—Berk. & C. E. Broome
B. & C.—Berk. & M. A. Curtis
B. & Cooke—Berk. & M. C. Cooke
B. & Mont.—Berk. & J. P. F. C. Montagne
B. & Rav.—Berk. & H. W. Ravenel
Berlese—A. N. Berlese
Bertol.—G. Bertolini
Biers—P. M. Biers
Big. & Guill.—R. Bigeard & H. Guillemin
Bijl—P. A. van der Bijl
Bizzoz.—G. Bizzozero
Bl.—C. L. Blume
Blackstone—J. Blackstone
Bl. & Nees—Bl. & T. F. L. Nees von Esenbeck
Błoński—F. Błoński
Boccone—P. Boccone
Boed.—K. B. Boedijn

Boehm.—G. R. Boehmer
Boisd.—J. A. Boisduval
Bolt.—J. Bolton
Bon.—H. F. Bonorden
M. Bon
Bond.—A. S. Bondarcev
Bond. & Boris.—Bond. & P. N. Borisov
Bond. & Kom.—Bond. & E. P. Komarova
Bond. & Ljub.—Bond. & L. V. Ljubarskij
Bond. & M. Bond.—Bond. & M. A. Bondarceva
Bond. & Parm.—Bond. & E. Parmasto
Bond. & S.—Bond. & R. Singer
M. Bond.—M. A. Bondarceva
Bong.—H. G. Bongard
Borisov—P. N. Borisov [Borissov]
I. Boršč.—I. G. Borščov [E. Borszczow]
Bosc—L. A. G. Bosc
Boud.—J. L. E. Boudier
Boud. & Fisch.—Boud. & E. Fischer
Bourd. & G.—Bourd. & A. Galzin
Bourd. & L. Maire—Bourd. & L. Maire
Boyce—J. S. Boyce
Branke—V. JU. fon Branke
Bref.—O. Brefeld
Bres.—G. Bresadola
Bres. & Cav.—Bres. & F. Cavara
Bresinsky & Stangl—A. Bresinsky & J. Stangl
Bret. & Niel—A. Le Breton & E. Niel
Breyne—J. Breyne (Lat., Breynius)
F. Brig.—F. Briganti
Britz.—M. Britzelmayr
Brongn.—A. T. de Brongniart
Brot.—F. de Avellar Brotero
Brotzm. & Gilb.—H. G. Brotzman & R. L. Gilbertson
J. G. Brown
Buchs—M. Buchs
Bull.—J. B. F. [or "Pierre"] Bulliard
Buller—A. H. R. Buller
Bundy—W. F. Bundy
Burt— E. A. Burt
Byl → Bijl
DC.—A. P. de Candolle
Cald.—L. Caldesi
S. Câm.—E. de Sousa da Câmara
Campb. & Dav.—W. A. Campbell & R. W. Davidson
Canf. & Gilb.—E. R. Canfield & R. L. Gilbertson
Cappelli—C. Cappelli

Carm.—D. Carmichael (1772–1827)
J. W. Carm.—J. W. Carmichael
Cartwr.—K. T. St G. Cartwright
Cartwr. & Findl.—Cartwr. & W. P. K. Findlay
Cash—E. K. Cash
Cejp—K. Cejp
Černj.—V. M. Černjajev [Czernajew]
Černý—A. Černý
Ces.—V. Cesati
Ces. & De-Not.—Ces. & C. De-Notaris
Chaill.—J. F. de Chaillet
Chev.—F. F. Chevallier
Chod. & Mart.—R. H. Chodat & C. E. Martin
C. M. Christ.—C. M. Christensen
M. P. Christ.—M. P. Christiansen
Clel.—J. B. Cleland
Clel. & Cheel—Clel. & E. Cheel
Clel. & Rodw.—Clel. & L. Rodway
F. Clem.—F. E. Clements
Clus.—C. Clusius [C. de l'Écluse]
Coker—W. C. Coker
Colenso—W. Colenso
Colm.—M. Colmeiro [y Penido]
Comes—O. Comes
Cooke—M. C. Cooke
Cooke & Ell.—Cooke & J. B. Ellis
Cooke & Harkn.—Cooke & H. W. Harkness
Cooke & Mass.—Cooke & G. E. Massee
Cooke & Q.—Cooke & L. Quélet
W. Cooke—W. B. Cooke
Corb.—L. Corbière
Corda—A. C. J. Corda
Cordier—F. S. Cordier
Corner—E. J. H. Corner
Cost. & Duf.—J. Costantin & L. Dufour
P. Cout.—A. X. Pereira Coutinho
Crag.—F. W. Cragin
Crouan—P. L. & H. M. Crouan
Cumino—U. Cumino
G. Cunn.—G. H. Cunningham
Currey—F. Currey
Curt.—W. Curtis
M. A. Curt.—M. A. Curtis
Dambl. & Moureau—J. Damblon & J. Moureau
Al. David—A[lix] David
Davids. & Campb.—R. W. Davidson & W. A. Campbell
R. W. Davids.—R. W. Davidson
DC.—A. P. de Candolle
Delastre—C. J. L. Delastre

Delle-Chiaje—S. Delle Chiaji

Del. & Mont.—A. R. Delile & J. P. F. C. Montagne

Demoulin—V. Demoulin

De-Not.—G. De-Notaris

Desm.—J. B. H. J. Desmazières

Desv.—A. N. Desvaux

Dicks.—J. J. Dickson

A. Dietr.—A. H. Dietrich

D. Dietr.—D. Dietrich

Dill.—J. J. Dillenius [= Dillen]

Ditm.—L. P. F. Ditmar

Doass. & Pat.—E. Doassans & N. F. Patouillard

B. O. Dodge

Doidge—E. M. Doidge

Domański—S. Domański

Dom. & Orl.—Domański & A. Orlicz

Donk—M. A. Donk

Doyer—C. M. Doyer

Dubois—F. N. A. Dubois

Duby—J. E. Duby

E. Duchesne—E. A. Duchesne

Dumée—P. Dumée

Dur. & Mont.—M. C. Durieu de Maisonneuve & J. P. F. C. Montagne

Duss—A. Duss

Eeden—F. W. van Eeden

Egeland—J. Egeland

Ehrenb.—C. G. Ehrenberg

Eliade—E. Eliade

Ell. & Ev.—J. B. Ell. & B. M. Everhart

Ell. & Gall.—J. B. Ell. & B. T. Galloway

Ell. & Harkn.—J. B. Ell. & H. W. Harkness

Ell. & Langl.—J. B. Ell. & A. B. Langlois

Ell. & Macbr.—J. B. Ell. & T. H. Macbride

Ell. & Sacc.—J. B. Ell. & P. A. Saccardo

J. B. Ell.—J. B. Ellis

Endl.—S. F. L. Endlicher

Englerth—G. H. Englerth

Enslin—J. C. Enslin

Erikss. & Strid—Jo. Erikss. & Å. Strid

Jo. Erikss.—J[ohn] Eriksson

Fairm.—C. E. Fairman

Falck—R. Falck

R. & O. Falck—Falck & O. Falck

Fallahyan—F. Fallahyan

Famintzin & Vor.—A. S. Famintzin & M. S. Voronin [M. Woronin]

Farinha—M. Farinha

Farinha & Rosado—Farinha & J. M. Rosado

Farl. & Burt—W. G. Farlow & E. A. Burt

Faull—J. H. Faull

Favre & Ruhlé— J. Favre & S. Ruhlé

Fay.—V. Fayod

Feltg.—J. Feltgen

Ferd. & Jørg.—C. C. F. Ferdinandsen & C. A. Jørgensen

Ferd. & W.—C. C. F. Ferdinandsen & Ø. Winge

Ferry—R. Ferry

Fic. & Sch.—H. D. A. Ficinus & C. Schubert

Fid.—O. Fidalgo

K. Fid.—M. E. P. K. Fidalgo

O. & K. Fid.—Fid. & M. E. P. K. Fidalgo

Forq.—L. Forquignon

Fr.—E. M. Fries

Fr. & Berk.—Fr. & M. J. Berkeley

R. Fr.—O. R. Fries

Fremr—V. Fremr

Frost—C. C. Frost

Fuck.—K. W. G. L. Fuckel

P. Gärtn.—P. G. Gärtner

P. Gärtn. & al.—P. G. Gärtner, B. Meyer, & J. Scherbius

Gaill.—A. Gaillard

Gard—M. Gard

Gasparrini—G. Gasparrini

Gaterau

Gaud.—C. Gaudichaud-Beaupré

Gaut.—L. M. Gautier

Gilb. & Lowe—R. L. Gilb. & J. L. Lowe

R. L. Gilb.—R. L. Gilbertson

Gillet—C. C. Gillet

Gillot—F. X. Gillot

Gillot & Luc.—Gillot & J. L. Lucand

Ginns—J. H. Ginns

Gled.—J. G. Gleditsch

J. F. Gmel.—J. F. Gmelin

Gonz.—R. Gonzáles Fragoso

Gouan—A. Gouan

Govi—G. Govi

Graff—P. W. Graff

Gramb.—E. Gramberg

S. F. Gray

Grev.—R. K. Greville

Grög.—F. Gröger

Güssow—H. T. Güssow

Guillaud & al.—J. A. Guillaud, L. Forquignon, & N. Merlet
Guillemot—J. Guillemot
Gunn.—J. E. Gunnerus
Haddow—W. R. Haddow
Haglund—Erik E. Haglund
Haller—A. von Haller
L. Hansen—L[ise] Hansen
Hara—K. Hara
Haracsi—L. Haracsi
Hard—M. E. Hard
Hariot—P. A. Hariot
Har. & Karst.—Hariot & P. A. Karsten
Har. & Pat.—Hariot & N. T. Patouillard
Harkn.—H. W. Harkness
Hartig—T. Hartig
R. Hartig—H. J. A. R. Hartig
Harz—C. O. Harz
Harzer—C. A. F. Harzer
Hazsl.—F. A. Hazslinszky
Hedgc. & Long—G. G. Hedgcock & W. H. Long
R. Heim
Hemmi—T. Hemmi
Hemmi & Ikeya—Hemmi & J. Ikeya
Hendr.—F. L. Hendrickx
B. Henn.—B. Hennig
P. Henn.—P. C. Hennings
Henriq.—J. A. Henriques
Heufl.—L. J. von Heufler, later L. von Hohenbühel, genannt Heufler zu Rasen und Perdonegg
Hill—J. Hill
Höhn.—F. X. R. von Höhnel
Hoffm.—G. F. Hoffmann
Holm → Holmskj.
Holmskj.—T. Holmskjold
Holterm.—C. Holtermann
Hook.—W. J. Hooker
Hook. f.—J. D. Hooker
Hornem.—J. W. Hornemann
House—H. D. House
Houtt.—M. Houttuyn
Hruby—J. Hruby
H. Huber
Hubert—E. E. Hubert
Huds.—W. Hudson
S. Hugh.—S. J. Hughes
Humb.—F. H. A. von Humboldt
Humphr. & Leus—C. J. Humphrey & S. Leus[-Palo]
Humphr. & Sigg.—C. J. Humphrey & P. V. Siggers

Hussey—A. M. Hussey
Igmándy—Z. Igmándy
Imaz.—R. Imazeki
Imaz. & Aosh.—Imaz. & K. Aoshima
Imaz. & Hongo—Imaz. & T. Hongo
Imaz. & Toki—Imaz. & S. Toki
Imbach—E. J. Imbach
Ingold—C. T. Ingold
Inz.—G. Inzenga
Istv.—G. Istvánffi de Csík-Madéfalvi (formerly, J. Schaarschmidt)
S. Ito
Jaap—O. Jaap
Jač.—A. A. Jačevskij [A. L. von/de Jaczewski]
Jacq.—N. J. von Jacquin
H. Jahn
Jel. & Tort.—M. B. Jelić & M. Tortić
Joach. & Dumée—L. Joachim & P. Dumée
Juel—H. O. Juel
Jullien
Jungh.—F. F. W. Junghuhn
Juss.—A. L. de Jussieu
Jørst.—I. Jørstad
Jørst. & Juul—Jørst. & J. G. Juul
O.K.—C. E. O. Kuntze
Kalchbr.—K. Kalchbrenner
Kalchbr. & Cooke—Kalchbr. & M. C. Cooke
Kallenb.—F. J. Kallenbach
P. Karst.—P. A. Karsten
Kartav.—N. T. Kartavenko, [N. T. Stepanova-Kartavenko]
C. H. Kauffm.—C. H. Kauffman
Kavina—K. Kavina
Kawam.—S. Kawamura
Keissl.—K. von Keissler
Kickx f.—J. J. Kickx (1803–64)
Killerm.—S. Killermann
L. Kirchn.—L. A. Kirchner
W. Kirschst.—W. Kirschstein
Kl.—J. F. Klotzsch
Y. Kobay.—Y. Kobayasi
Kögl—F. Kögl
Kön.—J. G. König [Kønig]
E. Komar.—E. P. Komarova
Konr.—P. Konrad
Konr. & M.—Konr. & A. Maublanc
Kotl.—F. Kotlaba
Kotl. & P.—Kotl. & Z. Pouzar
E. Krause—E. H. L. Krause
Kravc.—B. I. Kravcev
Kreisel—H. Kreisel

Kreis. & Jahn—H. Kreisel & H. Jahn
Krombh.—J. V. von Krombholz
Kühner—R. Kühner
Kumm.—P. Kummer
O. Kuntze → O.K.
Kunze—G. Kunze
L.—C. Linnaeus
Lam.—J. B. A. P. M. de Lamarck
Lapl.—M. C. de Laplanche
Lasch—W. G. Lasch
Laterr.—J. F. Laterrade
Latourr.—M. A. L. C. de Fleurieu de
  Latourrette
Laubert—R. Laubert
Lázaro—B. Lázaro é Ibiza
Lea—T. G. Lea
Lek—H. A. A. van der Lek
Lenz—H. O. Lenz
Letell.—J. B. L. Letellier
Lév.—J. H. Léveillé
Leyss.—F. W. von Leysser
Lib.—M. A. Libert
Liboš.—J. JA. Libošic [J. Liboschitz]
Lightf.—J. Lightfoot
Liljebl.—S. Liljeblad
Lind—J[ens] Lind
Lind. & Syd.—G. Lindau & P. Sydow
Lindblad—M. A. Lindblad
Lindr.—J. J. Lindroth
Linhart—G. Linhart
Link—J. H. F. Link
Litsch.—V. Litschauer
Litsch. & Lohw.—Litsch. & H. Lohwag
Ljub.—L. V. Ljubarskij
Lloyd—C. G. Lloyd
Lohw.—H. Lohwag
Lohw. & Follner—Lohw. & L. Follner
K. Lohw.—K. Lohwag
Lombard—F. F. Lombard
Lomb. & Gilb.—Lombard & R. L.
  Gilbertson
Long—W. H. Long
Long & Baxt.—Long & D. V. Baxter
Loud.—J. C. Loudon
Lour.—J. de Loureiro
Lowe—J. L. Lowe
Lowe & Gilb.—Lowe & R. L. Gilbert-
  son
Lowe & Lund.—Lowe & S. Lundell
Luc. & Gillot—L. Lucand & F. X.
  Gillot
Lübstorf—W. Lübstorf
F. Ludw.—F. Ludwig
Lumn.—S. Lumnitzer

Lundell—S. Lundell
Lund. & Pil.—Lundell & A. Pilát
Maas G.—R. A. Maas Geesteranus
Macbr.—T. H. Macbride (1848–1934)
Macrae & Aosh.—R. Macrae & K.
  Aoshima
P. Magn.—P. W. Magnus
Maheu—J. Maheu
Maire—R. C. J. E. Maire
Maire & Pol.—Maire & J. Politis
Maire & Wern.—Maire & R. G. Werner
L. Maire
Mal. & Bert.—Malenç. & R. Bertault
Malenç.—J. L. G. Malençon
Malinvaud—E. Malinvaud
Malkovský—K. M. Malkovský
Manc. & Sacc.—V. Mancini & P. A.
  Saccardo
Mang. & Pat.—L. A. Mangin & N. T.
  Patouillard
Mang. & Viala—L. Mangin & P. Viala
Maratti—G. F. Maratti
March. & Court.—L. Marchand & R.
  Courtois
L. March.—L. Marchand (fl. 1826–30)
Marcucci—M. Marcucci
Mart.—K. F. P. von Martius
C. E. Mart.—C. E. Martin
Martelli—U. Martelli
T. Martius
Mass.—G. E. Massee
Mattioli—P. A. Mattioli (Lat., Matthio-
  lus)
Mattirolo—O. Mattirolo
Maubl.—A. Maublanc
Mauri—E. Mauri
May—K. May
McCallum—A. W. McCallum
McGinty = Lloyd
McIlv.—C. McIlvaine
McKay—H. H. McKay
Mérat—F. V. Mérat [de Vaumartoise]
G. Meyer—G. F. W. Meyer
Mez—C. Mez
Mich.—P. A. Micheli
Michael—E. Michael
L. W. Mill.—L. W. Miller
A. Möll.—F. A. G. J. Möller
Moesz—G. von Moesz / Moesz G.
Mont.—J. P. F. C. Montagne
Mont. & Dur.—Mont. & M. C. Durieu
  de Maisonneuve
L. Morel—L. F. Morel
Morg.—A. P. Morgan

Moritzi—A. Moritzi
Moug.—J. B. Mougeot
Moug. & Nestl.—Moug. & C. G. Nestler
Moug. & Ferry—J. A. [= Antoine] Mougeot & R. Ferry
Moug. & Lév.—Moug. & J. H. Léveillé
Mounce & Macrae—I. Mounce & R. Macrae
O. F. Müll.—O. F. Müller
Müll. & Jahn—G. Müller & H. Jahn
Murašk.—K. E. Muraškinskij
Murr.—J. A. Murray
Murrill—W. A. Murrill
Mussat—E. Mussat
Nannf.—J. A. Nannfeldt
Nannf. & Du R.—Nannf. & G. E. Du Rietz
Nath.-W.—T. Nathorst-Windahl
Nees—C. G. D. Nees von Esenbeck
C. & T. Nees—Nees & T. F. L. Nees von Esenbeck
Nentien
Neuman—J. J. Neuman
Niemelä—T. Niemelä
Nikol.—T. L. Nikolajeva
Nobles—M. K. Nobles
Nobles & Frew—Nobles & B. P. Frew
Nocca & Balb.—D. Nocca & C. B. Balbis
Nohara—S. Nohara
Nyl. & Sæl.—W. Nylandei & T. Sælan
Oed.—G. C. Oeder
Oehm—G. Oehm
A. J. Olson
O.K.—C. E. O. Kunze
Opiz—F. M. Opiz (usually listed as P. M. Opiz)
Orl.—A. Orlicz
Otth—G. H. Otth
Otto—J. G. Otto
Oud.—C. A. J. A. Oudemans
Over.—C. van Overeem
C. & D. Over.—Over. & D. van Overeem-de Haas
Over. & Steinm.—Over. & A. Steinmann
Over. & Weese—Over. & J. Weese
Overh.—L. O. Overholts
Overh. & Engl.—Overh. & G. H. Englerth
Oveih. & L.—Overh. & J. L. Lowe
Panizzi—F. Panizzi
Paravicini—E. Paravicini
Park.-Rh.—A. F. Parker-Rhodes

Parm.—E. Parmasto
Passerini—G. Passerini
Pat.—N. T. Patouillard
Pat. & Gaill.—Pat. & A. Gaillard
Pat. & Har.—Pat. & P. A. Hariot
Pat. & Lag.—Pat. & N. G. Lagerheim
Paul.—J. J. Paulet
Payer—J. B. Payer
P. Beauv.—A. M. F. J. Palisot de Beauvois
A. Pears.—A. A. Pearson
Peck—C. H. Peck
Pegl.—D. N. Pegler
Pegl. & Gibs.—Pegl. & J. A. S. Gibson
Pegl. & Young—Pegl. & T. W. K. Young
Pegl. & Wat.—Pegl. & J. M. Waterston
Pers.—C. H. Persoon
Petch—T. Petch
Pfeiff.—L. Pfeiffer
Pilát—A. Pilát
Pil. & Lindtn.—Pilát & V. Lindtner
Pinto-L.—J. Pinto-Lopes
Plan.—J. J. Planer
Plowr.—C. B. Plowright
Poelt & Jahn—J. Poelt & H. Jahn
Poelt & Oberw.—J. Poelt & F. Oberwinkler
Poetsch—J. S. Poetsch
Poir.—J. L. M. Poiret
Pollini—C. Pollini
Poll. & Kauffm.—J. B. Pollock & C. H. Kauffman
Post—H. A. von Post
Pouz.—Z. Pouzar
Povah—A. H. W. Povah
Preuss—C. G. T. Preuss
Prilleux—E. E. Prilleux
Purt.—T. Purton
Quél.—L. Quélet
Quél. & Schulz.—Quél. & S. Schulzer von Müggenburg
Rab.—G. L. Rabenhorst
Radde & Walt.—G. F. R. Radde & A. Walter
Raddi—G. Raddi
Rafin.—C. S. Rafinesque[-Schmaltz]
Ravenel—H. W. Ravenel
Ray—J. Ray
Razzore—A. Razzore
G. F. Re
Rea—C. Rea
Reb.—J. F. Rebentisch
Reichenb.—H. G. L. Reichenbach

D. Reid—D. A. Reid
Reid & Austw.—D. A. Reid & P. K. C. Austwick
Rennerf.—E. Rennerfeldt
Retz.—A. J. Retzius
Richon—C. E. Richon
Rick.—A. Ricken
J. Rick
Riel—J. Riel
Ripart—J. B. Ripart
Risso—J. A. Risso
Robak—H. Robak
Rodw. & Clel.—L. Rodway & J. B. Cleland
D. P. Rog.—D. P. Rogers
Rog. & Mart.—D. P. Rog. & G. W. Martin
Rolland—L. L. Rolland
Roll-H.—F. Roll-Hansen
Romagn.—H. Romagnesi
Romell—L. Romell
Roques—J. Roques
Rostk.—F. W. T. Rostkovius
Rostr.—F. G. E. Rostrup
Roth—A. W. Roth
Roum.—C. Roumeguère
Roussel—H. F. A. de Roussel
Rox. Clem.—S. de Roxas [later, Rojas] Clemente y Rubio
Rubel—F. Rubel
Ryv.—L. Ryvarden
Sacc.—P. A. Saccardo
Sacc. & Berl.—Sacc. & A. N. Berlese
Sacc. & Cub.—Sacc. & G. Cuboni
Sacc. & Ell.—Sacc. & J. B. Ellis
Sacc. & Lind.—Sacc. & G. Lindau
Sacc. & Manc.—Sacc. & V. Mancini
Sacc. & D. Sacc.—Sacc. & D. Saccardo
Sacc.& Syd.—Sacc. & P. Sydow
Sacc. & Trav.—Sacc. & C. B. Traverso
Sacc. & Trott.—Sacc. & A. Trotter
Sarkar—A. Sarkar
Sart. & M.—A. Sartory & L. Maire
Saut.—A. E. Sauter
Schaeff.—J. C. Schaeffer
Schestunoff → Šestunov
Schlechtend.—D. F. L. von Schlechtendal
Schleich.—J. C. Schleicher
J. C. Schmidt (1793–1850)
Schoorel & Boed.—A. F. Schoorel & K. B. Boedijn
Schrad.—H. A. Schrader
Schrank—F. von Paula von Schrank

Schrenk—H. von Schrenk
Schrenk & Spauld.—Schrenk & P. Spaulding
A. Schröder
J. Schroet.—J. Schroeter
K. F. Schultz
S. Schulz.—S. Schulzer von Müggenburg
Schulz. & Bres.—S. Schulz. & G. Bresadola
Schum.—H. C. F. Schumacher
Schw.—L. D. von Schweini[t]z
Schwalb—K. J. Schwalb
Scop.—G. A. Scopoli
Seav.—F. J. Seaver
Secondat—J. B. de Secondat
Secr.—L. Secretan
Šestunov—N. Šestunov [Schestunoff]
Seyn.—J. de Seynes
Shirai—M. Shirai
Shope—P. F. Shope
Sibth.—J. Sibthorp
Siem.—W. Siemaszko
Siemss.—A. C. Siemssen
Siepm.—R. Siepmann
Siepm. & Zycha—Siepm. & H. Zycha
Sing.—R. Singer
Sin. & M. Bond.—JU. V. Sinadskij & M. A. Bondarceva
Škorič—V. Škorič
Sm.—J. E. Smith
A. L. Sm.—A. L. Smith
A. L. Sm. & Rea—A. L. Sm. & C. Rea
W. G. Sm.—W. G. Smith
Solenander—R. Solenander [cited by Pers. as "Solenaud."]
Sommerf.—S. C. Sommerfelt
Sorok.—N. V. Sorokin
Sow.—J. Sowerby
Speg.—C. L. Spegazzini
Spreng.—K. P. J. Sprengel
Stahlschmidt—C. Stahlschmidt
St-Am.—J. F. B. de Saint-Amans
Steinheim—C. Steinheim
Step.-Kart. → Kartav.
Sterb.—F. van Sterbeeck
Steud.—E. G. Steudel
Stev. & Cash—J. A. Stevenson & E. K. Cash
Steyaert—R. L. Steyaert
Strass.—P. Strasser
Strauss—F. K. J. von Strauss
Streinz—W. M. Streinz
Sumst.—D. R. Sumstine
Suza—J. Suza

Švarcm.—S. R. Švarcman
Svrček—M. Svrček
Sw.—O. P. Swartz
Syd.—P. Sydow
H. & P. Syd.—H. Sydow & P. Sydow
A. Teix.—A. R. Teixeira
Teix. & Rog.—A. Teix. & D. P. Rogers
Teng—S. C. Teng
Teramoto—T. Teramoto
Terracc.—T. Terracciano
Teston—D. Teston
Theiss.—F. Theissen
Theiss. & Syd.—Theiss. & H. Sydow
Thind & Ratt.—K. S. Thind & S. S. Rattan
Thore—J. Thore
Thüm.—F. K. A. E. J. von Thümen [von Gräfendorf]
Thunb.—C. P. Thunberg
Timm—J. C. Timm
Tochinai & Kamei—Y. Tochinai & S. Kamei
Tode—H. J. Tode
Torrend—C. Torrend
Tortič—M. [V.] Tortič
Tort. & Jel.—Tortič & M. B. Jelić
Tourn.—J. Pitton de Tournefort
Trapp.—J. E. van der Trappen
Tratt.—L. Trattinnick
Trav.—G. B. Traverso
Trog—J. G. Trog
Trott.—A. Trotter
L. Tul.—E. L. R. Tulasne
Ulbr.—E. Ulbrich
Underw.—L. M. Underwood
Vahl—M. Vahl
Vaill.—S. Vaillant
Vanin—S. I. Vanin
Velen.—J. Velenovský
Venturi—A. Venturi
Veull.—C. Veulliot
Vill.—D. Villars [also, Villar]
Vitt.—C. Vittadini
Viv.—D. Viviani

Vleugel—J. S. Vleugel
Voronich.—N. N. Voronichin [Woronichin]
Voronin—M. S. Voronin [M. Woronin]
Voronov—JU. N. Voronov [G. Woronow]
W. Voss
Vuyck—L. Vuyck
Wahl.—G. Wahlenberg
Wakef.—E. M. Wakefield
Wak. & Denn.— Wakef. & R. W. G. Dennis
Wak. & Pears.—Wakef. & A. A. Pearson
Wallr.—K. F. W. Wallroth
Warm.—J. E. B. Warming
Web.—G. H. Weber
G. F. Web.—G. F. Weber
Wehmeyer—L. E. Wehmeyer
Weinm.—J. A. Weinmann [J. A. Vejnman]
Weir—J. R. Weir
Westh.—G. C. A. van der Westhuizen
Wettst.—R. von Wettstein [Von Westersheim]
Willd.—C. L. Willdenow
Wint.—H. G. Winter
Wirtg.—P. W. Wirtgen
With.—W. Withering
Wittr.—V. B. Wittrock
Wojew.—W. Wojewoda
F. A. Wolf
J. E. Wright
Wulf.—F. X. von Wulfen
T. Wulff
Yamano—Y. Yamano
Yas.—A. Yasuda
Yen—H. C. Yen
Zaneveld—J. S. Zaneveld
Zant.—G. Zantedeschi
Zeller—S. M. Zeller
Zerova—M. J. Zerova
Zoll.—H. Zollinger

STRONGLY REDUCED BIBLIOGRAPHIC REFERENCES

(**AAp**), Atti R. Accad. pontaniana
(**AAR**), Atti I.R. Accad. Agiati [Rovereto]
(**AAt**), Actes Inst. bot. Univ. Athèn.
(**ABA**), An. Mus. nac. Hist. nat. B. Aires
(**Abc**), Acta bot. croat.
(**Abf**), Acta bot. fenn.
(**ABS**), Ark. Bot. [Sweden]
(**ABU**), Annln Bot. [ed. Usteri]
(**ABu**), Annls Jard. bot. Buitenz. (S. = supplement vol.)
(**ACE**), Atl. Champ. Eur. [ed. Kavina & Pilát] [Czech Ed.: Atl. Hub. evr.]
(**Afe**), Acta Soc. Fauna Fl. fenn.
(**AhG**), Abh. hall. naturf. Ges.
(**AHp**), Acta Horti petropol. / Trudy imp. S-Peterb. bot. Sada
(**AII**), Atti (R.) Ist. Incoragg. Sci. nat.
(**AIv**), Atti Ist. veneto Sci.
(**AJB**), Am. J. Bot.
(**AM**), Ann. Mag. nat. Hist.
(**Am**), Annls mycol.
(**AMo**), Ann. Missouri bot. Gdn
(**AMP**), Annls Mus. (natn.) Hist. nat., Paris
(**AmW**), Acta mycol. [Warszawa]
(**AN**), Am. Nat.
(**An**), Archs néerl. Sci. exact. nat.
(**AnH**), Ann. nat. Hist., Lond.
(**ANM**), Arch. Ver. Freunde Naturg. Mecklenb.
(**APe**), Ann. R. bot. Gdns Peradeniya
(**Apg**), Acta phytotax. geob., Kyoto
(**Aph**), Acta phytopath. Acad. Sci. hung.
(**ApJ**), Ann. phytopath. Soc. Japan
(**APo**), Acta Soc. Bot. Polon.
(**ArP**), Arch. Protistenk.
Arx **1970**, Gen. Fungi sporul. pure Cult.
A. & S. **1805**, Consp. Fung. nisk.
(**ASa**), An. Soc. ci. argent.
(**ASf**), Acta Soc. Sci. fenn.
(**ASL**), Annls Soc. bot. Lyon
(**ASl**), Atti Soc. ligust. Sci. nat. geogr.
(**ASn**), Annls Sci. nat. (Bot.)
(**ASv**), Atti Soc. ven.-trent. Sci. nat.
(**ATh**), Archf Theecult. Ned.-Indië
Atk. **1900**, Mushrooms: **1901**, Mushrooms, 2nd Ed.
(**AUL**), Annls Univ. Lyon
(**AW**), Annln naturh. Hofmus./Mus. Wien
(**BaB**), Boln Soc. argent. Bot.
(**BAC**), Bull. Dep. Agric. Dom. Can.
Badet **1934**, Contr. Étud. Hym. Fr.
(**BAg**), Ber. naturw. Ver. Augsburg

Balabán & Kotl. **1970**, Atlas dřevok. Hub

(**BAN**), Bull. Soc. Hist. nat. Afr. Nord

Barla **1859**, Champ. Prov. Nice

(**BAt**), Bull. Soc. Hist. nat. Autun

Batsch **1783**, Elench. Fung.: **1786**, Elench. Fung. Cont. 1; **1789**, Elench. Fung. Cont. 2

Batt. **1755**, Fung. arim. Hist.

Bauh. & Cherl. **H.**, Hist. Pl. univ. 3

C. Bauh. **1623**, πιναξ [Pinax]

Baumg. **1790**, Fl. lips.

(**Bba**), Ber. bayer. bot. Ges.

(**BBB**), Bull. Soc. r. Bot. Belg.

(**BbC**), Beih. bot. Cbl.

(**BBf**), Bull. Buffalo Soc. nat. Sci.

(**BbF**), Bull. Soc. bot. Fr.

(**BBr**), Bull. Jard. bot. Brux.

(**BBu**), Bull. Jard. bot. Buitenzorg & Bull. bot. Gdns Buitenzorg

(**BCb**), Bot. Cbl./Zbl.

(**BCó**), Boln Acad. nac. Ci. Córdoba

(**BCR**), Bot. Cbl. Deutschl. [ed. Rab.] [One volume published]

(**BdN**), (Amtl.) Ber. Vers. dt. Naturf. Ärzte

Bergeret **P.**, Phytonom. univ. [Dates uncertain.]

Berk. **1836**, Fungi (Sm., Engl. Fl. 5^2); **1860**, Outl. Br. Fungol.: **N.Z.**, Fungi *in* Hook. f., Bot. antarct. Voy. 2, Fl. Nov.-Zel.; **Ta.**, Fungi *in* Hook. f., Bot. antarct. Voy. 3, Fl. Tasm.

(**BFi**), Bidr. Känn. Finl. Nat. Folk

(**BFJ**), Bull. imp./Govt Forest Exp. Stn, Japan

(**BFY**), Bull. Sch. For. Yale Univ.

(**BG**), Bot. Gaz.

(**BGe**), Bull. Trav. Soc. bot. Genève (1879–1908) & Bull. Soc. bot. Genève II (1909–51)

Big. & Guill. **1909**, Fl. Champ. sup. Fr. [1]; **1913**, Fl. Champ. sup. Fr. 2

(**BIm**), Bres., Icon. mycol.

Birkf. & Hersch. **1961–8**, Morph.-anat. Bildt. Pilzk.

Bizzoz. **F.v.**, Fl. ven. critt.

(**BJ**), Bot. Jb.

(**BJG**), Bot. Jaarb. [Gent]

(**BK**), Bot. Közl.

(**BLa**), Ber. bot. Ver. Landshut

Blackstone **1746**, Specimen bot. Pl. Angl.

(**BlL**), Bull. mens. Soc. linn. Lyon

Bl. & Nees **1826**, Fungi iavan. *in* Nova Acta physoc.-med. Leop. Carol. 13 (1)

(**Bly**), Blyttia

(**BMa**), Bull. Soc. Sci. nat. (phys.) Maroc

(**BmF**), Bull. (trim.) Soc. mycol. Fr. (Atl. = Atlas)

(**BmI**), Bull. misc. Inf. Kew [Often erroneously cited as Kew Bull.]

(**BMs**), Bot. Mater. Inst./Otd. spor. Rast.

(**BMT**), Bot. Mag., Tokyo

(**BnH**), Beih. nova Hedw.

(**BNS**), Bull. New York St. Cab. / Bull. New York St. Mus. nat. Hist. / New York St. Mus. Bull. [Title varies]

(**BnS**), Ber. naturw. Ver. Schwab. u. Neub.

(**BnW**), Bijdr. natuurk. Wet.

(**Bo**), Bothalia

Boccone M.F., Mus. Fis.

Boehm. **1750**, Fl. Lips.

Bolt. **1788–91**, Hist. Fung. Halif.

(**BoN**), Bot. Notiser

Bond. **1953**, Trutov. Griby; **1956**, Prosobie Opred. Domav. Grib.

I. Boršć. [E. Borszcz.] **1856**, *in* Middendorff, Reise Sibir. **1** (2³)

Boud. **1904–11**, Ic. mycol. "1905–10" [Preliminary text 1904–9; final text 1911]

Bourd. & G. **1928**, Hym. Fr. "1927"

(**BOy**), Bull. Soc. Nat. Oyonnax

(**BPI**), Bull. Bur. Pl. Ind. U.S. Dep. Agric.

(**BPv**), Bull. Soc. Path. vég. Fr.

(**BPZ**), Bull. Pl. Dis. Div., Dep. sci. ind. Res., New Zeal.

(**Bra**), Bragantia

Bref. U., Unters. Gesammtgeb. Mykol.

Bres. **1926**, Sel. mycol. II *in* Studi tridentini II **6**; **F.m.**, Fungi mang. vel. Eur. media (1899); & II Ed. (1906); **F.t.**, Fungi trident.

Britz. **H.S.**, Hym. Südb. [As far as not published in periodicals]

(**BRo**), Bull. Soc. Amis Sci. nat. Rouen

(**Bro**), Brotéria (Bot.) 1902–32, (Ci. nat.) 1932–

Brot. **1804**, Fl. lusit. **2**

(**BSb**), Bolm Soc. broteriana

(**BSD**), Bull. Soc. Hist. nat. Doubs

(**BSe**), Boln R. Soc. esp. Hist. nat.

(**BSM**), Bull. Soc. imp. Nat. Moscou

(**BsM**), Biblphy syst. Mycol.

(**BSp**), Bull. Soc. portug. Sci. nat.

(**BT**), Bot. Tidsskr.

(**BTC**), Bull. Torrey bot. Cl.

(**BTS**), Bull. Tokyo Sci. Mus.

Bull. **1780–93**, Herb. Fr. [Plates]; **H.**, Hist. Champ. Fr. [Text]

(**BUS**), Bull. U.S. Dep. Agric.

(**BVP**), Bull. Div. Veg. Phys. Path. U.S. Dep. Agric.

(**BWb**), Bull. Washburn Lab. nat. Hist.

(**BZ**), Bot. Ztg

(**BŽ**), Bot. Ž.

DC. **1805**, Lam. & DC., Fl. franç., 3e Ed., **2**; **1815**, Lam. & DC., Fl. franç., 3e Ed., **5**

S. Câm. **1956**, Cat. syst. Fung. Lusit. **1**

Cappelli **1821**, Catal. Stirp. Regio Horto bot. taurinensi

Cartwr. & Findl. **1946**, Decay Timber; **1958**, Decay Timber, 2nd Ed.

(**ČcH**), Čas. čsl. Houb.

(**Cci**), Commentario Soc. critt. ital.

(**CDp**), C.M.I. Descr. pathog. Fungi Bact.

Chev. **1826**, Fl. gén. Env. Paris **1**; **1837**, Fung. Byss. Ill. **1**

(**CJB**), Canad. J. Bot.

(**CJR**), Canad. J. Res. (C)

Clel. **1934–5**, Toadst. Mushr. S. Austr. **1** (1934), **2** (1935)

Clus. **1601**, Fung. Pannon. obs. brev. Hist. [Reproduced by Istv. 1900]

(**ČM**), Česká Mykol.

(**ČnM**), Čas. národ. Mus. (Odd. přír.)

(**CoJ**), Collnea Bot. [ed. Jacq.]

Colm. **1867**, Enum. Cript. Esp. Port. **2** [Reprinted from Revta Prog. Ci. Exact. fis. nat. **17, 18**]

Cooke **I.**, Ill. Brit. Fungi

Cooke & Q. **1878**, Clav. Hym.

Corda **I.**, Ic. Fung.

Corner **1968**, Monogr. *Thelephora* (Beih. Nova Hedw. 27)

Cost. & Duf. **1900**, Nouv. Fl. Champ., 3e Ed.

P. Cout. **1919**, Eubasid. lusit.

(**Crf**), C.r. Ass. fr. Avanc. Sci.

Crouan **1867**, Fl. Finist.

(**CrP**), C.r. hebd. Séanc. Acad. Sci., Paris

(**CSg**), Comment. Soc. reg. Sci. gotting.

Cumino **1805**, Fung. Vallis Pisii Specim. *in* Mém. Acad. imp. Sci., Turin (Sci. phys.)
  1803–4 (Mém. prés.)

G. Cunn. **1965**, Polyp. New Zeal. (Bull. N.Z. Dep. sci. ind. Res. No. 164)

Curt. **F.l.**, Fl. londin. (F. = Fascicle)

(**DAb**), Dokl. Akad. Nauk belorussk. SSR

(**DAW**), Denkschr. math.-nat. Kl. K. Akad. Wiss.

(**DbA**), Dansk bot. Ark.

(**Dba**), Denkschr. bayer. bot. Ges.

Desm. **1823**, Cat. Pl. omiss.; **P.c.**, Pl. crypt. N. France & Pl. crypt. France

Dicks. **P.c.**, Fasc. Pl. crypt. Brit.

Dill. **1719**, Cat. Pl. Giss.; **1741**, Hist. Musc.

Donk **1933**, Rev. niederl. Hom.-Aphyll. 2 (& Meded. Nederl. mycol. Ver. 18–20 &
  Meded. bot. Mus. Herb. Rijks Univ. Utrecht No. 9) [Reprinted, Biblthca
  mycol. **21**. 1969]

(**DrG**), Denkschr. regensb. bot. Ges.

Dubois **1803**, Méth. Pl. Orléans [The "Nouvelle Edition" 1825 is an unchanged
  reprint with new title-pages]; **1833**, Méth. épr. Pl. Int. Fr., "Seconde Edition"
  [of Dubois 1803]

Duby **1830**, Bot. gallic. 2 [Sometimes DC. is cited as the author]

Dumée **N.A.**, Nouv. Atl. Champ.

Dur. & Mont. **1846–9**, *in* Dur., Expl. scient. Algérie, Bot. (Atl.)

(**EAT**), Eesti NSV Tead. Akad. Toim. (Biol. Seer.)

(**Eci**), Erb. critt. ital.

(**EFE**), Erdész. Faip. Egyet. Tud. Közl.

(**EFK**), Erdömérn. Föisk. Közl., Sopron

Ehrenb. **1818**, Sylv. mycol. berol.

Eliade **1965**, Consp. Macromic. România *in* Lucr. Grâd. bot. Buc. **1964–5**

J. B. Ell. **N.A.F.**, N. Am. Fungi [exs.]

Ell. & Ev. **N.A.F.**, N. Am. Fungi [exs.] II

(**EmB**), Encycl. méth. (Bot.) [By Lam., later, Poir.].

Erikss. & Strid **1969**, Stud. Aphyll. north. Finl. *in* Annls Univ. turku. "A, II: 40
  (Rep. Kevo Subartic Sta. 4)"

(**EtK**), Értek. természettud. Köreb. Magy. tudom. Akad.

Fallahyan **1964**, Étude cult. anat. Polyp. conidioph. Thèse Doct., Fac. Sci. Univ.
  Paris

Farl. & Burt **1929**, Ic. farlow.

(**Fb**), Fl. batava

(**Fd**), Fl. dan. [The fascicles (F.) are consecutively numbered throughout the series.]

Ferd. & Jørg. **1939**, Skovtraeernes Sygd. 2

Ferd. & W. **1943**, Mykol. EkskFlora

(**Ffg**), Fragm. flor. geobot.

(**Fgp**), Folia geobot. phytotax, bohemoslov.

(**Fic**), Fl. ital. crypt. (Fungi)

(**Fl**), Flora (1818–)

(**FpG**), Fl. polska (Grzyby) [First two volumes retroactively numbered]

(**FPJ**), Forsch. Geb. PflKrankh., [Kyoto, &c.] Japan

Fr. **1821**, Syst. Mycol. **1**; **1822**, Syst. mycol. **2** (1); **1832**, Syst. mycol. **3** (2) (Ind. =
Index, separately paged); **1838**, Epicr.; **1849**, Summ. Veg. Scand.; **1874**,
Hym. europ.: **A.S.**, Anteckn. Sver. väx. ätl. Svamp.; **E.**, Elench.; **F.s.**, Fl.
scanica; **I.**, Ic. sel. Hym.: **M.**, Monogr. Hym. Suec.; **O.**, Obs. mycol.; **Pr.**,
Fungi *in* J. G. C. Lehman, Pl. Preiss. **2**; **R.A.**, Reliq. Afz.; **S.S.**, Sver. ätl. gift.
Svamp.

(**FR**), Feddes Reprium Sp. nov. Reg. veg.

(**Fr**), Friesia

(**FsR**), Fl. sporov. Rast. SSSR / Fl. Pl. crypt. URSS

(**Fu**), Fungus

Fuck. **F.r.**, Fungi rhen. exs.

(**FwN**), Festschr. wetterau. Ges. Naturk. 1908

(**G**), Grevillea

(**GaS**), G. arcad. Sci.

Gaterau **1789**, Descr. Pl. Montauban

Gaud. **1827**, Bot. *in* Freycinet, Voy. Uranie

(**GbB**), Glasn. bot. Zav. Bašte Univ. Beogr.

(**GCh**), Gdnrs' Chron.

(**Gfl**), Gartenflora

Gillet **H.N.**, Nymén. Numéros Ordre Planches [1890]; **L.H.**, Liste Champ. Hym.
[1898]; **P.**, "Planches" to Champ. (Hym.) Fr.; **S.P.**, "Suites de planches"
to Champ. (Hym.) Fr. — The nos. cited for the plates are those given in L.H.
followed by those given in H.N.

Gled. **1753**, Meth. Fung.

J. F. Gmel. **1792**, C. Linn. Syst. nat., Ed. 13, **2** (2). "1791" & C. Linn. Syst. veg
[unnumbered Ed.] 2 1796

(**GN**), Glasgow Nat.

(**GöA**), Gött. Anz. gelehrt. Sachen

Gouan **1762**, Hort. reg. monsp.

Gramb. **P.H.**, Pilze Heim.

Grev. **S.**, Scott. crypt. Fl.

(**Gsk**), Glasn. skops. nauč. Društ.

(**GSP**), G. Sci. nat. econ. Palermo

(**GšP**), Glasn. šum. Pok.

(**GwB**), Gartenwelt, Berl.

(**H**), Hedwigia

Haller **1769**, Nomencl. Hist. Pl. ind. Helv.; **H.**, Hist. Stirp. indig. Helv. inch. **3**

Hara **1927**, List Jap. Fungi by M. Shirai, 3d Ed.

Hard **1908**, Mushr.

Hartig, **1833**, Abh. Verw. polycot. PflZelle Pilz- u. Schwammgeb.

R. Hartig **1874**, Wicht. Krankh. Waldb.; **1878**, ZersetzErsch. Holzes; **1882**, Lehrb.
Baumkrankh.

Harzer **1842–5**, Naturgetr. Abb. Pilze

R. Heim **1942**, Champ. destr. Bois Habit. *in* Circul. Inst. techn. Bâtim. Trav.
publ. (H) No. 1

B. Henn. **1927**, [Michael/Schulz] Führ. Pilzfr. **3**.

Henriq. **1881**, Contr. Fl. crypt. lusit.

(**HF**), HausschwForsch.

(**HJB**), Hook, J. Bot. (& Kew Gdn Misc.)

Hoffm. **1789**, Nomencl. Fung. **1**, Agarici [Except for Hoffm. 1790 no further parts
published]; **1790**, Nomencl. Fung. **1**, Agaricini, Cont. 1; **1796**, Deutschl. Fl. o.
bot. Taschenb. 2; **V.s.**, Veg. Herc. subterr.

Holmskj. **F.d.**, Beata ruris otia Fung. dan. imp. **1** (1790), **2** (1799, ed. Viborg) [Latin
text of vol. 1 (Coryph. Clav. Ram.) reprinted *in* 1796 (ABU Stück 17): 30–149, &

by Pers., Coryph. Clav. Ram. 1–119. 1797 with Praef. Ed. pp. iii–iv, Adnot.
pp. 120–130, in one vol. with Pers., Comm. Fung. clavaef. 1797 (consecutively
paged)]

Hook. **1822**, Fungi *in* Kunth, Syn. Pl. coll. Humb. & Bonpl.

Hornem. **1827**, Nom. Fl. dan.

Houtt. **1783**, Nat. Hist., tweede Deel, veertiende Stuk

Hubert **1931**, Outl. Forest Path.

Huds. **1762**, Fl. angl.; **1778**, Fl. angl., Ed. 2., 2

Humb. **1793**, Fl. friberg. Spec.

Hussey I., Ill. Brit. Mycol. [1], 2

(IDV), Issled. Prir. Dal'nego Vostoka

(IF), Indian Forester

(IFm), Icon. Fung. malay. [ed. Over. & Weese]

(Ih), Iheringia (Bot.)

Imaz. & Aosh. **1955**, Fung. Decay *in* Mem. scient. Inv. prim. For. Headw. River
Ishikari, Hokkaido, Japan "II-2"

Imaz. & Hongo **C.I.**, [Japanese title]. Col. Ill. Fungi Japan [1], 2.

Ingold **1953**, Disp. Fungi

Inz. **F.s.**, Funghi siciliani

(IO), Isis [ed. Oken]

(IPh), Indian Phytopath.

Istv. **1900**, Clusius-Codex

S. Ito **1955**, Mycol. Fl. Japan 2 (4)

(JAA), J. Arnold Arb.

Jač **O.G.**, Opred. Grib. (n.v.) [For fuller citation, see Bond. 1953: 693]

Jacq. **F.a.**, Fl. austriaca

(JAP), J. Acad. nat. Sci. Philadelphia

(JaR), J. agr. Res.

(JBD), J. Bot. [ed. Desv.], Paris

(JBL), J. Bot., Lond.

(JBM), J. Bot. [ed. Morot], Paris

(JCi), J. Cincinnati Soc. nat. Hist.

(JJB), J. Japan. Bot.

(JLS), J. Linn. Soc. (Bot.)

(JM), J. Mycol.

(JMS), J. Elisha Mitchell scient. Soc.

(Jna), Jb. nassau. Ver. Naturk.

(JPC), J. Phys. Chim. Hist. nat.

(JsC), Jber. schles. Ges. vaterl. Kult./Cult.

(JSL), Jorn. Sci. math. phys. nat., Lisboa

Jungh. **1838**, Praem. Fl. crypt. Javae *in* Verh. Batav. Genootsch. 17 [2]. 1839.

O.K. **1891**, Rev. Gen. Pl. 1, 2 [Consecutively paged]: **1898**, Rev. Gen. Pl. 3 (3)

Kalchbr. **1873–7**, [Hungarian title]. Icon. sel. Hym. Hung.

P. Karst. **1859**, Sydv. Finl. Polyp.; **1866**, Enum. Fung. Myxom. Lapp. orient.
lect., preprint from 1882 [!] (NfF 8); **1899**, Finl. Basidsv. (Florist. Handb.
Nybeg. utg. Soc. Fauna Fl. fenn. 1); **F.F.**, Fungi Fenn. exs.; **I.**, Ic. sel. Hym.
fenn. [Advance reprints from Acta Soc. Sci. fenn.]; **T.**, Tillägg [Supplements to
Krit. Öfvers. Finl. Basidsv. 1889 (BFi 48).]

Kartav [= Step.-Kart.] **1967**, Afillof. Griby Urala *in* Trudy Inst. Ekol. rast. život.,
Sverdlovsk No. 50.

Kawam. **1929**, [Japanese title]. Japan Fungi, 3rd Ed.; **I.** [Japanese title]. Icones
jap. Fungi.

(KB), Kew Bull. (1946–)

Killerm. **1928**, Unterkl. Eubasidii *in* Nat. PflFam., 2. Aufl., 6

(KnS), K. norske Vidensk. Selsk. Skr.

(KSN), K. svenska VetAkad. Skr. Naturskydds.

E. Komar. **1964**, Opred. trytov. Grib. Belorussii

Konr. & M. **I.**, Ic. sel. Fung.

E. Krause **B.r.**, Basidiom. rostock.; **M.B.**, Mecklenb. Basid.

Kravc. **1933**, Gribnye Bolez. sibir. Pichty [Publ. by Omskoe Bjuro Obščesta Kraevedenjja.]

Kreisel **1961**, Phytopath. Grosspilze Deutschl.: **1963**, Gatt. *Ganoderma* Deutschl. *in* Holzzerstörung Pilze, Int. Symp. Ebersw. 51–54. "1962".

Krombh. **1821**, Consp. Fung. escul.: **S.**, Nat. Abb. Beschr. Schw.

L. **1737**, Fl. lapp.: **1745**, Fl. suec.: **1753**, Sp. Pl.: **1755**, Fl. suec., Ed. 2: **1759**, Syst. Nat., Ed. 10, 2; **1763**, Sp. Pl., Ed. 2, 2

(LAC), (Justus) Liebig's Annln Chem.

Lam. **F.f.**, Fl. franç.

Lapl. **1894**, Dict. iconogr. Champ. sup.

Laterr. **1846**, Fl. bordel., 4e Ed.

Latourr. **1785**, Chloris lugd.

Lázaro **1917**, Polip. Fl. españ. [Reprint edition from 1916–7 (RMa 14–15)]

Lea **1849**, Cat. Pl. Cincinnati, Ohio.

Lenz **1831**, Nützl. schädl. Schwämme; **1840**, Nützl. schädl. Schwämme, 2. Aufl.; **1862**, Nützl. schädl. Schwämme, 3. Aufl. [Lind. & Syd., Thes. 1: 849. 1908 give "1861"]

Letell. **1826**, Hist. Descr. Champ.

Lév. **1855**, Icon. Champ. Paul.; **D.**, Obs. méd. Enum. Pl. Tauride *in* Demid., Voy. Russ. 2

Leyss. **1761**, Fl. halensis: **1783**, Fl. halensis, Ed. 2

(Li), Linnaea 1826–82

Lib. **P.A.**, Pl. crypt. Ard. [exs.]

Lightf. **1778**, Fl. scot. 2. "1777"

Liljebl. **1798**, Utk. svensk Fl., Upl. 2

Lind **1913**, Danish Fungi

Lind. & Syd. **T.**, Thes. Litt. mycol. lich.

Linhart **F.h.**, Fungi hung. [exs.] / Magy. Gomb. / Ungarns Pilze

Link **1804**, Bemerk. Reise Frankr. Span. u. Portugal: **1824–5**, *in* Linn. Sp. Pl., Ed. 4 cur. Willd., 6 (P. I [Hyphomycetes] 1824; P. II [Gymnomycetes], 1825).

(LJB), Lond. J. Bot.

(Ll), Lloydia

(LMW), Lloyd, Mycol. Writ. — Consisting of Mycol. Notes consecutively paged, Letters (**L.**) and various papers with titles of their own, e.g. **Ap.**, Syn. Sect. *Apus* Genus *Polyp.*; **F.**, Syn. Genus *Fomes*; **H.**, Syn. *Hexagona*; **M.**, Syn. Sect. *Microp.* &c.; **O.** Syn. Sect. *Ovinus* Polyp.; **O.S.**, Old Species: **P.I.**, Polyporoid Issue: **S.P.**, Syn. stip. Polyporoids: **S.S.**, Syn. stip. Stereums.

(LNF), Fungi exs. suec. [cur. Lund. & Nannf. Text also separately issued in fascicles, two fascicles corresponding to a century of Nos.]

(Lo), Lotos

Lobel **1581**, Kruydtboeck 2

Lohw. **1941**, Anat. Asco- u. Basidiom. [Handb. PflAnat. 6 (8)]

Lowe **1934**, Polyp. New York St. (Pil. sp.) (Techn. Publ. N.Y. St. Coll. For. Syracuse No. 41); **1942**. Polyp. New York St. (exc. *Poria*) (Techn. Publ. N.Y. St. Coll. For. Syracuse No. 66); **1946**, Polyp. New York St. *Poria* (Techn. Publ. N.Y. St. Coll. For. Syracuse No. 65): **1966**, Polyp. N. Am. *Poria* (Techn. Publ. St. Univ. Coll. For. Syracuse No. 90); **F.**, Polyp. N. Am. *Fomes* (Techn. Publ. St. Univ. Coll. For. Syracuse No. 80)

(LPB), Leafl. Philipp. Bot. [ed. Elmer]

Lumn. **1791**, Fl. poson.

(**LZS**), Lesn. Ž. S-Petersb.

(**M**), Mycologia (Gen. Ind. = General Index, Volumes 1–24)

P. Magn. **1905**, Pilze Tirol (Dalla Torre & Sarnth., Fl. Tirol 3)

(**MaJ**), Miscnea austr. [ed. Jacq.]

(**Mal**), Malpighia

Maratti **1822**, Fl. romana 2

Marcucci U., Unio itin. crypt. Sard.

Martelli **1886**, Florula bogos.

Mass. **B.F.**, Brit. Fung.-Fl.

Mattioli **1554**, Comm. Diosc. med. Mat. [The first Latin edition] (n.v.)

(**MB**), Magazin Bot. [Zürich; ed. Röm. & Ust.]

(**Mba**), Mitt. bayer. bot. Ges.

(**MBe**), Magazin Ges. naturf. Fr. Berl.

(**MBo**), Monti Boschi

(**MCS**), Meded. phytop. Lab. W. C. Scholten

Mérat **1821**, Nouv. Fl. Paris, 2e Ed., 1

Mez **1908**, Hausschw.

(**MF**), Mikol. Fitop.

(**Mfe**), Meddn Soc. Fauna Fl. fenn.

(**MH**), Mykol. Hefte

(**MHB**), Mém. Herb. Boissier

(**Mi**), Michelia

(**MiB**), Mitt. naturf. Ges. Bern

Mich. **1729**, Nova Pl. Gen.

Michael **F.P.**, Führ. Pilzk. (B. = Ausg. B)

(**MMb**), Mém. Soc. Émul. Montbéliard

(**MMH**), Mycol. MittBl., Halle

(**MmV**), Meded. Nederl. mycol. Ver.

(**MNM**), Mém. Soc. imp. Nat. Moscou

(**MnS**), Meddr norske SkogsforsVes.

(**MNY**), Mem. New York bot. Gdn

(**Mob**), Monographiae bot.

Mont. **1852**, Pl. cell. [Gay, Hist. Fis. Pol. Chile (Bot.) 7. "1850"]; **1856**, Syll. Crypt.;
    C., Bot., Pl. cell. *in* Sagra, Hist. Cuba

Moug. & Ferry **1887**, Champ. *in* Louis, Départ. Vosges, Fl. Vosges

Moug. & Nestl. **S.C.**—Stirp. Cryptog. Vogeso-Rhen.

(**MP**), Mykologia, Praha

(**MPa**), Mycol. Pap.

(**Msa**), Mitth. Ges. Salzburg. Landesk.

(**MSb**), Mems Soc. broter.

(**MSS**), Meddn St. SkogfInst.

(**MtK**), Mat(h). természettud. Közl.

Murašk. **1939**, Gorno-taežnye Trut. sibir. *in* Trudy omsk. sel'sk. Chozj. Inst. Kirova
    17; **1940**, Trut. sibir. II *in* Otd. Izd. omsk. Inst. sel'sk. Chozj. Kirova **17**.

Murr. **1784**, C. Linn. Syst. veg., Ed. 14

Murrill **S.P.**, South Polyp.: **W.P.**, West Polyp.

(**MVf**), Meddr Vestland. forstl. ForsStn

(**MWS**), Mem. Wernerian nat. Hist. Soc.

(**Myp**), Mycopath. Mycol. appl.

(**MZB**), Mag. Zool. Bot. [London]

(**NAF**), N. Am. Fl.

(**NaH**), Natur Heim., Münster

Nannf. & Du R. **1952**, Vilda Växt. Norden, Uppl. 2

(**NAu**), Nova Acta R. Soc. Scient. upsal.

(**NBe**), Notizbl. bot. Gart. Mus. Berl.

(**Nca**), Naturaliste canad.

Nees **1816**, Syst. Pilze: **1817**, Ueberbl. [Added to a re-issue of Nees 1816]

C. & T. Nees **1820**, Anhang u. Nachträge [to Bolt., Gesch. merkw. Pilze, which is a German translation of Bolt. 1788–91.]

Neuman **1914**, Polyp. Wisconsin *in* Bull. Wisconsin Geol. nat. Hist. Surv. **33**

(**NfF**), Notis. Sällsk. Fauna Fl. fenn. Förh.

(**NGi**), Nuovo Gior. bot. ital.

(**NH**), Nova Hedw.

Nikol. **1961**, Hydnaceae (Fl. Pl. crypt. URSS **6**)

(**NJB**), Neues J. Bot. [ed. Schrad.]

(**NMB**), Neues Magazin Bot.

(**NMN**), Nyt Mag. Naturvid. (1836–1934) & Nytt Mag. Naturvid. (1936–51)

(**NSn**), Nov. Sist. niz. Rast. / Nov. syst. Pl. non Vasc.

(**NwS**), Northw. Sci.

Nyl. & Sæl. **1859**, Herb. Mus. fenn.

(**NyM**), Nytt Mag. Bot.

(**NZL**), Naturw. Z. Land- u. Forstw.

(**ÖbZ**), Oesterr. bot. Z.

(**ÖfF**), Öfvers. finska Vet.-Soc. Förh.

Opiz **1852**, Seznam rostl. květ. české

Overh. **1929**, Res. Meth. Taxon. Hym. *in* Proc. int. Congr. Plant Sci. **2**; **1933**, Polyp. Pennsylv. *Polyporus* (Bull. Sch. Agric. Exp. Stn Pennsylv. St. Coll. No. 298): **1935**, Polyp. Pennsylv. St. Stn Pennsylv. St. Coll. No. 298): **1935**, Polyp. Pennsylv. II. Gen. *Cycl.* &c. (Bull. Sch. Agric. Exp. Stn Pennsylv. St. Coll. No. 316): **1942**, Polyp. Pennsylv. *Poria* (Bull. Sch. Agric. agric. Exper. Stn St. Coll. Pennsylv. No. 418): **1953**, Polyporac. U.S., Alaska, & Canada (Univ. Michigan Stud., Sci. Ser. **19**).

(**ÖVS**), Öfvers. K. VetAkad. Förh. [Stockholm]

(**PAA**), Proc. Am. Acad. Arts. Sci.

(**PAb**), Portug. Acta biol. (B)

(**PAP**), Proc. Acad. nat. Sci. Philadelphia

Parm. **1968**, Consp. Syst. Corticiac.; **1971**, [Russian title]. Lachnocladiac. Sov. Union "1970"; M.e., Eesti Seente Eks. / Mycoth. eston.

Pat. **1887**, Hym. Eur.; **1897**, Cat. rais. Pl. cell. Tunis; **1900**, Ess. tax. Hym.: T.a., Tab. anal. Fung.

Paul. **1812–35**, Traité Champ., Atl.; T., Traité Champ.

Payer **1850**, Bot. crypt.

(**PB**), Proc. Boston Soc. nat. Hist.

(**Pe**), Persoonia

Pers. **1800**, Comment. Schaeff. Ic. pictas: **1801**, Syn. Fung.; **1807**, Syn. Pl. **2**; **1818**, Traité Champ. comest.; **1822**, Mycol. europ. **1**; **1825** Mycol. europ. **2**; **1828**, Mycol europ. **3**; C., Comm. Fung. clavaef.; I.D. Ic. Descr. Fung.; I.p., Ic. pict. Fung.; O. Obs. mycol.; T., Tent. Disp. meth. Fung.

Petch **1923**, Diseases Tea Bush

(**Ph**), Phytopathology

(**Phm**), Phytomorphology

(**PIA**), Proc. Indiana Acad. Sci.

Pilát **1948**, Velen. Sp. nov. Bas. (Opera bot. čech. **6**)

(**PJS**), Philipp. J. Sci. & Philipp. J. Sci. (Bot.)

(**PK**), Pilz- u. Kräuterfr.

Plan. I.F., Ind. Pl. erfurt. Fung. add.

(**PMi**), Pap. Michigan Acad. Sci.

(**PNA**), Proc. K. Ned. Akad. Wet. (C).

(**PNP**), Pap. New Mexico Chapt. Phi Kappa Phi

(**PNW**), Proc. Linn. Soc. New S. Wales

Poelt & Jahn 1963-5, Mitteleurop. Pilze

Pollini **1824**, F. veron. **3**; **H.**, Horti Prov. veron. Pl. nov. *in* G. Fic. Chim. Stor. nat. **9**. 1816

(**PPA**), Proc. Pennsylv. Acad. Sci.

(**PrP**), Pacific rural Press

(**PTa**), Pap. Proc. R. Soc. Tasmania

Purt. **1817**, Bot. Descr. Br. Pl. Midl. **2**; **1821** App. Midl. Fl. [Often cited as Midl. Fl. 3]

(**QFA**), Q. Jl Florida Acad. Sci.

Quél. **1872**, Champ. Jura Vosges [Reprint from (MMb II **5**), with some minor alterations]; **1886**, Ench. Fung.; **1888**, Fl. mycol. Fr.

Rab. **F.e.**, Fungi europ. exs.; **Kl.** Klotzschii Herb. mycol.: **Kl. II**, Klotzschii Herb. mycol., Ed. nova

Raddi **1807**, Sp. nuove Funghi Firenze *in* Memorie Mat. Fis. Soc. ital., Modena

Ravenel **F.c.**, Fungi carol. exs.

Ray **1696**, Syn. meth. Stirp. brit., Ed. 2; **1724** Syn. meth. Stirp. brit., Ed. 3. [ed. Dill.); **H.**, Hist. Pl.

(**RBL**), Revta Biol., Lisb.

(**RBo**), Revue scient. Bourbonn.

(**RCL**), Revta Fac. Ci. Univ. Lisboa (C) II

(**Re**), Reinwardtia

G. F. Re **1827**, Fl. torin. 2 (2)

Rea **1922**, Brit. Basid.

Reb. **1804**, Prod. Fl. neomarch.

D. Reid **1955**, Aphylloph. Gastrom. Trist. da Cunha (Results Norw. scient. Exp. Tristan da Cunha No. 37)

Retz. **1779**, Fl. Scand. Prodr.; **1795**, Fl. Scand. Prodr., Ed. 2

(**RFJ**), Rep. Govt Forest Exp. Stn, Japan = (BFJ)

(**Rh**), Rhodora

(**Ri**), Rickia

Richon **1879**, Descr. Dess. Champ. 1 (n.v.); **1889**, Cat. rais. Champ. Marne

Rick. **1918**, Vadem. Pilzfr.; **1920**, Vadem. Pilzfr., 2. Aufl.

(**RjA**), Rad jugosl. Akad. Znan. Umjetn.

(**RKF**), Rab. Kryptog.-Fl. [So-called '2. Aufl.'], 1. Band, Pilze

(**RM**), Revue Mycol. [Paris] (S. = Suppl.)

(**Rm**), Revue mycol. [Toulouse]

(**RMa**), Revta (R.) Acad. Ci. ... Madr.

(**RMo**), Rep. Missouri bot. Gdn

(**RNS**), Rep. New York St. Mus. nat. Hist. [Title varies]

Rolland **1910**, Atl. Champ. Fr. [Date of issue of completed volume]

Romell **1907**, Säck-, Buk-, Hatt- o. Gélésv. *in* Krok & Almq., Svensk Fl., Uppl. **3**, 2; **1917**, Övrige Svampgr. *in* Krok & Almq., Svensk Fl., Uppl. **4, 2: F.s.**, Fungi exs. scand.

Rostr. **1902**, Plantepatologi

Roth **C.**, Catalecta bot.

Roum. **F.g.**, Fungi gall. exs.

Roussel **1806**, Fl. Calvados, 2e Ed.

Rox. Clem. **1864**, Plantas Titáguas *in* (RPC) **14** [Reprint cited; edited by Colm.: "Rojas"]

(**RPC**), Revta Progr. Ci. exact. fis. nat.

(**RTm**), Rep. Tottori mycol. Inst.

Sacc. **1915–6**, Hymeniales *in* Fl. ital. crypt., Fungi: **F.d.**, Fungi ital. autogr. delin.
Sart. & M. **1921**, Obs. mycol.
(**SB**), Sov. Bot.
(**Sbč**), Studia bot čech./čsl.
(**SbT**), Svensk bot. Tidskr.
(**Sbu**), Symb. bot. upsal.
(**SbW**), Sber. Akad. Wiss. Wien (Math.-nat. Kl., Abt. I)
Schaeff. **1763**, Fung. Bavar. nasc. Ic. 2; **1774**, Fung. Bavaria nasc. Ic. 4 (Ind.)
Schleich. **1821**, Cat. Pl. Helv., Ed. 4
Schrad. **1794**, Spic. Fl. germ.
Schrank **1789**, Baier. Fl. 2
J. Schroet. **1885–9**. Pilze Schles. 1 [Krypt.-Fl. Schles. 3 (1)]
K. F. Schultz **1806**, Prod. Fl. stargard.
S. Schulz. **1866**, Fungi *in* Verh. zool.-bot. Ges. Wien 16
Schum. **1803**, Enum. Pl. Saell. 2
Schw. **1822**, Syn. Fung. Carol. sup. *in* Schr. naturf. Ges. Leipz. 1; **1832**, Syn. Fung.
    Am. bor. *in* Trans. Am. phil. Soc. II 4
Schwalb **1891**, Buch Pilze
(**Sci**), Science, N.Y.
Scop. **1770**, Anni hist.-nat. 4 (= Annus IV. hist.-nat.); **1772**, Fl. carn., Ed. 2;
    P.s., Diss. Sci. nat. (Pl. subterr.)
(**Sčz**), Sb. čsl. Akad. zeměd. / Ann. Czech Acad. Agric.
Secondat **1785**, Mém. Hist. nat.
Secr. **M.**, Mycogr. suisse
Seyn. **F.**, Rech. Vég. inf. 1, Fistulines; **P.**, Rech. Vég. inf. 2, Polypores
(**SF**), Syll. Fung.
Sing. **1962**, Agaricales, 2nd Ed.: **M.a.**, Mycofl. austr. (Beih. nova Hedw. 29)
(**Slw**), Sylwan
(**SkT**), SkogsFör. Tidskr.
(**SnP**), Sb. nár. Mus. Praze (B) / Acta Mus. nat. Pragae (B)
Sommerf. **1826**, Suppl. Fl. lapp. ed. Wahl.
Sorok. **1892**, O nekot. bol. Vinogr. kavkazsk. Kraja
Sow. **1796–1815**, Col. Fig. Engl. Fungi
(**SPR**), Syll. Pl. nov., Ratisb.
Spreng. **1806**, Fl. halens. Tent. nov.; **1807**, Mant. I. Fl. halens.; **1827**, C. Linn.
    Syst. veg., Ed. 16, 4 (1)
St-Am. **1821**, Fl. agen.
Step.-Kart. → Kartav.
Sterb. **1675**, Theatr. Fung.; **1712**, Theatr. Fung., 2den Druck
Steud. **1824**, Nomencl. bot. Pl. crypt.
Stev. & Cash **1936**, New Fungus Names prop. Lloyd (Bull. Lloyd Libr. Mus. No. 35)
A. Stevenson **1961**, Cat. bot. Books Coll. Hunt 2 (2)
(**StP**), [Sturm (ed.)] Deutschl. Fl. (III. Abt., Pilze) (F. = Fasc.)
Švarcm. **1964**, Geterobaz. i avtobaz. Grivy (Fl. spor. Rast. Kazachst. 4)
(**SVH**), (K.) (svenska) Vet. Akad. Handl. & Nya Handl. svenska Vet. Akad. &
    Handl. svenska VetAkad., &c.
Sw. **1788**, Nova Gen. Sp. s. Prodr.
(**Sy**), Sydowia
Syd. **M.m.**, Mycoth. march.
(**SZF**), Schweiz. Z. Forstwes.
(**SZP**), Schweiz. Z. Pilzk.
(**Ta**), Taxon
(**TBS**), Trans. Br. mycol. Soc.
(**TEd**), Trans. Proc. bot. Soc. Edinb.

Terracc. **1872**, Relaz. itorno Peregr. Terra di Lavoro [1] (n.v.).

Thore **1803**, Essai Chloris Dép. Landes

Thüm. **M.u.**, Mycoth. univ.

Thunb. **1784**, Fl. iap.

Timm **1788**, Fl. megapol. Prod.

(**TlR**), Trudy lesn. opytn. Delu Ross. / Mitt. forstl. VersWes. Russl.

(**TLS**), Trans. Linn. Soc. Lond. (Bot.)

(**TmJ**), Trans. mycol. Soc. Japan

(**To**), Torreya

Tode **F.m.**, Fungi meckl.

Torrend **1940**, Polip. Bahia *in* Anais prim. Reun. sul-amer. Bot. "1938"; **F.e.**,
Fungi sel. exs. "Séries I–IV. N.os 1–100." 1910 [Printed sheet with Index]

Tourn. **1700**, Institutiones

Tratt. **1804–6**, Fungi austriaci ["Editio nova" 1830, text unaltered]

(**TrB**), Treb. Mus. Ci. nat., Barcelona

(**TSR**), Trudy bot. Inst. (Komar.) Akad. Nauk SSSR (ser. II, Spor. Rast.)

(**Ttk**), Trudy troitskosavsko-kjacht. Otd. imp. russk. geogr. Obšč.

Tul. **F.h.**, Fungi hypogaei. Monogr. Champ. hyp.

(**TUS**), Techn. Bull. U.S. Dep. Agric.

(**TÜT**), Tartu Riikl. Ülik. Toim.

(**UbZ**), Ukr. bot. Ž.

Vaill. **1727**, Bot. paris.

(**VBb**), Vesci Akad. Navuk Belarusk. SSR (Seryja bijal.) / Izv. Akad. Nauk belorussk.
SSR (Ser. biol.)

(**VBr**), Verh. bot. Ver. Prov. Brandenb.

(**VdF**), Vestn. dal'nevost. Fil. Akad. Nauk SSSR / Bull. far east. Branch Acad.
Sci. USSR

Velen. **1920–2**, České Houby

Vill. **1779**, Prosp. Hist. Pl. Dauph.; **1789**. Hist. Pl. Dauphiné 3 (2)

Vitt. **1832–5**, Descr. Funghi mang. Italia

Viv. **1834–8**, Funghi Italia

(**VW**), Verh. zool.-bot. Ges. Wien [If nothing to the contrary is stated then the
page numbers are those of the 'Abh.']

Wahl. **1812**, Fl. lapp.; **1820**, Fl. upsal.; **1826**, Fl. suec. 2

Wak. & Denn. **1950**, Common Br. Fungi

Wallr. **1833**, Fl. crypt. Germ. 2 (Bluff & Fingerh., Comp. Pl. germ. **4**)

Warm. **1890**, Handb. syst. Bot.

Web. **1787**, Suppl. Fl. hols.

Weinm. **1836**, Hym.- Gastro-m. ross.

Willd. **1787**, Fl. berol. Prodr.

With. **1776**, Bot. Arr. Veget. Gr. Brit. 2; **1792**, Bot. Arr. Br. Pl., 2d Ed., 3 (2);
1796, Arr. Brit. Pl., 3d Ed., **4**; **1801**, Syst. Arr. Br. Pl., 4th Ed., **4**

(**WPb**), Westfäl. Pilzbr.

(**WUS**), Washington Univ. Stud. (Part I, Scient. Ser.)

(**ZFJ**), Z. Forst- u. Jagdw.

(**ZgN**), Z. ges. Naturw.

(**ZP**), Z. Pilzk.

(**ZPf**), Z. Pilzfreunde

BIBLIOGRAPHY OF SPECIAL LITERATURE

The bibliography includes, *inter alia*, the titles of the 'special literature' cited on the preceding pages. The enumeration is the result of a sharp selection. With a very few exceptions the titles refer to publications that deal exclusively with separate European species or with groups of them, like genera, but then only if these treatments of genera or larger groups are not based more or less exclusively on non-European material. For instance, a monographic treatment of the Japanese species of *Grifola* will not be found listed; the same applies to Cunningham's "Polyporaceae of New Zealand." Publications with a wider scope than the 'polypores' also failed to qualify for admission. For instance, the section of "Die natürlichen Pflanzenfamilien" written by Killermann and containing besides the treating not only the 'Polyporaceae' but also various other groups is not included. A few exceptions were made, for instance for some works by Bourdot & Galzin, Overholts, and Lowe. Because it has been rigorously adapted to the limited scope of the present work the list is therefore far from world-wide.

The number of titles of special literature related to certain species has been strongly reduced in order to cut down the length of the bibliography. These species are *Fomes fomentarius, Heterobasidion annosum, Phaeolus schweinitzii, Phellinus igniarius,* and *Phellinus tremulae.*

It will be observed that many titles are replaced by mention of the language in which they are written, for instance, "[Russian title]." This is done for languages using many diacritical marks or an alphabet other than the 'Latin' (English) one, or where the title is written in national writing-symbols like Japanese. In these cases a second title is given in full, consisting of an English, French, or German translation, preferably one to be found in the paper itself or in the serial in which it was published. If such a translated title is copied from another source it appears in brackets ( ); if it is a translation for which I feel personally responsible it appears in square brackets [ ].

ANONYMUS [G. E. MASSEE?] (1892). *Trametes Trogii,* Berk. *In* Grevillea **21**: 45–46.
-――――― (1970). *Polyporus tuberaster* Fr. — Klumpen-Porling. *In* Mykol. MittBl., Halle **14**: 89–92.
ABBAYES, H. DES (1967). Le polypore du chêne-vert, *Hexagona nitida* Mont., en Vendée. *In* Bull. Soc. mycol. Fr. **83**: 354–355.
ADAMS, D. H. & L. F. ROTH (1967). Demarcation lines in paired cultures of *Fomes cajanderi* as a basis for detecting genetically distinct mycelia. *In* Can. J. Bot. **45**: 1583–1589 *1 pl.*
ALLEN, W. B. (1907). A note on *Trametes rubescens. In* Trans. Br. mycol. Soc. **2**: 161–163 *pl. 16 lower fs.*
ALMEIDA, M. G., C. C. RODRIGUES & N. J. TEIXEIRA (1964). Novos registos de colheitas de Polyporaceae em Portugal. *In* Bolm Soc. portug. Ci. nat. II **10**: 149–164.

AMBURGEY, T. L. (1967). Decay capacities of monokaryotic and dikaryotic isolates of *Lenzites trabea. In* Phytopathology **57**: 486–491 *3 fs.*

———— (1970a). Genetic control of basidiospore formation by isolates of *Lenzites trabea. In* Phytopathology **60**: 951–954 *3 fs.*

———— (1970b). Relationship of capacity to cause decay to other physiological traits in isolates of *Lenzites trabea. In* Phytopathology **60**: 955–960 *6 fs.*

AMES, A. (1913). A consideration of structure in relation to genera of the Polyporaceae. *In* Annls mycol. **11**: 211–253 *pls. 10–13.*

ANDERSON, B. A. (1931). The toxicity of water-soluble extractives of western yellow pine to *Lenzites sepiaria. In* Phytopathology **21**: 927–940 *2 fs.*

ANDREWS, S. R. (1955). Red rot of ponderosa pine. *In* Agric. Monogr. U.S.D.A. No. 23: ii & 34 pp. *19 fs. — Polyporus anceps* [= *Dichomitus squalens*].

———— (1971). Red rot of ponderosa pine. *In* Forest Pest Leafl. U.S.D.A. No. 123: 8 pp. *4 fs. — Polyporus anceps* [= *Dichomitus squalens*].

AOSHIMA, K. (1951). [Japanese title]. Studies on birch-wood-rotting fungus, *Fuscoporia obliqua* (Pers.) Aoshima, comb. nov. *In* Bull. Tokyo Univ. Forests No. 39: 185–207 *2 fs., pls. 2, 3.*

———— (1953a). Sexuality of *Elfvingia applanata (Fomes applanata). In* Nagaoa **3**: 4–11 *8 fs.*

———— (1953b). [Japanese title]. Butt-rot of *Abies mariesii* and *A. veitchii* caused by *Tyromyces balsameus* and *Fomitopsis annosa. In* J. Jap. For. Soc. **34**: 305–307 *4 fs.* (n.v.). — Cf. *in* Rev. appl. Mycol. **33**: 391. 1954.

———— (1954a). Germination of the basidiospores of *Elfvingia applanata* (Pers.) Karst. *(Fomes applanatus). In* Bull. Govt Forest Exp. Stn, Japan **67**: 5–18 *6 fs.*

———— (1954b). Decay of beech wood by the haploid and diploid mycelia of *Elfvingia applanata* (Pers.) Karst. *In* Phytopathology **44**: 260–265 *6 fs.*

———— (1963). [Japanese title]. A note on *Fomitopsis cytisina* (Berk.) Bond. et Sing. *In* J. Jap. For. Soc. **45**: 231–233.

———— (1967). Synopsis of the genus *Daedalea* Pers. ex. Fr. *In* Trans. mycol. Soc. Japan **8**: 1–4.

———— & H. FURUKAWA (1963). A note on *Grifola obducta* (Berk.) Aoshima et Furukawa, comb. nov. *In* Trans. mycol. Soc. Japan **4**: 91–93 *2 fs.*

———— & T. KOBAYASHI (1966). [Japanese title]. Notes on *Tyromyces albellus* (Peck) Bond. et Sing. and *Spongiporus lacteus* (Fr.) Aoshima, comb. nov. *In* Rep. Tottori mycol. Inst. **5**: 42–50 *2 fs., pls. 9, 10.*

ARCANGELI, G. (1878). Sulla *Fistulina hepatica* Fr. *In* Nuovo G. bot. ital. **10**: 369–374 *pl. 11.*

ARITA, I. and M. SAKAMOTO (1966). [Japanese title]. The mating system in some Hymenomycetes. I. The mating system in *Tyromyces albellus* and *Panus rudis. In* Rep. Tottori mycol. Inst. **5**: 37–41.

ATKINSON, G. F. (1908a). Observations on *Polyporus lucidus* Leys. and some of its allies from Europe and North America. *In* Bot. Gaz. **46**: 321–338 *5 fs., pl. 19.*

———— (1908b). On the identity of *Polyporus "applanatus"* of Europe and North America. *In* Annls mycol. **6**: 179–191 *pls. 2–4.*

AYE, D. (1931). Über Riechstoffe in Pilzen. *In* Arch. Pharm., Berl. **269**: 246–251. — Mainly of *Trametes suaveolens.*

BACCARINI, P. (1911). Sulla carie dell'*Acer rubrum* L. prodotta dalla *Daedalea unicolor* (Bull.) Fr. *In* Bull. Soc. bot. ital. **1911**: 100–104.

BAGCHEE, K. & B. K. BAKSHI (1951). *Poria monticola* Murr. on Chir (*Pinus longifolia* Roxb.) in India. *In* Nature, Lond. **167**: 824 *2 fs.*

BAILEY, H. E. (1941). Contributions to the biology of *Polyporus rheades* (Pers.) Fries. *In* Bull. Torrey bot. Cl. **68**: 198–201 *2 fs. — Inonotus dryophilus.*

BAKSHI, B. K. (1952). *Oedocephalum lineatum* is a conidial stage of *Fomes annosus*. *In* Trans. Br. mycol. Soc. **35**: 195 *pl. 7.*

———— (1957). Occurrence of *Polyporus squamosus* (Huds.) Fr. in India. *In* Indian Phytopath. **9**: 191–194 *(1)* & *4 fs.* "1956". — Perhaps not correctly named.

———— & T. G. CHROUDHURY (1961). Sexuality and interfertility studies in *Polyporus abietinus*, *P. pargamenus*, and *P. versatilis*. *In* Can. J. Bot. **39**: 997–999.

————, B. SINGH & S. GIBSON (1958). Occurrence of *Trametes sepium* in India. *In* Can. J. Bot. **36**: 603–606 *17 fs., 1 pl. (=fs. 1–5).*

BAMBERGER, M. & A. LANDSIEDL (1909). Zur Kenntnis des *Polyporus rutilans* (P.) Fr. (I. Mitteilung). *In* Sber. Akad. Wiss. Wien (Math.-nat. Kl., Abt. IIb) **118**: 457–458 & *in* Mh. Chem. **30**: 673–674.

BANERJEE, S. N. & A. SARKAR (1958). Formation of sporophores of *Ganoderma lucidum* (Leyss.) Karst. and *Ganoderma applanatum* (Pers.) Pat. in culture. *In* Indian J. mycol. Res. **2**: 80–82 *1 pl.* "1956".

———— & ———— (1959). Spore-forms in sporophores of *Ganoderma lucidum* (Leyss.) Karst. *In* Proc. Indian Acad. Sci. (B) **49**: 94–98 *1 f.*

BARBAS, P. (1957). Répartition et extension du *Phaeolus fibrillosus* (Karst.) B. and G. dans le nord des Vosges. *In* Revue Mycol. **22**: 291–292.

BARNETT, H. L. & V. G. LILLY (1949). The production of haploid and diploid fruit bodies of *Lenzites trabea* in culture. *In* Proc. W. Virginia Acad. Sci. **19**: 34–39 / W. Virginia Univ. Bull., Ser. 49, No. 8-I: 34–39.

BARRETT, D. K. (1970). *Armillaria mellea* as a possible factor predisposing roots to infection by *Polyporus schweinitzii*. *In* Trans. Br. mycol. Soc. **55**: 459–462 *1 f.*

BATKO, S. (1950). A note on the distribution of *Poria obliqua* (Pers.) Bres. *In* Trans. Br. mycol. Soc. **33**: 105–106 *pl. 8.*

BAXTER, D. V. (1924). The heart rot of black ash caused by *Polyporus hispidus* Fr. *In* Pap. Michigan Acad. Sci. **3**: 39–50 *1 f., pls. 8–14.*

———— (1925). *Fomes fraxineus* Fr. in culture. *In* Pap. Michigan Acad. Sci. **4** (1): 55–66 *pls. 3–8.*

———— (1935). Some resupinate polypores from the region of the Great Lakes. VI. *In* Pap. Michigan Acad. Sci. **20**: 273–282 *pls. 55–60.* — *Poria* [*Perenniporia*] *subacida*.

———— & W. E. MANIS (1939). *Polyporus ellisianus* (Murr.) Sacc. & Trott. and *Polyporus anceps* Pk. in culture: a study of isolates from widely separated forest regions. *In* Pap. Michigan Acad. Sci. **24**: 189–195 *2 fs., 3 pls.*

BAYLISS, J. S. (1908). The biology of *Polystictus versicolor* (Fries). *In* J. econ. Biol. **3**: 1–24 *pls. 1, 2.*

BECK [VON MANNAGETTA], G. VON (1888). *Poroptyche* nov. gen. Polyporeorum. *In* Verh. zool.-bot. Ges. Wien **38** (Verh.): 657–658 *(1) f.*

BELL, M. K. & J. H. BURNETT (1966). Cellulase activity of *Polyporus betulinus*. *In* Ann. appl. Biol. **58**: 123–130.

BENKERT, D. (1971). *Inonotus nidus-pici* Pilát und *Conocybe intrusa* (Peck) Sing., zwei für die Mykoflora der DDR neue Arten. [1. *Inonotus nidus-pici*]. *In* Feddes Reprium **81**: 645–646 *pl. 28.*

BENZONI, C. (1942). Sonderbare Abart des Zunderschichtporlings... *Fomes fomentarius* var. *lauri*. *In* Schweiz. Z. Pilzk. **20**: 152–154 *(2) fs.*

BERGSTÄDT, V. (1967). *Phellinus laevigatus* erstmals in der DDR gefunden. *In* Mykol. MittBl., Halle **11**: 53–55 *(1) f.*

———— (1970). *Tyromyces guttulatus* (Peck) Murr. *In* Mykol. MittBl., Halle **14**: 96–97.

BERLESE, A. N. (1889a). Note intorno al *Polyporus hispidus* del Fries, ed all' *Agaricum gelsis seu Moris etc.* Mich. Nov. Pl. Gen. 118, N. 7. *In* Nuovo G. bot. ital. **21**: 526–532.

———— (1889b). Ancora sul *Polyporus hispidus* del Fries e sull' *Agaricum Gelsis seu Moris etc.* Mich. Nova Pl. Gen. p. 118, n. 7. *In* Malpighia **3**: 367–371 *pl. 12.*

BIER, J. E. & D. C. BUCHANAN (1947). Relation of research in forest pathology to the management of second growth trees. I. *Poria weirii* root rot, an important disease affecting immature stands of Douglas-fir. *In* Br. Columbia Lumberm. **31** (2): 49–51, 64, 66 (n.v.). — Abstract *in* Rev. appl. Mycol. **26**: 432. 1947.

BIERS, P. M. (1922). Le *Polyporus (Ungulina) Inzengae* De Not., parasite du Peuplier. *In* Bull. Soc. Path. vég. Fr. **9**: 166–168.

BIRKFELD, A. (1965). Neue Funde der Braunen Borstentramete—*Trametes extenuata*—Dur. and Mont. *In* Mykol. MittBl., Halle **9**: 78–81 *(1) f.*

BIRKINSHAW, J. H., A. BRACKEN & W. P. K. FINDLAY (1944). Biochemistry of the wood-rotting fungi. 4. Metabolic products of *Trametes suaveolens* (Linn.) Fr. *In* Biochem. J. **38**: 131–132.

————, E. N. MORGAN & W. P. K. FINDLAY (1952). Biochemistry of the wood-rotting fungi. 7. Metabolic products of *Polyporus benzoinus* (Wahl.) Fr. [With:] Appendix. Antibacterial activity of the triterpenoid acid (polyporenic acid C) and of ungulinic acid, metabolic products of *Polyporus benzoinus* (Wahl.) Fr. by S. MARCUS. *In* Biochem. J. **50**: 509–517.

BJØRNEKAER, K. (1938). Undersøgelser over nogle danske Poresvampes Biologi med saerligt Hensyn til deres Sporefaeldning. Contributions to the knowledge—particularly the sporedischarge—of some Danish Polypores. *In* Friesia **2**: 1–41 *4 fs.*

BONDARCEV, A. S. (1912). [Russian title]. Pilze, gesammelt auf Stämmen verschiedener Baumgattungen in der Forstversuchs-Oberförsterei Brjansk. *In* Trudy lesn. opytn. Delu Ross. No. 37: (i) and 56 pp. *ill., 4 pls.*

———— (1924a). [Russian title]. On some southern species of Polyporaceae found in central and north Russia. *In* Bolezni Rast. **13**: 55–59.

———— (1924b). [Russian title]. [*Polyporus imberbis* (Bull.) Fr. as a parasite of trees]. *In* Bolezni Rast. **13**: 124–128.

———— (1934). [Russian title]. Polyporaceae des europäischen Teiles der Union der SSR und des Kaukasus. I. Die Gattungen *Fomes* und *Ganoderma*. *In* Trudy bot. Inst. Akad. Nauk SSSR (II) **2**: 485–532 *17 fs.*

———— (1936a). [Russian title]. Über das Vorkommen und die Verbreitung von *Polyporus destructor* (Schrad.) Fr. in der USSR. *In* Trudy bot. Inst. Akad. Nauk SSSR (II) **3**: 669–678 *4 fs.*

———— (1936b). [Russian title]. (Observations on the spore discharge of the polypore *Ganoderma applanatum* (Pers.) Pat.). *In* Sov. Bot. **1936** (6): 144–149.

———— (1940). [Russian title]. De fungis novis Polyporacearum. *In* Bot. Mater. Otd. spor. Rast. **5**: 17–23.

———— (1941). [Russian title]. De speciebus raris stirpis "lacteus" generis *Tyromycidis.* *In* Bot. Mater. Otd. spor. Rast. **5**: 100–106.

———— (1949a). [Russian title]. De fungo novo *Tyromyces kmetii* in USSR notula. *In* Bot. Mater. Otd. spor. Rast. **6**: 88–90.

———— (1949b). [Russian title]. Iterum de Polyporaceis in USSR minus cognitus. *In* Bot. Mater. Otd. spor. Rast. **6**: 90–95. — *Tyromyces albidus* and *T. floriformis.*

———— (1949c). [Russian title]. [Notes on the fungus *Polyporus buxi* on box]. *In* Bot. Ž. **34**: 296–298 *(1) f.*

————— (1950). [Russian title]. (New data on the polypore *Trametes odarota* Fr.). *In* Bot. Ž. **35**: 73–77 *(1) pl.*

————— (1953). Trutovye griby evropejskoj časti SSSR i Kavkaza. [Bracket fungi of the European part of the U.S.S.R. and of the Caucasus]. Moskwa and Leningrad, 1106 pp. *172 fs., 188 textpls., (1) colour-chart.* — English translation: The Polyporaceae of the European USSR and Caucasia. Jerusalem, v and 896 pp.

————— (1955). [Russian title]. *Phellinus conchatus* (Pers.) Quél. et formae eius. *In* Bot. Mater. Otd. spor. Rast. **10**: 187–196.

————— (1959). [Russian title]. On [the] systematical position of some species of the Polyporaceae. *In* Bot. Ž. **44**: 447–456 *5 fs.*

————— (1960). [Russian title]. [A fossil fungus of the genus *Ganoderma*]. *In* Bot. Ž. **45**: 1504–1506 *3 fs.*

————— (1963). [Russian title]. Species pro URSS rarae et novae Polyporacearum. *In* Bot. Mater. Otd. spor. Rast. **16**: 113–125 *7 fs.*

BONDARCEV, A. S. & B. I. KRAVCEV (1952). [Russian title]. De specie Polyporacearum in graminibus vegetabili. *In* Bot. Mater. Otd. spor. Rast. **8**: 121–124.

————— & L. V. LJUBARSKIJ (1938). [Russian title]. Pourriture du chêne de Mongolie causée par *Polyporus (Spongipellis) Litschaueri* (Lohw.) A. Bond. *In* Sov. Bot. **1938** (3): 121–125 *2 fs.* — Cf. Rev. appl. Mycol. **18**: 356. 1939.

————— & R. SINGER (1941). Zur Systematik der Polyporaceen. *In* Annls mycol. **39**: 43–65.

————— & ————— (1943). [Russian title]. A natural system of the pore fungi. *In* Sov. Bot. **1943** (1): 29–43. — Also known to me through an unpublished English translation made under the supervision of Dr. W. B. Cooke.

BONDARCEVA, M. A. (1959). [Russian title]. De forma nova *Phellini torulosi* (Pers.) Bourd. et Galz. in speciebus *Arbutus andrachne* L. et *A. unedo* L. inventa. *In* Bot. Mater. Otd. spor. Rast. **12**: 247–249.

————— (1961). [Russian title]. [A critical review of the most recent classifications of the family of the Polyporaceae]. *In* Bot. Ž. **46**: 587–593.

————— (1964). [Russian title]. Polyporacearum et Aporpiacearum regionis leningradensis species et formae novae. *In* Nov. Sist. niz. Rast. **1964**: 186–195.

————— (1970). [Russian title]. Species novae Polyporacearum. *In* Nov. Sist. niz. Rast. **6**: 142–146 *2 fs.*

————— & P. K. MICHALEVIČ (1968). [Russian title]. Polyporaceae pro Belorussia ignotae. *In* Nov. Sist. niz. Rast. **1968**: 142–144 *(1) f.*

BORISOV, P. N. (1940). [Russian title]. *Fomes igniarius* Fr. and some of its biological properties. *In* Sb. Trudy centr. naučno-issled. Inst. lesn. Chozj. **15**: 78–92 *6 fs.*

BORZINI, G. (1941). Primo contributo allo studio della possibilità di una coltivazione artificiale del "*Fomes officinalis*" (Will.) Fr. *In* Boll. R. Staz. Patol. veg., Roma II **21**: 221–234 *4 fs.*

BOSE, S. R. (1929). Artificial culture of *Ganoderma lucidum* Leyss. from spore to spore. *In* Bot. Gaz. **87**: 665–667 *1 f.* — Determination perhaps doubtful.

————— (1933). Abnormal spores of some *Ganoderma*. *In* Mycologia **25**: 431–434 *3 fs.*

————— (1934). Sexuality of *Polyporus ostreiformis* Berk. and *Polystictus hirsutus* Fr. *In* Cellule **42**: 249–266 *(1) pl.* — Reprinted *in* Kirtikar meml Vol. 1. 1942.

————— (1935). Cytology of secondary spore formation in *Ganoderma*. *In* Phytopathology **25**: 426–429 *2 fs.*

————— (1940). Effect of inversion of a small piece from the fruit-body of *Ganoderma lucidus* (Leyss.) Karst. growing *in situ* on the trunk of *Casuarina equisetifolia. In* Nature, Lond. **145**: 899–900 *3 fs.* — Fungus correctly named?

———— (1941). The nature of the colouring substances in coloured Polyporaceae. *In* Trans. natn. Inst. Sci. India **2**: 69–85 *pls. 1–4.* — Reprinted *in* Kirtikar meml Vol. **1**. 1942.

———— (1943a). Binucleate oidia on binucleate mycelium of *Polystictus hirsutus*, Fr. *In* Nature, Lond. **152**: 508–509.

———— (1943b). Moisture-relation as a determinant factor in the transformation of the basidia of certain Polyporaceae. *In* Mycologia **35**: 33–46 *8 fs.*

———— (1944). Importance of anatomy in systematics of Polyporaceae. *In* J. Indian bot. Soc. **23**: 153–157 *10 fs.*

———— (1952). Identity of *Polystictus* (=*Polyporus*=*Trametes*) *cinnabarinus* (Jacq.) Fr. with *Polystictus sanguineus* (L.) Fr. *In* Nature, Lond. **170**: 1020 *1 f.*

BOUDIER, J. L. E. (1887). Description de deux nouvelles espèces de *Ptychogaster* et nouvelle preuve de l'identité de ce genre avec les *Polyporus. In* J. Bot., Paris (ed. Morot) **1**: 7–12 *pl. 1.*

———— (1888). Note sur une forme conidifère curieuse du *Polyporus biennis* Bull. *In* Bull. Soc. mycol. Fr. **4**: lv–lix *pl. 3.*

———— (1899). Notes sur un cas de formation de chapeaux secondaires sur un pédicule de *Ganoderma lucidum. In* Bull. Soc. mycol. Fr. **15**: 311–312.

BOULTER, D. & A. BURGES (1955). Oxidases of *Polystictus versicolor. In* Experientia **11**: 188–189.

BOURDOT, H. & A. GALZIN (1925). Hyménomycètes de France (XI. Porés). *In* Bull. Soc. mycol. Fr. **41**: 98–255.

———— & ———— (1928). Hyménomycètes de France. Hétérobasidiés — Homobasidiés gymnocarpes. [V. Porés]. Sceaux, pp. 514–693. "1927".

BOURQUELOT, E. & H. HÉRISSEY (1895). Les ferments solubles du *Polyporus sulfureus* (Bull.). *In* Bull. Soc. mycol. Fr. **11**: 235–239.

BOYCE, J. S. (1924). An unusual infection of *Polyporus Schweinitzii* Fr. *In* Phytopathology **14**: 588.

BRAID, K. W. (1924). Some observations on *Fistulina hepatica* and hollow stag-headed oaks. *In* Trans. Br. mycol. Soc. **9**: 210–213.

BRANKE, V. JÜ. FON (1896). [Russian title]. [... *Fomes officinalis*, Fr. ...]. *In* Lesn. Ž. S.-Petersb., **1896**: 1191–1199 (n.v.).

BRATUS', V. M. (1959). [Ukrainian title]. [Tree species susceptible to *Trametes pini* (Thore) Fries [*Phellinus pini* (Thore ex Fries) Pilát]]. *In* Nauk. Praci Ukr. nauk.-dosl. Inst. Zach. Rosl. **1959**: 162–169.

———— & O. H. ČERNYCH (1966). [Ukrainian title]. Some data on the biology of *Fomitopsis annosa* (Fr.) Bond. et Sing. *In* Ukr. bot. Ž. II **23** (4): 96–103 *5 fs.*

———— & A. V. CYLJURYK (1964). [Ukrainian title]. The fungus *Phellinus tremulae* Bond. et Boriss. on dark-bark and green-bark forms of aspen. *In* Ukr. bot. Ž. II **21** (1): 90–97 *4 fs.*

BREFELD, O. (1888a). Untersuchungen aus dem Gesammtgebiet der Mykologie ... VIII. Heft: Basidiomyceten III.... [Die Gattung *Oligoporus*]. Pp. 114–142 *pl. 7 fs. 12–25, pl. 8.* "1889".

———— (1888b). Untersuchungen aus dem Gesamtgebiet der Mykologie. ... VIII. Heft: Basidiomyceten III. ... [Die Gattung *Heterobasidion*]. Pp. 154–184 *pls. 9–11.* "1889".

BRESINSKY, A. & J. STANGL (1968). Beiträge zur Revision M. Britzelmayrs "Hymenomyceten aus Südbayern" 7. Polyporaceae der Augsburger Umgebung. *In* Z. Pilzk. **34**: 71–79.

BRIGANTI, F. (1840). Descrizione di due nuove e rare specie di funghi della famiglia de' Porodermei. *In* Atti R. Ist. Incoragg. Sci. nat. Napoli **6**: 139–152 *(1) pl.* — *Polyporus calaber, P. nanus.*

———— (1847). Descrizione di un nuovo fungo del genere delle Dedalee, e del suo uso medico-economico. *In* Atti R. Ist. Incoragg. Sci. nat. Napoli **7**: 97–104 *(1) pl.* — *Daedalea hymenopus* (0).

BRODIE, H. J. (1936). The barrage phenomenon in *Lenzites betulina. In* Genetica **18**: 61–73 *pls. 1, 2.*

BROOKS, F. T. (1909). Notes on *Polyporus squamosus,* Huds. *In* New Phytol. **8**: 348–351 *f. 44.*

———— (1925). *Polyporus adustus* (Willd.) Fr. as a wound parasite of apple trees. *In* Trans. Br. mycol. Soc. **10**: 225–226.

BROTZMAN, H. G. & R. L. GILBERTSON (1967). *Tyromyces undosus* in North America. *In* Mycopath. Mycol. appl. **33**: 33–42 *6 fs.*

BROWN, J. G. (1930). Timber-rot in pepper-tree, *Schinus molle. In* Bull. agric. Exp. Stn Univ. Ariz. No. 132: 443–454 *fs. 13, 33, 8 pls.* — *Inonotus schini* (0).

BRUNNER, S. (1842). Einiges über den Stein-Locherpilz *(Polyporus tuberaster* Jacq. et Fries) und die Pietra fungaja der Italiener. *In* Neue Denkschr. allg. schweiz. Ges. ges. Naturw. **5**: 20 pp. *2 pls.* "1845". — Advance copy, 1842.

BRUNNTHALER, J. (1896). Über eine monströse Wuchsform von *Polyporus squamosus* (Huds.). *In* Verh. zool.-bot. Ges. Wien **46**: 435–437 *(1) f.*

BRUSHHABER, J. A. & S. F. JENKINS (1971). Mitosis and clamp formation in the fungus *Poria monticola. In* Am. J. Bot. **58**: 273–280 *29 fs.*

BUCHANAN, T. S. (1948). *Poria weirii*: its occurrence and behaviour on species other than cedars. *In* Northw. Sci. **22**: 7–12.

BUCHS, M. (1930a). Seltene schlesische Funde. I. *In* Z. Pilzk. **9**: 74–75 *pl. 4.* — *Polyporus [Grifola] frondosus.*

———— (1930b). Der Bergporling *Polyporus montanus* Quél. *In* Z. Pilzk. **9**: 8–9 *pl. 2.*

BUCHWALD, N. F. (1930). Tønder- eller Fyrsvampen (*Polyporus fomentarius* (L.) Fr.). Dens Naturhistorie, Historie or Anvendelse. The tinder fungus (*Polyporus fomentarius* (L.) Fr.), its natural history, history and application. *In* Meddr Foren. SvampekSk. Fremme **4**: 49–92 *4 fs.*

———— (1934). En sjælden Værtplante for Tøndersvamp. *Polyporus fomentarius* paa *Populus virginiana. In* Flora Fauna, Silkeborg **1934**: 34–37 *4 fs.*

———— (1938). Om Sporeproduktionens Størrelse hos Tøndersvampen *Polyporus fomentarius* (L.) Fr. On the size of the spore-production of the Tinder Fungus, *Polyporus fomentarius* (L.) Fr. *In* Friesia **2**: 42–69 *11 fs.*

———— (1941a). Lidt om Hymenoforets Variation hos Poresvampe og en ny Varietet af *Daedalea quercina* (L.) Pers., *D. quercina* var. *irpiciformis.* On the variation of the hymenophore of Polyporaceae and a new variety of *Daedalea quercina* (L.) Pers., *D. quercina* var. *irpiciformis* var. nov. *In* Friesia **2**: 161–165 *(1) f.*

———— (1941b). *Polyporus caesius.* En Poresvamp med blaat Sporestøv. *In* Friesia **2**: 186–187.

———— & S. HANSEN (1934). Om Fund of Tøndersvamp (*Polyporus fomentarius* (L.) Fr.) fra Postglacialtiden i Danmark. *In* Danm. geol. Unders. (IV. Række) **2** (11): 20 pp. *1 f.*

———— & E. HELLMERS (1946). Fortsatte Iagttagelser over Sporefældning hos Tøndersvamp (*Polyporus fomentarius* (L.) Fr.). Further observations on the spore-discharge of the true tinder fungus (*Polyporus fomentarius* (L.) Fr.). *In* Friesia **3**: 212–216 *1 f.*

———— & H. A. JØRGENSEN (1949). Er der Nogen Sammenhæng mellem Klimæt og Fremkomsten af Frugtlegemer hos *Polyporus dryadeus* (Pers.) Fr.?

Is there any relation between the climate and the appearance of fruit bodies in *Polyporus dryadeus* (Pers.) Fr.? *In* Friesia **3**: 381–387 *(1) f.*

BUCKLAND, D. C., A. C. MOLNAR & G. W. WALLIS (1954). Yellow laminated rootrot of Douglas fir. *In* Can. J. Bot. **32**: *69–81 fs., 3 pls.* — *Poria* [*Inonotus*] *weirii.*

BULATOV, P. K., M. P. BEREZINA & P. A. JAKIMOV (ed.) (1959). [Russian title]. [The "čaga" and its therapeutical use in stage IV cancer]. Leningrad, 334 pp. *23 fs.* (n.v.). — Contains *inter alia* A. S. BONDARCEV [The "čaga" and some of the most widespread polypores on birch], pp. 23–31, and O. P. NIZKOVSKAJA [Contributions to the biology of the causal agent of the "čaga" on birch], pp. 32–35. — Information taken from Rev. appl. Mycol. **40**: 190. 1961.

BULLER, A. H. R. (1906a). The biology of *Polyporus squamosus*, Huds., a timber-destroying fungus. *In* J. econ. Biol. **1**: 101–138 *fs. A–F, pls. 5–9.*

———— (1906b). The enzymes of *Polyporus squamosus*, Huds. *In* Ann. Bot. **20**: 49–59.

———— (1922a). Researches on fungi 2. ... [The monstrous fruit-bodies of *Polyporus rufescens*]. London, etc., 75–77 *fs. 23, 24.*

———— (1922b). Researches on fungi 2. ... [Chapter IV. The discharge of spores from certain Agaricineae and Polyporeae]. London, etc., 95–120 *fs. 32–42.*

———— (1922c). Researches on fungi 2. ... [Chapter V. *Fomes applanatus* and its spore-discharge period]. London, etc., 121–148 *fs. 43–50.*

BUTLER, E. J. (1909). *Fomes lucidus* (Leys.) Fr., a suspected parasite. *In* Indian Forester **35**: 514–518 *pl. 19.* — Specific determination uncertain.

CABRAL, R. V. DE GARCIA (1951). Anastomoses miceliais. Seu valor no diagnóstico das Poliporoses. *In* Bolm Soc. broter. II **25**: 291–364 *7 fs., 1 pl.*

CAMPBELL, A. H. & B. G. MUNSON (1936). Zone lines in plant tissues. III. The black lines formed by *Polyporus squamosus* (Huds.) Fr. *In* Ann. appl. Biol. **23**: 453–463 *pls. 18, 19.*

CAMPBELL, W. A. (1937). The cultural characteristics of *Fomes connatus*. *In* Mycologia **29**: 567–571 *2 fs.*

———— (1938). The cultural characters of the species of *Fomes*. *In* Bull. Torrey bot. Cl. **65**: 31–69 *128 fs.*

———— (1939). *Daedalea unicolor* decay and associated cankers of maple and other hardwoods. *In* J. For. **37**: 974–977 *1 f.*

———— & R. W. DAVIDSON (1938). A *Poria* as the fruiting stage of the fungus causing the sterile conks on birch. *In* Mycologia **30**: 553–560 *3 fs.* — *Poria* sp. [=*Inonotus obliquus*].

———— & ———— (1939). *Poria Andersonii* and *Polyporus glomeratus*, two distinct heart-rotting fungi. *In* Mycologia **31**: 161–168 *2 fs.*

———— & ———— (1941). Cankers and decay of yellow birch associated with *Fomes igniarius* var. *laevigatus*. *In* J. For. **39**: 559–560 *1 f.*

CAMPBELL, W. G. (1930). The chemistry of the white rots of wood. I. The effect on the wood substance of *Polyporus versicolor* (Linn.) Fr. *In* Biochem. J. **24**: 1235–1243.

———— (1931). The chemistry of the white rots of wood. II. The effect on wood substance of *Armillaria mellea* (Vahl) Fr., *Polyporus hispidus* (Bull.) Fr. and *Stereum hirsutum* Fr. *In* Biochem. J. **25**: 2023–2027.

———— (1932). The chemistry of the white rots of wood. III. The effect on wood substance of *Ganoderma applanatum* (Pers.) Pat., *Fomes fomentarius* (Linn.) Fr., *Polyporus adustus* (Willd.) Fr., *Trametes pini* (Brot.) Fr. and *Polyporus abietinus* (Dicks.) Fr. *In* Biochem. J. **26**: 1829–1838.

———— & S. A. BRYANT (1940). A chemical study of the bearing of decay by *Phellinus cryptarum* Karst. and other fungi on the destruction of wood by the death-watch beetle (*Xestobium rufovillosum* De G.). *In* Biochem. J. **34**: 1404–1414. — *Poria expansa*.

CARTWRIGHT, K. T. S. G. (1929). Notes on basidiomycetes grown in culture. I. The occurrence of abnormal fructifications in *Lenzites saepiaria*. *In* Trans. Br. mycol. Soc. **14**: 300–305 *4 fs*.

———— (1930). A decay of Sitka spruce timber, caused by *Trametes serialis*, Fr. A cultural study of the fungus. *In* Forest Prod. Res. Bull. [Res. Bull. Dep. scient. ind. Res. Forest Prod.] **4**: vi & 26 pp. *6 pls*. — *Fide* Nobles 1943 (CJR 21): 223–224 the fungus is *Poria microspora*.

———— (1932). Further notes on Basidiomycetes in culture. *In* Trans. Br. mycol. Soc. **16**: 304–307 *pls. 15, 16*. "1931".

———— (1937). A reinvestigation into the cause of "brown oak", *Fistulina hepatica* (Huds.) Fr. *In* Trans. Br. mycol. Soc. **21**: 68–83 *6 fs., pls. 1–7*.

———— (1940). Note on a heart rot of oak trees caused by *Polyporus frondosus* Fr. *In* Forestry **14**: 38–41 *3 fs*.

———— (1951a). A note on the occurrence of *Trametes hispida* in England, and its cultural characteristics. *In* Trans. Br. mycol. Soc. **33**: 332–334 *pls. 22, 23*.

———— (1951b). *Polyporus quercinus* on *Fagus silvatica*. *In* Trans. Br. mycol. Soc. **34**: 604–606 *pls. 29, 30*. — The fungus (K) is *Fomitopsis pinicola*.

CAVILL, G. W. K., P. S. CLEZY & J. R. TETAZ (1957). The structure of cinnabarin (Polystictin). *In* Proc. chem. Soc. **1957**: 346–347.

————, B. J. RALPH, J. R. TETAZ & R. L. WERNER (1953). The chemistry of mould metabolites. Part I. Isolation and characterisation of a red pigment from *Coriolus sanguineus* (Fr.). *In* J. chem. Soc. **1953**: 525–529.

ČERNÝ, A. (1959). [Czech title]. *Inonotus nidus-pici* Pilát, ein sehr schädlicher Parasit einiger Laubhölzer in der ČSR. *In* Sb. vys. Šk. zeměd. lesn. Brně (C) **1959**: 65–87 *7 fs*. — Cites previous literature, several items of which called the species *Poria obliqua*.

———— (1960). [Czech title]. *Phaeolus croceus* (Pers. ex Fr.) Pat. — A new polypore for Czechoslovakia. *In* Česká Mykol. **20**: 90–96 *5 fs*.

———— (1963a). [Czech title]. *Inonotus andersonii* (Ellis et Everhart) Černý comb. nov. — A new Polypore for Czechoslovakia. *In* Česká Mykol. **17**: 1–8 *7 fs*.

———— (1963b). [Czech title]. *Inonotus obliquus* (Pers. ex Fr.) Pilát, ein schädlicher Porling an *Fagus silvatica* L. und *Betula* spec. in der ČSSR. *In* Sb. vys. Šk. zeměd. lesn. Brně (C) **1963**: 133–148 *8 fs., (2) pls*. [=*fs. 5–8*].

———— (1965). [Czech. title]. Bionomie, Verbreitung und wirtschaftliche Bedeutung der Porlinge *Inonotus nidus-pici* Pilát und *Inonotus obliquus* (Pers. ex Fr.) Pilát in der ČSSR. *In* Sb. nár. Muz. Praze 21 (B): 157–244 *56 & 25 fs., 4 maps, Tabellen 3, 4*.

———— (1968). *Phellinus pilatii* sp. nov., ein sehr schädlicher Pilz an *Populus alba* L. und *Populus canescens* Smith. *In* Česká Mykol. **22**: 1–13 *11 fs*.

———— (1970). *Inonotus nidus-pici* Pilát, ein neuer sehr bedeutsamer Porling für Rumänien. *In* Revue roum. Biol. (Bot.) **15**: 399–418 *10 fs*.

CHIFFLOT, J. (1921). Un Champignon de 20 kilos. *In* Bull. Soc. mycol. Fr. **37**: 138–139. — *Polyporus* [*Laetiporus*] *sulphureus*.

CHILDS, T. W. (1937). Variability of *Polyporus schweinitzii* in culture. *In* Phytopathology **27**: 29–50 *3 fs*.

———— (1963). *Poria weirii* root rot. *In* Phytopathology **53**: 1124–1127 *2 fs*.

———— (1970). Laminated root rot of Douglas-fir in western Oregon and Washington. *In* Res. Pap. U.S.D.A. PNW-102: 27 pp. *9 fs*. — *Poria* [*Inonotus*] *weirii*.

———— & E. E. NELSON (1960). Laminated root rot of Douglas-fir. *In* Forest Pest Leafl. U.S.D.A. No. 48: 7 pp. *4 fs*. — *Poria* [*Inonotus*] *weirii*.

CHOW, C. H. & H. K. CHEN (1935). On the variation of *Ganoderma lucidum*, (Fries ex von Leysser) Karsten. *In* Bull. Fan meml Inst. Biol. **6**: 36–42 *pl. 2*.

CHRISTENSEN, C. M. (1940). Observations on *Polyporus circinatus*. *In* Phytopathology **30**: 957–963 *4fs*. — The fungus described is rather *Mucronoporus tomentosus?*

CHRISTISON, R. (1874). Notice of a remarkable *Polyporus* from Canada. *In* Trans. Proc. bot. Soc. Edinb. **12** [?]: 180–181. — *Agaricum officinale*.

CODINA VIÑAS [J. CODINA] (1925). A propos du *Polyporus tunetanus* Patouillard. *In* Bull. Soc. mycol. Fr. **41**: 82. — Determination of species doubtful (**0**).

COKER, W. C. (1946). The United States species of *Coltricia*. *In* Jl. E. Mitchell scient. Soc. **62**: 95–107 *pls. 17–22*.

COLEMAN, L. C. (1927). Structure of spore wall in *Ganoderma*. *In* Bot. Gaz. **83**: 48–60 *pl. 5*.

CONFERENCE [1962] and study tour on *Fomes annosus*. Scotland, June 1960. Int. Union Forest Res. Organiz., Sect. 24: Forest Prot. Firenze. 151 pp. *ill.*

———— [1970] on *Fomes annosus*, Proceedings of the Third International ∼, Aarhus, Denmark, July 29–Aug. 3. 1968. U.S.D.A. Forest Serv. Washington DC., iv and 208 pp. (n.v.). — Cf. Rev. Pl. Path. **50**: 571–572. 1971.

COOKE, M. C. (1878). Enumeration of *Polyporus*. *In* Trans. Proc. bot. Soc. Edinb. **13**: 131–159.

———— (1885–6). Praecursores ad monographia *Polyporum*. *In* Grevillea **13**: 80–87, 114–119. 1885; **14**: 17–21. 1885; **14**: 77–87, 109–115. 1886; **15**: 19–27, 50–60. 1886.

———— (1891). *Trametes* and its allies. *In* Grevillea **19**: 98–103.

COOKE, W. B. (1940). A nomenclatorial survey of the genera of pore fungi. *In* Lloydia **3**: 81–104.

———— (1949). Recent systems of polypore classification. *In* Lloydia **12**: 220–228.

———— (1960). The genera of pore fungi. *In* Lloydia **22**: 163–207. "1959".

COOL, C. (1920). *Trametes pini* (Brot.) Fr. Nieuw voor Nederland. *In* Nederl. kruidk. Arch. **29**: 126–128 *(1) f.*

CORNER, E. J. H. (1953). The construction of polypores—1. Introduction: *Polyporus sulphureus*, *P. squamosus*, *P. betulinus* and *Polystictus microcyclus*. *In* Phytomorphology **3**: 152–167 *13 fs*.

CORNU, M. (1876). Note sur le *Ptychogaster albus* Corda. *In* Bull. Soc. bot. Fr. **23**: 359–363.

COSTANTIN, J. (1894). Sur la culture du *Polyporus squamosus* et sur son *Hypomyces*. *In* Bull. Soc. mycol. Fr. **10**: 102–106.

———— (1895). Note sur la culture de la « pietra fungaia ». *In* Rev. gén. Bot. **7**: 433–437 *pl. 17*.

———— & L. DUFOUR (1923). Une maladie secondaire du Chêne causée par le *Polyporus (Phellinus) rubriporus*. *In* C. r. hebd. Séanc. Acad. Sci., Paris **177**: 806–809.

CUNNINGHAM, G. H. (1946). Notes on the classification of the Polyporaceae. *In* New Zeal. J. Sci. Techn. (A) **28**: 238–251 *10 fs*.

———— (1954). Hyphal systems as aids in identification of species and genera of the Polyporaceae. *In* Trans. Br. mycol. Soc. **37**: 44–50.

DA COSTA, E. W. B. & R. M. KERRUISH (1965). The comparative wood-destroying ability and preservative tolerance of monokaryotic and dikaryotic mycelia of *Lenzites trabea* (Pers.) Fr. and *Poria vaillantii* (DC. ex Fr.) Cke. *In* Ann. Bot. II **29**: 241–252 *9 fs*.

DAHNKE, W. (1957). Zweiter Beitrag zur Kenntnis der mecklenburgische Pilze: Porlinge und Leberpilze. *In* Arch. Freunde NatGes. Mecklenb. **3**: 33–43.

DANA, J. P. M. [1771]. Descriptio, & usus *Agarici*, seu *Boleti pellicei*. *In* Mélang. Phil. Math. Soc. r. Turin **4**: 161–167.

DARLEY, E. F. & C. M. CHRISTENSEN (1943). An unusual sporophore of *Trametes suaveolens* produced on artificially inoculated wood. *In* Phytopathology **33**: 328–330 *1 f.*

———— & ———— (1945). The culture designated Madison 517 identified as *Polyporus tulipiferus*. *In* Phytopathology **35**: 220–222 *1 f.*

DAVID, A. (1966). *Trametes ljubarskyi* Pilát polypore nouveau pour la flore européenne. *In* Bull. Soc. mycol. Fr. **82**: 504–511 *8 fs.*

———— (1967a). Caractères mycéliens de quelques *Trametes* (Polyporacees). *In* Naturaliste can. **94**: 557–572 *2 textpls.*

———— (1967b). *Lenzites reichardtii* Schulz., espèce nouvelle pour la flore française. *In* Bull. mens. Soc. linn. Lyon **36**: 155–163 *7 fs., 2 textpls.*

———— (1968). Caractères culturaux et comportement nucléaire dans le genre *Gloeophyllum* Karst. (Polyporaceae). *In* Bull. Soc. mycol. Fr. **84**: 119–126 *4 fs.*

———— (1969a). Caractères culturaux et cytologiques d'espèces du genre *Spongipellis* Pat. et affines. *In* Bull. mens. Soc. linn. Lyon **38**: 191–201 *6 fs., 1 textpl.*

———— (1969b). Caractères culturaux et cytologiques de quelques espèces rangées par Bourdot et Galzin et d'autres auteurs dans le genre *Phaeolus* (Polyporacées). *In* Naturaliste can. **96**: 211–224 *4 fs.*

———— (1971). Caractères mycéliens d'*Incrustoporia percandida* Malençon & Bertault. *In* Acta phytotax. barcin. **8**: 95–97 *(1) pl. (=f. 11).*

———— (1972). Caractères mycéliens de *Rigidoporus rivulosus* (Berk. et Curt.) comb. nov. *In* Bull. Soc. mycol. Fr. **87**: 415–419 *5 fs., 1 pl.*

DAVIDSON, R. W. & W. A. CAMPBELL (1943). Decay in mercantable black cherry on the Allegheny National Forest. *In* Phytopathology **33**: 965–985 *7 fs.*

————, C. M. CHRISTENSEN & E. F. DARLEY (1946). *Polyporus guttulatus* and *Ptychogaster rubescens*. *In* Mycologia **38**: 652–663 *3 fs.*

DAY, W. C., S. GOTTLIEB & M. J. PELCZAR (1953). The biological degradation of lignin. IV. The inability of *Polyporus versicolor* to metabolize sodium lignosulfate. *In* Appl. Microbiol. **1**: 78–81.

DECKER, B. (1970). *Polyporus tuberaster* Fr. — Klumpen-Porling. *In* Mykol MittBl., Halle **14**: 89–92.

DE GROOT, R. C. (1964). Color of the basidiospores of *Fomes pini*. *In* Mycologia **56**: 786–787.

———— (1965). Germination of basidiospores of *Fomes pini* on wood extract media. *In* Forest Sci. **11**: 176–180 *4 fs.*

DEMELIUS, P. (1917). Konidienbildung bei *Polyporus lucidus* Leyss. *(Ganoderma lucidum). In* Verh. zool.-bot. Ges. Wien **66**: 494–495.

DEMOULIN, V. (1967). Une nouvelle station française d'*Hexagona nitida* Mont. *In* Bull. Soc. mycol. Fr. **82**: 517–519 *1 f.* "1966".

DE VAY, J. E., S. L. SINDEN, F. L. LUKEZIC, L. F. WERENFELS & P. A. BACKMAN (1968). *Poria* root and crown rot of cherry trees. *In* Phytopathology **58**: 1239–1241 *1 f.* — *Poria ambigua* [*Oxyporus late-marginata*].

DION, W. M. (1952). Production and properties of a polyphenol oxidase from the fungus *Polyporus versicolor*. *In* Can. J. Bot. **30**: 9–21 *10 fs.* (graphs).

DÖRFELT, H. (1970). *Piptoporus betulinus* (Bull. ex Fr.) P. Karst. — Birkenporling— mit einer zweiten Röhrenschicht. *In* Mykol. MittBl., Halle **14**: 86–89 *2 fs.*

DOMAŃSKI, S. (1954). [Polish title]. [Observations on the biology of *Fomes igniarius* (Linn.) Fr. attacking white poplar *(Populus alba* L.)]. *In* Acta Soc. Bot. Polon. **23**: 589–616 *8 fs.* — English translation separately issued: Warzsawa, 20 pp. *10 fs. 1960d.*

——————— (1956). [Polish title]. A study of *Phellinus punctatus* (Fr.) Pilát with special reference to its cankerous activity on ash. *In* Acta Soc. Bot. Polon. **15**: 159–180 *12 fs.*

——————— (1959a). [Polish title]. Certaines espèces du genre *Poria* de la forêt vierge de Bialowieża en Pologne. *In* Monographiae bot. **8**: 153–169 *14 fs.*

——————— (1959b). [Polish title]. Deux rares porés: *Leptoporus lapponicus* (Rom.) Pil. et *Phaeolus alboluteus* (Ell. et Ev.) Pil. dans la forêt de Bialowieża en Pologne. *In* Monographiae bot. **8**: 171–181 *8 fs.*

——————— (1960a). [Polish title]. La morphologie des réceptacles de *Polystictus tomentosus* (Fr.) Karst. var. *circinatus* (Fr.) Sart. et Maire recueillis dans le Parc National à Ludwikowo. *In* Monographiae bot. **10**: 107–132 *9 fs.*

——————— (1960b). [Polish title]. Certains résultats des recherches sur *Gloeophyllum trabeum* (Pers. ex Fr.) Murr. *In* Monographiae bot. **10**: 133–146 *5 fs.*

——————— (1960c). [Polish title]. *Leptoporus albidus* (Schaeff. ex Secr.). Bourd. & Galz. et son diagnostic. *In* Acta Soc. Bot. Polon. **29**: 655–671 *9 fs., 1 tabl.*

——————— (1960d), see DOMAŃSKI, S. (1954).

——————— (1963). [Polish title]. Deux nouveaux genres des champignons de la groupe "*Poria* Pers. ex S. F. Gray". *In* Acta Soc. Bot. Polon. **32**: 731–739.

——————— (1964a). [Polish title]. *Tyromyces Lowei* (Pil. ex Pil.) Bond. en Pologne. *In* Fragm. flor. geob. **10**: 81–88 *3 fs.*

——————— (1964b). Révision de certaines espèces de champignons de la famille Polyporaceae. *In* Acta Soc. Bot. Polon. **33**: 167–178 *8 fs.*

——————— (1964c). [Polish title]. Wood-inhabiting fungi in Bialowieża virgin forests in Poland. I. *Poria subacida* (Peck) Sacc. and its diagnose. *In* Acta Soc. Bot. Polon. **33**: 661–678 *5 fs.*

——————— (1965a). [Polish title]. Wood-inhabiting fungi of Bialowieża virgin forest in Poland. III. *Cerioporiopsis placenta* (Fr. sensu J. Erikss.) Domański, its forms and their identification. *In* Acta Soc. Bot. Polon. **34**: 491–531 *10 fs.*

——————— (1965b). [Basidiomycetes: Aphyllophorales: Polyporaceae I, Mucronoporaceae I]. *In* Flora polska (Grzyby) **2**: 280 pp. *70 fs., 63 pls.*

——————— (1965c). [Polish title]. [Significance of microscopic structure of fruiting bodies in the taxonomy of polypores]. *In* Wiad. bot. **9**: 31–42.

——————— (1966a). [Polish title]. Wood-inhabiting fungi of Bialowieża virgin forest in Poland. IV. *Poria albidofusca*, sp. nov. and its diagnose. *In* Acta Soc. Bot. Polon. **35**: 461–475 *7 fs.*

——————— (1966b). [Polish title]. A comparative study of polyporoid fungi: *Polyporus dichrous* Fr. and *Polyporus pannocinctus* Romell. *In* Acta mycol. **2**: 151–168 *6 fs.*

——————— (1966c). [Polish title]. *Coriolellus malicola* (Berk. et Curt.) Murr. in Eurasia. *In* Acta Soc. Bot. Polon. **35**: 599–609 *4 fs.*

——————— (1968). [Polish title]. Wood-inhabiting fungi in Bialowieża virgin forests in Poland. V. *Trametella extenuata* (Dur. and Mont.) Domań. *In* Acta Soc. Bot. Polon. **37**: 125–144 *12 fs.*

——————— (1969a). [Polish title]. Wood-inhabiting fungi in Bialowieża virgin forests in Poland. VI. *Antrodia ramentacea* (Berk. and Br.) Donk. *In* Acta Soc. Bot. Polon. **38**: 57–68 *6 fs.*

——————— (1969b). [Polish title]. Wood-inhabiting fungi in Bialowieża virgin forests in Poland. VII. *Schizopora paradoxa* (Schrad. ex Fr.) Donk and its diagnose. *In* Acta Soc. Bot. Polon. **38**: 69–81 *5 fs.*

——————— (1969c). [Polish title]. Wood-inhabiting fungi in Bialowieża virgin forests in Poland. VIII. *Schizopora phellinoides* (Pilát) comb. nov. and its diagnose. *In* Acta Soc. Bot. Polon. **38**: 255–269 *7 fs.*

——————— (1969d). [Polish title]. Wood-inhabiting fungi in Bialowieża virgin forests in Poland. X. *Fibuloporia subvermispora* (Pilát) Domań., comb. nov. and its diagnose. *In* Acta Soc. Bot. Polon. **38**: 453–464 *7 fs.*

———— (1969e). [Polish title]. Wood-inhabiting fungi in Bialowieża virgin forests in Poland. XI. *Incrustoporia tschulymica* (Pilát) Domań. and its diagnose. *In* Acta Soc. Bot. Polon. **38**: 465–473 *4 fs.*

———— (1970a). [Polish title]. Wood-inhabiting fungi of Bialowieża virgin forests in Poland. IX. Further studies on *Ceriporiopsis placenta* (Fr. sensu J. Erikss.) Domań. *In* Acta Soc. Bot. Polon. **39**: 51–62 *4 fs.*

———— (1970b). [Polish title]. Wood-inhabiting fungi in Bialowieża virgin forests in Poland. XIII. Two species of *Diplomitoporus* Domań., gen. nov. *In* Acta Soc. Bot. Polon. **39**: 191–207 *9 fs.*

———— (1970c). [Polish title]. Wood-inhabiting fungi in Bialowieża virgin forests in Poland. XIV. *Coriolus hoehnelii* (Bres. in Höhn.) Bourd. & Galz. *In* Acta Soc. Bot. Polon. **39**: 521–530 *4 fs.*

———— (1970d). [Polish title]. Wood-inhabiting fungi in Bialowieża virgin forests in Poland. XV. *Polyporus dichrous* Fr. and *Polyporus pannocinctus* Romell in culture. *In* Acta Soc. Bot. Polon. **39**: 531–538 *5 fs.*

———— (1970e). [Polish title]. Wood-inhabiting fungi in Bialowieża virgin forests in Poland. XVI. *Coriolus foliaceo-dentatus* (Nikol.) Domański, comb. nov. *In* Acta Soc. Bot. Polon. **39**: 701–709 *4 fs.*

———— (1971). [Polish title]. Wood-inhabiting fungi in Bialowieża virgin forests in Poland. XVII. *Ceriporiopsis gilvescens* (Bres.) Domański. *In* Acta Soc. Bot. Polon. **40**: 295–303 *6 fs.*

———— & A. Orlicz (1966). [Polish title]. *Dichomitus campestris* (Quél.) comb. nov. in Poland. *In* Acta Soc. Bot. Polon. **35**: 627–636 *7 fs.*

———— & ———— (1967a). The taxonomy of *Polyporus megaloporus* Pers. *In* Int. Biodetn Bull. **3**: 79–80.

———— & ———— (1967b). [Polish title]. *Polyporus megaloporus* Pers. in the family Polyporaceae s. str. *In* Acta mycol. **3**: 51–62 *5 fs.*

———— & ———— (1967c). [Polish title]. A study on fungus *Ischnoderma corrugis* (Fr.) Domań. & Orlicz with special reference to structure of its carpophores. *In* Fragm. flor. geobot. **12**: 535–549 *5 fs.*

———— & ———— (1969). [Polish title]. A study of polypore *Irpex lacteus* (Fr. ex Fr.) Fr. *In* Acta mycol. **5**: 149–159 *6 fs.*

————, H. Orlos & A. Skirgiełło (1967). [Basidiomycetes: Aphyllophorales: Polyporaceae pileatae, Mucronoporaceae pileatae, Ganodermataceae, Bondarzewiaceae, Boletopsidaceae, Fistulinaceae]. *In* Flora polska (Grzyby) **3**: 398 & (i) pp. *92 fs., 29 pls.*

Donk, M. A. (1960). The generic names proposed for Polyporaceae. (The generic names proposed for Hymenomycetes—X). *In* Persoonia **1**: 173–302. — Reprinted *in* Biblthca mycol. **11**. 1968.

———— (1962). The generic names proposed for Polyporaceae. Additions and corrections. (The generic names proposed for Hymenomycetes—XIV). *In* Persoonia **2**: 201–210.

———— (1966a). *Osteina*, a new genus of Polyporaceae. *In* Schweiz. Z. Pilzk. **44**: 83–87.

———— (1966b). Notes on European polypores—I. *In* Persoonia **4**: 337–343.

———— (1967). Notes on European polypores—II. Notes on *Poria*. *In* Persoonia **5**: 47–130.

———— (1968). On *Cristella*. *In* Taxon **17**: 277–278.

———— (1969a). Notes on European polypores—III. Notes on species with stalked fruitbody. *In* Persoonia **5**: 237–263.

———— (1969b). Notes on European polypores—IV. On some species of *Ganoderma*. *In* Proc. K. Ned. Akad. Wet. (C) **72**: 273–282.

———— (1969c). On the typification of *Hexagonia Pollini* per Fr. *In* Taxon **18**: 663–666. — [Notes on European polypores—V].

———— (1971a). Notes on European polypores—VI. *In* Proc. K. Ned. Akad. Wet. (C) **74**: 1–24.

———— (1971b). Notes on European polypores—VII. *In* Proc. K. Ned. Akad. Wet. (C) **74**: 25–41.

———— (1971c). Notes on European polypores—VIII. *In* Persoonia **6**: 201–218.

———— (1971d). Multiple convergence in the polyporaceous fungi. *In* R. H. Petersen (ed.), Evol. higher Basidiom. pp. 393–422.

———— (1971e). Notes on European polypores—IX. *In* Proc. K. Ned. Akad. Wet. (C) **74**: 405–421.

———— (1972a). Notes on European polypores—X. *In* Proc. K. Ned. Akad. Wet. (C) **75**: 165–178.

———— (1972b). Notes on European polypores—XI. On some species of *Tyromyces*. *In* Proc. K. Ned. Akad. Wet. (C) **75**: 287–304.

———— (1973). Notes on European polypores—XII. *In* Proc. K. Ned. Akad. Wet. (C) 217–230.

DUMÉE, P. (1917). Notes de mycologie pratique (Suite) V. — Note sur les *Polyporus ulmarius* Sow. et *Polyporus fraxineus* Bull. *In* Bull. Soc. mycol. Fr. **33**: 28–32.

DURAND, J. (1924). Le *Trametes* du pin. *In* Revue Eaux Forêts 62: 59–61 *(2) pls.* — Cf. Rev. appl. Mycol. **3**: 494. 1924. — *Trametes [Phellinus] pini.*

DŽAFAROV, S. A. (1955). [Russian title]. [A new form of a parasitic fungus, *Ganoderma resinaceum* Boud. in Pat. forma *quercinum* f. n.]. *In* Dokl. Akad. Nauk. azerb. SSR **11** (2): 131–133 *(1) f.*

———— (1956). [Russian title]. [Fungi of the family Polyporaceae on tree-shrub species in the Lenkoran zone of the Azerbajdžan SSR]. *In* Izv. Akad. Nauk. azerb. SSR **1956** (5): 85–94.

———— (1957). [Russian title]. [Representatives of fungi from the genus *Ganoderma* on relic tree species in Talyš]. *In* Izv. Akad. Nauk. azerb. SSR **1957** (5): 117–134.

———— (1958). [Russian title]. [A new parasitic fungus on the chestnut-leaved oak]. *In* Lesn. Chozj. **11** (7): 42–43 *(1) f.* — *Ganoderma resinaceum* f. *quercinum.*

EBERT, P. (1940). Ein seltener Eggenpilz: *Irpex pachyodon* Fr. *In* Z. Pilzk. **19**: 12–13 *pl. 4 (3 lower fs.).*

———— (1967). *Trametes heteromorpha* (Fr.) Bres. im Erzgebirge. *In* Mykol. MittBl., Halle **11**: 37–40 *(1) f., (1) textmap.*

EDGERTON, C. W. (1907). The rate and period of growth of *Polyporus lucidus*. *In* Torreya **7**: 89–97. — *Ganoderma tsugae.*

EGELAND, J. (1914). Norske resupinate poresopper. *In* Nyt Mag. Naturvid. **52**: 123–171.

EISFELDER, I. & K. HERSCHEL (1966). *Agathomyia wankowiczi* Schnabl, die "Zitzengallenfliege" aus *Ganoderma applanatum. In* Westfäl. Pilzbr. **6**: 5–10 *7 fs.*

ELIADE, E. (1960). [Romanian title]. *Piptoporus quercinus* (Schrad.) Pilát—une espèce rare, nouvelle pour la mycoflore de la République populaire roumaine. *In* Comunle Acad. Rep. pop. rom. **10**: 743–746 *1 f.*

ELLIOTT, J. A. (1918). Wood-rots of peach trees caused by *Coriolus prolificans* and *C. versicolor. In* Phytopathology **8**: 615–617 *2 fs.*

ELLIS, J. B. & B. M. EVERHART (1889a). Some new species of hymenomycetous fungi. [*Mucronoporus* E. & E. A new genus of Polyporeae]. *In* Mycol. **5**: 28–29 *pl. 8.*

———— & ———— (1889b). *Mucronoporus* E. & E. *In* J. Mycol. **5**: 90–92.

ENSLIN, J. C. (1784). De *Boleto suaveolente* Linn. Erlangae, 32 pp. *(1) pl.* (Diss. inaug.).

———— (1798). Abhandlung über die Eigenschaften und den Gebrauch des wohlriechenden Weidenschwamms. Marburg, 88 pp. *1 pl.* (n.v.).

ERIKSSON, J. (1949). The Swedish species of the « *Poria vulgaris*-group ». *In* Svensk bot. Tidskr. **43**: 1–25 *6 fs., pls. 1–5.*

ETHERIDGE, D. E. (1955). Comparative studies of North American and European cultures of the root rot fungus, *Fomes annosus* (Fr.) Cooke. *In* Can. J. Bot. **33**: 416–428 *fs. 1, 2, 1 pl. (=f. 3).*

EYERDAM, W. J. (1953). A giant bracket fungus from southeastern Alaska. *In* Madroño **12**: 31–32. — *Fomes [Ganoderma] applanatus.*

FALCK, R. (1909). Die *Lenzites*-Fäule des Coniferenholzes. *In* HausschwForsch. **3**: ix–xxxii & 234 pp. *24 fs., 7 pls.*

———— & O. FALCK (1937). Die Ptychogasterfäule des Coniferenholzes. *In* Hausschw. Forsch. **12**: 1–62 & (i) *47 fs., 4 pls. (1–3, 3b).*

FALLAYAN, F. (1964). Étude culturale et anatomique de quelques Polypores conidiophorés à trame molle ou coriace. Paris, 89 pp. *15 and (8) textpls.* Thèse Fac. Sci. Univ. Paris, Sér. A No. 1066.

FARINHA, M. (1957). Contribuição para o estudo das Polyporaceae de Portugal. *In* Portug. Acta biol. (B) **6**: 4–25 *pls. 1–4*, 2 quadr.

———— (1964). Caracteres morfológicos das hifas dos himenóforos de Polyporaceae. *In* Portug. Acta biol. (B) **7**: 288–346.

———— & J. M. ROSADO (1952). Morphological, anatomical and cultural characters of *Polyporus ulicis* Boud. *In* Bolm Soc. broter. II **26**: 193–201 *43 fs., 2 pls.*

FAULL, J. H. (1916). *Fomes officinalis* (Vill.), a timber-destroying fungus. *In* Trans. R. Can. Inst. **11**: 185–209 *pls. 18–25.*

———— (1919). Pineapple fungus or enfant de pin or wabadou. *In* Mycologia **11**: 267–272. — *Fomes [Agaricum] officinalis.*

FAVRE, J. & S. RUHLÉ (1947). Deux champignons steppiques nouveaux pour la Suisse, *Polyporus rhizophilus* et *Disciseda circumscissa*. *In* Schweiz. Z. Pilzk. **25**: 57–61 *3 fs.*

FAYOD, V. (1889). Sopra un nuovo genere de Imenomiceti. *In* Malpighia **3**: 69–73 *3 fs.* — *Boletopsis.*

FERGUS, C. L. (1956). Some observations about *Polyporus dryadeus* on oak. *In* Pl. Dis. Reptr **40**: 827–829.

FERRY, R. (1894). Note sur *Poria contigua* (Pers.) Fr. *In* Revue mycol. **16**: 158–159 *pl. 150 f. 15.* — Description of spores does not agree.

FIDALGO, M. E. P. K. (1959). Note on *Lenzites cinnamomea* Fr. *In* Mycologia **50**: 753–756 *2 fs.* "1958".

———— (1962). The genus *Osmoporus* Sing. *In* Rickia **1**: 95–138 *fs. 1–9.*

———— (1968). Typification of the genus *Hexagona*. *In* Taxon **17**: 37–43.

FIDALGO, O. (1958a). The nomenclatural status of *Daedalea* Pers. ex Fr. and related genera. *In* Taxon **7**: 133–140.

———— (1958b). Studies on *Spongipellis borealis* (Fr.) Pat. *In* Mycopath. Mycol. appl. **10**: 1–18 *8 fs.*

———— (1959a). Studies on *Ptychogaster rubescens* Boud. the chlamydosporiferous form of *Polyporus guttulatus* Pk. *In* Mycologia **50**: 831–836 *4 fs.* "1958".

———— (1959b). Binomial combinations related to *Polyporus acanthoides* Fr. *In* Bull. Torrey bot. Cl. **86**: 130–136 *3 fs.* — *Abortiporus.*

———— (1964). Revisão do gênero *Heteroporus* Laz. emend. Donk. São Paulo. (vi) & 116 pp. *(10) pls.* (Tese, São Paulo). — Mimeographed. For printed edition, see O. Fidalgo *1969*.

———— (1967). Ultrasounds applied to the study of hyphal systems in Polyporaceae. *In* Mycologia **59**: 545–548.

———— (1969). Revision of the genus *Heteroporus* Láz. emend. Donk. *In* Rickia
4: 99–208 52 fs. — Printed edition of mimeographed paper of 1964.

FINDLAY, W. P. K. (1939). Note on an abnormal fungus on birch. *In* Trans. Br.
mycol. Soc. 23: 169–170 *pl. 3.* — *Inonotus obliquus.*

FISCHER, W. (1964). Bemerkenswerte Porlingfunde. Seltene Porlinge in Brandenburg.
In Mykol. MittBl., Halle 8: 47–48.

———— (1967). Die Zinnoberrote Tramete und ihr Vorkommen in der Niederlausitz.
In Niederlaus. florist. Mitt. 3: 48–49.

FISHER, E. (1934). Observations on *Fomes pomaceus* (Pers.) Big. & Guill. infecting
plum trees. *In* Trans. Br. mycol. Soc. 19: 102–113 *4 fs.*

FRIES, N. (1936). Über die Sexualität einiger Polyporaceen. *In* Svensk bot. Tidskr.
30: 355–361 *4 fs.*

———— & K. ASCHAN (1952). The physiological heterogeneity of the dikaryotic
mycelium of *Polyporus abietinus* investigated with the aid of micrurgical
technique. *In* Svensk bot. Tidskr. 46: 429–445 *6 fs., 1 pl.*

———— & L. JONASSON (1941). Über die Interfertilität verschiedener Stämme
von *Polyporus abietinus* (Dicks.) Fr. *In* Svensk bot. Tidskr. 35: 177–193
8 fs.

FRITZCHE, W. & K. HERSCHEL (1969). Beobachtungen an *Trametes extenuata*
Dur. et Mont. im Leipziger Raume. (Zusammenfassung von H. Jahn).
In Westfäl. Pilzbr. 7: 48–56 *7 fs.* "1968".

FURTADO, J. S. (1962). Structure of the spore of the Ganodermoideae Donk. *In*
Rickia 1: 227–241 *22 fs.*

———— (1965). Relation of microstructures to the taxonomy of the Ganodermoideae (Polyporaceae) with special reference to the structure of the
cover of the pilear surface. *In* Mycologia 57: 588–611 *27 fs.*

FÅHRAEUS, G. (1954). Further studies in the formation of laccase by *Polyporus
versicolor*. *In* Physiologia Pl. 7: 704–712. — Lists previous publications
on the subject.

GARD, M. (1922). L'apoplexie de la Vigne et les formes résupinées du *Fomes
igniarius* (L.) Fries. *In* Bull. Soc. Path. vég. Fr. 9: 22–28 *2 fs.*

GARREN, K. H. (1938). Studies on *Polyporus abietinus.* I. The enzyme-producing
ability of the fungus. *In* Phytopathology 28: 839–845. — II. The
utilization of cellulose and lignin by the fungus. *Ibid.* 28: 875–878.

GASPARRINI, G. (1842). Richerche sulla natura della Pietra fungaja e sul fungo
vi spopranasce. *In* Atti Accad. pontaniana 2: 197–254 *pls. 1–5.*

GAY, J. L., S. A. HUTCHINSON & J. TAGGART (1959). Spore discharge in the
Basidiomycetes. I. Periodicity of spore discharge by *Trametes gibbosa* Fr.
In Ann. Bot. II 23: 297–306 *3 fs., 2 pls.*

GIBSON, I. A. S. & J. TRAPNELL (1957). Sporophore production by *Polyporus
arcularius* Batsch ex Fr. in culture. *In* Trans. Br. mycol. Soc. 40:
213–220. — Correctly named?

GILBERTSON, R. L. (1954). *Polyporus montagnei* and *Cyclomyces greenei*. *In* Mycologia
46: 229–233 *9 fs.*

———— (1961). Polyporaceae of the western United States and Canada. I. *Trametes*
Fries. *In* Northw. Sci. 35: 1–20 *15 fs.*

GILLOT, F. X. (1909). Déformation coralloïde du *Polyporus umbellatus* Fr. *In* Bull.
Soc. mycol. Fr. 25: 64–65 *pl. 3.* — Certainly wrongly named.

GIRBARDT, M. (1955). Lebendbeobachtungen an *Polystictus versicolor* (L.). *In* Flora
142: 540–563 *26 fs.*

———— (1956). Schnallenanomalien bei *Polystictus versicolor* (L.). *In* Arch.
Mikrobiol. 23: 413–422 *7 fs.*

———— (1957). Der Spitzenkörper von *Polystictus versicolor* L. *In* Planta 50:
47–59 *6 fs.*

————— (1958). Über die Substruktur von *Polystictus versicolor* L. *In* Arch. Mikrobiol. **28**: 255–269 *16 fs.*

————— (1960). Licht- und elektronenoptische Untersuchungen an *Polystictus versicolor* (L.). I. Der Wassergehalt des Hyaloplasmas vegetativer Zellen. *In* Ber. dt. bot. Ges. **53**: 227–240 *2 fs., pls. 6, 7.*

————— (1961a). Licht- und elektronenmicroskopische Untersuchungen an *Polystictus versicolor*. II. Die Feinstruktur von Grundplasma und Mitochondrien. *In* Arch. Mikrobiol. **39**: 351–359 *17 fs.*

————— (1961b). Licht- und elektronenoptische Untersuchungen an *Polystictus versicolor* (L.). VII. Lebendbeobachtung und Zeitdauer der Teilung des vegetativen Kernes. *In* Exp. Cell Res. **23**: 181–194 *12 Abb., 1 f.*

————— (1962). Licht- und elektronenoptische Untersuchungen an *Polystictus versicolor*. III. Änderung der Grundplasmastrukturen während der Schnallenbildung. *In* Z. Naturf. **176**: 49–53 *fs. 1–2, (2) pls.* [*=fs. 3–7*]

GLASER, T. (1953). [Polish title]. Rarely met coralloid form of the sporophore *Fomes applanatus* (Pers.) Wallr. forma *coralloides*, f. nova. *In* Acta Soc. Bot. Polon. **22**: 805–810 *2 fs.*

GÖPFERT, H. (1968). *Phellinus tremulae* (Bond.) Bond & Boriss., der Espenfeuerschwamm in der Schweiz gefunden. *In* Schweiz. Z. Pilzk. **46**: 157–160 *2 fs.*

————— (1970). *Fomitopsis rosea*—der rosenrote Baumschwamm. *In* Schweiz. Z. Pilzk. **48**: 53–57 *fs. a–g.*

————— (1971). *Phellinus ferrugineofuscus* in der Schweiz gefunden. *In* Schweiz. Z. Pilzk. **49**: 33–36 *fs. a–f.*

GONZÁLES FRAGOSO, R. (1917). [Nota bibliográfica]. Lázaro e Ibiza (B.): Los Poliporáceos de la flora española. . . . *In* Boln R. Soc. esp. Hist. nat. **17**: 458–460.

GOOD, H. M. & W. SPANIS (1958). Some factors affecting the germination of spores of *Fomes igniarius* var. *populinus* (Neuman) Campbell, and the significance of these factors in infection. *In* Can. J. Bot. **36**: 421–437.

GOSSELIN, R. (1944). Studies on *Polystictus circinatus* and its relation to butt-rot of spruce. *In* Farlowia **1**: 525–568 *3 textpls.*

GOVI, G. (1968–9). Polyporaceae italiane. *In* Monte Boschi **19** (4): 35–50 *fs. 1–22, (1) fig., textpls. 1–3*; **19** (6): 31–44 *fs. 23–43, textpls. 4–6.* 1968; **20** (1): 51–60 *fs. 44–57, textpls. 7, 8*; **20** (3): 39–46 *fs. 58–77*; **20** (4): 55–64 *fs. 78–92, textpls. 9–12*; **20** (6): 41–49 *fs. 93–110, pls. 13–16.* 1969; **21** (3): 43–50 *pls. 17–19*; **21** (4): 45–54 *pls. 20–22*; **21** (6): 33–38 *pls. 23, 24.* 1970.

GRAFF, P. W. (1936). North American polypores.—I. *Polyporus squamosus* and its varieties. *In* Mycologia **28**: 154–170.

————— (1939). North American polypores.—II. *Polyporus biennis* and its varieties. *In* Mycologia **31**: 466–484.

GRAY, W. D. (1947). Physiological variation in isolates of *Polyporus schweinitzii* Fr. (Fungi; Basidiomycetes). *In* Trans. Kentucky Acad. Sci. **12**: 59–78 *1 f., 6 textpls.*

GRICIUS, A. (1967a). [Russian title]. Fungi of the genus *Phellinus* in the Lithuanian S.S.R. *In* Zinātn. Raksti Pētera Stučkas latv. valsts Univ. **74** (Bot.): 15–19. "1966".

————— (1967). [Russian title]. Resupinate polypores of the Lithuanian SSR. *In* Liet. TSR Mosklų Akad. Darb. (C) **2**: 65–85 *1 f.*

GRIFFIN, B. R. & C. L. WILSON (1967). Asexual nuclear behaviour and formation of conidia in *Fomes annosus*. *In* Phytopathology **57**: 1176–1177 *1 f.*

GRIPENBERG, J. (1951). Fungus pigments. I. Cinnabarin, a colouring matter from *Trametes cinnabarina* Jacq. *In* Acta chem. scand. **5**: 590–595.

————— (1958). Fungus pigments. VIII. The structure of cinnabarin and cinnabarinic acid. *In* Acta chem. scand. **12** (Part II): 603–610 *3 fs.* — Cites previous literature.

GUÉDÈS, M. (1970). La signification du stipe des carpophores d'après quelques Polyporales et Pleurotacées. *In* Nova Hedw. **18**: 271–281 *36 fs.* "1969".

GUÉGUEN, F. (1901). Sur une forme tératologique du *Ganoderma lucidum*. *In* Bull. Soc. mycol. Fr. **17**: 34–36 *(1) f.*

GÜSSOW, H. T. (1919). The Canadian tuckahoe. *In* Mycologia **11**: 104–110 *pls. 7–9.* — *Grifola tuckahoe* [=*Polyporus tuberaster*].

GUILLEMIN, F. (1923). Etudes sur les *Poria*. *In* Bull. Soc. Sci. nat. Saône-et-Loire II **23**: 21–33.

GUILLEMOT, J. (1894). Note sur les *Trametes hispida* Bagl. et *Trogii* Bk. *In* Bull. Soc. mycol. Fr. **10**: 73–74.

GUINIER, P. (1919). Une invasion de *Trametes Pini* Fr. dans une forêt de Pin maritime. *In* Bull. Soc. Path. vég. Fr. **6**: 48–51.

————— & R. MAIRE (1908). Sur l'orientation des réceptacles des *Ungulina*. *In* Bull. Soc. mycol. Fr. **24**: 138–140 *(2) fs.* — *Ungulina* [*Fomitopsis*] *pinicola*, *U.* [*Fomes*] *fomentaria*.

HAASE, J. (1961). Der Erreger der Gallen des Abgeflachten Porlings *Ganoderma applanatum* (Pers.) Pat. *In* Mycol. MittBl., Halle **5**: 18 *(2) fs.*

HADDOW, W. R. (1931). Studies in *Ganoderma*. *In* J. Arnold Arb. **12**: 25–46 *(1) f., pls. 29–30.*

————— (1938a). On the classification, nomenclature, hosts and geographical range of *Trametes Pini* (Thore) Fries. *In* Trans. Br. mycol. Soc. **22**: 182–193.

————— (1938b). The disease caused by *Trametes Pini* (Thore) Fries in white pine (*Pinus Strobus* L.). *In* Trans. R. Can. Inst. **22** (1): 21–80 *1 pl.* — For previous literature on the *Phellinus pini* complex.

————— (1941). On the history and diagnosis of *Polyporus tomentosus* Fries, *Polyporus circinatus* Fries and *Polyporus dualis* Peck. *In* Trans. Br. mycol. Soc. **25**: 179–190 *pl. 6.*

HAGSTRÖM, E. (1971). *Polyporus melanopus* and *P. picipes*—A taxonomic study. *In* Årsskr. Göteb. Svampkl. **1971**: 28–43 *5 fs.*

HANSEN, L. (1955). *Polyporus lentus* i Danmark. *In* Bot. Tidsskr. **52**: 170–171.

————— (1956). Two polyporaceous fungi with merulioid hymenophore. *In* Friesia **5**: 251–256 *(1) f.* — *Poria* [*Merulius*] *taxicola*, *Polyporus* [*Gloeoporus*] *dichrous*.

————— (1958). On the anatomy of the Danish species of *Ganoderma*. *In* Bot. Tidsskr. **54**: 333–352 *9 fs.*

————— (1960). *Lindtneria trachyspora*. A poriate corticiaceous fungus with coronate spores. *In* Bot. Tidsskr. **55**: 277–281 *4 fs.*

————— (1969). En ny poresvamp i Danmark: *Fibuloporia wynnei*. *In* Bot. Tidsskr. **64**: 242–243.

HARACSI, L. (1941). [Hungarian title]. Der zweigestaltige (schiefe) Holzschwamm: eine Gefahr der Zerreichen. [*Fomes obliquus* (Pers.) Fries]. *In* Erdész. Kisérl. **43**: 1–31 *(2) pls.* [=*10 fs.*]. — *Inonotus nidus-pici*, fide Černý.

————— & Z. IGMÁNDY (1956). [Hungarian title]. Das Vorkommen des Zerreichen-baumschwammes [*Xanthochrous obliquus* (Pers.) Bourd. and Galz.] an den Laubbäumen Ungarns. *In* Erdömérn. Föisk. Közl., Sopron **1956** (1): 73–87 *13 fs.* — *Inonotus nidus-pici*, fide Černý.

————— & ————— (1957). [Hungarian title]. Das Vorkommen des schwarzen Zerreichenzunderschwammes (*Xanthochrous obliquus* B. et G.) auf der gemeinen Esche. *In* Erdömérn. Föisk. Közl., Sopron **1957** (2): 85–96 *10 fs.* — *Inonotus nidus-pici*.

HARIOT, P. (1891). *Trametes hispida* Bagl. et *T. trogii* Ber[k]. *In* J. Bot., Paris (ed. Morot) **5**: 356.

HARMSEN, L. (1954). Om *Polyporus caesius* og *Ditiola radicata* som Tømmersvampe. *Polyporus caesius* and *Ditiola radicata* as timber fungi. *In* Bot. Tidsskr. **51**: 117–123 *11 fs.*

HARTIG, H. J. A. R. (1893). Die Spaltung der Oelbäume. *In* Forstl.-naturw. Z. **2**: 57–63 *pl. 4.* — "*Polyporus fulvus* var. *Oleae* Scop.".

HARZ, C. O. (1868). Beitrag zur Kenntniss des *Polyporus officinalis* Fries. *In* Bull. Soc. imp. Nat. Mosc. **41**: 3–40 *pls. 1, 2.* — Also published separately (Inaug. Diss.).

HASZLINZSKY, F. A. (1870). Der Nussschwamm als Farbenpflanze. *In* Öst. bot. Z. **20**: 77. — *Polyporus* [*Inonotus*] *hispidus.*

HEALD, F. D. (1906). A disease of the cotton-wood due to *Elfvingia megaloma*. *In* Rep. agric. Exp. Stn Nebraska **19**: 92–100 *pls. 1–4.*

HEDGCOCK, G. G. & W. H. LONG (1914). Heart-rot of oaks and poplars caused by *Polyporus dryophilus. In* J. agric. Res. **3**: 65–78 *pls. 8–10.*

HEIM, R. (1947). Sur les caractères des Polypores en culture artificielle. *In* C. r. hebd. Séanc. Acad. Sci., Paris **225**: 421–423.

————— (1962). Notules mycologiques sur la flore française. 9. Un Polypore australien parmi la flore parisienne (*Caloporus Hartmanni* Cooke). *In* Revue Mycol. **27**: 93–96. — (0).

————— (1963). L'organisation architecturale des spores de Ganodermes. *In* Revue Mycol. **27**: 199–211 *pls. 11–15.* "1962".

————— & R. LAMI (1950). La disparition des Peupliers de Paris sous les atteintes de *l'Ungulina Inzengae* (de Not.) Pat. *In* C. r. hebd. Séanc. Acad. Agric. Fr. **36**: 257–258.

HEINRICHER, E. & E. ELSLER (1910). *Pachyma Cocos* Fr. Ein interessanter Pilzfund aus Tirol. *In* Z. Ferdinand. Tirol III. Heft 54: 339–348 *(1) pl.* — Reprint seen.

HEMMI, T. & J. IKEYA (1939). [Japanese title]. Studies on *Lenzites gibbosa* (Pers.) Hemmi n. comb. causing wood-rot of deciduous trees. *In* Ann. phytopath. Soc. Japan **9**: 1–15 *pls. 1, 2.*

————— & S. KURATA (1931). Pathological studies on *Polyporus betulinus* (Bull.) Fr. *In* Forschn Geb. PflKrankh., Japan **1**: 206–224 *pls. 10–11.*

HENNIG, B. (1935). Merkwürdige Gallenbildung am Flachen Porling. *In* Natur Volk **65**: 543–547 *5 fs.*

HENNINGS, P. C. (1888). Über *Oligiporus rubescens* Bref. (=*Ptychogaster rubescens* Boud.). *In* Verh. bot. Ver. Brandenb. **30**: v–vii.

————— (1901). Über *Polyporus frondosus* (Fl. dan.) Fries, welcher aus einer sclerotiumartigen Knolle entstanden ist. *In* Verh. bot. Ver. Brandenb. **42**: xviii.

HENNINGSSON, B. (1965). Physiology and decay activity of the birch conk fungus *Polyporus betulinus* (Bull.) Fr. *In* Studia forest. suec. No. 34: 77 pp. *19 fs.*

HENRY, A. (1924). Larch Agaric. *In* Pharm. J. & Pharmacist **112**: 439–440. — *Agaricum officinale.*

HERINK, J. (1955). [Czech title]. [Sclerotia in *Polyporus umbellatus* (Pers. ex Fr.) Bond. et Sing.]. *In* Česká Mykol. **9**: 171–176 *(4) fs.*

HERRMANN, M. (1962). Die Verwendung des Echten Zunderschwammes—*Fomes fomentarius* (Fr.) Kickx, — einst und jetzt. *In* Mykol. MittBl., Halle **6**: 56–62. — Cites previous literature.

————— (1971). *Rigidoporus vitreus* (Pers. ex Fr.) Donk—ein resupinater Porling. *In* Mykol. MittBl., Halle **15**: 81–85 *3 fs.*

HERSCHEL, K. & M. HUTH (1970). Eine neue Pilzgalle in der deutschen Fauna. *In* Mykol. MittBl., Halle **13**: 100–103 *2 fs.* "1969".

HEUFLER, L. J. VON [=L. VON HOHENBÜHEL, genannt HEUFLER ZU . . .] (1870). Der *Fungus Laricis aureus* Matthiolis. *In* Öst. bot. Z. **20**: 193–199.

HILBORN, M. T. (1942). The biology of *Fomes fomentarius*. *In* Bull. Maine agric. Exp. Stn No. 409: 161–214 *17 pls.*

HILBORN, M. T. & D. H. LINDER (1939). The synonymy of *Fomes fomentarius*. *In* Mycologia **31**: 418–419.

HINTIKKA, V. (1970). First record of *Pycnoporellus albo-luteus* in N.W.-Europe. *In* Karstenia **11**: 33–34 *1 f.*

HIORTH, J. (1965). The phenoloxidase and peroxidase activities of two culture types of *Phellinus tremulae* (Bond.) Bond. et Boriss. *In* Meddr norske SkogforsVes. **20** / No. 75: 255–272 *5 fs.*

HIRT, R. R. (1926). Stratified sporophores of *Polyporus gilvus*. *In* Mycologia **18**: 111–113.

————— (1928). The biology of *Polyporus gilvus* (Schw.) Fries. *In* Bull. New York St. Coll. For., Syracuse **1** (1a): 47 pp. *2 fs., 11 pls.* (Techn. Publs No. 22).

————— (1930). *Fomes everhartii* associated with the production of sterile rimose bodies on *Fagus grandifolia*. *In* Mycologia **22**: 310–311 *pl. 25.*

————— (1932). On the biology of *Trametes suaveolens* (L.) Fries. *In* Bull. New York St. Coll. For., Syracuse **5** (1-c): 36 pp. *1 f. 7 textpls.* (Techn. Publ. No. 37).

————— (1949a). An isolate of *Poria xantha* on media containing copper. *In* Phytopathology **39**: 31–36 *2 fs.*

————— (1949b). Decay of certain hardwoods by *Fomes igniarius, Poria obliqua* and *Polyporus glomeratus*. *In* Phytopathology **39**: 475–480 *1 f.*

————— & H. HOPP (1942). Relation of tube layers to age in sporophores of *Fomes igniarius* on aspen. *In* Phytopathology **32**: 176–178 *1 f.*

————— & J. L. LOWE (1945a). *Poria microspora* in house timbers. *In* Phytopathology **35**: 217–218 *1 f.*

————— & ————— (1945b). *Polyporus versicolor* on Asiatic chestnut. *In* Phytopathology **35**: 574–575 *1 f.*

HÖHNEL, F. X. R. VON (1907). Fragmente zur Mykologie (III. Mitteilung . . .). [96. Über die sanguinolenten *Poria*-Arten Europas]. *In* Sber. K. Akad. Wiss. Wien (Math.-nat. Kl., Abt. I) **116**: 91–94.

————— (1907). Mykologisches. XVII. Über eine Krankheit der Feldahorne in den Wiener Donau-Auen. *In* Öst. bot. Z. **57**: 177–181. — *Poria [Inonotus] obliqua*.

HOLE, R. S. (1915). *Trametes Pini*, Fries, in India. *In* Indian Forest Rec. **5** (5): 159–184 *7 pls.*

HOPP, H. (1936). Appearance of *Fomes igniarius* in culture. *In* Phytopathology **26**: 915–917.

————— (1938). Sporophore formation by *Fomes applanatus* in culture. *In* Phytopathology **28**: 356–358 *1 f.*

HRUBY, J. (1931). *Melanopus (Polyporus) rhizophilus* (Pat.) in Mähren. *In* Öst. bot. Z. **80**: 72–73 *fig.*

HUBERT, E. E. (1929a). A butt-rot of balsam fir caused by *Polyporus balsameus* Pk. *In* Phytopathology **19**: 725–732 *3 fs.*

————— (1929b). A root and butt rot of conifers caused by *Polyporus circinatus* Fr. *In* Phytopathology **29**: 745–747.

HUMPHREY, C. J. & S. LEUS (1931). A partial revision of the *Ganoderma applanatum* group, with particular reference to its oriental variants. *In* Philipp. J. Sci. **45**: 483–589 *1 f., 36 pls.* [& mimeographed index, 5 pp., added to reprint].

————— & S. LEUS-PALO (1932). Studies and illustrations in the Polyporaceae. III. Supplementary notes on the *Ganoderma applanatum* group. *In* Philipp. J. Sci. **49**: 159–184 *12 pls.*

HYDE, J. M. & C. H. WALKINSHAW (1966). Ultrastructure of basidiospores and mycelium of *Lenzites saepiaria*. *In* J. Bact. **92**: 1218–1227 *12 fs.*

Høeg, O. A. and I. Jørstad (1938). *Ganoderma lucidum* in Norway. *In* K. norske Vidensk. Selsk. Forh. **10**: 201–202.

Igmándy, Z. (1953). [Hungarian title]. Schädigung von Zerreichen-Reinbeständen durch den schiefen Holzschwamm (*Fomes obliquus* (Pers.) Fries.) *In* Erdömérn. Föisk. Évk., Sopron 1951–2: 93–106 *3 fs*. — Date of publication, according to Rev. appl. Mycol. **36**: 72. 1957. — *Inonotus nidus-pici*.

————— (1955a). [Hungarian title]. Vorkommen und Schädigung des *Trametes Pini* (Thore) Fries in Ungarn. *In* Erdömérn. Föisk. Közl. 1954: 5–10 *(3) pls.* [=*6 fs.*].

————— (1955b). [Hungarian title]. Die Stockfäule-erregende Pilze der Zerreichen-Niederwälder. *In* Erdömérn. Föisk. Közl., Sopron **1955**: 131–147 *11 fs.*

————— (1956). [Hungarian title]. Contributions à la flore des Polyporaceae dans les Bassins carpathiques. [I and II]. *In* Bot. Közl. **46**: 306. — *Fomes [Agaricum] officinalis*, *Grifola ossea* [=*Osteina obducta*].

————— (1957a). [Hungarian title]. Contributions à la flore des Polyporaceae dans les Bassins Carpathiques. III. *Leptoporus litschaueri* (Lohwag) Pilát en Hongrie. *In* Bot. Közl. **47**: 101–103 *3 fs.*

————— (1957b). [Hungarian title]. L'occurrence et les dommages causés par le *Leptoporus irpex* (Schulz.) en Hongrie. *In* Erdömérn. Föisk. Közl., Sopron **1957** (1): 67–73 *5 fs.*

————— (1958). [Hungarian title]. Die Pilze Ödenburgs und seiner Umgebung. [I. Polyporaceae]. *In* Soproni Szemle **12**: 119–135 *14 fs.*

————— (1961). [Hungarian title]. A butt-rotting tinder [*Fomes fraxineus* (Fr.) Cooke] of the acacia (*Robinia pseudacacia* L.). *In* Erdészettud. Közl. **1961**: 69–77 *9 fs.*, *(2) pls.* (fs. 6–9).

————— (1962). [Hungarian title]. *Lenzites reichardtii* Schulzer. *In* Erdészettud. Közl. **1962**: 139–146 *7 fs. on (4) pls.*

————— (1963). Die holzzerstörenden Pilze der Robinie. *In* Lyer & Gillwald (ed.), Holzzerstörung Pilze 293–297 *2 textpls.* (on pp. 386, 387). "1962".

————— (1964a). [Hungarian title]. Vorkommen und Schädigung des Eichen-Rostporling's (*Xanthochrous cuticularis* (Bull.) Pat.) in Ungarn. *In* Mikol. Közl. **1964**: 38–45 *3 fs.*

————— (1964b). [Hungarian title]. Die holzzerstörenden Porlinge unserer Buchenbestände. *In* Erdész. Faip. Egyet. Tud. Közl. **1964**: 101–107 *8 fs.*

————— (1965). [Hungarian title]. Polypori Hungariae [I]. *In* Erdész. Faip. Egyet. Tud. Közl. 1–2: 203–222 *21 fs. (partly on 4 pls.).*

————— (1968a). [Hungarian title]. Die Porlinge Ungarns und ihre phytopathologische Bedeutung (Polyporei Hungariae) II. Teil. *In* Acta phytopath. Acad. Sci. hung. **3**: 221–239 *10 fs.*

————— (1968b). [Hungarian title]. Die Porlinge Ungarns und ihre phytopathologische Bedeutung (Polyporei Hungariae) III. *In* Acta phytopath. Acad. Sci. hung. **3**: 349–359 *7 fs.*, *(1) pl.*=*fs. 3, 6.*

————— (1970). Die Porlinge Ungarns und ihre phytopathologische Bedeutung (Polypori Hungaricae). IV. Teil. *In* Acta phytopath. Acad. Sci. hung. **5**: 279–301 *10 fs.* — *Phellinus.*

Igmándy, Z. & H. Pagony (1965). [Hungarian title]. Ein gefährlicher Kernfäulepilz der Silber- und Graupappelbestände in Ungarn. *In* Erdö **14**: 19–25 *6 fs.* — *Phellinus* sp.=*P. pilatii*, fide Černý, *1968*.

Ikediugwu, F. E. O., C. Dennis & J. Webster (1970). Hyphal interference by *Peniophora gigantea* against *Heterobasidion annosum*. *In* Trans. Br. mycol. Soc. **54**: 307–309 *pl. 26.*

Imazeki, R. (1951). *Cryptoderma Pini* (Brot. ex Fr.) Imazeki and *C. yamanoi* Imazeki. *In* Forschn Geb. PflKrankh., Japan **4**: 174–177.

———— (1957). [Japanese title]. Notes on sampling *I[s]chnoderma (Polyporus) resinosum*. *In* Trans. mycol. Soc. Japan 1 (6): 11–12.

———— & Y. KOBAYASI (1966). Notes on the genus *Coltricia* S. F. Gray. *In* Trans. mycol. Soc. Japan 7: 42–44 *2 fs.*

INGOLD, C. T. (1965). *Fomes fomentarius* on beech in Kent. *In* Trans. Br. mycol. Soc. 48: 81–82.

JAQUENOUD, M. (1968). Erstfund von *Fibuloporia wynnei* (Berk. and Br.) Bond. and Sing. in der Schweiz. *In* Schweiz. Z. Pilzk. 46: 25.

JACQUIOT, C. (1960). Contribution à l'étude de quelques espèces affines de la série des Igniaires. I. — *Phellinus robustus* Karst., *P. Hartigii* Allesch. and Schn., *P. fulvus* (Scop.) Pat. *In* Bull. Soc. mycol. Fr. 76: 83–106 *4 fs., pls. 1, 2.*

———— (1965). Contribution à l'étude de *Phaeolus alborubescens* B. et C. en culture. *In* C. r. 89e Congr. Socs sav. 1964: 304–305 *1 f.* (n.v.).

JAČEVSKIJ, A. A. [A. L. VON JACZEWSKI]. (1911). Bemerkungen zu der Mitteilung von P. Magnus über *Bresadolia caucasica* N. Schestunoff in der Hedwigia Band L, p. 101–104. *In* Hedwigia 50: 253–254 *1 f.*

JAHN, H. (1959). Ein merkwürdiger Flacher Porling. *In* Westfäl. Pilzbr. 2: 44–45 *(1) f.*

———— (1960). Zum Fund des Kupferroten Lackporlings (*Ganoderma pfeifferi* Bres.) bei Detmold. *In* Westfäl. Pilzbr. 2: 90–91. — *Ganoderma adspersum.*

———— (1961). Der Zonen-Porling *(Trametes zonata)*. *In* Westfäl. Pilzbr. 3: 10–12 *(1) f., (1) Bildbeilage.*

———— (1962). Der Espen-Feuerschwamm *(Phellinus tremulae)*, ein gefährlicher Feind der Espe. *In* Westfäl. Pilzbr. 3: 94–102 *4 fs. (1) Bildbeilage.*

———— (1963). Mitteleuropäische Porlinge (Polyporaceae s. lato) und ihr Vorkommen in Westfalen (unter Ausschluss der resupinaten Arten). *In* Westfäl. Pilzbr. 4: 144 pp. *7 fs., 66 Abb. (on textpls.).*

———— (1964). *Ganoderma europaeum* Steyaert, in der Schweiz gefunden. *In* Schweiz. Z. Pilzk. 42: 108–111 *(1) f.*

———— (1965a). *Pachykytospora tuberculosa* (DC. ex Fr.) Kotl. et Pouz. [= *Trametes colliculosa* (Pers.)] in Westfalen gefunden. *In* Westfäl. Pilzbr. 5: 77–79 *(1) f.*

———— (1965b). Entwickelung und Formen der Fruchtkörper beim Zunderschwamm, *Fomes fomentarius*. *In* Westfäl. Pilzbr. 5: 117–131 *9 fs.*

———— (1965c). *Inonotus polymorphus* (Rostk.) Bond. & Sing. in Westfalen gefunden. *In* Westfäl. Pilzbr. 5: 131–134 2 fs.

———— (1965d). Die *Phellinus robustus* var. *hippophaes* — *Ph. contiguus*-Ass., eine Pilzgesellschaft auf Sanddorn. *In* Westfäl. Pilzbr. 5: 139–141.

———— (1966a). Neue Funde von *Phellinus tremulae* in Mitteleuropa. *In* Z. Pilzk. 32: 30–32 *(1) f.*

———— (1966b). Richtigstellung zu einem angeblichen Fund von *Phellinus torulosus* (Pers.) Bourd. and Galz. in Deutschland. *In* Z. Pilzk. 32: 32–33. — *Ganoderma resinaceum.*

———— (1967a). Die resupinaten *Phellinus*-Arten in Mitteleuropa mit Hinweisen auf die resupinaten *Inonotus*-Arten und *Poria expansa* (Desm.) [= *Polyporus megaloporus* Pers.]. *In* Westfäl. Pilzbr. 6: 37–[124] *12 fs., 61 Abb. (on textpls.).*

———— (1967b). Zwei seltene Porlinge in Hessen gefunden: *Hapalopilus croceus* und *Buglossoporus quercinus*. *In* Westfäl. Pilzbr. 6: 145–152 *(4) fs.*

———— (1967c). *Trametes hoehnelii* (Bres.) und *Gloeoporus dichrous* (Fr.) als Nachfolger von *Inonotus*-Arten. *In* Westfäl. Pilzbr. 6: 159–162 *(1) f.*

———— (1968). Geotropisch verformter Porling an stehendem Stamm. *In* Westfäl. Pilzbr. 7: 12–13 *(1) f.* — *Fomitopsis pinicola.*

———— (1969). Die Gattung *Polyporus* s. str. in Mitteleuropa. *In* Schweiz. Z. Pilzk. 47: 218–227 *(1) f.*

————— (1970a). Ein resupinater Porling mit Nebenfruchtform: *Strangulidium rennyi* (B. and Br.) Pouz. mit *Ptychogaster citrinus* Romell. *In* Westfäl. Pilzbr. **8**: 13–16 *(1) f.*

————— (1970b). *Ceriomyces aurantiacus* Pat., eine Nebenfruchtform des Schwefelporlings *(Laetiporus sulphureus)*. *In* Natur Heim, Münster **30**: 85–88 *2 fs.*

————— (1971). Resupinate Porlinge, *Poria* s. lato, in Westfalen und im nördlichen Deutschland. *In* Westfäl. Pilzbr. **8**: 41–68 *13 fs.*

JAY, B. A. (1934). A study of *Polystictus versicolor*. *In* Bull. misc. Inf. Kew **1934**: 409–424 *5 fs., pls. 10, 11.*

JELIĆ, M. B. & M. TORTIĆ (1968). *Ischnoderma corrugis* (Fr.) Domań. et Orlicz, un nouvel élément dans la flore des macromycètes de Yougoslavie. *In* Glasn. bot. Zav. Bašte Univ. Beogr. II **3**: 233–237 *1 f., pls. 1–3.*

JOSSERAND, M. (1935). A propos des divers mécanismes assurant la libération des spores des champignons. La réaction d'un Polypore. *In* Bull. mens. Soc. linn. Lyon **4**: 74–76. — *Ganoderma applanatum.*

JUEL, H. O. (1914). Berichtigung über die Gattung "*Muciporus*". *In* Ark. Bot. **14** (1): 9 pp. *1 pl.*

JØRSTAD, I. & J. G. JUUL (1939). Råtesopper på levende nåletrær. I. (Fungi causing decay of living conifers. I.). *In* Meddr norske SkogsforsVes. 6/ Hft 22: 299–496 *38fs.*

KALANDRA, A. (1968). A contribution to the distribution and importance of *Phellinus pini* (Thore ex Fr.) Pil. var. *abietis* (P. Karst.) Pil. in the Carpathian Range, Czechoslovakia included. *In* Acta mycol. **4**: 275–277.

KALLENBACH, F. (1925). *Trametes cinnabarina* Jacq. (Zinnober-Tramete), *Polystictus hirsutus* Schrad. (striegeliger Porling) und *Lenzites tricolor* (schillender Blättling). *In* Z. Pilzk. **4**: 63–66.

————— (1932). *Ganoderma resinaceum* Boud.-Pat., ein seltener Lack-Porling. Der erste in Deutschland festgestellte Fund. *In* Z. Pilzk. **11**: *18 pls. 3, 4.*

————— (1933–4a). Gallen am flachen Porling *(Polyporus applanatus)*. *In* Z. Pilzk. **12**: 107–109 *pl. 17* 1933; **13**: 21–22, 42. 1934a.

————— (1934b). *Polyporus apalus* Lév., der Mehlstaub-Porling. *In* Z. Pilzk. **13**: 66–67 *pl. 10.* — *Oligoporus rennyi.*

————— (1934c). Ein riesiger Laubporling *(Polyporus frondosus)*. *In* Z. Pilzk. **13**: 22–24 *pls. 2, 3.* — *Meripilus giganteus, fide* K. Lohwag *1940.*

————— (1935–6a). Der weisse Porenschwamm *(Polyporus vaporarius)*. *In* Z. Pilzk. **14**: *pls. 5, 9, 10, 13, 14.* 1935; **15**: 14–15, 36–37, 73–75. 1936a. — Cf. J. Sponheimer *in* Z. Pilzk. **15**: 119–121.

[—————] (1936b–7). Ein gefährlicher Schädling an Stachel- u. Johannisbeersträuchern. *In* Z. Pilzk. **15**: *pl. 16.* 1936b; **16**: 11–12 *pl. 2.* 1937. — *Phellinus ribis.*

KARSTEN, P. A. (1859). Sydvesta Finlands Polyporeer. 47 pp.

————— (1879). Symbolae ad mycologiam fennicam. [b. Polyporaceae Fr.]. *In* Meddn Soc. Fauna Fl. fenn. **5**: 37–40. "1880".

————— (1881). Enumeratio Boletinearum et Polyporearum fennicarum, systemate novo dispositarum. *In* Revue mycol. 3 / No. 9: 16–19.

KARTAJEVSKAJA, N. I. (1928). [Russian title]. [Čaga (Contribution to the study of tree rots)]. *In* Trudy sib. Inst. sel'sk. Chozj. Lesov. / Mitt. forstl. Versuchswesen, Omsk 1 (6): 12 pp. *4 fs.*

KARTAVENKO, N. T. (1958). [Russian title]. *Fomitopsis officinalis* (Vill.) Bond. et Sing. in the « Island » pine-forests of the Trans-uralian forest-steppe. *In* Bot. Ž. **43**: 583–584.

————— (1961). [Russian title]. Species novae fungorum ad montes uralenses inventae. *In* Bot. Mater. Otd. spor. Rast. **14**: 189–196 *(1) f.*

KAUFFMAN, C. H. (1926). *Polyporus anceps* and *Polyporus immitis*. In Mycologia 18: 27–30 *pl. 5*.

————— & H. M. KERBER (1922). A study of the white heart-rot of locust caused by *Trametes robiniophila*. *In* Am. J. Bot. 9: 493–508 *3 fs*.

KEISSLER, K. VON (1907). Monströse Wuchsform von *Polyporus Rostkovii* Fr. *In* Annln KK. naturh. Hofmus. Wien 22: 143–144 *pl. 2*.

KENNEDY, L. L. & R. J. LARCADE (1971). Basidiocarp development in *Polyporus adustus*. *In* Mycologia 63: 69–78 *12 fs*.

KERRUISH, R. M. & E. W. B. DA COSTA (1963). Monocaryotization of cultures of *Lenzites trabea* (Pers.) Fr. and other wood-destroying Basidiomycetes by chemical agents. *In* Ann. Bot. II 27: 653–669.

KHAN, A. H. (1910). Root-infection of *Trametes pini* (Brot.) Fr. *In* Indian Forester 36: 559–562 *pls. 23, 24*.

KILLERMANN, S. (1919). Fund von *Polyporus montanus* Quélet in Bayern. *In* Hedwigia 61: 1–3 *pl. 1*.

————— (1927). Über zwei seltene Polyporaceen in Bayern. *In* Hedwigia 67: 125–130 *(1) f.* — *Polyporus xoilopus* [0] *P. wynnei*, *Poria mycorrhiza* [= *P. mollicula*].

KIRCHHEIMER, F. (1942). Über Reste von Zunderschwämmen aus der Braunkohlenzeit. *In* Z. Pilzk. 20: 85–91 *pls. 17–19*.

KIRCHMAYR, H. (1914). Über den Parasitismus von *Polyporus frondosus* Fr. und *Sparassis ramosa* Schäff. *In* Hedwigia 54: 328–337 *(2) fs*.

KIRSCHSTEIN, W. (1936). Ein schöner und eigenartiger Schimmelpilz. *In* Z. Pilzk. 15: 117–118 *pl. 15 f. 2.* — *Septocylindrium lindtneri* W. Kirschst. on *Ganoderma*.

KLINGEMANN, F. (1893). Über eine in der Natur vorkommende stickstoffhaltige Säure. *In* Justus Liebig's Annln Chem. 275: 89–91. — *Polyporus* [*Phellinus*] *igniarius* [?] on *Quercus* and *Salix*.

KLOTZSCH, J. F. (1833). De *Favolo*, genere Hymenomycetum a Friesio proposito. *In* Linnaea 8: 316–317 *pl. 5.* — *Favolus* [*Polyporus*] *boucheanus* (0).

KÖGL, F. (1926). Untersuchungen über Pilzfarbstoffe. V. Die Konstitution der Polyporsäure. *In* Justus Liebig's Annln Chem. 447: 78–85. — *Polyporus* [*Hapalopilus*] *rutilans*.

KOENIGS, J. W. (1960). *Fomes annosus*: a bibliography with subject index. *In* Occ. Pap. sth. Forest Exp. Stn U.S.D.A. No. 181: 35 pp.

KOMAROVA, E. P. (1956). [Russian title]. (Rare species of polypores found in the BSSR). *In* Vesci Akad. Navuk belarussk. SSR (Bial. Navuk) 2: 125–126 *5 fs*.

————— (1959a). [Russian title]. [The] new species *Tyromyces pseudohoehnelii* Bond. and Komarova. *In* Dokl. Akad. Nauk belorussk. SSR 3: 507–509 *1 f*.

————— (1959b). [Russian title]. Species rarae et formae novae Polyporacearum in Rossia alba inventae. *In* Bot. Mater. Otd. spor. Rast. 12: 249–257 *8 fs*.

————— (1959c). [Russian title]. [On the systematics of the pore fungi (fam. Polyporaceae)]. *In* Dokl. Akad. Nauk belorussk. SSR 3: 463–467.

————— (1960). [Russian title]. [New data on the pore fungus flora (fam. Polyporaceae) of the U.S.S.R.]. *In* Dokl. Akad. Nauk belorussk. SSR 4: 132–133 *1 f*.

————— (1961). [Russian title]. [On the problems of the systematics of the pore fungi (fam. Polyporaceae)]. *In* Sb. bot. Rabot 3: 32–42 *3 fs*.

————— (1964). Opredelitel' trutovych gribov Belorussii. [Key to the pore-fungi of White Russia]. Minsk, 344 pp. *92 fs., 44 textpls*.

KOPPE, F. (1956). Die Zitzengalle des Flachen Porlings in Westfalen. *In* Natur Heim., Münster 16: 7–9 *(1) f*.

KOTLABA, F. (1955a). [Czech title]. The Polyporaceae of the Soběslavská Blata. *In* Čas. národ. Mus. (Odd. přír.) 124: 145–150 *2 & 3 fs*.

———————— (1955b). [Czech title]. A new species of the Mycoflora of Czechoslovakia. — *Trametes subsinuosa* Bres. *In* Česká Mykol. **9**: 83–90 *(3) fs.*

———————— (1961). [Czech title]. Notes on the morphology of fruitbodies in the pore fungi (Polyporales). *In* Česká Mykol. **15**: 180–190 *(9) fs.*

———————— (1965a). [Czech title]. The boreal polypore *Phellinus ferrugineofuscus* (P. Karst.) Bourd. collected in Czechoslovakia. *In* Česká Mykol. **19**: 21–30 *pls. 1, 2.*

———————— (1965b). *Lenzites betulina* (L. ex Fr.) Fr.-Lupeník březovy. *In* Česká Mykol. **19**: 79–82 *pl. 57.*

———————— (1966). [Czech title]. What is *Polyporus sorbi* Velenovsky? *In* Česká Mykol. **20**: 184–188. — *Phellinus pomaceus.*

———————— (1968a). *Phellinus pouzarii* sp. nov. *In* Česká Mykol. **22**: 24–31 *(1) f. & fs. 1, 2.*

———————— (1968b). *Phellinus tricolor* (Bres.) comb. nov., a tropical relative of *Phellinus pilatii* Černý. *In* Česká Mykol. **22**: 174–179 *4 fs.*

———————— (1972). [Czech title]. Ecology and distribution of *Phellinus nigrolimitatus* (Romell) Bourd. & Galz. in Czechoslovakia. *In* Česká Mykol. **26**: 91–102 *3 fs., pls. 3, 4.*

KOTLABA, F. & Z. POUZAR (1956). [Czech title]. Polypori novi vel minus cogniti Čechoslovakiae I. [... — *Tyromyces albobrunneus* (Rom.) Bond.]. *In* Česká Mykol. **10**: 59–63 *(2) fs.*

———————— & ———————— (1957a). [Czech title]. Notes on classification of European pore fungi. *In* Česká Mykol. **11**: 152–170 *(2) fs.*

———————— & ———————— (1957b). [Czech title]. Polypori novi vel minus cogniti Cechoslovakiae II. *In* Česká Mykol. **11**: 214–224 *(4) fs.*

———————— & ———————— (1958). [Czech title]. Polypori novi vel minus cogniti Cechoslovakiae III. *In* Česká Mykol. **12**: 95–104 *(6) fs.*

———————— & ———————— (1959). [Czech title]. Polypori novi vel minus cogniti Cechoslovakiae IV. *In* Česká Mykol. **13**: 27–37 *(3) fs.*

———————— & ———————— (1963a). [Czech title]. A new genus of the polypores— *Pachykytospora* gen. nov. *In* Česká Mykol. **17**: 27–34 *1 f., pls. 1, 2.*

———————— & ———————— (1963b). [Czech title]. Three noteworthy polypores of the Slovakian Carpathians. *In* Česká Mykol. **17**: 174–185 *8 fs.* — *Pycnoporellus alboluteus, P. fibrillosus* [=*P. fulgens*], *Amylocystis lapponica.*

———————— & ———————— (1964a). [Czech title]. A study of *Tyromyces pannocinctus* (Romell) comb. nov. *In* Česká Mykol. **18**: 65–76 *pls. 5, 6.*

———————— & ———————— (1964b). [Czech title]. *Tyromyces gloeocystidiatus* Kotl. et Pouz. sp. nov. — A name for an old polypore. *In* Česká Mykol. **18**: 207–218 *4 fs., pls. 15, 16.*

———————— & ———————— (1965). [Czech title]. *Spongipellis litschaueri* Lohwag and *Tyromyces kmetii* (Bres.) Bond. et Sing., two rare polypores in Czechoslovakia. *In* Česká Mykol. **19**: 69–78 *pls. 5, 6.*

———————— & ———————— (1966a). [Czech title]. *Buglossoporus* gen. nov. — A new genus of polypores. *In* Česká Mykol. **20**: 81–89 *pls. 9, 10, 61.*

———————— & ———————— (1966b). [Czech title]. What is *Polyporus acanthoides* Bull. sensu Velenovsky? — with notes on the systematic position of *Polyporus croceus* (Pers.) ex Fr. *In* Česká Mykol. **20**: 97–104 *2 fs., pls. 11, 12.*

———————— & ———————— (1968a). [Czech title]. *Tyromyces balsameus* (Peck) Murrill in Bohemia. *In* Česká Mykol. **22**: 121–128 *3 fs.*

———————— & ———————— (1968b). [Czech title]. Some new data concerning *Phellinus tremulae* (Bond.) Bond. & Borisov. *In* Česká Mykol. **22**: 279–295 *5 fs., pls. 15, 70.*

———————— & ———————— (1969). *Inonotus rheades* (Pers.) Bond. et Sing. — Rezavec skořicový. *In* Česká Mykol. **23**: 163–170 *2 fs., pls. 8, 9, 73.*

————— & ————— (1970). Revision of the original material of *Phellinus sulphurascens* Pil., *Xanthochrous glomeratus* spp. *heinrichii* Pil., and *Polyporus rheades* Pers. (Hymenochaetaceae). *In* Česká Mykol. **24**: 146–152.

————— & ————— (1971): [Czech title]. *Ganoderma adspersum* (S. Schulz.) Donk—a species resembling *G. applanatum* (Pers. ex S. F. Gray) Pat. *In* Česká Mykol. **25**: 88–102 *5 fs.*

KRAMER, C. L. & D. L. LONG (1971). An endogenous rhythm of spore discharge in *Ganoderma applanatum*. *In* Mycologia **62**: 1138–1144 *4 fs.* "1970".

KREBS, G. (1961). Untersuchungen über die Fruchtkörperbildung, den Lack sowie Phenoloxydasen und Peroxydase bei *Ganoderma lucidum* (Leyss.) Karst. *In* Arch. Microbiol. **40**: 94–118 *10 fs.*

KREH, W. (1939). Die Gallen des flachen Porlings. *In* Aus der Heimat, Stuttg. **52**: 18–21 *fs. 3, 4, 7, pl. 8* [*fs. 1, 2, 5, 6*].

————— (1954). Über die Verbreitung der Zitzengallen des Flachen Porlings. *In* Z. Pilzk. No. 17: 17–18.

KREISEL, H. (1957). Zunderschwämme, *Fomes fomentarius* L. ex Fr., aus dem Mesolithikum. *In* Wiss. Z. Arndt-Univ. **6**: 299–301 *4 fs.*

————— (1960). *Ganoderma pfeifferi* Bres. ein wenig bekannter Lackporling. *In* Westfäl. Pilzbr. **2**: 85–89 *(1) f. & f. (lower) on p. 93.* — Sensu Kreisel = *G. adspersum.*

————— (1961). Die systematische Stellung der Gattung *Polyporus*. *In* Z. Pilzk. **26**: 44–47. "1960".

————— (1962). *Trametes extenuata* und *Trametes trogii* in Deutschland. *In* Ber. bayer. bot. Ges. **35**: 55–56 *2 fs.*

————— (1963a). Über *Polyporus brumalis* und verwandte Arten. *In* Feddes Reprium **68**: 129–138 *pls. 2, 3.*

————— (1963b). Die Gattung *Ganoderma* in Deutschland. *In* Lyr and Gillwald (ed.), Holzzerstörung Pilze, pp. 51–54. "1962".

————— (1968). Der Bergporling, *Bondarzewia montana*, in Thüringen. *In* Mykol. MittBl., Halle **12**: 22–23.

KŘÍŽ, K. (1964). [Czech title]. Die Verbreitung des Nördlichen Zinnoberschwammes, *Pycnoporus cinnabarinus* (Jacq. ex Fr.) Karst., in der Tschechoslowakei; ein weiterer Beitrag zur Kartierung der Makromyceten in Europa. *In* Česká Mykol. **18**: 129–143 *pls. 9, 10.*

KRULL, R. (1892). Über den Zunderschwamm *(Polyporus fomentarius)* und die Weissfäule des Buchenholzes. *In* Jber. schles. Ges. vaterl. Cult. **69** (II. Abth.): 131–133.

————— (1894). Über Infectionsversuche und durch Cultur erzielte Fruchtkörper des Zunderschwammes, *Ochroporus fomentarius* Schroet. *In* Jber. schles. Ges. vaterl. Cult. **71** (IIb. Abth.): 14–16.

KÜHNER, R. (1950a). Absence de boucles chez les Basidiomycetes de la série des Igniaires et comportement nucléaire dans le mycélium des *Hymenochaete* Lév. *In* C.r. hebd. Séanc. Acad. Sci., Paris **230**: 1606–1608.

————— (1950b). Comportement nucléaire dans le mycélium des Polypores de la série Igniaires. *In* C.r. hebd. Séanc. Akad. Sci., Paris **230**: 1687–1689.

————— (1950c). Absence de boucles et comportement nucléaire dans le mycélium de divers Homobasidiés. *In* C.r. hebd. Séanc. Acad. Sci., Paris **230**: 1888–1890.

KUZNECOVA-ZARUDNAJA, T. N. (1955). [Russian title]. De origine *Phellinus sp.* notula. *In* Bot. Mater. Otd. spor. Rast. **10**: 196–209. — "Čaga".

LA FUZE, H. H. (1937). Nutritional characteristics of certain wood-destroying fungi, *Polyporus betulinus* Fr., *Fomes pinicola* (Fr.) Cooke, and *Polystictus versicolor* Fr. *In* Plant Physiol. **12**: 625–646.

LAINE, L. (1968). Notes on the polypores (Polyporaceae) of Ahvenanmaa. *In* Karstenia **8**: 14–20. "1967".

————— & M. NUORTEVA (1970) Über die antagonistische Einwirkung der insektenpathogenen Pilze *Beauveria bassiana* (Bals.) Vuill. und *B. tenella* (Delacr.) Siem. auf den Wurzelschwamm (*Fomes annosus* (Fr.) Cooke). *In* Acta forest. fenn. **111**: 15 pp. *8 fs.*

LAMBINON, J. & G. MATHOT (1965). Structure anatomique et position systématique de *Coriolus pergamenus* (Fr.) Pat. *In* Lejeunia II No. 31: 7 pp. *(2) pls.*

LANGE, I. (1966). Das Bewegungsverhalten der Kerne in fusionierten Zellen von *Polystictus versicolor* (L.). *In* Flora (A) **156**: 487–497 *6 fs.*

LÁZARO É IBIZA, B. (1916–7). Los Poliporaceos de la flora española. (Estudio critico y descriptivo de los hongos de esta familia). *In* Rev. Acad. Ci. Madrid **14**: 427–464, 488–524, 574–592, 655–680, 734–759, 833–866. 1916; **15**: 87–120, 137–164, 209–232, 287–307. 1916; **15**: 369–384 *pls. 1–10.* 1917. Also: Madrid, 319 pp. *29 fs., 10 pls.* 1917. — Pages 299–319 of the book are not reprinted from the periodical.

LAZEBNÍČEK, J. (1962). [Czech title]. Zwei bemerkenswerte Funde von *Elfvingia applanata* (Pers. ex Wallr.) Karst. *In* Čas. slezsk. Muz. **11**: 41–46 *3 fs.*

LE BRETON, A. (1887). Une variété probable du *Polyporus obducens. In* Bull. Soc. Amis Sci. nat. Rouen III **23**: 31–34.

LEK, H. A. A. VAN DER (1921). Mycologische aanteekeningen. IV–V. (IV. *Polyporus tuberaster* in Nederland. — V. Een interessant exemplaar van *Polyporus giganteus*, bij Wageningen waargenomen]. *In* Meded. Nederl. mycol. Ver. **11**: 85–94 *pls. 1, 2.*

————— (1926). Mycologische aanteekeningen. VIII. *Polyporus Tuberaster* in Nederland. *In* Meded. Nederl. mycol. Ver. **15**: 131–132 *(1) pl.*

LEMBERG, R. (1952). Nitrogenous pigments from the fungus *Coriolus sanguineus (Polystictus cinnabarinus). In* Austr. J. exp. Biol. med. Sci. **30**: 271–278 *2 fs.*

LEONTOVYČ, R. (1956). [Czech title]. A rare discovery of *Ganoderma Pfeifferi* Bres. from Slovakia. *In* Česká Mykol. **10**: 99–102 *(2) fs.*

LEUCHS, J. C. (1832). Über die Fabrikation des Feuerschwammes oder Zunders. *In* J. techn. oekon. Chem. **14**: 434–438.

LIBERTA, A. E. (1966). On *Trechispora. In* Taxon **15**: 317–319.

LIESE, J. (1936). Beiträge zum Kiefernbaumschwammproblem. *In* Forstarchiv **12**: 37–48 *7 fs.* — Reprint seen.

LILLY, V. G. & N. L. BARNETT (1948). The inheritance of partial thiamine deficiency in *Lenzites trabea. In* J. agric. Res. **77**: 287–300 *2 fs.*

LINDEBERG, G. & G. FÅHRAEUS (1952). Nature and formation of phenol oxidases in *Polyporus zonatus* and *P. versicolor. In* Physiologia Pl. **5**: 277–283 *1 f.*

LINDROTH, I. I. (1904). Beiträge zur Kenntnis der Zersetzungserscheinungen des Birkenholz. *In* Naturw. Z. Land- u. Forstw. **2**: 393–406 *7 fs.* — "*Polyporus nigricans*", in part=*Phellinus nigricans*, in part=conks of *Inonotus obliquus.*

LIST, P. H. & H. G. MENSSEN (1959a). Basische Pilzinhaltsstoffe. 3. Mitteilung. Flüchtige Amine und Aminosäuren des Schwefelporlings, *Polyporus sulphureus* Bull. *In* Arch. Pharm., Berl. **292**: 21–28.

————— & ————— (1959b). Basische Pilzinhaltstoffe. 4. Mitteilung. Biogene Amine des Schwefelporlings, *Polyporus sulphureus* Bull. *In* Arch. Pharm., Berl. **292**: 260–271.

LJUBARSKIJ, L. V. (1960). [Russian title]. [The black birch fungus — čaga]. *In* Vopr. Geogr. Dal'n. Vostoka **16**: 231–248 *5 fs.* — *Inonotus obliquus.*

LLOYD, C. G. (1908–10a). Polyporoid Issue, No. 1. *In* Mycol. Writ. **3**: 1–16 *fs. 196–210.* 1908. [Including, *Polystictus* (section *Pelloporus*) pp. 1–14]. — No. 2. *Ibid.* 3: 17–32 *fs. 256–266.* 1909a. [Including, The genus *Favolus*, pp. 17–21 *fs. 256–258*]. — No. 3. *Ibid.* 3: 33–48 *fs. 357–373.* 1910a.

————— (1909b). A new genus, *Cyanosporus.* By N. J. McGinty. *In* Mycol. Writ. 3: 436.

————— (1909c). A new polyporoid genus. By N. J. McGinty. *In* Mycol. Writ. 3: 444 *f. 255.* — *Volvopolyporus.*

————— (1910b). A *Polyporus* that weeps. *In* Mycol. Writ. 3: 490–492 *f. 383.* — *Polyporus* [*Inonotus*] *dryadeus.*

————— (1911). Synopsis of the section *Ovinus* of *Polyporus.* Cincinnati, Ohio, pp. 71–94 *(1) portr., fs. 496–509* (and *in* Mycol. Writ. 3).

————— (1912). Synopsis of the stipitate polyporoids. Cincinnati, Ohio, ii pp. and pp. 95–208 *(1) portr., fs. 395–500* [and *in* Mycol. Writ. 3 and *in* Bull. Lloyd Libr. 20 (Mycol. Ser. 6)].

————— (1915a). Synopsis of the genus *Fomes.* Cincinnati, Ohio, pp. 209–288 *(1) portr., fs. 570–610* (and *in* Mycol. Writ. 4).

————— (1915b). Synopsis of the section *Apus* of the genus *Polyporus.* Cincinnati, Ohio, pp. 289–392 *(1) portr., fs. 631–706* (and *in* Mycol. Writ. 4).

LOHWAG, H. (1929). Mykologische Studien. III. *Xanthochrous cuticularis* (Bull.) Pat. *In* Arch. Protistenk. **65**: 321–329 *5 fs.*

————— (1930). Mykologische Studien. V. Zu *Xanthochrous cuticularis* (Bull.) Pat. und *Xanth. hispidus* (Bull.) Pat. *In* Arch. Protistenk. **72**: 420–432 *3fs., pls. 25–28.*

————— (1931a). Zur Ableitung von Polyporaceen über *Odontia. In* Annls mycol. **29**: 87–91 *1 f.*

————— (1931b). Mykologische Pilzstudien. VI. *Spongipellis Litschaueri* (=*Polyporus Schulzeri* Fr. sensu Bresadola). *In* Arch. Protistenk. **75**: 297–314 *2 fs., pls. 18, 19.*

————— (1931c). Zur Rinnigkeit der Buchenstämme. *In* Z. PflKrankh. **41**: 371–385 *5 fs.* — *Fomes fomentarius.*

————— (1935). Über eine Ahornkrankheit. *In* Cbl. ges. Forstw. **61**: 306–315. — *Poria* [*Inonotus*] *obliqua.*

————— (1936). Mykologische Studien. XI. *Poria obliqua* (Pers.) Bres. *In* Öst. bot. Z. **85**: 270–278 *6 fs.*

————— (1938). Der Lärchenporling. *In* Öst. Z. Pilzk. **2**: 18–22, 34–37. — Pp. 34–37, n.v.

————— (1943). Ein Pilz als Sanddornschädling. *In* Dt. Bl. Pilzk. **5**: 50–51. Mit "Ergänzung . . ." auf S. 58. — *Fomes* [*Phellinus*] *robustus.*

LOHWAG, H. & L. FOLLNER (1936). Die Hymenophore von *Fistulina hepatica. In* Annls mycol. **34**: 456–464 *5 fs.*

LOHWAG, K. (1935). Das mykologische Wachsfigurenkabinett und die "Pietra fungaja". *In* Öst. bot. Z. **84**: 210–218 *3 fs.*

————— (1937). *Fomes Hartigii* (Allesch.) Sacc. & Trav. und *Fomes robustus* Karst. *In* Annls mycol. **35**: 339–349 *2 fs.*

————— (1938a–40a). Verwachsungsversuche an Fruchtkörpern von Polyporaceen. (1. Teil.). *In* Biol. gen. **14**: 432–445 *5 fs.* 1938a. — II. *In* Annls mycol. **37**: 169–180 *7 fs.* 1939. — III. *Ibid.* **38**: 92–95 *2 fs.* 1940a.

————— (1938b). *Ganoderma resinaceum* Boud., Erreger einer characteristischen Fäule. *In* Cbl. ges. Forstwes. **64**: 258–260 *2 fs.*

————— (1940a). Untersuchungen über die Holzzerstörung durch *Fomes hartigii* (Allesch.) Sacc. et Trav. und *Fomes robustus* Karst. *In* Z. PflKrankh. PflSchutz **50**: 481–494 *7 fs.*

————— (1940b). Zur Anatomie des Deckgeflechtes der Polyporaceen. *In* Annls mycol. **38**: 401–452 *38 fs.*

————— (1940c). Der Laub-Porling in "Pilze der Heimat" [von Gramberg]. *In* Dt. Bl. Pilzk. II **2**: 23–24.

————— (1948). *Fomes officinalis* (Vill.) Neuman. *In* Schweiz. Z. Pilzk. **26**: 32–37 *(1) f.*

————— (1950). Bienenwabenfäule, hervorgerufen durch *Phellinus nigrolimitatus* (Romell) B. & G. *In* Int. Holzmarkt **1950** (Holzforsch., Ser. 3): 22–24 *3 fs.* — Reprint seen.

————— (1951). "Pignolifäule". *In* Int. Holzmarkt **42** (Mitt. öst. Ges. Holzforsch. 3, Folge 6): 2 pp. *1 f.* — Reprint seen.

————— (1955a). Wachstumversuche mit *Lenzites betulina* (L.) Fr. *In* Öst. bot. Z. **102**: 524–528 *2 fs.*

————— (1955b). *Fomes pinicola* (Sw.) Cooke = *Ungulina marginata* (Fr.) Pat., von der Mycelkultur zum Fruchtkörper. *In* Schweiz. Z. Pilzk. **33**: 69–74 *(1) f.*

————— (1956). Zitzenbildungen am Fruchtkörper von *Fomes annosus* (Fr.) Cooke. *In* Schweiz. Z. Pilzk. **34**: 74–75 *(1) f.*

————— (1957). Über die Abbauintensität des *Polyporus betulinus* (Bull.) Fr. *In* Beih. Sydowia **1**: 183–186.

————— (1960). *Poria obliqua* (Pers.) Bres. ein interessanter holzzerstörender Pilz. *In* Cbl. ges. Forstw. **77**: 52–56 *3 fs.*

————— (1964). Über die wirtschaftliche Bedeutung von *Gloeophyllum trabeum* (Pers. ex Fr.) Murr. (= *Lenzites trabea* Pers. Fr.) für Oesterreich. *In* Holzforsch. Holzverwert. **16**: 101–103.

————— (1965a). Birkenschwammschnitzereien aus der Steiermark. *In* Mitt. naturw. Ver. Steierm. **95**: 136–139 *pl. 3.*

————— (1965b). Zur Abbauintensität des Pilzes *Polystictus abietinus* (Dicks.) Fr. *In* Holz Roh- u. Werkstoff **23**: 1–2.

————— (1966). Der Spechtloch-Schillerporling (*Inonotus nidus-pici* Pilát). *In* Allg. Forstztg III **77**: (2) pp. *2 fs.* — Reprint seen.

LOMBARD, F. F., R. W. DAVIDSON & J. L. LOWE (1961). Cultural characteristics of *Fomes ulmarius* & *Poria ambigua*. *In* Mycologia **52**: 280–294 *4 fs.* "1960".

————— & R. L. GILBERTSON (1967). *Poria luteoalba* and some related species in North America. *In* Mycologia **58**: 827–845 *10 fs.*

LONG, W. H. (1913a). A preliminary note on *Polyporus dryadeus* as a root parasite on the oak. *In* Phytopathology **3**: 285–287.

————— (1913b). *Polyporus dryadeus*, a root parasite on the oak. *In* J. agric. Res. **1**: 239–250 *pls. 21, 22.*

————— (1916). Note on western red rot in *Pinus ponderosa*. *In* Mycologia **8**: 178–180. — See next title.

————— (1917). A preliminary report on the occurrence of western red rot in *Pinus ponderosa*. *In* Bull. U.S. Dep. Agric. No. 490: 8 pp. — *Polyporus* sp. = *Polyporus* [*Dichomitus*] *anceps, fide* Overholts.

————— (1930). *Polyporus dryadeus*, a root parasite on white fir. *In* Phytopathology **20**: 758–759.

————— (1945). *Polyporus farlowii* and its rot. *In* Lloydia **8**: 231–237 *4 fs.*

LOW, J. D. & R. J. GLADMAN (1960). *Fomes annosus* in Great Britain. An assessment of the situation in 1959. *In* Forest Rec., Lond. No. 41: 22 pp. *20 pls.*

LOWE, J. L. (1956). Type studies of the polypores described by Karsten. *In* Mycologia **48**: 99–125.

————— (1957a). Polyporaceae of North America. The genus *Fomes*. *In* Techn. Publs. St. Univ. Coll. For. Syracuse No. 80: 97 pp. *68 fs.*

————— (1957b). *Polyporus minusculoides* in America. *In* Pap. Michigan Acad. Sci. **42**: 37–39 *1 f.*

————— (1963). The polyporaceae of the world. *In* Mycologia **55**: 1–12.

———— (1966). Polyporaceae of North America. The genus *Poria. In* Techn. Publs. St. Univ. Coll. For. Syracuse No. 90: 183 pp. *159 fs.*

LOWE, J. L. & S. LUNDELL (1956). The identity of *Polyporus trabeus* Rostk. *In* Pap. Michigan Acad. Sci. 41: 21–25 *3 fs.*

LUDWIG, F. (1880a). *Ptychogaster albus* Corda, die Conidien-generation von *Polyporus Ptychogaster* n. sp. *In* Z. ges. Naturw. III 5: 424–432 *pls. 13, 14.*

———— (1880b). Einige interessante Pilzfunde. [(1). *Polyporus Ptychogaster* mihi]. In Verh. bot. Ver. Prov. Brandenb. 22: xiii–xiv. — See also F. Ludwig, op. cit. 31: ix. 1890.

———— (1882). *Polyporus agaricicola* Ludwig nov. spec. *In* Hedwigia 21: 145. — (0).

———— (1883). Über einige merkwürdige Löcherpilze (mit Rücksicht auf die Sporenverbreitung der Pilze, insbesondere der Hymenomyceten). *In* Z. Pilzfr. 1: 204–210 *pl. 9 f. 3.*

LUKACS, L. & J. ZELLNER (1933). Zur Chemie der höheren Pilze. (XXII. Mitteilung). Über *Ganoderma lucidum* Leiss. .... *In* Mh. Chem. 62: 214–216.

LUNDELL, S. & A. PILÁT (1936). Über *Polyporus Wynnei* Berk. & Br. eine für Schweden neue Art. *In* Svensk bot. Tidskr. 30: 229–233 *pls. 1, 2.* — Reproduced *in* Z. Pilzk. 16: 113–117 *pl. 15.* 1937.

LYR, H. (1963). Die Aufnahme von radioaktivem Pentachlorphenol durch Schüttelmycel von *Trametes versicolor. In* Lyr and Gillwald (ed.), Holzzerstörung Pilze, pp. 311–313.

MACDONALD, J. A. (1937). A study of *Polyporus betulinus* (Bull.) Fr. *In* Ann. appl. Biol. 24: 289–310 *22 fs., pls. 18, 19.*

———— (1938). *Fomes fomentarius* (Linn.) Gill. [*Ungulina fomentaria* (Linn.) Pat.] on birch in Scotland. *In* Trans. Proc. bot. Soc. Edinb. 32: 396–407 *3 fs., pls. 29, 30.*

MACRAE, R. (1967). Pairing incompatibility and other distinctions among *Hirschioporus [Polyporus] abietinus, H. fusco-violaceus,* and *H. laricinus. In* Can. J. Bot. 45: 1371–1398 *4 pls.*

———— & K. AOSHIMA (1967). *Hirschioporus [Lenzites] laricinus* and its synonyms: *L. abietis, L. ambigua, L. pinicola. In* Mycologia 58: 912–925 *21 fs.* "1966".

MADHOSINGH, C. (1962). Physiological studies on the *Pycnoporus* species. *In* Can. J. Bot. 40: 1073–1089 *2 fs.*

———— (1964). A seriological comparison of isolates of *Fomes roseus* and *Fomes subroseus. In* Can. J. Bot. 42: 1677–1683 *6 fs.*

———— (1967). Light and temperature effects on growth and sporophore formation of *Lenzites trabea. In* Can. J. Bot. 45: 525–530 *2 pls.*

MAGNUS, P. (1910). *Bresadolia caucasia* N. Schestunoff in litt., eine dritte *Bresadolia*art. *In* Hedwigia 50: 100–104 *pl. 2.*

MALENÇON, G. (1929). *Leucoporus Forquignoni* (Quél.) Pat. *In* Bull. Soc. mycol. Fr. 45 (Atl.): (1) p. *pl. 34.*

———— (1966). Le *Polyporus Peckianus* Cooke en Europe? *In* Bull. Soc. Nat. Oyonnax 16–18: 35–39 *1 f., (1) pl.*

MALKOVSKÝ, K. M. (1931). [Czech title]. Über den systematischen Wert von *Daedalea parasitica* Vel. *In* Mykologia, Praha 8: 66–77 *2 fs.*; and *in* Trav. mycol. tchécosl. No. 14: 12 pp. *2 pls.* — (0).

MANGIN, L. (1907). Note sur la croissance et l'orientation des réceptacles d'*Ungulina fomentaria. In* Bull. Soc. mycol. Fr. 23: 155–156 *(1) f.* on p. 158.

———— & N. T. PATOUILLARD (1922). Sur la destruction de charpentes au château de Versailles par le *Phellinus cryptarum* Karst. *In* C.r. hebd. Séanc. Acad. Sci., Paris 175: 389–394 *4 fs.* — *Poria expansa.*

MANION, P. D. & D. W. FRENCH (1968). Inoculation of living aspen trees with basidiospores of *Fomes igniarius* var. *populinus*. *In* Phytopathology **58**: 1302–1304 *2 fs.*

———— & ———— (1969). The role of glucose in stimulating germination of *Fomes igniarius* var. *populinus* basidiospores. *In* Phytopathology **59**: 293–296 *4 fs.*

MAŃKA, K. & T. STUBE (1952). [Polish title]. The birch-fungus *Poria obliqua* (Pers.) Bres. and its development on artificial media. *In* Acta Soc. Bot. Polon. **21**: 517–536 *6 fs.*

MARTELLI, U. (1889). Sul *Polyporus gelsorum* Fr. *In* Nuovo G. bot. ital. (& Bulletino Soc. bot. ital.) **21**: 292–295.

MARTINEZ, J. B. (1943). El Agárico blanco (« *Ungulina officinalis* » Pat.) en los cedrales del Marruencos español. *In* An. Jard. bot. Madrid **4**: 75–139 *11 textpls.*

MARTIN-SANS, E. (1924). Exemplaires remarquables de trois Polypores: *Cladomeris umbellata, Cladomeris sulfurea* et *Fomes lucidus*. *In* Bull. Soc. mycol. Fr. **40**: 186–188.

MARTIROSJAN, S. N. (1963). [Russian title]. [Some species of fungi of the genus *Tyromyces* Karst. new for the Armenian SSR]. *In* Izv. Akad. Nauk armjanskoj SSR (Biol.) **16** (12): 65–74 *6 fs.*

———— (1965a). [Russian title]. [Data on the study of representatives of fungi of the genus *Phellinus* Quél. in the Armenian SSR]. *In* Izv. Akad. Nauk armjanskoj SSR (Biol.) **17** (4): 77–87 *3 fs.*

———— (1965b). [Russian title]. [Data on the identification of Polyporaceae of Armenia]. *In* Izv. Akad. Nauk armjanskoj SSR (Biol.) **18** (2): 79–85.

MARTYN, J. (1744). An account of a new species of Fungus. *In* Phil. Trans. R. Soc. **43**: 263–264 *pl. 2 f. 1.* — Abnormal *Polyporus squamosus*.

MATTERS, C. N. (1955). Morphological characters of *Trametes cinnabarinus* (Jacq.) Fr., *Polyporus cinnabarinus* (Jacq.) Fr., and *Polystictus sanguineus* (Linn.) Mey. *In* Progr. Rep. C.S.I.R.O. Forest Prod. Lab., S. Melbourne [Div. Proj. P. 11, Sub-Proj. P. 11–12] No. 5: 1–22 *(4) pls.*

MATTIROLO, O. (1914). Notes sur l'histoire de la « Pierre à Champignons » (Pietra fungaia). *In* Bull. Soc. natn. Acclim. Fr. **59**: 585–601 *2 fs.* — To be consulted for previously published titles on *Polyporus tuberaster*.

MAYR, H. (1884). Zwei Parasiten der Birke, *Polyporus betulinus* Bull. und *Polyporus laevigatus* Fries. *In* Bot. Cbl. **19**: 22–29, 51–57 *pls. 1, 2.* — Determination of *P. laevigatus* doubtful.

McCRACKEN, F. I. & E. R. TOOLE (1969). Sporophore development and sporulation of *Polyporus hispidus*. *In* Phytopathology **59**: 884–885 *1 f.*

McKAY, H. H. (1959). Cultural basis for maintaining *Polyporus cinnabarius* and *Polyporus sanguineus* as two distinct species. *In* Mycologia **51**: 465–473 *2 fs.*

MEIER, W. (1922). *Polyporus mollis*. *In* Pilz- u. Kräuterfr. **5**: 166–167 *fs. 1, (2).* — *Tyromyces ptychogaster*.

MESSIKOMMER, E. (1950). Eine interessante mykologische Beobachtung am Fleischigzottigen Porling *(Polyporus hispidus)*. *In* Schweiz. Z. Pilzk. **28**: 207–208. — Guttation.

MEYER, H. (1936). Spore formation and discharge in *Fomes fomentarius*. *In* Phytopathology **26**: 1155–1156.

MIERSCH, J. (1965). *Tyromyces wynnei* (Berk. et Br.) Donk erstmalig in Mitteldeutschland gefunden. *In* Westfäl. Pilzbr. **5**: 100–102 *2 fs.*

MIKALAIKEVIČIUS, V. (1962). [Russian title]. Einige Angaben über ... *Phellinus tremulae* (Bond.) Bond. et Boriss. und *Phellinus igniarius* f. *betulae* Bond. emend. Boriss. .... *In* Bot. Issl. / Scr. bot., Tartu **2**: 201–210 *2 pls.*

MIZUMOTO, S. (1955?–7). [Japanese title]. Studies on *Lenzites abietina* Fr. and some of its allied species. I–XI. *In* J. Jap. For. Soc. **37**?–39. — N. v.; cited from Rev. appl. Mycol. **35**: 840 (III), **36**: 440 (VI), **37**: 124 (V, IX–XI).

MÖLLER, F. A. G. J. (1904). Über die Notwendigkeit und Möglichkeit wirksamer Bekämpfung des Kiefernbaumschwammes *Trametes Pini* (Thore) Fries. *In* Z. Forst- u. Jagdw. **36**: 677–715 *pls. 4–5*. — Review *in* Annls mycol. **3**: 113–115. 1905.

————— (1910). Der Kampf gegen den Kiefernbaumschwamm. *In* Z. Forst- u. Jagdw. **42**: 129–146.

————— (1914). Der Kampf gegen den Kiefern- und Fichtenbaumschwamm II. *In* Z. Forst- u. Jagdw. **46**: 193–208.

MOENS, J. (1971). *Ganoderma applanatum* en gallen. *In* Coolia **15**: 25–34 *3 fs.*

MOESZ, G. (1913). [Hungarian title]. Mykologische Mitteilungen. 1. Ein nord-afrikanischer Pilz im Grossen Alföld. *In* Bot. Közl. **12**: 231–232, (63)–(64).

MOLIN, N. (1957). Om *Fomes annosus* spridningsbiologi. A study on the infection biology of *Fomes annosus*. *In* Meddn St. SkogsforsknInst. **47** (3): 36 pp. *4 fs.*

MOLLIARD, M. (1909). Sur une forme hypochnée du *Fistulina hepatica* Fr. *In* Bull. Soc. bot. Fr. **56**: 553–556 *pl. 9.*

MONTAGNE, J. P. F. C. & M. C. DURIEU DE MAISONNEUVE (1876). Diagnose et description du *Lenzites (Suberosae) Warnieri*. *In* Actes Soc. linn. Bord. **31**: 203–204.

MONTGOMERY, H. B. S. (1936). A study of *Fomes fraxineus* and its effects on ashwood. *In* Ann. appl. Biol. **23**: 465–486 *5 fs., pl. 20*. — *Fomitopsis cytisina*.

MOROT, L. (1888). Note sur l'identité spécifique du *Polyporus abietinus* Fr. et de l'*Irpex fusco-violaceus* Fr. *In* J. Bot., Paris (ed. Morot) **2**: 30–32.

MORTON, H. L. & D. W. FRENCH (1970). Attraction toward and penetration of *Polyporus dryophilus* var. *vulpinus* basidiospores by hyphae of the same species. *In* Mycologia **62**: 714–720 *3 fs.*

MOUNCE, I. (1929). Studies in forest pathology. II. The biology of *Fomes pinicola* (Sw.) Cooke. *In* Bull. Dep. Agric. Dom. Can. II No. 111: 75 pp. *10 textpls.*

————— (1930). Notes on sexuality in *Fomes pinicola* (Sw.) Cooke, *Fomes roseus* (Fr.) Cooke, *Polyporus Tuckahoe* (Güssow) Sacc. et Trott., *P. resinosus* (Schrad.) Fr., *P. anceps* Peck, *Lenzites saepiaria* Fr., *Trametes protracta* Fr., and *T. suaveolens* (L.) Fr. *In* Proc. Can. phytopath. Soc. **1929**: 27–28.

—————, J. E. BIER & M. K. NOBLES (1940). A root-rot of Douglas fir caused by *Poria Weirii*. *In* Can. J. Res. (C) **18**: 522–533 *7 fs., 4 pls.*

————— & R. MACRAE (1936). The behaviour of paired monosporous mycelia of *Lenzites saepiaria* (Wulf.) Fr., *L. trabea* (Pers.) Fr., *L. thermophila* Falck, and *Trametes americana* Overh. *In* Can. J. Res. (C) **14**: 215–221 *1 pl.*

————— & ————— (1937). The behaviour of paired monosporous mycelia of *Fomes roseus* (Alb. & Schw.) Cooke and *Fomes subroseus* (Weir) Overh. *In* Can. J. Res. (C) **15**: 154–161 *12 fs., 1 pl.*

————— & ————— (1938). Interfertility phenomena in *Fomes pinicola*. *In* Can. J. Res. (C) **16**: 354–376 *1 f.*

MÜLLER, G. & H. JAHN (1966). Der Eschen-Baumschwamm, *Fomitopsis cytisina*, im Rheinland gefunden. *In* Westfäl. Pilzbr. **6**: 13–17 *fig.*

MÜLLER, C. J. (1872). Note on a British *Polyporus*. *In* J. Bot., Lond. **10**: 22–23. — *Polyporus [Hapalopilus] rutilans*, colour reaction with KOH solution.

MURAŠKINSKIJ, K. E. (1927). [Russian title]. (The larch bracket-fungus). Omsk, 20 pp. *2 fs.* — Cf. Rev. appl. Mycol. **6**: 521–522. 1927. — 'White agaric'.

MURRILL, W. A. (1903). A historical review of the genera of the Polyporaceae. *In* J. Mycol. **9**: 87–102.

————— (1916). A new genus of resupinate polypores. *In* Mycologia **8**: 56–57. — *Xanthoporia* [=*Inonotus*].

———— (1918). Murrill's and Saccardo's names on polypores compared. New York, i and 31 pp.

MYREN, D. T. & R. F. PATTON (1970). In vitro production of basidiospores of *Polyporus tomentosus*. *In* Phytopathology **60**: 911–912 *1 f*.

———— & ———— (1971). Establishment and spread of *Polyporus tomentosus* in pine and spruce plantations in Wisconsin. *In* Can. J. Bot. **49**: 1033–1040 *fs. 4, 5, 3 pls*.

NAGATOMO, I. (1931). [Japanese title]. Über das Verhalten von *Polyporus Schweinitzii* Fr. in Mischkulturen. *In* Forschn Geb. PflKrankh., Japan **1**: 192–205 *pls. 8, 9*.

NANNFELDT, J. A. (1956). *Polyporus hispidus* (Bull.) Fr. funnen på Oland. *In* Friesia **5**: 317–318.

———— (1967). Zur Erforschungsgeschichte von *Phellinus tremulae*. *In* Westfäl. Pilzbr. **6**: 130–134.

NEES VON ESENBECK, T. F. R. (1821). *Boleti fomentarii* Pers. varietas singularis; e fodinis Lithanthracum Leodinensis. *In* Nova Acta phys.-med. Acad. Caes. Leop.-Carol. **10**: 235–238 *pl. 18*. ——= ?

NEGER, F. W. (1905). Neue Beobachtungen an einigen auf Holzgewächsen parasitisch lebenden Pilzen. [I. *Irpex obliquus* (Schrad.) Fries, ein Wundparasit der Hainbuche]. *In* Festschr. Feier Forstlehranst. Eisenach pp. 86–? (n.v.).

NEGRUCKIJ, S. F. (1966). [Russian title]. Forma nova fungi *Fomitopsis annosa* (Fr.) Karst. in *Picea*. *In* Nov. Sist. niz. Rast. **1966**: 178–181 *2 fs*.

NELSON, E. E. (1967). Factors affecting survival of *Poria weirii* in small buried cubes of Douglas-fir heartwood. *In* Forest Sci. **13**: 78–84 *5 fs*.

NĚMEC, B. (1925a). Einiges über die Dorsiventralität der Fruchtkörper von Pilzen. *In* Stud. Pl. phys. Lab. Charles Univ., Prague **3**: 89–97 *13 fs*.

———— (1925b). Regenerationserscheinungen an *Lenzites separia* W. *In* Stud. Pl. physiol. Lab. Charles Univ., Prague **3**: 103–105 *(1) f*.

NEUHAUSER, K. S. & R. L. GILBERTSON (1971). Some aspects of bipolar heterothallism in *Fomes cajanderi*. *In* Mycologia **63**: 722–735 *3 fs*.

NICKOVSKAJA [NITZKOWSKAJA], O. P. (1963). Zur Biologie der sterilen Form des Pilzes *Inonotus obliquus* (Pers.) Pilát. *In* Lyr and Gillwald (ed.), Holzzerstörung Pilze pp. 41–42.

NICOLAS, G. (1926). Un nouvel hôte d'*Ungulina fraxinea* (Bull.). *In* Bull. Soc. mycol. Fr. **42**: 192–193.

NIEL, E. (1896). Observations sur le *Polyporus giganteus*, Pers. et le *P. acanthoides*, Bull. *In* Bull. Soc. mycol. Fr. **12**: 120–121.

NIEMELÄ, T. (1970). New data on *Albatrellus syringae* (Parmasto) Pouzar and *A. peckianus* (Cooke) Niemelä, n. comb. *In* Acta bot. fenn. **7**: 52–57 *5 fs*.

———— (1971). On Fennoscandian polypores. I. *Haploporus odorus* (Sommerf.) Bond. and Sing. *In* Annls bot. fenn. **8**: 237–244 *6 fs*.

———— (1972). On Fennoscandian polypores. II. *Phellinus laevigatus* (Fr.) Bourd. & Galz. and *P. lundellii* Niemelä, n. sp. *In* Annls bot. fenn. **9**: 41–59 *13 fs*.

NIKOLAJEVA, T. L. (1940). [Russian title]. Zur Monographie einiger Gattungen der Polyporaceen des europäischen Teiles der Soviet-Union und des Kaukasus *(Trametes, Daedalea, Lenzites)*. *In* Trudy bot. Inst. Akad. Nauk SSSR (II) **4**: 377–431 *42 fs*. "1938".

———— (1949). [Russian title]. De fungo novo generis *Irpicis*. *In* Bot. Mater. Otd. spor. Rast. **6**: 85–87 *3 fs*. — *Irpex* ["*Trametes*"] *foliaceo-dentata*.

———— (1953). [Russian title]. [Species of *Irpex* and *Hirschioporus* in the European part of the SSR and the Caucasus]. *In* Trudy bot. Inst. Akad. Nauk SSSR (II) **8**: 179–200 *13 fs*.

———— (1955). [Russian title]. [On the problem of the cause of "caga"]. *In* Bot. Ž. **40**: 233–237 *4 fs.*

NISIZAWA, K. (1955). Cellulose-splitting enzymes. V. Purification of *Irpex* cellulase and its action upon *p*-nitro-phenyl *β*-cellobioside. *In* J. Biochem., Tokyo **42**: 825–835 *4 fs.* — *Irpex lacteus.* This is only one of a long list of papers connected with this species.

NOBLES, M. K. (1943). A contribution toward a clarification of the *Trametes serialis* complex. *In* Can. J. Res. (C) **21**: 211–234 *79 fs., 4 pls. (=fs. 4–45)*.

———— (1958). Cultural characters as a guide to the taxonomy and phylogeny of the Polyporaceae. *In* Can. J. Bot. **36**: 883–926.

———— (1965). Identification of cultures of wood-inhabiting Hymenomycetes. *In* Can. J. Bot. **43**: 1097–1139.

———— (1971). Cultural characters as a guide to the taxonomy of the Polyporaceae. *In* R. H. Petersen (ed.), Evol. higher Basidiom. pp. 169–196 *1 f.*

NOBLES, M. K. & B. P. FREW (1962). Studies in wood-inhabiting Hymenomycetes. V. The genus *Pycnoporus* Karst. *In* Can. J. Bot. **40**: 987–1016 *98 fs., 2 pls. (=fs. 4–30)*.

NORDIN, I. (1967). [Swedish title]. The Swedish host species of *Fomes fomentarius* (L. ex Fr.) Kickx. *In* Friesia **8**: 32–50 *pls. 4–7.*

NOULET, J. B. (1871). Note sur le Polypore cinnabarin. *In* Mém. Acad. Sci. Toulouse VII **3**: 163–165.

NUORTEVA, M. & L. LAINE (1968). Über die Möglichkeiten der Insekten als Übertrager des Wurzelschwamms (*Fomes annosus* (Fr.) Cooke). *In* Annls entom. fenn. **34**: 113–135.

NUTMAN, F. J. (1929). Studies of wood-destroying fungi I. *Polyporus hispidus* (Fries). *In* Ann. appl. Biol. **16**: 40–64.

ÖHLMANN, J. (1960). Mykologische und Feuchtigkeitsuntersuchungen an einer vom Kiefernbaumschwamm befallenen Kiefer. *In* Arch. Forstw. **8**: 270–284 *4 fs.*

OEHM, G. (1933a). *Polyporus squamosus* (Huds.) Fr., seine Morphologie und physiologische Anatomie. (Beiträge zur Kenntnis der Hymenomyceten IV.). *In* Beih. bot. Cbl. (I) **51**: 101–158 *7 fs., pl. 1.*

———— (1933b). Zur entwickelungsgeschichtlichen Auffassung des Abschlussgewebes von *Polyporus squamosus* (Huds.) Fr. (Beiträge zur Kenntnis der Hymenomyceten, V.). *In* Beih. bot. Cbl. (I) **51**: 159–164 *4 fs.*

———— (1937a). Kritische Bemerkungen zur Nebensporenfrage bei höheren Pilzen, mit besonderer Berücksichtigung der Gattung *Ceriomyces*. (Beiträge zur Kenntnis der Hymenomyceten. VI.). *In* Beih. bot. Cbl. **37** (A): 250–263.

———— (1937b). Über den Einfluss des Lichtmangels auf *Polyporus squamosus* (Huds.) Fries. (Beiträge zur Kenntnis der Hymenomyceten. VII.). *In* Beih. bot. Cbl. (A) **57**: 264–266 *2 fs.*

———— (1937c). Über einen selbstständigen Nebenfruchtkörper von *Polystictus hirsutus*: [*Polystictus hirsutus* (Wulfen) Fries, status *Paramyces* m.] (Beiträge zur Kenntnis der Hymenomyceten. VIII.). *In* Beih. bot. Cbl. (A) **57**: 267–276 *3 fs., pl. 5.*

OLSON, A. J. (1941). A root disease of Jeffrey and ponderosa pine reproduction. *In* Phytopathology **31**: 1063–1077 *3 fs.* — *Cunninghamella* [*Oedocephalum*] *meineckella.*

ORLICZ, A. (1971). [Polish title]. *Tyromyces gloeocystidiatus* Kotl. et Pouz. and its identification. *In* Acta mycol. **7**: 15–26 *6 fs.*

ORŁOS, H. (1958). [Czech title]. De Polyporacearum sporificatione. *In* Česká Mykol. **12**: 200–204. — Czech with Russian summary.

———— (1961). [Polish title]. Studies on the ecological function of fungi of the family Polyporaceae in different forest types of the Bialowieża National Park. Studies on spore output in fungi of the family Polyporaceae. *In*

Prace Inst. badaw. Leśn. **193**: 3–100 *49 fs.*; **194**: 101–112. 1961 (n.v.). — Cf. Rev. appl. Mycol. **41**: 551. 1962.

OVEREEM, C. VAN (1925). *Laetiporus miniatus* (Junghuhn) van Overeem nov. comb. *In* Icon. Fung. malay. Heft 12: 6 pp. *pl. 12.* — (0).

OVERHOLTS, L. O. (1917). An undescribed timber decay of pitch pine. *In* Mycologia **9**: 261–270 *pls. 12, 13.* — *Polyporus [Skeletocutis] amorphus.*

———— (1946). The identity of *Poria monticola. In* Mycologia **38**: 674–676 *1 f.*

———— (1953). The Polyporaceae of the United States, Alaska, and Canada. Prepared for publication by J. L. Lowe. Ann Arbor, xiv and 466 pp. *frontisp., 132 pls.* (Univ. Michigan Stud., Sci. Ser. 19).

OWENS, C. E. (1936). Studies on the wood-rotting fungus, *Fomes Pini.* I. Variations in morphology and growth habit. *In* Am. J. Bot. **23**: 144–149 *textpls. 1–9.*
— II. Cultural characteristics. *Ibid.* **23**: 235–254 *1 f., textpls. 10–15.*

PARAVICINI, E. (1919). *Favolus europaeus.* Ein Schädling des Nussbaumes. *In* Schweiz. Z. Forstwes. **70**: 15–17.

PARKER-RHODES, A. F. (1955). The basidiomycetes of Skokholm Island. XIII. *Echinotrema clanculare* gen. et sp. nov. *In* Trans. Br. mycol. Soc. **38**: 366–368 *1 f.*

PARMASTO, E. (1956a). [Estonian title]. [Keys to the major polyporous fungi of the Estonian S.S.R.]. *In* Abiks Loodusevaat. No. 26: 71 pp. *68 fs., (3) pls. (=fs. 61–68).*

———— (1956b). [Estonian title]. On the biology of *Fomitopsis annosa* (Fr.) Karst. *In* Eesti NSV Tead. Akad. Toim. **5** (Biol. No. 3): 256–261 *2 fs.*

———— (1957). [Estonian title]. On the biology of *Inonotus obliquus* in the Estonian S.S.R. *In* Loodusuur. Seltsi Aastar. **50**: 203–208 *1 f.*

———— (1958). [Russian title]. Development and spore discharge of the fruit-bodies of Polyporaceae. *In* Eesti NSV Tead. Akad. Toim. **7** (Biol. No. 2): 83–93.

———— (1959a). [Russian title]. A new species of *Chaetoporus* (Fam. Polyporaceae). *In* Eesti NSV Tead. Akad. Toim. **8** (Biol. No. 2): 113–117 *3 fs., (2) pls.*

———— (1959b). [Russian title]. On the distribution of some rare species of Polyporaceae. *In* Eesti NSV Tead. Akad. Toim. **8** (Biol. No. 4): 226–278 *9 fs., (4) pls.*

———— (1959c). [Russian title]. De speciebus et formis novis Polyporacearum in RSS estonica inventis. *In* Bot. Mater. Otd. spor. Rast. **12**: 237–242 *6 fs., (1) pl. (=fs. 2, 3).*

———— (1959d). [Russian title]. Polyporaceae Reipublicae Estoniae S.S. *In* Trudy bot. Inst. Akad. Nauk SSSR (II) **12**: 213–273 *39 fs., (2) pls. (=fs. 14–16, 19).*

———— (1961). [Russian title]. On the place of *Chaetoporellus Šimani* (Pil.) Bond. within the Polyporaceae. *In* Eesti NSV Tead. Akad. Toim. **10** (Biol. No. 2): 118–122 *2 fs., (1) pl. (=f. 2).*

———— (1963). [Russian title]. Data on the fungus flora of the Komi A.S.S.R. *In* Tartu Riikl. Ülik. Toim. **136**: 103–129 *14 fs. in text & on (5) pls.*

———— (1967). [Russian title]. Polyporous fungi of the northern Soviet Union. *In* Mikol. Fitop. **1**: 280–286.

———— (1970). [Russian title]. On the dispersal of aphyllophoraceous fungi by basidiospores. I. Methods for investigation of polyporaceous fungi. *In* Eesti NSV Tead. Akad. Toim. **19** (Biol. No. 4): 355–361 *4 fs.*

PATOUILLARD, N. T. (1885). Contribution à l'étude des formes conidiales des Hyménomycètes; *Ptychogaster aurantiacus* Pat. sp. nov. *In* Revue mycol. **7**: 28–29 *pl. 50 f. 10.*

———— (1889a). Le genre *Ganoderma. In* [Bull.] Soc. mycol. Fr. **5**: 64–80 *pls. 10, 11.*

———— (1889b). Note sur la présence de basides à la surface du chapeau des Polypores. *In* [Bull.] Soc. mycol. Fr. **5**: 81–83. — *Phellinus* spp.

———— (1890). Sur la place du genre *Favolus*. *In* Bull. Soc. mycol. Fr. 6: xix–xxi.

PEGLER, D. N. (1964). A survey of the genus *Inonotus* (Polyporaceae). *In* Trans. Br. mycol. Soc. 47: 175–195 *2 fs.*

———— (1966). "Polyporaceae" — Part I. With a key to British genera. Reprint from "privately" published News Bull. Br. mycol. Soc. No. 26: 13–28 *28 fs.*

———— & T. W. K. YOUNG (1972). Reassessment of the Bondarzewiaceae (Aphyllophorales). *In* Trans. Br. mycol. Soc. 58: 49–58 *2 fs.*, *pls. 6–9.*

PELCZAR, M. J., S. GOTTLIEB & W. C. DAY (1950). Growth of *Polyporus versicolor* in a medium with lignin as the sole carbon source. *In* Archs. Biochem. 25: 449–451.

PENNINGTON, L. H. (1907). *Fomes pinicola* Fr. and its hosts. *In* Rep. Michigan Acad. Sci. 9: 80–82.

PERCIVAL, W. C. (1933). A contribution to the biology of *Fomes Pini* (Thore) Lloyd (*Trametes Pini* [Thore] Fries). *In* Bull. New York St. Coll. For. Syracuse 6 (1b): 72 pp. *21 fs.* (Techn. Publs No. 40).

PERLMAN, D. (1949). Studies on the growth and metabolism of *Polyporus anceps* in submerged culture. *In* Am. J. Bot. 36: 180–184.

PEŠEK, F. (1969). Distribution topographique d'isotope radioactif $^{40}_{19}$ K dans les chapeaux de *Trametes versicolor* et *Phellinus pomaceus* au point de vue de la présence comme des substances colorantes organiques, des résines et des substances aux qualités antibiotiques dans la famille Polyporaceae. *In* Friesia 9: 202–211 *3 fs.*

PETRAK, F. (1968). Alphabetisches Verzeichnis der von Fries in den Novae Symbolae Mycologicae und von Cooke in Grevillea XIV. bei der Gattung *Polystictus* Fr. eingereihten Polyporaceen. *In* Sydowia 20: 242–251.

PICCONE, A. (1876). Appunti sulla distribuzione geographica del *Polyporus Inzengae*, Ces. et DN. *In* Nuovo G. bot. ital. 8: 367–368. — Additional note: 'Qualche notizia del *Polyporus Inzengae* Ces. et Dntrs." *in* Nuovo G. bot. ital. 9: 155.

PILÁT, A. (1925). Revision der zentraleuropäischen resupinaten Arten der Gattung *Irpex* Fr. *In* Annls mycol. 23: 302–307.

———— (1926a). *Trametes Pini* (Brot.) Fries. — .... *In* Mykologia, Praha 3: 7–9 *(1) f.*, *(1) pl.*

———— (1926b). [Czech title]. *Trametes stereoides* (Fr.) Bres. en Bohême. *In* Mykologia, Praha 3: 103–105 *(1) f.*

———— (1926c). [Czech title]. *Xanthochrous cuticularis* (Bull.) Pat., espèce nouvelle en Bohême et en Moravie. *In* Mykologia, Praha 3: 120–122 *(2) fs.*

———— (1927). [Czech title]. ... *Polyporus officinalis* Vill. *In* Mykologia, Praha 4: 4–8 *(1) f.*

———— (1928). [Czech title]. Die Abwässerungs Kanäle im Hymenophor von *Xanthochrous hispidus* (Bull.) Pat. *In* Mykologia, Praha 5: 48–52 *2 fs.*

———— (1930a). Die Abwässerungskanäle in den Hymenophoren der Polyporaceen. *In* Z. Pilzk. 9: 118–124 *pl. 9.*

———— (1930b). [Czech title]. The polypore *Coriolus hirsutus* (Wulf.) Quél. as a parasite on fruit trees. *In* Ochr. Rostl. 10: 57–63 *5 fs.* (n.v.). — Cf. Rev. appl. Mycol. 10: 112–113. 1931.

———— (1931). Monographie der europäischen Polyporaceen mit besonderer Berücksichtigung ihrer Beziehungen zur Landwirtschaft. I. Teil. [Einleitung, *Caloporus*]. *In* Beih. bot. Cbl. (II) 48: 404–436 *6 fs.*, *pls. 13–15.*

———— (1933a). De *Poria aurea* Peck, specie americana in montibus Carpaticis orientalibus lecta. *In* Hedwigia 73: 31–33 *1 f.*, *pl. 1.* — See (0).

———— (1933b). Eine interessante Pilzinfektion des Weinberger Tunnels in Prag. *In* Annls mycol. 31: 59–72. — "*Leptoporus undatus*" [=*Rigidoporus* sp.].

———— (1934a). Monographie der europäischen Polyporaceen .... II. Teil: *Grifola* Gray und *Phaeolus* Pat. *In* Beih. bot. Cbl. (B) 52: 23–95 *11 fs.*, *pls. 2–15.*

——————— (1934b–5a). [Czech title]. De *Polyporello squamoso* (Huds.) Karst. et eius formis aberrantibus. *In* Čas. čsl. Houb. **14**: 102–106 *f. 43.* 1934b; **15**: 53–58 *fs. 34, 35, 143–145.* 1935a.

——————— (1935b). *Poria pearsonii* Pilát sp. n. *In* Trans. Br. mycol. Soc. **19**: 195–198 *pl. 3.*

——————— (1936a). Monographie der europäischen Polyporaceen .... III. Teil. [*Polyporellus*]. *In* Beih. bot. Cbl. (B) **56**: 1–82 *11 fs., pls. 1–8.*

——————— (1936b–42). Polyporaceae. [French edition]. *In* Atl. Champ. Eur. **3**: 1–96 *pls. 1–48.* 1936a; 97–176 *pls. 49–136.* 1937; 177–240 *pls. 137–168.* 1938a; 241–320 *pls. 169–208.* 1939; 321–360 *pls. 209–248.* 1940; 361–472 *pls. 249–304.* 1941; 473–624 *pls. 305–374.* 1942. — There is also a Czech edition. — and Cf. W. B. Cooke in Lloydia **22**: 207. 1960.

——————— (1936c). *Grifola Castaneae* (B. and G.) sur chêne en France. *In* Bull. Soc. mycol. Fr. **52**: 100–101 *pl. 1.*

——————— (1938). *Lindtneria* g.n., a new genus of the Phylacteriaceae with polyporoid hymenophore. *In* Stud. bot. čsl. **1**: 71–73 *1 f.*

——————— (1939). *Leptoporus Werneri* Pilát, Polyporacearum species nova marocana. *In* Stud. bot. čech. **2**: 61. — *Oxyporus* cf. *late-marginatus.*

——————— (1947). *Fomes cytisinus* (Berk.) Gill. an Edelkastanie. *In* Schweiz. Z. Pilzk. **25**: 161–163 *2 fs.*

——————— (1950). [Czech title].... (*Grifola umbellata* Pers. Pilát). *In* Česká Mykol. **4**:. 27–29 *(1) f.*

——————— (1952). [Czech title]. ... *Polyporellus rhizophilus* (Pat.) .... *In* Čas. národ. Mus. (Odd. příd.) **121**: 136–139 *3 fs.*

——————— (1965a). [Czech title]. ... *Amylocystis lapponica* (Romell) Bond. et Sing. *In* Česká Mykol. **19**: 9–10 *pl. 56.*

——————— (1965b). [Czech title]. *Bondarzewia montana* (Quél.) Sing. in urbe Praga lecta est. *In* Česká Mykol. **19**: 102–103 *(1) f., pl. 8.*

PINTO-LOPES, J. (1949a). On the Polyporaceae of Portugal. *In* Portug. Acta biol. (B) Vol. "Júlio Henriques": 211–217.

——————— (1949b). Algumas Poliporáceas da Serra do Gerez. *In* Bolm Soc. portug. Ci. nat. II **2**: 97–100.

——————— (1949c). Wood-rotting polypores of standing trees in Portugal. *In* Bolm Soc. broter. II **23**: 219–223.

——————— (1950a). Some new species of Polyporaceae recently recorded for Portugal. *In* Revta Fac. Ci. Univ. Lisb. II (C) **1**: 37–38.

——————— (1950b). Poliporoses e fungos da decomposição da madeira em Portugal. *In* Revta Fac. Ci. Univ. Lisb. II (C) **1**: 53–107 *10 pls.*

——————— (1952). "Polyporaceae". Contribuação para a sua bio-taxonomia. *In* Mem. Soc. broter. **8**: 5–215 *29 textpls.* — and Cf.: E. J. H. Corner (1954). [Bookreview of] "Polyporaceae" .... [With postscript by] J. Pinto-Lopes. In Trans. Br. mycol. Soc. **37**: 92–96.

——————— (1953a). Trois Polypores nouveaux pour la flore du Portugal. *In* Broteria (Ci. nat.) **22**: 78–79.

——————— (1953b). Polyporaceae de Portugal (excepto resupinadas). Revisão das colecções Portuguesas. *In* Revta Fac. Ci. Univ. Lisb. II (C) **3**: 157–238.

PINTO-LOPES, J. and M. FARINHA (1950). The presence or absence of clamp connections in the species of Polyporaceae. *In* Revta Fac. Ci. Univ. Lisb. II (C) **1**: 39–51.

PIRK, W. and R. TÜXEN (1957). Das Trametum gibbosae, eine Pilzgemeinschaft modernder Buchenstümpfe. *In* Mitt. flor.-soz. ArbGemeinsch. **6–7**: 120–126 *(1) f.*

PLUNKETT, B. E. (1958). Translocation and pileus formation in *Polyporus brumalis*. *In* Ann. Bot. II **22**: 237–249 *(1) pl.*

————— (1961). The change of tropism in *Polyporus brumalis* stipes and the effect of directional stimuli on pileus differentiation. *In* Ann. Bot. II **25**: 206–223 *5 fs., 1 pl.*

POETSCH, J. S. (1879). Neue Österreichische Pilze. *In* Öst. bot. Z. **29**: 289–291. — *Daedalea schulzeri, D. poetschii.*

POUZAR, Z. (1966a). Studies in the taxonomy of the polypores I. *In* Česká Mykol. **20**: 171–177.

————— (1966b). A new species of the genus *Albatrellus* (Polyporaceae). *In* Folia geobot. phytotax. **1**: 274–276 *pls. 5, 6.*

————— (1966c). Studies in the taxonomy of the polypores II. *In* Folia geobot. phytotax. **1**: 356–375.

————— (1967). Studies in the taxonomy of the polypores III. *In* Česká Mykol. **21**: 205–212.

————— (1971). Notes on taxonomy and nomenclature of *Ischnoderna resinosum* (Fr.) P. Karst. and *I. benzoinum* (Wahlenb.) P. Karst. (Polyporaceae). *In* Česká Mykol. **25**: 15–21.

————— (1972). Amyloidity in polypores I. The genus *Polyporus* Mich. ex Fr. *In* Česká Mykol. **26**: 82–90.

PRICE, S. R. (1913). On *Polyporus squamosus* Huds. *In* New Phytol. **12**: 269–281 *pl. 6.*

PRIEHÄUSSER, G. (1931). Über den Zunderschwamm *(Polyporus fomentarius)*. *In* Z. Pilzk. **10**: 115–119 *pls. 12, 14.*

PŘÍHODA, A. (1952). [Czech title]. The importance of *Phaeolus rutilans* (Pers.) Pat. in forestry. *In* Preslia **24**: 41–44 *2 fs.*

————— (1953a). [Czech title]. *In* Česká Mykol. **7**: 123–125 *(3) fs.* — On *Gloeophyllum trabeum.*

————— (1953b). [Czech title]. *(Trametes heteromorpha* (Fr.) Bres.). *In* Preslia **25**: 139–142 *(1) f.*

————— (1957). [Czech title]. Nocivitas *Fomitis pinicolae*. *In* Česká Mykol. **11**: 230–231.

PRILLEUX, E. (1889). La *Pachyma Cocos* en France. *In* Bull. Soc. bot. Fr. **36**: 433–436.

————— (1890). La *Pachyma Cocos* dans la Charente-Inférieure. *In* Bull. Soc. mycol. Fr. **6**: 95–97.

————— (1893). Sur le *Polyporus hispidus* (Bull.) Fr. *In* Bull. Soc. mycol. Fr. **9**: 255–259. — Review by R. Ferry *in* Revue mycol. **16**: 163–164 *pl. 150 f. 16.*

PRIOR, E. M. (1913). Contributions to a knowledge of the "snap-beech" disease. *In* J. econ. Biol. **8**: 249–263 *pls. 3, 4.* — *Polyporus [Bjerkandera] adustus.*

PRUTENSKAJA, M. (1965). [Russian title]. *In* Zašč. Rast. Vredit. Bolezn. **10** (9): 27–28 *(2) fs.* — On *Inonotus hispidus.*

RADZIJEVS'KYJ, H. H. (1960). [Ukrainian title]. (Little known fungi from the family Polyporaceae in the Ukraine). *In* Ukr. bot. Ž. **17**: 107–108 *2 fs.*

RAESTAD, R. (1941). The relation between *Polyporus abietinus* (Dicks. ex Fr.) Fr. and *Irpex fusco-violaceus* (Ehrenb. ex Fr.) Fr. *In* Nytt Mag. Naturvid. **81**: 207–231 *5 fs.*

RAHM, E. (1956). *Polyporus sulphureus*, Forma *conglobata* Pilát . . . . *In* Schweiz. Z. Pilzk. **34**: 137–138 *(1) f.*

RAUSCHERT, S. (1962). *Polyporus rhizophilus* Pat., ein für Deutschland neuer Steppenpilz. *In* Westfäl. Pilzbr. **3**: 53–59 *2 fs.*

RAZZORE, A. (1911). Un nuovo Poliporo resupinato. *In* Atti Soc. ligust. Sci. nat. geogr. **22**: 16–17. — *Poria suaveolens.*

REICHARDT, H. W. (1866). Miscellen. [6. Über das Vorkommen von *Polyporus Rostkowii* auf einer Panzerfregatte]. *In* Verh. zool.-bot. Ges. Wien **16**: 495–496.

REICHERT, I. & Z. AVIZOHAR (1939). An anatomical study of the fruit-body of the wood-rotting fungus *Ganoderma lucidum* (Leys.) Karst. in Palestine. *In* Palest. J. Bot. (Rehovot Ser.) **2**: 251–288 *1 f., pls. 11–14.*

RENNERFELDT, E. (1946). Om rotrötan (*Polyporus annosus* Fr.) i Sverige. Dess utbredning och sätt att uppträda. Über die Wurzelfäule (*Polyporus annosus* Fr.) in Schweden. Ihre Verbreitung und ihr Vorkommen unter verschiedenen Verhältnissen. *In* Meddn St. SkogsforsknInst. **35** (8): 88 pp. *35 fs.*

REZENDE-PINTO, M. C. DE (1940). Notas micologicas. [I. Algumas observações a propósito de *Poria viticola* Laz.]. *In* Broteria (Ci. nat.) **9**: 91–92 *pl. 1.*

————— (1942). Notas micologicas—III. *Poria viticola* (Schw.) Sacc., *Poria viticola* Laz., *Poria friesiana* Bres. *In* An. Fac. Ci. Pôrto **27**: 246–248.

RHOADS, A. S. (1918). The biology of *Polyporus pargamenus* Fries. *In* Bull. New York St. Coll. For. Syracuse **18** (5): 197 pp. *6 fs., 31 pls.* (Techn. Publs. No. 11).

RICEK, E. W. (1968). Ein Massenauftreten von Gallen an *Ganoderma applanatum* (Pers. ex Wallr.) Pat. *In* Sydowia **21**: 285–289 *(1) f.* "1967".

RILEY, C. G. (1952). Studies in forest pathology. IX. *Fomes igniarius* decay of poplar. *In* Can. J. Bot. **30**: 710–734 *16 fs., 3 pls.*

————— & J. E. BIER (1936). Extent of decay in poplar as indicated by the presence of sporophores of the fungus *Fomes igniarius* Linn. *In* For. Chron. **12**: 249–253 *4 fs.*

RISHBETH, J. (1950–1a, b). Observations on the biology of *Fomes annosus*, with particular reference to East Anglian pine plantations. I. The outbreaks of disease and ecological status of the fungus. *In* Ann. Bot. II **14**: 365–383 *1 f., pls. 12, 13.* 1950. — II. Spore production, stump infection, and saprophytic activity in stumps. *In* Ann. Bot. II **15**: 1–21 *pl. 1.* 1951a. — III. Natural and experimental infection of pines and some factors affecting severity of the disease. *In* Ann. Bot. II **15**: 221–246 *2 fs., pl. 12.* 1951b.

————— (1957). Some further observations on *Fomes annosus* Fr. *In* Forestry **30**: 69–89 *pl. 3, f. 3.*

————— (1959). Dispersal of *Fomes annosus* Fr. and *Peniophora gigantea* (Fr.) Massee. *In* Trans. Br. mycol. Soc. **42**: 243–260 *1 f.*

————— (1961). Inoculation of pine stumps against infection by *Fomes annosus*. *In* Nature, Lond. **191**: 826–827.

————— (1963). Stump protection against *Fomes annosus*. III. Inoculation with *Peniophora gigantea*. *In* Ann. appl. Biol. **52**: 63–77 *1 f., (1) pl.*

RITTER, G. (1964). *Piptoporus quercinus*—ein seltener Porling. *In* Mykol. Mitt. Bl., Halle **8**: 44–47.

ROBAK, J. (1932). Ein Polyporaceae mit tetrapolärer Geschlechtsverteilung. *Polyporus borealis* (Wahlenb.) Fr. (Vorläufige Mitteilung). *In* Svensk bot. Tidskr. **26**: 267–270.

ROBBINS, W. J. & A. HERVEY (1961). Light and the development of *Poria ambigua*. *In* Mycologia **52**: 231–247 *5 fs.* "1960".

RODRIGUES, C. C. (1970). Nova contribuição para o estudo das Polyporaceae di Portugal. *In* Bolm Soc. portug. Ci. nat. II **12**: 155–185 1 tabl. "1968–1969".

ROLL-HANSEN, F. (1940). Undersøkelser over *Polyporus annosus* Fr., særlig med henblikk på dens forekomst i Det sønnafjelske Norge. (Studies in *Polyporus annosus* Fr., especially in respect of its occurrence in Norway south of the Dovre Fell). *In* Meddr norske SkogforsVes. No. 24 / 7 (1): 100 pp. *33 fs.*

————— (1967). *Phellinus tremulae* (Bond.) Bond. & Boriss. and *Phellinus igniarius* (L. ex Fr.) Quél. on *Populus tremula* L. *In* Meddr norske SkogsforsVes. No. 85 **23**: 241–263 *10 fs.*

ROMAGNESI, H. (1944). Sur deux champignons nouveaux pour la flore française.

[I. — *Leptoporus ellipsosporus* (Pilát) nob.]. *In* Bull. Soc. mycol. Fr. **60**: 88–89.

ROMELL, L. (1911). Hymenomycetes of Lappland. First series (Polyporaceae). *In* Ark. Bot. **11** (3): 35 pp. *2 pls.*

———— (1912). Remarks on some species of the genus *Polyporus*. *In* Svensk bot. Tidskr. **6**: 635–644 *4 fs.*

———— (1916). Hvarifrån kommer det bruna pulvret å öfre sidan af *Polyporus applanatus* och andra *Ganoderma*-arter? From where originates the brown powder on the surface of *Polyporus applanatus*, *Pol. lucidus* etc.? *In* Svensk bot. Tidskr. **10**: 340–348.

———— (1926). Remarks on some species of *Polyporus*. *In* Svensk bot. Tidskr. **20**: 1–24.

ROSANOVA, M. A. (1925). [Russian title]. Sur la distribution du *Polyporus betulinus* Fr., *Fomes fomentarius* Fr. et du *Fomes igniarius* Fr. dans les forêts de bouleaux du district de Zvenigorod du gouvernement de Moscou. *In* Zašč. Rast. Vredit. **2**: 24–25.

ROSEN, H. R. (1927). A pink-colored form of *Polyporus sulphureus* and its probable relationship to root-rot of oaks. *In* Mycologia **19**: 191–194 *pls. 16, 17.*

ROSTKOVIUS, F. W. T. (1828–48). [*Polyporus*]. *In* Deutschl. Fl. (III. Abt., Pilze Deutschl.) **4**: 1–36 *pls. 1–16.* 1828; 37–68 *pls. 17–32.* 1830; 69–100 *pls. 33–48.* 1837; 101–132 *pls. 49–64.* 1838; Hefte 27–28: 1–48 *pls. 1–24.* 1848.

ROTHBERG, M. (1937). A cultural study of *Fistulina hepatica* (Huds.) Fries, isolated from decayed Jarrah (*Eucalyptus marginata* Sm.). *In* Proc. R. Soc. Victoria II **50**: 157–169 *2 fs., pls. 9, 10.* — Rather *Fistulina spiculifera* (Cooke) D. Reid 1963 (KB 17): 280.

ROUMEGUÈRE, C. (1881). Note sur le *Boletus ramosus* Bull. récemment trouvé en Belgique. *In* Revue mycol. **3** / No. 9: 3–4.

ROUTIEN, J. B. (1948). Hyphal proliferation through clamp-formation in *Polyporus cinnabarinus* Fr. *In* Mycologia **40**: 194–198 *1 f.*

RUBEL, F. (1778). *Agaricum officinale*. *In* Miscnea austr. (ed. Jacq.) **1**: 164–203 *pls. 20, 21.*

RUNGE, A. (1959). Die Fredeburger "Schwammklöpper". *In* Westfäl. Pilzbr. **2**: 8–9.

———— (1966). Die Verbreitung der Ochsenzunge (*Fistulina hepatica* Schff. ex Fr.) in Westfalen. *In* Natur Heim., Münster **26**: 118–121.

RYVARDEN, L. (1967). Flora over Norges kjuker. Flora of Norwegian non-resupinate pore fungi. *In* Blyttia **25**:137–216 *13 fs.*

———— (1968a). Flora over kjuker. Oslo, 96 pp. *19 fs.*

———— (1968b). The genus *Datronia* in Fennoscandia. *In* Svensk bot. Tidskr. **62**: 501–511 *5 fs.*

———— (1968c). New or interesting records of Norwegian polypores. *In* Nytt Mag. Bot. **15**: 267–270 *2 fs.*

———— (1969). The genus *Polyporus* s. str. in Norway. *In* Nytt. Mag. Bot. **16**: 151–157 *6 fs.*

———— (1970). New or interesting records of Norwegian polypores II. *In* Nytt Mag. Bot. **17**: 163–168 *2 fs.*

———— (1972). A note on the genus *Junghuhnia*. *In* Persoonia **7**: 17–21. 1972.

SACCAS, A. (1946). La polyporose du Groseillier due au *Xanthochrous ribis* (Schum.) Pat. Avec préface de R. Heim. *In* Cr. hebd. Séanc. Acad. Agric. Fr. **32**: 816–819.

SADEBECK, R. E. B. (1886). Über äussere Bedingungen für die Entwicklung des Hutes von *Polyporus squamosus*. *In* Bot. Cbl. **25**: 226–227.

SARGENT, F. R. LE (1901). Amadou touchwood, tinder or spunk; its history and uses. *In* Scient. Am. Suppl. **51**: 21227–21228 *3 fs.*

SARKAR, A. (1959a). Nuclear phenomena in the life cycle of *Ganoderma lucidum* (Leyss. ex Fr.) Karst. *In* Can. J. Bot. **37**: 59–64 *21 fs.*

———— (1959b). Studies in wood-inhabiting Hymenomycetes. IV. The genus *Coriolellus* Murr. *In* Can. J. Bot. **37**: 1251–1270 *116 fs. 2 pls. (=fs. 6–27)*.

———— (1959c). Developmental anatomy and differentiation of tissue systems of *Ganoderma lucidum* (Leyss. ex Fr.) Karst. *In Φυτον* [Phyton, Argentina] **13**: 89–104 *7 fs.*

———— (1959d). Spore-germination of *Ganoderma lucidum* (Leyss.) Karst. *In* Sci. Cult. **24**: 524–525 *1 f.*

SARTORY, A. & L. MAIRE (1920). *Poria terrestris* auct. *In* Annls Soc. linn. Lyon II **67**: 65.

———— & ———— (1923). Le *Polyporus tomentosus*, Fr. Le type, ses formes, ses variétés. *In* C. r. Ass. franç. Av. Sci. **46**: 773–783 *9 fs.*

———— & ———— (1924). Synopsis du genre *Fistulina*, Bull. 16 pp.

SCHELD, H. W. & J. J. PERRY (1970). Basidiospore germination in the wood-destroying fungus *Lenzites saepiaria*. *In* J. gen. Microbiol. **60**: 9–21.

———— & ———— (1970). Induction of dimorphism in the basidiomycete *Lenzites saepiaria*. *In* J. Bact. **102**: 271–277 *4 fs.*

SCHLESINGER, S. (1838). Untersuchungen des *Polyporus suaveolens*. *In* Reprium Pharm. **64**: 298–315.

SCHMID, G. (1934). Pietra fungaja. Ein mykologischer Briefwechsel Goethes. *In* Z. Pilzk. **13**: 71–80, 110–118 *pl. 13*, 140–151.

SCHMIEDER, J. (1886). Über die chemische Bestandteile des *Polyporus officinal*. (Agaricus alb. der Officinen). *In* Arch. Pharm., Berl. **24**: 641–668; and Inaug. Diss., Erlangen, 67 pp. — Cf. Sur la composition chimique du *Polyporus officinalis* (Fr.). Compte rendu par E. Bourquelot *in* [Bull.] Soc. mycol. Fr. **3**: 156–162. 1887.

SCHMITZ, H. (1925). Studies in wood decay V. Physiological specialization in *Fomes pinicola* Fr. *In* Am. J. Bot. **12**: 163–177 *pls. 8–10.*

———— & S. M. ZELLER (1919). Studies in the physiology of the fungi. IX. Enzyme action in *Armillaria mellea* Vahl, *Daedalea confragosa* (Bolt.) Fr., and *Polyporus lucidus* (Leys.) Fr. *In* Ann. Missouri bot. Gdn **6**: 193–200.

SCHORSTEIN, J. (1907). Über *Polyporus vaporarius* (Pers.). *In* Annls mycol. **5**: 46–48 *2 fs.* — *Poria* sp.

SCHREIBER, M. (1925). Über den Lärchen-Agaricus. *In* Cbl. ges. Forstw. **51**: 47–49.

SCHRENK, H. VON (1900). Two diseases of red cedar, caused by *Polyporus juniperinus* n. sp. and *Polyporus carneus* Nees. A preliminary report. *In* Bull. Div. Veg. Phys. Path. U.S. Dep. Agric. No. 21: 22 pp. *3 fs., 7 pls.* — *Polyporus carneus* [=*"Fomitopsis" cajanderi*].

———— (1901). A disease of the black locust (*Robinia Pseudoacacia* L.). *In* Rep. Missouri bot. Gdn **12**: 21–31 *pls. 1–3.* — "*Polyporus rimosus* Berk." [=*Phellinus robiniae*].

———— (1914). A trunk disease of the lilac. *In* Ann. Missouri bot. Gdn **1**: 253–262 *pls. 8, 9.* — *Polyporus* [*Coriolus*] *versicolor.*

SCHRÖDER, A. (1961). Ein neuer Fund des Kupferroten Lackporlings *(Ganoderma pfeifferi)* in Westfalen. *In* Westfäl. Pilzbr. **3**: 44–46 *(1) f.*

SCHULZ, R. (1917). Mitteilung über einige ungewöhnlich grosse Polyporaceen. *In* Verh. bot. Ver. Brandenb. **58**: 73–75.

SCHULZER VON MÜGGENBURG, S. (1874). Mykologischer Beitrag. *In* Verh. zool.-bot. Ges. Wien **24**: 451–452 *(1) f.* — "*Ceriomyces terrestris* nov. spec." [=*Abortiporus biennis*].

———— (1878). Mycologisches. *In* Flora **61**: 11–15. — On species of *Ganoderma.*

———— (1880a). Mykologisches. (Schluss.) *Daedalea polymorpha* Schlzr. olim

Ceriomyces terrestris. In Öst. bot. Z. **30**: 144–148 *(1) pl. — Abortiporus biennis.*

———— (1880b). Mykologisches. Die Doppelfructification des *Polyporus applanatus* P. *In* Öst. bot. Z. **30**: 321–323 *(1) f.*

———— (1880c). Mycologisches. *In* Flora **63**: 79–80. — *Polyporus observus* (0).

———— (1883). Le *Morchella rimosipes* DC. et le *Polyporus Sarrazini* nov. sp. *In* Revue mycol. **5**: 256–257.

SCHWANTES, H. (1962). Das Trametetum gibbosae Pirk. et Tx. in der Umgebung von Giessen. *In* Hess. flor. Briefe **11**: 25–27.

ŠEBEK, S. (1962). [Czech title]. A new find of *Polyporus rhizophilus* (Pat.) Sacc. in Bohemia. *In* Česká Mykol. **16**: 14–18 *(3) fs.*

SECONDAT, J. B. DE (1785). Mémoires sur l'histoire naturelle .... [Observations sur des champignons qui paroissent tirer leur origine d'une pierre]. Paris, pp. 37–40 *pls. 8–12.*

SEYNES, J. DE (1863). Polymorphisme des organes reproducteurs chez un *Fistulina*. *In* Bull. Soc. bot. Fr. **10**: 93–99 *pl. 2.*

———— (1867). Recherches sur quelques points de l'anatomie du genre *Fistulina*. *In* C. r. hebd. Séanc. Acad. Sci., Paris **64**: 426–429. — & cf. J. DE SEYNES (1867). On some points in the anatomy of the genus *Fistulina*. *In* Ann. Mag. nat. Hist. III **19**: 302–304.

———— (1874a). Recherches pour servir à l'histoire naturelle des végétaux inférieurs. I. Des Fistulines. Paris, v & 86 pp. *(3) fs., 7 pls.* — *Cf.* L. DEBAT: Rapport sur la monographie de M. de Seynes, intitulé: Recherches sur les Fistulines. *In* Annls Soc. bot. Lyon **3**: 32–35. Also, J. DE SEYNES: The conidia of *Fistulina*. *In* Grevillea **4**: 173–178. 1876.

———— (1874b). Note sur une monographie du genre *Fistulina* Bull. *In* Bull. Soc. bot. Fr. **21**: 191–194.

———— (1878). Les conidies du *Polyporus sulfureus* Bull., et leur développement. *In* C.r. hebd. Séanc. Acad. Sci., Paris **86**: 805–808.

———— (1879). Sur un appareil conidien (pycnide) du *Polyporus sulfureus* Bull. *In* C.r. Ass. fr. Avanç. Sci. **7**: 649–653 *pl. 10.*

———— (1888a). Recherches pour servir à l'histoire naturelle des végétaux inférieurs. II. Polypores. Paris, (i) & iv & 78 pp. *3 fs., 6 pls.*

———— (1888b). *Ceriomyces* et *Fibrillaria*. *In* Bull. Soc. bot. Fr. **35**: 124–127.

———— (1890). De la distribution des *Ceriomyces* dans la classification de Polypores. *In* Bull. Soc. bot. Fr. **37**: 109–112.

SHIGO, A. L. (1969). How *Poria obliqua* and *Polyporus glomeratus* incite cankers. *In* Phytopathology **59**: 1164–1165 *2 fs.*

———— & E. M. SHARON (1968). Discoloration and decay in hardwoods following inoculations with Hymenomycetes. *In* Phytopathology **58**: 1493–1498 *2 fs.* — Polypores.

SHIMANO, T., K. TAKI & K. GOTO (1953). [Japanese title]. Studies on the components of fungi I. On the constituents of *Polystictus versicolor* Fr. *In* A. Proc. Gifu Coll. Pharm. **3**: 43. — Chem. Abstr. **50**: 13183.

SHIRAI, M. (1905). On a medically, economically and vegetable-pathologically interesting fungus chu-ling (*Polyporus Chu-ling* nov. sp.) (Preliminary note). *In* Bot. Mag., Tokyo **19**: 91–92 *pl. 4.* — *Grifola umbellata.*

———— (1911). On a Japanese fungus probably identical with Pietra Fungaja (*Polyporus Tuberaster*) of Italy. *In* Miyabe-Festschr. (31)–(35) *pls. 1–4 (1, 2, 4, 5).* — Japanese.

SINADSKIJ, JU. V. & M. A. BONDARCEVA (1956). [Russian title]. (Little known bracket fungi on *Populus* and *Tamarix* and their importance in the Kara-Kalpak A.S.S.R.). *In* Bot. Ž. **41**: 1177–1183 *6 fs.*

———— & ———— (1960). [Russian title]. Nonnulae Polyporaceae in arboribus frondosis in valle inundato fluminis Ural. *In* Bot. Mater. Otd. spor. Rast. **13**: 221–232 *6 fs.*

————— & ————— (1962). [Russian title]. The bracket-fungi (Polyporaceae) of the game sanctuary « Krasnyj Les » (Krasnodar Territory). *In* Bot. Ž. **47**: 55–67 *3 fs.*

SINCLAIR, W. A. (1964). Root- and butt-rot of conifers caused by *Fomes annosus*, with special reference to inoculum dispersal and control of the disease in New York. *In* Mem. agric. Exp. Stn Cornell Univ. No. 391: 54 pp. *6 fs.* — With extensive bibliography.

SINGER, R. (1940). Notes sur quelques Basidiomycètes. VIe. Série. [1. Un genre nouveau de Polyporaceae]. *In* Revue Mycol. **5**: 3–4. — *Bondarzewia*.

————— (1944). Notes on taxonomy and nomenclature of the polypores. *In* Mycologia **36**: 65–69.

SINGH, S. (1924). The liability of deodar to the attack of *Trametes pini* (Brot.) Fr. in Lolab, Kashmir. *In* Indian Forester **50**: 361–365.

SISON, B. C., W. J. SCHUBERT & F. F. NORD (1958). On the mechanism of enzyme action. LXV. A cellulolytic enzyme from the mold *Poria vaillantii*. *In* Archs Biochem. Biophys. **75**: 260–272 *7 fs.*

ŠKORIČ, V. (1937). [Serbian title]. *Poria obliqua* (Pers.) Bres. Beitrag zur Biologie und Pathologie des Pilzes. *In* Glasn. šum. Pok. / Annls Exp. forest. **5**: 271–301 *8 fs., 4 pls.* — German summary. — The fungus treated is *Inonotus nidus-pici*, fide Černý.

SLEETH, B. & C. B. BIDWELL (1937). *Polyporus hispidus* and a canker of oaks. *In* J. For. **35**: 778–785 *2 fs.*

SMICKAJA, M. F. (1965). [Russian title]. Polyporaceae, gesammelt in den Buchenwäldern von Transkarpaten. *In* Probl. Izuč. Grib. Lisajn., Tartu pp. 73–77.

SMITH, E. C. (1930). *Trametes hispida* a destructive parasite in Apple orchards. *In* Mycologia **22**: 221–222 *pl. 26.*

SMITH, R. G. (1924). A chemical and pathological study of decay of the xylem of the apple caused by *Polystictus versicolor* Fr. *In* Phytopathology **14**: 114–118.

SNELL, W. H. (1921). Chlamydospores of *Fomes officinalis* in nature. *In* Phytopathology **11**: 173–175 *1 f.*

—————, W. G. HUTCHINSON & K. H. N. NEWTON (1928). Temperature and moisture relations of *Fomes roseus* and *Trametes subrosea*. *In* Mycologia **20**: 276–291 *2 fs., pl. 34.*

SOLOV'EV, F. A. (1927). [Russian title]. [A rot caused by the fungus *Polystictus triqueter* Fr.]. *In* Mater. Mikol. Fitopat. Ross. **6**: 295–300 *3 fs.*

SPAULDING, P. (1907). Heart rot of *Sassafras sassafras* caused by *Fomes ribis*. *In* Science II **26**: 479–480. — Cf. Overholts **1953**: 72.

————— (1911). The timber rot caused by *Lenzites sepiaria*. *In* Bull. Bur. Pl. Ind. U.S. Dep. Agric. No. 214: 46 pp. *3 fs., 4 pls.*

SREERAMULU, T. (1963). Observations on the periodicity in the air-borne spores of *Ganoderma applanatum*. *In* Mycologia **55**: 371–379 *3 fs.*

STAHLSCHMIDT, C. (1877). Über eine neue in der Natur vorkommende organische Säure. *In* Justus Liebig's Annln Chem. **187**: 177–197. — *Polyporus purpurascens* Stahlschm. [=*Hapalopilus rutilans*].

————— (1879). Beiträge zur Kenntniss der Polyporsäure. *In* Justus Liebig's Annln Chem. **195**: 365–372.

STANLEY-BROWN, J. (1891). Bernardinite: Is it a mineral or a fungus? *In* Am. J. Sci. III **42**: 46–50 *5 fs.* — *Polyporus* [*Agaricum*] *officinalis*.

STEVENS, N. E. (1912). *Polystictus versicolor* as a wound parasite of catalpa. *In* Mycologia **4**: 263–270 *pls. 74, 75.*

Stevenson, J. (1878). Mycological notes. *Ptychogaster albus*. *In* Scott. Natur. **4**: 250–252. — & *cf.* M. J. Berkeley *in* Gdnrs' Chron. II **8**: 753. 1877.

Steyaert, R. L. (1961a). Genus *Ganoderma* (Polyporaceae) taxa nova — I. *In* Bull. Jard. bot. Bruxelles **31**: 69–83.

————— (1961b). Note on the nomenclature of fungi and, incidentally, of *Ganoderma lucidum*. *In* Taxon **10**: 251–252.

————— (1967). Considérations générales sur le genre *Ganoderma* et plus spécialement sur les espèces européennes. *In* Bull. Soc. r. Bot. Belg. **100**: 189–211 *25 fs., pls. 1–4.*

Storck, R., M. K. Nobles & C. J. Alexopoulos (1971). The nucleotide composition of deoxyribonucleic acid of some species of Hymenochaetaceae and Polyporaceae. *In* Mycologia **63**: 38–49.

Strunk, C. (1963). Zur Frage des Aufbaues der Zellwand von *Polystictus versicolor*. *In* Lyr and Gillwald (ed.), Holzzerstörung Pilze pp. 127–130.

————— (1968). Zur Darstellung des Apicalporus bei *Polystictus versicolor*. *In* Arch. Mikrobiol. **60**: 255–261 *8 fs.*

Sumstine, D. R. (1907). *Polyporus pennsylvanicus* sp. nov. *In* J. Mycol. **13**: 137–138.

Suri, P. N. (1926). *Trametes pini* on Deodar in the Psaspa Valley, Burshahr state, Punjab. *In* Indian Forester **52**: 327–330 *pl. 7.*

Suvorov, P. A. (1963). [Russian title]. On the biology of *Laetiporus sulphureus* (Bull.) Bond. & Sing. *In* Bjull. mosk. Obšč. Ispyt. Prir. (Biol.) **68** (5): 66–77 *4 fs.*

————— (1966). [Russian title]. Studies of biological properties of *Fomes fomentarius* (Fries) Kic[k]x cultures from various trees. *In* Bjull. mosk. Obšč. Ispyt. Prir. (Biol.) **71** (1): 95–105.

————— (1967a). [Russian title]. Wood-destructive capability of the false tinder fungus *Fomes igniarius* (Fr.) Kic[k]x. *In* Mikol. Fitopat. **1**: 464–471 *2 fs.*

————— (1967b). Biological characteristics of *Fomes fomentarius*, found on spruce and birch. *In* Can. J. Bot. **45**: 1853–1857 *1 f.*

————— (1970). [Russian title]. Interaction between different strains of (*Fomes igniarius* (Fr.) Kickx) and other fungi at joint cultivation. *In* Mikol. Fitop. **4**: 13–18 *5 fs.*

Suza, J. (1935). Nová lokalita *Polyporellus rhizophilus* (Pat.) Pilát v Československu. *In* Čas. čsl. Houb. Mykol. Sb. **15**: 129–133 *fs. 74–75 (n.v.).*

Svrček, M. (1960). [Czech title]. *Fomitopsis rosea* (Alb. et Schw. ex Fr.) Karst. in vicinitate urbis Pragae. *In* Česká Mykol. **14**: 229–230.

Takemaru, T. & N. Fujioka (1969). [Japanese title]. The mating system of *Hirschioporus fusco-violaceus* (Fr.) Donk. *In* Rep. Tottori mycol. Inst. **7**: 59–63.

————— & A. Sano (1970). [Japanese title]. Multiple-allelomorph heterothallism in the tetrapolar basidiomycete *Coriolus versicolor* (Fr.) Quél. *In* Rep. Tottori mycol. Inst. **8**: 17–21.

Teixeira, A. R. (1956). Método para estudo das hifas do carpóforo de fungos polìporáceos. São Paulo, 23 pp. *15 fs.*

————— (1958a). Tipificação do gênero *Fomes* (Fries) Kickx. *In* Arq. Bot. Est. São Paulo **3**: 165–174.

————— (1958b). Studies on microstructure of *Laricifomes officinalis*. *In* Mycologia **50**: 671–676 *6 fs.*

————— (1961). Characteristics of the generative hyphae of polypores of North America, with special reference to the presence or absence of clamp-connections. *In* Mycologia **52**: 30–39 *14 fs.*

————— (1962a). Microestruturas do basidiocarpo e sistemática do gênero *Fomes* (Fries) Kickx. *In* Rickia **1**: 13–93 *(1) f., 4 textpls.*

————— (1962b). The taxonomy of the Polyporaceae. *In* Biol. Rev. **37**: 51–81 *2 fs.*

TESTON, D. (1953a, b). Étude de la différenciation des hyphes chez les Polypores dimidiés de la flore française. [I]. *In* Annls Univ. Lyon III (C) **7**: 11–23 *2 textpls.* 1953a — II. — Descriptions et figures. *In* Bull. Soc. Nat. Oyonnax **7**: 80–110 *10 textpls.* 1953b.

THOMS, H. & J. VOGELSANG (1907). Zur Kenntniss der Agaricinsäure. [I. Abhandlung]. *In* Justus Liebig's Annln Chem. **357**: 145–170. — Also for previous litterature.

TIKKA, P. S. (1934). Über die Stockfäule der Nadelwälder Nord-Suomis (-Finnlands). *In* Acta forest. fenn. **40**: 291–308 *3 fs., (2) pls.* — *Fomitopsis* [*Heterobasidion*] *annosa* and *Bjerkandera* [*Climacocystis*] *borealis.*

TOBLER, F. (1954). Flechtenähnliche Symbiose einer Polyporaceae mit Algen. *In* Ber. dt. bot. Ges. **67**: 406–409 *4 fs.* — *Coriolus* [*Cerrena*] *unicolor.*

TOOLE, E. R. (1955). *Polyporus hispidus* on southern bottomlands oaks. *In* Phytopathology **45**: 177–180 *2 fs.*

TORREND, C. (1910). *Trametes ochroleuca* (Berk.) Bres., v. *lusitanica* Torrend. *In* Bull. Soc. portug. Sci. nat. **4**: 35–37.

TORTIĆ, M. (1971). *Ganoderma adspersum* (S. Schulz.) Donk (=*Ganoderma europaeum* Steyaert) and its distribution in Jugoslavia. *In* Acta bot. croat. **30**: 113–118 *(2) pls.*

————— & M. B. JELIĆ (1969). Some interesting Macromycetes and their distribution in Jugoslavia. *In* Acta bot. croat. **28**: 379–386 *4 fs., 2 pls.*

TROTTER, A. (1946). Osservazioni sulla morfologia ed ecologia della "Pietra fungaia" *(Polyporus tuberaster).* *In* Ric. Ossni Divulg. fitopat. Capania Mezzogiorno **10**: 81–90 *4 fs.*

TUBEUF, K. VON (1906). Notizen über die Vertikalverbreitung der *Trametes Pini* und ihr Vorkommen an verschiedenen Holzarten. *In* Naturw. Z. Land- u. Forstw. **4**: 96–100, 328. — The species broadly conceived.

TULASNE, E. L. R. (1865). Note sur le *Ptychogaster albus* Co[rda]. *In* Annls Sci. nat. (Bot.) V **4**: 290–296; **15**: *pl. 12 fs. 1–4.* 1872.

ULBRICH, E. (1939a). Eine bisher unbekannte Gallenbildung des Weiden-Holz- schwammes (*Fomes salicinus* [Pers.] Fr.) und über die Gallen am Flachen Porling (*Ganoderma applanatum* [Pers.] Pat.). *In* Ber. dt. bot. Ges. **57**: 397–402 *1 f., pl. 19.*

————— (1939b). Über einen bemerkenswerten Fall von geotropischer Hymenial- Regeneration bei einer *Polystictus*-Art. *In* Ber. dt. bot. Ges. **57**: 241–246 *pl. 7.* — *Polystictus* [*Coriolus*] *versicolor.*

————— (1940). Wachstums-Beobachtungen an Fruchtkörpern einiger Polyporaceen und Boletaceen. *In* Notizbl. bot. Gart. Mus. Berl. **15**: 258–278.

————— (1941). Über die Gattung *Ceriomyces* Corda 1837 (*Ptychogaster* Corda 1838), die Ptychogasteraceae Falck 1939 und die *Ceriomyces-* (*Ptychogaster-*)Fäule an Nadelholz. *In* Notizbl. bot. Gart. Mus. Berl. **15**: 572–594.

————— (1942). Ein bemerkenswerter Fall von geotropischer Hymenialregeneration bei *Lenzites sepiaria* (Wulf.) Fr. *In* Notizbl. bot. Gart. Mus. Berl. **15**: 791–792.

UNDERWOOD, L. M. (1892). Connecting forms among polyporoid fungi. *In* Zoe **3**: 91–95.

USCUPLIC, M. & R. G. PAWSEY (1970). A selective medium for the isolation of *Polyporus schweinitzii. In* Trans. Br. mycol. Soc. **55**: 161–163 *pl. 19.*

VALADON, L. R. G. & R. S. MUMMERY (1969). A new carotenoid from *Laetiporus sulphureus. In* Ann. Bot. II **33**: 879–882 *1 f.*

VAN BAMBEKE, C. (1895). Note sur une forme monstrueuse de *Ganoderma lucidum* (Leys). *In* Bot. Jaarb. **7**: 93–116 *(3) fs. pls. 1, 2.*

————— (1902). Sur un exemplaire monstrueux de *Polyporus sulfureus* (Bull.) Fries. *In* Bull. Soc. mycol. Fr. **18**: 54–65 *pls. 2–4.*

———— (1906). Quelques remarques sur *Polyporus Rostkowii* Fr., espèce nouvelle pour la Flore belge. *In* Bull. Soc. r. Bot. Belg. **43**: 256–265 *pls. 1, 2.*

———— (1909). Sur *Polystictus cinnamomeus* (Jacq.) Sacc. et *Polystictus Montagnei* Fries. *In* Bull. Soc. r. Bot. Belg. **46**: 15–38 *1 pl.*

VANDENDRIES, R. (1934a). Les barrages sexuels chez *Lenzites betulina* (L.) Fr. *In* C. r. hebd. Séanc. Acad. Sci., Paris **198**: 193–195.

———— (1934b). Nouvelles recherches expérimentales sur les barrages sexuels de *Lenzites betulina* (L.) Fr. *In* Genetica **16**: 389–400 *pl. 3.*

———— (1934c). Contribution à l'étude de la sexualité dans le genre *Trametes*. *In* Bull. Soc. mycol. Fr. **50**: 98–110 *1 f., pls. 5, 6.*

———— (1936a). Les tendances sexuelles chez les Polyporés. I. *Leptoporus adustus* (Fr. ex Willd.) Quél. *In* Revue Mycol. **1**: 85–92 *1 f.*

———— (1936b). Les tendances sexuelles chez les Polyporés. II. *Leucoporus arcularius* (Batsch) Quél. *In* Revue Mycol. **1**: 181–190 *8 fs.*

———— (1936c). Les tendances sexuelles chez les Polyporés. III. *Leucoporus brumalis* (Fr. ex Pers.) Quél. *In* Revue Mycol. **1**: 294–302 *4 fs.*

———— (1937). Les tendances sexuelles des Polyporés. IV. *Melanopus squamosus* (Huds.) Pat. *In* Bull. Soc. mycol. Fr. **52**: 351–362 *13 fs.* "1936".

———— & H. J. BRODIE (1933). Manifestation de barrages sexuels dans le champignon tétrapolaire *Lenzites betulina* (L.) Fr. *In* Bull. Soc. r. Bot Belg. **65**: 109–111.

VAN DER BIJL, P. A. (1917). Heart rot of *Ptaeroxylon utile* (Sneezewood) caused by *Fomes rimosus* (Berk.). *In* Trans. R. Soc. S. Afr. **6**: 215–226 *pls. 39–44.*

VAN DER WESTHUIZEN, G. C. A. (1959). *Polyporus sulphureus*, a cause of heart-rot of *Eucalyptus saligna* in South Africa. *In* Jl S. Afr. For. Ass. **33**: 53–56 *1 f., 1 textpl.*

———— (1963). The cultural characters, structure of the fruit body, and type of interfertility of *Cerrena unicolor* (Bull. ex Fr.) Murr. *In* Can. J. Bot. **41**: 1487–1499 *fs. 1–14, 1 pl.* (=*fs. 15–22*).

VANTERPOOL, T. C. & R. MACRAE (1951). Notes on the Canadian tuckahoe, its occurrence in Canada and the interfertility of its perfect stage, *Polyporus tuberaster* Jacq. ex Fries. *In* Can. J. Bot. **29**: 147–157 *12 fs., 2 pls.* (=*fs. 1–9*).

VERRALL, A. F. (1937). Variation in *Fomes igniarius* (L.) Gill. *In* Techn. Bull. agric. Exp. Stn Univ. Minnesota No. 117: 41 pp. *13 fs.*

VESCO, G. DAL (1963). [Italian title]. Notes on the culture "in vitro" of upper Basidiomycetes. I. — "*Fistulina hepatica*". *In* Allionia **9**: 91–102 *5 fs.*

VESELÝ, R. (1957). [Czech title]. De *Fomite Pinicola* (Sw. ex Fr.) Gill. *In* Česká Mykol. **11**: 228–229 *pl. 28.*

VIMMER, A. (1929). [Czech title]. Diptères, parasites du genre *Polyporus* aux environs de Prague. *In* Mykologia, Praha **6**: 9–11 *(1) f.*

VORONICHIN, N. N. [N. N. Woronichin]. (1925). *Fomes torulosus* (Pers.) Lloyd und *Fomes Ephedrae* Woronich. in Transkaukasien. *In* Annls mycol. **23**: 295–301.

WAKEFIELD, E. M. (1915). *Fomes juniperinus* and its occurrence in British East Africa. *In* Bull. misc. Inf. Kew **1915**: 102–104 *(1) f.*

WALCH, H. (1967). Beiträge zur Physiologie der Gattung *Ganoderma* (Lackporlinge). Karlsruhe, Diss., Techn. Hochsch., (v) and 70 pp. *16 fs.*

———— (1968). Bildung und Nachweis von Phenoloxydasen bei einigen *Ganoderma*-Arten (Lackporlinge). *In* Arch. Mikrobiol. **60**: 314–325 *2 fs.*

WALCH, H. & H. KÜHLWEIN (1968). Zur Kenntnis der cellulolytische Aktivität in der Gattung *Ganoderma* (Lackporlinge). *In* Arch. Mikrobiol. **61**: 373–380 *3 fs.*

WALL, R. E. & J. E. KUNTZ (1964). Water-soluble substances in dead branches of aspen (*Populus tremuloides* Michx.) and their effects on *Fomes igniarius*. *In* Can. J. Bot. **42**: 969–977 *pl. 1.*

WALLIS, G. W. & G. REYNOLDS (1962). Inoculation of Douglas fir roots with *Poria weirii*. *In* Can. J. Bot. **40**: 637–645.

———— & ———— (1965). The initiation and spread of *Poria weirii* root rot of Douglas fir. *In* Can. J. Bot. **43**: 1–9 *2 pls.*

WEAN, R. E. (1937). The parasitism of *Polyporus schweinitzii* on seedling *Pinus strobus*. *In* Phytopathology **27**: 1124–1142 *3 fs.*

WEBSTER, H. (1899). A peculiar state of *Polyporus pergamenus*. *In* Rhodora **1**: 136–137.

WEIDNER, H. & F. SCHREMMER (1962). Zur Erforschungsgeschichte, zur Morphologie und Biologie der Larve von *Agathomyia wankowiczi* Schnabl, einer an Baumpilzen gallenerzeugenden Dipterenlarve. *In* Ent. Mitt. zool. StInst. zool. Mus. Hamb. **2**: 355–366 *5 fs., 2 pls.*

WEIR, J. J. (1914a). The cavities in the rot of *Trametes pini* as a home for hymenopterous insects. *In* Phytopathology **4**: 385.

———— (1914b). Two new wood-destroying fungi. *In* J. agric. Res. **2**: 163–167 *pls. 9, 10.* — *Fomes putearius* Weir [=*Phellinus nigrolimitatus*], *Trametes setosus* Weir [=*Phellinus viticola*].

———— (1921). *Polyporus dryadeus* (Pers.) Fr. on conifers in the Northwest. *In* Phytopathology **11**: 99.

———— (1923a). *Fomes roseus* (A. & S.) Cooke and *Trametes subrosea* nom. novum. *In* Rhodora **25**: 214–220.

———— (1923b). *Polyporus spraguei* Berk., cause of heart-rot. *In* Phytopathology **13**: 288.

———— (1927). Butt rot in *Diospyros virginiana* caused by *Polyporus Spraguei*. *In* Phytopathology **17**: 339–340.

WERESUB, L. K. (1967). Why the controversy over the type of *Cristella* Pat.? *In* Taxon **16**: 396–402 *6 fs.*

WEST, E. (1919). An undescribed timber decay of hemlock. *In* Mycologia **11**: 262–266. — Caused by *Ganoderma tsugae*.

WETTSTEIN, R. VON (1885). Über einen neuen *Polyporus* aus Niederösterreich. *In* Öst. bot. Z. **35**: 81–82. — *Polyporus laccatus* Kalchbr. apud Wettst.

WHITE, J. H. (1919). On the biology of *Fomes applanatus* (Pers.) Wallr. *In* Trans. R. Can. Inst. **12**: 133–174 *1 f., pls. 2–7.*

WHITNEY, R. D. (1962). Studies in forest pathology. XXIV. *Polyporus tomentosus* Fr. as a major factor in stand-opening disease of white spruce. *In* Can. J. Bot. **40**: 1631–1658 *ill., 7 pls.*

———— (1963). Artificial inoculation of small spruce roots with *Polyporus tomentosus*. In Phytopathology **53**: 441–443 *1 f.*

———— (1965). Mycorrhiza-infection trials with *Polyporus tomentosus* on spruce and pine. *In* Forest Sci. **11**: 265–270.

———— (1966a). Germination and inoculation tests with basidiospores of *Polyporus tomentosus*. *In* Can. J. Bot. **44**: 1333–1343 *fs. 1–7, 2 pls. (=fs. 8, 9).*

———— (1966b). Susceptibility of white spruce to *Polyporus tomentosus* in healthy and diseased stands. *In* Can. J. Bot. **44**: 1711–1716.

———— & W. P. BOHAYCHUK (1969). Storage of *Polyporus tomentosus* sporophores. *In* Can. J. Bot. **47**: 1489–1491 *1 pl.*

———— & ———— (1971). Germination of *Polyporus tomentosus* basidiospores on extracts from diseased and healthy trees. *In* Can. J. Bot. 4 : 699–703 *2 fs.*

WIEPKEN (1937). Ein merkwürdiger Fund. *In* Z. Pilzk. **16**: 25 *(1) f.* — Abnormal *Polyporus* [*Grifola*] *frondosus*.

WILLEMET, P. R. (1784). Reflexions botaniques et médinales sur la nature et les propriétés de l'agaric de chêne. *In* Nouv. Mém. Acad. Dijon (Sci. et Arts) 1784 (Sec. sém.): 85–95.

WILSON, C. L., J. C. MILLER & B. R. GRIFFIN (1967). Nuclear behavior in the basidium of *Fomes annosus*. *In* Am. J. Bot. **54**: 1186–1188 *18 fs.*

WILSON, M. (1927). The host plants of *Fomes annosus*. *In* Trans. Br. mycol. Soc. **12**: 147–149.

WINTERS, J. H., C. H. BECKER and W. M. LAUTER (1961). [Polish title]. A comparative phytochemical study of Polish and American *Poria obliqua* varieties. *In* Sylwan **105** (2): 15–25.

WOJEWODA, W. (1966). [Polish title]. *Ungulina corrugis* (Fr.) Bourd. & Galz., a species of the Polyporaceae family, new to Poland, found in Ojców National Park. *In* Fragm. flor. geobot. **12**: 513–517 *(2) pls.*

WOLF, E. (1891). *Trametes cinnabarina* und seine Zersetzung des Birkenholzes, nebst einem Beitrage zur Kenntnis des *Agaricus adiposus*. München, Inaug. Diss.

WOLF, F. A. (1941). A large fructification of *Polyporus sulphureus*. *In* Mycologia **33**: 136–137 *(1) f.*

WRIGHT, J. E. (1964a). Pseudoamyloid reaction in pore fungi. *In* Mycologia **56**: 692–695.

———— (1964b). Notes on *Poriae*. *In* Mycologia **56**: 785–786.

WULFF, T. (1909). Björktickan (*Polyporus betulinus* Fr.) och fnosktickan (*P. fomentarius* Fr.), ett par för björkskogen skadliga svampar. *In* SkogsvFör. Tidskr. **7** (Fackupps.): 1–14 *6 fs., pls. 1, 2.*

YAMANO, Y. (1931). On the morphology and physiology of *Fomes applanatus* (Fr.) Gill. and its allies. *In* Sci. Rep. Tôhoku Univ. (IV, Biol.) **6**: 199–236 *1 f., pls. 4–7.*

YDE-ANDERSEN, A. (1961a). *Polyporus Schweinitzii* Fr. i nåletræbevoksninger. *Polyporus Schweinitzii* Fr. in Danish conifer stands. *In* Friesia **6**: 347–355 *4 fs.*

———— (1961b). Om den årstidsbetingede variation i hyppigheden of størflade-infektioner med luftbårne *Fomes annosus*-sporer hos rødgran. *In* Dansk Skovforen. Tidsskr. **46**: 139–158 *1 f.*

ZAK, B. (1969). Characterization and classification of mycorrhizae of Douglas fir. I. *Pseudotsuga menziesii* + *Poria terrestris* (blue- and orange-staining strains). *In* Can. J. Bot. **47**: 1833–1840 *7 pls.*

ZANEVELD, J. S. (1941). Het vruchtlichaam van *Polyporus tuberaster* Jacq. ex Fr. in Nederland gevonden. *In* Fungus **13**: 1–5 *2 fs.*

ZELLER, S. M. (1916). Studies in the physiology of the fungi. II. *Lenzites saepiaria* Fries, with special reference to enzyme activity. *In* Ann. Missouri bot. Gdn **3**: 439–512 *pls. 8, 9.*

———— (1926). The brown-pocket heart rot of stone-fruit trees caused by *Trametes subrosea* Weir. *In* J. agric. Res. **33**: 687–693 *2 fs.*

ZEROVA, M. J. (1957). [Russian title]. *Polyporus rhizophilus* (Pat.) Sacc. and *Pleurotus eryngii* Fr. ex DC. var. *ferulae* Lanzi—interesting fungus species, new for the Ukrainian SSR, found in virgin steppes. *In* Ukr. bot. Ž. **14** (2): 69–71 *2 fs.*

ZIER, J. (1832). Über die Fabrication des Feuerschwammes. *In* J. tech. oekon. Chem. **13**: 463–473.

ZYCHA, H. (1964). Stand unserer Kenntnisse von der *Fomes annosus*-Rotfäule. *In* Forstarchiv **35**: 1–4. — Brief introduction to the phytopathology.

INDEX

The index contains (1) the accepted generic names of the "Check list" proper
and their synonyms, (2) the accepted specific (and infraspecific) epithets of the
same list and their binomial synonyms, (3) the generic names treated in the "Notes".
Misapplied names are not considered. The "List of omitted names" has not been
integrated into the index. To facilitate searching, the specific and infraspecific
epithets are followed by their first author, without the use of brackets. To save
space, however, the generic names are left out.